The SAGE
Handbook of
Remote Sensing

The SAGE Handbook of
Remote Sensing

Edited by
Timothy A. Warner,
M. Duane Nellis, and Giles M. Foody

Los Angeles | London | New Delhi
Singapore | Washington DC

First published 2009

SAGE Publications Ltd
1 Oliver's Yard
55 City Road
London EC1Y 1SP

SAGE Publications Inc.
2455 Teller Road
Thousand Oaks, California 91320

SAGE Publications India Pvt Ltd
B 1/I 1 Mohan Cooperative Industrial Area
Mathura Road
New Delhi 110 044

SAGE Publications Asia-Pacific Pte Ltd
33 Pekin Street #02-01
Far East Square
Singapore 048763

Library of Congress Control Number: 2008937505

British Library Cataloguing in Publication data

A catalogue record for this book is available from the British Library

ISBN 978-1-4129-3616-3

Typeset by CEPHA Imaging Pvt. Ltd., Bangalore, India
Printed in India at Replika Press Pvt Ltd
Printed on paper from sustainable resources

Dedication

This book is dedicated to the many pioneers in remote sensing, including:

Mike Barnsley

Jack Estes

Don Levandowski

Contents

Preface

Although the term remote sensing is about 50 years old, having been coined in 1958 by Evelyn Pruitt, a geographer at the US Office of Naval Research (Estes and Jensen, 1998), the subject matter covered by the field of remote sensing is vast. As a methodological approach, remote sensing has underpinnings in physics, mathematics, engineering, and computer science. Remote sensing plays an important role in the scientific, commercial, and national security arenas, and the applications of remote sensing extend from the Earth's atmosphere to the hydrosphere, cryosphere, biosphere, and lithosphere, as well as to the moon, planets, and asteroids.

The challenge in compiling a relatively comprehensive survey of such a vast field is evident in the fact that, to our knowledge, this book is the first comprehensive text in a quarter of a century. Our work follows in the tradition of the major series, *The Manual of Remote Sensing*, first published in two volumes by the American Society of Photogrammetry and Remote Sensing (ASPRS) in 1975, with a second edition in 1983. Notably, for the third edition, the idea of a single publication was abandoned and an apparently open-ended series was decided upon. As a result, six volumes in this series have already been published in the decade since 1997, with additional volumes planned.

For our book, we desired a single volume that provided as broad a view of the field as possible. Our aim was to give the reader a forward-looking perspective that also explained the developments that led to the current context. The chapters assume a basic background in remote sensing, but not necessarily in the specific topics covered. This book should therefore be particularly useful to professionals and advanced students who desire a systematic overview of the state of the art, as well as potential future challenges.

In addressing such a huge field we have by necessity had to be selective in our approach. Therefore, from the outset we limited our scope to the terrestrial Earth. By keeping this focus, we have been able to cover not only the traditional remote sensing applications, such as in soils, geology, and vegetation, but also the relatively new applications such as in the social sciences, biogeochemical modeling, and disaster monitoring.

The initial concept for this volume was developed in a 10-page outline, which was reviewed by 13 anonymous external reviewers. With advice and feedback from those reviewers, we recruited 33 authors to lead the individual chapters. Those lead authors recruited an additional 42 co-authors, resulting in a total of 75 authors. The chapters were reviewed by the editors as well as over 90 reviewers.

The book is organized in six major sections. Section I, an introduction, covers broad overarching issues, including remote sensing policy. Section II is a systematic treatment of the interaction of electromagnetic radiation with the terrestrial environment. This section provides a key background for the later section on remote sensing applications. The chapters are organized from short to long wavelength, specifically from the visible to microwave regions. Section III, on digital sensors and platforms, provides an overview of how the engineering of image acquisition influences image properties. The section includes chapters on sensor technology, as well as a series on satellite sensors, organized by relative spatial resolution. Separate chapters cover hyperspectral sensors, microwave sensors, airborne imaging, and airborne laser scanning (also known as lidar). Section IV covers remote sensing analysis, from design to implementation. This section covers both field work and image analysis issues, ending with a discussion on accuracy assessment. Section V, on remote sensing applications, comprises approximately one third of the book, and is organized in four subsections: (a) lithospheric sciences, (b) plant sciences, (c) hydrospheric and crysopheric sciences, and (d) global change and human environments. Section VI provides a short forward-looking summary of the book.

ACKNOWLEDGEMENTS

This book was only possible through the enthusiasm, cooperation, and support from a wide range of people. The role of the authors of course was central. In addition, the contribution of the external reviewers was particularly important in ensuring the highest quality for the chapters, and their role is gratefully acknowledged. We would also like to thank SAGE for unfailing encouragement and patience throughout this long process, especially Commissioning Editor Robert Rojek and Editorial Assistant Sarah-Jayne Boyd.

The external reviewers include:

Michael Abrams, NASA/Jet Propulsion Laboratory, USA

John Althausen, Lockheed Martin Corporation, USA

Hans-Erik Andersen, USDA Forest Service, USA

Paul Aplin, University of Nottingham, UK

Richard L. Armstrong, University of Colorado at Boulder, USA

Manoj Arora, IIT Roorkee, India

Gregory P. Asner, Carnegie Institution, USA

Paul Baumann, State University of New York, College at Oneonta, USA

Larry Biehl, Purdue University, USA

Michael P. Bishop, University of Nebraska-Omaha, USA

Janis L. Boettinger, Utah State University, USA

Zachary Bortolot, James Madison University, USA

Tomas Brandtberg, Uppsala University, Sweden

James B. Campbell, Virginia Tech, USA

Jocelyn Chanussot, Grenoble Institute of Technology, France

Jing M. Chen, University of Toronto, Canada

Shane Cloude, AEL Consultants, UK

Jeffrey D. Colby, Appalachian State University, USA

Kelley A. Crews, The University of Texas at Austin, USA

Paul Curran, Bournemouth University, UK

Mike A. Cutter, Surrey Satellite Technology Ltd, UK

Yongxin Deng, Western Illinois University, USA

Mark R. Drinkwater, European Space Agency, ESTEC, The Netherlands

Gregory Elmes, West Virginia University, USA

Robert Erskine, US Department of Agriculture, Agricultural Research Service, USA

Ian S. Evans, Durham University, UK

Steven Fassnacht, Colorado State University, USA

Pete Fisher, University of Leicester, UK

Igor Florinsky, Russian Academy of Sciences, Russia

Paul Frazier, University of New England, Australia

Bruce E. Frazier, Washington State University, USA

Mark Friedl, Boston University, USA

Joanne Irene Gabrynowicz, The University of Mississippi School of Law, USA

Kathleen Galvin, Colorado State University, USA

Jerry Griffith, University of Southern Mississippi, USA

Randy M. Hamilton, RedCastle Resources, Inc., USA

Ray Harris, University College London, UK

John C. Heinrichs, Fort Hays State University, USA

Geoffrey M. Henebry, South Dakota State University, USA

George F. Hepner, University of Utah, USA

Michael J. Hill, University of North Dakota, USA

Ryan Jensen, Brigham Young University, USA

Chris J. Johannsen, Purdue University, USA

John P. Kerekes, Rochester Institute of Technology, USA

Doug King, Carleton University, Canada

Philip Lewis, University College London, UK

Tim J. Malthus, University of Edinburgh, UK

Roger M. McCoy, University of Utah, USA

Greg McDermid, University of Calgary, Canada

Kyle McDonald, Jet Propulsion Laboratory, California Institute of Technology, USA

Assefa M. Melesse, Florida International University, USA

Gay Mitchelson-Jacob, Bangor University, UK

Mahta Moghaddam, The University of Michigan, USA

Keith Morris, Louisiana State University, USA

Atsushi Nara, Arizona State University, USA

Janet Nichol, The Hong Kong Polytechnic University, Hong Kong

Stuart Phinn, The University of Queensland, Australia

Leland E. Pierce, The University of Michigan, USA

Robert Gilmore Pontius Jr., Clark University, USA

Sorin Popescu, Texas A&M University, USA

Dale A. Quattrochi, NASA Marshall Space Flight Center, USA

R. Douglas Ramsey, Utah State University, USA

Jane M. Read, Syracuse University, USA

Ronald G. Resmini, George Mason University, USA

John Rogan, Clark University, USA

Kenton Ross, Science Systems and Applications, Inc., USA

Vincent V. Salomonson, University of Utah, USA

Michael E. Schaepman, Wageningen University, The Netherlands

James Schepers, US Department of Agriculture, Agricultural Research Service, USA

Edwin Sheffner, NASA – Ames Research Center, USA

Steve Stehman, SUNY College of Environmental Science and Forestry, USA

Douglas Stow, San Diego State University, USA

Gregg Swayze, US Geological Survey, USA

Richard M. Teeuw, University of Portsmouth, UK

Gerd Ulbrich, European Space Agency ESA/ESTEC, The Netherlands

Freek van der Meer, International Institute for Geo-Information Science and Earth Observation (ITC), and Utrecht University, The Netherlands

Jakob van Zyl, Jet Propulsion Laboratory, California Institute of Technology, USA

Wouter Verhoef, National Aerospace Laboratory NLR, The Netherlands

Eric Vermote, University of Maryland, and NASA Goddard Space Flight Center, USA

Chenghai Yang, US Department of Agriculture, Agricultural Research Service, USA

Stephen Yool, The University of Arizona, USA

Kenneth Young, University of Texas at Austin, USA

Howard A. Zebker, Stanford University, USA

Plus, an additional five reviewers who wished to remain anonymous.

REFERENCES

Estes, J. E., J. R. Jensen, and D. S. Simonett, 1980. Impacts of remote sensing on US Geography. *Remote Sensing of Environment*, 10: 43–80.

The Editors

Timothy A. Warner, West Virginia University

M. Duane Nellis, Kansas State University

Giles M. Foody, University of Nottingham

Notes on Contributors

EDITORS

Timothy A. Warner is Professor of both Geology and Geography at West Virginia University, in Morgantown, West Virginia, USA. He matriculated at the South African College Schools (SACS), has a B.Sc. (Hons.) in Geology from the University of Cape Town, and a Ph.D. from Purdue University in geological remote sensing. His research specialties include the spatial properties of images, high resolution remote sensing, and lidar. He has served as a founding board member and Secretary of AmericaView, and as Chair of the Remote Sensing Specialty Group (RSSG) of the Association of American Geographers (AAG). He received the 2006 *RSSG Outstanding Contributions Award*, and the 2006 *Boeing Award for Best Paper in Image Analysis and Interpretation* from the American Society of Photogrammetry and Remote Sensing. In 2007, he was a Fulbright Fellow at the University of Louis Pasteur, in Strasbourg, France. He serves on the editorial board of *Geographical Compass*.

M. Duane Nellis is Provost and Senior Vice President, as well as Professor of Geography, at Kansas State University. He has published over 100 articles and more than a dozen books and book chapters on various aspects of remote sensing and GIS applications to natural resources assessment, and other dimensions of rural geography. He is past president of both the Association of American Geographers (AAG), and the National Council for Geographic Education (NCGE). Nellis has also served as co-editor of the GIS/remote sensing journal *Geocarto International*. He has received numerous honors and awards including National Honors from the AAG, election as a fellow of the American Association for the Advancement of Science (AAAS), the Royal Geographical Society, and the Explorers Club. In addition, he is past Chair of the AAG Remote Sensing Specialty Group and received that groups, Outstanding Contributions Award. At Kansas State he received the University Outstanding Teaching Award, the Phi Kappa Phi Research Scholars Award, and the University Outstanding Advisor Award. Nellis completed his undergraduate degree at Montana State University and his masters and Ph.D. at Oregon State University. He started his academic career at Kansas State University, where he moved from Assistant to Full Professor, and where he served as the head of Geography Department and Associate Dean of Arts and Sciences. In 1997, he was named Dean of the Eberly College of Arts and Sciences at West Virginia University. He then returned to Kansas State University in 2004 as Provost.

Giles M. Foody completed B.Sc. and Ph.D. at the University of Sheffield in 1983 and 1986 respectively and is currently Professor of Geographical Information Science at the University of Nottingham. His main research interests lie at the interface between remote sensing, biogeography, and informatics. Topics of particular interest relate to image classification for land cover mapping and monitoring applications, addressing issues at scales ranging from the sub-pixel to global. His publication list includes seven books and more than 135 refereed journal articles as well as many conference papers. He currently serves as editor-in-chief of the *International Journal of Remote Sensing* and holds editorial roles with *Ecological Informatics* and *Landscape Ecology* as well as serving on the editorial boards of journals including *Remote Sensing of Environment*, *Geocarto International*, and *International Journal of Applied Earth Observation and Geoinformation*.

CHAPTER 1

Timothy A. Warner, M. Duane Nellis, and **Giles M. Foody** (see editors' biographical descriptions).

CHAPTER 2

Ray Harris is Emeritus Professor of Geography and a former Executive Dean of the Faculty of Social and Historical Sciences at University College London. In addition to UCL, his career has included periods at Logica UK plc, Software Sciences and the University of Durham. His research and teaching interests are in satellite Earth observation, science policy and in the Middle East. He has worked extensively with the European Space Agency, the European Commission, NASA, NOAA, JAXA and other space organizations on the applications of Earth observation data and on the development of data policy. He was a member of two International Council for Science (ICSU) panels on scientific data and information policy, and has worked on both GMES and GEOSS. For nearly 30 years he has worked on remote sensing applications in the UK, Kuwait, Oman, Jordan, Tunisia and Iran, focusing on agriculture, land degradation and land cover change. He is the author of four books on remote sensing and many journal articles, book chapters and reports.

CHAPTER 3

Willem J. D. van Leeuwen received the B.Sc. and M.Sc. degrees in soil science from the Wageningen University for Life Sciences, the Netherlands in 1985 and 1987 respectively, and the Ph.D. degree in soil and remote sensing science from the Department of Soil, Water, and Environmental Science, University of Arizona, Tucson, in 1995. He has been a research scientist and a member of the MODIS land science team, worked on global spectral vegetation index and albedo product and algorithm development, and is currently working on post wildfire vegetation recovery, land degradation, and vegetation community phenology research, employing remote sensing and geospatial tools. He also works on web-based decision support tools for natural resource managers. Since 2005, Dr. van Leeuwen is an Assistant Professor and has a joint appointment with the Department of Geography and Regional Development and the Office of Arid Lands Studies at the University of Arizona, Tucson, where he teaches classes in geographical field methods, biogeography, and remote sensing. He is a member of American Geophysical Union (AGU), American Society of Photogrammetry and Remote Sensing (ASPRS), Association of American Geographers (AAG), IEEE Geoscience and Remote Sensing Society (IGARSS), and the International Association for Landscape Ecology (IALE).

CHAPTER 4

Arthur P. Cracknell graduated in physics from Cambridge University, in 1961 and then did his D.Phil. at Oxford University on theoretical solid state physics. He has worked at Singapore University (now the National University of Singapore), Essex University, and Dundee University, where he became a professor in 1978. He retired from Dundee University in 2002 and now holds the title of emeritus professor there. He is currently working on various short-term contracts in several universities and research institutes in China and Malaysia.

After several years of research work in solid state physics, he turned his interests in the late 1970s to remote sensing and he has been the editor of the *International Journal of Remote Sensing* for over 20 years. His particular research interests in remote sensing include the extraction of the values of various geophysical parameters from satellite data and the correction of remotely-sensed images for atmospheric effects. He and his colleagues and research students have published around 300 research papers and he is the author or co-author of several books, both on theoretical solid state physics and on remote sensing. He also pioneered the teaching of remote sensing at postgraduate level at Dundee University.

Doreen S. Boyd received the B.Sc. degree in geography from the University of Wales, UK, in 1992 and the Ph.D. degree from the University of Southampton, UK, in 1996. She is currently an Associate Professor with the School of Geography, University of Nottingham, UK, having held Lectureships at Manchester, Kingston, and Bournemouth Universities. Between 2004 and 2006, she held the position of senior research leader in Research and Innovation at Ordnance Survey, Great Britain. Her main research interests are in the remote sensing of terrestrial ecosystems and the monitoring of environmental change. She serves on the editorial board of *Visual Geosciences* and the advisory board of the *Journal of Maps*. She is the co-founder and co-chair of the laser scanning and LiDAR special interest group of the Remote Sensing and Photogrammetry Society.

CHAPTER 5

Dale A. Quattrochi is currently employed by NASA as a Geographer and Senior Research Scientist at the George C. Marshall Space Flight Center in Huntsville, Alabama. He holds a B.S. (1973, Geography) from Ohio University, an M.S. (1978, Geography/Remote Sensing) from the University of Tennessee, and a Ph.D. (1990, Geography/Remote Sensing) from the University of Utah. Dr. Quattrochi's research interests focus on the application of thermal remote sensing data for analysis of heating and cooling patterns across the landscape as they impact local and regional environments. Much of his work has been on applying thermal infrared data to analysis of the urban heat island effect. He is also working on the application of remote sensing data to public health effects, such as asthma and respiratory distress. Additionally, he is also conducting research on the use of geospatial statistical techniques, such as fractal analysis, to multiscaled remote sensing data. Dr. Quattrochi is the recipient of numerous awards, including the NASA Exceptional Scientific Achievement Medal, NASA's highest science award, which he received for his research on urban heat islands and remote sensing. He is also a recipient of the Ohio University College of Arts and Sciences distinguished Alumni Award.

Jeffrey C. Luvall is a NASA Senior Research Scientist at Marshall Space Flight Center. He holds a B.S. (1974, Forestry) an M.S. (1976, Forest Ecology) from Southern Illinois University, Carbondale, IL, and a Ph.D. (1984, Tropical Forest Ecology) from the University of Georgia, Athens, GA. His current research involves the modeling of forest canopy energy budgets using airborne thermal scanners. These investigations have resulted in the development of a Thermal Response Number (TRN), which quantifies the land surface's energy response in terms of $kJ\,m^{-2}\,°C^{-1}$ and can be used to classify land surfaces in regional surface budget modeling by their energy use. A logical outgrowth of characterizing surface energy budgets of forests is the application of thermal remote sensing to quantify the urban heat island effect. One important breakthrough is the ability to quantify the importance of trees in keeping the city cool. His current research involves alternate mitigation strategies to reduce ozone production through the use of high albedo surfaces for roofs and pavements and increasing tree cover in urban areas to cool cities.

CHAPTER 6

Mahta Moghaddam received a B.S. degree (with highest distinction) from the University of Kansas, Lawrence, in 1986 and M.S. and Ph.D. degrees from the University of Illinois, Urbana-Champaign, in 1989 and 1991, respectively, all in electrical and computer engineering. She is an Associate Professor of Electrical Engineering and Computer Science at the University of Michigan, where she has been since 2003. From 1991 to 2003, she was with the Radar Science and Engineering Section, NASA Jet Propulsion Laboratory (JPL) in Pasadena, CA. She has introduced innovative approaches and algorithms for quantitative interpretation of multichannel SAR imagery based on analytical inverse scattering techniques applied to complex and random media. She has also introduced a quantitative approach for data fusion by combining SAR and optical remote sensing data for nonlinear estimation of vegetation and surface parameters. She has led the development of new radar instrument and measurement technologies for subsurface and subcanopy characterization. Dr. Moghaddam's research group is engaged in a variety of research topics related to applied electromagnetics, including the development of advanced radar systems for subsurface characterization, continental scale wetlands mapping with SAR, mixed-mode high resolution medical imaging techniques, and smart sensor webs for remote sensing data collection and validation.

CHAPTER 7

John P. Kerekes received his B.S., M.S., and Ph.D. degrees in electrical engineering from Purdue University, West Lafayette, Indiana, USA, in 1983, 1986, and 1989, respectively. From 1983 to 1984, he was a member of the technical staff with the Space and Communications Group, Hughes Aircraft Co., El Segundo, California, USA. From 1989 to 2004, he was a technical staff member with Lincoln Laboratory, Massachusetts Institute of Technology, Lexington, Massachusetts, USA. In 2004, he became an Associate Professor in the Center for Imaging Science, Rochester Institute of Technology, Rochester, New York, USA. He is a senior member of the Institute of Electrical and Electronics Engineers (IEEE) and a member of Tau Beta Phi, Eta Kappa Nu, American Geophysical Union, American Meteorological Society, American Society for Photogrammetry and Remote Sensing, and SPIE. Since 2000 he has served as an Associate Editor of the *IEEE Transactions on Geoscience and Remote Sensing*. He was the General co-chair of the 2008 International Geoscience and Remote Sensing Symposium (IGARSS'08) held in Boston, Massachusetts, USA. His research interests include the modeling and analysis of remote sensing system performance in pattern recognition and geophysical parameter retrieval applications.

CHAPTER 8

Thierry Toutin, educated both in France and Canada, received his final diploma, the Dr.-Ing. degree, in geodetic sciences and remote sensing from the Ecole Nationale des Sciences Géographiques of the Institut Géographique National, Paris, France, in 1985. After a few years in the Canadian private industry, he has worked since 1988 as senior research scientist at the Canada Center for Remote Sensing, Natural Resources Canada. He currently develops mathematical tools and prototype systems for stereoscopy, radargrammetry, interferometry, and the chromostereoscopy, using a broad range of Earth observation data (airborne and spaceborne; VIR and SAR; fine to coarse resolution). In recent years, he has focused mainly on 3D physical models and their generalization to fine spatial resolution optical imagery (SPOT5, EROS, IKONOS, Quickbird, Formosat, etc.). His main fields of interest are 3D modeling and reconstruction, interactive feature extraction, cartographic applications of Earth observation data, and the integration of multisource data.

CHAPTER 9

Samuel N. Goward is Professor of Geography at the University of Maryland. He has been involved in land remote sensing since the early 1970s. One primary research focus area has been automated processing and analysis of regional and global data sets from AVHRR and Landsat. From 1997 to 2002 he served as the Landsat Science Team leader and was recently selected to serve as a member of the new USGS/NASA Landsat Science Team. He also continues to work with the NASA Landsat Project Science Office to develop operational concepts including the long-term acquisition plan. Currently his research is carried out under the North American Carbon Program, in association with NASA and US Forest Service colleagues, seeking to improve forest dynamics analysis with Landsat time-series data. Earlier he also worked with the NASA Stennis Space Center, to evaluate commercial sources of land remote sensing data, including the IKONOS and QuickBird. Over the last decade he served as co-chair of the advisory committee for the USGS National Satellite Land Remote Sensing Data Archive (NSLRSDA) at USGS EROS and continues on the editorial board of *Remote Sensing of Environment*. Among several honors, he has recently been awarded the USGS *John Wesley Powell Award* (2006) and the USGS/NASA *William T. Pecora Award* (2008) for contributions to the Landsat Mission.

Terry Arvidson has been part of the Landsat program since 1979, from pre-proposal phases through on-orbit operations, from developer and tester to operations engineer and project manager. Currently, she is a manager of sustaining engineering for Landsat 7, and supports both the USGS and NASA/GSFC. Ms. Arvidson serves as the liaison between the satellite operations team and the Landsat Science Project Office. She has been an active member of the international Landsat Ground Station Operators Working Group since the 1980s. Ms. Arvidson managed the development of the Landsat-7 Long-Term Acquisition Plan (LTAP), working with the science community on specialized requirements for land covers such as glaciers and reefs, and maintaining the LTAP databases. She continues to interface with the science community on scheduling and operations issues and in support of Drs. Goward and Williams on the Landsat Science Team. Ms. Arvidson has researched the

Landsat historical archive for the Landsat Legacy project, including internationally-held archives, and participated in oral history interviews and document preservation. She has published numerous articles on the LTAP and the Landsat archive history, and co-edited, with Drs. Goward and Williams, a PE&RS special issue on Landsat 7. Ms. Arvidson has a B.Sc. degree from the University of Maryland.

Darrel L. Williams serves as Associate Chief of the Hydrospheric and Biospheric Sciences Laboratory within the Earth Sciences Division at NASA's Goddard Space Flight Center. He also serves as the Project Scientist for the Landsat 5 and 7 missions currently in orbit, and is entrusted with ensuring the scientific integrity of these missions. Prior to his more recent roles in science management, his remote sensing research involved the development of enhanced techniques for assessing forest ecosystems worldwide. He has authored ~100 publications in the field of quantitative remote sensing and served on the editorial board of the *International Journal of Remote Sensing* throughout the 1990s. Dr. Williams has received numerous prestigious awards such as the NASA Medal for Outstanding Leadership (1997), NASA's Exceptional Service Medal (2000), and the 'Aviation Week and Space Technology 1999 Laurels Award' for outstanding achievement in the field of *Space* in recognition of his science leadership of the Landsat 7 mission. Recently Dr. Williams received an 'Outstanding Alumni Award' from the School of Forest Resources at the Pennsylvania State University. Additional awards have been bestowed by the US Department of Agriculture, the US Department of the Interior, and the American Society of Photogrammetry and Remote Sensing.

Richard Irish accepted a position, in 1993, with Science Systems Applications, Inc., to work on NASA's Landsat-7 program at the Goddard Space Flight Center. There, he developed the cloud cover recognition algorithm used for Landsat-7, created the Calibration Parameter File used for radiometric and geometric processing and updates, and defined the standard Landsat-7 distribution product, now an international exchange standard that is used world wide. Mr. Irish continues his work within NASA's Biospheric Sciences Branch on the Landsat program. His research endeavors include developing cloud shadow discrimination and multiscene merging algorithms for the TM, ETM+, and LDCM missions. He is also the Landsat-7 science liaison to the user community. He wrote and maintains the frequently visited Landsat-7 Science Data Users Handbook web site.

James R. Irons is the Associate Chief of the Laboratory for Atmospheres, NASA Goddard Space Flight Center (GSFC). He is also the NASA Landsat Data Continuity Mission (LDCM) Project Scientist. Prior to 2007, Dr. Irons worked for 28 years as a physical scientist in the Biospheric Sciences Branch, NASA GSFC where he served as the Landsat-7 Deputy Project Scientist beginning in 1992. Dr. Irons' career has been devoted to advancing the science and practice of land remote sensing. His research has focused on applying Landsat data to land cover mapping. His research has also encompassed the characterization and understanding bi-directional reflectance distribution functions (BRDFs) for land surfaces, particularly plant canopies and soil. He is the principal or co-author of 35 peer reviewed journal articles and two book chapters. Dr. Irons received his B.Sc. degree in environmental resources management in 1976 and the M.Sc. degree in agronomy in 1979 from the Pennsylvania State University. He received his Ph.D. degree in agronomy in 1993 from the University of Maryland.

CHAPTER 10

Christopher Owen Justice received his Ph.D. in geography from the University of Reading, UK. In 1978 he came to NASA's Goddard Space Flight Center as a National Academy of Sciences post-doctoral fellow. In 1981 he took a fellowship position at ESA ESRIN and in 1983 he returned to the Goddard Space Flight Center to work with AVHRR data on land studies and helped form the GIMMS Group with Compton Tucker and Brent Holben. Since 2001 he has been a professor and research director in the Geography Department of the University of Maryland. He is a team member and land discipline chair of the NASA Moderate Imaging Spectroradiometer (MODIS) Science Team and is responsible for the MODIS Fire Product and helped develop the MODIS Rapid Response System. He is a member of the NASA NPOESS Preparatory Project (NPP) Science Team. He is co-chair of the GOFC/GOLD-Fire Implementation Team, a project of the Global Terrestrial Observing System (GTOS), and a member of the Integrated Global Observation of Land (IGOL) Steering Committee and leader of the GEOSS Agricultural Monitoring Task. He is on the Strategic Objective Team for USAID's Central Africa Regional Project for the Environment. He is a Co.I.

on the USGS Landsat Science Team. He is Program Scientist for the NASA Land Cover Land Use Change Program. His current research is on land cover and land use change, the extent and impacts of global fire, global agricultural monitoring (with the US Department of Agriculture, Foreign Agricultural Service, and the GIMMS group at Goddard Space Flight Center), and their associated information technology and decision support systems.

Compton James Tucker III received his B.S. degree in biology in 1969 from Colorado State University. After working in two banks and realizing banking was not his calling, he returned to Colorado State University and received his M.S. degree in forestry in 1973 and his Ph.D. degree, also in forestry, in 1975. He came to NASA's Goddard Space Flight Center as a National Academy of Sciences post-doctoral fellow in late 1975. Since 1977 he has been a physical scientist and leader of the GIMMS group at NASA's Goddard Space Flight Center. In the mid 1970s he contributed to the sensor configuration of Landsat's thematic mapper instrument. He has been a pioneer in demonstrating the utility of coarse-resolution remote sensing using AVHRR and similar data for large-scale vegetation studies exploiting temporal information. Currently he is using satellite data to study climatically-coupled hemorrhagic fevers, global primary production including agricultural monitoring, tropical deforestation and habitat fragmentation, and glacier variations from the 1970s to the present. Since 2005 he has worked for NASA at the Climate Change Science Program Office in the areas of land use and land cover change and climate and worked to prioritize satellite and *in situ* observations for climate research.

CHAPTER 11

Douglas A. Stow is a Professor of Geography at San Diego State University (SDSU) and specializes in remote sensing. He received B.A., M.A., and Ph.D degrees in Geography from the University of California, Santa Barbara. His remote sensing studies focus on land cover change analyses with emphases on Mediterranean-type and Arctic tundra ecosystems, and urban areas. He is the co-director of the Center for Earth Systems Analysis Research and doctoral program coordinator. Stow is currently the P.I. for a NASA REASoN project on integration of remote sensing and decision support systems for international border security. He has also served as P.I. for several state and local agency contracts, and as a co-investigator on numerous NASA, NSF, and NIH grants. He is the author or co-author of over 100 refereed publications and 35 conference proceedings papers, mostly on remote sensing topics.

Lloyd L. Coulter has worked as a staff researcher in the Department of Geography at San Diego State University, since November 1998. He specializes in remote sensing and image processing. Mr. Coulter has served as technical lead on several projects using fine spatial resolution imagery for detecting changes in southern California native habitat and for mapping such things as invasive plants, urban irrigated vegetation, urban canyon fire hazards, and land use. Mr. Coulter is also the operator of an ADAR 5500 airborne digital multispectral camera system owned and operated by the Department of Geography. He has several years of experience in airborne digital image acquisition and post-processing.

Cody A. Benkelman is the lead engineer at Mission Mountain Technology Associates, which provides remote sensing, image processing, and geographic information systems services. He served as lead engineer and co-founder of Positive Systems, Inc., developing multispectral airborne imaging systems and image processing software. Mr. Benkelman also served as principal investigator and project manager on numerous NASA R&D projects, focused on development of image co-registration software (SBIR Phase I and Phase II, 2004–2006), multispectral data acquisition for the EOS Science Data Buy Program (1997–2001) and imaging system design and development (Earth Observation Commercialization and Applications Program, 1993). Mr. Benkelman was awarded peer-reviewed certification as a 'Mapping Scientist in Remote Sensing' by the American Society for Photogrammetry and Remote Sensing (ASPRS), certification number RS144, effective 10/6/03. He received his M.S. degree in electrical engineering from the University of Colorado in 1987 and a B.S. in physics from Montana State University in 1981.

CHAPTER 12

Michael E. Schaepman is full Professor in remote sensing at the University of Zurich in Switzerland and adjoint Professor in Geo-Information science with special emphasis on remote sensing at Wageningen

University (WU) in The Netherlands. His specialization is in quantitative, physical based remote sensing using imaging spectrometers and multiangular instruments. He pays particular attention on the retrieval of land surface variables in vegetated areas. After obtaining M.Sc. (1994) and Ph.D. (1998) degrees from the University of Zurich (CH) in geography and remote sensing, he spent part of his post doctorate at the University of Arizona (College of Optical Sciences, Tucson, AZ) before being appointed full chair in Wageningen in 2003 and scientific manager in 2005, and full chair in Zurich in 2008 respectively. He serves as Chairman of the ISPRS WG VII/1 on Physical Modeling and has significantly contributed to the further development of imaging spectroscopy over recent years, namely to ESA missions such as LSPIM, SPECTRA, FLEX and APEX. Michael E. Schaepman has co-authored more than 300 scientific publications (>60 peer reviewed papers).

CHAPTER 13

Josef Martin Kellndorfer's research focuses on the monitoring and assessment of terrestrial and aquatic ecosystems using geographic information systems (GIS) and remote sensing technology. He studies land-use, land cover change and their links to the carbon cycle with a focus on climate change at a regional and global scale. With his scientific findings he strives to support environmental policy decisions at the global scale, and is involved in supporting the UNFCCC negotiations on 'Reducing Emissions from Deforestation and Degradation' (REDD). Dr. Kellndorfer has been principal and co-investigator on numerous projects involving imaging radar technology. His current research activities include a NASA-funded project to generate the first high-resolution above-ground biomass and carbon dataset of the United States based on the integration of space shuttle radar and optical satellite imagery, as well as research on forest monitoring using the new class of space-borne imaging radar satellites like ALOS//PALSAR, EnviSat, Radarsat, and TerraSAR-X. Before joining the Woods Hole Research Center, Dr. Kellndorfer was a research scientist with the Radiation Laboratory in the Department of Electrical Engineering and Computer Science at the University of Michigan. Dr. Kellndorfer holds a diploma degree in physical geography, computer science, and remote sensing, and a doctorate in geosciences from the Ludwig-Maximilians-University in Munich, Germany. Dr. Kellndorfer is a senior member of the IEEE Geoscience and Remote Sensing Society.

Kyle C. McDonald is a Research Scientist in the Water and Carbon Cycles Group of JPL's Science Division. He received the Bachelor of Electrical Engineering degree (co-operative plan with highest honors) from the Georgia Institute of Technology, Atlanta, Georgia in 1983, the M.S. degree in numerical science from Johns Hopkins University, Baltimore, Maryland, in 1985, and the M.S. and Ph.D. degrees in electrical engineering from the University of Michigan, Ann Arbor, Michigan, in 1986 and 1991, respectively.

Dr. McDonald has been employed in the Science Division, Jet Propulsion Laboratory, California Institute of Technology, Pasadena, since 1991, and is currently a Research Scientist in the Water and Carbon Cycles Group. He specializes in electromagnetic scattering and propagation, with emphasis on microwave remote sensing of terrestrial ecosystems. His research interests have primarily involved the application of microwave remote sensing techniques for monitoring seasonal dynamics in boreal ecosystems, as related to ecological and hydrological processes and the global carbon and water cycles. Recent activities have included development of radar instrumentation for measuring sea ice thickness from airborne platforms. Dr. McDonald has been a Principal and co-investigator on numerous NASA Earth Science investigations. He is a member of NASA's North American Carbon Program (NACP) science team, NSF's Pan-Arctic Community-wide Hydrological Analysis and Monitoring Program (Arctic-CHAMP) Science Steering Committee, and the ALOS PALSAR Kyoto and Carbon Initiative science panel.

CHAPTER 14

Juha Hyyppä received his Master of Science, the Licentiate in Technology, and the Doctor of Technology degrees from the Helsinki University of Technology (HUT), Faculty of E.Eng., all with honors, in 1987, 1990, and 1994, respectively. He has been Professor and Head of the Department at the Finnish Geodetic Institute since 2000. He has docentship in space technology especially in remote sensing (HUT, E.E., 1997–), in laser scanning (HUT, Surveying, 2004–), and in remote sensing of forests (Helsinki University, 2005–).

He has been Earth Observation Programme Manager at National Funding Agency Tekes, responsible for the coordination of national and international (ESA and EU) remote sensing activities of Finland, Finnish adviser to ESA Earth Observation Programme Board, and to ESA Potential Participant Meetings (1994–1995), coordinator of the Design Phase of the National Remote Sensing Programme (1995), President of EuroSDR Com II (information extraction) 2004–2010, co-chair to ISPRS WG III/3 2004–2008, Vice-President of ISPRS Com VII 2008–2012, and Principal Investigator in ESA/NASA Announcement of Opportunity studies and coordinator for more than 10 international research projects. His references are represented by over 200 scientific/technical papers (more than 100 refereed papers). His personal hobby is the development of retrieval methods for laser-assisted individual tree based forest inventory together with Finnish industry.

Wolfgang Wagner received the Dipl.-Ing. degree in physics and the Dr.techn. degree in remote sensing, both with excellence, from the Vienna University of Technology (TU Wien), Austria, in 1995 and 1999 respectively. He received fellowships to carry out research at the University of Bern, Atmospheric Environment Service Canada, NASA Goddard Space Flight Center, European Space Agency, and the Joint Research Centre of the European Commission. From 1999 to 2001 he was with the German Aerospace Agency. In 2001 he was appointed Professor for Remote Sensing at the Institute of Photogrammetry and Remote Sensing of TU Wien. Since 2006 he has been the head of the institute. In the period 2008–2012 he is the president of ISPRS Commission VII (Thematic Processing, Modeling and Analysis of Remotely Sensed Data). His main research interests lies in geophysical parameter retrieval techniques from remote sensing data and application development. He focuses on active remote sensing techniques, in particular scatterometry, SAR and airborne laser scanning. He is a member of the Science Advisory Groups for SMOS and ASCAT and committee Chair of the EGU Hydrologic Sciences Sub-Division on Remote Sensing and Data Assimilation. Since December 2003 he has been the coordinator of the Christian Doppler Laboratory for 'Spatial Data from Laser Scanning and Remote Sensing'.

Markus Hollaus, born in 1973, finished his studies of land and water management and engineering at the University of Natural Resources and Applied Life Sciences (BOKU), Vienna, in March 2000. During his studies he received a fellowship to study at the Norwegian University of Science and Technology (NTNU) in Trondheim, Norway. From 2001 to 2003 he was a research scientist at the Institute of Surveying, Remote Sensing and Land Information at the BOKU. He was involved in several remote sensing and GIS projects with the focus on land use/cover classification and change. From 2004 to 2008 he was research scientist at the Institute of Photogrammetry and Remote Sensing (TU – Vienna) and also worked for the Christian Doppler Laboratory on 'Spatial Data from Laser Scanning and Remote Sensing'. He received the Dr.techn. (Ph.D.) degree in November 2006 with the thesis 'Large scale applications of airborne laser scanning for a complex mountainous environment'. Since 2009 he is university assistant at the Institute of Photogrammetry and Remote Sensing at the TU Vienna. The focus of his work is the derivation and modeling of vegetation parameters from airborne laser scanner data and aerial photographs and the classification of 3D point clouds using full-waveform airborne laser scanner data.

Hannu Hyyppä received his Master of Science, the Licentiate in Technology, and the Doctor of Technology degrees from the Helsinki University of Technology (HUT), Faculty of Civil Engineering, in 1986, 1989, and 2000, respectively. He has a docentship at HUT. Currently, he is post-doctoral fellow of the Academy of Finland in the Department of Surveying, at the Institute of Photogrammetry and Remote Sensing, Helsinki University of Technology. Previous employment include Research Fellow, Research Scientist and Assistant Coordinator, part-time R&D director of DI_Ware Oy, part-time president of UbiMap Oy, Project Manager, Development and Planning Engineer at Consulting Company Plancenter Ltd, Assistant, Senior Assistant, and Junior Fellow of the Academy of Finland and Research Scientist at the Laboratory of Road and Railway Engineering of the Department of Civil Engineering and Surveying at the HUT. He has 18 years of work experience in the research of civil and environmental engineering and geoinformatics. His references are represented by over 100 publications in the fields of civil and environmental engineering and geoinformatics, including more than 20 scientific refereed publications. His interests include the use of laser scanning and geoinformatics in new applications in built environment.

CHAPTER 15

Gabriela Schaepman-Strub obtained her Ph.D. degree in natural sciences from the University of Zurich, Switzerland, in 2004. In 2001, she was a guest researcher at the Department of Geography, Boston University. She obtained a post-doctoral fellowship for prospective researchers from the Swiss National

Science Foundation in 2005 and was an external post-doctoral fellow of the European Space Agency (2005–2007) at Wageningen University, the Netherlands. She is currently affiliated with the Nature Conservation and Plant Ecology Group and the Centre for Geo-information at Wageningen University. Her experience include performing and analyzing field spectrometer and goniometer measurements of vegetation canopies, reflectance product terminology, albedo analysis of tundra areas in Northern high latitudes, and plant functional type related analysis in highly dynamic (e.g., floodplain) and vulnerable (e.g., peatland) ecosystems. Her main interests lie in linking advanced vegetation products with dynamic vegetation models, and investigating remote sensing based land surface albedo products for climate modeling applications.

Michael E. Schaepman (see Chapter 12).

John Martonchik obtained the Ph.D. degree in astronomy from the University of Texas at Austin, in 1974. He joined NASA's Jet Propulsion Laboratory in 1972 and is currently in the Multi-angle Imaging element of the Earth and Space Sciences Division with the title of Research Scientist. His experience include analyzing telescopic and spacecraft observations of planetary atmospheres, laboratory and theoretical studies of the optical properties of gaseous, liquid, and solid materials, and development and implementation of 1- and 3-dimensional radiative transfer and line-by-line spectroscopy algorithms for studies of planetary atmospheres and Earth tropospheric remote sensing. He has been involved in several NASA Land Processes programs including Remote Sensing Science, FIFE, and BOREAS and is presently the Aerosol/Surface product algorithm scientist for the EOS MISR experiment.

Thomas Painter is Assistant Professor of Geography and Director of the Snow Optics Laboratory at the University of Utah, Salt Lake City. He is also Affiliate Research Scientist with the National Snow and Ice Data Center and Western Water Assessment of the University of Colorado, Boulder. His research focuses on radiative, hydrologic, and climatic forcings of dust and soot in snow and ice, alpine surface radiation, multispectral and hyperspectral remote sensing of snow physical properties, snowmelt hydrology, snow radiative properties, integration of remote sensing and distributed snow models, dust source mapping, and robotic goniometry. He is currently a member of the GOES-R cryosphere team, developing the fractional snow cover algorithm for the next generation geostationary satellite. His research on radiative and climate effects of dust in snow has been the subject of stories on National Public Radio, Reuters, The Weather Channel, and myriad articles in the domestic and international media. He is a member of the AGU Cryospheric Executive Committee and the AGU Hydrology Remote Sensing Technical Committee. His memberships in professional organizations include the American Geophysical Union, the European Geophysical Union, International Glaciological Society, and the Western Snow Conference.

Stefan Dangel obtained his Ph.D. degree in physics from the University of Zurich in 1997, specializing in quantum optics and nonlinear dynamics of pattern formation. His research interests include nonlinear wave propagation in low frequency seismology with applications for the oil and gas industry as well as spectro-directional effects, BRDF retrieval for field and laboratory goniometer measurements and goniometer measurement intercomparison in the field of remote sensing. He has contributed to ESA's SPECTRA mission as principal investigator for the development of a SPECTRA, end-to-end simulator. He also obtained a Master's degree in music. His current focus is on teaching mathematics, physics and bassoon.

CHAPTER 16

Freek van der Meer has an M.Sc. in structural geology and tectonics of the Free University of Amsterdam (1989) and a Ph.D. in remote sensing from Wageningen Agricultural University (1995) both in the Netherlands. He started his career at Delft Geotechnics (now Geodelft) working on geophysical processing of ground penetrating radar data. In 1989 he was appointed lecturer in geology at the International Institute for Aerospace Surveys and Earth Sciences (ITC in Enschede, the Netherlands) where he has worked to date in various positions (presently Professor and Chairman of the Earth Science Department). His research is directed toward the use of hyperspectral remote sensing for geological applications. In 1999, Dr. van der Meer was appointed full professor at the Delft University of Technology. In 2004, Dr. van der Meer was appointed adjunct professor at the Asian Institute of Technology in Bangkok (Thailand). In 2005 he was appointed Professor in Geological Remote Sensing at the University Utrecht. Professor van der Meer published over 100 papers in international journals, authored more than 150 conference papers and reports, has supervised over 50 M.Sc. projects and graduated eight Ph.D. candidates. He is the past chairman of the Netherlands Society for Earth Observation and Geoinformatics, chairman of the special interest group geological remote sensing of EARSeL, member of the Royal Netherlands Academy of Sciences, Associate Editor for *Terra Nova*, editor for the *International Journal of Applied Geoinformation Science and Earth Observation*, editor for the *Netherlands Journal of Geosciences*, and editor of the *Remote Sensing and Digital Image Processing Series* of Springer.

Harald van der Werff received his M.Sc. degree in geology from Utrecht University. Thereafter he worked as a researcher at the German Space Organization DLR in Oberpfaffenhofen in the spectroscopy group led by Andreas Mueller. In 2001 he joined ITC as a Ph.D. candidate working on the development of spectral-spatial contextual image analysis techniques. He received his Ph.D. in 2006 from the University of Utrecht on a thesis entitled 'Knowledge based remote sensing of complex objects'. To date Dr. van der Werff works as an Assistant Professor at ITC. His research interests are on (geological) hyperspectral remote sensing and on the integration of spectral and spatial information of remotely sensed images. Current research is on airborne detection of hydrocarbon spills from pipelines and geological interpretation of hyperspectral data (OMEGA, CRISM) from Mars by segmentation and landform analysis.

Steven M. de Jong is Professor in Physical Geography with emphasis on land degradation and remote sensing at the Faculty of Geosciences of Utrecht University since 2001. From 1998 to 2001 he was head of the Centre for Geo-information and Remote Sensing of Wageningen University. De Jong is chairman of the research school Centre for Geo-ecological Research (ICG) and research director of Physical Geography, Utrecht. From 1995 to 1996 he worked as a visiting scientist at NASA's Jet Propulsion Laboratory in Pasadena and conducted research to applications of NASA's Airborne Visible Infrared Imaging Spectrometer (AVIRIS). In 1997, 1998, and 2001 de Jong was Principle Investigator of several experimental campaigns investigating the usefulness of imaging spectrometers (DAIS7915, HyMap) for environmental applications in France and Spain. From 1998, to 2001 he was leader of a project investigating the use of SPOT-XS and IKONOS imagery for urban mapping in Burkina Faso. In 1994 he completed his Ph.D. thesis 'Soil Erosion Modelling using Hyperspectral Images in Mediterranean Areas'. De Jong is a member of the editoral board of *Remote Sensing and Digital Image Processing book series* (Kluwer) and of the *International Journal of Applied Earth Observation and Geo-information* (Elsevier).

CHAPTER 17

Chris J. Johannsen is a Professor Emeritus of Agronomy and Director Emeritus of the Laboratory for Applications of Remote Sensing (LARS) at Purdue University. His research has related to remote sensing and GIS applications for precision farming, soil pattern influences on reflectance, spatial-spectral-temporal resolution impacts and land degradation. He is co-editor of a book titled *Remote Sensing for Resource Management*, contributor to 16 book chapters and author or co-author of over 260 papers and articles. He served as International President of the Soil and Water Conservation Society in 1982–1983. Dr. Johannsen was responsible for the collection of ground reference information at LARS (1966–1972), continued research involving uses of reference information at the University of Missouri – Columbia (1972–1984) and resumed research, education, and outreach responsibilities for LARS as Director (1985–2003). He has received much recognition for his work including Fellow of five professional societies. Recently, he received the prestigious Hugh Hammond Bennett Award from the SWCS for his work on spatial technologies relating to studying land degradation.

Craig S. T. Daughtry is a Research Agronomist in the USDA-ARS Hydrology and Remote Sensing Laboratory in Beltsville, Maryland, USA. His research has focused on measuring and modeling the spectral reflectance of crops and soils. Daughtry joined the Laboratory of Applications of Remote Sensing (LARS) at Purdue University in 1976 and made significant advancements in integrating remotely sensed data into crop growth and yield models. After joining ARS in 1987, he has developed innovative techniques for measuring optical properties of leaves, increasing sampling efficiency, and managing the spatial variability of crops and soils. He also pioneered the use of fluorescence and shortwave infrared technologies to estimate crop residue cover for quantitatively assessing conservation tillage practices and tracking carbon sequestration. He is author or co-author of over 180 papers and articles. He has served on various committees of American

Society of Agronomy and editorial boards of *Photogrammetric Engineering and Remote Sensing and Agronomy Journal.*

CHAPTER 18

James W. Merchant is Professor in the School of Natural Resources, University of Nebraska-Lincoln (UNL) and is Director of UNL's Center for Advanced Land Management Information Technologies (CALMIT). Dr. Merchant received a B.A. in geography from Towson University, Baltimore, Maryland, and both the M.A. and Ph.D. in geography from the University of Kansas. His research has focused upon (1) development of strategies for large-area land cover characterization using digital multispectral satellite data, (2) spatial and contextual analysis of digital images, and (3) applications of geographic information systems in management of natural resources. Dr. Merchant was recipient of the 1999 Outstanding Contributions Award presented by the Nebraska GIS/LIS Association and the 1998 Outstanding Achievements Award conferred by the Remote Sensing Specialty Group of the Association of American Geographers. In 1997 he was honored with the John Wesley Powell Award that recognizes significant achievements in contributing to the research of the US Geological Survey. From 2000–2007 Dr. Merchant served as Editor of *Photogrammetric Engineering and Remote Sensing*, the journal of the American Society for Photogrammetry and Remote Sensing (ASPRS).

Sunil Narumalani is a Professor in the School of Natural Resources, and Associate Director of the Center for Advanced Land Management Information Technologies (CALMIT), University of Nebraska, Lincoln (UNL). He received his Ph.D. in geography from the University of South Carolina in 1993. Dr. Narumalani teaches courses in remote sensing (digital image analysis), introductory and advanced geographic information systems. His research focuses on the use of remote sensing for the extraction of biophysical information from satellite data and aircraft multispectral scanner systems, integration of geospatial data sets for ecological and natural resources mapping and monitoring, and the development of new image analyses techniques. Some of Dr. Narumalani's recent research has been on using remote sensing and GIS for the assessment of coral reefs and seagrasses off the coast of Florida and in the Caribbean. Over the past several years he has also been involved with projects pertaining to homeland security and military applications of geospatial technologies including the development of workshops for military intelligence units, integration of geospatial technologies for the National Guard, and initiating operational geographic databases for the Nebraska Emergency Management Agency (NEMA). Dr. Narumalani is also the Geography Program Coordinator at UNL.

CHAPTER 19

John R. Jensen is a Carolina Distinguished Professor in the Department of Geography at the University of South Carolina (USC). He majored in physical geography, cartography, and remote sensing at California State University, Fullerton, 1971 (B.A.); Breghan Young University, 1972 (M.A.); and UCLA, 1976 (Ph.D.). While at UCLA, he was trained in photogrammetry at Aero Service, Inc. In 1977, he accepted a professorship at the University of Georgia. In 1981, he went to USC and helped in developing the Ph.D. in GIScience. Dr. Jensen has mentored 60 M.S. and 28 Ph.Ds. His research focuses on: (a) remote sensing of wetland resources and water quality, (b) development of algorithms to classify land cover and detect change, and (c) the development of remote sensing-assisted decision support systems. Dr. Jensen was President of ASPRS in 1996. He has published >120 remote sensing articles. He was a co-author of ASPRS' *Manual of Remote Sensing* (1st and 2nd editions) and *Manual of Photographic Interpretation* (1997). He co-authored *Geographic Information for Sustainable Development in Africa* (2002) published by the National Academy Press. His textbooks *Remote Sensing of the Environment* (2007) and *Introductory Digital Image Processing* (2005) are used throughout the world. He received the ASPRS *SAIC John E. Estes Teaching Award* in 2004.

Jungho Im received his B.S. in 1998 in oceanography from Seoul National University, an M.C.P. in 2000 in environmental management from Seoul National University, and his Ph.D. in 2006 in geography with Dr. Jensen at the University of South Carolina. From 2006 to 2007, he worked as a post-doctoral research scientist in the Center for GIS and Remote Sensing, Department of Geography, University of South Carolina. In 2007, he became an Assistant Professor in the Environmental Resources and Forest

Engineering, State University of New York College of Environmental Science and Forestry, Syracuse, New York, USA. He is a member of the Association of American Geographers (AAG) and American Society for Photogrammetry and Remote Sensing (ASPRS). His research interests include Geographic Information Systems (GIS), GIS-based modeling, digital image processing, and environmental remote sensing.

Perry Hardin is currently an Associate Professor of Geography at Brigham Young University. He received his Ph.D. in geography from the University of Utah in 1989 where his dissertation focused on statistical classification of Landsat imagery. He has authored several journal papers related to nonparametric classification methods, confusion matrix analysis, and the use of neural networks in remote sensing. His current research interest is the use of neural networks to estimate biophysical and socioeconomic parameters (e.g., leaf area index, population data) in urban areas where calibration ground data is unavailable. Dr. Hardin served for two years on the editorial board of *Photogrammetric Engineering and Remote Sensing* and as chair of the Publications Committee for the American Society of Photogrammetry and Remote Sensing.

Ryan R. Jensen is an Associate Professor in the Department of Geography at Brigham Young University. Before this, he was an Assistant and then Associate Professor in the Department of Geography, Geology, and Anthropology where he served as the Director of the Center for Remote Sensing and Geographic Information Systems and as the Associate Director for Forest Research in the Center for State Park Research.

Dr. Jensen received his B.S. (cartography and geographic information systems) and M.S. (geography) from Brigham Young University. He received his Ph.D. from the University of Florida in geography with a minor in botany and a concentration in interdisciplinary geographic information systems. His research interests include using remote sensing and GIS to study biogeography and landscape patterns. He currently has active research programs in urban forestry, fire ecology in the southeastern (United States) coastal plain, and hyperspectral remote sensing.

CHAPTER 20

Shunlin Liang received his Ph.D. degree from Boston University. He was a post-doctoral research associate in Boston University from 1992 to 1993, and Validation Scientist of the NOAA/NASA Pathfinder AVHRR Land Project from 1993 to 1994. He joined the University of Maryland in 1993 and currently is a professor.

His present research interests focus on radiative transfer modeling, inversion of environmental information from satellite observations, spatio-temporal analysis of remotely sensed data, integration of numerical models with different data from various sources (i.e., data assimilation), and remote sensing applications to agriculture, weather and climate, and carbon and water cycles.

He is a principal investigator of numerous grants and contracts from NASA, NOAA, and other funding agencies. He is an Associate Editor of the *IEEE Transactions on Geoscience and Remote Sensing*, a member of several satellite science teams (e.g., MODIS, MISR, ASTER, EO1) of NASA and other space agencies, and co-chairman of the International Society for Photogrammetry and Remote Sensing Commission VII/I on Fundamental Physics and Modeling. He is an author of about 100 peer-reviewed journal papers and the book entitled *Quantitative Remote Sensing of Land Surfaces* (2004).

CHAPTER 21

Stephen V. Stehman has been a Biometrician in the Department of Forest and Natural Resources Management at the State University of New York College of Environmental Science and Forestry since 1989. He received his Ph.D. in biometry from Cornell University, an M.S. in statistics from Oregon State University, and a B.S. in biology from Penn. State University. His research activity has focused on the theory and practical application of rigorous sampling strategies for assessing map accuracy.

Giles M. Foody (see editors' biographical descriptions).

CHAPTER 22

Yongxin Deng is an Assistant Professor in the Department of Geography, Western Illinois University, teaching GIS and physical geography courses. He earned his Ph.D. in the Department of Geography, University of Southern California in 2005, focusing on geographical information science. His earlier experience in China includes a Bachelor's degree in physical geography in 1986 from Xinjiang University, a Master's degree in soil science in 1991 from the Chinese Academy of Sciences, and six ensuing years' (1991–1997) research at the Chinese Academy of Sciences, mostly conducting environmental impact assessment and natural resource evaluations in arid northwestern China. His current research interests include quantitative and classification methods in GIS, digital terrain analysis, mountain landscapes, urban landscape modeling, fuzzy logic, and spatial scale.

CHAPTER 23

Xianfeng Chen was born in Kurla, China, in 1963. He studied physical geography in the Department of Geography, Xinjiang University, China, from 1982 to 1986, and graduated with a B.S. degree in 1986. For the next three years he studied geomorphology at the graduate school of the Chinese Academy of Sciences, receiving an M.S. degree in 1989. He was recruited as a research scientist by Xinjiang Institute of Geography, Chinese Academy of Sciences, in 1989. Beginning in the summer of 1997, he spent a year at West Virginia University as a visiting scholar, working with Dr. Tim Warner. After his return to China, he was involved in several research projects in remote sensing and GIS. In 2000, he began the Ph.D. program in geology at West Virginia University. His dissertation, 'Integrating Hyperspectral and Thermal Remote Sensing for Geological Mapping,' was supervised by Dr. Tim Warner. After graduation in 2005, he joined the faculty of Slippery Rock University of Pennsylvania. His research interests include geological applications of hyperspectral and thermal remote sensing, land use and land cover studies, and environmental modeling with GIS and remote sensing.

David J. Campagna is a Remote Sensing Geologist with over 20 years of industry and academic experience. He is currently the Chief Science Officer of SkyTruth, Inc., a non-profit remote sensing group and serves as Adjunct Faculty in the Department of Geology and Geography at West Virginia University. He also consults for worldwide investigations of unconventional reservoirs, naturally fractured reservoirs, and structurally complex terrains. Dr. Campagna has previously worked with Advanced Resources International, Unocal Exploration, and Petro-Hunt in both domestic and international operations. He holds a Ph.D. in structural geology from Purdue University.

CHAPTER 24

James B. Campbell, PhD, has devoted his career to applications of aerial survey and remote sensing to studies of land use, geomorphology, and soils. Since 1997 he has served as co-director of Virginia Tech's Center for Environmental Analysis of Remote Sensing (CEARS). He is author of numerous technical articles and several books; his recent research addresses human impact upon coastal geomorphology and the history of aerial survey. Employment: Professor, Department of Geography, Virginia Tech, Blacksburg, VA.

CHAPTER 25

Michael A. Wulder received his B.Sc. (1995) degree from the University of Calgary and his M.E.S. (1996) and Ph.D. (1998) degrees from the Faculty of Environmental Studies at the University of Waterloo. On graduation, he joined the Canadian Forest Service of Natural Resources Canada at the Pacific Forestry Centre, in Victoria, British Columbia, as a Research Scientist. Dr. Wulder's research interests focus on the application of and integration of remote sensing and GIS to address issues of forest structure and function. He is an Adjunct Professor in the Department of Geography at the University of Victoria and the Department of Forest Resources Management of the University of British Columbia. Dr. Wulder is also a member of the GOFC-GOLD Land Cover Implementation Team and the USGS/NASA Landsat Science Team.

Joanne C. White received her B.Sc. (1994) and M.Sc. (1998) degrees in geography from the University of Victoria. She has worked in the fields of remote sensing and GIS, in a forestry context, for over 13 years. Joanne has been employed by federal, provincial, and private forest agencies, and has experience in operational, strategic, and research-oriented environments. She joined the Canadian Forest Service of Natural Resources Canada at the Pacific Forestry Centre, in Victoria, British Columbia, in 2003 as a Spatial Analyst.

Nicholas C. Coops received his B.App.Sc. (1991) and Ph.D. (1996) degrees from the Royal Melbourne Institute of Technology in Melbourne, Australia. He then worked as a research scientist at the CSIRO Australia Research Institution in the Forestry and Wildlife Divisions, in Canberra and Melbourne, from 1994 to 2004. In 2004, Dr. Coops accepted a faculty position at the University of British Columbia in Vancouver, British Columbia. He is the Canadian Research Chair in remote sensing and the editor of the *Canadian Journal of Remote Sensing*. His research interests focus on the theoretical development and application of remote sensing technologies to vegetation studies.

Stephanie Ortlepp received her B.Sc.F. (1999) and her M.Sc. (2003) degrees from the Faculty of Forestry at the University of British Columbia. After graduation, she worked for Pacific Geomatics Ltd. as an image processing specialist for over two years, before joining the Canadian Forest Service of Natural Resources Canada at the Pacific Forestry Centre, in Victoria, British Columbia in 2007 as a Monitoring Analyst.

CHAPTER 26

M. Duane Nellis (see editors' biographical descriptions).

Kevin P. Price is a Professor of Geography and served as the Associate Director of the Kansas Applied Remote Sensing (KARS) Program for 15 years at the University of Kansas (KU). While at KU, he has also retained a courtesy appointment in the Environmental Studies Program. He recently accepted a new position at Kansas State University (KSU), and will have a joint appointment in Agronomy and Geography from August of 2008. Given his agricultural academic training and interests, he looks forward to working at KSU, which is a well-respected agricultural school. He has been the recipient of national awards for service within the remote sensing community and served as the Associate Chair and Chair of the Remote Sensing Specialty Group of the Association of American Geographers. He also served on two National Research Council (NRC) committees that addressed issues of sustainable development in Africa and environmental satellite data utilization. His academic training includes a Ph.D. in geography (specialty in biogeography, remote sensing, and GIS) at the University of Utah, and B.S. and M.S. degrees in range science at Brigham Young University. Before coming to KU, he was an Assistant Professor in Geography at Utah State University where he also served as an Adjunct Professor in the Departments of Forestry, Range Science, Fisheries and Wildlife, and Landscape Architecture. Dr. Price's research has focused on the use of tools in the GISciences to study 'natural' and anthropogenically driven forcings that influences ecosystem (including agro-ecosystem) dynamics. He has conducted research throughout most regions of the world, and is the author or co-author of over 230 publications and an investigator on 75 research grants and contracts.

Donald Rundquist is a Professor with the School of Natural Resources, Institute of Agriculture and Natural Resources, University of Nebraska-Lincoln (UNL). He also serves as Director of the Center for Advanced Land Management Information Technologies (CALMIT). Rundquist holds a Ph.D. in geography from UNL (1977). He has been conducting research in and teaching courses on the subject of remote sensing since the early 1970s. His research involves high spectral and spatial resolution remote sensing of both cropland-vegetation canopies and surface waters, field techniques in support of remote-sensing campaigns, and airborne imaging spectrometry.

CHAPTER 27

Samantha Lavender has over 15 years research experience, and is Reader in Geomatics at the University of Plymouth (UoP), England, and Managing Director of ARGANS Limited. Her research interests have focused primarily on the quantitative remote sensing of water bodies using their color signature. Often called ocean color, it is used to quantify concentrations and behavior of what is dissolved or suspended by modeling the optical properties. Research extends from the movement of sediments in the coastal zone to phytoplankton in the open ocean. Both topics link to wider issues, such as climate change, as remote sensing is an important monitoring tool. Through the UoP Geomatics research group this links into Geographical Information Systems and coastal zone management. Recent community activities have also included the NERC Centre for observation of Air–Sea Interactions and fluXes (CASIX) Centre of Excellence and ESA GlobColour project; demonstrating an Earth Observation based service. She is also a council member for the Remote Sensing and Photogrammetric Society and, at an international level, Chair of the International Society for Photogrammetry and Remote Sensing Working Group VIII.6 (Coastal Zones Management, Ocean Colour and Ocean State Forecasting) and a committee member on the International Ocean Colour Coordinating Group.

CHAPTER 28

Jeff Dozier, Professor of Environmental Science and Management at the University of California, Santa Barbara, teaches and does research in the fields of snow hydrology, Earth system science, remote sensing, and information systems. He has pioneered interdisciplinary studies in two areas: one involves the hydrology, hydrochemistry, and remote sensing of mountainous drainage basins; the other is in the integration of environmental science with computer science and technology. In addition, he has played a role in development of the educational and scientific infrastructure. He founded UCSB's Donald Bren School of Environmental Science and Management and served as its first Dean for six years. He was the Senior Project Scientist for NASA's Earth Observing System in its formative stages when the configuration for the system was established. Professor Dozier received his B.A. from California State University, Hayward in 1968 and his Ph.D. from the University of Michigan in 1973. He is a Fellow of the American Geophysical Union, the American Association for the Advancement of Science, and the UK's National Institute for Environmental eScience. He is also an Honorary Professor of the Chinese Academy of Sciences, a recipient of the NASA Public Service Medal, and one of two winners of the 2005 Pecora Award from the Department of Interior and NASA.

CHAPTER 29

Gregory P. Asner's is a Staff Scientist in the Department of Global Ecology at the Carnegie Institution for Science. He is also a professor in the Department of Environmental Earth Systems Science at Stanford University. His scientific research centers on how human activities alter the composition and functioning of ecosystems at regional scales. Asner combines field work, airborne and satellite mapping, and computer simulation modeling to understand the response of ecosystems to land use and climate change. His most recent work includes satellite monitoring of deforestation, selective logging and forest disturbance throughout the Amazon Basin, invasive species and biodiversity in tropical rainforests, and climate effects on tropical forest carbon dynamics. His remote sensing efforts focus on the use of new technologies for studies of ecosystem structure, chemistry and biodiversity in the context of conservation, management, and policy development. He directs the Carnegie Airborne Observatory, a new airborne laser and hyperspectral remote sensing platform designed for regional assessments of the carbon, water, and biodiversity services provided by ecosystems to society.

Scott V. Ollinger is an Associate Professor at the University of New Hampshire with appointments in the Institute for the Study of Earth Oceans and Space and the Department of Natural Resources. His specialties are in ecosystem ecology and biogeochemistry with emphasis on basic ecological processes and interactions with human-induced environmental stressors. His research interests include: carbon and nitrogen cycling in forests; factors affecting carbon assimilation and storage by ecosystems; ecological effects of climate change, rising CO_2 and air pollution; and regional patterns of climate and atmospheric pollutant deposition in the Northeastern US. His work typically involves equal measures of field studies, remote sensing, and ecosystem modeling. At present, he is a principal investigator with the North American Carbon Program, an Associate Editor for Biogeochemistry and he serves or has served on numerous science advisory boards and steering committees. He also teaches courses in forest ecology and biogeochemistry. Dr. Ollinger received his Ph.D. from the University of New Hampshire in 2000.

CHAPTER 30

Janet Nichol is a Physical Geographer, specializing in biogeography and remote sensing. She obtained her B.Sc. from Queen Mary College London University, M.A. from the Institute of Arctic and Alpine Research at the University of Colorado, and Ph.D. from the University of Aston in Birmingham, England. She has lectured and conducted research in the United Kingdom, Nigeria, Singapore, Ireland, and Hong Kong and is currently a Professor in the Department of Land Surveying and GeoInformatics at the Hong Kong Polytechnic University. Her main research interests are in the application of remote sensing techniques to environmental assessment and monitoring, including the urban heat island, air quality, urban environmental quality, ecological mapping and evaluation, and the assessment of climate change impacts. She has published widely on these topics and is a reviewer for journals specializing in remote sensing, planning, and environmental issues.

CHAPTER 31

Kelley A. Crews is at the University of Texas at Austin where she is an Associate Professor of Geography and the Environment and Director of the GIScience Center, and is a faculty research associate of the Population Research Center, Center for Space Research, Environmental Science Institute, and Lozano Long Institute of Latin American Studies. She is currently both Chair of the Remote Sensing Specialty Group and member of Honors B Committee of the AAG, and Deputy Director of the Awards Committee for the ASPRS. She serves on the advisory committee for NASA's only *socioeconomic* data and applications center housed at CIESIN of the Earth Institute at Columbia University. She is currently a member of the editorial boards of *Geocarto International*, *Geography Compass*, and *Southwestern Geographer*, and has served on review panels for NSF and SSRC. She recently co-edited a "Focus" section of *Professional Geographer* and also has a co-edited book with Kluwer Academic Publishers on expanding the socio-spatial frontiers of GIScience. Her publications span outlets focusing on remote sensing (*Photogrammetric Engineering and Remote Sensing*; *International Journal of Remote Sensing*), environment and policy (*Agriculture, Ecosystems, and Environment*; *Environmental Management*), and population–system interactions (*Population Research and Policy Review*; *Urban Ecosystems*). Her research focuses on the global tropics, with current projects in the Andes–Amazon Corridor and the Okavango Delta, Botswana, and funded by grants from NASA and NSF.

Stephen J. Walsh is at the University of North Carolina at Chapel Hill and is Professor of Geography, member of the Ecology Curriculum, and Research Fellow of the Carolina Population Center where he is faculty advisor to the Spatial Analysis Unit. He has served as Chair of the Geographic Information Systems and the Remote Sensing Specialty Groups of the AAG. He was awarded the Outstanding Contributions Award and Medal from the Remote Sensing Specialty Group of the AAG and National Research Honors for Distinguished Scholarship from the AAG, and was elected into the American Association for the Advancement of Science. He is currently a member of the editorial boards of *Annals of the Association of American Geographers*, *Plant Ecology*, and *Geocarto International*, and has served on review panels for the NSF and NIH. He has co-edited special remote sensing and GIS issues in the *Journal of Vegetation Science*, *Geomorphology*, *GeoForum*, and *Photogrammetric Engineering and Remote Sensing*. He has co-edited a series of three books for Kluwer Academic Publishers and has published over 170 articles and book chapters on the research and practice of digitally integrated spatial science. His current research is conducted in Thailand, the Ecuadorian Amazon, the Galapagos Islands, and the Mountains of the American West and is funded by grants from NASA, NIH, NSF, and the USGS.

CHAPTER 32

Richard Teeuw is Senior Lecturer in applied geomorphology and remote sensing in the School of Earth and Environmental Sciences at the University of Portsmouth, England. Richard has extensive experience of mapping natural resources and geohazards in Africa, working as a consultant for the overseas development agencies of Canada (CIDA), Germany (GTZ), and Japan (JICA), as well as various mineral exploration organizations. Current interests focus on using low-cost satellite imagery for mapping geohazards and vulnerability, leading to disaster risk reduction, with research projects in SE Spain, Turkey, Costa Rica, and the Caribbean. Currently Chair of the Geological Remote Sensing Group (http://www.grsg.org), Richard has also edited books on the remote sensing of geohazards and the uses of geoinformatics for fieldwork.

Paul Aplin is an Associate Professor in Geographical Information Science at the University of Nottingham, UK. He specializes in environmental remote sensing, with principal research interests in the development of innovative approaches for land cover classification, the influence of scale of observation on image analysis and the application of spatial approaches for ecological investigation. He is currently engaged as Chairman of the Remote Sensing and Photogrammetry Society and Book Series Editor of the International Society for Photogrammetry and Remote Sensing.

Nicholas McWilliam has been developing information management and mapping applications in humanitarian disaster response with the UK-bsaed NGO MapAction since 2003, through research and training projects and emergency-response missions. Most recently he worked for the UN Joint Logistics Centre in South Sudan. Before that he was a lecturer in GIS for geography and life sciences, and worked for British Antarctic Survey's mapping centre. His first GIS use was modeling large mammal populations in Tanzania, and he has remained involved in National Parks mapping in East Africa. He co-edited the Royal Geographical Society's fieldwork manual *GIS, GPS and Remote Sensing* and regularly runs GIS workshops with the RGS for student research projects.

Toby Wicks is the European Engagement Manager at EuroGeographics, the association representing Europe's National Mapping and Cadastral Agencies. He is motivated by the use of geographic information to better coordinate response to sudden-onset disasters. Previously Toby has held positions at Ordnance Survey Great Britain, the World Health Organisation, and Spot Image. Toby has a Ph.D. in remote sensing, is a trustee and council member of the UK's Remote Sensing and Photogrammetry Society, a Fellow of the Royal Geographical Society and has been a volunteer for the UK-based NGO MapAction since 2004.

Matthieu Kervyn is a young Belgian Geoscientist. He holds a Master's degree in Geography from the Université Catholique de Louvain. Since 2004, he developed a Ph.D. in volcanology within the Mercator and Ortelius Research Centre for Eruption Dynamics at Ghent University, under the supervision of Dr. Gerald G.J. Ernst, and Prof. P. Jacobs. His research interests focus on the use of low cost remote sensing, analogue and numerical modeling to gain insights on the relationships between eruption dynamics, volcano growth, and associated hazards. Matthieu has been involved in assessing hazards and developing low cost remote sensing monitoring at several African volcanoes, including Oldoinyo Lengai, Tanzania, and Mt. Cameroon.

Gerald G. J. Ernst is a Geohazard Researcher interested in understanding how eruptions and volcanoes work and in applications to poverty alleviation and to hazard assessment at volcanoes in the developing world, especially across Central Africa. Exploring new applications of remote sensing including low cost approaches and combining this with analogue modeling is a key focus of interest. After 12 years of training, researching, and lecturing in volcanology and geological fluid dynamics at the Department of Earth Sciences, University of Bristol, UK, Dr. Ernst has joined the Belgian NSF (Flanders) in 2003 and is now working toward establishing the Mercator and Ortelius Research Centre for Eruption Dynamics – a school for volcanology research and an analogue modeling laboratory at the University of Ghent, Belgium. Dr. Ernst has published approximately 30 peer-reviewed articles related to volcanology, analogue modeling, or volcano remote sensing and co-edited the first textbook on *Volcanoes and the Environment* (CUP, 2005). In recent years, he has been developing initiatives working with African colleagues to develop the capacity for volcano research and monitoring in sub-tropical developing countries. He has received four prizes in recognition of his efforts so far including the 2002 Golden Clover Prize from the *Fondation Belge de la Vocation*, a foundation patronized by HM Queen Fabiola of Belgium. He trained over 50 students through supervision of research projects. Two former students he helped train are now award-winning volcano scientists.

CHAPTER 33

Timothy A. Warner (see editors' biographical descriptions).

Abdullah Almutairi is an Assistant Professor of Geography at Emam Muhammad Bin Saud University, in Riyadh, Kingdom of Saudi Arabia (KSA). He has a Bachelor's degree in Geography from the Emam

Muhammad Bin Saud University. His received his graduate education in geography with a specialization in remote sensing at West Virginia University, where he earned a Master's degree in 2000 and a Ph.D. in 2004. He has served as Chair of the Geography Department at Emam Muhammad Bin Saud University. He is a member of the Social Geographical Society and GIS Club, and has served as a consultant to the KSA Ministry of High Education. His research specialties include the spatial properties of images, high resolution remote sensing, and the use of remote sensing and GIS in planning.

Jong Yeol Lee is Research Fellow of the Geospatial Science Research Centre at the Korea Research Institute for Human Settlements (KRIHS), an affiliated research institute of the Office of the Prime Minster of the Republic of Korea. He graduated with a B.S. in geography from the College of Education at Seoul National University. He has a Ph.D. from West Virginia University in geography with specialization in remote sensing. He has served as the director of the Geospatial Science Research Centre at KRIHS, and is an Adjunct Professor at the University of Seoul. He serves on the board of directors of the Korean Geography Society, and is also a member of the executive board of the Korean Society of Remote Sensing. His specialty is high spatial resolution image analysis, integration of remote sensing and GIS, and regional analysis in regional studies.

CHAPTER 34

Giles M. Foody, Timothy A. Warner, and **M. Duane Nellis** (see editors' biographical descriptions).

Introduction

Remote Sensing Scale and Data Selection Issues

Timothy A. Warner, M. Duane Nellis,
and Giles M. Foody

Keywords: remote sensing data, scale, spectral scale, spatial scale,
temporal scale, radiometric scale.

INTRODUCTION

Remote sensing can be termed a mature discipline, in the sense that the underlying physical principles are well understood, and applications are beginning to appear in operational contexts spanning a diverse array of applications. In addition, the supporting technology has evolved to the extent that image acquisition, field work, and digital analysis are today much more sophisticated than in the early days of analog imaging, computer mainframe-based processing, and qualitative analysis. However, with the wide range of remotely sensed data that is now available, the rapid and continued advances in the power and storage capacity of modern desktop computers, and the sophistication of the many software packages available, remote sensing is far from a static field. Indeed, the last decade has seen the development of commercial fine resolution remote sensing from space (Toutin, in this volume), the exponential growth of lidar (also known as airborne laser scanning) (Hyyppä et al., in this volume), and the increasing sophistication and automation of image processing, to name just a few examples. This rapid evolution of remote sensing technology suggests that there is a need for a periodic and relatively comprehensive review of the field of remote sensing. This book is an attempt to address that need.

In this introductory chapter we lay the groundwork for a theme that is common throughout many of the chapters in this book, namely, the trade-offs and issues that should be considered in selecting data for a specific problem. For example, in Chapter 25 Wulder et al. consider data selection within the context of vegetation characterization, and in Chapter 31, Crews and Walsh review data selection from the perspective of social scientists. This introductory chapter provides a broad perspective on this important topic.

Ironically, selecting data is today more challenging than in the past, a consequence of the wide range of data currently available. In the past, few remotely sensed data sets were available, and consequently the properties of the available data tended to determine the nature of the problems that could be addressed. Thus, an important part of early remote sensing research using the Earth Resources Technology Satellite (ERTS, later renamed Landsat) was simply to ask the question, 'What can we do with these new data?' Today, we have a vast array of data to select from in remote sensing, and so a new problem has emerged – how do we optimize the data characteristics that we use, so that the data will most effectively address a particular application or research problem? It should thus be clear that the definition of an optimal data set is entirely dependent on the aims of the project for which the data are intended.

Adding to the complexity of choosing data attributes are three related issues. Firstly, there are fundamental physical and engineering trade-offs that limit the nature and detail of the data that can be collected using an imaging system (Kerekes, in this volume; Figure 1.1). These constraints help explain the design choices made in satellite-borne sensors, and likewise need to be considered by those planning their own custom acquisitions of aerial imagery (Stow, in this volume).

A second issue that makes selecting the appropriate data for a project complex is that, just as too little data will likely reduce quality of the analysis, data with too much detail may also have a negative effect (Latty et al. 1985). It is intuitive that too much spatial detail can be burdensome for a computer-based analysis, and the same principle applies to

other components of imagine information, including the spectral, radiometric, and temporal scales of the data. For example, Hughes (1968) showed that an excessive number of spectral bands can lead to lower classification accuracy, an observation that is known as the *Hughes phenomenon* (Swain and Davis 1978).

The last issue, perhaps the most important of the three, is the need to match the scale of the analysis to the scale of the phenomena under investigation (Wiens 1989). Inferences drawn from an analysis at one spatial scale are not necessarily valid at another scale, an issue known in ecology as *cross-level ecological fallacy* (Robinson 1950, Alker 1969). In geography, the dependence of observed patterns on how data are aggregated is known as *the modifiable areal unit problem* (MAUP, Openshaw and

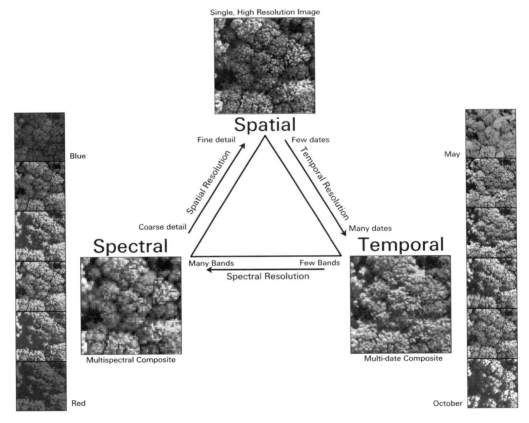

Figure 1.1   Given a limited bandwidth for image acquisition, storage, and communication, trade-offs have to be made regarding the spatial, spectral, and temporal scale of the imagery that can be acquired. Radiometric scale (not shown) is also important. (See the color plate section of this volume for a color version of this figure).
Source: Figure reproduced from T. Key, T. Warner, J. McGraw, and M. A. Fajvan, 2001. A comparison of multispectral and multitemporal imagery for tree species classification. *Remote Sensing of Environment* 75: 100–112.

Taylor 1979, Openshaw 1983, 1984). The MAUP has two components (Jelinsky and Wu 1996):

- The scale problem, which focuses on how results may vary as the size of the aggregation units (pixels, in the typical remote sensing analysis) varies.
- The zoning (or aggregation) problem, which focuses on how the results may vary as the shape, orientation and position of the units vary, even as the number of aggregation units is held constant.

In remote sensing, attention has usually focused on the MAUP scale problem, and less attention has been applied to the zoning problem (for an exception, see Jelinsky and Wu 1996), because most pixels are assumed to represent a similar, approximately square shape. However, NOAA Advanced Very High Resolution Radiometer (AVHRR) Global Area Coverage (GAC) data is produced by aggregating a linear-oriented subset of finer scale Local Area Coverage pixels (Justice and Tucker, in this volume), thus potentially opening the GAC data to zoning problems. Clearly, both scale and zoning MAUP problems are potentially present when ancillary vector-derived data are used in a remote sensing analysis (Merchant and Narumalani, in this volume).

Woodcock and Strahler (1987) provide a useful remote sensing conceptual framework that categorizes images based on the size of the pixels relative to objects in the scene. Thus an H-resolution image has pixels small enough to resolve objects or phenomena of interest in the scene. In contrast, in an L-resolution image, the pixels are too large to resolve the individual objects. However, most scenes have objects at a variety of scales, and therefore it may be more useful to refer to H- and L-resolution image elements, both of which are likely to be present in any one image (Ferro and Warner 2002).

Central to the ideas presented so far is the concept of scale (Quattrochi and Goodchild 1997, Walsh et al. 1997, 2003, Marceau and Hay 1999, Spiker and Warner 2007). Landscape ecology recognizes scale as having two attributes: grain and extent (Turner et al. 2001). Although there are numerous definitions of these terms, for our purposes we will define *grain* as the finest level of measurement, the degree of detail, or the sampling unit. An example of grain is the instantaneous field of view (IFOV) of the sensor, which in turn is related to the ground sampling distance or ground resolution element, depending on the context. (Although pixel size is not as precise a term, for simplicity we will use it to represent the concept of ground sampling distance in this chapter.) *Extent* can be defined as the range over

which measurements are made, for example, the area represented in an imaged scene. Grain and extent tend to be inversely related, simply because the total amount of data that can be collected is usually constrained.

Even though the examples given here draw on image spatial properties, the term scale is often also applied to the three other attributes of image data already referred to, namely the spectral, radiometric, and temporal properties. Although scale is a common thread in this chapter, it is important to note that it is not the only attribute that is important in selecting data to address a particular problem.

The remainder of this chapter is organized in seven major sections. Following this general introduction, we discuss factors that influence the optimal characteristics of each of the four major types of image properties: spatial, spectral, radiometric, and temporal. We then present some examples of the interactions and trade-offs between the individual types of major image properties, before considering some broader, more general issues. In the concluding sections, we look to the future to discuss challenges and opportunities on the horizon.

SELECTING IMAGES WITH OPTIMAL SPATIAL PROPERTIES

Scale and image spatial properties

The concept of scale is particularly useful for discussing image spatial properties (Cao and Lam 1997, Marceau and Hay 1999). For example, the section in this book on satellite-borne sensors is partly organized along the lines of pixel size. Thus, we have chapters on fine (Toutin, in this volume), moderate (Goward et al., in this volume), and coarse spatial resolution (Justice and Tucker, in this volume) sensors. However, the challenges that the authors of these chapters faced, both in arriving at these terms, and in using them consistently, suggests that meaning of scale varies greatly depending on the focus of the analysis, and perhaps also the historical context of the time. Thus, despite its name, the Advanced Very High Resolution Radiometer (AVHRR), with 1.1 km pixels, is grouped in this book with coarse resolution sensors. The Landsat Enhanced Thematic Mapper Plus (ETM+), which we treat as a moderate resolution sensor, has also been termed a fine spatial resolution sensor by some. Adding complexity is the fact that many satellite-borne sensors have bands of differing spatial resolution. For example ASTER acquires data in three bands with 15 m pixels, six bands with 30 m pixels, and five bands with 90 m pixels. It is apparent that spatial resolution of modern satellite sensors fall

Table 1.1 Image spatial resolution categories

Pixel size (m)	Spatial resolution	Example satellite-borne sensors
<1	very fine	WorldView
1–10	Fine	IKONOS
10–100	Moderate	ASTER, AWIFS, ETM+, MSS, SPOT
100–1000	Coarse	MODIS, MERIS
>1000	Very coarse	AVHRR, GOES, METEOSAT

along a continuum, and therefore attempts to label sensors by simple spatial resolution descriptors is inherently arbitrary. Nevertheless, to minimize confusion, we have attempted throughout this book to standardize as far as possible on the terms summarized in Table 1.1.

Although often used interchangeably, *spatial resolution* and *pixel size* are not strictly speaking equivalent. This is because pixel size refers to the sampling frequency, and not the ground resolution element or sampling area. Thus, for example, the Landsat MultiSpectral Scanner (MSS) oversampled data along the scan line, producing pixels that are smaller than the ground resolution element. In addition, spatial resolution is dependent on the spectral radiometric properties of both the object being resolved, and the background against which it is being resolved. Generally, a higher spectral radiometric contrast between an object and its background will result in a higher apparent spatial resolution. At the one extreme, an object with no contrast against the background is not resolvable, irrespective of its size. At the other extreme, it is potentially possible to detect the presence of a single, bright object that is much smaller than a pixel, as long as the object is surrounded by a much darker background. However, for this latter example, it is not normally possible to predict where in the pixel that object occurs, so in that sense, the resolution is ultimately limited by the pixel size. Nevertheless, because of mixed pixels, and the low contrast of most Earth scenes, objects generally need to be multiple times the size of a single pixel before they are large enough to be discerned as distinct spatial features.

A more precise way of specifying resolution is the modulation transfer function (MTF). This is a specification of how contrast in the scene is represented in ('transferred to') the image. To measure MTF, a test signal of multiple bars of defined contrast, and varying spatial frequency (width of the bars), is imaged, normally in a laboratory setting. The contrast in the resulting image, at each of the various spatial frequencies, is then measured as a proportion of the original contrast. A similar measure is the point spread function (PSF), which characterizes how a point signal is blurred when it is measured by the sensor (Huang et al. 2002).

Blurring results from the effects of the atmosphere, the sensor optics and electronics, and image resampling. Because of blurring, the information in a pixel usually includes a component from neighboring pixels (Zhang et al. 2006). Huang et al. (2002) have shown how modeling of the PSF can be used to reduce this adjacency effect, and thus improve the overall fidelity of the image.

In real images, quantifying spatial resolution requires identification and exploitation of natural boundaries between features in the image. Tarnavsky et al. (2004) used the full-width-half-maximum (FWHM) of the line spread function (LSF), derived from the study of the edges of objects in the image, to compare the spatial fidelity of scanned aerial film, and digital aerial images.

Image spatial extent and pixel size are generally inversely related. Thus, spatial resolution generally limits the potential extent of the scene. For example, it is possible to collect a global set of near cloud-free Landsat 7 ETM+ imagery, with 30 m pixel size, on a seasonal basis (Goward et al., in this volume). However, MODIS with 250 m visible and near infrared (NIR) pixels, can provide weekly global composites of nearly cloud-free imagery (Justice et al. 2002). In contrast, despite almost a decade of data collection by multiple commercial companies, there is as yet no fine spatial resolution global data set.

Choosing an optimal spatial scale

What is the optimal spatial resolution for a particular project? As already mentioned, it is important to clarify the interpretation objective of a project, before this question can be addressed. If the aim is to map the location of discrete objects, or the overall spatial patterns in an image data set, then methods that estimate optimal resolution based on finding the pixel size with the maximum local variation have been shown to be very effective. For example, Woodcock and Strahler (1987) related the graph of local variation plotted against pixel size to the average size of objects in an image. Variograms, which characterize the variability between measurements as a function of distance between those measurements (Jupp et al. 1989), have a particularly rich theoretical underpinning (Matheron 1971, Journel and Huijbregts 1978, Jupp et al. 1988). Variograms have been used to identify optimal distances between field measurements and the optimal pixel size (Hyppänen 1996, Atkinson and Curran 1997). An alternative measure, lacunarity, which is based on fractal theory, is useful for identifying multiple scales in an image (Butson and King 2006).

If the aim is to map the size and spatial extent of individual objects or regions, then it is important to have a pixel size much smaller than the distance calculated for optimal sampling, as described above. However, if the resolution becomes too fine, unwanted spatial detail will likely be resolved in the image, and, at least using conventional image analysis techniques, classification accuracy may be lower (Latty et al. 1985). On this basis, the optimal resolution has been defined as the scale that minimizes variance within the classes to be mapped (Marceau et al. 1994). An important consequence of this definition is that the optimal scale is therefore likely to be class-dependent (Marceau et al. 1994).

Hengl (2006) provides a thorough overview of the issues associated with choosing an optimal scale. He recommends a scale that is a compromise between *the coarsest legible scale*, which respects the scale and properties of the dataset; and *the finest legible scale*, which preserves at least 95% of the object or scene variability (Hengl 2006). McCloy and Bøcher (2007) extend Woodcock and Strahler's (1987) local variance concept to show how a graph of average local variance (AVL) can help predict a scale that minimizes within class variance, and thus optimizes the accuracy of subsequent classifications.

Image geometric properties

Another issue that should be considered in selecting data is the quality of the georeferencing to a cartographic projection. High quality georeferencing is generally expensive. For an image acquired from a nadir-viewing sensor, a simple polynomial warp that does not include terrain correction may be sufficient, and if local map control at a sufficient scale is available, can be applied routinely. Topographically induced image distortion increases with increasing angle away from nadir, as does the distortion of the shape and size of the pixel. Thus, with sensors that have a pointing capability, the view angle is an important variable to consider in selecting data. However, the increasing sophistication and availability of automated photogrammetric software makes it potentially possible for non-specialists to generate high quality orthorectifications, although the procedure remains relatively complex.

The quality of the image geometric properties is particularly important for multi-temporal analysis. Even a 0.2 pixel misregistration can cause as much as 10% error in the estimate of the change in spectral values, depending on the heterogeneity of the scene (Townshend et al. 1992). The quality of georeferencing is also important for change detection derived from object-based classification. In object-based classification, pixels are first grouped into so-called image objects, which are then classified as a single unit (Jensen et al., in this volume). In a series of experiments on the effects of

misregistration on object-based change detection, Wang and Ellis (2005) found change detection error increased with increased positional error, increased landscape heterogeneity, and finer change detection resolution (the local region over which change is identified). The relationships between these variables were summarized using regression, and then used to calculate an optimal change detection resolution, based on a desired degree of accuracy (Wang and Ellis 2005).

SELECTING IMAGES WITH OPTIMAL SPECTRAL PROPERTIES

Scale and image spectral properties

When the concept of scale is applied to spectral properties, *spectral grain* can be used to refer to the wavelength interval, or width, of the spectral bands. Multispectral sensors, with a coarse spectral grain, have bands that span hundreds to thousands of nm. The *spectral extent* can be used to describe the spectral wavelength region encompassed by the bands (e.g., many optical sensors operate in the visible and near-infrared spectral region), and the total number of bands. The definition of hyperspectral data, which usually emphasizes the number, width and contiguity of the spectral bands (Schaepman, in this volume), thus encompasses the concepts of both spectral grain and extent.

The specific location and width of spectral bands can be very important for subsequent analysis. For example, Teillet et al. (1997) show that normalized difference vegetation index (NDVI) values are not necessarily comparable between satellites with different spectral properties, even if the data are atmospherically corrected and radiometrically calibrated. The width and location of the red band used in the NDVI calculation is particularly important, and should ideally be less than 50 nm wide (Teillet et al. 1997). Thus the spectral grain of Envisat Medium Resolution Imaging Spectrometer (MERIS) appears to be more appropriate for NDVI work than either the Landsat TM or SPOT HRV sensors (Teillet et al. 1997).

The choice between using multispectral and hyperspectral data has important ramifications for the range of information extraction routines that are appropriate for subsequent analysis. Multispectral analysis techniques tend to use data from within the scene to develop empirical models and classifications. Obtaining sufficient reliable within-scene training data can be a major challenge with multispectral analyses. In addition, the spectral separability of the classes of interest may be limited with multispectral data.

Hyperspectral analysis techniques often employ methods that are not premised on requiring in-scene knowledge. For example, hyperspectral methods may employ theoretical biophysical models, or draw on spectral libraries for classification (Chen and Campagna, in this volume). Spectral libraries consist of high quality spectra, usually acquired under laboratory conditions, which are assumed to represent material classes over wide areas. A number of extensive mineralogical spectral libraries are available in the public domain (for example, Clark et al. 2003); more recently an urban land cover library has been developed (Herold et al. 2003). The availability of spectral libraries for vegetation tends to be more limited, because of the phenological and environmental variation in vegetation properties limit the generalization that can be achieved. One of the difficulties in exploiting library spectra is that scaling from small laboratory samples and field spectrometer measurements to pixels, is complex (Baccini et al. 2007).

Choosing the optimal spectral bands

In the early days of digital image processing of remotely sensed data, limited computing power made it attractive to select only the most useful bands for classification. This constraint has largely fallen away with the steady improvement in computing power. Nevertheless band reduction is still often desirable, especially as advances in sensor technology enable data acquisition in more bands. The Hughes phenomenon (Hughes 1968, Warner and Nerry 2008), which has already been referred to above, is assumed to result from the increased number of parameters needed to characterize the distributions of training samples as the number of bands increases. The effect of the Hughes phenomenon is most likely classifier-dependent, and indeed, support vector machines are thought to be less susceptible to this problem (Melgani and Bruzzone 2004).

The simplest way of selecting bands is to use knowledge of the spectral properties of interest. For example, in a vegetation application one might select bands from the visible, NIR, and short wave infrared (SWIR) to sample spectral regions influenced by vegetation pigments, leaf structure, and moisture status, respectively (van Leeuwen, in this volume). In geological applications, one might use spectral libraries to identify the wavelengths associated with important diagnostic absorption features of the minerals and rocks of interest (Chen and Campagna, in this volume).

A variety of automated and statistical approaches have been proposed for selecting optimal subsets of image bands that carry the most information (Serpico and Moser 2007). One assumption common to many band selection methods is that highly correlated bands are redundant (Wiersma

and Landgrebe 1980, Miao et al. 2007). Using the statistical method of principle component analysis (PCA) (Jensen 2005), the axes of multidimensional data can be rotated so that an n-band original data set is transformed to n new orthogonal and uncorrelated bands. The new bands are normally ordered according to the proportion of the original variance each new band explains. This strategy generally works very well, with the first few principle components carrying most of the information, and the remaining, low variance components generally dominated by noise. PCA is one of the most widely-used general image analysis techniques, having applications that go well beyond data compression and band selection. The minimum noise fraction (MNF) transformation (Green et al. 1988), typically applied to hyperspectral data, is a cascaded sequence of PCA transformations in which the noise is isolated and removed.

Despite the robustness of PCA, it is important to be aware that this method uses correlated variance as a surrogate measure for information. In situations where the signal of interest is not correlated across bands, but is instead isolated in a narrow spectral absorption feature, PCA will not be so useful. In addition, although highly correlated bands are likely somewhat redundant, they may nevertheless contain non-redundant information that can be very useful for separating subtle spectral differences (Warner and Shank 1997).

An alternative to this focus on covariance is data transformations and band selections that specifically enhance the spatial patterns in the resulting images. The spatial analog to PCA is multivariate spatial correlation (MSC) (Wartenberg 1985), which can be used to transform and compress image data (Warner 1999). Comparisons of the autocorrelation of ratios of image bands have also been used to select individual bands, and combinations of bands (Warner and Shank 1997). This autocorrelation-based method of selecting bands has been found not only to increase classification accuracy, but also to result in classifications that have higher autocorrelation, and thus potentially more clearly defined spatial patterns (Warner et al. 1999).

Data fusion

Data fusion has been defined as:

> a formal framework in which [there] are expressed means and tools for the alliance of data originating from different sources. It aims at obtaining information of greater quality; the exact definition of 'greater quality' will depend upon the application. (Wald 1999: 1191)

Pohl and Van Genderen (1998) note that data fusion can take place at three different levels in the image processing chain of analysis:

1 At the *pixel level*, by combining raw image bands of different sources.
2 At the *feature level*, by segmenting the images to identify image objects, and combining the different images in the context of each image object.
3 At the *decision level*, where each image is first analyzed separately, and then the derived information is combined.

The attributes of the data that are combined through data fusion could potentially cover any individual or combinations of the four attributes of scale: spatial, spectral, radiometric, and temporal, as well as a combination of imagery with ancillary data (Pohl and Van Genderen 1998). In this section, which focuses on image spectral properties, the discussion will be limited to attempts to increase the information content of a data set by combining images of disparate wavelengths at the pixel level (Briem et al. 2002). Subsequently, in the section on interactions between the different scale components, pan-sharpening using multi-spatial resolution data fusion will also be discussed.

The underlying rational for multi-wavelength data fusion is that different wavelength regions may respond to different physical phenomena. Thus, for example, a combined analysis of optical and synthetic aperture radar imagery potentially can provide information about vegetation type, biomass, structure, and water content (Hill et al. 2005).

Similarly, combining hyperspectral VNIR and SWIR with multispectral thermal infrared (TIR) data may allow the incorporation of temperature or emittance variations in discrimination between land cover units. For mineral mapping, SWIR bands often provide an ability to discriminate clays, whereas multispectral thermal bands are valuable for separating silicate minerals (Chen and Campagna, in this volume; Chen et al. 2007a). However, the benefits of combining these disparate wavelength regions varies greatly with classification method used (Chen et al. 2007b), and for some classifiers, the accuracy may actually decline when disparate data are combined. This suggests that a suitable approach for mineral discrimination may sometimes be an expert system that adapts to the spectral pattern of each pixel to draw on different classifiers, using different wavelength intervals, to classify each pixel independently.

The fusion of VNIR and SWIR data with multispectral thermal data also holds promise for classification in the urban environment, especially for the discrimination of different roof and road materials. In a study of Strasbourg, France, it was found that various combinations of four to six broad bands from the visible, NIR and SWIR, together with six multispectral TIR bands, resulted in higher classification accuracy than with using 71 hyperspectral visible, NIR, and SWIR bands (Warner and Nerry 2008). Unfortunately, there are currently no planned medium or high spatial resolution thermal satellite-based sensors, and therefore opportunities to exploit data fusion with TIR may remain limited.

SELECTING IMAGES WITH OPTIMAL RADIOMETRIC PROPERTIES

Scale and image radiometric properties

Radiometric resolution is arguably as important as spatial, spectral, and temporal resolution, yet does not seem to receive as much attention as the other image attributes. When scale is applied to radiometric properties, *grain* refers to the fineness of the division between successive brightness levels the sensor measures. *Extent* refers to the range of brightness levels over which the sensor can differentiate changes in radiance. A sensor with a rather unusual radiometric extent is the Operational Linescan System (OLS), which is flown aboard the Defense Meteorological Satellite Program (DSPM). The OLS is particularly sensitive to a range of low light levels, which makes it possible to detect illumination at night from street lights (Henderson et al. 2003) and other sources of illumination, such as fires and flares.

The number of bits over which the signal is quantized can serve as an indicator of the radiometric grain. An eight-bit resolution (2^8, or 0–255 DN values) has been until recently a common choice, partly because this data range corresponds to the underlying structure of computer data storage. Nevertheless, it is important to consider the range of radiometric values actually filled (Malila 1985), as well as the noise in the data. Thus, radiometric grain is perhaps more usefully characterized as the minimum radiance change that can be detected reliably. This change can be measured in radiance units, or as the signal-to-noise ratio. The latter measure is normally defined as the mean signal divided by the standard deviation of the noise. Atkinson et al. (2007) have demonstrated the utility of using land-cover-specific variograms to estimate the signal-to-noise ratio based on the relative variance of both the signal and noise. This land-cover-specific measure of radiometric grain emphasizes the importance of the scene context in interpreting measures of noise.

Over time the radiometric range of data quantization from available sensors has increased notably. Tarnavsky et al. (2004) have shown that scanned color infrared aerial photographs have more noise than Airborne Data Acquisition and Registration (ADAR) 5500 multispectral images, which are acquired using digital cameras. The original Landsat MSS sensor recorded just six bits of data, although the data for the first three bands were scaled non-linearly to provide an effective seven-bit range (Goward et al., in this volume). In contrast, Landsat TM data is quantized over eight bits. Malila (1985) used an analysis of entropy to show the importance of this radiometric improvement in increasing the information content compared to the improvement in the number, width and location of the spectral bands. On the other hand, Narayanan et al. (2000) suggest that TM imagery can potentially be compressed to as few as only four bits per pixel, and still produce classifications that are similar in accuracy to the original eight-bit data.

The commercial high resolution sensors of IKONOS, Quickbird and OrbView are all quantized with 11-bit data (Toutin, in this volume). Nevertheless, purchasers of these data sets are offered degraded 8-bit versions of the data, perhaps reflecting legacy software or limited hardware and software available to some purchasers. Based on the personal experience of the authors, one of the advantages of the higher radiometric resolution of the commercial sensors appears to be the increased information content in dark areas of the images, especially shadows.

Radiometric normalization and calibration

Many image analysis procedures can be undertaken with images in DN format. However, some change detection techniques and most biophysical transformations (e.g. vegetation indices) require normalization or calibration to radiance units or equivalent reflectance (Teillet et al. 1997, Song et al. 2001). For example, conversion to reflectance is particularly important for hyperspectral data, especially if the imagery is to be classified using spectral libraries (Chen and Campagna, in this volume). In comparing radiance and reflectance measurements between sensors, and between field spectrometers and remote imaging devices, it is particularly important to define and consider the geometric arrangement of the illuminating energy and the observing sensor. Schaepman-Strub et al. (in this volume) provide a comprehensive review of the terminology and the relationships between different types of spectral measurements.

Conversion to reflectance requires information about the spectral sensitivity of the sensor, as well as both solar illumination and atmospheric transmission and scattering. The effect of topography on illumination may be calculated if a sufficiently detailed digital elevation model is available

(Warner and Chen 2001). However, the bidirectional reflectance distribution function (BRDF), or dependence of reflectance on the geometry on the illumination and observation (Schaepman-Strub et al., in this volume), varies between different materials, and thus if a single BRDF model is used to normalize topographic variations in an area of varying land cover properties, the calculated reflectances may have cover-dependent errors.

SELECTING IMAGES WITH OPTIMAL TEMPORAL PROPERTIES

Scale and image temporal properties

The application of the concept of scale to image temporal properties is somewhat more complex than in the spatial and spectral domains. Normally, an image is acquired in a single, very short period of time, which might be referred to as the temporal scale extent. If only one image is considered, the grain and extent are identical. On the other hand, the concept of temporal scale is very useful for discussing multitemporal image archives, as well as for characterizing change detection and time series analyses. The temporal *extent* of an archive is quite straightforward, and is the overall period of time covered. However, the temporal *grain* can potentially refer to two different attributes. In the case of a series of individual images, the grain might be the period *between* the image acquisition dates. However, for coarse resolution data, single bands are often generated on a pixel by pixel basis from multiple sequential images, using algorithms that minimize the effects of cloud. For such data, the final image represents a multitemporal composite, where each pixel has been individually selected from the images acquired during the compositing period (Holben 1986). Thus, at least for multitemporal composited data, grain could also refer to *the period of time over which the image data have been integrated*. For example, a compositing period of a week or a month is often used to generate some image data products (Justice et al. 2002).

Cloud-free multitemporal composites have been found to be particularly useful for characterizing the annual pattern of ecosystem response to annual weather patterns (Loveland et al. 1995). For example, the date of onset of greenness, total integrated greenness over time, and maximum greenness, have been used to classify different land cover classes. By extending such studies over multiple years, apparent changes in climate have been observed, including an earlier spring greenup at high latitudes (Myneni et al. 1997, Delbart et al. 2006). However, Schwartz et al. (2002) have cautioned that the integration of data over a week or longer periods can result in uncertainty and bias in the phenological trends identified.

For change detection studies, the temporal extent of the available image archive constrains the period over which change can be observed. Thus, the Landsat TM and ETM+ sensors provide a particularly important long term data set, with a temporal extent of over 25 years (Goward et al., in this volume). The temporal extent of change detection studies can be extended back to 1972 by using Landsat MSS imagery, and for some areas, to as early as 1960, by using declassified CORONA imagery, although the latter are mostly digitized black and white film. However, for change detection studies, images from different sensors should be used with caution, because it can be challenging to differentiate between real changes in the scene, and changes in the sensors.

The grain, or revisit period of the sensor, also constrains the potential differentiation of events within the period studied. However, the actual availability of cloud free imagery is usually some small fraction of what might be assumed based on only the sensor revisit time.

Image acquisition frequency

Finding recent imagery tends to be an important consideration for some applications. Procedures for satellite data collection vary greatly between the nadir viewing sensors, such as Landsat ETM+, and pointable satellites, a category which includes all fine resolution sensors, such as IKONOS and WorldView. For nadir-viewing sensors, the operators usually attempt to acquire and archive all images on a systematic basis, at least when the satellite is within sight of a receiving station. Landsat ETM+ is unique in that the operators have a policy of acquiring multiple global data sets on a regular basis (Goward et al., in this volume). For pointable satellites, image acquisition is prioritized based on requests from customers, who pay a premium for tasking the satellite. Thus, archive imagery is only available over limited areas, and new acquisitions may be delayed depending on the priorities of the operator. These same issues tend to apply to other sensors that have only a limited acquisition capability, such as ASTER and HYPERION.

Obtaining images of the appropriate season is also important. This is particularly true of vegetation studies, where the timing of phenological events such as leaf out and senescence may be as valuable as spectral information (Key et al. 2001).

Geostationary satellites, such as European EuMetSat's Meteosat Second Generation (MSG) satellites and the planned US National Polar-orbiting Operational Environmental Satellite

System (NPOESS) satellites, offer the greatest potential for high frequency of coverage. For example METOSAT-9 acquires full disk images of Earth every 15 minutes, and in rapid scanning mode, where only part of the Earth disk is imaged, images can be acquired even more frequently. The trade-off with geostationary sensors is the comparatively low spatial resolution, for example 1–3 km pixels at the sub-satellite point for the METEOSAT Spinning Enhanced Visible and Infrared Imager (SEVIRI) instrument. Nevertheless, this high temporal frequency of acquisition opens the possibility for completely new remote sensing applications associated with highly dynamic phenomena, such as modeling the growth and development of individual fires (Umamaheshwaran et al. 2007).

Airborne sensors (Stow et al., in this volume) can provide high spatial resolution as well as complete user-control of acquisition timing, including not just the date, but even time of day. In practice, however, mobilization and operational costs may limit the degree to which the user can achieve this flexibility.

Acquisitions for time-critical events

Time-critical applications of remote sensing include disaster response (Teeuw et al., in this volume) and precision agriculture (Nellis et al., in this volume) support. When timing is critical, pointable sensors clearly have advantage over nadir viewing sensors in that they have a shorter potential revisit period.

For disaster response, a rapid delivery of analyzed imagery requires a series of expedited responses, starting with emergency tasking of the satellite, pre-preprocessing by the satellite operator, and internet-based data delivery. Following receipt of the data, the analyst may need to perform additional georeferencing work before interpretation can be done. Because time is normally very limited, relatively routine or simple methods are necessary.

The fact that rapid response requires some advance planning and organization is demonstrated by the establishment of the International Charter on Space and Major Disasters (International Charter 2007, Harris, in this volume, Teeuw et al., in this volume). This agreement, initiated by the French, European, and Canadian space agencies in 2000, now includes the space agencies of six other countries, and additional agreements with commercial satellite operators. The charter provides for 24-hour availability of a single point of contact for requesting emergency remote sensing support. In France, the organization Service Régional de Traitement d'Image et de Télédétection (SERTIT) has been contracted by the French space agency,

CNES, to provide 24-hour availability of image analysts (SERTIT 2005). SERTIT places its image map products on a website, for free download (http://sertit.u-strasbg.fr/documents/RMS_page_garde/RMS_page_garde.htm).

Of course, data currency is a concern not just in disaster response, but in all applications studying dynamic phenomena. Satellite images typically require preprocessing by the data provider prior to being made available to the user. Additional bottlenecks may occur in the distribution, although internet access to the data can overcome this problem.

INTERACTIONS BETWEEN DIFFERENT COMPONENTS OF SCALE

So far, the discussion has been limited to each of the different components of scale: spatial, spectral, temporal, and radiometric. However, clearly, these components are linked. For example, if image acquisition is constrained by the rate at which data are stored and transmitted, then increasing one type of resolution (such as spectral resolution), will necessarily require changes to other types of resolution (such as spatial resolution) (Figure 1.1). The Compact Airborne Spectral Imager (CASI), manufactured by ITRES of Canada, is a good example of an instrument that is designed to have maximum flexibility within the constraints of data acquisition trade-offs. CASI is a programmable sensor, in which the operator chooses the number, width, and location of spectral bands prior to image acquisition. Because longer integration times are needed as the number of bands imaged increases, there is an inverse relationship between the number of bands and the spatial resolution for this sensor (ITRES 2007).

Alternatively, it is possible in some instances to overcome the spatial-spectral constraint described above by employing pan sharpening, in which data fusion is used to combine high spatial resolution, panchromatic (i.e., single band) images with comparatively low spatial resolution, multispectral images (Alparone et al. 2007). Pan sharpening has become increasingly important since the SPOT sensors popularized the concept of acquiring simultaneous high spatial resolution panchromatic data to complement a lower spatial resolution multispectral data set, and this design approach has been followed for a number of subsequent sensors, including ETM+ (Goward et al., in this volume), IKONOS, and QuickBird (Toutin, in this volume). The aim of pan sharpening is quite simple: to incorporate the spatial detail from the panchromatic image, and the spectral information from the multispectral images. The challenge, however, is to ensure that the combined data set maintains a spectral balance such that when the images are displayed as a color composite, the colors of the sharpened images are similar to the original, low spatial resolution multispectral data set (Alparone et al. 2007). This challenge is particularly great if the panchromatic band is poorly correlated with the individual multispectral bands (Gross and Schott 1998, Price 1999).

Pohl and Van Genderen (1998) provide a comprehensive review of pan sharpening methods. Alparone et al. (2007) empirically compared eight different methods, and found that multiresolution analysis, incorporating for example wavelets or Laplacian pyramids to characterize the spatial dependence of DN values on scale, generally outperformed component substitution, in which some transformed component of the multispectral data set, such as the first principal component, is replaced by the panchromatic data. In particular, the two methods found to have the best results both take into account physical models of the image formation, namely the modulation transfer function (Alparone et al. 2007). Wang et al. (2005) use a theoretical framework, which they term general image fusion, to compare the different methods, and conclude that the optimal method is multiresolution analysis-based intensity modulation. Pan sharpening using spectral mixture analysis also shows promise, especially for hyperspectral imagery (Gross and Schott 1998).

There are other complex interactions between the different types of resolution. Malila (1985) has found that, although the increased number and range of spectral bands of TM compared to MSS provide a great deal more information as indicated by studies of entropy, if both TM and MSS had been quantized at just five bits, the information content of the two sensors would have been approximately equal.

Key et al. (2001) compared the value of multiple spectral bands with multiple image dates for classifying individual deciduous trees species. Their study showed that a single, optimally chosen, multispectral image acquired during peak autumn colors resulted in relatively high classification accuracy. However, multiple dates of single band imagery could provide a similar high accuracy. This finding suggests that if the spatial resolution of multispectral imagery is too coarse, panchromatic imagery, which typically has a higher spatial resolution, may be substituted, if multiple dates can be obtained (Key et al. 2001). For example, the current highest spatial resolution from commercial satellites is provided by the WorldView-1 sensor, launched in 2007. WorldView-1 provides imagery with 0.5 m pixels, but only panchromatic data, with no multispectral bands (DigitalGlobe 2007).

OTHER ISSUES

Additional, broader issues should be considered in selecting image data sets. Data cost, particularly for the new commercial sensors, can be high. However, the commercial providers generally make a distinction between new acquisitions, which require tasking the satellite, and existing images in the companies' archives, charging a premium for the former. Commercial image licensing agreements may constrain sharing the data with others, even in the same organization. Thus purchasers should consider the long-term use of imagery, and consider paying extra to have more flexible use of the data. One of the major advantages of US government data, including Landsat TM, ETM+, and Terra and Aqua MODIS data, is not only the very economical price, but the absence of constraints on data sharing (Harris, in this volume). Indeed, large internet archives of US satellite imagery are available for free downloading (Table 1.2).

A second major issue relates to data volume. Large volumes of data can strain computer storage and processing capacity. Although this issue is far less significant today compared to when early sensing systems such as the Landsat MSS were launched, it is still important for projects that cover relatively large geographic areas, or use multiple dates of images.

In addition to improvements in computer hardware, software has also advanced considerably since the early 1970s. Early programs, typically running on main-frame computers, often were based on command-line program initiation. Today, remote sensing packages typically have graphical user-interfaces, and even semi-automated 'wizards' that help guide the less sophisticated users. Furthermore, there are now specific programs for advanced analysis such as for photogrammetry and hyperspectral classification. On the other hand, the development of software that integrates remote sensing analysis and GIS analysis has been more mixed (Merchant and Narumalani, in this volume).

FUTURE CHALLENGES AND OPPORTUNITIES

It is evident that the number and diversity of satellite-borne sensors will only grow in future years, especially as the commercial satellite sector grows, and additional nations launch and operate their own satellite programs. Thus, the challenges, and opportunities, in selecting data to address specific problems, will also likely grow. Some specific trends can be observed with regards to image spatial and spectral properties, as well as the availability of relatively new types of image data.

With regards to spatial resolution, it appears for the moment that ~0.5 m is the smallest pixel size of space-borne imagery that will be available to non-government users, due to security issues. Thus, the operating licenses for both Worldview-1 (Digital-Globe 2007) and the planned GEOEYE-1 (GeoEye 2007) limit the spatial resolution of imagery that is sold to the general public to 0.5 m.

In terms of spectral properties, one likely future development is finally to achieve operational hyperspectral imaging from space. For the user, space-based hyperspectral imagery should be more economical than contracting for airborne hyperspectral data. An operational satellite-borne hyperspectral system will also remove the geographical constraints of the narrow swath of the experimental satellite-based Hyperion hyperspectral sensor. Once these financial and geographical barriers are removed, hyperspectral analysis may enter the mainstream, especially if there is continued improvement in the ease of use of hyperspectral software analysis tools. Nevertheless, limits on the signal-to-noise and spatial resolution for space-based hyperspectral sensors may ensure that aerial hyperspectral imaging will continue to play an important role for some time to come.

Another area of likely future importance, and challenge to users, will be greater integration of diverse wavelength regions and characteristics, including hyperspectral VNIR and SWIR,

Table 1.2 *Sources of free imagery*

Facility	Example data	URL[1]
Global Land Cover Facility, University of Maryland	TM, MSS, MODIS, ASTER	http://glcf.umiacs.umd.edu
AmericaView	Landsat	http://glovis.texasview.org
USGS EROS	Landsat	http://edc.usgs.gov/products/satellite/landsat_ortho.html
USGS-NASA DataPool	ASTER, MODIS	http://lpdaac.usgs.gov/datapool/datapool.asp
Boston University Climate and Vegetation Group	AVHRR, MODIS	http://cliveg.bu.edu/modismisr/products/products.html
Boston University Land Cover and Land Cover Dynamics	MODIS	http://duckwater.bu.edu/lc/datasets.html

[1] URLs current as of January 2008.

hyperspectral thermal, and multi-wavelength, fully polarimetric radar. The integration of lidar with multispectral and hyperspectral imagery seems a particularly promising area (Bork and Su 2007).

Relatively new types of data will also likely become more available, although once again the general exploitation of these data may be dependent on the development of easy to use software. Polarization information, currently used mainly with microwave wavelengths, holds promise for improved image analysis of optical wavelengths (Zallat et al. 2004). Multi-angular imaging, already available from the Multiangle Imaging SpectroRadiometer (MISR) experimental satellite, allows characterization and exploitation of BRDF information (Armston et al. 2007, Jovanovic et al. 2007). One particularly interesting application of BRDF information is for mapping wetlands by exploiting the distinctive and strong angular reflection signature of water compared to other surface types. This approach has been shown to be effective for discriminating inundated areas with emergent vegetation, open water, and non-inundated areas (Vanderbilt et al. 2002). One strength of this approach is that as the pixel size increases, the accuracy of unmixing the proportions of these cover types tends to increase (Vanderbilt et al. 2007), making the method particularly effective for global-scale hydrological modeling.

In conclusion, remote sensing has advanced greatly since the early 1970s and since the beginnings of regular satellite Earth observations with the ERTS/Landsat MSS sensor. The many advances in remote sensing technology have themselves brought new challenges, as exemplified by issues such as the Hughes Phenomenon (Hughes 1968). Although many of these challenges can be addressed through innovative research, one area outside the control of most individual scientists is the general area of remote sensing policy (Harris, in this volume). For example, despite the importance of data continuity in global change studies, there unfortunately seems to be a lack of political will, at least in the United States, to support an aggressive, long term strategy to ensure data continuity for moderate resolution imaging. This problem has recently been highlighted by the difficulties associated with the Landsat Data Continuity Mission (Goward et al., in this volume). Despite these difficulties, remote sensing offers a powerful, objective, and consistent tool for studying the earth, from local to global scales. The value of remote sensing is demonstrated by the growing number of research studies, and the increasing use of remote sensing in operational environments. The chapters that follow give insight into the many facets and key issues of this rapidly developing subject.

ACKNOWLEDGMENTS

John Althausen (Lockheed Martin Corporation), Yongxin Deng (University of Western Illinois University), Gregory Elmes (West Virginia University), and Douglas Stow (San Diego State University), provided valuable suggestions for improvement of this chapter.

REFERENCES

Alker, H. R., 1969. A typology of ecological fallacies. In M. Dogan and S. Rokkan (eds), *Quantitative Ecological Analysis in the Social Sciences*. The MIT Press, Cambridge, MA, USA.

Alparone, L., L. Wald, J. Chanussot, C. Thomas, P. Gamba, and L. M. Bruce, 2007. Comparison of pansharpening algorithms: Outcome of the 2006 GRS-S data-fusion contest. *IEEE Transactions on Geoscience and Remote Sensing*, 45(10): 3012–3021, doi 10.1109/TGRS.2007.904923.

Armston, J. D., P. F. Scarth, S. R. Phinn, and T. J. Danaher, 2007. Analysis of multi-date MISR measurements for forest and woodland communities, Queensland, Australia. *Remote Sensing of Environment*, 107: 287–298.

Atkinson, P. M. and P. J. Curran, 1997. Choosing an appropriate spatial resolution for remote sensing investigations. *Photogrammetric Engineering and Remote Sensing*, 63(12): 1345–1351.

Atkinson, P. M., I. M. Sargent, G. M. Foody, and J. Williams, 2007. Exploring the geostatistical method for estimating the signal-to-noise ratio of images. *Photogrammetric Engineering and Remote Sensing*, 73(7): 841–850.

Baccini, A., M. A. Friedl, C. E. Woodcock, and Z. Zhu, 2007. Scaling field data to calibrate and validate moderate spatial resolution remote sensing models. *Photogrammetric Engineering and Remote Sensing*, 73(8): 945–954.

Bork E. W. and J. G. Su, 2007. Integrating LIDAR data and multispectral imagery for enhanced classification of rangeland vegetation: A meta analysis. *Remote Sensing of Environment*, 111: 11–24.

Briem, G. J., J. A. Benediktsson, and J. R. Sveinsson, 2002. Multiple classifiers applied to multisource remote sensing data. *IEEE Transactions on Geoscience and Remote Sensing*, 40(10): 2291–2299.

Butson, C. R. and D. J. King, 2006. Lacunarity analysis to determine optimal extents for sample-based spatial information extraction from high-resolution forest imagery. *International Journal of Remote Sensing*, 27(1): 105–120.

Cao, C. and N. S.-N. Lam, 1997. Understanding the scale and resolution effects in remote sensing and GIS. In D. A. Quattrochi and M. F. Goodchild (eds), 1997. *Scale in Remote Sensing and GIS*. Lewis Publishers, Boca Roton, FL, USA, Chapter 3, pp. 57–72.

Chen, X. and D. Campagna, in this volume. Geological applications. Chapter 22.

Chen, X., T. A. Warner, and D. J. Campagna, 2007a. Integrating visible, near infrared and short wave infrared hyperspectral and multispectral thermal imagery for geologic mapping: simulated data. *International Journal of Remote Sensing*, 28(11): 2415–2430.

Chen, X., T. A. Warner, and D. J. Campagna, 2007b. Integrating visible, near-infrared and short-wave infrared hyperspectral and multispectral thermal imagery for geological mapping at Cuprite, Nevada. *Remote Sensing of Environment*, 110: 344–356.

Clark, R. N., G. A. Swayze, R. Wise, K. E. Livo, T. M. Hoefen, R. F. Kokaly, and S. J. Sutley, 2003. *USGS Digital Spectral Library splib05a*. U.S. Geological Survey, Open File Report 03-395. (http://pubs.usgs.gov/of/2003/ofr-03-395/ofr-03-395.html, last date accessed 24 September 2007).

Crews, K. and S. J. Walsh, in this volume. Remote sensing and the social sciences. Chapter 31.

Delbart, N., T. Le Toan, L. Kergoat, and V. Fedotova, 2006. Remote sensing of spring phenology in boreal regions: A free of snow-effect method using NOAA-AVHRR and SPOT-VGT data (1982–2004). *Remote Sensing of Environment*, 101: 52–62.

DigitalGlobe 2007. Worldview-1. http://www.digitalglobe.com/about/worldview1.html (last date accessed 14 October 2007).

Ferro, C. J. and T. A. Warner, 2002. Scale and texture in digital image classification. *Photogrammetric Engineering and Remote Sensing*, 68(1): 51–63.

GeoEye 2007. GeoEye imagery products: GEOEYE-1. http://www.digitalglobe.com/about/worldview1.html (last date accessed 14 October 2007).

Goward, S. N., T. Arvidson, D. L. Williams, R. Irish, and J. Irons, in this volume. Moderate spatial resolution optical sensors. Chapter 9.

Green, A. A., M. Berman, P. Switzer, and M. D. Craig, 1988. A transformation for ordering multispectral data in terms of image quality with implications for noise removal. *IEEE Transactions on Geoscience and Remote Sensing*, 26: 65–74.

Gross, H. N. and J. R. Schott, 1998. Application of spectral mixture analysis and image fusion techniques for image sharpening. *Remote Sensing of Environment*, 63: 85–94.

Harris, R. in this volume. Remote sensing policy. Chapter 2.

Henderson, M., E. T. Yeh, P. Gong, C. Elvidge, and K. Baugh, 2003. Validation of urban boundaries derived from global night-time satellite imagery. *International Journal of Remote Sensing*, 24(3): 595–609.

Hengl, T., 2006. Finding the right pixel size. *Computers and Geosciences*, 32: 1283–1298. doi: 10.1016/j.cageo.2005.11.008

Herold, M., M. Gardner, and D. A. Roberts, 2003. Spectral resolution requirements for mapping urban areas. *IEEE Transactions on Geoscience and Remote Sensing*, 41(9): 1907–1919.

Hill, M. J., C. J. Ticehurst, J-S. Lee, M. R. Grunes, G. E. Donald, and D. Henry, 2005. Integration of optical and radar classifications for mapping pasture type in Western Australia. *IEEE Transactions on Geoscience and Remote Sensing*, 43(7): 1665–1681. doi 10.1109/TGRS.2005.846868.

Holben, B., 1986. Characteristics of maximum-value composite images from temporal AVHRR data. *International Journal of Remote Sensing*, 7(11): 1417–1434.

Huang, C., J. R. G. Townshend, S. Liang, S. N. V. Kalluri, and R. S. DeFries, 2002. Impact of sensor's point spread function on land cover characterization: assessment and deconvolution. *Remote Sensing of Environment* 80(2) 203–212.

Hughes, G. F., 1968. On the mean accuracy of statistical pattern recognizers. *IEEE Transactions on Informational Theory*, IT-14: 55–63.

Hyppänen, 1996. Spatial autocorrelation and optimal spatial resolution of optical remote sensing data in boreal forest environment. *International Journal of Remote Sensing*, 17(17): 3441–3452.

Hyyppä, J., W. Wagner, M. Hollaus, and H. Hyyppä, in this volume. Airborne laser scanning. Chapter 14.

International Charter, 2007. International charter: Space and major disasters. http://www.disasterscharter.org (last date accessed: September 24, 2007).

ITRES, 2007. Flight planning flexibility. http://www.itres.com/Flight_Planning_Flexibility (last date accessed October 14, 2007).

Jelinsky, D. E. and J. Wu, 1996. The modifiable areal unit problem and implications for landscape ecology. *Landscape Ecology*, 11(3): 129–140.

Jensen, J. R., 2005. *Introductory Digital Image Processing: A Remote Sensing Perspective*. Prentice Hall, Upper Saddle River, NJ, USA.

Jensen, J. R., J. Im, P. Hardin, and R. R. Jensen, in this volume. Image classification. Chapter 19.

Journel, A. G. and C. J. Huijbregts, 1978. *Mining Geostatistics*. Academic Press, London.

Jovanovic, V., C. Moroney, and D. Nelson, 2007. Multi-angle geometric processing for globally geo-located and co-registered MISR image data. *Remote Sensing of Environment*, 107(1–2): 22–32.

Jupp, D. L. B., A. H. Strahler, and C. E. Woodcock, 1988. Autocorrelation and regularization in digital images. I. Basic theory. *IEEE Transactions on Geoscience and Remote Sensing*, 26(4): 463–473.

Jupp, D. L. B., A. H. Strahler, and C. E. Woodcock, 1989. Autocorrelation and regularization in digital images. II. Simple image models. *IEEE Transactions on Geoscience and Remote Sensing*, 27(3): 247–258.

Justice, C. O. and C. J. Tucker, in this volume. Coarse spatial resolution optical sensors. Chapter 10.

Justice, C. O., J. R. G. Townshend, E. F. Vermote, E. Masuoka, R. E. Wolfe, N. Saleous, D. P. Roy, and J. T. Morisette, 2002. An overview of MODIS Land data processing and product status. *Remote Sensing of Environment*, 83: 3–15.

Kerekes, J. P., in this volume. Optical sensor technology. Chapter 7.

Key, T., T. Warner, J. McGraw, and M. A. Fajvan, 2001. A comparison of multispectral and multitemporal imagery for tree species classification. *Remote Sensing of Environment*, 75: 100–112.

Latty, R. S., R. Nelson, B. Markham, D. Williams, D. Toll, and J. Irons, 1985. Performance comparison between information extraction techniques using variable spatial resolution data. *Photogrammetric Engineering and Remote Sensing*, 51(9): 1459–1470.

Loveland, T. R., J. W. Merchant, J. F. Brown, D. O. Ohlen, B. C. Reed, P. Olson, and J. Hutchinson, 1995. Seasonal land-cover regions of the United States. *Annals of the Association of American Geographers*, 85(2): 339–355.

Malila, W. A., 1985. Comparison of the information contents of Landsat TM and MSS data. *Photogrammetric Engineering and Remote Sensing*, 51(9): 1449–1457.

Marceau, D. J. and G. J. Hay. 1999. Remote sensing contributions to the scale issue. *Canadian Journal of Remote Sensing*, 25(4): 357–366.

Marceau D. J., D. J, Gratton, R. A. Fournier, and J. P. Fortin, 1994. Remote-sensing and the measurement of geographical entities in a forested environment. 2. The optimal spatial-resolution. *Remote Sensing of Environment*, 49(2): 105–117.

Matheron, G., 1971. *The Theory of Regionalized Variables and its Applications*. Cahiers du Centre de Morphologie Mathematique de Fontainbleau, No. 5.

McCloy, K. R. and P. K. Bøcher, 2007. Optimizing image resolution to maximize the accuracy of hard classification. *Photogrammetric Engineering and Remote Sensing*, 73(8): 893–903.

Melgani, F. and L. Bruzzone, 2004. Classification of hyperspectral remote sensing images with support vector machines. *IEEE Transactions on Geoscience and Remote Sensing*, 42: 1778–1790.

Merchant, J. W. and S. Narumalani, in this volume. Integrating remote sensing and geographic information systems. Chapter 17.

Miao, X., P. Gong, S. Swope, R. Pu, R. Carruthers, and G. L. Anderson, 2007. Detection of yellow starthistle through band selection and feature extraction from hyperspectral imagery. *Photogrammetric Engineering and Remote Sensing*, 73(9): 1005–1015.

Myneni, R. B., C. D. Keeling, C. J. Tucker, G. Asrar, and R. R. Nemani, 1997. Increased plant growth in the northern high latitudes from 1981 to 1991. *Nature*, 386: 698–702.

Narayanan, R. M., T. S. Sankaravadivelu, and S. E. Reichenbach, 2000. Dependence of information content on gray-scale resolution. *Geocarto International*, 15(4): 15–27.

Nellis, M. D., K. Price, and D. Rundquist, in this volume. Remote sensing of cropland agriculture. Chapter 25.

Openshaw, S. 1983. The modifiable areal unit problem. *Concepts and Techniques in Modern Geography* 38. Geo Books, Norwich, UK.

Openshaw, S. 1984. Ecological fallacies and the analysis of areal census data. *Environment and Planning A*, 6: 17–31.

Openshaw, S. and P. J. Taylor. 1979. A million or so correlation coefficients: three experiments on the modifiable areal unit problem. In R. J. Bennett (ed.), *Statistical Applications in the Spatial Sciences*. Pion, London, UK.

Pohl, C. and J. L. Van Genderen, 1998. Multisensor image fusion in remote sensing: concepts, methods and applications. *International Journal of Remote Sensing*, 19(5): 823–854. doi: 10.1080/014311698215748.

Price, J. C. 1999. Combining multispectral data of differing spatial resolution. *IEEE Transactions on Geoscience and Remote Sensing*, 37(3): 1199–1203.

Quattrochi, D. A. and M. F. Goodchild (eds), 1997. *Scale in Remote Sensing and GIS*. Lewis Publishers, Boca Roton, FL, USA.

Robinson, W.S. 1950. Ecological correlations and the behavior of individuals. *American Sociological Review*, 15: 351–357.

Schaepman, M. E., in this volume. Imaging spectrometers. Chapter 12.

Schaepman-Strub, G., M. E. Schaepman, J. V. Martonchik, T. H. Painter, and S. Dangel, in this volume. Terminology of radiometry and reflectance – from concepts to measured quantities. Chapter 15.

Schwartz, M. D., B. C. Reed, and M. A. White, 2002. Assessing satellite-derived start-of-season measures in the conterminous USA. *International Journal of Climatology*, 22: 1793–1805. doi: 10.1002/joc.819.

Serpico, S. B. and G. Moser, 2007. Extraction of spectral channels from hyperspectral images for classification purposes. *IEEE Transactions on Geoscience and Remote Sensing*, 45(2): 484–495.

SERTIT 2005. SERTIT: Introduction. http://sertit.u-strasbg.fr/english/en_presentation.htm (last date accessed September 24 2007).

Song, C., C. E. Woodcock, K. C. Seto, M. Pax Lenney, and S. A. Macomber, 2001. Classification and change detection using Landsat TM data: When and how to correct atmospheric effects? *Remote Sensing of Environment*, 75: 230–244.

Spiker, S. and T. A. Warner, 2007. Scale and spatial autocorrelation from a remote sensing perspective. In J. Gattrell and R. Jensen (eds.), *Geo-Spatial Technologies in Urban Environments*. Springer-Verlag, Heidelberg, pp. 197–213.

Stow, D. A., L. L. Coulter, and C. A. Benkelman, in this volume. Airborne digital multispectral imaging. Chapter 11.

Swain, P. and S. M. Davis, 1978. *Remote Sensing: The Quantitative Approach*. McGraw Hill, New York.

Tarnavsky, E., D. Stow, L. Coulter, and A. Hope, 2004. Spatial and radiometric fidelity of airborne multispectral imagery in the context of land-cover change analyses. *GIScience and Remote Sensing*, 41(1): 62–80.

Teeuw, R., P Aplin, N. McWilliam, T. Wicks, K. Matthieu, and G. Ernst, in this volume. Hazard assessment and disaster management using remote sensing. Chapter 32.

Teillet, P. M., K. Staenz, and D. J. Williams, 1997. Effects of spectral, spatial and radiometric characteristics on remote sensing vegetation indices of forested regions. *Remote Sensing of Environment*, 61: 139–149.

Toutin, T., in this volume. Fine spatial resolution optical sensors. Chapter 8.

Townshend, J. R. G., C. O. Justice, C. Gurney, and J. McManus, 1992. The impact of misregistration on change detection. *IEEE Transactions on Geoscience and Remote Sensing*, 30(5): 1054–1060.

Turner, M. G., R. H. Gardner, and R. V. O'Neill, 2001. *Landscape Ecology in Theory and Practice: Pattern and Process*. Springer, New York, USA.

Umamaheshwaran, R., W. Bijker, and A. Stein, 2007. Image mining for modeling of forest fires from Meteosat images. *IEEE Transactions on Geoscience and Remote Sensing*, 45(1): 246–253. doi 10.1109/TGRS.2006.883460.

Vanderbilt, V. C., G. L. Perry, G. P. Livingston, S. L. Ustin, M. C. Diaz Barrios, F.-M. Bréon, M. M. Leroy, J.-Y. Balois, L. A. Morrissey, S. R. Shewchuk, J. A. Stearn, S. E. Zedler,

J. L. Syder, S. Bouffies-Cloche, and M. Herman, 2002. Inundation discriminated using sun glint. *IEEE Transactions on Geoscience and Remote Sensing*, 40(6): 1279–1287.

Vanderbilt, V. C., S. Khanna, and S. L. Ustin, 2007. Impact of pixel size on mapping surface water in subsolar imagery. *Remote Sensing of Environment*, 109: 1–9.

van Leeuwen, W. J. D., in this volume. Visible, near-IR and shortwave IR spectral characteristics of terrestrial surfaces. Chapter 3.

Wald, L., 1999. Some terms of reference in data fusion. *IEEE Transactions on Geoscience and Remote Sensing*, 37(3): 1190–1191.

Walsh, S. J., A. Moody, T. R. Allen, and D. G. Brown, 1997. Scale dependence of NDVI and its relationship to mountainous terrain. In D. A. Quattrochi and M. F. Goodchild (eds.), *Scale in Remote Sensing and GIS*. Lewis Publishers, Boca Raton, pp. 27–55.

Walsh, S. J., L. Bian, S. McKnight, D. G. Brown, and E. S. Hammer, 2003. Solifluction steps and risers, Lee Ridge, Glacier National Park, Montana, USA: a scale and pattern analysis. *Geomorphology*, 55: 381–398.

Wang, H. and E. C. Ellis, 2005. Image misregistration error in change measurements. *Photogrammetric Engineering and Remote Sensing*, 71(9): 1037–1044.

Wang, Z., D. Ziou, C. Armenakis, D. Li, and Q. Li, 2005. A comparative analysis of image fusion methods. *IEEE Transactions on Geoscience and Remote Sensing*, 43(6): 1391–1402. doi 10.1109/TGRS.2005.846874

Warner, T. A., 1999. Analysis of spatial patterns in remotely sensed data using multivariate spatial correlation. *Geocarto International*, 14(1): 59–65.

Warner, T. A. and M. Shank, 1997. An evaluation of the potential for fuzzy classification of multispectral data using artificial neural networks. *Photogrammetric Engineering and Remote Sensing*, 63(11): 1285–1294.

Warner, T. and X. Chen, 2001. Normalization of Landsat thermal imagery for the effects of solar heating and topography. *International Journal of Remote Sensing*, 22(5): 773–788.

Warner, T. A. and F. Nerry, 2008. Does a single broadband or multispectral thermal data add information for classification of visible, near- and shortwave infrared imagery of urban areas? *International Journal of Remote Sensing*, in press.

Warner, T. A., K. Steinmaus, and H. Foote, 1999. An evaluation of spatial autocorrelation-based feature selection. *International Journal of Remote Sensing*, 20(8): 1601–1616.

Wartenberg, D., 1985. Multivariate spatial correlation: a method for exploratory geographical analysis. *Geographical Analysis*, 17(4): 263–283.

Wiens, J.A. 1989. Spatial scaling in ecology. *Functional Ecology*, 3: 385–397.

Wiersma, D. J. and D. A. Landgrebe, 1980. Analytical design of multispectral sensors. *IEEE Transactions on Geoscience and Remote Sensing*, GE-18: 180–189.

Woodcock, C. E. and A. H. Strahler, 1987. The factor of scale in remote sensing. *Remote Sensing of Environment*, 21: 311–332.

Wulder, M., J. C. White, S. Ortlepp, and N. C. Coops, in this volume. Remote sensing for studies of vegetation condition: theory and application. Chapter 25.

Zallat, J., C. Collet, and Y. Takakura, 2004. Clustering of polarization-encoded images. *Applied Optics*, 43: 283–292.

Zhang, P., J. Li, E. Olson, T. J. Schmit, J. Li, and W. P. Menzel, 2006. Impact of point spread function on infrared radiances from geostationary satellite. *IEEE Transactions on Geoscience and Remote Sensing*, 44(8): 2176–2183.

Remote Sensing Policy

Ray Harris

Keywords: global policies, access, government, pricing, geospatial data, United States, Europe, India, Japan.

INTRODUCTION

Policy: (1) a course or principle of action adopted or proposed by an organization or an individual; (2) prudent or expedient conduct or action. Origin: from the French *police*–bill of lading, contract of insurance. *Oxford English Dictionary*

Act so as to produce the greatest good for the greatest number. *The Principle of Utility*, Jeremy Bentham (1748–1832)

The history of satellite remote sensing has so far shown a commendable although unconscious example of Bentham's *Principle of Utility*, or *Utilitarianism* (Harte and North 2004). Satellite remote sensing missions have mainly been general purpose satellite missions, such as Landsat, designed to capture environmental data that can be used by anyone with the knowledge and technical capability to do so. Sometimes policy development has preceded this technical capability and sometimes followed it, but remote sensing policy has in general been characterized by utilitarianism, that is the greatest good for the greatest number.

Remote sensing policy is mainly written by governments in some form, be it through national legislation as in the USA or through national representation in international organizations, such as the European Space Agency (ESA) or Eumetsat. In that sense remote sensing is often an extension of national policy. The USA is a good case in point. Each fiscal year a report is sent to the US Congress entitled *Our Changing Planet*

(CCSP 2006). The report, which summarizes a great deal of US remote sensing research and applications, describes the activities and the plans of the Climate Change Science Program that was established under the US Global Change Research Act of 1990 and the US Climate Change Research Initiative established by the US President in 2001. While the report has the appearance of a science progress report it has a foundation in national government policy that is part of a wider US policy landscape, for example on science, national security and industry privatization. The report *Our Changing Planet* is transmitted each year to the US Congress by the Secretaries of State for Commerce and for Energy, both political appointments, together with the Director of the Office for Science and Technology Policy of the Executive Office of the President. Where governments appear not to be directly involved in remote sensing policy they still have a responsibility for regulation or licencing. Fine spatial resolution missions such as DigitalGlobe or IKONOS operate with a licence issued by the US government which in turn relays the national commitments it has entered into as well as reflecting national government priorities.

Government influence is therefore important in understanding remote sensing policy. As different national governments around the world have different political complexions, and indeed political complexions that change over time, so remote sensing policies are different. The remote sensing policy debate is especially contentious within the Group on Earth Observation (GEO) initiative to the extent that the term 'data policy', a key part of remote sensing policy, is deliberately avoided and only the term 'data sharing' is allowed (Achache 2006). This book examines a very wide variety of remote sensing data, techniques to process the data and applications of the data. This chapter looks at remote sensing policy, trying to disentangle the ways in which the organizations responsible for providing the data examined in later chapters have come to their different views. This chapter will concentrate on data policy because this is where remote sensing policy has the greatest impact on access to and use of remote sensing data, while commenting on wider policy concerns where appropriate. The chapter opens by examining remote sensing policy agreements reached at the global scale, and then goes on to examine the policies developed in the USA and Europe as major organizational actors in remote sensing. A review of selected national policies is used to highlight differences in approach to policy, for example India's very clear concern with national security, before the conclusion points to critical tensions such as pricing policy and the overall sustainability of remote sensing.

Whether in the public sector or the private sector Bentham's *Principle of Utility* is without doubt an unconscious characteristic of remote sensing. One could go even further along the road of utilitarianism and argue that the comments of the Roman senator Cicero are also applicable to twenty-first century remote sensing (Oxford 1981):

Salus populi suprema est lex
[The good of the people is the chief law]

GLOBAL SCALE REMOTE SENSING POLICY

United Nations principles on remote sensing

On 3 December 1986 the United Nations (UN) reached agreement on the *UN Resolution Relating to the Remote Sensing of the Earth from Outer Space* (Jasentuliyana 1988, von der Dunk 2002). This Resolution contains 15 principles on remote sensing that were agreed as a compromise between the perspective of state territorial sovereignty and the principle of the freedom to use outer space that is embodied in the Outer Space Treaty.[1] Those nations that had satellite remote sensing capability both wanted and needed freedom to capture remote sensing data for any and all parts of the Earth. Some of those nations that lacked a satellite remote sensing capability wanted to control access to the outer space above their territory in much the same way as they controlled the air space above their territory (Harris and Harris 2006). This approach of control

envisaged the concept of ownership extending to a limitless distance above a nation's territory and would have invited all organizations that orbited remote sensing spacecraft to seek the permission of each and every country to allow orbital passes over their country. The compromise between the points of view of open access and control was the agreement of the 15 UN Principles on Remote Sensing. While all 15 principles are relevant to this book, four principles are particularly important.

Principle I. For the purposes of these principles with respect to remote sensing activities: (a) The term 'remote sensing' means the sensing of the Earth's surface from space by making use of the properties of electromagnetic waves emitted, reflected or diffracted by the sensed objects, for the purpose of improving natural resources management, land use and the protection of the environment.

This first principle (of which only part (a) of five parts is reproduced here) provides a definition of the scope of the later principles. At the time of the agreement the UN principles were thought to apply to civil remote sensing only of the land surface, but since 1986 the term 'protection of the environment' has taken on a much wider meaning because of the concerns about climate change (IPCC 2007) and it is now difficult to identify which elements of the Earth system (ocean, ice, atmosphere, land) fall outside the scope of the protection of the environment. Principle I may therefore now be considered very wide in scope.

Principle IV. Remote sensing activities shall be … carried out for the benefit and in the interests of all countries, irrespective of their degree of economic or scientific development, and stipulates the principle of freedom of exploration and use of outer space on the basis of equality. These activities shall be conducted on the basis of respect for the principle of full and permanent sovereignty of all States and peoples over their own wealth and natural resources, with due regard to the rights and interests, in accordance with international law, of other States and entities under their jurisdiction. Such activities shall not be conducted in a manner detrimental to the legitimate rights and interests of the sensed State.

This principle strikes at the core of the dilemma noted above: the freedom of the use of outer space by those nations equipped to do so and the sovereignty that nations have over their own territory and resources.

Principle XII. As soon as the primary data and the processed data concerning the territory under its jurisdiction are produced, the sensed State shall have access to them on a non-discriminatory basis and on reasonable cost terms.

Principle XII means that space-faring nations cannot keep the remote sensing data collected by their space missions to themselves, and answers in part the question posed by Principle IV. Principle XII allows a state sensed by a remote sensing satellite to have access to the data collected by the satellite under three conditions: as soon as the data are produced; on a non-discriminatory basis; and on reasonable cost terms. None of these three conditions of access is tightly defined: the balance of issues around these three terms is discussed at length by Frans von der Dunk in Harris (2002).

Principle XIV. *... States operating remote sensing satellites shall bear international responsibility for their activities and assure that such activities are conducted in accordance with the provisions of the Treaty and the norms of international law, irrespective of whether such activities are carried out by governmental or non-governmental entities or through international organizations to which such States are parties. This principle is without prejudice to the applicability of the norms of international law on State responsibility for remote sensing activities.*

The UN principles were agreed between states, so are private companies and other organizations exempt? Principle XIV covers both governmental and non-governmental entities which brings private companies and other organizations into the scope and the legitimacy of the UN Principles. This principle therefore covers the authority of governments to grant licences to private companies such as DigitalGlobe and to participate in international organizations such as Eumetsat.

United Nations Charter Space and major disasters

As well as drawing up the set of 15 Principles the United Nations has been active in arranging major meetings of all UN member states to discuss the opportunities offered by the use of outer space. There have been three such major meetings called UNISPACE conferences. At the third UNISPACE conference in Vienna in 1999 the ESA and the French Space Agency (CNES) launched the idea of a UN Charter on Space and Major Disasters. The basic idea of the Charter is to provide a unified system of space data acquisition and delivery to those affected by natural or man-made disasters through the mechanism of authorized users. The UN Charter has two major objectives.

- Supply during periods of crisis, to states or communities whose population, activities or property are exposed to an imminent risk, or are already victims, of natural or technological disasters, data providing a basis for critical information for the anticipation and management of potential crises.
- Participation, by means of this data and of the information and services resulting from the exploitation of space facilities, in the organization of emergency assistance or reconstruction and subsequent operations.

Following the lead given by Europe other members have joined the Charter, namely Canada, India, Japan, the US National Oceanic and Atmospheric Administration (NOAA), US Geological Survey (USGS) and the participants in the Disaster Management Constellation of small satellites (Algeria, Nigeria, Turkey and the United Kingdom).

When a disaster strikes an authorized user can contact a single point to request satellite remote sensing data acquisition. The space agency members of the Charter then work together to plan image acquisitions and provide data of the disaster location to the authorized users free of all charges. Each year there are approximately 20–30 activations of the Charter, acquiring data of, for example, floods in Indonesia, a typhoon in the Philippines, an oil slick off the coast of Lebanon, an earthquake in Pakistan, the 2004 tsunami in the Indian Ocean and forest fires in Portugal. Further discussion on the use of remote sensing in disaster applications is given by Teeuw et al. (in this volume).

World Meteorological Organisation Resolution 40

A second policy related to remote sensing that is global in nature was that agreed by the World Meteorological Organisation (WMO) in 1995. At the Twelfth Meteorological Congress in Geneva in June 1995 the WMO passed 41 resolutions covering a wide range of its activities from the use of the Portuguese language to the Global Climate Observing System (WMO 1995). One of these resolutions, Resolution 40, states the WMO policy for exchanging meteorological data including remote sensing meteorological data. The policy applies to all 187 WMO member states and so is a policy that is global in reach. WMO Resolution 40 has at its core:

As a fundamental principle ..., WMO commits itself to broadening and enhancing the free and unrestricted international exchange of meteorological and related data and products.

Resolution 40 then provides advice to WMO member states on the practice of the resolution (WMO 1995):

> Members shall provide on a free and unrestricted basis essential data and products which are necessary for the provision of services in support of the protection of life and property and the well-being of all nations, particularly those basic data and products as … required to describe and forecast accurately weather and climate, and support WMO programmes.

Annexes to the resolution provide advice on how to implement the basic ideas of free and unrestricted access. The meteorological community has always practiced relatively unrestrictive exchange of weather data and this principle is followed through to cover meteorological remote sensing data. Resolution 40 is important in remote sensing policy because it provides a clear statement of one community's view of how remote sensing data should be regarded. There is an implicit assumption that meteorological remote sensing data are a public good (Samuelson 1954, Pearce 1995, Longhorn and Blakemore 2004, Miller 2007), an attractive idea in the development of the information society but an idea not without its problems such as financing the systems that deliver the data.

International Council for Science

In 2004, the International Council for Science (ICSU) published a report that is essentially a guidance document for science data policy (ICSU 2004). It covers all science data and information, identifying especially remote sensing data and biomedical data as exemplars of massive data sets that are presenting new challenges to science. The ICSU recommendations cover the roles of the public and private sectors in the production of scientific data and information, data rescue and safeguarding, interoperability, dissemination, intellectual property rights and funding. For remote sensing policy a key issue can be summed up in the recommendation on professional data management (ICSU 2004):

> The panel recommends that ICSU play a major role in promoting professional data management and that it foster greater attention to consistency, quality, permanent preservation of the scientific data record, and the use of common data management standards throughout the global scientific community.

This book is concerned with the acquisition, treatment and use of remote sensing data, yet these data are more than just digits. They are information resources about the state of the Earth, resources that are important in understanding the Earth both now and in the future. Professional approaches to data management will improve the access to remote sensing data and will improve the opportunity to gain a greater scientific and operational return on the large investments involved. The International Polar Year and the Electronic Geophysical Year (both having a focus on 2007–08) have been stimulated by the ICSU policy ideas in developing their own policies and frameworks for data, including policies on the legacy that will be left in the form of professionally archived data sets.

All users of remote sensing data benefit from improvements in policy definition because it means that the conditions of access are explicit and known. Initiatives such as Global Monitoring of Environment and Security (GMES) and Global Earth Observation System of Systems (GEOSS) increasingly rely on data that are robust and have a known pedigree, which in turn means that professional data management and clear data policies become essential to progress in the field of remote sensing.

UNITED STATES

The United States has the most developed and the most formal approach to remote sensing policy. The major national initiatives that incorporate remote sensing are frequently passed as national laws and are then subject to regular, formal review. The US has an overall law on access to all data produced by the Federal government. This is the Paperwork Reduction Act of 1995[2] that was made operational in the Office of Management and Budget (OMB) Circular A-130. The Act is relevant to remote sensing because it mandates that all data produced by the Federal government, including remote sensing data produced by federal agencies, should be provided to users with no restrictions and no copyright protection. This means that when a user acquires, for example, a Landsat ETM+ or a MODIS digital image then this data set can be provided to other users free of any copyright restrictions. By contrast this is not the case for SPOT (Satellite Pour l'Observation de la Terre) data which cannot be copied freely to other users by the initial purchaser. The general approach in the US is that the tax payer has paid once for remote sensing data and so to maximize the value of the data then they should be provided to as many users as can benefit from them with low barriers to use.

The US Climate Change Science Program (CCSP) was created in 2002 as a combination and integration of two government policy actions, the Global Change Research Act of 1990,[3] which was approved by Congress, and the Climate Change Research Initiative which was established by

President Bush in 2001. This formal arrangement for global change science explains in part why the US National Aeronautics and Space Administration (NASA) is so well funded for its remote sensing work compared to other countries. In Financial Year 2005, NASA received US$292 million for research and US$972 million for space-based observations as part of the Climate Change Science Program (CCSP 2006). Presidential directives have long been important for remote sensing in the US. The development of fine spatial resolution commercial systems, such as IKONOS, depended on the US government's agreement for commercial operators to launch remote sensing sensors with pixel sizes of around 1 m. For example, in April 2003 a National Security Presidential Directive provided guidance on licencing commercial satellite sensors of fine spatial resolution, and indicated how the US government would access these data and how the data would fit in with US foreign policy (Anon 2003).

Landsat missions have been exemplars of the strong role that government plays in US remote sensing (Goward et al., in this volume). Since its inception in 1972 as the Earth Resources Technology Satellite (ERTS), the Landsat program has had involvement from NASA, NOAA, the Department of Defense and the private sector (CES 1995). The Land Remote Sensing Policy Act of 1992[4] directed that when a successor was considered to Landsat 7 'preference should be given to the development of such a system [namely the Landsat follow-on] by the private sector'. This broad approach, called the Landsat Data Continuity Mission (LDCM), had its own policy approved by NASA and the US Geological Survey (USGS) in 2002 (Gillespie 2005) that focused on the supply of data to the user community rather than necessarily on the provision of a satellite. Long term provision of Landsat sensor data has been the subject of a further law, Public Law 102-555, that established the National Satellite Land Remote Sensing Data Archive (NSLRSDA). The law stated that:

> It is in the best interest of the United States to maintain a permanent, comprehensive Government archive of global Landsat and other remote sensing data.

The NSLRSDA is located at the EROS Data Center and provides an extensive archive of remote sensing data in the United States. An interesting policy element of the archive is that all organizations in the US that wish to dispose of a large archive of remote sensing data must first offer the archive to the NSLRSDA, which in turn may or may not choose to accept it.

NOAA implements US government policy by providing much of its remote sensing data free of all charges. The basic model is that the data themselves are free: any costs that are incurred above the provision of the basic data are in principle met by the user. Thus, data that are provided through online web access or by direct broadcast from NOAA satellites are free of charge because no extra costs fall on NOAA, while large amounts of data that require special archive access or are transcribed to CD are provided at the cost of fulfilling a user request, also termed COFUR. The COFUR concept is a common part of remote sensing data policy in the US and elsewhere as it plays out the policy of no restrictions on distribution of data produced by the Federal government while avoiding limitless financial commitments through excessive user demand. For example, Landsat 7 data are provided at COFUR, approximately US$600 per scene. The data from the Advanced Spaceborne Thermal Emission and Reflection Radiometer (ASTER) flying on the NASA Terra platform was originally provided completely free of charge, but in August 2002 a charge of US$55 per scene was introduced for ASTER data (increased in 2006 to US$85) to bring the data provision in line with NASA's COFUR data policy.

EUROPE

European Space Agency

The main ESA remote sensing satellite so far in the twenty-first century has been Envisat, launched on 1 March 2002. The predecessors to Envisat were the ERS-1 and ERS-2 satellites. The data policy for ERS-1 was not agreed by ESA member states until well after the launch of ERS, so for the Envisat mission ESA was determined to plan ahead on policy. During the period 1997–1999 ESA and the member states developed a data policy and an implementation plan to formalize the conditions under which Envisat sensor data are made available. The Envisat data policy was subsequently retrofitted to apply to the ERS missions for the last few years of the ERS satellites' lifetime. Taken together, the Envisat data policy and its accompanying implementation document is probably the fullest statement of an Earth observation data policy yet produced. The objectives of the Envisat data policy are to maximize the beneficial use of Envisat data and to stimulate a balanced development of science, public utility and commercial applications, consistent with the Envisat mission objectives (Harris 1999, 2003). The Envisat data policy recognizes two categories of use of the data, namely (ESA 1998):

Category 1 use. Research and applications development use in support of the mission objectives, including research on long term issues of

Earth system science, research and development in preparation for future operational use, certification of receiving stations as part of the ESA functions, and ESA internal use.

Category 2 use. All other uses which do not fall into Category 1 use, including operational and commercial use.

An important element here is the word *use* and not *user*. A user, for example a national meteorological service, can have many different uses (e.g., research, operations, commercial) but it is the use criterion that is the objective assessment and the use criterion that fits with the objectives of the Envisat data policy.

ESA is normally responsible for the distribution of category 1 use data. Such distribution is typically to Principal Investigators. ESA delegates the distribution of category 2 use data to distributing entities appointed by ESA, although ESA retains the intellectual property rights in the standard Envisat products. The distributing entities, currently the companies EMMA and SARCOM, then sell Envisat standard products and value-added products that they produce themselves in the open market, thus contributing to the development of a market for Envisat data.

European Commission

The European Commission has been active in creating policies that implicitly include remote sensing data through directives either on the environment or on general data bases. The European Council Directive 90/313/EEC of 7 June 1990 (EurLex 2006a) defined the terms of access to environmental information held by public bodies with the main objective being freedom of access to the data. The Directive mandates European Union (EU) Member States to ensure that public authorities make available information on the environment, including remote sensing data, to any natural or legal person and that the charge for supplying the information must not exceed a reasonable cost. This is somewhat similar to the COFUR concept used in the United States, although the term 'reasonable cost' is not as clear as the term COFUR. A replacement Directive 2003/4/EC of the European Parliament and Council entered into force on 14 February 2003 (EurLex 2006b). This new directive ensures that environmental information is systematically available and disseminated to the public. The environmental information available includes the following.

- Data on activities affecting the environment
- Environmental impact studies and risk assessments

- Reports on the state of the environment
- Environmental authorizations and agreements

The Directive mandates that the environmental information must be available no later than one month after the receipt of a request, but that 'all information held by public authorities relating to *imminent threats* to human health or the environment is *immediately disseminated* to the public likely to be affected' (my emphases).

By contrast the Directive on the legal protection of databases, Directive 96/9/EC of 11 March 1996 (EurLex 2006c), is designed to afford an appropriate and uniform level of protection of databases to secure remuneration to the maker of the database. There is some scope for conflict between the Environmental Information Directive and the Database Directive. The first is targeted at public information and the second is targeted at private sector information: indeed the Database Directive was stimulated by concern for protection of the music industry as the industry became increasingly digital. However, even though the Database Directive was intended for the digital music industry it is also potentially applicable to the digital remote sensing sector. This is what lawyers call 'black letter law' where the written word is definitive, and not conditioned by the intentions of the original authors.

The EU and the ESA have gone through a series of steps along the road of creating a common European Space Policy (Peter 2007). The EU and ESA agreed to publish a White Paper and then a Framework Agreement (EC 2003) on how to collaborate and these discussions were held at ministerial level in Europe. The broad balance of the European Space Policy is for:

1 the EU to focus on space-based applications that contribute to the achievement of its policies, particularly Galileo for navigation and GMES for environment and security, and to provide the optimum regulatory environment for space activities in Europe;
2 the ESA to focus on space science, space exploration and on the development of the basic tools that allow access to space and to the exploitation of space for the benefit of European citizens.

Eumetsat

Eumetsat is primarily responsible for the Meteosat series of geostationary meteorological satellites. In addition, Eumetsat launched its first polar orbiting satellite on 19 October 2006. Eumetsat has had an explicit interest in the policy for its data since 1989 when it organized the First Eumetsat Workshop on Legal Protection of Meteorological Satellite Data

(Eumetsat 1991). Starting in 1998 and continuing to 2006 Eumetsat has defined and later refined its policy on access to Meteosat data. The core of the Eumetsat data policy is a set of 12 agreed policy statements that define access conditions for the national meteorological services (NMS) of Eumetsat member states, the national meteorological services of non-member states, education and research users, and all other users including commercial broadcasters. Eumetsat fulfils its obligations under WMO Resolution 40 through its policy statements and through its implementing rules, especially rule 4:

> Eumetsat shall make its Six-hourly Meteosat Data, its WEFAX Data, the Meteosat Derived Products and the data offered through its Meteosat Internet Service available to all users world-wide on a free and unrestricted basis as 'Essential' Data and Products in accordance with WMO Resolution 40 (Cg-XII). (Eumetsat 2006)

Eumetsat recognizes a special category of use in education and research and makes special provision for such use, requesting though that 'all results obtained [from the research] are openly available at delivery costs only, without delay linked to commercial objectives, and that the research itself is submitted for open publication' (Eumetsat 2006). The policy defines a concept that is essentially the same as COFUR in the USA, although it is named by Eumetsat as 'without charge'. This term is intended to mean without charge for the data, but explicitly includes the cost of distribution media, documentation, software licences, delivery or transmission, direct labour costs and any other direct costs.

The technical way in which Eumetsat controls its data is through encryption, a concept that may be more widely applicable in remote sensing (Harris and Browning 2005). The six-hourly Meteosat data are available without encryption. All other data are normally encrypted and can be decrypted only through a decryption key provided to recognized

users by Eumetsat. When disasters or emergencies occur Eumetsat makes Meteosat data available without charge for a limited period.

Eumetsat acknowledges in its policy the concept of an ability to pay. It distinguishes poor countries that have a Gross National Income (GNI) per capita of below or equal to US$3500 per annum and grants to the national meteorological service of these countries Meteosat data without charge for official duty use. For countries with a GNI per capita above US$3500 per annum the national meteorological services are charged a fee for hourly, half hourly and quarter hourly Meteosat data. Table 2.1 shows a selection of the charges based on these rules.

Eumetsat is moving beyond a data policy for its Meteosat satellites because it now has responsibility for the Eumetsat Polar System (EPS). EPS is not only a means of sharing responsibility for the provision of global meteorological satellite data, EPS also carries instruments for NOAA, in particular the highly successful Advanced Very High Resolution Radiometer (AVHRR). This sharing has led to discussions between Eumetsat and NOAA on the policy for EPS data (Ernst 2004). There is broad agreement on sharing of data between Eumetsat and NOAA, with the access dependent on the respective policies of Eumetsat and the USA. However, there are two challenges. First, in a situation of crisis or war NOAA may ask Eumetsat to authorize selective denial of critical data from US instruments on Eumetsat satellites to an adversary of the US or its allies. Second, there is debate around access control to data from EPS on an instrument by instrument basis to account for the different policy positions found in the USA and Europe. This policy debate illustrates the close linkage between remote sensing policy and remote sensing instruments and their data.

Comparisons – USA and Europe

Both the USA and Europe are committed to maximizing the use of remote sensing data. However,

Table 2.1   Annual fees applicable to national meteorological services of non member states of Eumetsat for official duty use of Meteosat data. Period 2007–08. Gross National Income per capita data are drawn from the World Bank.

Country	GNI per capita US $	Hourly data € 000	Half-hourly data € 000	Quarter-hourly data € 000
Brazil	2,710	0	0	0
Burundi	100	0	0	0
Canada	23,930	60	80	100
Cyprus	12,320	60	80	100
Egypt	1,390	0	0	0
Israel	16,020	60	80	100

Source: Eumetsat 2006

the policies differ. The US view is that the maximum value can be gained from government investment in space by encouraging free and open access to federally-produced data, while at the same time encouraging the private sector to be involved in remote sensing through a licencing structure. The European view is that while governments need to be involved in remote sensing during its early years as a way of stimulating a new, high technology sector, governments want an exit that encourages other organizations to take over the operations of remote sensing missions in the long term. The maximum value to the tax payer is achieved in the European perspective through the growth of operational environmental services using remote sensing in both the private and public sectors, building on the research and development investments by governments in remote sensing systems but at the same time encouraging the transition of the funding to the users who are ultimately in the best position to judge the value of remote sensing information.

These two views are echoed around the world. Two common opinions are often heard. First, remote sensing should be freely available and free of charge. Second, the government departments that fund remote sensing missions seek to recoup all or part of their investment in those missions through sales of the remote sensing data. These two sentiments are largely incompatible.

NATIONAL CONTRASTS

Security

India has the most coherent remote sensing programme on Earth. It is coherent in that it has a staged structure for spatial resolution from the sensors with a 2 km pixel size on the INSAT geostationary meteorological satellites down to the 1 m pixel size of Cartosat. In addition, Indian remote sensing satellites have consistent wavelength bands across many missions that allow for inter-comparison of data at different spatial scales. As with the USA, India has a formal remote sensing policy adopted by the government, the ISRO EOS Policy 01:2001 (ISRO 2001). The policy affirms that there is a national commitment to remote sensing in support of public policy objectives such as improvement of agriculture, regional development and poverty alleviation, and in that sense remote sensing is in direct support of the public good. The government of India owns the data from its remote sensing satellites and gives a licence to use the data to users, a model comparable to that used in Europe. Where the remote sensing policy in India stands out is in its explicit recognition of security, noting that the 'security situation of the country

[is of the] utmost importance'. The Indian policy defines three categories of data type in relation to security.

1 Remote sensing data with a pixel size larger than 5.8 m is available on a non-discriminatory basis and on request.
2 Data with a pixel size in the range 5.8 m to 1 m is actively screened to identify sensitive areas.
3 Data with a pixel size less than 1 m is subject to a formal clearance procedure and is reviewed in detail before access is granted.

A concern for security in remote sensing is not uncommon in many countries. Remote sensing data are often seen as a form of spying, especially at very fine spatial resolution. In the Sultanate of Oman, for example, areas of the country regarded as militarily sensitive, such as barracks, ports and the sultan's palaces, are censored and blacked out of aerial photographs, even those aerial photographs used by government ministries. The development of remote sensing from space at very fine spatial resolution means that data with a pixel size comparable to low altitude aerial photography are openly available, revealing all the barracks, ports and palaces in the case of Oman. Malaysia plans to launch its Razaksat remote sensing satellite in 2007 or 2008. Razaksat has a sensor with a 2.5 m pixel size. Data from Razaksat will be screened by an internal security organization before distribution because the government policy is to regard as sensitive any image data with a pixel size less than 5 m. In India Google Earth gave rise to security concerns. In *The Times of India* on 26 December 2005 President Abdul Kalam declared serious security concerns that Indian army headquarters, other defence establishments such as the Rashtrapati Bhavan and the Parliament House were readily visible on Google Earth images. An expert group was formed by the Indian government to establish India's formal position on access to Google Earth. Subsequent to this national security concern by politicians, the Indian Army's 3 Corps, based at Rangapahar, decided to subscribe to *Google Earth Pro* to gain access to remote sensing data of the northern states of India to assist with counter-insurgency operations (Anon 2006). While Google Earth is free to any user, Google Earth Pro costs US$400 for a licence and provides additional geographic data of a higher quality.

Military and intelligence requirements have always been important stimuli to remote sensing in the USA. While the country recognizes open access to data for both scientific and commercial reasons, security remains a concern for both domestic and foreign policy. An interesting illustration is the case of Israel. The Kyl–Bingaman Amendment to the U.S. National Defense Authorization Act of 1997[5] requires the operators of very fine spatial resolution sensors to degrade the spatial resolution of their data over Israel to no better than the resolution available from other commercial sources (see also the discussion on fine spatial resolution remote sensing by Toutin (in this volume)). This restriction has been called 'shutter control'. The Kyl–Bingaman Amendment translates into a limit on spatial resolution of 2 m pixel size: data of Israel may be collected at pixel sizes smaller than 2 m but can only be distributed at a 2 m pixel size or larger. The US retains the right under its legislation to restrict access to any remote sensing data from US government or commercial sources at times of national crisis.

Japan and the public good

Japan shares with India an extensive remote sensing programme, one that is broadly focused on the public good. The Japanese space agency JAXA has two core objectives for its remote sensing missions: to contribute to the protection of the Earth-environment system and to contribute to sustainable development. Japan has two main policy approaches in remote sensing. First, the scientific approach where its remote sensing data are used in national and international science projects. Second, a public benefit approach in which JAXA cooperates with operational agencies in Japan such as the Geographical Survey Institute, the Meteorological Agency and the Fisheries Agency to exploit the research investment in remote sensing by JAXA. An example of the public benefit approach can be seen in the name of the operational geostationary meteorological satellite mission MTSAT. The MT part of the name refers to the Japanese Ministry of Transport which is the institutional home for the Meteorological Agency. There is some similarity with Europe on remote sensing policy in that research and development is funded by one agency and when remote sensing systems move to an operational status then they fall under the responsibility of another organization that has an operational responsibility.

Geospatial data in Australia, New Zealand, and Canada

Australia and New Zealand have well-developed policies on geospatial data, of which remote sensing data form a part. In August 2000, the Australian government established an Interdepartmental Committee on Spatial Data Access (IDC 2001). This Interdepartmental Committee produced four main recommendations on access to geospatial data, of which the first two are particularly important for remote sensing.

1 Provide fundamental spatial data free of charge over the Internet, and at no more than the marginal cost of transfer for packaged products and full cost of transfer for customized services, without any copyright licence restrictions on commercial value-adding. Fundamental spatial data sets … will be identified in a public schedule.
2 Develop an Internet-based public access system, within the framework of the Australian Spatial Data Infrastructure. Agencies will be responsible for maintaining their own data access and management systems, but must comply with an agreed set of standards.

These ideas share common ground with the US policy approach in reducing barriers to data access. One objective of the Australia and New Zealand Land Information Council (ANZLIC) is to maximize the benefits of geospatial data by providing data at marginal cost of transfer so that (ANZLIC 1999):

… the community has easy, efficient and equitable access to spatial data in an environment where technology requirements, data formats, institutional arrangements and contractual conditions do not inhibit use.

In New Zealand, the safeguarding of geospatial data is formalized through the Public Records Act and implemented by a specific government department known as Archives New Zealand. The approach taken by Archives New Zealand is comparable to the National Satellite Land Remote Sensing Data Archive in the USA: if a government department wishes to dispose of a data set, including a remote sensing data set, then it must act within a legal framework to offer the data set to the archive.

GeoConnections in Canada is the organization that is implementing Canadian national policy on geospatial data. A guide developed in GeoConnections on best practice in the dissemination of government spatial data in Canada (Werschler and Rancourt 2003) noted the problems posed by the variety of policies at federal and state level:

… the data dissemination and licensing frameworks used to promote, extend and support the use of government geographic data generally have not kept pace with developments in technical capacity and growing user demand. … The variety of terms of use, fee structures, source acknowledgment and termination clauses used by federal departments makes it difficult to optimise the use of government geographic data.

An interesting development here is the view of copyright. The concept of copyright is often regarded as a restriction, but GeoConnections promotes the view that copyright can be used to protect or even promote data integrity and quality, a way of stating a brand or quality seal on data and information products. At the present stage of maturity of remote sensing such an approach has merit for helping to build a sound foundation for the exploitation of remote sensing data. For example, the Normalized Difference Vegetation Index (NDVI) is a commonly used product in remote sensing applications (Cracknell 1996, Mather 2004), but there is no agreement on which specific wavelengths of data are used to create an NDVI product beyond a combination of visible and near infrared wavebands, but the data used will affect the resulting NDVI product and its use in applications projects. Copyright could be used as a way of defining the input variables in an NDVI product, the processing steps followed and the quality of the output product.

DISCUSSION AND CONCLUSIONS

Pricing policy

An implicit theme in the development of policy in remote sensing is the question *who pays?* Open access and no cost for remote sensing data carries with it the implication that some organization,

typically a government department, will pay for the remote sensing mission, the launch, the ground segment and the operations. As long as the government maintains funding then open and free access can be maintained. Conversely, if there is a change in government policy then funding may dry up. An alternative approach is for the user to pay. The success of the user-pays model depends on the ability and appetite of the user organization to pay and therefore on the value of the remote sensing data to the user (Miller 2007). Remote sensing data are objectively neither expensive nor cheap. A 100 euro image may be expensive to one user if it has no information of value, but cheap to another user if it allows that organization to be more effective in what it does. In considering remote sensing pricing policy there are normally seven models that are used, and the arguments for and against these models are summarized in Table 2.2 (Harris 2002).

Sustainability of remote sensing

The pricing policy debate is wider than an argument about who pays for remote sensing data. One aspect of many remote sensing policies is to achieve sustainable systems in the long term. For example, a contribution envisaged by the Group on Earth Observations (Christian 2005, Lautenbacher 2006) is to:

… realize a future wherein decisions and actions for the benefit of humankind are informed via coordinated, comprehensive and sustained Earth observations and information.

Table 2.2 Pricing models for remote sensing data

Pricing model	Main characteristics
Free data for all users	Encourages sharing and extensive use, and is simple to administer. Maintains the costs on the supplier and fails to recognize the value of the data through the pricing mechanism.
Marginal cost price to all users	Termed COFUR in the USA and 'without charge' by Eumetsat. Encourages active selection by the user and avoids large deficits consequent upon large data requests. Maintains the core costs on the supplier and is expensive to administer.
Market price for all users	Can produce surpluses that can be reinvested in future missions. May restrict the use of data because of high prices.
Full, commercial price	Recovers all investment costs and can be sustainable. May be too expensive to create a healthy market for remote sensing data.
Two tier prices	Market prices for all except for a preferred group (usually research) that pays a marginal cost price. Recognizes differing abilities to pay and focuses on value for most users. May be open to misuse and is difficult to administer.
Access key pricing	Encrypted data are provided for free and the charge is for a decryption key. Encourages wide dissemination and at the same time focuses attention on the role of the remote sensing data in a user organization. Could involve very large volumes of data and restricts access to those with suitable technology.
Information content pricing	Products are sold on the basis of the information not on the basis of data volume. Maintains a focus on the value of the data to the user. Challenges the remote sensing community on how to measure and attribute value.

The question then becomes how best to achieve sustainable remote sensing? One route is to encourage those remote sensing systems that have become mature and for which there is a user base to make a transition from the research domain to the user domain. This has an important shift in the type of funding: a shift from funding in a sponsorship mode that is seeking to experiment, test and evaluate remote sensing systems to funding in an operational mode where the information provided meets an operational need that can be given a value. The operational user can equally be in the public or the private sector, for example a government environment department or a private forestry company. A second route is to regard remote sensing as a legitimate responsibility of government in collecting information that can be used for the greatest good for the greatest number. ESA uses this concept when referring to Envisat as providing a *Health Check* on planet Earth. This route is comparable to building road infrastructure that operational road users (public buses, private cars, company lorries) then use. A problem with remote sensing policy currently is that these two routes are often muddled with conflicting voices making conflicting demands.

The sustainability of remote sensing has been the focus of attention of several major initiatives, none of them so far successful in achieving operational remote sensing systems.

- The Centre for Earth Observation (CEO) of the European Commission provided funding to bring remote sensing scientists, technologists and users together to build environmental information systems that could form the basis of operational remote sensing. The CEO adopted and expanded a model developed in the UK by the British National Space Centre (BNSC), the Applications Development Programme.
- NASA provided funding in its Earth Observations Commercialization Applications Program (EOCAP) to stimulate a wider use in the private sector of remote sensing data from NASA missions (Macaulay 1995).
- ESA and the European Commission are collaborating on the GMES initiative. GMES focuses on a short list of remote sensing applications and provides funding to technologists and users to explore the use of remote sensing data in the expectation that users will be persuaded of the value of the remote sensing data and continue to pay for the data once GMES has finished. The GEOSS has similar objectives to GMES but is global in scope (Macaulay 2005).

Given that remote sensing can be technically used for all of the applications envisaged in these environmental programmes, applications such as crop monitoring, forest mapping, urban expansion measurement and natural disaster management, why has remote sensing not become truly operational, with the notable exception of meteorology? Satellite communications has moved from the research domain to the operational domain, so why not remote sensing? As this chapter has shown, remote sensing policy drives access to data and drives the opportunities to develop remote sensing missions. More analysis of how remote sensing policy influences remote sensing systems will help tease out how the remote sensing science and technology described in this book can take further steps into a sustainable, operational mode. Alongside this challenge there are also policy challenges in aligning international remote sensing policy, particularly the UN Principles on Remote Sensing, with new technology that employs pixel sizes of less than 0.5 m, improving technical access to data through international agreement on data formats and standards, ensuring that remote sensing data are safeguarded on a century time scale, and developing institutional capacity to use the data.

It is clear that remote sensing can be used for the greatest good for the greatest number. Remote sensing needs to take further policy steps to maximize its utility.

NOTES

1 The full title of the Outer Space Treaty is the *Treaty on Principles Governing the Activities of States in the Exploration and Use of Outer Space, Including the Moon and Celestial Bodies*, adopted 19 December 1966, opened for signature 27 January 1967, entered into force 10 October 1967 (von der Dunk 2002).

2 US Public Law 104-13, 109 Stat 163.

3 US Public Law 101-606, 104 Stat. 3096–3104.

4 US Public Law 102-155, section 401.

5 US Public Law 104-21.

REFERENCES

Achache, J., 2006. Implementation of GEO data sharing principles – challenges and opportunities, *Scientific Data and Knowledge within the Information Society*, CODATA 06 Conference, Beijing, China, October 2006.

Anon., 2003. *US Commercial Remote Sensing Policy*, National Security Presidential Directive 27, Washington DC, USA.

Anon., 2006. *GIS News*, May 2006.

ANZLIC, 1999. Policy Statement on Spatial Data Management: Towards the Australian Spatial Data Infrastructure, *Australia and New Zealand Land Information Council*, Canberra, Australia.

CCSP, 2006. *Our Changing Planet*, The U.S. Climate Change Science Program for Fiscal Year 2006, Washington DC.

CES, 1995. *Earth Observations from Space: History, Promise and Reality*, Committee on Earth Studies, Space Studies Board, National Research Council, Washington DC, 310pp.

Christian, E., 2005. Planning for the Global Earth Observation System of Systems (GEOSS), *Space Policy*, 21(2): 105–110.

Cracknell, A. P., 1996. *The Advanced Very High Resolution Radiometer (AVHRR)*, Taylor and Francis, London.

EC, 2003. *Space: a New European Frontier for an Expanding Union. An Action Plan for Implementing the European Space Policy*, European Commission, Brussels, 11 November 2003, COM(2003) 673.

Ernst, K., 2004. *The New Eumetsat Data Policies*, European Centre for Space Law, Paris, 12 March 2004, Eumetsat document EUM/LAD/VWG/04/205, Darmstadt, Germany.

ESA, 1998. *Envisat Data Policy*, European Space Agency ESA/PB-EO(97)57 rev.3, Paris, 19 February 1998.

Eumetsat, 1991. *First Eumetsat Workshop on Legal Protection of Meteorological Satellite Data*, Darmstadt, Germany, ISBN 92-9110-004-8, 310pp.

Eumetsat, 2006. *Eumetsat Data Policy*, Darmstadt, Germany, 28pp.

EurLex, 2006a. *Council Directive 90/313/EEC of 7 June 1990 on the Freedom of Access to Information on the Environment*, available at http://eur-lex.europa.eu/LexUriServ/, accessed 20 July 2006.

EurLex, 2006b. *Directive 2003/4/EC of the European Parliament and of the Council of 28 January 2003 on Public Access to Environmental Information and Repealing Council Directive 90/313/EEC*, available at http://eur-lex.europa.eu/LexUriServ/, accessed 20 July 2006.

EurLex, 2006c. *Directive 96/9/EC of the European Parliament and of the Council of 11 March 1996 on the Legal Protection of Databases*, available at http://eur-lex.europa.eu/LexUriServ/, accessed 20 July 2006.

Gillespie, A. R., 2005. *The Evolution of Remote-sensing Technologies and their Use in Developing Countries*, Master of Environmental Studies thesis, University of Waterloo, Ontario, Canada, 173pp.

Goward, S. N., T. Arvidson, D. L. Williams, R. Irish, and J. Irons, in this volume. Moderate Spatial Resolution Optical Sensors. Chapter 9.

Harris, A. and R. Harris, 2006. The need for air space and outer space demarcation, *Space Policy*, 22: 3–7.

Harris, R., 1999. The new ERS and Envisat data policies, *From Data to Information*, Remote Sensing Society, Reading, Conference Proceedings, pp. 341–345.

Harris, R., 2002. *Earth Observation Data Policy and Europe*, A. A. Balkema, Lisse, The Netherlands.

Harris, R., 2003. Earth observation and principles on data, in Harrison, C. and J. Holder (eds) *Law and Geography. Current Legal Issues volume 5*, Oxford University Press, Oxford, pp.539–555.

Harris, R. and R. Browning, 2005. *Global Monitoring: the Challenges of Access to Data*, UCL Press – Cavendish Publishing Ltd., London, 229pp.

Harte, N. and J. North, 2004. *The World of UCL*, UCL Press, London.

ICSU, 2004. *Scientific Data and Information*, Report of the CSPR Assessment Panel, International Council for Science, Paris, 42pp.

IDC, 2001. *A Proposal for a Commonwealth Policy on Spatial Data Access and Pricing*, Commonwealth Interdepartmental Pricing Committee on Spatial Data Access and Pricing, Canberra, Australia.

IPCC, 2007. *Fourth Assessment Report*, Intergovernmental Panel on Climate Change, WMO and UNEP, Geneva, Switzerland.

ISRO, 2001. *Remote Sensing Data Policy*, ISRO EOS 01:2001, Indian Space Research Organisation, Bangalore, India, 4pp.

Jasentuliyana, N. 1988. United Nations Principles on Remote Sensing, *Space Policy*, 4(4): 281–284.

Lautenbacher, C. C., 2006. The Global Earth Observation System of Systems: science serving society, *Space Policy*, 22(1): 8–11.

Longhorn, R. and M. Blakemore, 2004. Re-visiting the valuing and pricing of digital geographic information, *Journal of Digital Information* 4(2). http://journals.tdl.org/jodi/issue/view/18 (Last accessed January 26 2008).

Macaulay, M. K., 1995. NASA's Earth observations commercialization application program, *Space Policy*, 11(1): 53–65.

Macaulay, M. K., 2005. Is the vision of the Earth summit realizable? *Space Policy*, 21(1): 29–40.

Mather, P. M., 2004. *Computer Processing of Remotely Sensed Data*. Wiley, Chichester, 3rd edition.

Miller, L., 2007. *The Social and Economic Value of Earth Observation Data*, PhD thesis, University of London.

Oxford, 1981. *The Concise Oxford Dictionary of Quotations*, Oxford University Press, Oxford, 464pp.

Pearce, D., 1995. *Capturing Environmental Value. Blueprint 4*, Earthscan Publications Ltd, London, 212pp.

Peter, N., 2007. The EU's emergent space diplomacy, *Space Policy*, 23(2): 97–108.

Samuelson, P., 1954. The pure theory of public expenditure, *Review of Economics and Statistics* 36(4): 387–389.

Toutin, T., in this volume. Fine Spatial Resolution Optical Sensors. Chapter 8.

Teeuw, R., P. N. Aplin, McWilliam, T. Wicks, K. Matthieu, and G. Ernst, in this volume. Hazard Assessment and Disaster Management using Remote Sensing. Chapter 32.

von der Dunk, F., 2002. United Nations Principles on Remote Sensing and the user, in Harris R., 2002. *Earth Observation Data Policy and Europe*, A. A. Balkema, Lisse, The Netherlands, pp.29–40.

Werschler, T. and J. Rancourt, 2003. *The Dissemination of Government Geographic Data in Canada*. Guide to best practice, version 1.0, GeoConnections, Ottawa, Canada.

WMO, 1995. *Twelfth World Meteorological Congress, Abridged Final Report with Resolutions*, WMO No. 827, Secretariat of the World Meteorological Organisation, Geneva, 162pp.

Electromagnetic Radiation and the Terrestrial Environment

Visible, Near-IR, and Shortwave IR Spectral Characteristics of Terrestrial Surfaces

Willem J. D. van Leeuwen

Keywords: spectral signatures, reflectance, atmosphere, spectral response, minerals, soils, rocks, vegetation, non-natural features.

INTRODUCTION

Reflectance spectra have been used since the beginning of the remote sensing era to obtain information about the composition of materials and features on the Earth's surface. Traditionally, most operational- as well as research-based spaceborne sensors only acquire relatively broad bandwidth multispectral data. Nevertheless, the development of hyperspectral sensors has demonstrated that the information content of high spectral resolution data is particularly useful for identifying surface mineralogy. Since imaging spectroscopy has important commercial applications in mineral resource exploration, many nations and private companies have developed airborne and satellite hyperspectral systems (e.g., Airborne Visible/Infrared Imaging Spectrometer (AVIRIS) (NASA 2007); Hymap (Integrated Spectronics 2007); Compact Airborne Spectrographic Imager (CASI) (ITRES 2007); Hyperion (NASA 2003); Probe-1 (EarthSearch Sciences 2006)). Hyperspectral data has also been shown to be useful in retrieving biogeochemical data from vegetation and soils. Natural resource management applications (e.g., forestry, agriculture, global climate change, and variability) generally rely on relatively frequent, timely

and synoptic multispectral satellite information. Because multispectral sensors have a limited number of relatively broad bands, many narrow reflectance and absorption features of surface materials are not detectable due to spectral undersampling and averaging or smoothing. In general, as the complexity of spectral data increases, more information is potentially available about the identity, concentration, proportion, and spatial distribution of the constituents of the objects imaged, particularly when analyzed with powerful analytical processing techniques that have been developed over the last two decades (Singh, 1989; Schowengerdt, 1997).

Visible, near-infrared, and shortwave infrared electromagnetic radiation is reflected, absorbed, and transmitted in different ways by both biotic and abiotic terrestrial surfaces. Each of these terrestrial surfaces can be composed of pure substance and mixtures, including minerals, soils, snow, water, and living organisms, all of which can exhibit diagnostic spectral features at different wavelengths. Consequently, many terrestrial features exhibit specific spectral characteristics that can be used to extract information from remotely sensed data. This chapter will provide an overview of the optical interactions of sunlight with terrestrial surface materials and features with a focus on spectral reflectance. The spectral reflectance characteristics for several representative terrestrial materials will be highlighted and evaluated in terms of how their spectral response changes with respect to sensor characteristics.

The spectral wavelength (λ) range of interest for this chapter is between 0.4 and 2.5 μm; this range is usually subdivided into the visible (0.4–0.7 μm), near-infrared (NIR, 0.7–1.1 μm), and shortwave infrared (SWIR, 1.1–2.5 μm). Laboratory spectral measurements often use a tungsten lamp for illumination of minerals, rocks, vegetation, and man-made materials and features to acquire and characterize their spectral 'signatures'. The United States Geological Survey (USGS) (http://speclab.cr.usgs.gov/spectral-lib.html; Clark et al. 2003, 2007) and NASA Jet Propulsion Laboratory (JPL) (http://speclib.jpl.nasa.gov/) on-line spectral libraries are an excellent resource for examining spectral signatures of terrestrial surface materials. The sun, atmosphere, and Earth surface spectral interactions often result in slightly different spectral characteristics compared to laboratory spectra due to signal degradation by the atmosphere and limited solar irradiance beyond 2.5 μm. Many satellite sensors like ASTER, MODIS and MERIS have spectral channels that are suitable

for atmospheric as well as terrestrial observations. From the perspective of wanting to discriminate or identify terrestrial spectral signatures, these atmospheric bands can be used to correct apparent or at-sensor radiance or reflectance values to surface reflectance. Such corrections ultimately improve our ability to perform quantitative remote sensing applications like change detection, materials identification, global albedo quantification, and vegetation time-series analysis related to climate variability and change.

PRINCIPLES OF ELECTROMAGNETIC INTERACTIONS (0.4–2.5 μM) AT THE EARTH'S SURFACE

Intermittent atmospheric absorption windows due primarily to O_2, O_3, H_2O, CO_2, and CH_4 gases as well as scattering and absorption by air molecules and aerosols will affect all outdoor reflectance measurements that use the Sun as an energy source (Figure 3.1). The fraction of incident solar radiation reflected by a particular surface material at a given wavelength (i.e., spectral reflectance) is closely linked to the materials' physical (for example density, surface

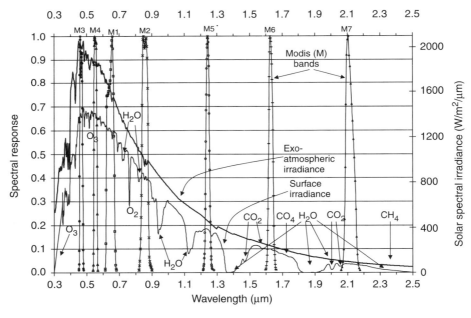

Figure 3.1 Top of the atmosphere (TOA) spectral solar irradiance and an example of the solar spectral irradiance at the Earth's surface after absorption and scattering by the atmospheric components water vapor, aerosols, ozone, carbon dioxide, methane, and oxygen. Strategically placed spectral bands for MODIS are distributed over the atmospheric windows to avoid strong atmospheric effects.

roughness, structure), chemical (for example content and/or concentration of chlorophyll, water, iron oxide, mineral composition, organic matter), and biological properties (for example photosynthetic activity, senescence, vegetation phenology). Knowing the spectral response function is important for calibrating how much spectral irradiance (E_λ; in W m$^{-2}\mu$m^{-1}) hits the top of the atmosphere (TOA), determining how much down-welling spectral irradiance arrives at the surface, and calibrating the amount of spectral upwelling radiance (L_λ; in W m^{-2} sr$^{-1}\mu$m^{-1}) that travels back through the atmosphere and gets measured by a sensors' detectors. The linear equation (1) describes how each channels' digital number (DN) counts can be converted to spectral irradiance values based on channel specific calibration coefficients a and b. E_λ depends on the Sun–Earth distance d in astronomical units (Teillet et al. 2001).

$$L_\lambda = aDN + b \qquad (1)$$

$$\rho^{app} = \pi L_\lambda / d^2 E_\lambda \; \cos\theta_s \qquad (2)$$

The top of the atmosphere or apparent reflectance (ρ^{app}; unitless) is computed based upon Equation (2) and can be expressed in terms of surface reflectance and atmospheric correction parameters to derive the atmospherically corrected surface reflectance (Eqn (3)). For a perfectly diffusing (Lambertian), homogeneous and flat surface the following equation has been developed (Vermote et al. 1997):

$$\rho^{app}(\theta_v, \theta_s, \varphi) = \tau_g(\theta_v, \theta_s)[\rho^i(\theta_v, \theta_s, \varphi) \\ + \tau_{tw}(\theta_v, \theta_s)\rho^s/(1 - S\rho^s)] \quad (3)$$

where ρ^{app} is the apparent reflectance, ρ^i is the intrinsic atmospheric reflectance due to Rayleigh and aerosol scattering, ρ^s is the surface reflectance of a ground target, θ_v is the sensor view angle at the surface relative to nadir, θ_s is the solar view angle at the surface relative to nadir, φ is the relative azimuth angle between view, and solar azimuth angles, τ_g is the gaseous transmittance, τ_{tw} is the two-way atmospheric transmittance due to Rayleigh and aerosol scattering, and S is the Spherical albedo of the atmosphere due to Rayleigh and aerosol scattering.

The spectral bandwidth, sampling frequency, and signal-to-noise-ratio of multispectral (e.g., Landsat, MODIS, VIIRS, AVHRR) and hyperspectral (AVIRIS, EO-1) spectroradiometers varies significantly among sensors and greatly affects the ability to resolve narrow absorption and reflectance features. The spectral response functions or bandpass profiles and their spectral resolution combined with sensor design and data downlink rates are important determinants of how well and frequently a sensor can be used for global to regional ecosystem monitoring or for geological and biogeochemical applications (Swayze et al. 2003).

This chapter includes both laboratory and field acquired spectral measurements. Laboratory spectra are continuous while field-based measurements generally display a gap in the strong water absorption bands at 1.4 and 1.9 μm. Data from spectral libraries provided by the U.S. Geological Survey (Clark et al. 2003) and NASA Jet Propulsion Laboratories' 'ASTER spectral library' (NASA 1999[1]) have been utilized along with measurements recently acquired with a full range spectroradiometer, FieldSpec®3, manufactured by Analytical Spectral Devices Inc. (ASD).

Absorption and scattering processes

This section provides a brief description of the basic principles that govern the electromagnetic interactions with Earth's surface material and features between 0.4 and 2.5 μm. The radiation budget equation as governed by the law of conservation of energy, states that the fractional reflectance (ρ) transmittance (τ) and absorption (α) characteristics for each target on the Earth's surface sums to 1 ($1 = \rho + \tau + \alpha$). Figure 3.2 shows that plants have low reflectance and transmittance and high absorptance in the visible wavelengths, high transmittance and reflectance and relatively low absorptance in the NIR, and moderate to high absorptance in the SWIR regions, especially in the water absorption bands at 1.45, 1.9, and 2.4 μm for plants.

Electromagnetic radiation will interact at the atomic and molecular level of terrestrial features, resulting in spectral responses that are a function of electronic, vibrational, and rotational processes. Hunt (1977, 1982); Burns (1993); and Farmer (1974) provide significant insights into the principles of the complex interactions of photons and materials. Clark (1999) discusses scattering properties of materials and other related topics.

Factors affecting spectral reflectance signatures

There are a wide range of factors that create uncertainties in spectral reflectance signatures. The following list of model parameters, system characteristics, and environmental factors associated with the calculation of reflectance are often taken into account to minimize reflectance data uncertainties, enhance interpretation, and permit multitemporal analysis:

- spectral response function,
- vicarious calibration,

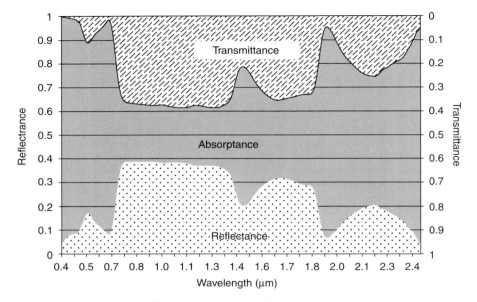

Figure 3.2   Herbaceous leaf reflectance and transmittance spectral signatures and resulting absorptance.
Source: **Based on data from Asner et al. 1998.**

- radiance values,
- atmospheric parameterization of radiative transfer correction algorithms,
- field of view and sun-view geometry,
- topography,
- satellite orbital drift,
- variable pixel resolutions across swaths,
- cloud cover and sub pixel clouds,
- snow, dust, and soil background,
- vegetation cover characteristics,
- co-registration,
- algorithm selection rules,
- time interval,
- scale and aggregation,
- and metadata.

Bidirectional reflectance distribution function (BRDF)

The BRDF is wavelength specific and yields the reflectance of a surface feature as a function of illumination and viewing geometry in units of steradians (sr; Eqn (4)). Reflectance hotspots occur when satellite view and sun angle are aligned. Specular reflectance properties (solar incidence similar to the angle of exitance) are characteristic of flat and smooth surfaces. Perfectly diffuse (Lambertian) reflectance rarely exists, both soils and vegetation canopies tend to exhibit varying levels of anisotropy depending on solar illumination and view angles (Kimes 1983). Generally,

measurements made in the backscatter direction (sensor view and solar illumination in the same direction) have higher reflectance values than in the forward scatter direction. Backscatter and forward scatter view and sun angle interactions with non-Lambertian Earth surfaces results in bidirectional distributed reflectance observations that depend on spectral band characteristics, surface roughness, shadows, topography, vegetation geometry, and volume scattering. BRDF models (e.g., Walthall et al. 1985, Roujean et al. 1992) can be parameterized with sufficient bidirectional reflectance factor (BRF; Eqn (5)) observations after which reflectance data can be normalized to nadir view angle or to derive albedo. The BRF relates the reflectance values from a target surface to the reflectance values that would be observed from a Lambertian surface (e.g., spectralon panel) at the target location. The BRF is generally measured in the field. Additional information on BRDF and other reflectance terms is given by Schaepman-Strub et al. (in this volume).

$$\mathrm{BRDF} = \mathrm{L}(\theta_v, \theta_s, \varphi, \lambda)/E(\lambda)(sr^{-1}) \quad (4)$$
$$\mathrm{BRF} = \pi\,\mathrm{BRDF} \quad (5)$$

The wavelength dependency of the BRDF (Eqn (5)) will alter Earth's surface spectral signature when data are collected at different sun/view geometries. In remote sensing science, the BRDF is needed for the correction of view and illumination angle effects for deriving albedo, land cover classification and change detection, atmospheric correction, image compositing and mosaicking, and many other applications.

Time-series of spectral signatures

Time-series of daily or bi-weekly discrete snapshots of local, regional or global regions are commonly used to analyze spatio-temporal trends and dynamics in natural and non-natural (urban) resources. Spectral signatures that are used in time-series analysis to determine change or variability of particular phenomena are acquired throughout the year during the satellite overpass times. Therefore, all terrestrial spectral reflectance data are affected by seasonal changes (or 'noise') in sun angle, and variable atmosphere and sensor perturbations. For example, applications which are time sensitive that examine the relationship between climate and vegetation phenology, land cover change, or global crop production often incorporate these variables into their analysis to account for seasonal trends that are not related to landscape dynamics. Separating the chronologic and random 'noise' in time-series data from an ecosystem's time-series spectral signature is often a challenge, but can be accomplished through normalization schemes or by conducting season-to-season comparisons.

Spatial mixtures of spectral signatures

The total net point-spread-function (PSF) (Schowengerdt 1997) of the sensor's optics, detectors, electronics and movement has a blurring effect on the spatial responsivity and affect the fidelity of the spectral signatures measured by satellite sensors especially when adjacent pixels have significantly contrasting spectral characteristics. Scattering of reflected signatures in the atmosphere and the PSF will degrade the clarity of remotely sensed imagery because most pixel values are actually representations of the pixel *with* contributions from the adjacent areas. Spatial features like texture, edges, and other patterns will become spectrally less separable due to the mixing of spectral responses that overlap. Spectral signature(s) of a pixel representing the landscape can be pure (e.g., mineral or soil) or a mixture of two or more landscape components (e.g., soil, rock, grass, shrub, tree). Woodcock and Strahler (1987) called the former representation a simple scene or pixel and the latter a complex scene or mixed pixel. Landscape spectral heterogeneity and the spatial resolution of the sensors determine which spectral signatures are represented in each pixel. Woodcock and Strahler (1987) demonstrated with geostatistical analysis that pixels that are smaller than the size of the landscape components are needed for representing the spectral signatures of simple scenes/pixels; pixels that are larger than the size of the components in the scenes are represented by complex mixtures of the unique spectral landscape component signatures present in the scene.

CONTINUITY OF MULTI-SENSOR SPECTRAL SIGNATURES

Although multi-sensor continuity can be obtained by using cross-sensor translation equations, the interactions between the spectral characteristics of surface components, sensor-specific spectral band characteristics and atmospheric scattering and absorption windows will introduce uncertainty due to insufficient knowledge about the atmospheric conditions that affect the signal of the Earth's surface at the time of data acquisitions (van Leeuwen et al. 2006).

Multispectral and hyperspectral data will have different information content because fine spectral reflectance and absorptance characteristics can be identified with hyperspectral data while multispectral data generally have levels of spectral resolution that do not permit such recognition of unique materials and features. However, several narrow band sensors like MERIS in the visible, and ASTER in the SWIR, have greater potential to identify a specific minerals' electronic and vibrational features and subtle movement of vegetation's red edge, a feature that is discussed later in this chapter.

Uncertainty arises when multi-sensor reflectance data are used in the production of time-series reflectance imagery with similar but divergent variables and parameters. Spatial and temporal continuity and consistency (i.e., artifacts, seasonal, inter-annual, and trend analysis) are most important for seasonal resource monitoring applications. For example, remotely sensed vegetation phenology observations are useful integrative measures for tracking long term climate change and variability. The usefulness of a reflectance product is greatly improved when some quality assurance indicators (flags) associated with the reflectance products (e.g., metadata about data source and processing algorithms) are included and consequently become part of the quality assurance data for up- or downstream products.

In order to demonstrate the effects that inherent sensor characteristic (spectral band pass and channel location) have on the spectral reflectance of Earth features, a numerical experiment was conducted to simulate sensor specific reflectance data. QuickBird, IKONOS, ASTER, LANDSAT-7, LANDSAT-5, IRS, SPOT-5, MODIS, VIIRS, AVHRR-14, AVHRR-18, and MERIS spectral response functions data were used in the analysis (van Leeuwen et al. 2006). For the IRS

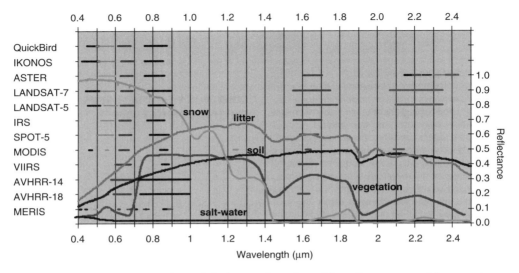

Figure 3.3   Spectral bandwidths and their wavelength positions for a range of frequently used sensors (QuickBird, IKONOS, ASTER, LANDSAT-7, LANDSAT-5, IRS, SPOT-5, MODIS, VIIRS, AVHRR-14, AVHRR-18, and MERIS). The bandwidth is represented by the full width at half-maximum (FWHM) spectral response. Soil, snow, green leaf, and senescent brown leaf spectral signatures are shown to illustrate the importance of the position of the instrument band-passes as they affect the reflectance response of a surface consisting of a mixture of contributing components (e.g., soil, vegetation, water, NPV). The bright soil is an Alfisol soil developed under temperate forests of the humid mid-latitudes. (See the color plate section of this volume for a color version of this figure).
Source: Based on data from NASA, 1999[1].

and VIIRS sensors we used the full-width half-maximum (FWHM) values to convolve leaf and soil reflectance data. The spectral response functions (not shown) and band widths vary widely among these sensors as can be seen in Figure 3.3 and Table 3.1. Vegetation and soil spectra were simulated using hyperspectral measurements of a bright soil from the ASTER spectral library (NASA 1999[1]), and herbaceous leaf spectra (reflectance and transmittance; Figure 3.2) from Asner et al. (1998). The sensor spectral response data were used to calculate weighted soil and leaf spectra for each sensor's specific spectral channels (e.g., red and NIR). The sensor specific spectral bandwidths are shown in Figure 3.3. The resulting leaf reflectance and transmittance and soil reflectance data were used to simulate sensor-specific canopy reflectance data for a range of leaf area index (LAI) values. The scattering by arbitrarily inclined leaves (SAIL) model (Verhoef 1984; 1998) was then used to simulate the spectral reflectance values for a range of LAI values (LAI = 0, 0.1, 0.3, 0.5, 1.0., 1.5, 2.0, 3.0, 4.0, 5.0) and sensor specific spectral channels (red, NIR) at a constant nadir view angle and 40° sun angle. For each sensor, the normalized difference vegetation index (NDVI) was calculated from these red and NIR

reflectance values to demonstrate the effect of sensor specific spectral response functions on both reflectance and derived products like the NDVI (see Table 3.1). The output from the SAIL model allows us to estimate the effects of spectral bandwidths and response of the different sensors on the NDVI.

The reflectance and derived $NDVI_{sensor}$ values for a range of LAI values vary significantly among the different sensors (Table 3.1). Simulated NDVI values for soil (LAI = 0) varied from 0.10 (IKONOS and ASTER) to 0.15 (AVHRR), while the highest LAI values resulted in simulated NDVI values between 0.69 (AVHRR-14) and 0.77 (VIIRS and MERIS).

These simulation experiments highlight the fact that the sensor specific spectral band response curves in combination with the vegetation and soil spectral signatures cause sensor-dependent reflectance and NDVI responses with respect to the range of LAI values. Hence, caution should be taken when multi-sensor and instrument reflectance data are used, compared, aggregated, mixed, or interpreted. The relationship of biophysical properties and spectral reflectance data measured with each sensor will also vary with different land cover types (Myneni et al. 1997).

Table 3.1   Full width of the spectral bands at half of the maximum spectral response for the blue, red, NIR and SWIR bands, and corresponding range of NDVI values (for a herbaceous cover and bright soil) for QuickBird, IKONOS, ASTER, LANDSAT-7, LANDSAT-5, IRS, SPOT-5, MODIS, VIIRS, AVHRR-14, AVHRR-18, and MERIS

Sensor	Full width half maximum sensor spectral bands						NDVI
	Blue (nm)	Green (nm)	Red (nm)	NIR (nm)	SWIR-1 (nm)	SWIR-2 (nm)	
QuickBird	446–512	494–590	614–682	755–874			0.1230 – 0.7350
IKONOS	421–516	506–596	632–698	757–853			0.1002 – 0.7088
ASTER		512–601	628–691	754–859	1607–1702	2146–2184	0.1084 – 0.7471
						2187–2228	
						2240–2287	
						2298–2369	
						2363–2431	
LANDSAT–7	441–514	519–601	630–692	771–898	1547–1749	2065–2346	0.1220 – 0.7638
LANDSAT–5	452–518	438–610	626–693	776–905	1567–1784	2097–2349	0.1194 – 0.7638
IRS		520–590	620–680	770–860	1550–1700		0.1244 – 0.7571
SPOT–5		499–588	619–686	782–883	1585–1683		0.1214 – 0.7237
MODIS	456–475	544–564	620–670	837–876	1230–1254	2086–2140	0.1462 – 0.7577
					1616–1644		
VIIRS			600–680	846–885	1580–1683		0.1287 – 0.7819
AVHRR–14			575–705	720–1000			0.1571 – 0.7249
AVHRR–18			588–680	732–995	1577–1637		0.1502 – 0.6888
MERIS	407–417	505–515	615–625	750–757			0.1215 – 0.7700
	437–447	555–565	660–670	758–765			
	485–495		677–685	773–780			
			703–713	855–875			
				880–890			
				895–905			

SPECTRAL SIGNATURES OF EARTH'S SURFACE FEATURES

The spectral characteristics and phenomena of both abiotic and biotic surfaces like soils, rocks, minerals, water, vegetation, water, snow, and man-made structures will be highlighted, and are essential for employing multiple sensors to explore and monitor Earth's surface resources. Figure 3.3 shows examples of spectral signatures for green vegetation, non-photosynthetic vegetation (NPV), an Alfisol (Alfisols are well-developed soils with a subsurface clay horizon and are normally found under temperate forests of the humid midlatitudes), snow, and water. All five signatures have unique spectral reflectance features and shapes that can be mapped or exploited in the analysis of hyperspectral or multispectral imagery. Remote sensing applications and analysis techniques that take advantage of these pure spectral signatures are numerous. Not only can data be analyzed with spectral libraries, new spectral signatures can be identified, extracted, and visualized as well. Analysis techniques that use this kind of spectral information include color transforms, spectral derivatives, spectral angle mapper (SAM) (Kruse et al.

1993), linear multiple endmember mixture models (Smith et al. 1985; Huete 1986; van Leeuwen et al. 1997), spectral ratio techniques; spectral continuum removal (Clark and Roush 1984), and spectral feature fitting (Clark et al. 2003). Color transforms convert three-band red, green, blue (RGB) images to other color space such as the hue, saturation, value (HSV), or the USGS Munsell color chart (Kruse and Raines 1994). Linear mixture models often use a factor analysis methodology to determine the relative abundance of surface components that are represented in multispectral or hyperspectral imagery based on the components' spectral characteristics (Adams and Gillespie 2006). The spectral signatures or reflectance values of a pixel are often assumed to be a linear combination of the reflectance signature of each pure 'endmember' present within the pixel. It should be noted that scattering processes in the visible and NIR are generally non-linear but the approximation has been used extensively. Spectral unmixing results are highly dependent on the selected endmembers. Spectral angle mapper (SAM; Kruse et al. 1993) determines the spectral similarity between two spectra by calculating the angle between the spectra of the pixels and reference spectra. The spectral feature fitting algorithm is an absorption feature methodology that uses scaled reference spectra to compute the abundance of these features in the image spectra after the continuum is removed from both data sets.

Water and snow spectral signatures

Water absorbs most radiation in the longer wavelength and has only low reflectance in the shorter visible wavelengths (Figure 3.3). Water naturally appears blue or blue-green due to stronger scattering at these wavelengths. Water reflectance values will be higher if suspended sediments or dissolved organic or inorganic materials are present in the upper layers of the water body (Dekker et al. 1992). However, shallow clear water and water with suspended sediment can be easily confused since these two phenomena will appear spectrally similar. Water with algae or phytoplankton and high chlorophyll concentrations will reflect the green and absorb more of the blue wavelengths causing the water to appear greener. Snow on the other hand has very high reflectance values in the visible wavelength with rapidly decreasing reflectance values toward longer wavelengths (Figure 3.3). The reflectance properties of snow depend on the size distribution of ice grains, the amount of liquid water inclusions, as well as the amount of solid (e.g., dust) and soluble (e.g., salt) impurities (Dozier et al. 1988; Dozier 1989). Snow has a small absorption feature at 0.90 μm and strong absorption features at 1.03, 1.25, and 1.50 μm. The contrast between SWIR and NIR reflectance values for snow is very strong. Spectra of clouds often appear to be similar to spectra of snow (Gao et al. 1998). Reflectance values are generally higher for water- and ice-clouds than for snow. Snow and clouds can best be distinguished at the 1.6 μm wavelength. Snow has very low reflectance, while the reflectance of cirrus and optically thick clouds remains high at 1.6 μm (Dozier 1989). Water and snow surfaces have spectrally distinct features which sets them apart from those of soils and vegetation.

Vegetation spectral signatures

Actively growing plants exhibit a significant contrast between strong absorption in the red and high reflectance in the near-infrared (NIR) regions of the spectrum. The amount of absorption in the red and reflectance in the near-infrared varies with both the type of vegetation and the density or vigor of the plants. The natural change of a green leaf as it senesces is shown in Figure 3.4. We see that a healthy green leaf (spectrum c) has very low reflectance values in the blue and red due to strong chlorophyll *a* and *b* absorptions (centered

around 0.65 μm which accounts for 60–75% of the absorption and causes plants to appear green), carotenoids absorptions (centered around 0.43 μm; accounts for 25–35% absorption), pigment absorptions (xanthophylls absorb visible radiation strongly in the 0.35–0.5 μm region, causing visually attractive autumn colors in many mid-latitude countries). We observe very high reflectance values in the near-infrared with two water absorption features around 0.985 and 1.2 μm. Between the highly absorbing visible region and the highly reflective NIR is the so-called *red-edge* at around 0.72–0.76 μm (Horler et al. 1983). Shifts in the *red-edge* are known to be caused by stresses related to plant nutrient and water availability (Collins et al. 1983). The SWIR reflectance properties of leaves are strongly affected by water absorption and lignin and cellulose concentrations. Senescent leaves have several absorption (2.1 and 2.3 μm) and reflectance peak features (2.0, 2.2 and 2.4 μm) in the SWIR, while healthy green leaves generally have a reflectance maximum at 2.2 μm. The SWIR reflectance for NPV leaves (spectrum a in Figure 3.4) is generally higher than the SWIR reflectance of live green (c) or yellow (b) leaves.

At the individual leaf scale, leaf pigment concentrations (chlorophyll and carotenoids), leaf water content and leaf structure cause variations in leaf reflectance, transmittance and absorption signatures (Gates et al. 1965; Sinclair et al. 1971; Woolley 1971; Walter-Shea et al. 1992). Leaf pigments in healthy green leaves cause strong absorption of sunlight, which results in low reflectance and transmittance in the visible parts of the spectrum (0.4–0.7 μm). The near-infrared is minimally absorbed by green leaves, although certain photosynthetically active plants like cacti tend to manifest some strong absorption features in the NIR (spectrum d in Figure 3.4). Reflectance and transmittance properties in the NIR are mostly a function of leaf structure (mesophyll structure, leaf thickness, external hairs or thorns) and to a lesser extent leaf water content. The leaf mesophyll structure affects the intra-leaf scattering in both the visible and near-infrared (NIR) since it determines the mean path length of incident radiation through the leaf. A longer path length will improve the photon capture by plant pigments and decrease the reflectance and transmittance in the visible part of the spectrum. Since the NIR has low absorptance, this scattering mechanism will increase the reflectance in the NIR. The transmission will decrease with thicker leaves for both the NIR and SWIR. Dicotyledonous and monocotyledonous leaves have different reflectance and transmittance properties because of variability in their mesophyll structure (Sinclair et al. 1971).

Dicotyledonous leaves (e.g., trees) may differ markedly in the structure of their adaxial (leaf face) and abaxial (leaf back) sides, showing

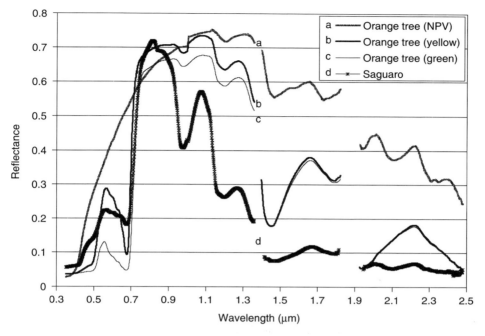

Figure 3.4 Leaf spectral signatures measured with an ASD spectroradiometer. Green, yellow, red, and dry senescent leaves show the effect of pigments, mesophyll, and water content on spectral signatures. The two plants are broad leaf orange tree (*Citrus sinensis*), and Saguaro cactus (*Carnegiea gigantean*).

corresponding differences in their reflectance and transmittance spectral signatures. Monocotyledons (e.g., wheat, grass) generally have very similar adaxial and abaxial leaf scattering properties (Gates et al. 1965; Woolley 1971). Reflectance differences between vegetation and senescent vegetation can be attributed to histological and optical properties (van Leeuwen and Huete 1996). Senescence of plant components occurs during or after plant maturity or can be caused by stress factors like lack of water and nutrients or extreme temperatures. These natural and environmental factors can cause changes in leaf pigmentation concentration and composition (Sanger 1971) and changes in leaf structure (more air-filled intercellular leaf space). Senescence and decomposition of leaves will finally cause a breakdown of all pigments. The internal leaf structure will collapse, and different leaf structures, mainly consisting of cellulose, will be left in different stages of decomposition, changing color from yellow/red/brown to brown/gray and black.

Variability in the bidirectional spectral properties at the canopy and landscape scale (Asner 1998) are affected by sun-sensor geometry, canopy structure (e.g., LAI and leaf angle distribution), bidirectional plant component scattering properties (Jacquemoud and Baret 1990), bidirectional soil reflectance properties, and bidirectional diffuse and direct irradiation (Colwell 1974). Therefore, the spectral response at the leaf scale is significantly different from the spectral response at the canopy and landscape scale.

Figure 3.5 shows SAIL modeled canopy spectral signatures for a range of LAI values and an Alfisol soil background. The data are based on the herbaceous leaf reflectance and transmittance data shown in Figure 3.2, the soil spectral data from Figure 3.3, and a spherical leaf angle distribution. Figure 3.5 shows that when the canopy density increases, as indicated by increasing LAI values, the associated red reflectance values decrease, the NIR reflectance values increase, and the SWIR reflectance values decrease (Figure 3.5). Absorption bands in the soil dominated spectral signature (LAI = 0.01) are clearly visible at low LAI values, but are obscured by the vegetation signature as the LAI increases. Vegetation canopy absorption bands in the red and SWIR wavelength are increasingly prominent at higher LAI values.

Dry and wet soil signatures

Soil optical properties have been measured and analyzed in the laboratory and with airborne

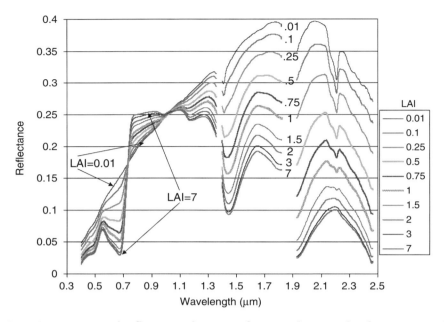

Figure 3.5   Canopy spectral reflectance signatures for LAI values ranging from 0.01 to 7 based on soil and leaf spectra and a simulation using the SAIL model.

sensors and field-based radiometric experiments. It should be noted that the focus is on reflective properties of soils and that the sensible depth is about two times the effective diameter of the particle size according to a modeling study by Liang (1997). Soil reflectance properties are governed by a combination of biogeochemical elements, moisture conditions at the surface and geometrical-optical scattering (Hapke 1981; Baumgardner et al. 1985; Ben-Dor et al. 1999; Huete, 2004). Iron oxides and hydroxides (Fe–O, Fe–OH) in most soils cause steep decreases in reflectance toward the blue wavelengths. The visible and near-infrared regions are characterized by pronounced spectral absorption features associated with red ferric iron (hematite; 0.7 and 0.87 μm) and a ferrous iron absorption feature near 1.0 μm (goethite). Other mineralogical forms of iron can cause weaker absorptions between 0.4 and 0.55 μm. Reflectance spectra of moist soils include absorption bands at 1.4 and 1.9 μm. Iron oxide absorptions also occur in the SWIR and can overwhelm the water absorption bands in this region. Minerals that have OH (e.g., illite, montmorillonite), CO_3 (e.g., calcite), or SO_4 (e.g., gypsum) exhibit absorption features in the 1.3–2.5 μm region. Clays with hydroxyls absorb near 1.4 and 2.2–2.4 μm. (Baumgardner et al. 1985; Mulders, 1987; Clark et al. 1990). Organic material has absorption features in the 1.45, 1.7 (C–H overtone stretch), 2.1 and 2.3 μm bands (Nagler et al. 2000).

Soils are composites of a number of abiotic and biotic constituents that make it complicated to distinguish many soil spectral signatures. Based on work by Condit (1970); Stoner and Baumgardner (1981); and Baumgardner et al. (1985), five distinct spectral curves represent most soil groups (Figure. 3.6):

1   organic dominated soils with high organic matter (OM) content and low iron content have a concave (upward) shape from 0.5 to 1.3 μm (Figure. 3.6 spectrum g, 11.3% clay, 8.4% OM);
2   iron dominated soils with high iron content and lower amounts of organic matter have reflectance curves that show evidence of absorption features at 0.4–0.5, 0.75, 0.9 and 1.0–1.1 μm (spectrum h, 7.0% free Fe, 17.7% clay, and 0.5% OM);
3   minimally altered soils with both low organic matter (OM) and low iron content that have a convex (upward) curve shape from 0.5 to 1.3 μm (spectrum c, quartz sand);
4   organic matter affected soils with lower organic matter content (spectrum f, 14.9% clay, 0.5% OM), not yet fully decomposed, and low iron content have concave (upward) curves from 0.5 to 0.75 μm and convex (upward) curves from 0.75 to 1.3 μm, and resembles spectra of senescent leaves; and

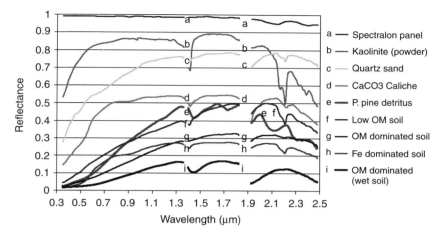

Figure 3.6 Kaolinite, sand, iron, and organic matter (OM) affected and dominated soils as well as dry and wet soils and detritus spectral signatures based on outdoor ASD measurements (Tucson AZ; March, 2007). Spectralon reflectance data are shown in this and subsequent reflectance figures because a Spectralon panel was used as a reference calibration source.

5 iron affected soils with low organic matter content and medium iron content have slight ferric iron absorption at 0.7 μm, and stronger Fe absorption at 0.9 μm, especially within kaolinitic soils.

Several additional noteworthy features of soil reflectance are shown in Figure 3.6. The spectrum of the $CaCO_3$-rich soil (spectrum d; 9.0% clay, 0.0% OM, and 3.0 $CaCO_3$) is very bright with weak absorption features. The spectrum of the organic forest floor has a concave shape in the visible region typical of organic matter dominated soils, but also displays several absorption bands in the NIR and SWIR regions (spectrum e). At 2.2 μm, soils may have an absorption band (spectra h and f) while detritus manifests a reflectance peak (spectrum e). Organic matter dominated soils could conceal these features and thereby making the spectrum flat in the SWIR (spectrum g). High soil moisture content reduces soil brightness in the visible and NIR (spectrum h) and strongly absorbs incident light in the SWIR region resulting in much lower reflectance values than usually present in dry soils (spectrum g).

Mineral and rock spectral signatures

With the advent of multispectral remote sensing, geoscientists have been able to utilize the reflectance characteristics of rocks and minerals to discriminate rock and soil types based upon general mineral assemblages and to identify specific mineral constituents based upon characteristic absorption features in the visible, NIR and SWIR

regions of the electromagnetic spectrum. The most common rock and soil forming minerals (e.g., quartz, feldspars) do not exhibit distinctive spectral features in the visible to the SWIR wavelength range. However, many minerals have characteristic absorption features that permit their direct identification. The pioneering work of Graham Hunt of the U.S. Geological Survey documented the reflectance spectra of a wide range of minerals (Hunt and Salisbury 1970; 1971; Hunt et al. 1971a,b, 1972; Hunt 1977) and rocks (Hunt et al. 1973a,b,c) and ultimately led to the development of hyperspectral sensors (Goetz et al. 1985). As previously noted, spectral libraries are now available for a large variety of minerals and rocks through the USGS (Clark et al. 2007) and NASA (1999) spectral libraries.

Characteristic absorption features in the visible and near-infrared are associated with electronic transitions and charge transfer processes. Generally, these absorption features are crystal field effects in transition elements (Fe, Cr, Co, Ni) (van Der Meer and De Jong 2001). In the shortwave infrared, vibrational processes associated with the presence of H_2O, hydroxyl (OH), carbonate, and sulfates give rise to deep and often diagnostic absorption features.

The ability to map surface geology and mineralogic alteration associated with hydrothermal ore deposits was first documented by Abrams et al. (1977). This and similar initial studies utilized airborne scanners with broad spectral bands in the visible and shortwave infrared that permitted the differentiation of rock alteration zones based upon the general presence of mineral absorption features in the blue/green and at 2.2 μm. With the launch of Landsat-4 and the Thematic Mapper (TM) and the subsequent Enhanced Thematic Mapper Plus (ETM+) with a relatively broad band at 2.2 μm allowed for the recognition of the presences of clay and carbonate/sulfate minerals (Goetz and Rowan 1981). Ultimately, higher spectral resolution sensors such as ASTER and the AVIRIS hyperspectral sensor, which have the capability to discern narrow diagnostic absorption features, began to be used for geologic and alteration mapping (e.g., Goetz and Rowan 1981; Marsh and McKeon, 1983; Kruse 1988; Kruse et al. 1990; Rowan et al. 2003, 2006).

Figure 3.7 displays a range of characteristic mineral spectra found in the NASA spectral library (NASA 1999). Though quartz (spectrum e)

displays no distinctive absorption features, the other mineral samples presented here provide a sample of the range of characteristic absorption features. Jarosite (spectrum i), an iron (Fe_3+) sulfate, exhibits both strong blue/green absorption as well as an electronic absorption feature between approximately 0.8–1.1 μm, characteristic of all ferrous iron minerals. Gypsum (spectrum a), a common low-temperature mineral found in soil ($CaSO_4 \cdot 2H_2O$), exhibits distinctive absorption features in the 1.4 and 1.9 μm water related absorption bands due to its significant water content, and a broad absorption doublet at 2.2 μm (spectrum a). Kaolinite (spectrum f) and muscovite (spectrum d) are common 'clay' minerals that are also present in hydrothermal alteration zones often associated with economic mineral deposits. Muscovite

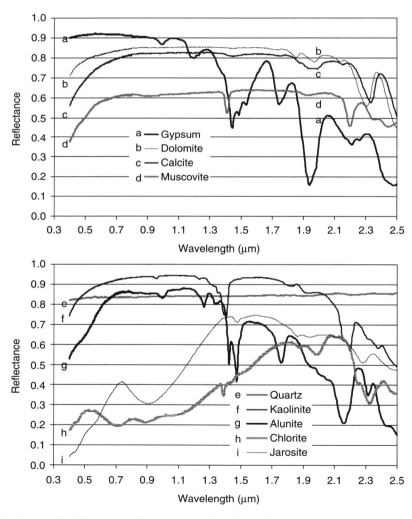

Figure 3.7 Spectral reflectance of representative minerals.
Source: Based on data from NASA, 1999[1].

(spectrum d) exhibits a single strong absorption at 2.2 μm while both kaolinite (spectrum f) and alunite (spectrum g) exhibit characteristic 'doublet or triplet' absorption bands in the 2.2 μm region. The phyllosilicate chlorite (spectrum h) also displays a characteristic absorption feature centered near 2.3 μm. Calcite (spectrum c) and dolomite (spectrum b) display distinctive absorption features near 2.35 and 2.32 μm, respectively. Calcite (spectrum c; $CaCO_3$) has a relatively weak band at 2.15 μm and a much stronger feature at 2.34 μm. Dolomite (spectrum b; $CaMg(CO_3)_2$) because of the presence of magnesium, has these same bands shifted to slightly shorter wavelengths (2.13 and 2.32 μm).

When these minerals appear as constituents of rocks they can often be recognized even in moderate quantities (~few weight percentage). In Figure 3.8 the spectral curves for both calcite (spectrum a) and limestone (spectrum c) are displayed and there is a clear match in the absorption features at 1.99, 2.15, and 2.34 μm. The andesite (an igneous volcanic rock), spectrum d, has only weak absorption in the 2 to 2.5 μm region as its composition (primarily plagioclase plus pyroxene and/or hornblende) do not contain spectral absorption features at these wavelengths. The sandstone (spectrum b) sample has a reddish color due to the presence of a Fe-oxide or Fe-hydroxide and the spectra shows a characteristic steep fall off in the

blue/green and a small shoulder at 0.90 μm. The sample also displays the characteristic kaolinite doublet at 2.17 and 2.20 μm. Volcanic rock (spectrum e) and obsidian (spectrum f) are relatively dark rocks with the least number of absorption features.

Spectral signatures of non-natural material and features

Non-natural or constructed material and features are becoming more prevalent as human population increases and more cities and roads are constructed, destructed, modified, and modernized throughout the world. Building materials can consist of one or multiple minerals, chemicals, inorganic and organic materials, paints and plastics, and these can be transformed into many shapes, structures, textures, and colors. It is clearly impossible to represent the spectral characteristics of all man-made materials here. Therefore, the spectral characteristics of selected synthetic materials and man-made materials and features are presented in this section (Figure 3.9). Knowledge about the spectral characteristics can aid urban change detection and characterization (e.g., mapping of roofing materials, roads). Bright and dark colored features like concrete/cement (spectrum d) and asphalt roads (spectrum c) that represent impervious surfaces

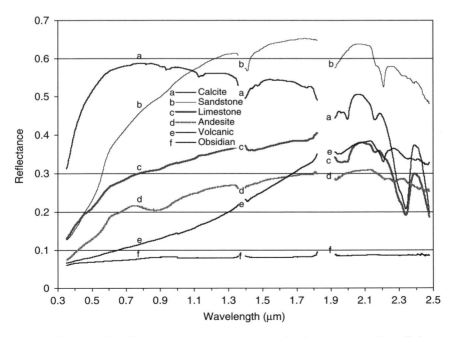

Figure 3.8   Rock spectral reflectance measurements acquired under natural sunlight conditions with an ASD spectroradiometer.

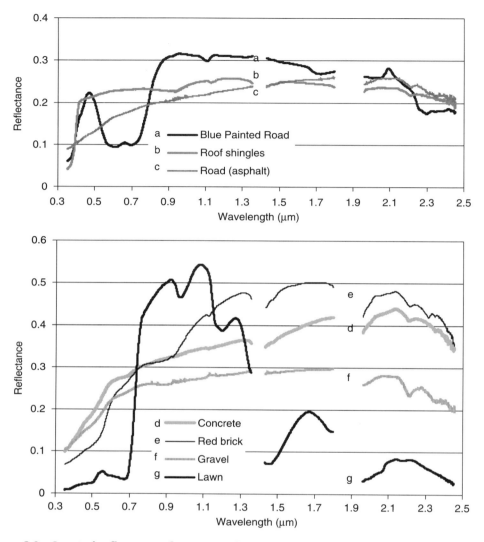

Figure 3.9 Spectral reflectance of non-natural or constructed features found in an urban environment.

have fairly discernaible spectral characteristics. Blue-painted signs on the road (spectrum a) have high reflectance values in the blue wavelength and lower in the other visible wavelengths. Red bricks (spectrum e) and gravel (spectrum f) have a high red reflectance value that increases into the NIR. Bright white roofs (spectrum b) have relatively similar reflectance values for all visible wavelengths. A manicured grass lawn (spectrum g) has a strong vegetation signature, low in the visible and SWIR wavelengths, and high in the NIR. Many spectral characteristics identified in previous sections can be used to identify the components that are likely constituting man–made or non-natural features.

Remote sensing imagery can be used to direct disaster response and to find fire hazards in urban or built-up settings. Cedar shake roofs or tar roofs are more susceptible to ignition from wild fires than tile or metal roofs and can be mapped using a spectral image classification approach (Herold et al. 2003a). Clark et al. (2005) used AVIRIS data to assess the asbestos content of the World Trade Center dust. Recent urban remote sensing models use spectral unmixing algorithms based on vegetation, soil and impervious surface spectral endmembers (Ridd 1995), spectral classification methods (Herold et al. 2003a), and spatiotemporal fragmentation analysis (Herold et al. 2003b) to understand spatial and temporal urban growth patterns and processes.

SUMMARY AND FUTURE DIRECTIONS

This chapter gave a short overview and highlighted some of the visible, NIR and SWIR spectral characteristics of terrestrial surfaces. Although much is known about the spectral characteristic of terrestrial surfaces, it is clear that the amount of information we can extract from airborne and satellite data is often obscured by atmospheric aerosols, clouds, coarse spatial and spectral resolutions, infrequent observations and discontinuity among past and current sensors. However, with the recent development and launch of higher resolution and better calibrated sensors, coupled with improved spectral, spatial, and temporal analysis techniques, more quantitative information is being exploited by industrial users, resource managers, and decision makers, for the benefit of society.

Sun synchronous multispectral sensors have only limited capacity to resolve and inventory all the spectral information of the Earth's surface, but they do allow for continuous monitoring of the Earth's resources as long as capable sensors continue to be built and launched without interruption.

Calibrated airborne hyperspectral sensors do have the spectral resolution necessary to better characterize the Earth's surface. However, they are very expensive to operate and data from these systems are not readily available for many areas across the globe. More spectral data with a diverse range of spatial and temporal resolutions needs to be available to further advance the field of remote sensing and imaging spectroscopy applications in the next decade.

More tools and sensors are being developed that integrate remotely sensed data, GIS, and global positioning system technologies and provide a more holistic approach to address challenges in earth systems science.

Spectral analysis techniques and spectral libraries have been developed and incorporated into commercial of-the-shelf software. A wealth of information can be found in the ASTER and USGS spectral libraries. The increasing availability of improved and user friendly data analysis systems and tools, and more accessible data formats that provide access to these data to a broader range of user communities further promotes the use and understanding of spectral information for exploring and managing Earth's environment.

Our future calls for increased focus on the synergistic use of passive (multispectral, hyperspectral, and thermal infrared) and active (e.g., lidar, microwave) sensor systems to augment data exploitation, expand our knowledge, and further our understanding of Earth systems.

ACKNOWLEDGMENTS

I am indebted to Dr. Stuart E. Marsh (Arizona Remote Sensing Center, University of Arizona) for his contribution to the minerals and rocks section of this chapter and grateful for his edits and thoughtful advice. I am also indebted to Dr. Craig Rasmussen (Soil Water and Environmental Science Department, University of Arizona) for providing soil samples and their properties. Many thanks to Drs. Tim Warner and Gregg Swayze, and an anonymous reviewer for their valuable comments on an earlier version of this chapter.

NOTE

1 Spectra are reproduced from the ASTER Spectral Library through the courtesy of the Jet Propulsion Laboratory, California Institute of Technology, Pasadena, California. Copyright © 1999, California Institute of Technology. All rights reserved.

REFERENCES

Abrams, M. J., R. Ashley, L. Rowan, L. Goetz, and A. Kahle, 1977. Mapping hydrothermal alteration in the cuprite mining district, Nevada using aircraft scanner data for the spectral region 0.46 to 2.36 microns. *Geology*, 5: 713–718.

Adams, J. B. and A. R. Gillespie, 2006. *Remote Sensing of Landscapes with Spectral Images: A Physical Modeling Approach.* Cambridge University Press, Cambridge, UK.

Asner, G. P., 1998. Biophysical and biochemical sources of variability in canopy reflectance. *Remote Sensing of Environment*, 64: 234–253.

Asner, G. P., C. A. Wessman, D. S. Schimel, and S. Archer, 1998. Variability in leaf and litter optical properties: implications for canopy BRDF model inversions using AVHRR, MODIS, and MISR. *Remote Sensing of Environment*, 63: 200–215.

Baumgardner, M. F., L. F. Silva, L. L. Biehl, and E. R. Stoner, 1985. Reflectance properties of soils. *Advances in Agronomy*, 38: 1–44.

Ben-Dor, E., J. R. Irons, and G. Epema, 1999. Soil reflectance, In A. N. Rencz (ed.), *Remote Sensing for the Earth Sciences.* John Wiley & Sons, Inc., New York, pp. 111–188.

Burns, R. G., 1993. Origin of electronic spectra of minerals in the visible-near infrared region. In C. M. Pieters and P. A. J. Englert (eds), *Remote Geochemical Analysis: Elemental and Mineralogical Composition*, Cambridge University Press, Cambridge, UK pp. 3–29.

Clark, R. N., 1999. Spectroscopy of rocks and minerals, and principles of spectroscopy. In A. N. Rencz (ed.), *Manual of Remote Sensing, Volume 3, Remote Sensing for the Earth Sciences*, John Wiley and Sons, New York, Chapter 1, pp. 3–58.

Clark, R. N. and Roush, T. L., 1984. Reflectance spectroscopy: Quantitative analysis techniques for remote sensing applications. *Journal of Geophysical Research*, 89(B7):6329-6340.

Clark, R. N., A. J. Gallagher, and G. A. Swayze, 1990. Material absorption band depth mapping of imaging spectrometer data using the complete band shape least-squares algorithm simultaneously fit to multiple spectral features from multiple materials. In *Proceedings of the Third Airborne Visible/Infrared Imaging Spectrometer (AVIRIS) Workshop*, JPL Publication 90-54, pp. 176–186.

Clark, R. N., G. A. Swayze, K. E. Livo, R. F. Kokaly, S. J. Sutley, J. B. Dalton, R. R. McDougal, and C. A. Gent, 2003. Imaging spectroscopy: Earth and planetary remote sensing with the USGS Tetracorder and expert systems. *Journal of Geophysical Research*, 108(E12): 5131, pp. 5–1 to 5–44. doi:10.1029/2002JE001847.

Clark, R. N., G. A. Swayze, T. M. Hoefen, R. O. Green, K. E. Livo, G. Meeker, S. Sutley, G. Plumlee, B. Pavri, C. Sarture, J. Boardman, I. Brownfield, and L. C. Morath, 2005. Environmental mapping of the World Trade Center area with imaging spectroscopy after the September 11, 2001 attack. In Gaffney, J. S. and Marley, N. A. (eds.), *Urban Aerosols and Their Impacts: Lessons Learned from the World Trade Center Tragedy*, American Chemical Society Book Series 919, Oxford University Press.

Clark, R. N., G. A. Swayze, R. Wise, E. Livo, T. Hoefen, R. Kokaly, S. J. Sutley, 2007. USGS digital spectral library splib06a. U.S. Geological Survey, Digital Data Series 231, http://speclab.cr.usgs.gov/spectral.lib06. Last accessed December 25, 2007.

Collins, W., S-H. Chang, G. Raines, F. Canney, and R. Ashley, 1983. Airborne biogeophysical mapping of hidden mineral deposits. *Economic Geology and the Bulletin of the Society of Economic Geologists*, 78: 737–749.

Colwell, J. E., 1974. Vegetation canopy reflectance. *Remote Sensing of Environment*, 3: 175–183.

Condit, H.R. 1970. The spectral reflectance of American soils. *Photogrammetric Engineering*, 36: 955–966.

Dekker, A. G., T. J. Malthus, M. M. Wijnen, and E. Seyhan, 1992. The effect of spectral bandwidth and positioning on the spectral signature analysis of inland waters. *Remote Sensing of Environment*, 41: 211–225.

Dozier, J., 1989. Spectral signature of alpine snow cover from the Landsat Thematic Mapper. *Remote Sensing of Environment*, 28: 9–22.

Dozier, J., R. E. Davis, A. T. C. Chang, and K. Brown, 1988. The spectral bidirectional reflectance of snow. *Spectral Signatures of Objects in Remote Sensing*, 87–92, European Space Agency, Paris.

EarthSearch Sciences Inc., 2006. The Probe-1 Hyperspectral Instrument. http://www.earthsearch.com/index.php?sp=10, last accessed December 24, 2007.

Farmer, V. C. (ed.), 1974. *The Infra-Red Spectra of Minerals.* Mineralogical Society, London.

Gao, B.-C., W. Han, S.-C. Tsay, and N. F. Larsen, 1998. Cloud detection over Arctic region using airborne imaging spectrometer data during the daytime. *Journal of Applied Meteorology*, 37: 1421–1429.

Gates, D. M., H. J. Keegan, J. C. Schleter, and V. R. Weidner, 1965. Spectral properties of plants. *Applied Optics*, 4: 11–20.

Goetz, A. F. H. and L. C. Rowan, 1981. Geologic remote sensing. *Science*, 211: 781–791.

Goetz, A .F. H, G. Vane, J. E. Solomon, and B. N. Rock, 1985. Imaging Spectroscopy for Earth Remote Sensing. *Science*, 228: 1147–1153.

Hapke, B., 1981. Bidirectional reflectance spectroscopy. 1. Theory. *Journal of Geophysical Research*, 86: 3039–3054.

Herold, M., M. E. Gardner, and D. A. Roberts, 2003a. Spectral resolution requirements for mapping urban areas. *IEEE Transactions on Geoscience and Remote Sensing*, 41: 1907–1919.

Herold, M., N. C. Goldstein, and K. C. Clarke, 2003b. The spatiotemporal form of urban growth: measurement, analysis and modeling. *Remote Sensing of Environment*, 86: 286–302.

Horler, D. N. H., M. Dockray, and J. Barber, 1983. The red edge of plant leaf reflectance. *International Journal of Remote Sensing*, 4: 273–288.

Huete, A. R., 1986. Separation of soil–plant spectral mixtures by factor analysis, *Remote Sensing of Environment*, 19: 237–251.

Huete, A. R., 2004. Remote sensing of soils and soil processes, In: S. Ustin, (ed.), *Remote Sensing for Natural Resources Management and Environmental Monitoring: Manual of Remote Sensing*, 3rd edn. Vol. 4, John Wiley & Sons, Inc., New York, pp. 1–48.

Hunt, G. R., 1977. Spectral signatures of particulate minerals in the visible and near-infrared. *Geophysics*, 52: 501–513.

Hunt, G. R., 1982. Spectroscopic properties of rocks and minerals, In R. S. Carmichael (ed.) *Handbook of Physical Properties of Rocks, Volume I.* CRC Press, Boca Raton, pp. 295–385.

Hunt, G. R. and J. W. Salisbury, 1970. Visible and near-infrared spectra of minerals and rocks, I, Silicate Minerals. *Modern Geology*, 1: 283–300.

Hunt, G. R. and J. W. Salisbury, 1971. Visible and near-infrared spectra of minerals and rocks, II, Carbonates. *Modern Geology*, 2: 23–30.

Hunt, G. R., J. W. Salisbury, and C. J. Lenhoff, 1971a. Visible and near-infrared spectra of minerals and rocks, III, Oxides and hydroxides. *Modern Geology*, 2: 195–205.

Hunt, G. R., J. W. Salisbury, and C. J. Lenhoff, 1971b. Visible and near-infrared spectra of minerals and rocks, IV. Sulphides and sulphates. *Modern Geology*, 3: 1–14.

Hunt, G. R., J. W. Salisbury, and C. J. Lenhoff, 1972. Visible and near-infrared spectra of minerals and rocks, V. Halides, arsenates, vanadates, and borates. *Modern Geology*, 3: 121–132.

Hunt, G. R., J. W. Salisbury, and C. J. Lenhoff, 1973a. Visible and near-infrared spectra of minerals and rocks, VI. Additional silicates. *Modern Geology*, 4: 85–106.

Hunt, G. R., J. W. Salisbury, and C. J. Lenhoff, 1973b. Visible and near-infrared spectra of minerals and rocks, VII. Acidic igneous rocks. *Modern Geology*, 4: 217–224.

Hunt, G.R., J. W. Salisbury, and C. J. Lenhoff, 1973c. Visible and near-infrared spectra of minerals and rocks, VIII. Intermediate igneous rocks. *Modern Geology*, 4: 237–244.

Integrated Spectronics, 2007. Integrated Spectronics. http://www.intspec.com, last accessed December 24, 2007.

ITRES, 2007. Compact Airborne Spectrographic Imager – CASI, 2007. http://www.itres.com/CASI_1500, Last accessed December 24, 2007.

Jacquemoud, S. and F. Baret, 1990. PROSPECT: a model of leaf optical properties, *Remote Sensing of Environment*, 34: 75–91.

Kimes, D. S. 1983. Dynamics of directional reflectance factor distributions for vegetation canopies. *Applied Optics*, 22(9): 1364–1372

Kruse, F. A., 1988. Use of Airborne Imaging Spectrometer data to map minerals associated with hydrothermally altered rocks in the northern Grapevine Mountains, Nevada and California: *Remote Sensing of Environment*, 24(1): 31–51.

Kruse, F.A. and Raines, 1994. A technique for enhancing digital color images by contrast stretching in Munsell color space. In: *Proceedings of the ERIM Third Thematic Conference*, Environmental Research Institute of Michigan, Ann Arbor, MI, 1994, pp. 755–760.

Kruse, F. A., K. S. Kierein-Young, and J. W. Boardman, 1990. Mineral mapping at Cuprite, Nevada with a 63 channel imaging spectrometer: *Photogrammetric Engineering and Remote Sensing*, 56(1): 83–92.

Kruse, F. A., A. B. Lefkoff, J. B. Boardman, K. B. Heidebrecht, A. T. Shapiro, P. J., Barloon, and A. F. H. Goetz, 1993. The Spectral Image Processing System (SIPS) – Interactive visualization and analysis of imaging spectrometer data. *Remote Sensing of the Environment*, 44: 145–163.

Liang, S., 1997. An investigation of remotely sensed soil depth in the optical region. *International Journal of Remote Sensing*, 18(16): 3395–3408.

Marsh, S. E. and J. B. McKeon, 1983. Integrated analysis of high-resolution field and airborne spectroradiometer data for alteration mapping. *Economic Geology*, 78: 618–632.

Mulders, M. A., 1987. *Remote Sensing in Soil Science*. Elsevier, Amsterdam.

Myneni, R. B., R. R. Nemani, and S. W. Running, 1997. Estimation of global leaf area index and absorbed PAR using radiative transfer model. *IEEE Transactions on Geoscience and Remote Sensing*, 35: 1380–1393.

Nagler, P. L., C. S. T. Daughtry, and S. N. Goward, 2000. Plant litter and soil reflectance. *Remote Sensing of Environment*, 71: 207–215.

NASA, 1999. Aster spectral Library – http://speclib.jpl.nasa.gov/, Last accessed March 22, 2007.

NASA, 2003. Earth Observing – 1: Hyperion. http://eo1.gsfc.nasa.gov/Technology/Hyperion.html, Last accessed July 26, 2007.

NASA, 2007. AVIRIS. http://aviris.jpl.nasa.gov, Last accessed December 24, 2007.

Ridd, M .K., 1995. Exploring a V-I-S (Vegetation-Impervious Surface-Soil) model for urban ecosystems analysis through remote sensing: comparative anatomy for cities. *International Journal of Remote Sensing*, 16(12): 2165–2186.

Roujean, J.-L., M. Leroy, P.-Y. Dechamps, 1992. A bidirectional reflectance model of the earth's surface for the correction of remote sensing data. *Journal of Geophysical Research*, 97(D18): 20455–20468.

Rowan, L. C., S. J. Hook, M. J. Abrams, and J. C. Mars, 2003. Mapping hydrothermally altered rocks at Cuprite Nevada using Advanced Spaceborne Thermal Emission and Reflection Radiometer (ASTER), a new satellite imaging system. *Economic Geology*, 98: 1019–1027.

Rowan, L. C., R. G. Schmidt, and J. C. Mars, 2006. Distribution of hydrothermally altered rocks in the Reko Diq, Pakistan mineralized area based on spectral analysis of ASTER data. *Remote Sensing of Environment*, 104: 74–87.

Sanger, J. E., 1971. Quantitative investigations of leaf pigments from their inception in buds through autumn coloration to decomposition in falling leaves. *Ecology*, 52(6): 1075–1089.

Schaepman-Strub, G., M. E. Schaepman, J. V. Martonchik, T. H. Painter, and S. Dangel, in this volume. Radiometry and reflectance: From terminology concepts to measured quantities. Chapter 15.

Schowengerdt, R. A., 1997. *Remote Sensing: Models and Methods for Image Processing* (Second Edition) Academic Press, San Diego, California.

Sinclair, T. R., R. M. Hoffer, and M. M. Schreiber, 1971. Reflectance and internal structure of leaves from several crops during a growing season. *Agronomy Journal*, 63: 864–868.

Singh, A., 1989. Digital change detection techniques using remotely-sensed data. *International Journal of Remote Sensing*, 10(6): 989–1003.

Smith, M. O., P. E. Johnson, and J. B. Adams, 1985. Quantitative determination of mineral types and abundance from reflectance spectra using principal components analysis. Proceedings of the 15th Lunar and Planetary Science Conference, Part 2. *Journal of Geophysical Research*, 90: C797–C808.

Stoner, E. R. and M. F. Baumgardner, 1981. Characteristic variations in reflectance of surface soils. *Soil Science Society of America Journal*, 45: 1161–1165.

Swayze, G. A., R. N. Clark, A. F. H. Goetz, T. G. Chrien, and N. S. Gorelick, 2003. Effects of spectrometer band pass, sampling, and signal-to-noise ratio on spectral identification using the Tetracorder algorithm, *J. Geoph. Research (Planets)*, 108(E9): 5105.

Teillet, P. M., J .L. Barker, B. L. Markham, R. R. Irish, G. Fedosejevs, and J. C. Storey, 2001. Radiometric cross-calibration of the Landsat-7 ETM+ and Landsat-5 TM sensors based on tandem data sets. *Remote Sensing of Environment*, 78: 39–54.

van Der Meer, F. D. and DeJong, S. M., 2001. *Imaging Spectrometry: Basic Principles and Prospective Applications*. Kluwer Academic Publishers, Boston.

van Leeuwen, W. J. D. and A. R. Huete, 1996. Effects of standing litter on the biophysical interpretation of plant canopies with spectral indices. *Remote Sensing of Environment*, 55: 123–138.

van Leeuwen, W. J. D., A. R. Huete, C. L. Walthall, S. D. Prince, A. Begué, and J. L. Roujean, 1997. Deconvolution of remotely sensed spectral mixtures for retrieval of LAI, fAPAR and soil brightness. *Journal of Hydrology*, 188–189: 697–724.

van Leeuwen, W. J. D., B. Orr, S. Marsh, and S. Herrmann, 2006. Multi-sensor NDVI data continuity: uncertainties and implications for vegetation monitoring applications. *Remote Sensing of Environment*, 100(1): 67–81.

Verhoef, W., 1984. Light scattering by leaf layers with application to canopy reflectance modeling: the SAIL model. *Remote Sensing of Environment*, 16: 125–141.

Verhoef, W., 1998. Theory of radiative transfer models applied in optical remote sensing of vegetation canopies. Unpublished PhD thesis, Wageningen Agricultural University, The Netherlands.

Vermote, E., D. Tanré, J. L. Deuzé, M. Herman, and J. J. Morcrette, 1997. Second simulation of the satellite signal in the solar spectrum: An overview. *IEEE Transactions on Geoscience and Remote Sensing*, 35(3): 675–686.

Walter-Shea, E. A., B. L. Blad, C. J. Hays, M. A. Mesarch, D. W. Deering, and E. M. Middleton, 1992, Biophysical properties affecting vegetative canopy reflectance and absorbed photosynthetically active radiation at the FIFE site, *Journal of Geophysical Research*, 97(D17): 18, 925–18, 934.

Walthall, C. L., J. M. Norman, J .M. Welles, G. Campbell, and B. L. Blad, 1985. Simple equation to approximate the bi-directional reflectance from vegetative canopies and bare soil surfaces. *Applied Optics*, 24: 383–387.

Woodcock, C. E. and A. H. Strahler, 1987. The factor of scale in remote sensing, *Remote Sensing of Environment*, 21: 311–332.

Woolley, J. T., 1971. Reflectance and transmittance of light by leaves. *Plant Physiology*, 47: 656–662.

Interactions of Middle Infrared (3–5 µm) Radiation with the Environment

Arthur P. Cracknell and Doreen S. Boyd

Keywords: middle infrared radiation, reflectance, emissivity, brightness temperature, Advanced Very High Resolution Radiometer, clouds, fires, hot spots, vegetation.

INTRODUCTION

In passive remote sensing systems, of which infrared systems are an example, the remote sensing instrument detects electromagnetic radiation which falls on it, but which was generated by some other source. At ultraviolet, visible and near infrared wavelength (below about 1 µm wavelength) this radiation is almost universally reflected solar radiation. For far infrared radiation (above about 10 µm wavelength), frequently referred to as thermal infrared radiation, the radiation is emitted from the surface of the target. For middle infrared radiation (MIR; 3–5 µm wavelength), the subject of this chapter, the radiation received by a remote sensing system during daytime comprises substantial components of both reflected solar radiation and emitted radiation from the target under observation; at night there is only the emitted radiation.

The measured MIR radiance by a sensor is often expressed as a brightness temperature in band i:

$$L_{bb,i}\left(T_{B,i}\right) = L_{i,A} + L_{i,B} + L_{i,C} + L_{i,D} + L_{i,E} + L_{i,F} \tag{1}$$

where $T_{B,i}$ is the measured brightness temperature at top of the atmosphere. The contributions to the MIR radiance (A to F) on the right-side of this equation can be described as follows:

(A) Direct solar reflection:

$$L_{i,A} = \tau_{i\uparrow}\,\rho_{b,i}\left(\theta_v,\,\varphi_v;\,\theta_s,\,\varphi_s\right)\tau_{i\downarrow}E_i\left(\theta_s,\,\varphi_s\right)\cos\left(\theta_s\right) \tag{2}$$

where $\tau_{i\uparrow}$ is the atmospheric transmission (gaseous absorption and scattering) along the target-to-sensor path in the direction θ_v, φ_v, $\rho_{b,i}\left(\theta_v,\,\varphi_v;\,\theta_s,\,\varphi_s\right)$ is the bidirectional reflectance distribution function (BRDF), $E_i(\theta_s,\,\varphi_s)$ is the incoming irradiance and $\tau_{i\downarrow}$ is the atmospheric transmission (gaseous absorption and scattering) along the Sun-to-target path in the direction θ_s, φ_s. Hereafter, radiation with a propagation direction toward the target (or the sensor) is called downward (upward) or downwelling (upwelling) radiation and noted with a down (up) arrow as a subscript.

(B) Atmospheric reflection due to scattering in the atmosphere:

$$L_{i,B} = \rho_{atm,i}E_i\left(\theta_s,\,\varphi_s\right) \tag{3}$$

where $\rho_{atm,i}$ is the reflectance of the atmosphere itself.

(*C*) Solar reflection after atmospheric scattering:

$$L_{i,C} = \tau_{i\uparrow}\rho_{hd,i}E_{scat\downarrow,i} \qquad (4)$$

where $\rho_{hd,i}$ is the surface hemispheric directional reflectance and $E_{scat\downarrow,i}$ is the total solar irradiance scattered by the atmosphere toward the target.

(*D*) Surface thermal emission:

$$L_{i,D} = \tau_{i\uparrow}\varepsilon_i\,(\theta_v,\,\varphi_v)\,L_{bb,i}\,(T_s) \qquad (5)$$

where $\varepsilon_i(\theta_v,\,\varphi_v)$ the surface emissivity in the observation direction θ_v, φ_v in band *i* and $L_{bb,i}\,(T_s)$ is the Planck distribution function and T_s is the temperature of the surface. Thermal emission can be larger than the direct solar reflection depending principally on the radiant temperature of the surface. These are the two major contributions to the measured MIR radiance.

(*E*) Atmospheric emission toward the sensor:

$$L_{i,E} = L_{bb,i}\,\left(T_{atm\uparrow}\right) \qquad (6)$$

where the atmosphere is considered as a black body with an equivalent upward brightness temperature $T_{atm\uparrow}$.

(*F*) Atmospheric emission toward the target reflected by the surface:

$$L_{i,F} = \tau_{i\uparrow}\rho_{hd,i}L_{bb,i}\,\left(T_{atm\downarrow}\right) \qquad (7)$$

where $T_{atm\downarrow}$ is the equivalent downward atmospheric brightness temperature.

The proportions of the contributions *A* to *F* differ across the environment, both spatially and temporally. To derive meaningful measurements of the environment (i.e., variables such as surface temperature and surface reflectivity (and its related emissivity)) Equation (1) must be inverted. This is not a straightforward task, nevertheless progress has resulted in two main approaches being used to process the total MIR radiation signal (Boyd and Petitcolin 2004). One approach emphasizes the solar components of the MIR radiance but tends to have poor descriptions of the thermal contributions (e.g., Roger and Vermote 1998), and the other approach emphasizes the separation of land surface emissivity and temperature, with some focus on the estimation of the MIR reflectance (e.g., Goïta and Royer 1997).

Despite the problem of the complex signal recorded by sensors at MIR wavelengths, there are some properties of this radiation that make it particularly suitable for studying certain environmental phenomena. This realization has come about as a result of using the data acquired in AVHRR (Advanced Very High Resolution Radiometer) channel 3 (3.5–3.9 μm) which has

been available for some 30 years, providing a strong heritage of MIR interactions with the environment. Though there currently are now a number of sensing instruments used by the Earth observation community that record MIR radiation, either already in operation or planned (Table 4.1), for many years the main source of middle infrared data from Earth-observing satellites was from the AVHRR series and only since the 1990s have data from the geostationary GOES Imager and Meteosat SEVIRI sensors and polar orbiting ATSR/AATSR, MODIS, HSRS and VIRS (tropical regions) sensors, among others, been available for use (Table 4.1). Consequently this chapter predominantly reviews the major contribution made by the AVHRR to remote sensing in middle infrared wavelengths, but also provides examples on how other sensors have built on this foundation. An overview of the AVHRR sensor is provided in the section below; additional information about AVHRR can be found in Justice and Tucker (in this volume) and Cracknell (1997).

The first AVHRR was flown on the NOAA polar-orbiting meteorological satellite TIROS-N, which was launched in October 1978. This instrument had four wavebands located in the visible (channel 1; 0.58–0.68 μm), near infrared (NIR) (channel 2; 0.725–1.00 μm), MIR (channel 3; 3.55–3.93 μm) and thermal infrared (TIR) (channel 4; 10.30–11.30 μm) parts of the spectrum. Since then, there has been a continuous succession of AVHRRs flown on the successors to TIROS-N, from NOAA-6 onwards (Table 4.1). The AVHRR/2 had five wavebands, with one extra to that carried on AVHRR/1 at TIR wavelengths (channel 5; 11.50–12.50 μm). The current AVHRR/3 has six wavebands, with three located in the visible, NIR and shortwave infrared (SWIR) (channel 3A: 1.58–1.64 μm). The SWIR band is operational during the day-time overpasses. The remaining three wavebands are located in the MIR (channel 3B: 3.55–3.93 μm, used for night-time operation) and TIR parts of the spectrum. All subsequent satellites in the series NOAA-K/L/M will have six channels. The NOAA polar-orbiting spacecraft constitute an operational system so that a continuous global data set is available from 1978 to the present day and will continue to be available from the NOAA series, and from its successors in the National Polar-orbiting Operational Environmental Satellite System (NPOESS) for the foreseeable future (see, for instance, Hutchison and Cracknell 2006).

The AVHRR was designed primarily for meteorological purposes, including an ability to estimate sea surface temperatures (SST) which is an important input to numerical weather forecasting models. Channels 3 (3B), 4 and 5 of the AVHRR are provided with onboard calibration facilities and this means that the data from channel 4 and channel 5 can be used to estimate SST accurately.

Table 4.1 Examples of past, current and planned satellite sensors which acquire radiation at middle infrared (3.0–5.0 μm) wavelengths

Instrument	Organization	Orbit type	Spatial resolution of MIR waveband (m)	Platform	Name and spectral locations of wavebands at MIR wavelengths (μm)
AATSR	ESA	Sun-sync	1000	Envisat	1. 3.55–3.93
AIRS	NASA	Sun-sync	13500	AQUA	5. 3.74–4.61
ASTR	ESA	Sun-sync	1000	ERS-1 ERS-2	1. 3.55–3.93
AVHRR/1 AVHRR/2	NOAA	Sun-sync	1000	NOAA-9 to NOAA-14	3. 3.55–3.93
AVHRR/3	NOAA	Sun-sync	1000	NOAA-15 to NOAA-18	3B. 3.55–3.93 (night-time operation)
GLI	NASDA	Sun-sync	1000	ADEOS-II	30. 3.55–3.88
HSRS	DLR	Sun-sync	370	BIRD	1. 3.40–4.20
ILAS-II	NASDA		1000	ADEOS-II	2. 3.00–5.70
IMAGER	NOAA	Geo-stationary	1000	GOES-8 GOES-10 GOES-11 GOES-12	2. 3.80–4.00
JAMI	JAXA	Geo-stationary	4000	MTSAT-IR	IR1. 3.50–4.00
MODIS	NASA	Sun-sync	1000	Terra Aqua	20. 3.66–3.84 21. 3.93–3.99 22. 3.93–3.99 23. 4.02–4.08 24. 4.43–4.50 25. 4.49–4.55
MTI	USDE	Near polar–sun-sync	20	MTI	J. 3.49–4.10 K. 4.85–5.05
MVISR	China	Sun-sync	1100	Fengyun 1C Fengyun 1D	3.55–3.93
OCTS	JAPAN	Sun-sync	700	ADEOS	9. 3.55–3.88
SEVIRI	EUMETSAT	Geo-stationary	4800	MSG	IR 3.9. 3.48–4.36
SOUNDER	NOAA	Geo-stationary	8000	GOES	13. 3.68–3.82 14. 3.95–4.01 15. 4.10–4.17 16. 4.43–4.48 17. 4.50–4.55 18. 4.55–4.59
VIRS	NOAA/NASA/NASDA	Equatorial	2000	TRMM	3. 3.55–3.93
VIIRS	NASA-NOAA-DoD	Sun-sync	742	NPP NPOES	M12. 3.61–3.79 M13. 3.98–4.13

This is commonly done using a two-channel algorithm to eliminate atmospheric effects. During the night, when channel 3 (MIR) receives only emitted middle infrared data, its calibrated channel-3 data can be used in conjunction with the channel-4 and channel-5 data in a three-channel algorithm to determine a more accurate value of the sea surface temperature than would be obtained by using only channel-4 and channel-5 data. This three channel SST algorithm approach has been adopted by other sensors which record radiation in the MIR and TIR spectral regions with good effect (e.g., ADEOS II GLI; Sakaida et al. 2006). The use of MIR radiation for land surface temperature (LST) estimation, though based on the same principles as that of SST estimation, is more difficult as for the land the emissivity shows large variations (Becker and Li 1990), according to the nature of the surface, whereas the value of the emissivity of the sea is known and is very close to 1.0 and for most practical purposes is taken to be equal to 1.0.

There have been enormous successes achieved with AVHRR data in fields far removed from the original meteorological purpose (see Cracknell 1997).

However, apart from a few notable exceptions, channel-3 (MIR) data have not shared in this success. There are a few reasons for this. The first is the quality of the data. AVHRR data in channels 1, 2, 4 and 5 is of very good quality. However, channel-3 (MIR) images often have a very pronounced herring-bone pattern which is noise or an artifact. This is worse in data from some instruments in the series than from others; it can, however, be removed by using an appropriate filtering technique. The second reason for the relative unpopularity of channel-3 (MIR) data is that in the daytime the radiation received in channel-3 is a mixture of emitted radiation and reflected solar radiation. This makes the interpretation of the data more complicated than is the case for the data from the other channels. Nonetheless, MIR radiation is suitable for Earth observation, particularly in endeavours such as the discrimination of cloud types, the study of small intense heat sources on land and in the mapping and monitoring of the land surface.

CLOUD CLASSIFICATION

Channel-3 (MIR) data plays an integral part in multi-spectral classification schemes for clouds. The scanners on the early meteorological satellites had very few spectral bands, commonly just one visible band and one thermal infrared band. With the images from these satellites clouds could be seen, either from their albedos or temperatures, and weather systems could be followed in a way that was previously impossible before the advent of satellite meteorology. Experienced analysts could identify different types of clouds from their shape and context in satellite images. However, at that stage automatic pattern recognition algorithms were not well developed and the development of a multi-spectral classification scheme, based only on the individual pixel values was not feasible. With the advent of the AVHRR and its five spectral channels (wavebands) that changed. First of all, various workers developed schemes for cloud screening of AVHRR images, i.e., detecting and removing pixels that were affected by cloud (Bell and Wong 1981, Inoue 1985, Scorer 1989). This led to the development of classification schemes which could be applied on a pixel-by-pixel basis and which were based on the differences of the reflectivity (albedo) and of emissivities and temperatures of different types of clouds in the different spectral channels (bands) (Liljas 1987). Now multi-spectral analysis has become a fundamental technology used in the meteorological analysis of remotely-sensed satellite data.

The middle infrared has proved to be very useful in discriminating between snow and cloud cover.

Ice is very absorbent at middle infrared wavelengths so that its reflectivity (or albedo) is very low. Therefore, clouds composed of ice crystals, as well as sea ice and snow on the ground, have very low reflectivity and can be detected in a daytime channel-3 image. The absorption coefficient of water at this wavelength is much lower, and thus the reflectivity is much higher than for ice. Therefore, it was believed that accurate snow maps could be generated from daytime AVHRR channel-3 data if the emitted component in the observed radiance could be isolated and removed from the reflected (solar) component. Although initial attempts proved successful in differentiating between snow and water clouds, the presence of cirrus (ice) clouds was problematic (Allen et al. 1990). Subsequently, a different approach was taken to separate the emitted and reflected (solar) components of the channel-3 (MIR) signal in snow and cloud discrimination by producing the channel-3 (MIR) albedo image (Hutchison et al. 1997a). This new approach simultaneously enhanced the signature of thin cirrus clouds and suppressed the signature of snow, making it possible (1) to differentiate water clouds from snow, (2) to differentiate snow from ice (cirrus) clouds, and (3) to differentiate water clouds from thin cirrus clouds, although some ambiguity exists for cirrus clouds with larger optical depths (Hutchison et al. 1997b). In fact, the procedure even allowed snow to be manually classified in the presence of overcast thin cirrus clouds using only signatures in the solar channels rather than longwave infrared channels (Hutchison and Locke 1997). Enhanced algorithms which use MIR radiation acquired by the more recently launched sensors have proved successful. For example, the properties of upper level ice clouds, optical depth and effective size can be derived using the Visible-Infrared Solar-Infrared Split-Window Technique (VISST) based on TRMM VIRS data (Minnis et al. 2007). The new NPOESS VIIRS also promises an increase in the accuracy of cloud classifications (Hutchinson and Cracknell 2006).

SMALL INTENSE SOURCES OF HEAT

A unique capability of MIR data is in the study of small intense sources of heat. Such sources of heat are indicative of both unmanaged fires (e.g., wildfires or accidental fires) and managed fires (e.g., agricultural, peat-bog burning), as well as industrial activities (e.g., steel works and oil refineries). Satellite remote sensing improves the spatio-temporal sampling of hot spots affording the ability to monitor these dynamic events. Thus, early developments in using MIR radiation acquired by the AVHRR sensors for the purpose of hot spot detection have been fundamental, but were quite unforeseen.

A fire that generates a large smoke plume is very often clearly visible in channel-1 AVHRR data. However, the particles in dust clouds and smoke plumes are often so small that they do not reflect strongly at MIR wavelengths. Thus in channel-3 (MIR) the smoke plume will, most likely, be invisible but the site of the fire itself will probably show up very clearly. Moreover, fires that are considerably smaller than the AVHRR sensor's Instantaneous Field of View (IFOV) can be quite apparent in channel-3 (MIR) data. Examples include gas flares, blast furnaces and agricultural straw fires. The ability of the channel-3 (MIR) data to detect sub-pixel size high temperature sources, or 'hot spots', has been well documented in studies of gas flares, forest and range fires, and steel production plants (e.g., Muirhead and Cracknell 1984, Matson et al. 1987). Hot spots in channel-3 (MIR) imagery can usually be distinguished from pixel noise or highly reflective cloud (Muirhead and Cracknell 1984) by their characteristic point-spread pattern or, if necessary by comparison with other AVHRR channels.

The principle of what is involved in fire detection with AVHRR data is explained by Matson et al. (1987). At typical room, land or sea temperatures, say ~300 K, the Planck distribution function peaks in the vicinity of channel-4 and channel-5 wavelengths (i.e., 10–12 μm). For a flame temperature, say 900–1000 K, the peak moves to the vicinity of channel-3 (MIR) wavelengths, i.e., 3.5–3.9 μm (Figure 4.1). Kennedy et al. (1994) give some numbers to illustrate this. For a pixel corresponding to an IFOV of 1.1 km by 1.1 km, containing a single surface element of uniform temperature (300 K) the average radiant energy in channels 3 and 4 would be 0.442 and 9.68 W m^{-2} μm^{-1} sr^{-1}, respectively, assuming an emissivity of 1 and no attenuation of the signal by the atmosphere. For a pixel containing two equal area elements with different temperatures, a fire at 800 K and the non-fire background of 300 K, the average radiant energy for channels 3 and 4 would be 670 and 49.9 W m^{-2} μm^{-1} sr^{-1}, respectively. The radiance received by channel 3 increases by a factor of about 1500, compared to a factor 5 for channel 4. It is this differential response which is the basis for AVHRR fire detection. Figure 4.2 shows a typical plot of channel-3 (MIR) and channel-4 brightness temperatures over two high-temperature sources. Typical temperature differences between the two channels over land surfaces in general are usually about 1–2 K. Target 1

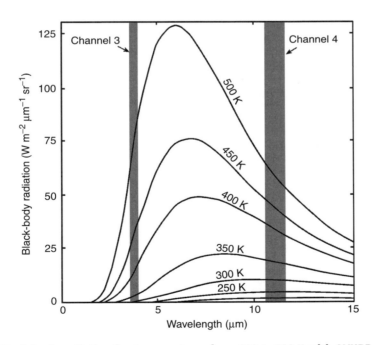

Figure 4.1 **Black body radiation for temperatures from 200 to 500 K with AVHRR channel-3 and channel-4 wavelengths indicated.**
Source: Copyright Taylor & Francis Ltd, http://www.informaworld.com, reprinted by permission of the publisher from Matson, M., G. Stephens and J. Robinson, 1987. Fire detection using data from the NOAA-N satellites. International Journal of Remote Sensing 8: 961–970.

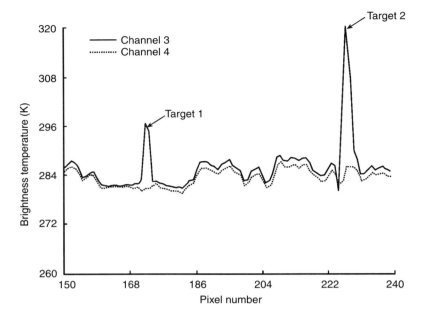

Figure 4.2   A typical plot of channel-3 (MIR) and channel-4 (TIR) brightness temperatures over two high-temperature sources located in Idaho. Target 1 is a small controlled forest fire and target 2 is a phosphorus plant.
Source: Copyright Taylor & Francis Ltd, http://www.informaworld.com, reprinted by permission of the publisher from Matson, M., G. Stephens and J. Robinson, 1987. Fire detection using data from the NOAA-N satellites. International Journal of Remote Sensing 8: 961–970.

in Figure 4.2 is a small controlled forest fire and target 2 is a phosphorus plant. At these sources, the channel-3 (MIR) brightness temperatures are 16.2 and 33.9 K higher than the corresponding channel-4 brightness temperatures. These targets are smaller than the IFOV and so these two higher temperatures correspond to pixel averages and do not give the hot target's temperature directly. One can, however, use an approach developed by Dozier (1981) and applied to fires by Matson and Dozier (1981) to estimate both the area and temperature of the hot target.

In the absence of an atmospheric contribution or attenuation, the upwelling radiance sensed by a downward-pointing radiometer is given by:

$$L(T) = \frac{\int_0^\infty \varepsilon_\lambda B(\lambda, T)\phi(\lambda)d\lambda}{\int_0^\infty \phi(\lambda)d\lambda} \qquad (8)$$

where $B(\lambda, T)$ is the Planck distribution function, $\phi(\lambda)$ is the spectral response of the detector, λ is the wavelength and the target area is assumed to be at a uniform temperature, T, and ε_λ, is the emissivity, to allow for the target not being a black body. For most Earth surfaces ε_λ is relatively independent of λ over the range of an AVHRR channel,

so one can drop the λ subscript and take ε outside the integral in equation (8). We suppose that the channel-3 (MIR) and channel-4 (TIR) radiances are $L_3(T)$ and $L_4(T)$, respectively, as functions of blackbody temperature (i.e., $\varepsilon = 1$) and that the inverse functions of energy, E, are $L_3^{-1}(E)$ and $L_4^{-1}(E)$, respectively, for channels 3 and 4. For temperatures from 100 to 1000 K the expression in Equation (8) was calculated numerically by Dozier (1981) using $\phi(\lambda)$ for the AVHRR on NOAA-6. The resulting values for $L_3(T)$ and $L_4(T)$ are shown in Figure 4.3. The shift in the peak of the Planck function toward shorter wavelengths, as temperature increases, causes the radiant contribution to channel 3 to increase more rapidly, at higher temperatures, than that to channel 4.

Suppose that we now have a mixed pixel corresponding to a hot target at temperature T_t which occupies a proportion p of the IFOV (where $0 < p < 1$) and a background temperature T_b which occupies the remaining fraction $(1 - p)$ of the IFOV. Strictly speaking, we should specify the location of the hot target within the IFOV and use the point spread function for the AVHRR, but we ignore that just now. The brightness temperatures T_j ($j = 3, 4$) sensed by the AVHRR in channels 3 and 4 will be, in the absence of an atmospheric

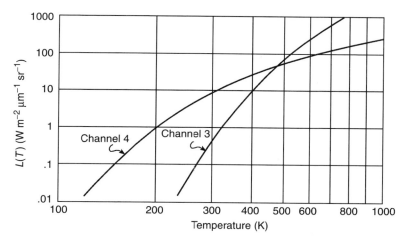

Figure 4.3   The upwelling radiance detected in channel 3 and channel 4 of the AVHRR on NOAA-6.
Source: Reprinted from Remote Sensing of Environment Vol. 11, J. Dozier, A method for satellite identification of surface temperature fields of subpixel resolution, 221–229, Copyright (1981), with permission from Elsevier.

contribution or attenuation (and at night so that there is no reflected solar radiation in channel 3);

$$T_j = L_j^{-1}[pL_j(T_t) + (1 - p)L_j(T_b)] \qquad (9)$$

where $j = 3, 4$. The background temperature can be estimated reasonably accurately from nearby pixels. This means that in the two equations (9) there are just two unknowns, p, the fraction of the IFOV occupied by the hot target, and T_t, the target temperature.

Using the theory described above with the data shown in Figure 4.2, Matson et al. (1987) were able to estimate the areas and temperatures of the two high-temperature targets causing the two peaks. For the two sources in Figure 4.3 they found the area and temperature were 0.28 ha and 430 K for target 1 and 1.7 ha and 483 K for target 2. As evidenced by the calculated target sizes and the detection of the phosphorus plant, target 2, it does not need a 1.1 km square target to cause a response in channel 3 (a 1.1 km square has an area of $((1.1 \times 10^3)^2/10^4)$ ha = 121 ha). Thus we see an example which shows that small sub-IFOV size high-temperature sources, such as fires, can be detected by channel 3. Very hot and small targets such as waste gas flares from offshore oil platforms have been detected as well as small fires from straw burning (see below). The question arises as to what is the minimum size of the subresolution-scale high temperature sources which can be detected; a discussion of this question is given in Section 6.2.4 of Cracknell (1997).

It should be noted that the theory leading up to Equation (9) neglects atmospheric effects and, of

course, assumes that there is no reflected solar radiation. It, therefore, applies only to night-time data, since during the day there is a significant reflected solar radiation contribution to the radiance detected in channel 3 of the AVHRR.

A problem which arises is that the radiation from a fire is very intense; the contrast between fires and normal Earth temperatures is thus so sharp that fires easily saturate sensors instrumented for normal Earth surface observation and of course then Equation (9) cannot be used. Yet another problem is that the AVHRR does not provide continuous data gathering; only providing one set of instantaneous data a few times each day. When active fires, as opposed to post-fire char or scars, are being observed, measurements pertain to the instant of observation, not to the fire's average condition or ultimate magnitude. Thus, measurements made of a fire will vary considerably, both over its life-cycle and with the variations of the wind and the fuel bed. Moreover, fires typically manifest a strong diurnal rhythm, varying over time and space. Consequently, fire measurements from a sensor platform with a fixed hour of observation may have strong biases that are difficult to eliminate. This limitation can be overcome by use of the MIR radiation acquired by the geostationary GOES Imager and Meteosat SEVIRI sensors (Roberts et al. 2005). These sensors offer the advantage of frequent observations of the Earth's surface, yet despite their coarse spatial resolution are able to detect hot spots, though the spectral contrast between hot spots pixels and surrounding pixels is reduced. Indeed, MIR radiation acquired by geostationary satellites has been used in algorithms to routinely detect active fires, for example, the Automated Biomass Burning Algorithm (ABBA) and its successor Wildfire ABBA (Prins and Menzel 1994, Prins et al. 1998).

The theory developed by Dozier (1981) outlined above was for mixed pixels but included no analysis of AVHRR data for fires. Dozier's theory was applied to AVHRR data for fires by Matson and Dozier (1981) and later extended by Wan (1985) to arbitrarily dimensioned pixel fields overlaid by an N-layered absorbing and multiply scattering atmosphere. Matson et al. (1984) examined fire appearances in AVHRR imagery, and studied the correspondence between fires sighted with HRPT (high resolution picture transmission) data and fire sightings by fire control agencies in the western United States. Agricultural, slash and burn and range improvement fires were dominant; a few wildfires, a peat-bog fire, gas flaring and mine smelts were also recorded. Matson et al. (1987) showed channel-3 (MIR) images of examples of fires in various parts of the world.

Another example of the observation of small intense sources of heat with data from channel 3 of the AVHRR is related to the identification and location of gas flares on oil and gas platforms in the North Sea (Muirhead and Cracknell 1984). Using data from channel 3 of the AVHRR, one can identify gas flares which show up as pixels with very high temperature values. These pixels occur sometimes as isolated pixels, sometimes as a cluster of 2 or 3 pixels but never more than about 5 or 6 pixels. It is not that a flare is several kilometres in size but rather that its effect is seen in several pixels because its temperature is so high and because of the point spread function of the AVHRR. It was possible to show that very nearly every flare in each scene could be assigned unambiguously to a platform that could be held responsible for it.

The possible use of AVHRR channel-3 data in relation to post-harvest straw burning by farmers has been studied by Muirhead and Cracknell (1985) who used channel-3 AVHRR data to investigate the extent of straw and stubble burning across Britain in the summer of 1984 as a result of a large surplus of straw. The results obtained from data for various days showed that a voluntary agreement by farmers about when to burn straw was largely being followed.

In other parts of the world it is forest fires which have been widely studied using MIR data. Early work on detecting sub-pixel-size fires had made use of the 1.1 km resolution HRPT or LAC (local area coverage) data. However, the relatively high-resolution data is only available if there is an appropriate direct readout ground receiving station or if the data capture has been scheduled for tape-recording on board the spacecraft. There are thus some areas of the world for which the high-resolution data has not been obtained on a regular and frequent basis. Operational global coverage is, however, available with the lower resolution GAC (global area coverage) data and these data were used by Malingreau et al. (1985) to examine the catastrophic Indonesian fires of 1983 in Kalimantan, Borneo, in Indonesia. GAC data were used because the LAC (HRPT) data was not available and despite an initial expectation that the GAC resolution (4 km) would be too coarse for observing forest fires, many fires were in fact observed. Matson et al. (1987) also demonstrated that the spatial resolution of GAC data is not too coarse for that data to be used to study forest fires.

Forest fire monitoring in north America has been studied by Flannigan and Vonder Haar (1986). Satellite observations were compared to daily fire reports from the Canadian Forest Service based on aerial reconnaissance. In total, 355 fires were observed. A fire's duration appeared to be as important as, or even more important than, its size in determining its probability of being observed. Less than 0.1% of all fires saturated the sensor. The probability of identification increased rapidly for the first three passes following the outbreak of a fire and then tapered off. The Matson–Dozier algorithm was used to evaluate the sizes and temperatures of fires in the non-discarded pixels; the derived estimates were then compared with data in the fire reports. Calculated temperatures tended to be more indicative of smouldering than flaming combustion. Where more than five of the eight pixels contiguous to a target pixel showed signs of fire, it was assumed that the entire pixel was burnt.

A serious set of fires in the Yellowstone National Park (USA) in September 1988 was monitored using channel-1 and channel-3 (MIR) AVHRR data by Hastings et al. (1988). A number of workers have studied fires in the Amazon Basin in Brazil in connection with the problem of tropical deforestation. Matson and Holben (1987) applied the Matson–Dozier algorithm and made NDVI computations for all hotspots (defined as being all pixels for which the channel-3 (MIR) brightness temperature exceeded 307 K) encountered in a 3° by 6° area from a single LAC image centred near Manaus in the Amazon Basin. Of the 169 hotspots observed 143 (85%) saturated the sensor. The remaining 26 fires were analyzed to determine their size and temperature. These 26 fires had areas that ranged from 0.27 to 9.04 ha, and temperatures that ranged from 376 to 609 K. The computed values seemed reasonable but no ground data were available to check the results. Typical observations of fires in Amazonia suggest that colonists and slash-and-burn cultivators generally burn 1–20 ha at a time. Smaller fires predominate. Flaming combustion occupies a large fraction, but usually less than one half, of the burn site during the hour or two of intense burning. Further work on the Amazon

Basin was carried out by Malingreau and Tucker (1988) and Setzer et al. (1988).

An unusual example of the observation of hot spots in channel-3 (MIR) AVHRR data was provided by Dousset et al. (1993). Hundreds of fires were set alight in Los Angeles in a civil disturbance following the verdict in a court case on 29 April 1992. In the channel-3 (MIR) AVHRR image shown by Dousset et al. from 03.47 PDT on 30 April 1992, approximately 10 hours after the start of the riots, an exceptionally large thermal anomaly was visible. It extended over more than 85 km^2 in south central Los Angeles where, on average, three new fires were started every minute during the three hours preceding the capture of the image.

The issue of saturation of channel 3 is very important (Setzer and Verstraete 1994). Channel 3 supposedly saturates for Earth targets at about 47°C, or 320 K and also with sunglint in water bodies (Khattak et al. 1991) or sunlight reflected from some types of exposed soils (Grégoire et al. 1993). Calculations based on the use of Wien's displacement law predict that a fire of about 30 m by 30 m in extent and near the centre of a channel-3 (MIR) IFOV should saturate it. However, this expectation is not supported by available measurements. For instance, large fires in forests and grasslands, sometimes occupying areas many times the size of the IFOV of the AVHRR have been associated repeatedly with non-saturated values of channel 3 (for references, see Setzer and Verstraete 1994). This discrepancy between theory and practice is intriguing. Nevertheless, subsequent sensor developments have considered the potential of signal saturation and consequently, the EOS MODIS 'fire' Band 21 (3.929–3.989 μm) is designed to perform radiometric measurements over a very hot target without saturation (Petitcolin and Vermote 2002). There are a number of global fire-related products now offered from MODIS (see more in Justice et al. 2002); this is yet another example of the useful heritage provided by the AVHRR sensor.

Building on the use of the AVHRR sensor in the detection of hot spots, there has been an increasing amount of research on the possibility of provision of information on the properties of hot spots, particularly wildfires. Properties such as fire radiative energy which is a direct function of fuel biomass combustion, is necessary for the understanding of atmospheric chemistry and climate radiative forcing processes. Fire radiative power (FRP) estimates are integrated over time to provide fire radiative energy information and can be directly remotely sensed using MIR radiation. MIR radiation acquired by both polar-orbiting EOS MODIS, BIRD HSRS sensors and geostationary GOES Imager and Meteosat SEVIRI sensors has been employed to do this in a number of ways.

One has used the EOS-MODIS MIR 'fire' channel (band 21) in the algorithm:

$$FRP(MODIS) = 4.34.10\text{-}19(TMIR - Tb.MIR) \quad [W\,m^{-2}] \quad (10)$$

where TMIR and Tb.MIR respectively are the radiative brightness temperatures of the fire pixel and the neighbouring nonfire background (K) recorded in the MODIS MIR channel (Kaufman et al. 1998a, b). However, this approach will overestimate FRP fires over 1600 W m^{-2} since the coefficients were not designed for these fires (Wooster et al. 2005). The MIR radiance method suggested by Wooster et al. (2003) is based on the same principles as that of the MODIS method; however, it is designed for use by any sensor with a MIR spectral band. It is most suitable for situations where fire fractions are potentially large (Wooster et al. 2005) and, once parameterized, has been used successfully to estimate FRP of fires observed by the BIRD HSRS sensor a microsatellite designed for the study of active fires (Wooster et al. 2003, Siegert et al. 2004) and the geostationary Meteosat SEVIRI sensor (Roberts et al. 2005). The MIR radiance method has a potential disadvantage in that it underestimates radiative emissions from fires cooler than 650 K.

LAND SURFACE ANALYSIS

A significant contribution that satellite remote sensing makes to Earth observation is the ability to acquire information about the land surface at regional to global scales and at high temporal resolution. As such the MIR radiation acquired over the land surface has proved to be useful in the estimation of land cover temperatures and this has been carried out using MIR radiation acquired by a number of sensors in addition to the AVHRR. For instance, MIR radiation is used in split-window algorithms to overcome problems of atmospheric water vapour effects (e.g., Sun and Pinker 2003 using GOES Imager). Related to this work has been the recent interest in determining spectral emissivity in the middle infrared spectral region (e.g., Payan and Royer 2004 using MODIS and ASTER, Peres and DaCamara 2005 using METEOSAT SEVIRI, Petitcolin et al. 2002 using MODIS) for improved land surface characterization.

It can be argued that the most important component of the land surface is that of vegetation. However, there has been relatively limited activity in the use of MIR radiation for the study of vegetation, even though there are many attributes of the MIR spectral region that lend themselves to remote sensing of vegetation, including the small scan angle effects apparent at MIR wavelengths, the very low variation in solar irradiance in the MIR spectral region and the lower temperature estimate error due to emissivity uncertainty relative to that at longer thermal infrared wavelengths (Salisbury and D'Aria 1994). Moreover, there are many successes in the use of MIR radiation for the study of a range of vegetative environments. These successes have facilitated the ability to conduct long-term environmental monitoring. The use of channel-3 (MIR) data acquired by the NOAA AVHRR instruments allows for decades of accurately calibrated data because on-board thermal infrared calibration techniques can be applied to data acquired in channel-3 (MIR) (see Cracknell 1997 for instance). The progressive development of the methods for processing the radiation acquired by instruments in the middle infrared wavelength spectral region has allowed the informed use of the unique information provided by this part of the spectrum for measuring and monitoring the vegetation (Boyd and Petitcolin 2004). This work has also been underpinned by laboratory-based radiometric studies and again it is the AVHRR sensor that provides much of the methodological foundation and experience-base for this activity.

STUDYING VEGETATIVE ENVIRONMENTS USING MIR RADIATION

In areas where vegetation attains high biomass and/or the atmosphere has high aerosol loadings, the use of radiation at visible and near infrared wavelengths, which are commonly used when studying vegetation, has been somewhat unsuccessful. This has prompted the use of MIR radiation. A major focus of employment of MIR radiation in vegetation studies to date has been for low-latitude vegetation monitoring and the study of ecosystem change and dynamics. Divisions between major vegetation types and eco-biological zones have been detected using MIR radiation (Laporte et al. 1995) and other studies have demonstrated that MIR radiation provides a relatively large spectral contrast (compared to that in visible and NIR wavelengths) between mature forest and other land cover (Achard and Blasco 1990). A number of studies have noted temporal variation in MIR radiation measured from tropical vegetation (e.g., Achard and Estreguil 1995) and attributed it to the coupled effect of seasonal changing climatic conditions and vegetation phenological characteristics. This variation has been shown to be more accurate than the NDVI for characterizing changes in density and vigour of the vegetation in West Africa. Grégoire et al. (1993) also noted the potential of using MIR radiation alongside the NDVI for the definition of new land cover classes, particularly

at continental and global scales. Channel-3 (MIR) AVHRR data has also been used in conjunction with data from other channels to study forest clearing, in which burning plays a part, in various remote areas (Tucker et al. 1984, Malingreau et al. 1989). In these cases channel-3 (MIR) data is only part of the data used in a classification scheme.

Building on the success of these studies, researchers have explored the use of MIR radiation for the mapping and estimation of biophysical properties of forests. Theoretically, an inverse relationship exists between leaf area (and other related biophysical properties) and MIR radiation (Boyd and Petitcolin 2004). Empirical relationships have been derived between forest biophysical properties (e.g., tree density, basal area, forest cover and regeneration age) of forests and radiation in MIR wavelengths, both alone and within vegetation indices. Of note is that using MIR improved the accuracy with which these vegetation properties were estimated (e.g., Foody et al. 1996, Lucas et al. 2000).

There is an important issue, however, that may compromise the use of MIR radiation for the study of vegetation; at and beyond the regional scale the emitted radiation component of the signal may be subject to additional confounding variables, rather than representing the intrinsic properties of the surface itself (Kaufman and Remer 1994, Boyd and Duane 2001). Emitted radiation, though related to vegetation canopy properties, may also be influenced by varying localized atmospheric conditions, such as wind speed and air vapour conductance, site specific factors, such as topography and aspect, and soil moisture conditions. Indeed, these factors have been related more strongly to emitted radiation from vegetation canopies than to their properties such as basal area and tree density. It may be preferable, therefore, to remove the emitted component and use only the reflected component of the MIR radiation (Kaufman and Remer 1994) in the study of vegetation at regional to global scales. This has been made possible by the evolution of the methods by which MIR reflectance component can be retrieved from the MIR radiation signal (see Introduction, above).

The most common application of the retrieved MIR reflectance from the total MIR radiation signal for the study of vegetation has been in the mapping, estimation and monitoring of vegetation properties (e.g., leaf area) and their changes (as a function of fire or drought). Studies of the former have demonstrated that MIR reflectance shows promise for the discrimination of different vegetation types (Holben and Shimabukuro 1993, Goïta and Royer, 1997), estimation of total and leaf biomass of several forest ecosystems (Boyd et al. 1999, 2000) and monitoring intra- and inter-annual changes in vegetation (Boyd and Duane 2001). In each of these studies, MIR reflectance has been more useful than the more commonly-used normalized difference vegetation index (NDVI) for describing the vegetative environment under study. This information about the vegetative environment has also been incorporated with that provided by NIR radiation in a vegetation index, VI3 (NIR–MIR/NIR+MIR), in order that the MIR reflectance becomes less sensitive to bidirectional reflectance and subpixel clouds (Kaufman and Remer 1994). The study of vegetation change through drought-induced ENSO events has also advocated the use of MIR reflectance in a vegetation index; the Ts/VI3 (Boyd et al. 2006). The measurement and monitoring of change in the environment as a result of fire can also benefit from the use of MIR reflectance. MIR reflectance has been recommended for mapping burn scars in a range of vegetative environments. In this regard, MIR reflectance (from the NOAA AVHRR) has been used within an index, such as the VI3 (Barbosa et al. 1999, Roy et al. 1999, Vafeidis and Drake 2005) and the GEMI3, an empirical modification of the global monitoring index, GEMI (Barbosa et al. 1999, Pereira 1999). Each of these indices demonstrated a strong spectral contrast between the burn scar and surrounding vegetation. Again, it has been this sort of experimental work that has led to the delivery of datasets such as the MODIS burned area product (Roy et al. 2002).

CONCLUSION

Since the launch of the first NOAA AVHRR in 1978 remotely sensed data acquired at MIR wavelengths have been available for the measurement of key properties about the environment at coarse spatial resolution and fine temporal resolution. These remotely sensed data acquired at MIR wavelengths are now complemented by similar data acquired by a suite of other satellite sensors, yet there has been a seeming under use of these data, probably a result of the complicated data collected in this part of the spectrum. However, as this chapter has highlighted, much work on the processing of the MIR radiation signal has meant that researchers have been able to use the information in the MIR radiation signal to estimate properties of the environment, such as cloud type, hot spots, land surface temperature and vegetation type and biomass.

Earth observation using radiation at MIR wavelengths should continue for the foreseeable future and work should continue on refining, both how MIR radiation data are processed and how they are optimally applied to the study of the environment. As demonstrated by Henebry (2006) this work should be assisted by the airborne and field based sensors that are now available for use (e.g., MASTER, BOMEM MR154 FT-spectroradiometer). It has been argued that due to spectral position of MIR radiation, there will be the opportunity to take advantage of scientific advances focusing on both solar reflection and thermal emission from the environment (Boyd and Petitcolin 2004). To realize fully the potential of remote sensing for Earth observation, the data that are routinely collected in *all* wavebands, including the undervalued MIR, should be considered.

REFERENCES

Achard, F. and F. Blasco, 1990. Analysis of seasonal evolution and mapping of forest cover in West Africa with use of NOAA AVHRR HRPT data. *Photogrammetric Engineering and Remote Sensing*, 56: 1359–1365.

Achard, F. and C. Estreguil, 1995. Forest classification of Southeast Asia using NOAA AVHRR data. *Remote Sensing of Environment*, 54: 198–208.

Allen, R. C., P. A. Durkee and C. H. Wash, 1990. Snow/cloud discrimination with multispectral satellite measurements. *Journal of Applied Meteorology*, 29: 994–1004.

Barbosa, P. M., D. Stroppiana, J.-M. Gregoire and J. M. C. Pereira, 1999. An assessment of vegetation fire in Africa (1981–1991): burned areas, burned biomass, and atmospheric emissions. *Global Biogeochemical Cycles*, 13: 933–950.

Becker, F. and Z.-L. Li, 1990. Temperature independent spectral indices in thermal infrared bands. *Remote Sensing of Environment*, 32: 17–33.

Bell, J. J. and M. C. Wong, 1981. The near-infrared radiation received by satellites from clouds. *Monthly Weather Review*, 109: 2158.

Boyd, D. S. and W. J. Duane, 2001. Exploring spatial and temporal variation in middle infrared reflectance measured from the tropical forests of west Africa. *International Journal of Remote Sensing*, 22: 1861–1878.

Boyd, D. S. and F. Petitcolin, 2004. Remote sensing of the terrestrial environment using middle infrared radiation 3.0–5.0 μm. *International Journal of Remote Sensing*, 25: 3343–3368.

Boyd, D., G. M. Foody and P. J. Curran, 1999. The relationship between the biomass of Cameroonian tropical forests and radiation reflected in middle infrared wavelengths 3.0–5.0 μm. *International Journal of Remote Sensing*, 20: 1017–1023.

Boyd, D. S., T. E. Wicks and P. J. Curran, 2000. Use of middle infrared radiation to estimate leaf area index of a boreal forest. *Tree Physiology*, 20: 755–760.

Boyd, D. S., P. C. Phipps and G. M. Foody, 2006. Dynamics of ENSO drought events on Sabah rainforests observed by NOAA AVHRR. *International Journal of Remote Sensing*, 27: 2197–2219.

Cracknell, A. P., 1997. *The Advanced Very High Resolution Radiometer*. Taylor and Francis, London.

Dousset, B., P. Flament and R. Bernstein, 1993. Los Angeles fires seen from space. *EOS Transactions, American Geophysical Union*, 74: 33–38.

Dozier, J., 1981. A method for satellite identification of surface temperature fields of subpixel resolution. *Remote Sensing of Environment,* 11: 221–229.

Flannigan, M. D. and T. H. Vonder Haar, 1986. Forest fire monitoring using NOAA satellite AVHRR. *Canadian Journal of Forest Research,* 16: 975–982.

Foody, G. M., D. S. Boyd and P. J. Curran, 1996. Relationships between tropical forest biophysical properties and data acquired in AVHRR channels 1–5. *International Journal of Remote Sensing,* 17: 1341–1355.

Goïta, K. and A. Royer, 1997. Surface temperature and emissivity separability over land surface from combined TIR and SWIR AVHRR data. *IEEE Transactions on Geoscience and Remote Sensing,* 35: 718–733.

Grégoire, J. M., A. S. Belward and P. Kennedy, 1993. Dynamics of signal saturation with channel 3 of the AVHRR (Advanced Very High-Resolution Radiometer) – Major handicap or information source for environmental monitoring of the Sudan and Guinea. *International Journal of Remote Sensing,* 14: 2079–2095.

Hastings, D., M. Matson and A. H. Horvitz, 1988. AVHRR Catalog. *Photogrammetric Engineering and Remote Sensing,* 54: 1469–1470.

Henebry, G. M., 2006. Mapping human settlements using the mir-IR: advantages, prospects, and limitations. In Q. Weng and Q. Quattrochi (eds), *Urban Remote Sensing.* CRC Press.

Holben, B. N. and Y. E. Shimabukuro, 1993. Linear mixture model applied to coarse spatial resolution data from multispectral satellite sensors. *International Journal of Remote Sensing,* 14: 2231–2240.

Hutchison, K. D. and A. P. Cracknell, 2006. *Visible Infrared Imager Radiometer Suite. A New Operational Cloud Imager.* CRC, Taylor and Francis, Boca Raton.

Hutchison, K. D. and J. K. Locke, 1997. Snow cover identification through cirrus-cloudy atmospheres using daytime AVHRR imagery. *Geophysical Research Letters,* 24: 1791–1794.

Hutchison, K. D., B. J. Etherton and P. C. Topping, 1997a. Validation of automated cloud top phase algorithms: distinguishing between cirrus clouds and snow in *a-priori* analyses of AVHRR imagery. *Optical Engineering,* 36: 1727–1737.

Hutchison, K. D., B. J. Etherton, P. C. Topping and A. H. L. Huang, 1997b. Cloud top phase determination from the fusion of signatures in daytime AVHRR imagery and HIRS data. *International Journal of Remote Sensing,* 18: 3245–3262.

Inoue, T., 1985. On the temperature and effective emissivity determination of semitransparent cirrus clouds by bi-spectral measurements in the 10 micron window region. *Journal of the Meteorological Society of Japan,* 63: 88–99.

Justice, C. O. and C. J. Tucker, in this volume. Coarse Spatial Resolution Optical Sensors. Chapter 10.

Justice, C. O., L. Giglio, S. Korontzi, J. Owens, J. T. Morisette, D. Roy, J. Descloitres, S. Alleaume, F. Petitcolin and Y. Kaufman, 2002. The MODIS fire products. *Remote Sensing of Environment,* 83: 244–262.

Kaufman, Y. J. and L. A., Remer, 1994. Detection of forests using MID-IR reflectance: An application for aerosol studies. *IEEE Transactions on Geoscience and Remote Sensing,* 32: 672–683.

Kaufman, Y. J., R. G. Kleidman and M. D. King, 1998a. SCAR-B fires in the tropics: Properties and remote sensing from EOS-MODIS. *Journal of Geophysical Research-Atmospheres,* 103 (D24): 31955–31968.

Kaufman, Y. J., C. O. Justice, L. P. Flynn, J. D. Kendall, E. M. Prins, L. Giglio, D. E. Ward, W. P. Menzel and A. W. Setzer, 1998b. Potential global fire monitoring from EOS-MODIS. *Journal of Geophysical Research-Atmospheres,* 103 (D24): 32215–32238.

Kennedy, P. J., A. S. Belward and J. -M. Grégoire, 1994. An improved approach to fire monitoring in West Africa using AVHRR data. *International Journal of Remote Sensing,* 15: 2235–2255.

Khattak, S., R. A. Vaughan and A. P. Cracknell, 1991. Sunglint and its observation in AVHRR data. *Remote Sensing of Environment,* 37: 101–116.

Laporte, N., C. Justice and J. Kendall, 1995. Mapping the dense humid forest of Cameroon and Zaire using AVHRR satellite data. *International Journal of Remote Sensing,* 16: 1127–1145.

Liljas, E., 1987. Multispectral classification of cloud, fog and haze. In R. A. Vaughan (ed.), *Remote sensing applications in meteorology and climatology.* Reidel, Dordrecht. Chapter 16, pp. 301–319.

Lucas, R., M. Honzak, G. M. Foody, P. J. Curran and D. T. Nguele, 2000. Characterising tropical forest regeneration in Cameroon using NOAA AVHRR data. *International Journal of Remote Sensing,* 21: 2831–2854.

Malingreau, J. P. and C. J. Tucker, 1988. Large scale deforestation in the southeastern Amazon Basin. *Ambio,* 17: 49–55.

Malingreau, J. P., G. Stephens and L. Fellows, 1985. Remote sensing of forest fires: Kalimantan and North Borneo in 1982–3. *Ambio,* 14: 314–315.

Malingreau, J. P., C. J. Tucker and N. Laporte, 1989. AVHRR for monitoring global tropical deforestation. *International Journal of Remote Sensing,* 10: 855–867.

Matson, M. and J. Dozier, 1981. Identification of subresolution high temperature sources using a thermal IR sensor. *Photogrammetric Engineering and Remote Sensing,* 47: 1311–1318.

Matson, M. and B. N. Holben, 1987. Satellite detection of tropical burning in Brazil. *International Journal of Remote Sensing,* 8: 509–516.

Matson, M., S. R. Schneider, B. Aldridge and B. Satchwell, 1984. Fire detection using the NOAA-Series satellites. Technical Report NESDIS 7, NOAA, Washington, DC.

Matson, M., G. Stephens and J. Robinson, 1987. Fire detection using data from the NOAA-N satellites. *International Journal of Remote Sensing,* 8: 961–970.

Minnis, P., J. P. Huang, B. Lin, Y. H. Yi, R. F. Arduini, T. F. Fan, J. K. Ayers and G. C. Mace, 2007. Ice cloud properties in ice-over-water cloud systems using Tropical Rainfall Measuring Mission (TRMM) visible and infrared scanner and TRMM Microwave Imager data. *Journal of Geophysical Research-Atmospheres,* 112 (D6): Art. No. D06206.

Muirhead, K. and A. P. Cracknell, 1984. Identification of gas flares in the North Sea using satellite data. *International Journal of Remote Sensing,* 5, 199–212.

Muirhead, K. and A. P. Cracknell, 1985. Straw burning over Great Britain detected by AVHRR. *International Journal of Remote Sensing*, 6: 827–833.

Payan, V. and A. Royer, 2004. Spectral emsivity of northern land cover types derived with MODIS and ASTER sensors in MWIR and LWIR. *Canadian Journal of Remote Sensing*, 30: 150–156.

Pereira, J. M. C., 1999. A comparative evaluation of NOAA/AVHRR vegetation indexes for burned surface detection and mapping. *IEEE Transactions on Geoscience and Remote Sensing*, 37: 217–226.

Peres, L. F. and C. C. DaCamara (2005) Emissivity maps to retrieve land-surface temperature from MSG/SEVIRI. *IEEE Transactions on Geoscience and Remote Sensing*, 43: 1834–1844.

Petitcolin, F. and E. Vermote, 2002. Land surface reflectance, emissivity and temperature from MODIS middle and thermal infrared data. *Remote Sensing of Environment*, 83: 112–134.

Petitcolin, F., F. Nerry and M-P. Stoll, 2002. Mapping directional emissivity at 3.7 μm using a simple model of bi-directional reflectivity. *International Journal of Remote Sensing*, 23, 3443–3472.

Prins, E. M. and W. P. Menzel, 1994. Trends in South-American biomass burning detected with the Goes Visible Infrared Spin Scan Radiometer Atmospheric Sounder from 1983 to 1991. *Journal of Geophysical Research–Atmospheres* 99 (D8): 16719–16735.

Prins, E. M., J. M. Feltz, W. P. Menzel and D. E. Ward, 1998. An overview of Goes-8 diurnal fire and smoke results for scar-B and 1995 fire season in South America. *Journal of Geophysical Research–Atmospheres*, 103 (D24): 31821–31835.

Roberts, G., M. J. Wooster, G. L. W. Perry, N., Drake, L. -M. Rebelo and F. Dipotso, 2005. Retrieval of biomass combustion rates and totals from fire radiative power observations: Application to Southern Africa using Geostationary SEVIRI imagery. *Journal of Geophysical Research–Atmospheres*, 110 (D21): D21111.

Roger, J. C. and E. F. Vermote, 1998. A method to retrieve the reflectivity signature at 3.75 μm from AVHRR data. *Remote Sensing of Environment*, 64: 103–114.

Roy, D. P., L. Giglio, J. D. Kendall and C. O. Justice, 1999. Multi-temporal active-fire based burn scar detection algorithm. *International Journal of Remote Sensing*, 20: 1031–1038.

Roy, D. P., P. E. Lewis and C. O. Justice, 2002. Burned area mapping using multi-temporal moderate spatial resolution data – a bi-directional reflectance model-based

expectation approach. *Remote Sensing of Environment*, 83: 264–287.

Sakaida, F., K. Hosoda, M. Moriyama, H. Murakami, A. Mukaida and H. Kawamura, 2006. Sea surface temperature observation by global imager (GLI)/ADEOS-II: Algorithm and accuracy of the product. *Journal of Oceanography*, 62: 311–319.

Salisbury, J. W. and D. M. D'Aria, 1994. Emissivity of terrestrial materials in the 3–5 μm atmospheric window. *Remote Sensing of Environment*, 47: 345–361.

Scorer, R. S., 1989. Cloud reflectace variations in channel-3 (MIR). *International Journal of Remote Sensing*, 10: 675–686.

Setzer, A. W. and M. M. Verstraete, 1994. Fire and glint in AVHRR's channel 3: a possible reason for the non-saturation mystery. *International Journal of Remote Sensing*, 15: 711–718.

Setzer, A. W., M. C. Pereira, A. C. Pereira and S. A. O. Almeida, 1988. Relatorio de Atividades do Projeto IBDF-INPE 'SEQE' – Ano 1987. INPE-4534-RPE/565, Inst. Nacional de Pesquisas Espacias, 12.201, São Jos, dos Campos, SP, Brazil.

Siegert, F., B. Zhukov, D. Oertel, S. Limin, S. E. Page, and J. O. Rieley, 2004. Peat fires detected by the BIRD satellite. *International Journal of Remote Sensing*, 25: 3221–3230.

Sun, D. L. and R. T. Pinker, 2003. Estimation of land surface temperature from a Geostationary Operational Environmental Satellite (GOES-8). *Journal of Geophysical Research–Atmospheres* 108 (D11): 4326.

Tucker, C. J., B. N. Holben and T. E. Goff, 1984. Intensive forest clearing in Rondonia, Brazil, as detected by satellite remote sensing. *Remote Sensing of Environment*, 15: 255–261.

Vafeidis, A. T. and N. A. Drake, 2005. A two-step method for estimating the extent of burnt areas with the use of coarse-resolution data. *International Journal of Remote Sensing*, 26: 2441–2459.

Wan, Z.-M., 1985. Land surface temperature measurement from space. PhD thesis, Department of Geography, University of California, Santa Barbara.

Wooster, M. J., B. Zhukov and D. Oertel, 2003. Fire radiative energy for quantitative study of biomass burning: derivation from the BIRD experimental satellite and comparison to MODIS fire products. *Remote Sensing of Environment*, 86: 83–107.

Wooster, M. J., G. Roberts, G. L. W. Perry and Y. J. Kaufman, 2005. Retrieval of biomass combustion rates and totals from fire radiative power observations: FRP derivation and calibration relationships between biomass consumption and fire radiative energy release. *Journal of Geophysical Research–Atmospheres*, 110 (D24): D24311.

Thermal Remote Sensing in Earth Science Research

Dale A. Quattrochi and Jeffrey C. Luvall

Keywords: thermal infrared remote sensing, surface energy balance, thermal radiation, emissivity, atmospheric correction, evaporation, evapotranspiration, urban heat island.

INTRODUCTION

In this chapter, we focus on elucidating the principles of thermal infrared (TIR) remote sensing for observation of Earth surface characteristics and for research relating to analysis of biophysical Earth processes, in particular landscape characterization and measurement of land surface processes. We will draw on selected case studies that exemplify how TIR data are essential for developing a better understanding, and more robust models, of land–surface energy balance interactions.

Because the basic theory of TIR remote sensing is well developed in the literature (see for example, Sabins 1978, Chapter 5, Price 1989, Jensen 2000, Chapter 8), it will be covered only briefly here as a foundation for the remaining parts of the chapter. Additional background on the instrumentation, optics, and engineering related to TIR sensors is also covered extensively in the literature; a particularly thorough overview of TIR instrumentation may be found at the Omega.com website (Omega 2007). Another important area of TIR research relates to geology, including volcanology and mineral exploration. For space reasons, these latter topics cannot be covered here, but additional information can be found at The Geological Remote Sensing Group web site at http://www.grsg.org/ and in the TIR Chapter in Gupta's (2003) book, *Remote Sensing Geology.*

A BRIEF HISTORY OF THERMAL INFRARED REMOTE SENSING

Thermal remote sensing at its beginnings was shielded in the cloak of secrecy due to its strategic advantage as a tool for military uses. Although the astronomer Sir Frederick William Herschel (1738–1822) discovered the TIR portion of the electromagnetic spectrum in 1800, it was not until 1880 that S. P. Langley developed a device that could measure temperature variations with accuracy. His instrument, called the bolometer, could measure temperature variations of $1/10,000\,°C$ (Jensen 2000). Primitive radiation thermometers were employed during World War I to detect men and even aircraft from a distance of 120 m. In the 1930s, Germany developed the Kiel system for discriminating between bombers and night fighters (Jensen 2000). Great Britain and the United States also developed TIR surveillance techniques during World War II, and the single most important development in TIR technology was the invention of the detector element during the war (Jensen 2000). Detectors convert incoming radiant and thermal energy into an analog electrical signal. This was a vast improvement over what was previously available in the form of only thermal photographs.

The first modern radiation thermometers were not available until after World War II. Early detectors consisted of lead sulfide photodetectors and were the first widely used detectors in industrial radiation thermography. Now there are very fast detectors consisting of mercury-doped germanium (Ge:Hg), indium antimonide (In:Sb) and other substances that are very sensitive to TIR radiation (Jensen 2000). Migration of TIR technology from the military sector, which had developed very sensitive satellite and aircraft remote sensing systems, began in the 1960s. In 1968, the U.S. government declassified the production of TIR remote sensing systems that did not exceed a specific spatial resolution and temperature sensitivity (i.e., prior to 1968, thermal data could be collected but at a very coarse resolution) (Jensen 2000). The first civilian satellite remote sensing data (including thermal data at a very coarse resolution) were collected for weather observation in 1960 onboard the TIROS weather satellite. The United States (U.S.) National Aeronautics and Space Administration (NASA) launched the Heat Capacity Mapping Mission (HCMM) in 1978 for Earth remote sensing observations. HCMM obtained 600 m × 600 m spatial resolution TIR (10.5–12.6 μm) scenes during both the day-time and night-time. In the same year, NASA launched the Coastal Zone Color Scanner (CZCS), which also included a TIR sensor, to monitor sea-surface temperature. The U.S. National Oceanic and Atmospheric Administration (NOAA) began collecting thermal data with the Advanced Very High Resolution Radiometer (AVHRR) onboard the Polar Orbiting Environmental Satellites (POES), beginning with TIROS-N in 1978. The latest genesis of the AVHRR sensor still collects TIR data today at spatial resolutions of 4 km × 4 km and 1.1 km × 1.1 km (Justice and Tucker, in this volume).

Perhaps the most widely used TIR measurements collected from satellite for Earth observations has been Landsat Thematic Mapper (TM) data (Goward et al., in this volume). The Landsat 4 and 5 TM sensors were launched by NASA in 1982 and 1984 respectively, and collected TIR data at 120 m × 120 m spatial resolution over the range 10.4–12.5 μm. Landsat 7 was launched in 1999 carrying the Enhanced Thematic Mapper Plus (ETM+) sensor, for which the TIR data spatial resolution was increased to 60 m × 60 m. As a complement to the Landsat TM and TM+ TIR sensors, NASA launched the Advanced Spaceborne Thermal Emission and Reflection Radiometer (ASTER) on the Terra Earth observing satellite platform. ASTER, which was built for the Japanese Space Agency, collects data at a 90 m × 90 m spatial resolution over five channels between 8.12 and 11.65 μm.

In addition to spaceborne TIR sensors, NASA has also developed a number of aircraft-mounted thermal instruments that have been used for Earth science research (Table 5.1). The Thermal Infrared Multispectral Scanner (TIMS) sensor is a six-channel instrument that collects data entirely within the thermal spectrum for six spectral channels from 8.2 to 11.60 μm. As a further evolution of the TIMS, the Advanced Thermal and Land Applications Sensor (ATLAS) was developed by NASA. The ATLAS sensor includes the same six TIR spectral channels as the TIMS, and also includes nine other spectral channels in the visible and middle TIR reflective range. The advantage of these sensors is that unlike satellite platforms that are placed in a set orbit, aircraft-mounted instruments can collect data at different altitudes thereby providing TIR data at different spatial scales to study specific Earth processes, such as monitoring geological phenomena (e.g., volcanoes), biological and physiological interactions (e.g., evapotranspiration, vegetation energy balances), landscape and agricultural characteristics (e.g., soil moisture, evaporation), forest energy exchanges (e.g., forest surface temperature distributions), and a variety of energy flux interactions (e.g., urban heat island dynamics, surface energy balances) (Quattrochi and Luvall 1999, Quattrochi et al. 2000).

The German Aerospace Center (DLR) has developed a 79-channel Digital Airborne Imaging Spectrometer (DAIS 7915). This sensor covers the spectral range from the visible to the TIR. DAIS 7915 has six TIR channels ranging from 8.5 to 12.3 μm, and one channel at 4.3 μm (Strobl et al. 1996). More recently, the DLR has supported the development of the Airborne Reflective Emissive Spectrometer (ARES), with 30 TIR bands (Müller et al. 2005).

A BASIC PRIMER ON THERMAL INFRARED RADTIATON THEORY AND RELATED PRINCIPLES

Thermal radiation

All bodies above absolute zero radiate energy. The amount of energy and the wavelengths of this energy are emitted is a function of its blackbody temperature. Planck's law gives the intensity radiated by a blackbody as a function of frequency (or wavelength).

$$L_{BB\lambda} = \frac{C_1}{\lambda^5 \pi \left[e^{C_2/\lambda T} - 1 \right]} \qquad (1)$$

where $L_{BB\lambda}$ is spectral blackbody radiance (in $W\,m^{-2}sr^{-1}$; i.e. energy per unit area, per unit wavelength, per unit solid angle), λ is the wavelength, T is the temperature of blackbody (Kelvin), C_1 is the first fundamental physical constant characterizing blackbody radiation,

Table 5.1   Characteristics of several airborne multispectral visible and thermal scanners

Sensor	FOV[1] (°)	IFOV[2] (mrad)	Bandwidth (nm)	Number of channels	Radiometric resolution (bits)	Calibration
DAIS[3]	9.00	3.3	450–1050	32	15	VIS, SWIR, and TIR on ground laboratory calibration facility
			1500–1800	8		
			1900–2450	32		
			3000–5000	1		TIR-Onboard ambient and heated blackbodies
			8700–12315	6		
MIVIS[4]	±35.53	2.0	430–830	20	12	On ground VIS
			1150–1550	8		Daedalus AB532 optical bench
			1980–2450	64		SWIR and TIR – blackbody flat plate
			8210–12700	10		
MASTER[5]	±42.96	2.5	400–13000	50	16	VIS–SWIR Laboratory Integrating sphere
						MIR–TIR 2 on-board blackbodies
ATLAS[6]	37.00	2.00	450–12200	15	8	VIS–onboard integrating radiance sphere
						Mir–TIR– N₂ cooled blackbodies
TIMS[7]	30.00	2.5	8000–12000	6	8	TIR–N₂ cooled blackbodies

[1] Field of View.
[2] Instantaneous Field of View.
[3] Digital Airborne Imaging Spectrometer 7915.
[4] Multispectral Infrared Visible Imaging Spectrometer.
[5] MODIS/ASTER Airborne Simulator.
[6] Advanced Thermal and Land Applications Sensor.
[7] Thermal Infrared Multispectral Scanner.

3.7415×10^{-16} W m^{-2}, C_2 is the second fundamental physical constant characterizing blackbody radiation, 0.0143879 m K, and ε_λ is the surface spectral emissivity at a given wavelength λ.

The dominant wavelength (λ_{max}) for a blackbody at a specific temperature is determined by Wien's Displacement Law:

$$\lambda_{max} = \frac{k}{T} \qquad (2)$$

where T is the blackbody temperature (Kelvin) and k is the constant 2898 μm K.

The TIR wavelengths are ideally suited for observing the Earth's surface (Table 5.2). The peak spectral exitance for most of the Earth's surfaces and ecosystems are within the range of the TIR region. High temperature features such as fires and volcanoes are better observed using the mid-infrared.

The Stefan–Boltzmann Law of thermal radiation gives the energy emitted by a blackbody (per unit time) at an absolute temperature of T. The long-wave energy emitted from a surface ($L \uparrow$) is dependent on surface temperature:

$$L \uparrow = \varepsilon \left[\sigma T^4 \right] \qquad (3)$$

Table 5.2   The relationship between a blackbody temperature and peak spectral exitance

Surface	Temperature (K)	Wavelength (μm) of peak spectral exitance
Sun	6000	0.483
Volcano	1350	2.147
Forest fire	1000	2.898
Urban areas	315	9.200
Forest	300	9.660
Earth	288	10.063
Ice	273	10.615

where ε is emissivity, σ is the Stefan–Boltzman constant (5.7×10^{-8} W m^{-2} K^{-4}) and T is the land surface temperature (Kelvin). Emissivity is calculated as:

$$\varepsilon = \frac{S_r}{S_b} \qquad (4)$$

where S_r is the measured surface radiance and S_b is the blackbody radiance at the same temperature.

Emissivity is a ratio of the actual energy emitted by an object to the energy an ideal blackbody would emit at the same temperature. Most objects are 'gray' bodies, i.e., they have an emissivity that is less than one and as defined have a constant emissivity over all wavelengths. However, such an ideal 'gray' body does not exist in practice. Many geologic mineral deposits and soils exhibit different emissivities at different wavelengths. These wavelength-dependent emissivities have been the basis for mapping geologic formations using thermal data (Kahle 1987, Hook et al. 1992). Changes in ε will affect the apparent radiant temperatures obtained from those surfaces. It is difficult to measure emissivity using most satellite sensors because very few such sensors measure more than a single broadband thermal image (8–12 μm). Spectral variation in emissivity can be measured using a multispectral thermal instrument. Figure 5.1 demonstrates ε differences from several natural surfaces measured with the airborne ATLAS instrument (Airborne Thermal and Land Applications Sensor). Any two thermal channels plotted gives a scattergram that shows that: (1) radiance increases with higher temperature and higher emissvity; (2) any material for which ε_λ is constant will plot as a line segment in the scattergram, the segment's position is a function of T; (3) all such materials will plot as coaxial segments; (4) vegetation has nearly uniform ε in the thermal bands; (5) sensor gain and offset change the slope and scale of the data; (6) sensor noise causes dispersion from an ideal line; and (7) departure from the line defined by spectral gray materials must be caused by variation in ε between channels. The wavelength variation of soil emissivity channels is evident in Figure 5.1 from the non-coaxial plotting of the points. This is in contrast to the metal surfaces that have low emissivity, but plot along the same trend line. This last mentioned attribute can be observed in a multispectral false color composite of the image data, where different thermal channels are assigned to the red, green, and blue display bands. This produces a false-color image which is gray where ε is relatively similar between the different bands, and colors of various hues where ε varies with wavelength.

Atmospheric corrections and data calibration

Accurate calibration of thermal date from either a satellite or an aircraft based sensor requires the integration of sensor calibration and characteristics with a determination of atmospheric radiance. Atmospheric radiance must be accounted for in order to obtain calibrated surface temperatures. Although the bandpasses for satellite and aircraft thermal channels used for surface temperature mapping fall within the atmospheric window for atmospheric long wave transmittance (8.0–12.0 μm), the maximum transmittance across

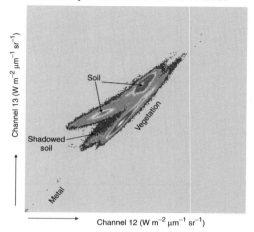

Emissivity differences in natural surfaces

Figure 5.1 Emissivity differences in natural surfaces as indicated from a scatter plot of two thermal infrared channels of ATLAS (Advanced Thermal and Land Applications Sensor) data. Frequency of occurrence of a particular combination of energy values in the original image is indicated by a particular gray shade in the figure.

this region is only about 80% (Figure 5.2). Typical broadband thermal sensors cover 8–12 μm; some airborne and space-borne sensors such as the ATLAS have several bands within this region. The amount of atmospheric radiance in the atmospheric window is mostly dependent on the atmospheric water vapor content, although there is an ozone absorption band around 9.5 μm. Radiosondes, which measure atmospheric water vapor, temperature and pressure, are sometimes launched concurrently with image acquisitions. The atmospheric profiles obtained from these radiosondes are then incorporated into an atmospheric modeling program such as the MODTRAN4 model for calculation of atmospheric radiance (Berk et al. 1999). The output from MODTRAN4 is combined with calibrated spectral response curves for each band, and blackbody radiance information, recorded during the flight. For example, for ATLAS processing, a combination of software is used, including the public domain image processing/remote sensing package ELAS (Beverly and Penton 1989), and a series of custom programs, WATTS and ENERGY (Graham et al. 1986, Rickman et al. 2000) , to produce data calibrated to physical units of energy.

Other techniques have been used for calibration verification of TIR data. For example, Hook et al. (2007) used instrumented buoys located in Lake Tahoe on the California–Nevada border and

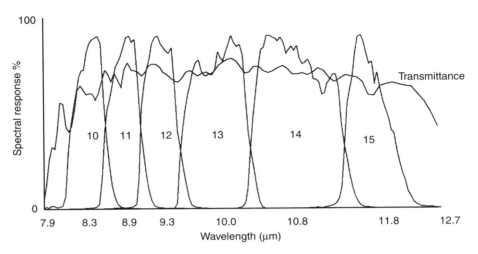

Figure 5.2   Graph of thermal atmospheric transmittance and ATLAS thermal spectral responses. Numbers in the graph represent ATLAS band numbers, those below the graph indicate the band centers in micrometers.

ground-based radiance methods to validate the absolute radiometric calibration of ASTER and MODIS data. Their approach is predicated on the use of a homogeneous isothermal surface (lake surface temperatures) for the calibration of the satellite data to give at-sensor radiance, L_s, by wavelength (λ) in the mid and TIR region (Hook et al. 2007):

$$L_{s\lambda} = \left[\varepsilon_\lambda L_{bb\lambda}(T) + (1 - \varepsilon_\lambda)L_{sky\,\lambda}\right]\tau_\lambda + L_{atm\,\lambda} \quad (5)$$

where ε_λ is the surface spectral emissivity at a given wavelength, $L_{bb\lambda}(T)$ is the spectral radiance from a blackbody at surface temperature T, $L_{sky\,\lambda}$ is the spectral downwelling radiance on the surface from the atmosphere, τ_λ is the spectral atmospheric transmittance, and $L_{atm\,\lambda}$ is the at sensor spectral path radiance from the atmosphere including emission and scattering.

The removal of atmospheric effects defines surface radiance by wavelength as:

$$L_\lambda = \varepsilon_\lambda L_{BB\lambda}(T) \quad (6)$$

Basic thermal equations related to evapotranspiration and surface energy budgets

Surface temperature is a major factor in the surface energy budget. Use of energy terms in modeling surface energy budgets allows the direct comparison of various land surfaces encountered in a landscape, from vegetated (forest and herbaceous) to non-vegetated (bare soil, roads, and buildings) (Oke 1987). The partitioning of energy budget

terms depends on the surface type. In natural landscapes, the partitioning is dependent on canopy biomass, leaf area index, aerodynamic roughness, and moisture status, all of which are influenced by the development stage of the ecosystem. In urban landscapes, coverage by man-made materials substantially alters the surface energy budget.

The Penman–Monteith equation (Monteith 1973) can be used to illustrate these factors. It combines those environmental factors and plant factors that are important in determining the latent heat flux (evapotranspiration), LE, from a surface:

$$LE = \left(\frac{s(R_n - G) + \left[\rho_a C_\rho D/R_a\right]}{s + \gamma\left[1 + R_c/R_a\right]}\right) \quad (7)$$

where R_n is net all wave radiation at the surface, G is energy storage into or out of the surface, s is the slope of the saturation vapor–pressure relationship, ρ_a is the density of the air, C_ρ is the specific heat of air, D is the vapor-density deficit of the air, γ is the psychrometric constant, R_a is aerodynamic resistance, and R_c is canopy resistance (stomatal resistance/leaf area index).

The surface temperature T_s is dependent on the partitioning of the energy received at the surface into sensible and latent heat.

$$T_s = T_a + \frac{R_b}{C_\rho}(R_n - LE) \quad (8)$$

where T_a is air temperature and R_b is boundary layer (aerodynamic) resistance.

Luvall and Holbo (1989, 1991) present the Thermal Response Number (TRN), a remote sensing technique for describing the surface energy budget within a forested landscape. This procedure treats changes in surface temperature as an aggregate response of the dissipated thermal energy fluxes (latent heat and sensible heat exchange, and conduction heat exchange with biomass and soil). The TRN is therefore directly dependent on surface properties (canopy structure, amount and condition of biomass, heat capacity, and moisture). A time interval of 15–30 minutes between remote sensing over flights of the same area using the Thermal Infrared Multispectral Scanner (TIMS) for selected forested landscapes has revealed a measurable change in forest canopy temperature due to the change in incoming solar radiation. Surface net radiation integrates the effects of the non-radiative fluxes, and the rate of change in forest canopy temperature presents insight on how non-radiative fluxes are reacting to radiant energy inputs. The ratio of net radiation to change in temperature can be used to define a surface property referred to as the TRN, with units of kJ m^{-2} K^{-1}:

$$\text{TRN} = \sum_{t_1}^{t_2} Rn\Delta t / \Delta T \qquad (9)$$

where:

$$\sum_{t_1}^{t_2} Rn\Delta t \qquad (10)$$

represents the total amount of net radiation (Rn) for that surface over the time period between flights ($\Delta t = t_2 - t_1$) and ΔT is the change in mean temperature of that surface.

The mean spatially averaged temperature for the surface elements at the times of imaging is estimated from:

$$T = \frac{\sum T_p}{n} \qquad (11)$$

where each T_p is the temperature of a pixel in the thermal image and n is the number of pixels of the surface element.

The TRN provides an analytical framework for studying the effects of surface thermal response for large spatial resolution map scales that can be aggregated for input to coarser scales as needed by climate models. The importance of TRN is that: (1) it is a functional classifier of land cover types; (2) it provides an initial surface characterization for input to various climate models; (3) it is a physically based measurement; and (4) it can be determined completely from a pixel by pixel measurement or for a polygon from a landscape feature which represents a group of pixels. The TRN can be used as an aggregate expression of both environmental energy fluxes and surface properties such as forest canopy structure and biomass, age, and physiological condition; urban structures and material types.

Thermal energy theory as applied to ecological thermodynamics

Thermal remote sensing can provide environmental measuring tools with capabilities for measuring ecosystem development and integrity. Recent advances in applying principles of nonequilibrium thermodynamics to ecology provide fundamental insights into energy partitioning in ecosystems. Ecosystems are nonequilibrium systems, open to material and energy flows, which grow and develop structures and processes to increase energy degradation. More developed terrestrial ecosystems will be more effective at dissipating the solar gradient (degrading its exergy[1] content).

Thermal energy theory has been developed as a result of research to uncover principles of ecological development (see Kay 2000 for an overview). The research in ecological thermodynamics has focused on linking physics and systems sciences with biology, and especially linking the science of ecology with the laws of thermodynamics. Such research follows on the observation that similar developmental processes are observed in ecosystems across a range of scales, for example from small laboratory microcosms, to prairie grass systems, to vast forest systems and ocean plankton systems. Such similar phenomenology has long suggested underlying processes and rules for the development of ecological patterns of structure and function (Odum 1969). Furthermore, recent advances in nonequilibrium thermodynamics coupled with the investigation of self-organizing phenomena in quite different types of systems (from simple convection cell systems to forested ecosystems), have revealed that all self-organizing phenomena (including ecosystem development) involve similar processes, processes which are mandated by the second law of thermodynamics. This conclusion, as discussed below, provides a basis for a quantitative description of ecosystem development (Kay 1991, Kay and Schneider 1992a, b, Schneider and Kay 1993, 1994a, b, Regier and Kay 1996, Fraser and Kay 2004).

The study of self-organization phenomena in thermodynamic systems that are open to energy and/or material flows, and which reside in quasi-stable states some distance from equilibrium (Nicolis and Prigogine 1977), is the basis for this theory. Both non-living self-organizing systems (like convection cells, tornadoes, and lasers), and living self-organizing systems (from cells to ecosystems), are dependent on exergy (high quality energy) fluxes, from outside sources, to sustain their self-organizing processes. These processes are maintained by the destruction of the exergy, conversion of the high quality energy flux into a flux of lower quality forms of energy. Consequently, these processes increase the entropy of the larger 'global' system, in which the self-organizing system is embedded. Crucial insights into the dynamics of self-organizing systems can be gained from examining the role of the second law of thermodynamics in determining these dynamics.

Using recent advances in thermodynamics, specifically exergy or second law analysis, the second law of thermodynamics can be extended so as to apply to nonequilibrium regions and processes. In this circumstance, systems can be described in terms of the exergy fluxes setting up gradients (e.g., temperature and pressure differences in classical thermodynamic systems). With the establishment of these gradients, the system is no longer in equilibrium. The system responds to these imposed gradients, by self-organizing in a way which resists the ability of the exergy fluxes to establish gradients, and hence move the system further away from equilibrium. More formally, the thermodynamic principle (the restated second law) which governs the behavior of systems states that, as systems are moved away from equilibrium, they will utilize all avenues available to counter the applied gradients. As the applied gradients increase, so does the system's ability to oppose further movement from equilibrium (Schneider and Kay, 1994a). As a system self-organizes it will become more effective at exergy utilization.

Kay and Schneider (1994) have focused on the application of this thermodynamic principle to the science of ecology. Ecosystems are viewed as open thermodynamic systems with a large gradient impressed on them by the exergy flux from the sun. Ecosystems, according to the restated second law, develop in ways which systematically increases their ability to degrade the incoming solar exergy, hence counteracting the sun's ability to set up even larger gradients.

Thus it should be expected that more mature ecosystems will degrade the exergy they capture more completely than a less developed ecosystem. The degree to which incoming solar exergy is degraded is a function of the surface temperature of the ecosystem (see Fraser and Kay 2004 for details). If a group of ecosystems receives the same amount of incoming radiation, we would expect that the most mature ecosystem would reradiate its energy at the lowest quality level and thus would have the lowest surface temperature.

Terrestrial ecosystem's surface temperatures have been measured using TIMS. Luvall and his coworkers (Luvall and Holbo 1989, 1991, Luvall et al. 1990) have documented ecosystem energy budgets, including tropical forests, mid-latitude varied ecosystems, and semi-arid ecosystems.

These data show that within a given biome type, and under similar environmental conditions (air temperature, relative humidity, winds, and solar irradiance), the more developed the ecosystem, the cooler its surface temperature and the more degraded the quality of its reradiated energy. In the context of nonequilibrium thermodynamic theory, three hypotheses about ecosystem development can be developed that relate to the use of thermal remote sensing measures (Kay and Schneider 1994, Schneider and Kay 1994a, Schneider and Sagan 2005):

1 The ratio (Rn/K^*) of net all-wave radiation to net short-wave radiation (K^*), received at the surface, will be larger for more developed ecosystems and these systems will have a lower surface temperature.
2 Spatial variation of surface temperature (T) will be less for more developed ecosystems. The spatial variation can be indexed using the beta index (Holbo and Luvall 1989). The more developed ecosystems will have a larger beta index.
3 More developed ecosystems will exhibit a smaller temperature change in response to a given amount of energy input (net radiation). This can be measured using TRN (Luvall and Holbo 1989). The more developed ecosystems should have a larger TRN.

The studies discussed above suggest that ecosystems develop structure and function that degrades the quality of the incoming energy more effectively, that is they degrade more exergy. Furthermore, those studies show that analysis of airborne collected radiated energy fluxes is a valuable tool for measuring the energy budget and energy transformations in terrestrial ecosystems. Given the hypothesis, formulated from first principles of nonequilibrium thermodynamics of self-organizing systems, that a more developed ecosystem degrades more exergy, the ecosystem temperature, Rn/K^*, beta index, and TRN are excellent candidates for indicators of ecological integrity. The potential for these methods to be used for remote sensed ecosystem classification and ecosystem health/integrity evaluation is apparent.

LANDSCAPE CHARACTERIZATION AND MEASUREMENT OF SURFACE ENERGY FLUXES FROM THERMAL INFRARED REMOTE SENSING DATA

TIR remote sensing data can provide important measurement of surface energy fluxes and temperatures that are integral to understanding landscape characteristics, such as those related to biophysical processes and responses across the landscape, and interactions between the land and the lower atmosphere. The two most fundamental ways that TIR data can be applied to characterization of landscape processes and the responses that result from them are: (1) through measurement of surface temperatures as related to specific landscape and biophysical components; and (2) through relating surface temperatures with energy fluxes for specific landscape phenomena or processes. The interaction of solar, atmospheric, and terrestrial radiation at the Earth's surface that ultimately drives energy balance fluxes across the landscape may be stated as:

$$R_{net} = K \downarrow (1 - \alpha) + L \downarrow - L \uparrow$$
$$= LE + H + P_s \pm G \qquad (12)$$

where R_{net} is the net radiation energy balance, $K \downarrow$ is the incoming short-wave radiation received at the surface (both direct and diffuse) [at night the term $K \downarrow (1 - \alpha)$ is zero], α is the albedo of the surface (i.e., reflectivity), $L \downarrow$ is the downward longwave atmospheric radiation incident on the surface, $L \uparrow$ is the upward longwave radiation emitted by the surface, G is heat flux into and out of the ground or other surfaces, H is the sensible heat transfer between atmosphere and ground, LE is the heat loss by evaporation from the surface (or plant cover) or gain by condensation (i.e., dew or frost formation), and P_s is heat production or heat rejection from man-made sources, including human and animal metabolism (Landsberg 1981).

Although it is conceivable that remote sensing technology can contribute to the measurement of all the parameters given in Equation (12), the factor $L \uparrow$ is the most amenable to measurement as a separate flux, particularly from TIR data. Fundamental to the characterization of landscape processes using TIR data is the measurement of longwave thermal energy emitted from land surface features. Measurement of the magnitude and direction distribution of longwave thermal energy provides quantification of a basic input to the energy budget of the landscape ecosystem. Understanding how thermal energy is partitioned across a landscape, and the dynamics and variation in surface temperatures emanating from various landscape elements (forest, crops, built-up surfaces), is essential to defining the overall mechanisms that govern land–atmosphere interactions. The energy budget dynamics that drive these interactions over landscapes may be defined as: (1) the coupling of extant energy balances with the environment; (2) the level of energy balances with the environment; (3) the kinds of energy transformations that occur, especially those that are biologically controlled; and (4) the mix of energy outputs that can be regarded as yields. Variations in the magnitude

of surface thermal energy for a specific surface or surfaces often affect the density, dynamics, and importance of other energy fluxes linked to specific landscape characteristics (e.g., evapotranspiration, nutrient cycling) (Miller 1981, Quattrochi and Luvall 1999).

In order to better understand the exchange of heat and moisture between the land surface and the lower atmosphere, it is important to quantify the components of the surface energy balance in a distributed fashion across the landscape (or at the 'landscape scale'). TIR remote sensing data can provide spatially distributed information on a number of fundamental land surface characteristics and state variables that control the surface energy balance. In conjunction with meteorological measurements recorded at the near-surface (e.g., from meteorological instrument towers) and using a relatively simple model, aircraft- and satellite-based TIR remote sensing data can be used to create a representation (i.e., map) of spatially distributed surface energy balance components over a landscape. The simplest form of the energy balance equation (assuming there is no advection of energy into the area (e.g., by surface wind) may be expressed by:

$$R_{net} = G + H + LE \qquad (13)$$

where R_{net} refers to the net radiation balance, G refers to the soil heat flux (i.e., the energy used to warm the near-surface soil layers), H is the sensible heat flux (the energy used to transfer heat from the surface to the atmosphere), and LE relates to the latent heat flux (the energy used to transfer water vapor from the surface to the atmosphere).

TIR remote sensing data obtained from both satellite and aircraft have been used extensively in surface energy research to measure surface temperatures over a wide range of environmental conditions. For example, TIR data have been used to measure thermal emissions of complex terrain (Goward et al. 1985), nocturnal temperature patterns and delineation of cold-prone areas (Chen et al. 1979, 1982), characterization of desert terrain (Peterson et al. 1987), and boreal forest assessment (Sellers et al. 1995). More recently, TIR data obtained from satellite and aircraft sensors have been used to derive radiometric surface temperatures of biophysical components, as well as for estimating surface energy fluxes across the landscape (Gillies and Temesgen 2004, Humes et al. 2004). A brief discussion of these investigations is useful to elucidate the role that TIR data can play in both measuring land surface energy fluxes and assessing the spatial distribution of energy fluxes across the landscape.

Humes et al. (2004) illustrate the utility of TIR data for mapping spatially distributed information on a number of key land surface characteristics and state variables that control surface energy balances. In their study, Landsat TM thermal data were used in conjunction with a relatively simple 'snapshot' model to compute spatially distributed values of net radiation, soil heat flux, sensible heat flux, and latent heat flux over the Little Washita Experimental Watershed located in south-central Oklahoma, USA. The key information provided by remotely sensed data in the model included surface temperature, land cover type, and estimates of vegetation density. Hence, Landsat TIR data were important for quantifying the components of the surface energy balance at a landscape-scale to better understand the exchange of heat and moisture between the land surface and the lower atmosphere.

Gillies and Temesgen (2004) focus on the derivation of biophysical variables from thermal and multispectral remote sensing data (in this case from the Thematic Mapper Simulator airborne sensor). TIR remote sensing data were coupled with a Soil-Vegetation-Atmosphere-Transfer (SVAT) model to derive biophysical variables. The SVAT model was then used to derive estimates of vegetation cover and surface radiant temperature using what is called the 'triangle method'. The triangle method combines measurements of surface radiant temperature (T_0) and reflectance in the red and near-TIR portions of the electromagnetic spectrum. The reflectance measurements are used to calculate the Normalized Difference Vegetation Index (NDVI). NDVI is then plotted as a function of T_0 to evaluate the relationship between these two variables, as well as providing an overlaying index of moisture availability to establish a 'warm edge' and a 'cold edge' index (Figure 5.3). Thus, the triangle technique within the overall scheme of an SVAT model (wherein TIR data are a critical input to the model) is a technique for inversely deriving biophysical variables relating to fractional vegetation cover, moisture availability, evapotranspiration, and sensible heat flux.

MEASUREMENT OF EVAPOTRANSPIRATION, EVAPORATION, AND SOIL MOISTURE FROM THERMAL REMOTE SENSING DATA

One of the most significant applications of TIR remote sensing data has been in inferring evapotranspiration (ET), evaporation, and soil moisture of vegetated surfaces. A good review of TIR attributes for measurement of ET, evaporation, and soil moisture is given by Moran and Jackson (1991). Some of the more important early work on using TIR remote sensing data to estimate soil moisture and evaporation was performed by Idso, Reginato, and Jackson (see, e.g., Idso et al. 1975,

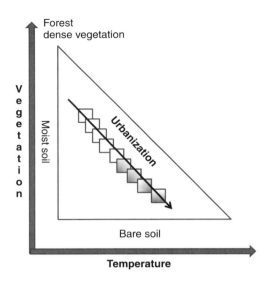

Figure 5.3 Schematic triangle showing the boundary values for zero fractional vegetation cover (bare ground: the triangle's base) and 100% vegetation cover (forest dense vegetation: near the triangle's upper vertex). Zero moisture availability is indicated on the right-hand side of the triangle) and the line of field capacity (moist soil: the left-hand side of the triangle). The arrow represents the migration of a pixel over time in response to urbanization (adapted from Carlson, 1995).

1976, Reginato et al. 1976, 1985, Jackson 1986). These studies verified that thermal remote sensing data, in conjunction with *in situ* measurements, can be used to measure, or at least quantitatively infer, evaporation and soil water content. Reginato et al. (1976) estimated soil water content for bare soil from TIR data and air temperatures. Also, Reginato et al. (1985) calculated evapotranspiration by combining remote sensed reflected solar radiation and surface temperatures with ground station meteorological data (i.e., incoming solar radiation, air temperature, windspeed, and vapor pressure) to calculate net radiation and sensible heat flux. They found that evapotranspiration could be adequately evaluated using a combination of remotely sensed and ground-based meteorological data, and that maps of evapotranspiration for relatively large areas could be made using airborne (and potentially satellite) sensors.

Data from the Heat Capacity Mapping Mission (HCMM) have been used to study evaporation rates. Price (1980) used HCMM satellite data and a coupled numerical model to infer 24 hour mean evaporation rates and diurnal heat capacity as controlled by variation in soil moisture. In a related work, Price (1982) used HCMM data obtained for day-time and night-time in conjunction with the results from numerical simulation, to relate satellite observations of surface temperature to surface evaporation, and the quantity of thermal inertia and diurnal heat capacity present for an agricultural area in eastern Washington State. The values derived from HCMM data were found to be in close agreement with evaporation measurements made on the ground within the study area.

Data from the NOAA AVHRR family of sensors have also been used to estimate and model evapotranspiration and soil moisture for vegetated areas. TIR data from the AVHRR sensor were employed by Taconet et al. (1986) in developing a methodology as input to a numerical vegetation/soil/surface-atmosphere model that includes parameterization of energy transfers within vegetation canopy. This investigation showed that a single surface temperature measured near midday using AVHRR data is sufficient for obtaining surface energy fluxes and canopy over dense vegetation. Results from this study also appear to closely simulate those obtained using ground-level measurements of surface fluxes.

AVHRR data have also been used in conjunction with the Normalized Difference Vegetation Index (NDVI) to infer and model soil moisture. For example, Ottlé and Vidal-Madjar (1994) utilized calculations of NDVI and TIR data to predetermine land surface parameters and landscape characteristics and to estimate the surface energy budget, and consequently, evaporation. These estimates of NDVI can also be coupled with hydrologic models to parameterize evaporation based on land cover (primarily vegetation) and to derive seasonal changes in landscape evaporation.

More recently, Carlson et al. (2004) demonstrated the potential of TIR remote sensing data for monitoring rapid soil drying and its implications for the derivation of quantitative estimates of soil moisture and surface energy fluxes. The problems related with associating correct levels or depths for soil moisture estimates are particularly acute when comparing estimates made via *in situ* measurements and those derived through analysis of remote sensing data. Carlson et al.'s (2004) research provides methods for quantifying soil water content and surface radiant temperature using both *in situ* and TIR remote sensing data, as well as to study the interrelationships between vegetation and surface energy fluxes.

In summary, it is apparent that TIR remote sensing data are important in providing measurements of surface temperatures for use in modeling of landscape–atmosphere energy exchanges to estimate evapotranspiration, evaporation, and soil moisture. Additionally, multitemporal TIR data collected from satellites are paramount for parameterizing surface moisture conditions and in

developing better simulations of landscape soil moisture and evaporation conditions over multiple space and time scales.

URBAN HEAT ISLAND CHARACTERIZATION USING THERMAL INFRARED REMOTE SENSING DATA

In Chapter 30, *Remote Sensing of Urban Areas* (Nichol, in this volume), the application of satellite thermal remote sensing data for assessing urban heat islands is described in detail. In this section, as a precursor to that discussion, we provide some background on the potential for using high spatial resolution airborne TIR data for analysis of thermal patterns across the urban landscape. As discussed in Chapter 30, the urban landscape is extremely heterogeneous. Although we can obtain good estimates of urban surface temperatures via satellite data, the spatial resolution of TIR satellite data (e.g., 60 m for Landsat ETM+, and 90 m for ASTER) is not adequate for measuring surface temperatures from the discrete array of land cover types (e.g., pavement, rooftops, automobile parking lots), the collective surface radiation characteristics of which drive the development of the urban heat island (UHI) effect. Thus, in order to make empirical observations and to derive quantitative measurements of the discrete surfaced energy balance characteristics of individual urban land cover types that force development of the UHI, it is necessary to obtain TIR data at an urban scale; i.e., at a spatial scale that permits discrimination of individual surface types that are ubiquitous in urban areas, such as asphalt, concrete and rooftops.

The utility of high spatial resolution TIR data for analysis of urban surface thermal characteristics has been documented in several publications by Quattrochi, Luvall and others (Quattrochi and Ridd 1994, 1998, Lo et al. 1997, Quattrochi et al. 2000, Gluch et al. 2006, Gonzalez et al. 2007). ATLAS data have been collected over a number of cities in the United States and its territories including Atlanta, Georgia, Salt Lake City, Utah, Sacramento, California, Baton Rouge, Louisiana, and San Juan, Puerto Rico. Because the ATLAS is mounted on an aircraft, data can be collected at relatively high spatial resolutions (e.g., 10 m pixels) for deriving accurate surface temperature measurements and thermal energy balance fluxes of discrete urban surfaces that comprise the urban landscape. Examples of the daytime TIR data from the ATLAS are given in Figure 5.4 for the Louisiana State University campus in Baton Rouge, and in Figure 5.5 for San Juan and Atlanta. It can be seen from the images (particularly in Figure 5.5 (bottom) that with high spatial analysis of thermal responses, it is possible to empirically identify different urban

Figure 5.4   Daytime ATLAS images of the Louisiana State University campus showing a natural color image on the left and a thermal image on the right, both at 5 m spatial resolution. (See the color plate section of this volume for a color version of this figure).

surface features, such as rooftops, pavements, and vegetation. The methodology for calculating surface temperatures from high spatial resolution TIR data is outlined in the references noted above and will not be expounded upon here (i.e., Quattrochi and Ridd 1994, 1998, Lo et al. 1997, Quattrochi et al. 2000, Gluch et al. 2006, Gonzalez et al. 2007). Figures 5.4 and 5.5 illustrate the complexity of thermal responses across the city landscape. With high spatial resolution TIR data, therefore, it is possible to derive thermal responses for individual surface types that collectively comprise the urban land covers that ultimately affect both the spatial pattern and dynamics of the UHI across the city at a microscale. Thus, we can relate these discrete surface temperatures with other thermophysical properties of individual surfaces such as surface thermal conductivity, heat capacity, thermal diffusivity, and moisture availability to develop a model of the 'thermal building blocks' extant across the urban landscape. In turn, this is extremely useful for developing a better quantitative and empirical understanding of the spatial and thermal energy dynamics that force development of the UHI phenomena for different urban areas in different climatic settings, and with differing urban morphologies.

THE FUTURE OF TIR REMOTE SENSING

It is a paradox that given the great utility of TIR Earth remote sensing, that the future prospects for new TIR Earth remote sensing systems, particularly in the United States, are somewhat cloudy. At this time, the Landsat Data Continuity Mission (LDCM) as a follow-on to Landsat 7, has no plans for a TIR band as is currently available on the Landsat ETM+. The U.S. National Research Council (NRC) has recently published a report entitled *Earth Science and Applications from Space: National Imperatives for the Next Decade and Beyond* (NRC 2007) that is being used as a guideline by NASA for planning future Earth science remote sensing satellite instruments. The report recommends only one satellite mission that can provide TIR data at spatial resolutions useable for land remote sensing. This mission, called the Hyperspectral Infrared Imager (HyspIRI), is still in the planning stages, and has its heritage in the imaging spectrometer Hyperion on the NASA EO-1 platform launched in 2000 and in ASTER, the Japanese multispectral and TIR instrument flown on Terra. At present, HyspIRI is conceptually configured as a multispectral TIR sensor with seven thermal bands between 7.5–12 μ m and one band at 3–5 μm for thermal measurement of forest fires and volcanoes, with a 60 m spatial resolution in all bands. HyspIRI is planned for launch in 2013–2016.

Another satellite platform that is in the developmental stage that will carry a TIR sensor is the National Polar Orbiting Environmental Satellite System (NPOESS). NPOESS is a satellite system for use in monitoring global environmental

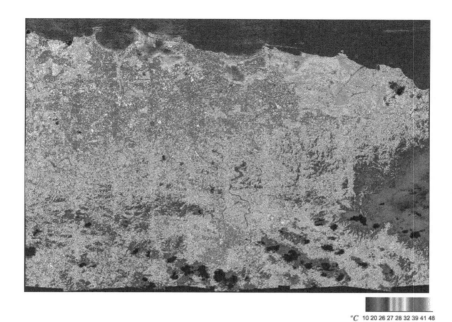

°C 10 20 26 27 28 32 39 41 48

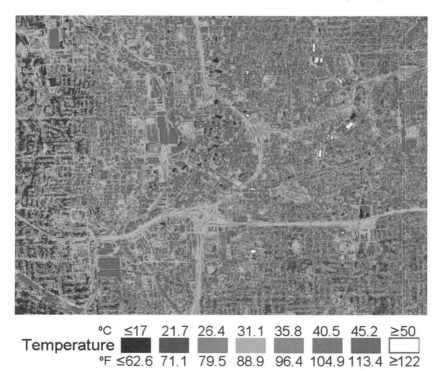

°C	≤17	21.7	26.4	31.1	35.8	40.5	45.2	≥50
Temperature								
°F	≤62.6	71.1	79.5	88.9	96.4	104.9	113.4	≥122

Figure 5.5   Daytime 10 m spatial resolution ATLAS color density sliced thermal images at 10 m spatial resolution of the eastern end of Puerto Rico (top) and of the Atlanta, Georgia central business district (bottom). (See the color plate section of this volume for a color version of this figure).

conditions, and to collect and disseminate data related to: weather, atmosphere, oceans, land, and near-space environment. The NPOESS program is managed by the U.S. Department of Commerce, the U.S. Department of Defense, and NASA. One of the instruments on the NPOESS platform, the Visible/Infared Imager/Radiometer (VIIRS),will have four thermal bands from 8.5–12.0 μm, with a spatial resolution of 400 m. NPOESS is currently set to be launched around 2013.

At this time, therefore, the future of TIR remote sensing is potentially promising, albeit somewhat uncertain. There are new TIR sensors that will be coming on-line in the future, but there will be also be a gap in TIR data as current satellites age and become non-functional. Landsat 7 has line scanning and image correction problems that affect all bands and make data from this satellite less than optimal (Goward et al., in this volume). The MODIS and ASTER sensors on the NASA Terra platform are now nine years old and we cannot realistically expect them to be continuously operational until the time the HyspIRI and NPOESS platforms are launched. Additionally, outside of TIR bands on future U.S. and European geostationary weather satellites, a survey of planned remote sensing satellite launches indicates a paucity of Earth observation satellites that will collect TIR data useful for analysis of Earth land surface temperatures (Bakker 2003).

Despite the uncertainty in the future of Earth remote sensing satellite platforms that will have TIR capabilities, there are several TIR airborne sensors that can potentially be used to bridge this gap in satellite thermal instruments (Table 5.1). Airborne sensors have historically been used as a development stage for space based thermal imagers such as the MASTER (Hook et al. 2001), which was used for both MODIS (Moderate Resolution Imaging Spectroradiometer) and ASTER satellite development. One of the virtues of airborne scanners is they can collect data at varying spatial resolutions, unlike satellite systems. At least for the interim, therefore, we may have to rely on these TIR instruments to provide data on land surface thermal responses to augment the current suite of satellite-borne sensors (i.e., Landsat ETM+, MODIS, ASTER), and to fill the gap in future satellite TIR instruments.

NOTE

1. In thermodynamics, the exergy of a system is the maximum work available through any process that brings the system into equilibrium with a heat reservoir (environment). Exergy is the energy available for use. See Fraser and Kay (2004) for a discussion of exergy in an ecological context.

REFERENCES

Bakker, W., 2003 Earth Observation Satellite Launch Table. http://www.itc.nl/~bakker/launch-table.html. Last accessed January 22, 2008.

Berk, A., G. P. Anderson, P. K. Acharya, J. H. Chetwynd, L. S. Bernstein, E. P. Shettle, M. W. Matthew, and S. M. Adler-Golden, 1999. *MODTRAN4 Users Manual*. U.S. Air Force Research Laboratory, Space Vehicles Directorate, Air Force Material Command, Hanscom Air Force Base, MA.

Beverly, A. M. and P. G. Penton, 1989. *ELAS, Earth Resources, Laboratory Applications Software, Volume II, User Reference*. NASA Earth Resources Laboratory Report No. 183. NSTL, MS.

Carlson, T. N., 1995. *Soil-Vegetation-Atmosphere Transfer (SVAT) Modeling/Remote Sensing. NASA-EOS Report, 1995*. Earth System Science Center, Pennsylvania State University, http://dbwww.essc.psu.edu/reports/annual_95/svat.html (last accessed January 22, 2008).

Carlson, T. N., D. A. J. Ripley, and T. J. Schmugge, 2004. Rapid soil drying and its implications for remote sensing of soil moisture and the surface energy fluxes. In D. A. Quattrochi and J. C. Luvall (eds), *Thermal Remote Sensing in Land Surface Processes*. CRC Press, Boca Raton, FL. Chapter 6: 185–204.

Chen, E., L. H. Allen, Jr., J. F. Bartholic, R. G. Bill, Jr., and R. A. Sutherland, 1979. Satellite-sensed winter nocturnal temperature patterns of the Everglades agricultural area. *Journal of Applied Meteorology*, 18: 992–1002.

Chen, E., L. H. Allen, Jr., J. F. Bartholic, and J. F. Gerber, 1982. Delineation of cold-prone areas using nighttime SMS/GOES thermal data: Effects of soils and water. *Journal of Applied Meteorology*, 21: 1528–1537.

Fraser, R. and J. J. Kay, 2004. Exergy analysis of ecosystems: Establishing a role for thermal remote sensing. In D. A. Quattrochi and J. C. Luvall (eds), *Thermal Remote Sensing in Land Surface Processes*. CRC Press, Boca Raton, FL. Chapter 9: 283–360.

Gillies, R. and B. Temesgen, 2004. Coupling thermal infrared and visible satellite measurements to infer biophysical variables at the land surface. In D. A. Quattrochi and J. C. Luvall (eds), *Thermal Remote Sensing in Land Surface Processes*. CRC Press, Boca Raton, FL. Chapter 5: 160–184.

Gluch, R., D. A. Quattrochi, and J. C. Luvall, 2006. A Multiscale approach to urban thermal analysis. *Remote Sensing of Environment*, 104:123–132.

Gonzalez, J. E., J. C. Luvall, D. L. Rickman, D. Comarazamy, and A. J. Picon, 2007. Urban heat island identification and climatologic analysis in a coastal, tropical city: San Juan, Puerto Rico. In Q. Weng and D. A. Quattrochi (eds), *Urban Remote Sensing*. CRC Press, Boca Raton, FL. Chapter 11: 223–268.

Goward, S. N., T. Arvidson, D. L. Williams, R. Irish, and J. Irons, in this volume. Moderate Spatial Resolution Optical Sensors. Chapter 9.

Goward, S. N., G. D., Cruickshanks, and A. S. Hope, 1985. Observed relation between thermal emission and reflected spectral radiance of a complex vegetated surface. *Remote Sensing of Environment*, 18: 137–146.

Graham, M. H., B. G. Junkin, M. T. Kalcic, R. W. Pearson, and B. R. Seyfarth, 1986. *Earth Resources Laboratory*

Applications Software. Revised January 1986. NASA Earth Resources Laboratory Report No. 183. NSTL, MS.

Gupta, R. P., 2003. *Remote Sensing Geology*. Springer, Cambridge, MA.

Holbo, H. and J. Luvall, 1989. Modeling surface temperature distributions in forest landscapes. *Remote Sensing of Environment*, 27: 11–24.

Hook, S. J., Gabell, A. R., Green, A. A., and P. S. Kealy, 1992. A Comparison of Techniques for Extracting Emissivity Information from Thermal Infrared Data for Geologic Studies. *Remote Sensing Environment*, 42: 123–135.

Hook, S. J., J. J. Myers, K. J. Thome, M. Fitzgerald, and A. B. Kahle, 2001. The MODIS/ASTER Airborne Simulator (MASTER) – A New Instrument for Earth Science Studies. *Remote Sensing of Environment*, 76: 93–102.

Hook, S. J., R. G. Vaughan, H. Tonooka, and S. G. Schladow, 2007. Absolute Radiometric In-Flight Validation of Mid Infrared and Thermal Infrared Data From ASTER and MODIS on the Terra Spacecraft Using the Lake Tahoe, CA/NV, USA, Automated Validation Site. *IEEE Transactions Geoscience and Remote Sensing*, 45: 1798–1807.

Humes, K., R. Hardy, W. P. Kustas, J. Prueger, and P. Starks, 2004. High spatial resolution mapping of surface energy balance components with remotely sensed data. In D. A. Quattrochi and J. C. Luvall (eds), *Thermal Remote Sensing in Land Surface Processes*. CRC Press, Boca Raton, FL. Chapter 3: 110–132.

Idso, S. B., R. D. Jackson, and R. J. Reginato, 1975. Estimating evaporation: A technique adaptable to remote sensing science. *Science*, 189: 991–992.

Idso, S. B., R. D. Jackson, and R. J. Reginato, 1976. Compensating for environmental variability in the thermal inertia approach to remote sensing of soil moisture. *Journal of Applied Meteorology*, 15: 811–817.

Jackson, R. D., 1986. Soil water modeling and remote sensing. IEEE *Transactions on Geoscience and Remote Sensing*, GE-24 (1): 37–46.

Jensen, J. R., 2000. *Remote Sensing of the Environment: An Earth Resource Perspective*. Prentice Hall, Upper Saddle River, NJ.

Justice, C. O. and C. J. Tucker, in this volume. Coarse Spatial Resolution Optical Sensors. Chapter 10.

Kahle, A. B. 1987. Surface emittance, temperature, and thermal inertia derived from Thermal Infrared Multispectral Scanner (TIMS) data for Death Valley, California. *Geophysics*, 52: 858–874.

Kay, J. J., 1991. A nonequilibrium thermodynamic framework for discussing ecosystem integrity. *Environmental Management*, 15(4): 483–495.

Kay J., 2000. Ecosystems as self-organizing holarchic open systems: narratives and the second law of thermodynamics. In: S. Jorgensen and F. Miller (eds) *Handbook of Ecosystem Theories and Management*. CRC Press, London.

Kay, J. J. and E. D. Schneider, 1992a. Thermodynamics and measures of ecosystem integrity. In: D. H. McKenzie, D. E. Hyatt, and V. J. Mc Donald (eds). *Ecological Indicators, Volume 1. Proceedings of the International Symposium on Ecological Indicators*, Fort Lauderdale, Florida, Elsevier, New York, 456 pp.

Kay, J. J. and E. D. Schneider, 1992b. Thermodynamics and measures of ecosystem integrity. In D. H. McKenzie, D. E. Hyatt, and V. J. McDonald (eds). *Ecological Indicators, Volume I, Proceedings of the International Symposium on Ecological Indicators*. Fort Lauderdale, FL. Elsevier, Amsterdam, pp. 159–182.

Kay, J. and E. D. Schneider, 1994. Embracing complexity, the challenge of the ecosystem approach. *Alternatives*, 20(3): 32–38.

Landsberg, H. E., 1981. *The Urban Climate*. Academic Press, New York.

Lo, C. P., D. A. Quattrochi, and J. C. Luvall, 1997. Application of high-resolution thermal infrared remote sensing and GIS to assess the urban heat island effect. *International Journal of Remote Sensing*, 18: 287–304.

Luvall, J. C. and H. R. Holbo, 1989. Measurements of short-term thermal responses of coniferous forest canopies using thermal scanner data. *Remote Sensing of Environment*, 27: 1–10.

Luvall, J. C. and H. R. Holbo, 1991. Thermal remote sensing methods in landscape ecology. In M. G. Turner and R. H. Gardner (eds), *Quantitative Methods in Landscape Ecology*. Springer-Verlag, New York.

Luvall, J. C., D. Lieberman, M. Lierberman, G. S. Hartshorn, and R. Peralta, 1990. Estimation of tropical forest canopy temperatures, thermal response numbers, and evapotranspiration using an aircraft-based thermal sensor. *Photogrammetric Engineering and Remote Sensing*, 56(10): 1393–1401.

Miller, D. H., 1981. *Energy at the Surface of the Earth*. Academic Press, New York.

Monteith, J. L., 1973. *Principles of Environmental Physics*. Edward Arnold, London, 241 pp.

Moran, M. S. and R. D. Jackson, 1991. Assessing the spatial distribution of evapotranspiration using remotely sensed inputs. *Journal of Environmental Quality*, 20: 725–737.

Müller, A. M., R. Richter, M. Habermeyer, S. Dech, K. Segl, and H. Kaufmann, 2005. Spectroradiometric requirements for the reflective module of the airborne spectrometer ARES. *IEEE Geoscience and Remote Sensing Letters*, 2(3): 329–332.

NRC, 2007. *Earth Science and Applications from Space: National Imperatives for the Next Decade and Beyond*. National Research Council, The National Academies Press, Washington, DC, 400 p.

Nichol, J., in this volume. Remote Sensing of Urban Areas. Chapter 30.

Nicolis, G. and I. Prigogine, 1977. *Self-Organization in Nonequilibrium Systems*. J. Wiley & Sons, New York, 345 pp.

Odum E., 1969. The strategy of ecosystem development. *Science*, 164: 262–270.

Oke, T. R., 1987. *Boundary Layer Climates*. John Wiley & Sons, New York, 372 pp.

Omega, 2007. Transactions in measurement and control. Volume 1 – Non-contact temperature measurement. http://www.omega.com/literature/transactions/volume1/ trantocvol1.html (last accessed July 12 2007).

Ottlé, C. and D. Vidal-Madjar, 1994. Assimilation of soil moisture inferred from infrared remote sensing in a hydrological model over the HAPEX-MOBILHY region. *Journal of Hydrology*, 158: 241–264.

Peterson, G. W., K. F. Connors, D. A. Miller, R. L. Day, and T. W. Gardner, 1987. Aircraft and satellite remote sensing of desert soils and landscapes. *Remote Sensing of Environment,* 23: 253–271.

Price, J. C. 1980. The potential of remotely sensed thermal infrared data to infer surface soil moisture and evaporation. *Water Resources Research*, 16(4): 787–795.

Price, J. C. 1982. Estimation of regional scale evapotranspiration through analysis of satellite thermal-infrared data. *IEEE Transactions on Geoscience and Remote Sensing*, GE-20(3): 286–292.

Price, J.C., 1989. Quantitative aspects of remote sensing in the thermal infrared. In G. Asrar (ed.), *Theory and Applications of Optical Remote Sensing*. John Wiley & Sons, New York. Chapter 15: 578–603.

Quattrochi, D. A. and J. C. Luvall, 1999. Thermal infrared remote sensing for analysis of landscape ecological processes: methods and applications. *Landscape Ecology*, 14: 577–598.

Quattrochi, D. A. and M. K. Ridd, 1994. Measurement and analysis of thermal energy responses from discrete urban surfaces using remote sensing data. *International Journal of Remote Sensing*, 15: 1991–2022.

Quattrochi, D. A. and M. K. Ridd, 1998. Analysis of vegetation within a semi-arid urban environment using high spatial resolution airborne thermal infrared remote sensing data. *Atmospheric Environment*, 32(1): 19–33.

Quattrochi, D. A., J. C. Luvall, D. L. Rickman, M. G. Estes, Jr., C. A. Laymon, and B. F. Howell, 2000. A decision support information system for urban landscape management using thermal infrared data. *Photogrammetric Engineering and Remote Sensing* 66(10): 1195–1207.

Regier, H. A. and J. J. Kay., 1996. An heuristic model of transformations of the aquatic ecosystems of the great lakes–St. Lawrence river basin. *Journal of Aquatic Ecosystem Health*, 5(1): 3–21.

Reginato, R. J., S. B. Idso, J. F. Vedder, R. D. Jackson, M. B. Blanchard, and R. Goettelman, 1976. Soil water content and evaporation determined by thermal parameters obtained from ground-based and remote measurements. *Journal of Geophysical Research*, 8(9): 1617–1620.

Reginato, R. J., R. D. Jackson, and P. J. Pinter, Jr., 1985. Evapotranspiration calculated from remote multispectral and ground station meteorological data. *Remote Sensing of Environment*, 18: 75–89.

Rickman, D. L., J. C. Luvall, and S. Schiller, 2000. An algorithm to atmospherically correct visible and thermal airborne imagery. Workshop on Multi/Hyperspectral Technology and Applications, Redstone Arsenal, AL.

Sabins, S. F. Jr., 1978. *Remote Sensing: Principles and Interpretation*. W. H. Freeman and Company, San Francisco.

Schneider, E. D. and J. J. Kay, 1993. Exergy degradation, thermodynamics and the development of ecosystems. *Proceedings of an International Conference on Energy Systems and Ecology*, Cracow, Poland, 10 pp.

Schneider, E. D. and J. J. Kay, 1994a. Life as a manifestation of the second law of thermodynamics. *Mathematical Computer Modeling*, 19(6–8): 25–48.

Schneider, E. D. and J. J. Kay, 1994b. Complexity and thermodynamics: towards a new ecology. *Futures*, 24(6): 626–647.

Schneider, E. D. and D. Sagan, 2005. *Into the Cool; Energy Flow, Thermodynamics, and Life*. 378 pp. The University of Chicago Press.

Sellers, P. J., F. Hall, H. Margolis, B. Kelly, D. Baldocchi, G. den Hartog, J. Cihlar, M. Ryan, B. Goodson, P. Crill, K. J. Ranson, D. Lettenmaier, and D. E. Wickland, 1995. The Boreal Ecosystem-Atmosphere study (BOREAS): An overview and early results from the 1994 field year. *Bulletin of the American Meteorological Society*, 76 (9): 1549–1577.

Strobl, P., R. Richter, F. Lehmann, A. Mueller, B. Zhukov, and D. Oertel, 1996. Preprocessing for the Airborne Imaging Spectrometer DAIS 7915. *SPIE Proceedings*, 2758: 375–382.

Taconet, O., T. Carlson, R. Bernard, and D. Vidal-Madjar, 1986. Evaluation of a surface/vegetation parameterization using satellite measurements of surface temperature. *Journal of Climate and Applied Meteorology*, 25: 1752–1767.

6

Polarimetric SAR Phenomenology and Inversion Techniques for Vegetated Terrain

Mahta Moghaddam

Keywords: Synthetic Aperture Radar (SAR), polarimetry, radiative transfer, coherent scattering, inverse scattering, classification, vegetation.

INTRODUCTION

Characterizing and monitoring the state of the Earth's vegetation cover for carbon and water cycle studies is of increasing importance in the face of evident global climate change. Radar remote sensing has long been recognized as a key component of an effective Earth observing system, due to the strong relationships of radar backscattering coefficient (sometimes also called radar backscatter) with vegetation geometric and compositional properties. Recognizing the sensitivity of radar measurements to vegetation variables, many radar instruments have been flown on airborne and spaceborne platforms for synoptic observations of vegetation cover. In particular, synthetic aperture radar (SAR) systems have been developed, which offer high resolution and allow for great flexibility in measurement parameters. Kellndorfer and McDonald (in this volume) provide a comprehensive review of passive and active microwave sensors.

A number of associated radar scattering models have been developed, which predict the backscattering cross-sections for different frequencies and polarizations for terrain covered with shrubs, forests, agricultural crops, or no vegetation at all.

Using these scattering models, sensitivity analyses have been carried out to better understand SAR backscatter due to various terrain covers, and subsequently retrieve information about the areas imaged. The focus of this chapter is on polarimetric SAR and related models. Another major radar modality is interferometric SAR (InSAR), where two antennas are used to receive the backscattered waves, and the images are formed from the interference of the two received signals. InSAR has been shown to be a very powerful technique for topography and vegetation height mapping (Treuhaft et al. 1996, Cloude and Papathanassiou 1998, Truehaft and Siqueira 2000). This topic would require its own dedicated chapter to be treated in sufficient depth. It is therefore not discussed here.

The information retrieval process, which is the ultimate goal of any remote sensing device, consists of two major analysis components for radars: the forward model and the inverse model. The forward model, or the scattering model, is as introduced in the previous paragraph: given the description of a particular vegetated scene, a forward scattering model predicts the radar measurements. The inverse model, on the other hand, relies on the data alone to estimate variables describing the scene and ideally uses no prior knowledge of the scattering scene parameters. The same physical relationships captured in the forward model are at work in the inverse model and it follows that the inverse models have the forward model at their core.

This chapter is devoted to a discussion of each of these two closely related topics. The forward scattering models of vegetation and underlying ground are presented first, followed by the inverse scattering models. The former have a longer and more diverse history, not only because they are the prerequisite for the latter, but also because they can be more systematically treated. For each topic, the existing classes of solutions will be presented, with some highlights of formulation strategies and results of application to synthetic or actual data. A vast body of work is available on these topics, especially for the forward scattering models. It is therefore noted that this chapter is merely a brief survey of these models. A list of references is provided for further reading.

RADAR SCATTERING MODELS

For the purpose of calculating radar scattering coefficients, vegetated terrain is typically modeled as a two-layer discrete random medium situated on a continuous random rough surface representing the ground. Figure 6.1 depicts the typical electromagnetic wave scattering geometry. The waves are incident at an arbitrary angle with respect to the vertical. The vegetation could consist of two layers, one representing woody stems (tree trunks, for example) and modeled as nearly vertical cylinders, and another representing the crown layer and modeled as randomly oriented smaller cylinders, perhaps with a bi-modal size distribution, and containing small disks or needles to represent foliage. For short vegetation or crops, the depth of the woody layer can be set to zero. For some crop types such as corn, the highly oriented stalks warrant the scattering medium to be represented as a uniaxial crystal, resulting in large differences between polarization responses. The signal can follow a number of paths as it interacts with vegetation and ground and before it is received back at the radar. The figure shows the paths, which include (1) direct backscatter from the crown (branch) layer, (2) direct backscatter from the ground, (3) double-bounce scattering between branches and ground or the opposite path (B–G and G–B), (4) double-bounce scattering between stems and ground or the opposite path G–T and T–G), and (5) backscattering from the ground to branches to ground (G–B–G). In reality, there are also other paths through which the signal travels before returning to the radar receiver, which include multiple scattering within the crown layer, multiple scattering within the stem layer, and multiple scattering between the stem and crown layers. Higher-order contributions include such mechanisms as crown–ground–crown and stem–ground–stem, and so on, and are most evident in cross-polarized scattering. In general, however, interactions besides the main five listed above are usually quite small. The total backscattering coefficient is calculated by summing the contributions from each of these mechanisms.

In both of the layers, the cylinders are of finite length, and could be of varying size scales compared to the wavelength. Each vegetation layer both scatters and attenuates the signals. Generally speaking, longer wavelengths are attenuated less than the shorter wavelengths, and may also be scattered to a smaller degree from the vegetation volume if they are significantly larger than vegetation components.

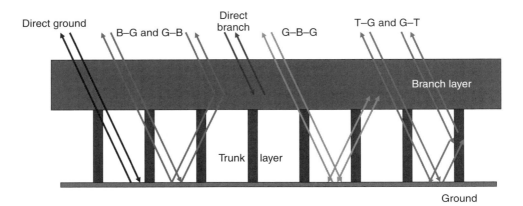

Figure 6.1   Geometry of electromagnetic scattering from vegetated terrain. (See the color plate section of this volume for a color version of this figure).

Scattering models of vegetation

There are two categories of approaches for calculating the radar backscatter from vegetation. Both approaches use essentially the same scattering geometry and vegetation model as shown in Figure 6.1. A detailed comparison of the two approaches is given in Saatchi and McDonald (1997).

The first approach uses vector radiative transfer theory (RT) to formulate solutions for the radar backscattering coefficients (Ulaby et al. 1990, Ishimaru 1997, Tsang et al. 1985). The RT theory formulates the transport of energy through a discrete random medium whose particles can scatter, absorb, or emit radiation. In the first-order RT, considered by most of the existing vegetation models, multiple interactions between particles are ignored. An exception is the model developed by Karam et al. (1992), where the second-order effects are also included. Since RT tracks the transport of energy and not the propagation of wave fields, it cannot properly account for coherent wave effects. Coherent effects are due to numerous specular reflections added exactly in-phase, which enhance scattering. The RT-based models, therefore, tend to underestimate the total radar backscatter by an amount proportional to the coherent wave contribution. The coherent effects are prominent in the specular interactions of crown-ground and trunk-ground.

The second approach starts from Maxwell's equations and proceeds to derive the full wave scattering matrix for each of the random media depicted in Figure 6.1. First, the scattering matrix of a single finite cylinder of arbitrary orientation is derived based on vector scattering theory. The scattering from the ensemble is then calculated by integrating over an arbitrary, but realistic, probability density function for the size and orientation of scatterers within a given volume of vegetation. The interactions between vegetation layers can be accounted for by cascading the scattering matrices of the relevant layers. Additionally, attenuation due to each layer is calculated from the forward scattering matrix, and is included in the calculation of the backscattered signal from the layer below it. The inclusion of the layer attenuation is also referred to as the distorted Born approximation, since it modifies the single-scattering (Born approximation) solution by allowing the waves to travel in an equivalent, lossy medium, which is more realistic than traveling in free space. Among these models are those by Durden et al. (1989), Chauhan and Lang (1991), Wang et al. (1993), Lin and Sarabandi (1999), Israelsson et al. (2000), Picard et al. (2003), and Thirion et al. (2006).

Although the two-layer vegetation is the most widely used model, most real landscapes have vegetation that is of either mixed species or the same species at different stages of growth. Therefore, there is a need to model multilayer vegetation. Recently, one such model has been developed, which uses the vector RT formulation (Liang et al. 2005). The scattering geometry is shown in Figure 6.2, where it is seen that an arbitrarily large number of layers can be considered, each with a distinct distribution function for the scatterers. This model further includes a tapered crown that can extend into any number of crown layers.

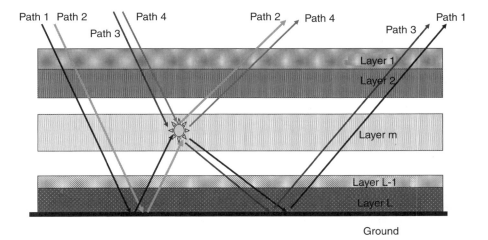

Figure 6.2 Multilayer model geometry used in the Liang et al. (2005) vector RT model, showing the four possible interactions for each layer of vegetation: (1) ground-layer m-ground, (2) ground-layer m, (3) same as two but opposite path, (4) direct from layer m. (See the color plate section of this volume for a color version of this figure).

Comparison with actual SAR data from the JPL airborne SAR (AIRSAR) at a forested site in Australia, with abundance of ground reference data, suggests that this more realistic forest model can predict the radar backscatter more accurately than the traditional two-layer model. A parallel model of an N-layer vegetated medium using Maxwell-based scattering has yet to be developed.

In efforts to make the vegetation scattering models more realistic, several three-dimensional (3D) scattering simulators have also been developed recently. These models replace the planarly layered vegetation with a full 3D representation, and allow linkage to vegetation growth models to impose natural and realistic constraints on the vegetation physical make-up (Sun and Ranson 1995, Woodhouse and Hoekman 2000).

Scattering models of a single, slightly rough, ground surface

An integral part of a forest scattering model is the model for the underlying ground. As seen from Figures 6.1 and 6.2, there are several ways in which the scattering from ground contributes to the total radar backscatter: direct ground, crown-ground, and stem-ground. A random rough surface can be defined in terms of its root-mean-square (rms) surface height, the correlation length, and its permittivity. Various statistical representations have been suggested for typical, naturally occurring rough surfaces, which include Gaussian and exponential distributions. If there are multiple crown layers, each interacts with the ground and adds to the total backscattered signal.

Since analytical solutions for scattering from general random rough surfaces do not exist, approximate solutions are often used to account for the direct ground scattering term. These solutions are typically derived for the small roughness and/or small slope regimes, and include the small perturbation method (SPM), the Kirchhoff approximation (KA), and the geometric optics approximation, which is closely related to the KA. These techniques have been well developed and are treated in depth in several textbooks (Tsang et al. 1985, Ishimaru 1997, Kong 2000). For the crown-ground and the trunk-ground double-bounce terms, the roughness effect of ground is captured via an exponential correction term to the Fresnel specular reflection coefficient (Ishimaru 1997).

For higher radar frequencies and in the presence of substantial vegetation, the contribution of the ground surface (direct or double-bounce) is not as marked as that for the lower frequencies, since the canopy attenuation masks the backscattered waves. For sparse or no vegetation, on the other hand, ground scattering becomes quite important at higher frequencies. At lower frequencies, where canopy attenuation is much smaller even if the vegetation is quite dense, ground double-bounce scattering is an important (often dominant) effect. On the other hand, direct ground scattering is less important due to the lower backscattering coefficients at lower frequencies if the ground is assumed to be only slightly rough, i.e., rms height of less than 0.05λ. Therefore, the choice of the ground scattering model is most critical in (1) higher frequencies and sparse vegetation and (2) lower frequencies and dense vegetation. The above two cases and recent model developments to address them will be discussed later in the next subsection. Other than these limiting cases, the most popular choice for the underlying ground of a forested region has been the SPM, both in the RT-based and the full wave vegetation scattering models.

Scattering models of general roughness and layered ground

When the roughness of the underlying ground surface cannot be regarded as only slightly rough, analytical methods listed previously no longer hold, and semi-analytical or numerical techniques have to be enlisted. These techniques rely on direct solutions of Maxwell's equations. A recent popular method has be the Integral Equation Method (IEM), initially proposed by Fung (1994) and later extended and improved in several ways (e.g., Hsieh et al. 1997, Hsieh and Fung 1999). Another, more accurate (but also more computationally complex) method is where the rough surface is randomly and numerically generated, then discretized into fine samples, and the total solution is found by Monte Carlo simulations of the scattered waves for a large number of surface realizations. Method of moments (MoM) is the most widely used frequency domain technique for this purpose (Tsang et al. 1994, Kapp and Brown 1996, and many others), along with various acceleration techniques used to reduce the computational complexity of MoM (e.g., El-Shenawee et al. 2001, Moss et al. 2006). In both of these classes of approaches, a major concern is error induced by truncating the numerical simulation grid. Incident wave tapering is typically used to minimize the diffraction effects from the edges of the truncated grid. This treatment is only realistic if it can represent an actual radar system, whose antenna beam could illuminate an area on the ground that is many tens of wavelengths. Therefore, the numerical simulation regions generally span 30–40 wavelengths.

In all known radar forest scattering models, the ground under the vegetation canopy has been assumed to be a single rough surface. With the recent interest and prospects for availability of lower-frequency radars in the ultra-high frequency (UHF) and very-high frequency (VHF) bands, the validity of a single rough surface model for the forest floor has to be revisited. Due to the large depth of penetration at UHF and VHF, the waves can travel well inside the vegetation layer and into the ground surface, then scatter from subsurface layers, even for dense forests. It is therefore necessary that the radar scattering model also capture these effects. There are two scattering contributions to consider: (1) direct ground return, and (2) double-bounce scattering between trunks or crown and ground.

The direct backscatter from ground with subsurface layers can be treated with both analytical and numerical techniques, much like the single-layer ground. Naturally, the complexity of the solution is increased for the multilayer ground. For this reason and due to the novelty of the problem, there are far fewer published techniques available on this topic. On the analytical side, limiting assumptions have been made such as a single slightly rough interface on top of or embedded in a layered medium (Fuks and Voronovich 2000). Two ways of simplifying the analysis have been by ignoring multiple scattering between the rough boundaries (Nghiem et al. 1995) and by using the reduced Rayleigh equations to eliminate scattered fields inside the layered medium in the case of two independent rough boundaries (Soubret et al. 2001).

A recent solution based on the small perturbation assumption offers a desirable mix of accuracy and computational efficiency for low-frequency radars (Tabatabaeenejad and Moghaddam 2006). Fields in each region are represented as the summation of up-going and down-going waves, with their amplitudes found by simultaneously matching the boundary conditions. The boundary conditions are imposed up to the first order, as in any other first-order SPM. However, the resulting equations are solved in the far field without any further approximations. Consequently, this method intrinsically takes into account multiple scattering processes between the boundaries, all of which are considered rough simultaneously. This is a distinguishing factor from the previous methods in that it can be extended to higher orders, because it does not rely on the assumption that each rough boundary contributes to the solution independently of the other boundaries, which is only true in the first-order solution. This technique has been applied to an arbitrary number of rough layers.

The numerical models for treating subsurface ground layers are also few. Their advantage over the analytical models is that they are not, in principle, restricted to the small roughness or small slope regimes, though computational complexity and accuracy may ultimately limit their utility. Moss et al. (2006) have developed an accelerated MoM to solve the two-layer rough surface problem using tapered wave illumination. Due to the added edge effects from the second layer and the interaction between the layers, the length of the simulated

surface for the two-layer problem has to be larger than that of the single surface, on the order of 50 wavelengths. Another more recent solution is based on the extended boundary condition method (EBCM) and scattering matrix technique (Kuo and Moghaddam 2007), which can be used to solve an arbitrary number of rough layers and permittivity (moisture) profiles between each two rough interfaces. The reflection and transmission matrices of rough interfaces are constructed using EBCM. The permittivity profiles are modeled as stacks of thin dielectric layers. The interactions between the rough interfaces and stratified dielectric profiles are taken into account by applying the generalized scattering matrix technique. The scattering coefficients are obtained by combining the powers computed from the resulting periodic plane wave modes of the overall system. This method could be considered advantageous to MoM due to its numerical efficiency and that it does not rely on the tapered illumination assumption.

Sensitivity of radar measurements to vegetation and ground variables

Regardless of the approach used to formulate the scattering from vegetation, the quantities used to parameterize the models are the following or a subset thereof:

- Trunks: height, radius size distribution, distribution of inclination angle from vertical, spatial density, and permittivity (or dielectric constant).
- Branches: layer height, length distribution, radius distribution, orientation angle distribution, spatial density, and permittivity. If a model contains a bimodal distribution of branches, then each of the preceding parameters will have two entries. Leaf distributions may be included here.

The behavior of radar backscatter in the presence of vegetation depends on the specific value of each of the above attributes, and certainly on measurement parameters such as frequency, polarization, and incidence angle. It has long been observed that radar backscatter manifests a 'saturation' effect as the vegetation biomass represented by the above parameters increases, in that the value of backscatter does not change measurably as biomass increases (Dobson et al. 1992, Imhoff 1995a). Higher frequencies reach saturation biomass faster than lower frequencies, but the specific saturation point is a strong function of the vegetation canopy structure and not just the value of biomass alone (Moghaddam et al. 1994, Imhoff 1995b, Woodhouse 2006). At C-band, saturation has been observed for biomass values of $4–6\,kg/m^2$, whereas for P-band saturation could take place for 10–15 kg/m^2 of biomass.

To model the ground surface for frequencies that do not penetrate significantly into the ground, the quantities of interest are:

- top soil permittivity (moisture content)
- rms surface height
- surface correlation length.

For lower frequencies, where subsurface layers contribute to the backscattering coefficients, the following variables need to be included in addition to the above:

- number of subsurface layers
- depth of each layer
- surface rms height and correlation length for each layer
- permittivity (moisture content) of each layer.

The complex nature of dependencies of the radar backscatter on scene attributes increases with increasing number of layers of vegetation and ground. Figures 6.3–6.5 show calculations of co-polarized (HH) radar backscatter at C-band

(5.6 cm), L-band (24 cm), and P-band (68 cm) for two different forest types defined in Table 6.1 as a young jack pine (YJP) and an old jack pine (OJP) stand. Each figure shows the polarimetric contributions of individual scattering mechanism, as calculated from a vector wave scattering model (Durden et al. 1989, Moghaddam and Saatchi 1995). The YJP forest has short trees with dense crowns and a rough ground covered with debris. The OJP forest has tall trees, a sparse crown layer, and smoother ground covered with lichen. The calculations show the prominent role of trunk-ground contribution (TG–HH in the figures) in the OJP forest, especially at the two lower frequencies of L–and P-band. In contrast, the YJP forest scattering is dominated by backscattering from the crown layer (B–HH), especially at the C-band.

Figure 6.6 shows the expected depth of sensitivity of reflected microwaves (i.e., penetration depth or 'skin depth') from soil at L-band, UHF, and VHF ranges. The one-way penetration depth could reach tens of centimeters to several meters depending on the moisture content and soil type. The penetration depth is defined as the distance over which the wave amplitude decays to $1/e$ (-8.7 dB) of its original value. At any frequency, the loss tangent of the soil medium, i.e., the ratio of the imaginary to real parts of the dielectric constant,

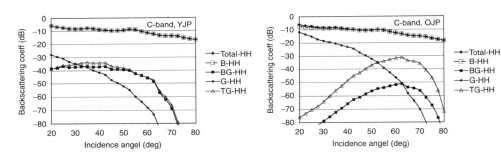

Figure 6.3   Simulations of HH radar backscatter for YJP (left) and OJP (right) of Table 6.1 at C-band.

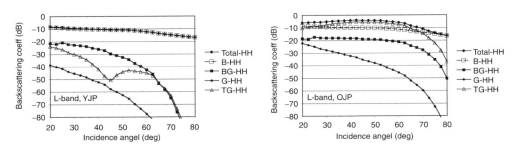

Figure 6.4   Simulations of HH radar backscatter for YJP (left) and OJP (right) of Table 6.1 at L-band.

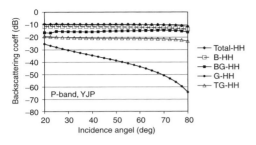
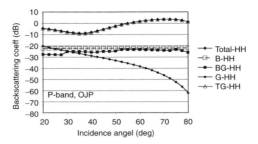

Figure 6.5 Simulations of HH radar backscatter for YJP (left) and OJP (right) of Table 6.1 at P-band.

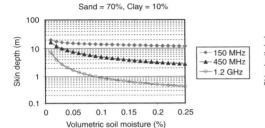
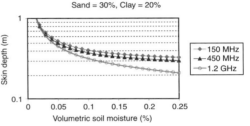

Figure 6.6 Penetration depth at L-band (1.2 GHz), UHF (450 MHz), and VHF (150 MHz) for different soils. (See the color plate section of this volume for a color version of this figure).

Table 6.1 Some tree and soil parameters for two boreal forest stands used as examples throughout this chapter

Stand type	Density #/m^2	Crown layer height (m)	Trunk height (m)	Diameter (cm)	Primary branch density (#/m^3)	Primary branch length (m)	Soil permittivity
YJP	1.0	2.8	3.8 ± 0.8	4.0 ± 1.4	30	0.8 ± 0.22	(8,1.5)
OJP	0.3	9.0	15.1 ± 3.0	13.0 ± 4.9	7	0.7 ± 0.18	(5,1)

determines the loss encountered by the waves as they travel through the medium and is a function of soil texture and moisture content. For the same soil type and frequency, higher moisture content results in higher loss tangent, and therefore smaller penetration depth.

Figure 6.7 shows possible coherent wave scattering paths at lower frequencies within a forest, whose ground has one or more subsurface layers (Moghaddam and Tabatabaeenejad 2005). The waves are specularly reflected from the stems, reflect from but also transmit through the ground, and go through multiple reflections and transmissions within the ground layers before reaching the radar receiver. The total effect can be captured through a generalized reflection coefficient (instead of the Fresnel reflection coefficient). Figure 6.8 shows the significant contributions of coherent

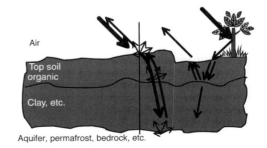

Figure 6.7 Coherent wave scattering paths between trees, surface, and subsurface. These paths contribute significantly to radar backscatter at lower frequencies. (See the color plate section of this volume for a color version of this figure).

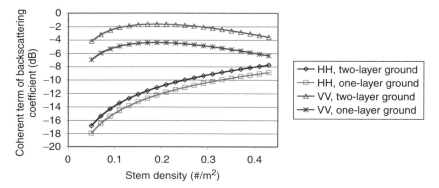

Figure 6.8   Effect of stem-subsurface multiple scattering at VHF, showing the potentially large errors committed if the effect of the subsurface interface is not included in the calculation of the coherent backscattering coefficient. The effect is more pronounced for the VV polarization. (See the color plate section of this volume for a color version of this figure).

wave multiple scattering for different vegetation covers, assuming a two-layer ground. For lower frequencies, this effect cannot be ignored. In fact, it may be used to derive important and reliable information about subsurface layers.

RADAR INVERSION MODELS

Radars are primarily sensitive to the strength and geometric distribution of dielectric properties of the targets in a scene. Dielectric constants are strong functions of their water content. The relationship between dielectric constant and soil moisture, for example, has been studied extensively and determined for various frequencies and soil textures (e.g., Dobson et al. 1985, Peplinski et al. 1995a, b). Likewise, the dielectric constant of vegetation has been shown to be a strong function of its moisture content (El-Rayes and Ulaby 1987). The relation between microwave measurements and dielectric constant and soil and vegetation geometry can be studied via the radiative transfer or wave scattering models discussed above. In the 'forward' mode, these models predict the radar backscatter measurements given the scene properties, and in the 'inverse' mode they are used to estimate the scene properties from radar scattering measurements. We will provide an overview of the latter in this section.

Inversion vs. Classification

Several researchers in recent years have addressed the problem of relating radar observations to the bio- and geo-physical parameters of forests (van Zyl 1989, Dobson et al. 1992, Le Toan et al. 1992, Pierce et al. 1994, Ranson and Sun 1994, Cloude and Pottier 1996, Haddad et al. 1996, Moghaddam and Saatchi 1999, Saatchi and Moghaddam 2000, Moghaddam et al. 2000, and many others). Their works can be grouped into two categories: classification algorithms and quantitative estimation/inversion of scene parameters.

Classification algorithms, as the name suggests, aim to assign the unknowns to discrete classes as opposed to estimating their value quantitatively (e.g., van Zyl 1989, Cloude and Pottier 1996, Freeman and Durden 1998). For example, forest biomass may be classified into large ranges such as <50 tons/ha, between 50 and 100 tons/ha, and >100 tons/ha.

Classification algorithms can generally be divided into supervised and unsupervised. Supervised classifications are those where a reference data set for the scene variables, the so-called ground-truth, is used along with the measured radar data to derive classification rules based on their mathematical dependencies. The physical dependencies of the measurements and the scene variables are not entered into the classification scheme in any quantitative fashion. Decision-tree algorithms are among this group (Hess et al. 1995, 1997). Recently, statistically based decision tree algorithms have shown great promise for large and complex data sets where a large number of classes are to be identified with closely related spectral properties (Breiman 2001, Gislason et al. 2006, Whitcomb et al. 2007). Unsupervised classification techniques group radar measurements based on their expected behavior derived from scattering models. In this regard, unsupervised classification is rooted in the physics of the scattering problem, and hence can take more advantage of information contained in measurements (Cloude and Pottier 1997, Freeman and Durden 1998, Lee et al. 1999, 2004).

The second general category of efforts to retrieve vegetation parameters from radar observations has concentrated on direct retrieval of target parameters from radar scattering data using empirical and/or theoretical models called quantitative estimation/ inversion. For example, biomass retrieval has been addressed in many studies (Dobson et al. 1992, Le Toan et al. 1992, Ranson and Sun 1994, Saatchi and Moghaddam 2000) using mainly empirical regression models. The regression curves are generated by making radar measurements at available frequencies, incidence angles, and polarizations, then plotting the data against unknown variables of interest and fitting the data to a suitable curve. These models do not fully take advantage of the information contained in the physical scattering process that relates the measurements to the unknowns. Nevertheless, they are quite useful and numerically very efficient. In another group of methods, scattering models have been incorporated to construct and train neural networks to estimate forest scattering parameters (Chua 1993, Pierce et al. 1994, Benediktsson and Sveinsson 1997, Wang and Dong 1997). These methods are more flexible and universal, but since they involve many unknown parameters, training times are long, and must be repeated for every new scene. Finally, another set of techniques has been proposed to quantitatively invert the forest variables directly from radar data using the scattering models. The implementation of this type of technique is more difficult than classification and empirically based techniques, but more specific and higher accuracy retrievals can be made, which have a wide range of validity. A later subsection is devoted to this topic.

Empirically based inversion algorithms

An effective method for inversion of canopy and ground surface parameters has been empirically derived relationships between SAR measurements and the unknown parameters, with varying degrees of guidance from the theoretical scattering models. Some examples were given in the previous subsection. Another example is the work of DuBois et al. (1995) that derived an empirical analytical model relating polarimetric L-band SAR measurements to soil moisture and soil surface roughness. Using this regression-based model, they were able to estimate soil moisture with about 4% accuracy. Oh et al. (1992) and Shi et al. (1997) also derived an empirical model and estimation algorithm for soil moisture. Similar approaches have also been used widely for estimating vegetation parameters such as biomass (Le Toan et al. 1992, Dobson et al. 1995, Saatchi and Moghaddam 2000). In particular, low frequencies such as VHF have been used to map forest biomass for high stem volume forests

(Israelsson et al. 1997, Smith and Ulander 2000). These techniques are accurate and efficient in their geographic region of validity, and hence could serve the purpose of monitoring the dynamics of a variable of interest quite well. Their limited spatial range of validity, however, generally prevents their applicability to other locations.

Direct scattering inversion

An alternate and complementary approach to both classification and empirically based estimation algorithms is that based on direct scattering model inversion. The advantage is that not only is quantitative information about scattering attributes of terrain directly obtained, but also the algorithm is not simply developed for a fixed area as in the case of empirical inversion models.

A successful inversion scheme has been developed based on the electromagnetic scattering model for each of the scattering mechanisms present in vegetated terrain (Moghaddam and Saatchi 1999, Moghaddam et al. 2000). In this approach, the dominant scattering mechanisms are identified first (e.g., using a classification algorithm). Therefore, the number of unknowns can be reduced. Theoretical scattering models are then used in the forward mode to obtain parametric closed-form models relating the unknowns of interest to radar measurements at various frequencies and polarizations. A nonlinear estimation algorithm is finally employed to retrieve the unknowns. This approach is fast, and it provides physical insight into the retrieval problem. A major application of this type of inversion, where a number of parameters can be retrieved while keeping some other parameters fixed, is in monitoring temporal change in the canopy status with respect to the unknown parameters. For example, the parameters of interest could be the real and imaginary parts of dielectric constant of crown layer components, which are surrogates for moisture content. This inversion algorithm can be summarized as follows:

1 Derive scattering-based closed form models that relate the normalized radar backscatter cross-section for multiple frequencies and polarizations to the complex dielectric constant of branch layer components. These models could be generated for where the radar backscatter is dominated by a single scattering mechanism, such as the branch layer volume scattering, or mixed mechanisms (for example trunk-ground double bounce in addition to branch layer volume scattering). The closed form models are derived by numerically simulating a large number of radar backscatter values over a

wide range of forest parameter values, then fitting a multidimensional polynomial to the simulated data. The dimension of the polynomials is the same as the number of unknowns in the problem, and their order is determined by minimizing the fitting error.

2 Given the parametric models, a nonlinear optimization scheme is used to estimate the unknowns from available SAR data (Moghaddam et al. 2000). An iterative estimation algorithm is typically used, which includes provisions for treating the statistical properties of the data and the unknowns, as well as the ill-conditioned nature of the problem. Both local and global optimization algorithms have been considered. While the local optimizations are numerically efficient and usually convergent for low-dimensional problems, the global optimization algorithms are far superior in their accuracy and convergence properties for higher dimensional problems. However, the latter suffer from poor numerical efficiency since they often require a very large number of iterations (Tabatabaeenejad and Moghaddam 2007).

When several radar frequencies such as C-band, L-band, P-band, and VHF are available, more than one scattering mechanism is at play. One mechanism possibly dominates at a given frequency but not at others. For example, for a dense forest canopy, crown layer scattering may dominate at C-band, but trunk-ground mechanism may dominate at L- and P-bands. At VHF, trunk-ground, and trunk-subsurface will dominate. One general approach for inversion would be to start from the highest frequencies and successively characterize and 'remove' the upper layers of canopy and/or ground (Moghaddam 2001).

This multi-tier approach can be further described as follows: The forest canopy is conceptualized as a layered medium consisting of a branch and leaf layer with a specified random collection of disks and cylinders, a trunk layer with randomly located nearly vertical cylinders that may or may not extend into the branch layer, and an underlying rough ground with possible subsurface layers. Depending on the depth and scattering characteristics of each layer, different frequencies and polarizations may be used to obtain quantitative information about that layer.

The approach is to first estimate the unknowns of the top-most layer with the highest frequencies available. The reasons are that the higher frequencies such as C-band are generally attenuated as they go through the top branch layer and do not carry much backscattering information from the layers below, and that due to their shorter wavelengths, they are more sensitive to the smaller scatterers usually present in the upper layers of the canopy. The algorithm described in Moghaddam and Saatchi (1999) can be used to estimate relevant unknowns within this layer. Allometric equations for the species under study can be used to relate canopy parameters (e.g., height and trunk diameter) to reduce the number of unknowns to that suitable for estimation with the available number of data channels.

With the branch and leaf layer specified, a forest scattering model (e.g., Durden et al. 1989) can be used to simulate the backscattering contribution of the layer at lower frequencies and remove that contribution from the total measured backscattered signal. The remaining signal is due to the double-bounce and ground scattering mechanisms. Unless the canopy is very sparse or parts of the surface are exposed, there is generally very little ground backscattering contribution. The lower frequencies, such as P-band, can be used in the algorithm described in Moghaddam et al. (2000) to derive trunk and ground characteristics using the double-bounce mechanism. Again, the number of unknowns can be reduced by utilizing canopy-specific allometric equations. Figure 6.9 shows an example of applying this technique to the YJP and OJP stands of Table 6.1.

If the ground is exposed to a degree that rough-surface scattering dominates the measurements, all frequencies can be used, each suitable for the derivation of different characteristics of the ground. Lower frequencies will be more appropriate for estimation of dielectric constant (moisture) of the surface and subsurface layers, whereas the higher frequencies will be more sensitive to roughness properties.

At each stage, be it the solution to volume scattering, the double-bounce, or ground scattering, the inverse problem is stated as deriving the set of unknowns that best fits the observations given the respective (polynomial) scattering model. Hence, the basics of the estimation algorithm are the same in all cases. In the examples shown above, a nonlinear optimization technique is used with an iterative preconditioned conjugate gradient algorithm at its core. Solutions are found within a few iterations given a tolerable error condition. Depending on the unknowns involved and the data channels used for each mechanism, the associated covariances used to describe their statistical characteristics have to be carefully defined in each case. Data covariances are calculated from the SAR data directly, whereas unknown variable covariances have to be assumed *a priori*.

There are two other groups of techniques that must be noted with regards to inversion of canopy and ground parameters. One group is based on Bayesian estimation (Haddad et al. 1996, Paloscia et al. 2005), and aims to find the probability that an unknown belongs to a range of values given

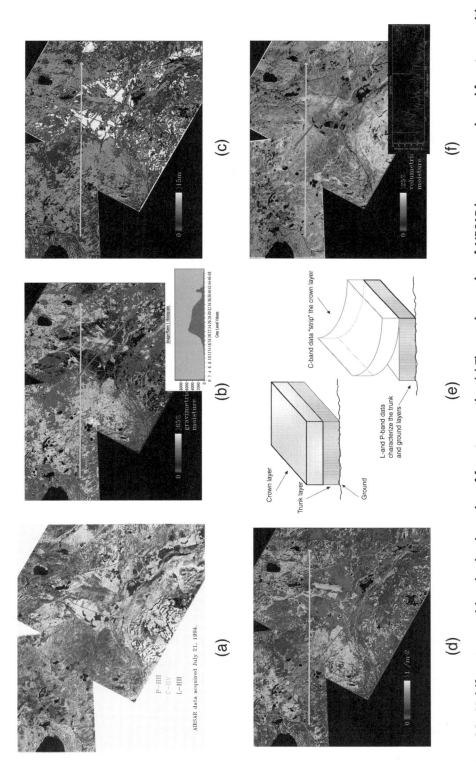

Figure 6.9 Multifrequency multimechanism inversion of forest properties. (a) Three-color overlay of AIRSAR imagery over a boreal forest area with mixed species. (b) Inversion of crown layer moisture from C-band. (c) Inversion of crown layer (and stand) height from C-band. (d) Inversion of stem density from C- and L-band. (e) Conceptual representation of removing the top layer using highest frequency. (f) Inversion of subcanopy soil moisture, after simulation of crown layer backscatter at L- and P-bands, then subtracting from total and using remainder to estimate soil moisture. (See the color plate section of this volume for a color version of this figure).

the scattering model and the radar measurements. As such, this approach has similarities to both classification and direct inversion. Finally, there is the semi-empirical approach to inversion (e.g., Saatchi and Moghaddam 2000), where some of the scene unknowns are found through empirical relationships derived from the data themselves, and some remaining unknowns are found from direct inversion.

The usefulness of each of the above techniques to extract information from radar imagery depends on the specific application at hand, the complexity of the scene, and the available data. While the direct quantitative inversion techniques are perhaps the most desirable, they are analytically and computationally quite demanding. Hence, they may not yet be possible or practical in some situations. Classification techniques, therefore, may sometimes be the best available approach as an initial analysis tool.

REMAINING CHALLENGES

SAR data are rich with information about the geometry and material make-up of the imaged scenes, especially if used for studying the time evolution and dynamics of our environment. Although significant progress has been made in the past few decades in understanding and retrieving information from SAR images, there are still outstanding challenges, especially with regards to recent and upcoming SAR systems and those currently under conceptual development. In the area of forward modeling, advances need to be made to achieve more realistic depictions of natural terrain, man-made targets, and their combinations. This is especially true for lower-frequency systems, which are extremely useful for observing densely forested regions and subsurfaces. To this end, further advances are needed in the development of coherent scattering models for complex and multilayered scenes. In the area of inversion, quantitative approaches need to be further explored and applied to more complex environments and multiple frequencies. Since a large number of iterations are often needed for finding an optimum solution for the inversion algorithms when the number of unknowns is large, improving numerical efficiency and developing new estimation techniques is urgently needed. Signal-target sensitivities need to be more clearly quantified and used to design the next-generation SAR systems. Finally, since the inversion algorithms rely on absolute levels of backscattering coefficients, system calibration is a critical issue in their success. Therefore, more robust and accurate calibration strategies need to be implemented.

REFERENCES

Benediktsson, J. A. and J. R. Sveinsson, 1997. Feature Extraction for Multisource Data Classification with Artificial Neural Networks, *International Journal of Remote Sensing*, 18(4): 727–740.

Breiman, L., 2001. Random forests. *Machine Learning*, 45: 5–32.

Chauhan, N. and R. Lang, 1991. Radar modeling of a boreal forest. *IEEE Transactions on Geoscience and Remote Sensing*, 29: 627–638.

Chua, H. T., 1993. An artificial neural network for inversion of vegetation parameters from radar backscattering coefficients, *Journal of Electromagnetic Waves and Applications*, 7: 1075–1092.

Cloude, S. R. and C. Papathanassiou, 1998. Polarimetric SAR interferometry. *IEEE Transactions on Geoscience and Remote Sensing*, 36(5): 1551–1565.

Cloude, S. R. and E. Pottier, 1996. A review of target decomposition theorems in radar polarimetry. *IEEE Transactions on Geoscience and Remote Sensing*, 34(2): 498–518.

Cloude S. R. and E. Pottier, 1997. An entropy based classification scheme for land applications of polarimetric SAR. *IEEE Transactions on Geoscience and Remote Sensing*, 35(1).

Dobson, M. C., F. Ulaby, M. Hallikainen, and M. El-Rayes, 1985. Microwave dielectric behavior of wet soil. Part II. Dielectric mixing models. *IEEE Transactions on Geoscience and Remote Sensing*, 23(1): 35–46.

Dobson, M. C. F. T. Ulaby, Christensen 1992. Dependence of radar backscatter on conifer forest biomass. *IEEE Transactions on Geoscience and Remote Sensing*, 30: 412–415.

Dobson, M. C., F. T. Ulaby, L. E. Pierce, T. L. Sharik, K. M. Bergen, J. Kelldorfer, J. R. Kendra, E. Li, Y. C. Lin, A. Nashashibi, K. Sarabandi, and P. Siqueira, 1995. Estimation of Forest Biophysical Characteristics in Northern Michigan with SIR-C/X-SAR. *IEEE Transactions on Geoscience and Remote Sensing*, 33(4): 877–895.

DuBois, P., J. van Zyl, and T. Engman, 1995. Measuring soil moisture with imaging radars. *IEEE Transactions on Geoscience and Remote Sensing*, 33(4): 915–926.

Durden, S. L., J. J. van Zyl, and H. A. Zebker, 1989. Modeling and observation of the radar polarization signature of forested areas. *IEEE Transactions on Geoscience and Remote Sensing*, 27: 290–301.

El-Rayes, M. and F. Ulaby, 1987. Microwave dielectric spectrum of vegetation. I – Experimental observations. II – Dual dispersion model. *IEEE Transactions on Geoscience and Remote Sensing*, 25: 541–557.

El-Shenawee, M., C. Rappaport, E. Miller, and M. Silevitch, 2001. Three-dimensional subsurface analysis of electromagnetic scattering from penetrable/PEC objects buried under rough surfaces: Use of the steepest descent fast multipole method (SDFMM). *IEEE Transactions on Geoscience and Remote Sensing*, 39: 1174–1182.

Freeman, A. and S. L. Durden, 1998. A three-component scattering model for polarimetric SAR data. *IEEE Transactions on Geoscience and Remote Sensing*, 36(3): 963–973.

Fuks, I. M. and A. G. Voronovich, 2000. Wave diffraction by rough interfaces in an arbitrary plane-layered medium. *Waves in Random Media*, 10: 253–272.

Fung, A. K., 1994. *Microwave Scattering and Emission Models and Their Applications*. Artech House, Norwood, MA.

Gislason, P. O., J. A. Benediktsson, and J. R. Sveinsson, 2006. Random forests for land cover classification. *Pattern Recognition Letters*, 27(4): 294–300.

Haddad, Z., P. Dubois, and J. van Zyl, 1996. Bayesian estimation of soil parameters from radar backscatter data. *IEEE Transactions on Geoscience and Remote Sensing*, 34(1): 76–82.

Hess, L. L., J. M. Melack, S. Filoso, and Y. Wang, 1995, Delineation of inundated area and vegetation along the Amazon floodplain with the SIR-C synthetic aperture radar. *IEEE Transactions on Geoscience and Remote Sensing*, 33: 896–904.

Hess, L. L., J. M. Melack, E. M. L. M. Novo, C. C. F. Barbosa, and M. Gastil, 1997. Dual-season mapping of wetland inundation and vegetation for the central Amazon basin. *Remote Sensing of Environment*, 87: 404–428.

Hsieh, C. Y. and A. K. Fung, 1999. Application of an extended IEM to multiple surface scattering and backscatter enhancement. *Journal of Electromagnetic Waves and Applications*, 13(1): 121–136.

Hsieh, C. Y., A. K. Fung, G. Nesti, A. J. Sieber, and P. Coppo, 1997. A further study of the IEM surface scattering model. *IEEE Transactions on Geoscience and Remote Sensing*, 35(4): 901–909.

Imhoff, M., 1995a. Radar backscatter and biomass saturation: Ramifications for global biomass inventory. *IEEE Transactions on Geoscience and Remote Sensing*, 33(2): 511–518.

Imhoff, M., 1995b. A theoretical analysis of the effect of forest structure on synthetic aperture radar backscatter and the remote sensing of biomass. *IEEE Transactions on Geoscience and Remote Sensing*, 33(2): 341–352.

Ishimaru, A., 1997. *Wave Propagation and Scattering in Random Media*. IEEE Press, New York.

Israelsson, H., L. M. H. Ulander, J. L. H. Askne, J. E. S. Fransson, P.-O. Frolind, A. Gustavsson, and H. Hellsten, 1997. Retrieval of forest stem volume using VHF SAR. *IEEE Transactions on Geoscience and Remote Sensing*, 35(1): 36–40.

Israelsson, H., L. Ulander, T. Martin, and J. Askne, 2000. A coherent scattering model to determine forest backscattering in the VHF-Band. *IEEE Transactions on Geoscience and Remote Sensing*, 38(1): 238–248.

Kapp, D. and G. Brown, 1996. A new numerical method for rough-surface scattering calculations. *IEEE Transactions on Antennas and Propagation*, 44(5).

Karam, M., A. Fung, R. Lang, and N. Chauhan, 1992. A microwave scattering model for layered vegetation. *IEEE Transactions on Geoscience and Remote Sensing*, 30(4): 767–784.

Kellndorfer, J. and K. McDonald, in this volume. Active and Passive Microwave Systems. Chapter 13.

Kong, J. A., 2000. *Electromagnetic Wave Theory*. EMW, New York.

Kuo, C. H. and M. Moghaddam, 2007. Electromagnetic scattering from multilayer rough surfaces separated by media of arbitrary dielectric profiles for remote sensing of soil moisture. *IEEE Transactions on Geoscience and Remote Sensing*, 45(2): 349–367.

Lee, J. S., M. Grunes, T. Ainsworth, D. Li-Jen, D. Schuler, and S. Cloude, 1999. Unsupervised classification using polarimetric decomposition and the complex Wishart classifier. *IEEE Transactions on Geoscience and Remote Sensing*, 37(5): 2249–2258.

Lee, J. S., M. R. Grunes, E. Pottier, and L. Ferro-Famil, 2004. Unsupervised terrain classification preserving polarimetric scattering characteristics. *IEEE Transactions on Geoscience and Remote Sensing*, 42(4): 722–731.

Le Toan, T., A. Beaudoin, J. Riom, and D. Guyon, 1992. Relating forest biomass to SAR data. *IEEE Transactions on Geoscience and Remote Sensing*, 30: 403–411.

Liang, P., M. Moghaddam, L. Pierce, and R. Lucas, 2005. Radar backscattering model for multilayer mixed species forests. *IEEE Transactions on Geoscience and Remote Sensing*, 43(11).

Lin, Y. and K. Sarabandi, 1999. A Monte Carlo coherent scattering model for forest canopies using fractal generated trees. *IEEE Transactions on Geoscience and Remote Sensing*, 37(1): 440–451.

Moghaddam, M., 2001. Estimation of comprehensive forest variable sets from multiparameter SAR data over a large area with diverse species. *Proc. IGARSS'2001*, Sydney, Australia.

Moghaddam, M. and S. Saatchi, 1995. Analysis of scattering mechanisms in SAR imagery over boreal forest: Results from BOREAS'93. *IEEE Transactions on Geoscience and Remote Sensing*, 33: 1290–1296.

Moghaddam, M. and S. Saatchi, 1999. Monitoring tree moisture using an estimation algorithm applied to SAR data from BOREAS. *IEEE Transactions on Geoscience and Remote Sensing*, 37(2): 901–916.

Moghaddam, M. and A. Tabatabaeenejad, 2005. Coherent model for VHF scattering from mixed forests on multilayer rough ground. *IEEE–APS*, Washington DC.

Moghaddam, M., S. Durden, and H. Zebker, 1994. Radar measurement of forested areas during OTTER. *Remote Sensing of Environment*, 47(2): 154–166.

Moghaddam, M., S. Saatchi, and R. Cuenca, 2000. Estimating subcanopy soil moisture with radar. *Journal of Geophysical Research – Atmospheres*, 105(D11): 14899–14911.

Moss, C. D., T. M. Gezegorcyk, H. C. Han, and J. A. Kong, 2006. Forward–backward method with spectral acceleration for scattering from layered rough surfaces. *IEEE Transactions on Antennas Propogation*, 54(3): 1006–1016.

Nghiem, S., R. Kwok, S. H. Yueh, J. A. Kong, C. C. Hsu, M. A. Tassoudji, and R. T. Shin, 1995. Polarimetric scattering from layered media with multiple species of scatterers. *Radio Science*, 30(4): 835–852.

Oh, Y., K. Sarabandi and F. T. Ulaby, 1992. An empirical model and an inversion technique for radar scattering from bare soil surfaces. *IEEE Transactions on Geoscience and Remote Sensing*, 30(2): 370–381.

Paloscia, S., P. Pampaloni, S. Pettinato, P. Poggi, and E. Santi, 2005. The retrieval of soil moisture from Envisat/ASAR data. *EARSeL EProceedings*, 4: 44–51.

Peplinski, N. R., F. T. Ulaby, and M. C. Dobson 1995a. Dielectric properties of soils in the 0.3–1.3 GHz Range. *IEEE Transactions on Geoscience and Remote Sensing*, 33(3).

Peplinski, N. R., F. T. Ulaby, and M. C. Dobson, 1995b. Corrections to 'Dielectric Properties of Soils in the

0.3–1.3 GHz Range'. *IEEE Transactions on Geoscience and Remote Sensing*, 33(6).

Picard, G., Le Toan, T., and F. Mattia, 2003. Understanding C-band radar backscatter from wheat canopy using a multiple-scattering coherent model. *IEEE Transactions on Geoscience and Remote Sensing*, 41(7): 1583–1591.

Pierce, L. E., K. Sarabandi, and F. T. Ulaby, 1994. Application of an artificial neural network in canopy scattering inversion. *International Journal of Remote Sensing*, 15: 3263–3270.

Ranson, K. J. and Q. Sun, 1994. Mapping biomass of a northern forest using multifrequency SAR data. *IEEE Transactions on Geoscience and Remote Sensing*, 32: 388–396.

Saatchi, S. and K. McDonald, 1997. Coherent effects in microwave backscattering models for forest canopies. *IEEE Transactions on Geoscience and Remote Sensing*, 35(4): 1032–1044.

Saatchi, S. and M. Moghaddam, 2000. Estimation of crown and stem water content and biomass of boreal forest using polarimetric SAR imagery. *IEEE Transactions on Geoscience and Remote Sensing*, 38(2): 697–709.

Shi, JC., J. Wang, A. HSU, P. O'Neill, and E. T. Engman, 1997, Estimation of bare surface soil moisture and surface roughness parameter using L-band SAR image data, *IEEE Transactions on Geoscience and Remote Sensing*, 35(3), pp. 1254–1266.

Smith, G. and L. Ulander, 2000. A model relating VHF-band backscatter to stem volume of coniferous boreal forest. *IEEE Transactions on Geoscience and Remote Sensing*, 38(2): 728–740.

Soubret, A., G. Berginc, and C. Bourrely, 2001. Backscattering enhancement of an electromagnetic wave scattered by two-dimensional rough layers. *Journal of the Optical Society of America A*, 18: 2778–2788.

Sun, Q. and K. J. Ranson, 1995. A three-dimensional radar backscatter model of forest canopies. *IEEE Transactions on Geoscience and Remote Sensing*, 33(2): 372–382.

Tabatabaeenejad, A. and M. Moghaddam, 2006. Bistatic scattering from layered rough surfaces. *IEEE Transactions on Geoscience and Remote Sensing*, 44(8): 2102–2115.

Tabatabaeenejad, A. and M. Moghaddam, 2007. Inversion of a layered rough surface model: maximizing the number of retrievable parameters for the design of future subsurface sensing radar systems. *Proc. IGARSS'07*, Barcelona, Spain.

Thirion, L., E. Colin, and C. Dahon, 2006. Capabilities of a forest coherent scattering model applied to radiometry, interferometry, and polarimetry at P-band and L-band. *IEEE Transactions on Geoscience and Remote Sensing*, 44(4): 849–862.

Treuhaft, R. and P. Siqueira, 2000. Vertical structure of vegetated land surfaces from interferometric and polarimetric SAR. *Radio Science*, 35: 141–177.

Treuhaft, R., S. Madsen, M. Moghaddam, and J. van Zyl, 1996. Vegetation characteristics and underlying topography from interferometric data. *Radio Science*, 31: 1449–1495.

Tsang, L., R. T. Shin, and J. A. Kong, 1985. *Theory of Microwave Remote Sensing.*, Wiley Interscience, New York.

Tsang L., C. H. Chan, K. Pak, H. Sanganni, A. Ishimaru, and P. Phu, 1994. Monte Carlo simulations of large-scale composite random rough-surface scattering based on the banded-matrix iterative approach. *Journal of the Optical Society of America A*, 11(2): 691–696.

Ulaby, F., K. Sarabandi, K. McDonald, M. Witt, and M. C. Dobson, 1990. Michigan microwave canopy scattering model. *International Journal of Remote Sensing*, 11: 1223–1253.

van Zyl, J. J., 1989. Unsupervised classification of scattering behavior using radar polarimetry data. *IEEE Transactions on Geoscience and Remote Sensing*, 27: 36–45.

Wang, Y. and D. Dong, 1997. Retrieving forest stand parameters from SAR backscatter data using a neural network trained by a canopy backscatter model. *International Journal of Remote Sensing*, 18(4): 981–989.

Wang, Y., J. Day, and G. Sun, 1993. Santa Barbara microwave backscattering model for woodlands. *International Journal of Remote Sensing*, 14(8): 1146–1154.

Whitcomb, J., M. Moghaddam, K. McDonald, J. Kellndorfer, and E. Podest, 2007. Wetlands map of Alaska using L-band radar satellite imagery. *Remote Sensing of Environment*, in review.

Woodhouse, I, 2006. Predicting backscatter-biomass and height-biomass trends using a macroecology model. *IEEE Transactions on Geoscience and Remote Sensing*, 44(4): 871–877.

Woodhouse, I. and D. Hoekman, 2000. Radar modeling of coniferous forest using a tree growth model. *International Journal of Remote Sensing*, 21(8): 1725–1737.

Digital Sensors and Image Characteristics

7

Optical Sensor Technology

John P. Kerekes

Keywords: optical sensors, multispectral imagers, hyperspectral imagers.

INTRODUCTION

Optical imaging sensors are a key technology in the field of remote sensing. Nearly all applications of remote sensing rely on optical imagery as a primary data source for analysis. The interpretation of visible images is as intuitive for remote sensing analysts as looking at a photograph. This comfort and reliance on optical imagery comes naturally as the very first remote sensing instrument was the human visual system. The human eye collects light in the visible optical spectrum and records the intensity with rods and cones on the retina. Optical remote sensing systems in use today operate with many similar components to the human visual system but provide dramatically richer information through higher resolution and additional phenomenology available by sensing in spectral regions beyond what the human eye can see.

The main function of electro-optical (EO) imaging sensors is to collect incident electromagnetic (EM) radiation and convert it to a stored representation useful for remote sensing analysis. These sensors operate in the optical region of the EM spectrum traditionally defined as radiation with wavelengths between 0.4 and 15 μm. This range spans the visible (VIS, 0.4–0.7 μm), the near infrared (NIR, 0.7–1.1 μm), the shortwave infrared (SWIR, 1.1–2.5 μm), the midwave infrared (MWIR, 2.5–7.5 μm), and the long wave infrared (LWIR, 7.5–15 μm) spectral regions.

While it is possible to build a single sensor spanning this entire range, it is more common to find sensors (or at least sensor subsystems) limited to one or two of these regions.

Optical imaging sensors resolve and sample the incident EM field spatially, spectrally, radiometrically, and temporally producing an image of the radiance emanating from the scene. The resolution and accuracy with which these four domains are sampled determine the image characteristics and ultimately the quality and utility of the recorded image. While many applications still rely primarily on the visual interpretation of the imagery, modern remote sensing systems make precise radiometric measurements of the incident radiance and quantitative applications are on the rise.

This chapter provides a general introduction to optical imaging sensors. The sensor system elements which convert the scene radiance field to a digital image are first described. Metrics are then presented which quantify the performance of sensors in several domains. The chapter concludes with a discussion of remaining challenges and new directions in optical sensor technology.

SENSOR SYSTEMS

All optical imaging sensors have in common the basic components illustrated in Figure 7.1.

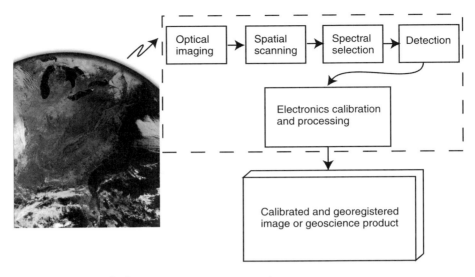

Figure 7.1   Electro-optical sensor system components.

While other chapters in this handbook describe details for specific satellite or airborne sensors, this chapter describes these elements generically and provides references for the interested reader to pursue. The following subsections trace the flow of the incident electromagnetic waves as they are focused by the *optical imaging* subsystem, spatially sampled by the *spatial scanning* subsystem creating a two-dimensional image, spectrally filtered by the *spectral selection* devices, converted to an electrical signal by the *detectors* and then *processed* to a final image by the *electronics* and the *calibration* algorithms to an accurate representation of the scene with proper radiometric units and geometric orientations. Further details on sensor systems may be found in the texts by Wolfe and Zissis (1985), Wyatt (1987), Hobbs (2000), and Schott (2007).

Optical imaging

The first element in an optical imaging system is the telescope. The primary functions of the optical telescope are to gather enough light to make an accurate measurement and to focus that light onto the focal plane to form a sharp image (Fischer and Tadic 2000). While many modern visible cameras and early astronomical telescopes use refractive optics (a series of lenses aligned on a common optical axis with light refracted, or bent, as it passes through), these systems suffer from geometrical and chromatic aberrations, and have a limited useful spectral range due to transmittance losses of glass in spectral regions beyond the visible and near infrared. Most modern remote sensing systems use reflective optics (mirrors) in a standard

Cassegrain or the Ritchey–Chretien modified configuration which uses parabolic (or hyperbolic) mirrors co-aligned, but with openings in the mirror to allow the focused light to pass through. While some light is lost in this arrangement due to the obscuration of the secondary mirror, the reflective approach can be used over a wide spectral range.

In modern optical imaging systems with solid state detectors there are three key parameters that dominate in defining the imaging capability of a well-designed system: the entrance aperture size, the focal length, and the diameter of the field stop. The aperture size is generally defined by the diameter of the primary mirror, the focal length by the placement and shape of the mirrors, and the size of the field stop is usually the linear dimension of an individual detector element on the focal plane.

The size of the aperture is important in that it limits the total light collected as well as achievable spatial resolution. The *Rayleigh criterion* states through Equation (1) that the closest distance d_{Rayleigh} between two point sources on the ground that can be distinguished is determined by the size of the aperture, D_A, the wavelength of light λ, and the height (or slant range) H of a sensor relative to the ground. This limit is due to diffraction of light waves (Born and Wolf 1999) and is a measure of the maximum potential ground resolution of the system:

$$d_{\text{Rayleigh}} = 1.22 \frac{\lambda}{D_A} H \qquad (1)$$

The size of the field stop, or detector element, can also determine ground resolution. The angular *Instantaneous Field of View (IFOV)* θ_{IFOV} is defined by the ratio of the field stop diameter (or detector linear size) D_F divided by the focal length f. The resolution size of the pixel on the ground is given by equation (2):

$$d_{\text{GIFOV}} = \frac{D_F}{f}H = \theta_{\text{IFOV}}H \qquad (2)$$

When reported as a linear distance on the ground the term *Ground Instantaneous Field of View (GIFOV)* is commonly used. A well designed optical system will not be diffraction limited but limited by the detector field stop, and have $d_{\text{GIFOV}} \gg d_{\text{Rayleigh}}$ at the longest wavelength collected by the sensor. This is equivalent to having the diffraction blur circle much larger than the detector field stop. It is important for the user of remote sensing imagery to be aware that some systems have a spatial resolution that varies with spectral band due to their design and diffraction effects.

Spatial scanning

Early photographic remote sensing systems captured an entire frame, or *field-of-view (FOV)*, at once with the optical system projecting the scene onto the film. With the advent of individual solid state detectors which collect a single measurement known as a picture element, or *pixel*, corresponding to a single IFOV projected onto the ground, some form of scanning was necessary to form a complete two-dimensional image. Detector technology has continued to advance with detectors combined first to form linear arrays and then later into two-dimensional arrays. These different detector formats use various scanning techniques to form images, each placing different requirements on focal plane complexity and platform stability. The following discussion assumes a single spectral band system; spectral imaging systems which measure a vector of spectral information for each pixel are addressed in the next section.

Figure 7.2 shows three basic scanning techniques. *Line scanners* use an individual detector and thus were the form of many early solid state imaging systems. The across-track pixels, or lines of an image, are formed by sweeping the projected view of the detector across the ground with a rotating mirror. The down-track pixels, or columns, are formed by the forward movement of the sensor platform. The simplicity of the focal plane and potential for a wide FOV of these types of systems is traded-off with platform stability since changes in the platform's orientation relative to the earth will lead to non-uniform spacing of the pixels. For example, it is quite common for aircraft to be buffeted by winds and to roll slightly relative to the flight-line. With a mirror rotating at a constant velocity

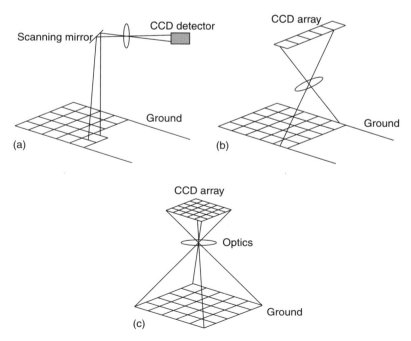

Figure 7.2   Scanning techniques imaging systems: (a) line scanner, (b) pushbroom, (c) framing camera.

the pixels then correspond to non-uniform sampling of the ground. A variation of this technique is known as a *whiskbroom scanner* which uses multiple detectors arranged in a line parallel to platform motion and thus collects several lines (rows) of the image simultaneously. The advantage of this mode is the detectors can dwell longer for each sample and increase the signal-to-noise ratio. The disadvantage is the additional cost of the detectors and the additional processing necessary to compensate for variations in detector responsivity.

The development of linear arrays of hundreds of detectors enabled imaging sensors to operate in a *pushbroom scanner* mode. These types of systems generally do not require the use of a moving mirror as the linear array images a line across the entire field of view. Thus, one row of an image is collected all at once, while the platform motion provides the down track sampling or columns of the image. An advantage here is the significantly longer dwell time for the detectors and the resultant improvement possible in signal-to-noise ratio. Another advantage is the across-track sampling is done by the fixed spacing of detectors in the array, thereby eliminating the nonuniform sampling possible with scanning systems. The disadvantages include the cost of the arrays, the limited field of view since the optics must image the full swath, and the added processing and calibration to compensate for variations in detector responsivity.

A third technique for scanning requires no moving mirror or platform motion and just like the original film cameras images the scene all at once. *Framing cameras* use two-dimensional detector arrays in the focal plane to image the entire swath and the down-track columns simultaneously. The advantage here is the enormous gain in available dwell time, as well as the ability to image from a stationary platform such as an aerostat or geosynchronous satellite. Also, since an entire image is collected at once, platform instability is much less an issue and the images have high geometric fidelity. The disadvantages include the potentially high cost of the array and the processing associated with non-uniformity correction as with the other detector arrays. While framing cameras do not require a moving mirror to collect an image, some systems include a movable mirror to point the array to various areas within the sensor's *field of regard (FOR)*. This allows a system with modest single frame coverage to collect imagery over a wide area and build a *mosaic* of images into one large image.

Spectral selection

So far we have just considered imaging systems that collect a single image with a given spectral bandpass. There is rich information in the phenomenology of materials associated with their variation in spectral response across wavelengths (Solé et al. 2005). An obvious example is the value of color compared to black and white photography. Spectral imaging systems fall into two categories: *multispectral imagers (MSI)* which collect imagery in a handful (roughly three to ten) irregularly and often widely spaced spectral bands while *hyperspectral imagers (HSI)* collect imagery in many tens to hundreds of contiguous narrow spectral bands. MSI systems typically have spectral resolution $\lambda/\Delta\lambda \sim 10$ (the ratio of the channel central wavelength λ to the spectral bandwidth $\Delta\lambda$) while HSI systems have $\lambda/\Delta\lambda \sim 100$. In both cases the images of the various bands are usually spatially co-registered; that is, a given pixel index in each band image samples the same location in the scene. For visualization purposes these spectral images are often displayed as an image cube, with a single band image on the top and the depth provided by the spectral dimension.

The simplest multispectral systems use separate co-aligned optical apertures and focal planes with a different pass band filter on each focal plane. Figure 7.3 shows an example of a multiband sensor using this approach. The Wildfire Airborne Sensor Program (WASP) sensor developed at the Rochester Institute of Technology has a color RGB imager co-aligned with SWIR, MWIR, and LWIR cameras. Alternatively, a single aperture can be used with beam splitters to feed multiple focal planes with filters. These types of systems are most common for instruments with multiple bands spanning a wide spectral range such as the full optical remote sensing spectrum from 0.4 to 15 μm. An issue with the use of separate apertures and focal planes is it is extremely difficult to obtain and maintain alignment between the separate optical paths and the data usually suffer from misregistration artifacts.

There are several spectral selection techniques used in hyperspectral imaging systems. They can be grouped into two categories based on the type of focal plane and spatial scanning technique with which they can be used. *Dispersive* methods spread the spectrum out spatially and require a linear (or two-dimensional) array and can be used with line scanner, whiskbroom or pushbroom scanning systems. *Temporal* methods collect the spectrum over time, and thus can be used with either single detector elements in an across track scanning mode or two-dimensional arrays in a framing camera mode.

Dispersive hyperspectral systems often use either a *grating* or a *prism*. A diffraction grating is a series of closely spaced parallel thin slits that cause incident white light to be reflected or transmitted at an angle proportional to wavelength. The slits break up the incident collimated beam into many light beams in which monochromatic

Figure 7.3   WASP sensor in laboratory and placement in aircraft (insert).

light will constructively interfere only in a specific direction. The full bandpass of the incident light forms a 'rainbow' on the detector array thus achieving spectral sampling. This grouping is repeated at integer multiples of the diffraction angle leading to multiple bands or orders of the diffraction pattern. Since energy from multiple wavelengths can overlap at the same diffraction angle, order sorting filters are often required to ensure only the desired wavelength is measured by a given detector. The NASA Atmospheric Infrared Sounder (AIRS) used a grating for its spectral dispersion (Aumann et al. 2003).

Prisms achieve spectral dispersion through wavelength dependence of the index of refraction of transmissive materials. Generally, the index of refraction varies proportionally to the frequency of light, or inversely with wavelength. Thus the shorter wavelengths of an incident beam will be refracted at a higher angle to normal than will the longer wavelengths. This leads to a spreading of the spectrum which can then be sampled across a detector array. The HYDICE airborne HSI system used a prism for its spectral dispersion (Rickard et al. 1993).

Alternatives to the spatial dispersion techniques are methods which sample the spectrum over time. The simplest of these uses a rotating filter wheel to sequentially sample the scene in multiple bands. For moving or scanning platforms, this sampling must occur fast enough that the scene appears stationary. A variation on this use of filters is to align strips of filters across a detector array such that platform motion allows the scene to be sequentially sampled in the multiple wavelengths. However, this latter technique places stringent requirements

on platform stability and orientation to ensure the multispectral images sample the exact same spot on the ground with each filter.

In the past decade, new types of tunable filters have been applied to the spectral selection task eliminating the need for a mechanical moving wheel or platform motion. These systems include *Acousto-Optical Tunable Filters (AOTF)* (Denes et al. 1998) or *Liquid Crystal Tunable Filters (LCTF)* (Stevenson et al. 2003). These devices work on the principle of selective transmission of light through an optical material in which acoustic waves of varying frequency (AOTF) or varying voltage (LCTF) are applied to selectively refract the light and allow only a small wavelength range to pass through. These types of systems have the advantage of being able to rapidly select the wavelength of light sensed and thus enable tunable spectral sensing, but generally suffer from low throughput and limited angular fields of view.

An *interferometer* is another temporal spectral selection technique that is particularly well suited to very high spectral resolution ($\lambda/\Delta\lambda$ $\sim$ 1,000–10,000) systems. These instruments collect an interferogram, the result of interfering the incident light beam with a delayed version of itself, which then must be digitally processed to obtain the optical spectrum. The most common interferometer is a Michelson *Fourier Transform Spectrometer (FTS)* which is configured to collect the Fourier transform of the spectrum by moving a mirror to change the optical path difference between two mirrors and causing constructive and destructive interference for the incident light as measured by the detector on the focal plane (Beer 1992). The interferogram, sampled over time, is then processed through an inverse Fourier transform to obtain the desired optical spectrum.

Detection

While the earliest remote sensing systems used photographic film to record images (Peres 2007), nearly all modern systems use solid state semiconductor detectors to convert the incident optical radiation into an electrical signal. Initially these detectors were single pixel devices and required scanning to create an image, but as the technology has matured detectors have evolved into one-dimensional linear and two-dimensional area arrays.

Semiconductor detectors operate as *photon* detectors by converting incident photons into photoelectrons producing a measurable voltage or current (Dereniak and Boreman 1996). The effectiveness by which detectors perform this conversion can be characterized by their *quantum efficiency* which is the ratio of the number of photoelectrons produced per incident photon (usually expressed as a percent), or their *responsivity* which is the ratio of the voltage (or current) produced to the incident optical power. In order to be detected an incident photon's energy must be sufficient to move electrons in the detector to the conduction band. The energy required to do this is referred to as the *band gap* (Wolfe and Zissis 1985). Since the energy of a photon is inversely related to its wavelength, the band gap of a given material sets an upper limit on the wavelength response of a detector.

Detectors are made from various materials depending on the wavelength band they are designed to measure (Norton 2003). Systems that image in the visible and near infrared (up to $1 \mu m$) generally will use silicon (Si) detector arrays. Silicon is a popular choice because of its relatively low cost (driven by the consumer imaging market), typically high efficiency and room temperature operation. At longer wavelengths indium gallium arsenide (InGaAs) is commonly used with responsivity spanning 0.8 to 2.8 μm. Indium antimonide (InSb) is also used at these wavelengths and beyond (0.3–5.5 μm), but requires cryogenic cooling to 77 K while InGaAs can operate at 220 K which is achievable with passive radiators in space or thermal electric cooling. Mercury cadmium telluride (HgCdTe) is a flexible material that can be made to be responsive across a reasonably wide band anywhere within a large part of the optical spectrum (0.7–15 μm) by adjusting the relative concentrations of Cd and Te during fabrication. However, it also requires cooling to 77 K to be efficient. Arsenic-doped silicon (Si:As) is another flexible material that can be made responsive from the short wave infrared to the far infrared (2.5–25 μm) but requires liquid helium cooling to 10 K to be efficient.

Random detector noise mechanisms can be lumped into two categories: *photon noise* and *fixed noise*. Photon noise results from fluctuations in the arrival rate of photons and is inherent in any detection process. Photon noise obeys *Poisson* statistics with a standard deviation that is proportional to the square root of the incident radiance and thus will be larger for higher signals. Fixed noise sources within a detector do not depend upon the incident radiance and thus can be characterized by a constant standard deviation. Examples of fixed noise sources include random movement of charge carriers in the material due to thermal effects and noise associated with the read out of the photoelectrons. The total noise in a detector is the root-sum-square of photon and fixed noise levels. System designers strive for a condition known as *Background Limited Performance (BLIP)* where the total noise is dominated by the photon noise resulting from the incident flux on the detector. Note that this flux is the total from the scene plus any instrument self emission that makes its way onto the detector.

Electronics, calibration, and processing

The electrical signal produced by the detector is usually amplified and then converted to a binary number by an *analog-to-digital (A/D)* converter. The number of bits in the A/D converter determines the radiometric resolution of the sensor. While the earliest Landsat sensors (*circa* 1972) had just six bits of resolution leading to $2^6 = 64$ possible output levels, most modern sensors have ten or more bits ($2^{10} = 1024$ output levels). Once the signal is in digital counts, it can be stored, transmitted, or processed without additional noise corrupting the measurement. The digital data are either transmitted in real-time to a ground station or stored on-board the platform for later transmission or retrieval.

The data usually undergo further processing before distribution to users. This processing can include *radiometric calibration, geometric registration* and *rectification*, or even processing to a geophysical parameter. The details of this processing are system specific but generic descriptions follow here.

Radiometric calibration is the process whereby digital counts from the sensor are calibrated to physical radiometric units such as spectral radiance (e.g., $W m^{-2} sr^{-1} \mu m^{-1}$) (Chen 1997). Often ancillary data files are also produced with information about the quality of the data such as bad pixel locations, sensor noise levels, and estimates of the spectral and radiometric calibration accuracy. Geometric registration is the step where images from multiple spectral bands are co-aligned so a specific pixel index corresponds to the same physical location on the Earth's surface. Rectification is a regridding of the image so that it is arranged in a rectilinear array.

PERFORMANCE METRICS

In this section we address quantitative metrics that describe the performance of optical remote sensing instruments from the point of view of a remote sensing analyst or user of the data. In particular, the discussion includes metrics for spatial, spectral, and radiometric resolutions, along with geometric accuracy. In most remote sensing applications it is these characteristics that determine the usefulness of remote sensing imagery. For more detail on optical sensor performance metrics refer to the text by Holst (2005).

Spatial resolution

A primary measure of performance of any imaging system is the spatial resolution achieved. There are a number of ways that resolution can be characterized, and are often erroneously used interchangeably. Earlier in this chapter we defined the Rayleigh criterion for determining resolution. However, the achieved resolution will depend on the complete system including ground processing. The spatial resolution of an optical system is most properly reported in angular units (e.g., milliradians) which can then be converted to a physical distance (e.g., meters or kilometers) by multiplying by the height above or slant range to the Earth's surface.

The simplest metric for remote sensing image resolution is *pixel size* or the linear dimension on the ground corresponding to the length of one side of a hypothetical square representing the area measured in a given pixel. In many situations this simplistic view adequately characterizes image resolution, but it can be important to recognize that reality is much more complicated.

A more accurate description of image resolution can be derived from the system *Point Spread Function (PSF)*. The PSF describes the two-dimensional normalized distribution of intensity measured on the focal plane from an ideal point source located in the scene. Figure 7.4 presents a notional PSF projected onto the ground for a hypothetical imaging system. Note the system PSF is often circularly, or at least elliptically, symmetric.

Another accurate measure of the spatial resolution is the *Ground Resolved Distance (GRD)* which is the linear width of the PSF at a specified level, commonly selected as the Full Width at Half Maximum (FWHM). For non-circularly symmetric PSFs this can be the geometrical average between the across track and down track widths. The PSF used here should be the total system PSF

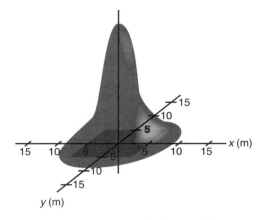

Figure 7.4   Point spread function of an optical system projected onto the ground. Shown underneath the PSF is a projection of a typical pixel size on to the ground.

which is a convolution of PSFs due to the aperture, optical elements, field stop (detector), and detector sampling effects, among others. These effects can also be characterized by their *Modulation Transfer Function (MTF)*, which is the Fourier transform of the PSF. The system MTF then becomes the product of component MTFs.

For systems imaging through a significant part of the atmosphere such as from a satellite, atmospheric scattering from aerosols, molecules and turbulence can degrade spatial resolution. In fact, it is common to include an atmospheric PSF when performing system PSF calculations. A net result of these atmospheric effects is the scattering of radiance reflected from nearby surfaces *outside* the geometrical projection of the sensor's IFOV *into* the measurement for that pixel (Schott 2007). This is known as the atmospheric *adjacency effect* and is most significant at wavelengths below 1 μm and in hazy atmospheres (Kaufman 1984). An example of this effect can be observed in an image when a band of anomalously high near infrared values is found on the edge of a dark asphalt highway surrounded by lush vegetation.

A metric that is commonly used as a surrogate for true resolution is the *Ground Sample Distance (GSD)*. GSD properly refers to the distance on the ground between the centers of the pixels in the image as collected on the sensor focal plane. Remote sensing systems are often designed with the pixel spacing approximately equal to the optical resolution (Fiete 1999). However, some systems are spatially oversampled providing closely spaced, but blurry, pixels, leading to this metric being an optimistic value for the true resolution. GSD is also relevant when displaying an image on a computer monitor. If the sensor acquired data with different GSDs in the across track and down track directions, the image will not have the correct aspect ratio (height to width) unless resampled to a uniform grid.

Spectral resolution and accuracy

Spectral performance can be characterized with many metrics. The specific term *spectral resolution* is often defined as the unitless ratio $\lambda/\Delta\lambda$. The normalized response versus wavelength for a given spectral channel is known as the *spectral response function (SRF)* and in general is a composite of spectrally selective effects in the optics, the spectral selection elements and the detector spectral responsivity. For broadband or multispectral imaging systems, the SRF is often approximated by a rectangular function and the spectral bandwidth $\Delta\lambda$ is defined as the width. For SRFs that do not have a flat response a criterion such as 90% of the peak may be used in defining this width.

For hyperspectral systems with SRFs that are best modeled by triangles or Gaussian shapes, the FWHM is commonly used to define the bandwidth. For interferometric systems whose unapodized response function is a $\sin(x)/(x)$ curve, the distance from the central peak to the first zero crossing is a more appropriate representation of the spectral bandwidth. (Note that interferometers often report spectral measurements as a frequency with units of wavenumbers, or cm^{-1}. A handy conversion relationship is a wavelength λ in μm is equal to 10,000 divided by the frequency ν in cm^{-1}.) For imaging spectrometry or hyperspectral systems with many narrow contiguous spectral channels, the *spectral sampling interval*, or spacing between adjacent spectral channel centers, is also a parameter of interest.

Another spectral metric is *spectral calibration accuracy*, which is the accuracy (in percent) of knowledge of the central wavelength of each spectral channel (can also be applied to knowledge of the spectral bandwidth). Another effect is *spectral misregistration* which is the result when a single pixel index in an image cube represents radiance from slightly different areas on the ground from spectral channel to channel. Spectral registration accuracy is often reported as a percentage of the integrated spatial response area for a pixel in a given band that overlaps the response area for another spectral band.

A spectral metric that is particular to dispersive spectral imaging systems using a two-dimensional focal plane is *spectral smile*. This refers to an artifact in which the central wavelength of a given spectral channel varies across the spatial dimension of the detector array. Smile is usually measured as a percent deviation from a nominal wavelength and will vary across the array. The name arises from a plot of locations on the array for a constant center wavelength which can trace out an arc, or a smile, across the array. This can necessitate that physics-based algorithms be applied differently across the columns of an image due to their requirement of accurate wavelength knowledge.

Radiometric resolution and accuracy

Radiometric resolution refers to how fine a difference in incident spectral radiance can be measured by the sensor. We discussed earlier the number of digital bits available in the A/D conversion step as a metric for radiometric resolution. However, this is more precisely termed *precision* (pun intended!). *Accuracy* is how well a measurement of a parameter compares to its true value and can be characterized by long-term deterministic calibration errors or short-term random noise errors.

The accuracy to which data can be calibrated to a physical quantity such as in-band spectral radiance (W m^{-2} sr^{-1} μm^{-1}) is referred to as *absolute radiometric accuracy*. This is usually measured as the long term average difference between the reported and the true radiance for a measurement of a calibration source, and can generally be thought of as a deterministic or constant error source.

It is common to report random *sensor noise* in terms of the equivalent change in spectral radiance at the input aperture of the sensor required to yield the same standard deviation of a noise-only output signal. This *Noise Equivalent Delta Radiance (NEΔL)* represents the short term accuracy with which a radiometric measurement can be made by the sensor. For thermal infrared sensors, this is often converted to *Noise Equivalent Delta Temperature (NEΔT)* corresponding to the smallest detectable change in temperature of a black body source in the scene. For sensors operating in the visible or reflective solar part of the spectrum, this term can be converted to *Noise Equivalent Delta Reflectance (NE$\Delta\rho$)* which represents the smallest equivalent change in surface reflectance measurable by the sensor. In all these cases the scene conditions and instrument settings need to be specified for the reported values to be meaningful.

As an example of the relative magnitude of various noise sources in a typical instrument consider Figure 7.5. Here we have plotted the NEΔL for the HYDICE hyperspectral imager (Rickard et al. 1993) as calculated by a design spreadsheet. (Other HSI sensors have similar performance.) This example is for a case with high scene radiance. Except for regions of atmospheric absorption and at the extreme ends of the spectrometer wavelength range, the dominant noise mechanism in this example is the photon noise from the scene. However, in this example we also see the noise in the electronics circuitry is nearly comparable to the

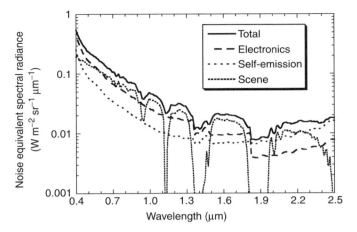

Figure 7.5   Total noise together with three major contributing noise sources for an example hyperspectral imager.

scene noise. The photon noise due to the thermal self emission of the instrument is also seen to be significant, particularly at the longest wavelengths.

Another radiometric noise metric is the signal-to-noise ratio (SNR) defined on a spectral channel-by-channel basis as

$$\mathrm{SNR}_\lambda = \frac{\bar{S}_\lambda}{\sigma_\lambda} \qquad (3)$$

where $\bar{S}_\lambda$ is the measured mean signal, and σ_λ is the measured standard deviation for the mean signal level, each for a given spectral channel λ and for the case of viewing a uniform scene.

A random error source that can occur in linear or two-dimensional detector arrays is residual error from array nonuniformity corrections which is often referred to as *pattern noise*. All detector arrays suffer from a variation in responsivity from detector to detector. Usually a correction map is determined from measurements in the laboratory and used to perform a *flat-fielding* of the array. However, these corrections often do not entirely eliminate the responsivity variations. This can happen because the responsivity of the detector elements may change from the laboratory to the field (or during a collection) due to temperature sensitivity or bias voltage variations. Since one does not know *a priori* where on the focal plane a given object or location on the earth will be imaged, or what the exact magnitude of the error will be, this error source can be considered to be random in nature.

Most noise sources can be considered statistically independent from each other and thus can be added in quadrature. (Adding in quadrature refers to taking the square root of the sum of the terms squared). It can also be a reasonable assumption

that noise is spectrally uncorrelated in multispectral or hyperspectral instruments. However, there are some noise effects that can be correlated across spectral channels. For example, there may be cross-talk between detectors which would lead to the photon noise being correlated across spectral channels. These effects are very instrument specific, but the user should be aware that they do exist.

Geometric accuracy

Geometric accuracy refers to both the fidelity of reproducing the spatial relationships on the ground accurately in the image, and the knowledge of the corresponding geographic location (e.g., latitude/longitude) of each pixel in the image. Geometrical errors can be reported as a percent of a pixel for relative errors or directly in meters. The field of *photogrammetry* (McGlone 2004) is devoted to the science and technology of measuring spatial relationships on the ground using remote sensing images. While the scope of this subject is beyond what can be discussed in this chapter, we include a brief review here.

Many of the geometric inaccuracies in remote sensing imagery are simply due to the geometric realities of the scene and the imaging process and are not due to any errors inherent in the instrument. For line scanning systems (or wide field of view pushbroom or framing arrays) which scan off nadir, the geometric projection of the detector IFOV will cover a larger ground area at the edges of the image than at nadir. Often these systems sample at uniform angular intervals across-track. This results in increased sample spacing (in terms of the distance on the ground) at the edges of the swath. Off-nadir viewing also can lead to imaging the side of buildings or terrain features, thereby displacing the measurement from its true horizontal location.

The process of converting an image with nonuniform pixel sampling to a uniform rectilinear grid is termed *rectification*. Often images will also be corrected such that the off nadir effects are compensated through a process known as *orthorectification* (Falkner and Morgan 2002). *Ortho-images* appear as if every pixel was viewed from directly above with compensation for all vertical relief displacement.

Some inaccuracies are the result of effects or errors in the sensor or platform. Optical aberrations in the instrument can lead to *pincushion* or *barrel* distortions which displace uniform intervals on the ground into nonuniform spacing in the imagery reminiscent of the objects with these names (Figure 7.6). These effects are most significant with framing array sensors. Often they can be calibrated out using a geometrical *camera model* of the sensor, but there can be model inaccuracies that lead to residual geometrical errors in the image.

Random changes and inaccurate knowledge of the orientation of the sensor and its platform will also lead to geometrical errors. Airborne sensors in particular are frequently buffeted by winds and experience changes in their pitch, yaw, or roll orientation. For line scanners, these effects can lead to severe nonuniform spacing in the ground projection of the pixels in both across track and down track directions. Although less of a problem, this is also an issue with pushbroom scanners. For framing arrays, these effects affect the entire frame and are visible when comparing sequential frames, but only affect the orientation of the entire frame, not from pixel to pixel.

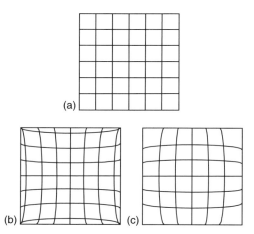

Figure 7.6   Effects on regularly spaced features for (a) ideal image, (b) pincushion distortion, and (c) barrel distortion.

Images are often also *georegistered* which is the process by which the image is resampled and aligned in proper orientation with a standard map projection. For nadir-viewing framing array images and satellite pushbroom images this usually can be accomplished with a *rotation, scaling, and translation (RST)* transformation. Ground control points can be selected in the image, identified with their true locations in the map projection, and then used to derive the RST parameters. However, it can be significantly more complicated and difficult with airborne line scanners due to severe nonuniform sampling and random (due to winds) orientation. A spatially adaptive transform is necessary with best results achieved when accurate sensor position and orientation information is available from an onboard location and orientation unit and combined with an accurate sensor model to project each pixel to its true ground location. Figure 7.7 presents an example of an original unprocessed airborne line scanner image along with the results after rectification and georegistration (Lach and Kerekes 2007).

CHALLENGES AND NEW DIRECTIONS IN OPTICAL SENSOR TECHNOLOGY

Advances in remote sensing technology have occurred in the past through a combination of a 'pull' from users needs and a 'push' from new technology developments. The future will be no different and some current trends offer insight into new needs and capabilities to be expected.

Starting from the user's perspective, we are currently experiencing a rapidly growing commercial market for high resolution color satellite and airborne imagery driven by consumer and civil government mapping applications. This demand for imagery is creating a need for more data providers as well as increased requirements for the accurate geopositioning of the data. Among earth science and environmental applications, we see a call for more capable and flexible sensors, but also with lower data cost and more timely data delivery.

From the technology side, the emerging field of nanotechnology is impacting nearly all technology areas and remote sensing will be no different. In particular, micro electrical and mechanical systems (MEMS) technology has the promise of miniaturizing and expanding the capability of imaging systems and spectral selection devices. In a related area, continued research into new materials will lead to more efficient, more capable and lower cost detectors.

A topic that crosses over between user applications and technology is the growing understanding of remote sensing as a *science*. This refers

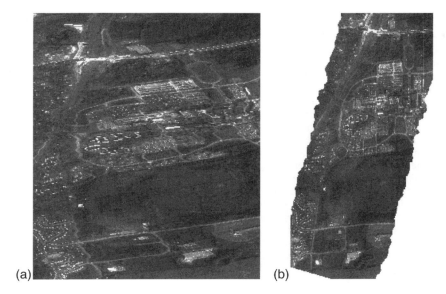

(a)   (b)

Figure 7.7   Effects of registration: (a) unprocessed airborne line scanner image and (b) after rectification and georegistration.

to the understanding and ability to quantitatively predict (without extensive analysis) what information might be extracted from a specific remote sensing system or image in a given application. While this has been studied in the context of single band images used for visual interpretation, the application to machine-processed spectral imagery has the potential of leading to a theory of optimal design and operation of remote sensing systems and the resultant cost savings and data efficiencies.

While many aspects of optical sensor technology will no doubt see development we have chosen to highlight below a few of those mentioned: MEMS, detectors, geopositioning hardware and software, and quantitative metrics for remotely sensed spectral image quality.

Micro electrical and mechanical systems

The miniaturization of electrical, mechanical, and optical devices could lead to pervasive impacts in optical remote sensing technology. One impact is obviously the decrease in size for components, but a potentially greater advantage is the flexibility and adaptability afforded by MEMS devices. As an example, work by Newman et al. (2006), is exploring the use of digital micromirror devices as a *grating electromechanical system (GEMS)* to form a tunable imaging spectrometer. By selectively addressing individual gratings in the device, they have demonstrated rapid adaptation of a broad

band imager to a spectral imager, all without moving parts. The impact of MEMS and nanotechnology is further discussed in the next two sub sections.

Detector technology

While the overall high system costs of satellite and advanced research sensors allow the use of relatively expensive detector arrays as needed, wide proliferation of remote sensing technology is limited by high cost or insufficient performance. Silicon detector arrays have benefited tremendously from the high volume commercial markets for digital cameras, and have led to many vendors offering airborne imaging systems operating in the visible and near infrared. But there are many applications that require sensing beyond the 1 μm cut-off of silicon. Beyond the increased cost of the detector material for these wavelengths, they also often require active cooling to achieve useful efficiencies which drives up the size and cost of the instrument. New generations of detector materials such as extended InGaAs and short wave HgCdTe offer promise of improved performance with less cooling requirements (Yoon et al. 2006). Also devices such as microbolometers offer the promise of improved thermal infrared detectors without heavy cooling requirements. As an example of the convergence of technology trends, Tezcan et al. (2003) reported the use of nanotechnology to construct low-cost uncooled microbolometer arrays.

Geometric positioning technology

While photogrammetrists and surveyors have always been concerned with the geometrical accuracy of remotely sensed images, it has been less of an issue in the broader remote sensing community. However, three emerging trends have placed increased emphasis on geometric accuracy. One is the growing commercial market for remote sensing imagery in mapping and navigation applications where the imagery has to be properly registered to a map projection. This is true not only for nadir viewing imagery, but also for the emerging 3D mapping market where consumers are being provided with perspective views from the ground or air in the visualization of map displays. Another closely related trend is the increasing use and availability of high resolution imagery. As the pixel sizes decrease there is a greater need for accurate geolocation. The third trend driving more emphasis on geopositioning is the increased use of multiple images of a given site collected by the same sensor over time (*multitemporal*) or collected by many different sensors (*multisensor*). When analyzing the images from multiple sensors it is important that they be geometrically aligned in order to fully extract the synergistic benefits of the multiple data sets.

New geopositioning technology includes advances in hardware as well as processing software. Most systems now use signals from the *Global Positioning System (GPS)* to know where they are in three-dimensional space and time, and *Inertial Navigation Systems (INS)* to know their orientation (Farrell and Barth 1998). MEMS technology in particular is enabling dramatic reductions in the size of INS units, while the consumer and auto markets are driving down the size and cost of GPS receivers. These reductions in size and cost will enable further deployment of airborne sensing systems in small aircraft or unmanned aerial vehicles (UAVs). However, while smaller size sometimes is associated with lower accuracy, advances in the processing software can improve the accuracy and thus will also be an important part of future developments in geolocation systems.

Remotely sensed spectral image quality

This chapter has addressed several objective sensor performance metrics which describe how well various components in the imaging system perform. Taken together, these metrics provide measures of *image fidelity* for a system. That is, they help describe how accurately the sensor reproduces the scene. A related, but distinct, topic is that of *image utility*. In the remote sensing context, image utility addresses how useful a given image might be for a particular remote sensing application. While fidelity may be a component of utility, it is not the whole story. Utility is determined by whether the image has the necessary resolutions (radiometric, spatial, spectral, and temporal) and coverage to capture the scene phenomena of interest for a given application. Taken together, image fidelity and utility form the basis for the broader concept of *image quality*.

In the context of panchromatic or single-band imagery, the concept of image quality has been formalized through the development of the *National Image Interpretability Rating Scale (NIIRS)* and its quantitative representation the *General Image-Quality Equation (GIQE)* (Leachtenauer et al. 1997). The GIQE combines several component performance metrics into a single number that captures the overall quality of an image. Originally developed for military applications, these metrics also have application to civilian and scientific uses (Leachtenauer and Driggers 2001). The NIIRS scale includes both image fidelity through parameters such as SNR as well as utility in that it places images at a specific level (0 through 9) based on image analysis tasks (e.g., distinguish a bomber from a fighter aircraft) performed by a human observer. There was an extension of the NIIRS to multispectral imagery, but it also was based on visual interpretation of images displayed as three-band color composites. While there have been initial efforts (Kerekes et al. 2005), no generalization of this approach to spectral image quality has emerged. A key distinction that makes this extension a challenge is the requirement for machine processing of spectral imagery since it has dimensionality (number of spectral bands) exceeding the three bands that can be displayed on a monitor's red, green, and blue channels for visual interpretation.

It remains a goal of remote sensing system research to achieve a generalized objective measure of the quality of a spectral image. This is a complex topic as it very much depends on how and what information the user is interested in extracting from the image. However, in practice, the designers and operators of remote sensing systems are most interested in optimizing the ability of the system to provide the desired scientific information from the scene. Thus, if one could develop such a quality metric it would be very useful in making sensor design choices, scheduling of data collections and indexing image databases. While the specifics of such a metric would vary from application to application, it would be most beneficial if developed within a generalized framework with common parameters. Clearly this is a challenging task, but if achieved, a remote sensing spectral image quality metric would be valuable across the field and thus is a goal worthy of pursuit.

REFERENCES

Aumann, H., M. Chahine, C. Gautier, M. Goldberg, E. Kalnay, L. McMillin, H. Revercomb, P. Rosenkranz, W. Smith, D. Staelin, L. Strow, and J. Susskind, 2003. AIRS/AMSU/HSB on the Aqua Mission: design, science objectives, data products, and processing systems. *IEEE Transactions on Geoscience and Remote Sensing*, 41(2): 253–264.

Beer, R., 1992. *Remote Sensing by Fourier Transform Spectrometry*. Wiley & Sons, Hoboken, New Jersey.

Born, M. and E. Wolf, 1999. *Principles of Optics*, 7th edition. Cambridge University Press, Cambridge, United Kingdom.

Chen, H., 1997. *Remote Sensing Calibration Systems: An Introduction*. Deepak Publishing, Hampton, Virginia.

Denes, L., M. Gottlieb, and B. Kaminsky, 1998. Acousto-optic tunable filters in imaging applications. *Optical Engineering*, 37(4): 1262–1267.

Dereniak, E. and G. Boreman, 1996. *Infrared Detectors and Systems*. John Wiley & Sons, New York.

Falkner, E. and D. Morgan, 2002. *Aerial Mapping: Methods and Applications*. CRC Press, Boca Raton, Florida.

Farrell, J. and M. Barth, 1998. *The Global Positioning System and Inertial Navigation: Theory and Practice*, McGraw-Hill, Columbus, Ohio.

Fiete, R., 1999. Image quality and λFN/p for remote sensing systems. *Optical Engineering*, 38(7): 1229–1240.

Fischer, R. and B. Tadic, 2000. *Optical System Design*, McGraw-Hill, Columbus, Ohio.

Hobbs, P., 2000. *Building Electro-Optical Systems: Making it All Work*. John Wiley & Sons, New York.

Holst, G., 2005. *Electro-Optical Imaging System Performance*, SPIE Press, Bellingham, Washington.

Kaufman, Y., 1984. Atmospheric effect on spatial resolution of surface imagery, *Applied Optics*, 23(19): 3400–3408.

Kerekes, J., A. Cisz, and R. Simmons, 2005. A comparative evaluation of spectral quality metrics for hyperspectral imagery. *SPIE*, 5806: 469–480.

Lach, S. and J. Kerekes, 2007. Multisource data processing for semi-automated radiometrically-correct scene simulation, IEEE GRSS/ISPRS Joint Workshop on Remote Sensing and Data Fusion over Urban Areas, Paris, France.

Leachtenauer, J. and R. Driggers, 2001. *Surveillance and Reconnaissance Imaging Systems: Modeling and Performance Prediction*. Artech House, Inc., Norwood, Massachusetts.

Leachtenauer, J., W. Malila, J. Irvine, L. Colburn, and N. Salvaggio, 1997. General image-quality equation: GIQE, *Applied Optics*, 36(32): 8322–8328.

McGlone, C. (ed.), 2004. Manual of Photogrammetry. American Society of Photogrammetry and Remote Sensing, Bethesda, Maryland.

Newman, J., M. Kowarz, J. Phalen, P. Lee, and A. Cropper, 2006. MEMS programmable spectral imaging system for remote sensing, *SPIE* vol. 6220.

Norton, P., 2003. Detector focal plane array technology. In: R. Driggers (ed.), *Encyclopedia of Optical Engineering*. Marcel Dekker Inc., New York. Chapter X: 320–348.

Peres, M. (ed.), 2007. *The Focal Encyclopedia of Photography: Digital Imaging, Theory and Applications, History, and Science*, 4th edition. Focal Press, Burlington, Massachusetts.

Rickard, L., R. Basedow, E. Zalewski, P. Silverglate, and M. Landers, 1993. HYDICE: an airborne system for hyperspectral imaging, *SPIE* vol. 1937: 173–179.

Schott, J., 2007. *Remote Sensing: The Image Chain Approach*, 2nd edition, Oxford University Press, New York.

Solé, J., L. Bausá and D. Jaque, 2005. *An Introduction to the Optical Spectroscopy of Inorganic Solids*. John Wiley & Sons, New York.

Stevenson, B., W. Kendall, C. Stellman, and F. Olchowski, 2003. PHIRST Light: A liquid crystal tunable filter hyperspectral sensor. *SPIE* vol. 5093: 104–113.

Tezcan, D., S. Eminoglu, and T. Akin, 2003. A low-cost uncooled infrared microbolometer detector in standard CMOS technology. *IEEE Transactions on Electron Devices*, 50(2): 494–502.

Wolfe, W. and G. Zissis (eds), 1985. *The Infrared Handbook*. Environmental Research Institute of Michigan, Ann Arbor, Michigan.

Wyatt, C., 1987. *Radiometric System Design*. Macmillan Publishing, New York.

Yoon, W., M. Dopkiss, and G. Eppeldauer, 2006. Performance comparisons of InGaAs, extended InGaAs, and short-wave HgCdTe detectors between 1 μm and 2.5 μm, **SPIE** vol. 6297.

8

Fine Spatial Resolution Optical Sensors

Thierry Toutin

Keywords: spacecraft, orbits, camera, CCD.

INTRODUCTION

For a long time spaceborne remote sensing, especially sensors with fine spatial resolution, remained in the military domain (Aplin et al. 1997). In 1986, a breakthrough was realized with the launch of first French satellite Satellite Pour l'Observation de la Terre (SPOT) carrying sensors with spatial resolution as fine as 10 m. Later on to re-enforce the presumed United States (US) leadership in civil satellites, the US Land Remote Sensing Policy Act of 1992 allowed the commercialization of satellites carrying up to 1 m resolution sensors, and subsequently IKONOS, the first commercial satellite (0.81 m resolution), was successfully launched in September 1999. Because more than a dozen commercial satellites have now been launched within 0.5–10 m resolution range, a better distinction between these sensors is required: fine spatial resolution (FSR) for 1–10 m and very fine spatial resolution (VFSR) for 1 m and below.

This chapter first addresses generalities on the technology of FSR/VFSR optical sensors with their associated space platforms. It thus focuses on specific FSR/VFSR spaceborne sensors in orbit in 2009 with their available image data types, products and accuracy.

SPACE PLATFORMS AND ORBITS

This section mostly focuses on Earth observation (EO) spacecraft, artificial satellites orbiting the Earth, carrying FSR/VFSR optical sensors. EO satellites obey the celestial mechanical laws as defined by Newton and Kepler for an unperturbed trajectory (Keplerian orbit) and by Gauss and Lagrange for a perturbed trajectory (osculatory orbit) (Escobal 1965, Centre National d'Études Spatiales 1980). A number of perturbations (Earth surface irregularities, atmospheric drag, etc.) slowly change the Keplerian orbit based on the two-body attraction of Newton's law into an osculatory orbit (Centre National d'Études Spatiales 1980). Information on orbits is often needed and orbital models are used depending of their utility and required accuracy (Bakker 2000):

1   to calculate the satellite location on its osculatory orbit to compute Earth coordinates of scanned pixels, requiring high accuracy (metres) over a small time frame (seconds);
2   to predict when the satellite will pass over a specific area, requiring low accuracy (km) but over a long time frame (days).

Many orbital models have been developed since 1960 using the same mechanical laws with Gaussian/Lagrange equations but the differences between the orbital models are mainly in the number and types of perturbations and the techniques to integrate them. As defined and adapted by the North American Aerospace Defense Command, Simplified General Perturbations (SGP), SGP4 and most-accurate SGP8 are the orbital models to be used for low and near Earth satellites (orbital period less than 225 min and altitude less than 6,000 km), while SDP4 and most-accurate SDP8 for highly-elliptic orbits are for deep-space satellites (orbital period greater than 225 min and altitude over 6,000 km). Other orbital models also exist to fulfil different requirements or for specific satellites.

An important factor for an artificial satellite is its final orbit because each orbit has associated advantages and disadvantages. The orbit altitude determines the sensor resolution, for example, DigitalGlobe decided to reduce the QuickBird-2 orbit altitude to achieve 0.61 m resolution instead of the 1 m resolution originally envisaged, without changing sensor characteristics. The orbit inclination determines the percentage of Earth that can be imaged (mainly in the highest latitudes), while other orbit variables determine the repeat cycle and the forelap depending on latitude. Consequently, orbit selection and sensor characteristics are closely related. Most, if not all, of the commercial EO spacecrafts have near-Earth, retrograde, quasi-circular, quasi-polar, geosynchronous and sunsynchronous orbits.

Near-Earth orbits (more than 300 km altitude) are high enough to reduce the atmospheric drag. A sensor aboard a near-Earth orbit satellite should have higher performances (optics and detectors) to achieve the same high-level details of the Earth than a sensor in a low-Earth orbit satellite. Retrograde orbits with 90–180° inclination are westward-launched orbits, which require extra fuel to compensate for the Earth's rotation, but they provide the only solution for obtaining sunsynchronous orbits. Quasi-circular orbits avoiding large changes in altitude enable images with similar scales to be acquired, which is desirable for EO. Quasi-polar orbits with 90–100° inclination enable sensors to image the entire Earth, including most of the poles. Because geosynchronous orbits have a repeating ground track, they have an orbital period that is an integer multiple of the Earth's sidereal rotation period. This integer multiple is called the repeat cycle. The special case of a geosynchronous orbit that is circular and directly above the Equator is called a geostationary orbit. The satellite track on the ground, also called path, is kept fixed within certain limits related to orbit maintenance accuracy. Paths are artificially divided into squared scenes at regular intervals, generating rows. An example of path-row system is the World Reference System of the Landsat satellites. The main advantage is that the satellite follows a fixed pattern on the Earth, which is desirable for operational EO systems. Sunsynchronous orbits enable satellites to pass overhead at the same local solar time and thus to acquire images with almost-identical illumination lighting conditions. While the comparison of multi-date images is easier with a sunsynchronous orbit, this approach has some disadvantages, especially as variations of illumination reveal different structural details. In addition, sunsynchronous orbits require retrograde orbits and strict relationships between orbital parameters (mainly inclination and height), which must be preserved during the satellite's lifetime.

Military spy spacecrafts generally have completely different orbit characteristics depending on their mission goals and their typically short lifetime (few weeks). Their orbits are:

1 low-Earth (100–300 km) to increase sensor resolution and detail quality on the ground;
2 prograde, less than 90°-inclination, to reduce fuel at launch;
3 elliptical to offer better details of Earth at the perigee;
4 rarely polar, with 70–80° inclination, because latitudes over 70° do not contain many primary targets (so far!); and
5 not geosynchronous and not sunsynchronous because they do not necessarily require to view the same area during their short-life mission and with the same illumination condition.

FINE SPATIAL OPTICAL SENSORS

This section reviews the basic technology of fine spatial optical sensors (e.g., film-frame/panoramic cameras and push-broom/digital-frame sensors), and summarizes and compares their general characteristics.

Film frame and panoramic cameras

Film frame and panoramic cameras acquiring basic photography in one or more spectral bands are historically one of the oldest types of equipment used for performing space-based Earth observations. The panoramic camera is an extension of the frame camera with a rotating lens system scanning the terrain perpendicular to the satellite displacement (e.g., the US KH-1/4 and Russian KVR-1000). Film frame cameras with return of the film to Earth for development and analysis were used for intelligence gathering since 1959. They were first declassified in 1987 for FSR and 1992 for VFSR by the USSR, and in 1995 by the US. The main advantages of film cameras are a large image format (mainly when using a panoramic camera) with high information density and geometric integrity and a well-known conical geometry. However, the disadvantages, such as the analog output, long process of film transmission and pre-processing, limited spectral coverage, difficult calibration process, as well as the limited ground coverage and repeat period, are more constraining, and therefore less desirable than current digital and geospatial information technologies. While photographic cameras and film recovery practices were abandoned a few decades ago, these images provide unique archives and are still a cheap source of historical data for EO and monitoring.

The main elements in a camera (normal and panoramic) are (i) the optics with their lenses and filters, (ii) the diaphragm and the shutter, (iii) the exposure control system, (iv) the film (black and white (B/W) and colour) with its magazine, and (v) the camera body by itself (Bonneval, 1972, Slama et al. 1980, Morain and Budge 1998). In fact, what determines the spatial resolution of the camera are primarily the resolving power (in line pairs) of the film, the quality of the optics and the aperture of the shutter. The main qualities of the optics are a high resolving power, a large field of view (FOV), a strong and equal brilliance in the full FOV, a low curvature of field and low aberrations and distortions. One of the most useful tools for the evaluation of the resolution performance of an optical system is the optical or modulation transfer function method developed by Schade (1948), because it provided a mathematical basis for the construction of a single-type of target applicable to all kind of systems (Slama et al. 1980).

The spectral characteristics of a camera system are determined by the film and the filters. Because normal emulsion is sensitive to blue light, the film can be sensitized to have an extended spectral sensitivity into the green (orthochromatic emulsion), red (panchromatic emulsion) and near-infrared (near-infrared (NIR) sensitive emulsion). In addition, colour-glass or coated glass filters can be used to limit the sensitivity to a specific spectral range (Bakker 2000).

Push-broom and digital frame sensors

Push-broom scanners are systems that capture an image using a linear array that record simultaneously each picture element along a line without the use of electromechanical components (Kerekes, in this volume). Because the push-broom scanner operates using a linear array of solid semiconductive elements without any mechanical scanning mirror, their mechanical reliability is very high.

Charge coupled devices (CCD) are mostly adopted for linear array sensors, which are therefore called linear CCD sensors or CCD cameras. All elements are thus at fixed distances from each other and the recorded images will not have geometric distortions caused by irregular movement of a scanner. The first civilian use of CCD-based push-broom scanner with five 1200-detector elements on EO satellites was the High Resolution in the Visible (HRV) sensor of the French SPOT-1 launched in 1986. The push-broom scanner has an optical lens, through which a line image is detected simultaneously, perpendicular to the flight direction (conical geometry) and the second dimension of the image is generated by the satellite displacement along its orbit (cylindrical geometry) (Toutin 2001). The qualities of optical lenses are the same as those required for cameras. Apart from these geometric advantages, there are also some radiometric advantages (larger dynamic range and very narrow spectral bands, more stable and easier calibration and digital output) but some disadvantages (limited spectral sensitivity beyond about 2.5 μm and non-identical detector characteristics). The radiometric differences generate vertical banding in recorded images and require calibration coefficients for each detector. Nevertheless, CCD push-broom scanners with 5,000–10,000 linear CCD sensors are the most common, if not the only used sensor for acquiring FSR/VFSR images from space.

Digital frame cameras are a recent two-dimensional (2D) matrix extension of push-broom scanners and can be used as a replacement of a film frame camera with a longer life expectancy. They combine some advantages of both sensors: the instantaneous acquisition of a camera over a large area, the digital nature of a push-broom scanner and the geometric integrity of both sensors. On the other hand, it also shares some disadvantages: the limited CCD spectral sensitivity and the calibration of each CCD in the matrix, but without band striping. The first use of CCD-based frame camera was on the Russian RESURS-DK launched 15 June 2006 (Anshakov and Skirmunt 2000).

General characteristics of FSR/VFSR sensors

Different technologies are used by push-broom scanners to increase the performance of CCD linear array sensors (Toutin 2004a). Some agile scanners (QuickBird, OrbView, etc.), which continuously rotate when acquiring images, have forward and reverse scanning to increase their imaging capability. North-to-South rotation is also used to achieve a smaller ground sampling distance (GSD) (QuickBird) or to increase the line integration time (EROS), but this approach generates large nonlinear distortions in the imaging geometry.

Time Delay and Integration (TDI) uses rectangular CCD chips with many lines, whose signals are summed up to generate an image line in order to increase the line integration time for a better image quality (QuickBird, IKONOS). Even with the development of the 5,000–10,000 element detector, multiple identical CCDs are still needed, either butted with some overlap (QuickBird, IRS) and/or staggered (SPOT5-HRG, OrbView), in order to achieve a large FOV and small GSD. In the staggered technique, at least two CCDs are shifted a half of a pixel relative to each other. Both the butted and the staggered techniques require some geometric and radiometric preprocessing by the satellite operator to generate an 'ideal virtual focal plane' (Toutin 2004a). Because multiband (MB) CCDs (with filters to narrow the spectral range) are usually identical to panchromatic (Pan) CCDs the larger GSD of MB images is achieved by averaging pixels (generally four pixels) in the column direction and by increasing the integration time in the line direction (generally by four). This technique produces better quality images than using CCDs with a larger GSD.

Due to the satellite movement, push-broom scanners, acquiring images line by line, generate more distortions in the imaging geometry than frame cameras (film or digital), which acquire 'rectangle' images instantaneously. The FOV of film frame cameras (10–20° equivalent to 70–160 km on the ground) is larger than those of push-broom scanners (1–5° for 10–60 km) and digital frame cameras (5° for 30 km). The main drawbacks of film frame cameras are, however, their hard-copy nature and limited life time.

Most of these sensors have stereo capability achieved by off-nadir pointing: (1) across track from two different orbits by rotating a mirror (e.g., SPOT HRV) or by rolling the satellite (e.g., IRS-1C/D, Kompsat), (2) along and below track from the same orbit using fixed fore and aft sensors (e.g., TK-350; SPOT5-HRS), or (3) along track from the same orbit at 360°-azimuth by rotating the satellite in the three axes (all agile satellites). The across-track and agile viewing capability increases the repetitive coverage of areas, while the forward and reverse scanning of agile satellites increases the stereo capability. In addition, the simultaneous along-track same-date stereo-data acquisition gives advantages in terms of instantaneous stereo acquisition and limited radiometric variations between the two images versus the across-track multi-date stereo-data acquisition (Toutin 2001).

In terms of radiometry, the film camera with its spectral sensitivity and resolving power should provide better image quality than the digital scanners (spectral sensitivity of up to about 2.5 μm). However, the film quality of Russian cameras has been questioned because numerous archived images were poorly processed during original production

and also scratched. On the other hand, some problems have been noticed on images acquired from push-broom scanners: non-homogeneous and non-additive noise, which increases with intensity; banding/staircase effects; white dotted lines; strong reflection and saturation; spilling in scan direction due to specular reflection; and other artefacts, including ghosting of moving objects, edge sharpening and bright horizontal and vertical lines. Some of these radiometric effects are related to the new technologies of agile satellites that have been used to increase their capability (time delay and integration, scanning modes) and sensor resolution (smaller GSD, larger integration time), as well as the preprocessing of raw data to generate the 'ideal virtual' focal plane.

REVIEW OF FSR/VFSR SENSORS AND IMAGE PRODUCTS

Film frame and panoramic cameras

CORONA was a codeword name of a series of US military reconnaissance satellites used for photographic surveillance from June 1959 until May 1972, with the first successful photographs acquired 18 August 1960 (McDonald 1995, GlobalSecurity.org 2007) (Table 8.1). The satellites were retro actively designated in 1962 as Keyhole, KH-1 to KH-4, KH-4A and KH-4B, with the incrementing number indicating changes in the instrumentation, such as single-panoramic (KH-1 to KH-3) to double-panoramic cameras (KH-4/4A/4B) with 30° along-track stereo-capability. All panoramic cameras used special 70 mm film and scanned at ±35° across track with image-motion compensation. The quality and spatial resolution of cameras improved during the duration of the program from 12.2 m (KH-1) to 1.8 m (KH-4B). Likewise, the mission life extended from one day (KH-1) to 19 days (KH-4B) (Smith 1997). A spin-stabilized attitude control system and other communications and control systems were included in the Agena vehicle. During nine successful missions, the single panoramic cameras KH-1, KH-2 and KH-3 collected 1,400 scenes in 1960, 7,200 scenes in 1960–61 and nearly 10,000 scenes in 1961–62, respectively. The stereo panoramic cameras KH-4, KH-4A and KH-4B during 85 successful missions collected nearly 102,000 scenes in 1962–63, 518,000 scenes in 1963–69 and 189,000 scenes in 1967–72, respectively. Occasionally one camera malfunctioned or was programmed to halt, while its mate continued to operate, thus reducing the number of stereo-scenes. A very limited amount of colour test film was attached to the end of KH-4B missions. All the scenes together roughly cover 1.1–1.4 billion km^2 of the Earth's surface

Table 8.1   Characteristics of CORONA KH1—4/GAMBIT KH-7 film frame panoramic cameras of the USA

Sensor	KH-1–3	KH-4	KH-4A/-4B	KH-7
Operation period	1960–1962	1962-1963	1964–1972	1963–1967
Mean altitude (km)	165–460	165—460	185/150	122–140
Orbit inclination (°)	80–84	80—84	80–84	(88–89)*
Focal length (mm)	609.6	609.6	609.6	∼ 960
Ground coverage (km)	15× 210 – 42× 580		17× 232 14× 188	22× 750
Ground resolution (m)	4.6–7.5	7.5	2.7/1.8	1.2/0.6
Scale	1 : 275, 000 to 1 : 760, 000		1 : 305/247, 000	1: 50, 000
Forelap/overlap	No	60%	Up to 100%	Variable

* KH-7 orbit inclination is approximate.

Table 8.2   Characteristics of film and digital frame cameras of USSR/Russia

Sensor name	MK-4	KFA-1000	KFA-3000	KVR-1000	DK
Sensor type	Film	Film	Film	Film	Digital
Spacecraft	RESURS-F2	RESURS-F1/F2/F3	RESURS-F3	Kosmos	RESURS
Operation period	1988–1993	79–93/88–93/93–94	1993–1994	1981–2000	2006–
Mean altitude (km)	260 (230)*	260 (230)*	270	220	350 (or 510)
Orbit inclination (°)	82.3; (72.9)*	82.3; (72.9)*	82.3; 72.9	70.4	64.8 (or 70.4)
Swath	160	72/122/217	35	160	28.3
Focal length (mm)	300	1000	3000	1000	?
Frame size (mm)	180× 180	300× 300	300× 300	180× 720	(4.7× 28.3)
Spectral bands	Six from 460–900 nm	Pan: 570–810 nm	510–760 nm	570–810 nm	2 VIR and
		Colour 570–810 nm			1 near IR
Ground spatial	6–8 m	Pan: 4–6 m	Pan: 2 m	Pan: 2 m	Pan:<1 m
resolution (m)		Colour: 6–10 m			MB: 2–3 m
Scale	1 : 900, 000	1 : 250, 000	1 : 80, 000	1 : 220, 000	NA
Forelap/overlap	60%	60%		10%	No

* Starting May 1991, a new profile was chosen with a mean altitude of 230 km and orbital inclination of 73°.

(McDonald 1997). CORONA KH-4B stereo data can be used for three-dimensional (3D) topographic mapping (Altmaier and Kany 2002) but, the B/W data is of limited utility for thematic applications.

GAMBIT was the program name for the first VFSR USA intelligence satellite, designated KH-7, and which complemented the broad area coverage of the Corona systems (Table 8.1). The KH-7 system operated on 34 successful missions from 1963 to 1967 and was designed mainly to obtain technical measurements of military targets. Initially, KH-7 imagery attained spatial resolutions of roughly 1.2 m, increasing to 0.6 m by 1966. Simultaneously, mission length expanded from one day to eight days (McDonald 1995). The first KH-7 imagery was acquired pointing at nadir, while the later KH-7 sensor pointing capacity compares favourably with the QuickBird and IKONOS systems of 35 years later!

CORONA and GAMBIT were officially classified until 1992, but images acquired between 1960 and 1980 were declassified on 22 February 1995 in two phases: on 1996 880,000 images (1959 to 1972) and in 2002 50,000 images (1963 to 1980). Scanned CORONA and GAMBIT

photographs are available from the US Geological Survey (USGS) Earth Resources Observation and Science Data Center (http://edc.usgs.gov/products/satellite/declass1.html). An agreement of the imaged countries is sometimes necessary; some related issues of data policy are discussed in Harris (in this volume).

USSR RESURS-F satellites have collected images of the Earth's surface in visible and near-infrared spectral bands with different medium-spatial resolution cameras (e.g., KATE-200) since 1962, but with FSR sensors, such as KFA-1000/3000 in the 1970s. The RESURS-F system based on the recoverable Vostok capsule consists of four spacecraft (Table 8.2) (Lukashevich 1994):

- RESURS-F1 spacecraft was equipped with two large format stereoscopic cameras KFA-1000 acquiring spectrozonal stereoscopic images at a scale of 1:250,000, and three KATE-200 15–30 m resolution frame cameras, with one camera linked to a stellar camera to provide simultaneous star backgrounds for precise geographical location determination. The RESURS-F1 program achieved 50 23-day successful missions with only two launch failures.

- RESURS-F1M, a modified F1 system in 1997, carried three KFA-1000 cameras, which acquired pictures on two-layer colour spectrozonal film, and one panchromatic camera KATE-200 capturing B/W pictures.
- RESURS-F2 spacecraft was equipped with a MK-4 camera system, which acquired multiband stereoscopic imagery recorded on three separate panchromatic films at a scale of 1:900,000. Three bands can be selected from the six available. Although RESURS-F2 had two on-board stellar cameras to increase orientation accuracy information, they were not always operated, thus limiting the cartographic capabilities without ground control. Eight missions were conducted during 1988–1993.
- RESURS-F3 spacecraft was virtually identical to the Resurs-F1 but carried the KFA-3000 panchromatic frame camera system acquiring B/W imagery at scale of 1:70, 000 to 1:90,000 with 2–3 m resolution. Two missions were flown during July–August 1993 and June 1994.

The USSR Space Mapping System 'KOMETA' on Kosmos consisted of stereoscopic cameras TK-350 and a FSR panoramic camera KVR-1000 (±20° across-track) (Figure 8.1), integrated with on-board equipment (two star positioning cameras, laser altimeter, navigation sensors, Doppler systems and synchronizing devices) for precise external orientation determination. The spatially-degraded 2 m GSD images acquired by the KVR-1000 camera have two other designations, KWR-1000 and DD-5, which may stand for 'Digital Degradation by a factor of 5' (Gupta 1994); in this condition, the GSD of the raw data would be 0.4 m. The Kosmos satellite operates in near-circular orbit at an average altitude of 220 kilometres and has a longer 45-day orbital duration with a total film capacity approximately equivalent to 10.5 million km². There have been 20 missions starting in 1981, and with the latest in 2000. The panchromatic KVR-1000 images (490–590 nm) are sold in both film (40 × 40 km) and 8-bit digital format (4 × 4 km up to 40 × 40 km). The KOMETA system allows production of Level-3 elevation models and 2 m spacing orthorectified images for 1:25,000 scale topographic and digital maps (Lavrov 1996, 2000).

Since 1991, most of the Russian satellite sensor images can be delivered from the archives on-line by the company SOVINFORMSPUTNIK (http://www.sovinformsputnik.com/thecomp.html). Different products are available: raw imagery (film and digitized copies) and orthorectified imagery with or without the use of ground control

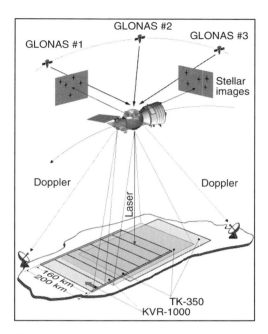

Figure 8.1   The space mapping system 'KOMETA' with stereo TK-350 camera, panoramic KVR-1000 camera, two star positioning cameras, laser altimeter, GLONASS navigation sensors and Doppler systems (adapted from Lavrov 1996).

points (GCPs). However, the film quality has been questioned, especially for the archived images, which were scratched and poorly processed during original production. In addition starting in 1992, Russia has acquired a huge and unique archive of 1-m resolution digital satellite images, which were gathered by Russian intelligence panchromatic cameras with a larger swath than other camera systems (Kirillov 2001). The digital orthorectified imagery are delivered with 15 m positioning accuracy without ground control and 1–2 m with global positioning system (GPS) GCPs, which is equivalent to a map standard of 1:10,000 scale. The archives contain data of substantial areas of the US, Europe and Asia as well as selected areas in South America and the Far East.

Push-broom scanners

In this section, a brief summary of the distinctive characteristics for each push-broom system currently operating in 2009 is given, including satellite name, orbit characteristics and sensor properties (Table 8.3), as well as image products and processing (Table 8.4).

Table 8.3 Summary of existing VFSR/FSR push-broom scanners

Platform Sensor	Country Year	Height (km)/ Inclination (°)	GSD (m) / Nb of CCDs	MB / GSD (m)	FOV (°) / Swath (km)	Field of regard (°)	Revisit time (days)	Stereo B/H	Nb. of bits
IRS-1C/1D PAN	India 1995/97	817/98.6	5.8/12,288	None	4.9/70	±26 across	5	Across Up to 1	6
IKONOS-2 OSA/TDI	USA 09/1999	680/98.2	0.8[A]/13,816	4/4	0.93/11	45[A] (at 360°)	2–3.5	Agile Variable	11
Kompsat-1 EOC	Korea 12/1999	685/98.13	6.6/2,592	None	1.4/17	±45[C] across	3	Across Up to 1.1	8
EROS A1 PIC/TDI	Israel 12/2000	480/97.4	1.8[B]/7,800	None	1.5/13[B]	45 (at 360°)	2–4	Agile Variable	11
QuickBird-2 BHRC60/TDI	USA 10/2001	450/52	0.61/27,568	4/2.44	2.12/16	45 (at 360°)	1–3	Agile Variable	11
SPOT-5 HRG	France 05/2002	822/98.7	(5/3.5)/ 12,000[D]	4/10	4.2/60	±27 across	3–6	Across Up to 1	8
Orbview-3 OHRIS	USA 06/2003	470/97	1/8,000	4/4	0.97/8	50 (at 360°)	1–3	Agile Variable	11
Formosat-2 RSI/TDI	Taiwan 05/2004	891/99.14	2/12,000	4/8	1.5/24	45 (at 360°)	1	Agile Variable	8
Cartosat-1 PAN (2)	India 05/2005	618/97.87	2.5/12,288	None	2.16/30	±26 across	5	Along 0.62	10
Beijing-1 CMT	China 10/2005	686/98.2	4/(6 000)	None	2/24	±30 across	4	Across Up to 1.1	8
TopSat AOC/TDI	UK 10/2005	686/98.2	2.5/6,000	3/5	1.2/15	±30 across	4	Across Up to 1.1	11
ALOS PRISM (3)	Japan 01/2006	692/98.16	2.5/14,000	None	2/35	±1.5 across	46	Along 0.5/1.0	8
EROS B PIC-2/TDI	Israel 04/2006	~ 500/97.4	0.7/20,000	None	0.8/14	45 (at 360°)	1–3	Agile Variable	10
Kompsat-2 MSC	Korea 07/2006	685/98.13	1/15,000	4/4	1.3/15	30 along 56 across	2	Agile Variable	10
Cartosat-2 PAN	India 01/2007	635/97.92	0.8/12,000	None	0.59/9.6	45 (at 360°)	1–4	Agile Variable	10
WorldView-1 PAN/TDI	USA 09/2007	496/97.2	0.5/35,000	8/2	2.12/17.6	45 45 (at 360°)	2–6	Agile Variable	11
CBERS-2B HRC	China–Brazil11/2007	778/98.5	2.5/10,368	None	2.1/27	Few	5	Along Up to 1	8

Notes:
A The spatial resolution at nadir is 0.81 m; the field of regard can be up to 60° but with 2-m resolution.
B 1-m resolution but with 6.25-km swath is obtained by applying "over-sampling".
C For cartographic mapping, up to 30° across-track viewing is only used.
D Two staggered CCD lines are used to achieve the 3.5-m GSD.

Table 8.4 Processing levels of available data products of VFSR/FSR sensors

System	Processing Level				
	Raw data without any correction; orbit oriented	Radiometric correction; orbit oriented	Radiometric and geometric corrections; orbit oriented	Radiometric and geometric corrections; map oriented	Radiometric and geometric corrections with control data; map oriented	Radiometric and geometric corrections with control data and elevation (DTM); map oriented
SPOT1–5	0A	1A	1B	2A	2B	3
IRS1 C/D		1A	1B	2A		3
IKONOS-2				Geo Standard	Reference Pro	Precision Precision Plus
EROS A		1A	1B			Special request
Kompsat-1	1A	1R	1GR		1GC.P	1GC.D
QuickBird-2		Basic	Standard			Ortho DG/DOQQ
OrbView-3		Basic				Ortho
Formosat-2		1A	1B			3
Cartosat-1	0A	0B	1SYS		2GCP	3A/B DEMA/B
Beijing-1						
Topsat	0	1A		2A		
ALOS	1A	1B1	1B2	1B2	2B	3
EROS B		1A	1B			Special request
Kompsat-2		1A	1B	2A		3
Cartosat-2				User request		Precision/High-Precision
WorldView-1		Basic	Standard			Ortho DG/DOQQ

The two Indian Remote Sensing satellites (IRS-1B/1C) (http://www.nrsa.gov.in/satellites/IRS_satellites.html), launched in 1995 and 1997 with 5.8 m Pan sensors, were the first FSR satellites with a resolution finer than that of the 10 m SPOT-HRV. Three similar CCD linear arrays were used to achieve a 70 km swath. Because their alignment was far outside the system accuracy, a simultaneous calibration/rectification process based on ground control was necessary to achieve GSD positioning accuracy (Cheng and Toutin 1998). The processing levels provided are 1A/B, 2A and 3 (Table 8.4) supplied in the Landsat Ground Station Operations Working Group (LGSOWG) format, georeferencing Tagged Image format (GeoTIFF), hierarchical data format (HDF) and Fast format. IRS Pan data can be used for 3D topographic mapping and updating at 1:50, 000–100,000 but is of limited utility for thematic applications.

IKONOS (http://www.geoeye.com/), derived from the Greek word 'image', was the first VFSR commercial agile (up to 60°) satellite using declassified military technologies. The sensor simultaneously collects Pan images (450–900 nm) and MB images (blue 450–520 nm, green 520–600 nm, red 630–690 nm, NIR 760–900 nm), which greatly improves the pan-sharpening process. The Pan GSD is 0.81 m at nadir and 2 m at 60°-viewing, but image GSD is processed at 1 m. Each band combines three butted CCD linear arrays (for increasing the swath) and TDI with up to 32 CCD linear arrays (for increasing image quality), but generally 13 are utilized. The reverse and forward scans also use two different CCD groups. The revisit frequency of the satellite is 2.9 days at 1 m resolution and 1.5 days at 1.5 m resolution. Unfortunately, SpaceImaging, now GeoEye, does not supply raw data and sensor information, which is regarded as proprietary, but instead provides a parametric model for orthorectification (Grodecki and Dial 2003). The processing levels are Geo or Geo Ortho Kit (2A), Reference/Pro (2B) and Precision/PrecisionPlus (3). The differences between each 2B or 3 products are the positioning accuracy. The data utilization can be summarized as follows: linear elements with Pan, location of houses with MB and their shape with Pan-sharpened; vegetation mapping up to a scale of 1:10,000 using MB data (Pan-sharpened data has limitations for vegetation stratification); and, bathymetry mapping of shallow water with the blue channel.

The Earth Observing Camera (EOC) of Kompsat-1 (KOrean MultiPurpose SATellite; http://www.kari.re.kr/) has a single push-broom scanner (Pan 510–730 nm). With the body pointing method, Kompsat-1 has ±45° across-track field of regard (FOR), but up to 30° roll-tilting, which is only used for mapping purposes. Based on Kompsat-1, Kompsat-2, the last VFSR Korean satellite launched on 28 July 2006, has improved

capabilities: higher spatial resolution (0.7 m) and attitude control, larger across-track FOR and smaller revisit rate. Kompsat-2 has a camera with five bands (Pan 500–900 nm, blue 450–520 nm, green 520–600 nm, red 630–690 nm, NIR 760–900 nm) with TDI using up to level 32, and FOR with the body pointing method is ±56° across-track and ±30° along-track. The CCD line rate can be changed in the range from 7,100 to 2,200 during stereo imaging to reduce image degradation due to smear effects. It is capable of acquiring up to 7,500 images daily, equivalent of 1.7 million km^2 of coverage. All processing levels, except 2B for Kompsat-1, are provided. The sensor model is, however, supplied for Kompsat-2 inside the SPOT-Image DIctionary MAintenance Programs (DIMAP) format. The applications of Kompsat-1 mainly include land and ocean mapping at medium scales, while those with Kompsat-2 are more equivalent to other VFSR 1 m Pan and 4 m MB sensors.

EROS A1 (http://www.imagesatintl.com/) carries an asynchrone sensor, meaning that the satellite ground track is longer than the imaged ground area (Figure 8.2). Consequently, EROS continuously rotates from South to North to increase its line integration time and image quality. The nadir GSD is 1.8 m but increases up to 2.5 m off-nadir, generating large distortions between stereo-images (Figure 8.3, Toutin 2004b). Recently, modifications of the primary sensor by applying 'over-sampling' techniques carried out by the manufacturer have already doubled the system's performance: 1.4 m resolution but with reduced image coverage. Slightly larger and similar in appearance to EROS A, EROS B has superior capabilities, including a larger CCD camera and selectable TDI configuration, a Pan resolution of 0.70 m, an improved pointing accuracy using a star sensor, a larger on-board recorder and a faster data communication link. EROS data are provided as Basic Scene, Vector Scene, Stereo (Triplet and Hypersampled Scene only for EROS A) in all processing levels. EROS A can be utilized as cost saving alternative if only panchromatic data are required and also to pan-sharpen other existing MB. EROS B utilization can be summarized as: shape of houses and linear elements clearly visible for 3D topographic mapping and updating (1:10–25,000), 3D city buildings, visualization and simulation; vegetation mapping and monitoring with pan-sharpened image data as texture may allow separation of some species; water analysis utilizing blue channel and off-nadir viewing for avoiding wave reflection.

The QuickBird satellites are the first in a constellation of spacecraft that DigitalGlobe™ (http://www.digitalglobe.com/) is developing to offer VSFR imagery. Because of the satellite speed and line integration time, the 0.61 m GSD is achieved with South-to-North body rotation, which introduces some non-linearity distortions. The sensor simultaneously collects data in five bands (Pan 450–900 nm, blue 450–520 nm, green 520–600 nm, red 610–680 nm, NIR 790–900 nm). Each band combines six butted CCD linear arrays (for the 16.5-km swath) and TDI with up to 32 CCD linear arrays (for increasing image quality). The Pan, MB, Pan+MB, 3-band colour or Pan-sharpened images

are provided as *Single Area* (16.5× 16.5 km) and *Strip* (16.5× 165 km) with different processing levels: Basic (1A), Standard (2A), Ortho Ready Standard (2B) and Ortho (3) supplied in GeoTIFF 1.0 or New Industry Text Format (NITF) 2.0/2.1 by special delivery services or ftp. The data utilization is the same as EROS B due to their similar sensor characteristics.

The two SPOT5 HRG sensors (http://www. cnes.fr/) have 5 m spatial Pan resolution. However, using a viewing system (two staggered CCD lines offset of 0.5 pixel across-track and 3.5 pixels along track) and CNES-licensed processing software the Supermode spatial resolution is finer than 3 m (Baudoin 1999). With a spectral beam splitter, HRG simultaneously collect one Pan (490–600 nm) and four coregistered MB bands (green 490–610 nm, red 610–690 nm, NIR 780–890 nm, SWIR 1580–1750 nm), but the SWIR band has 20 m spatial resolution. In addition, the two 10 m resolution SPOT5 HRS sensors provide ±20° along-track stereoscopic Pan data (490–690 nm) with image GSD of 10 m in pixel by 5 m in line due to a resampling for improving elevation parallax measurements. The stereo acquisition mode provides 120 km swath images with lengths up to 600 km. All processing levels of data products are available by default in the DIMAP format, but the old Committee on Earth Observing Satellites (CEOS) format is maintained for the lowest levels. The spectrum of data utilization is very broad (3D topographic/thematic mapping, monitoring, disaster management, surveillance) due to its different sensors and resolutions.

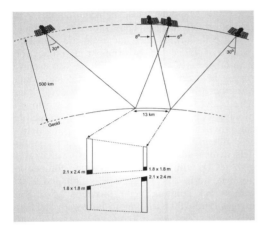

Figure 8.2   Geometry of acquisition of EROS with the body rotation due to its asynchrone system and its impact on pixel spacing (Toutin 2004b).

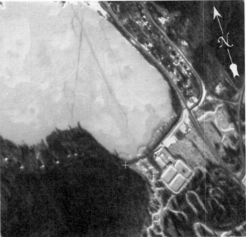

Figure 8.3   Result of the EROS body rotation on the geometry of stereo images (512 by 512 pixels). Note the shape and size variations of the lake, snowmobile tracks, and roads (Toutin 2004b). EROS Images © and courtesy ImageSat Intl., 2002.

OrbView-3's imaging instrument (www. orbimage.com) uses two staggered CCD lines shifted by 0.5 pixels in both directions and over-sampling to achieve 1 m GSD. However, this 1 m GSD is not exactly equivalent to real 1 m projected GSD without over-sampling. Without TDI, the imaging time has also to be increased by a continuous body-pointing movement to improve the image quality. Even though the instrument and the focal plane are different compared to IKONOS, the images have similar characteristics: five bands (Pan 450–900 nm, blue 450–520 nm, green 520–600 nm, red 625–695 nm, NIR 760–900 nm) with the same GSD, although the swath at 7 km is smaller. The instrument's storage capacity is limited to 4 GB. Three processing levels are only provided for Pan or MB images: Basic (1A), Basic Enhanced™ with satellite modelling, Geo (2A) and Ortho (3). Other highest products are recently offered: DEM/DSM, Thematic/Feature Map™. The data utilization is similar to other VFSR data.

The RSI of Formosat-2 (http://www.nspo. org.tw/) has five bands (Pan 450–900 nm, blue 450–520 nm, green 520–600 nm, red 630–690 nm, NIR 760–900 nm) with 2 m spatial resolution and a body pointing capability of ±45° in the roll and pitch axis providing a cross-track observation capability of 968 km about nadir for event/disaster monitoring. No TDI/butted/staggered CCD lines are necessary for this FSR sensor due to its limited 24 km swath width, when compared to SPOT5 HRG. Three processing levels are provided (1A, 1B and 3 with user-provided GCPs and digital terrain model) by ftp in 24 hours or on CD-ROM. The utilization is mainly dedicated for monitoring sites, detecting changes (infrastructures, land) and surveillance (humanitarian efforts, targets, moving objects) due to its one-day revisit time. 3D topographic mapping and updating at 1:10,000 scale can also been realized with the across-track capability.

Cartosat-1 (http://www.nrsa.gov.in/) has a 12 K, 7 μm, 35 μm staggered CCD for each of the two Pan (500–750 nm) sensors (nadir and backward). Cartosat-1 also carries a Solid State Recorder with a capacity of 120 GB. In order to ensure along-track stereo imaging the spacecraft body is steerable (yaw manoeuvering of the satellite platform) to compensate the Earth rotation effect. For any given latitude, stereo imaging can also be achieved by mounting the payloads at appropriate yaw angle with respect to each other. A combination of fixed mounting, catering to stereo acquisition requirements for Indian latitudes and a yaw manoeuvring for other regions with minimum power consumption was adopted. When no stereo acquisition is performed, the spacecraft can be manoeuvered such that the image strips will fall side by side so that wider swath images are obtained by the two cameras. All processing levels are available, as well as DEM and 2D–3D image-map products that are supplied in CEOS super structure and IRS fast formats with Cartosat specific changes. The utilization is mainly dedicated for 3D topographic mapping and updating at 1:10,000 scale.

Beijing-1 and Topsat (http://www.sstl.co.uk/) launched 27 October 2005 are the latest satellites in the Disaster Monitoring Constellation (DMC) to complement four other DMC satellites launched in 2002 and 2003 by Algeria, Nigeria, Turkey and the UK. All are operational in a 686 km low Earth orbit. The China Mapping Telescope acquired only 4 m resolution Pan images. The Chinese Government will use it during emergencies to aid decision-making of the central government in conjunction with other DMC satellites. It will be also used to survey land resources, for geological and water resources research, to monitor floods and winter wheat, and for urban planning. TopSat's primary mission is to demonstrate that a microsatellite can provide fine spatial resolution imaging delivered directly from the satellite to ground terminals in near-real time. The scanner, which uses eight CCDs with TDI, forward motion compensation with up-to-20° pitch compensation, and ±30° across-track pointing angle, provides 2.5 m Pan images (500–700 nm) over 15 km swath and 5-m MB images (blue 400–500 nm, green 500–600 nm, red 600–700 nm) over a 10 km swath.

The Panchromatic Remote-sensing Instrument for Stereo Mapping (PRISM) of ALOS, launched 24 January 2006 (http://www.eorc.jaxa.jp/) is a Pan radiometer (520–770 nm) with a 2.5 m spatial resolution consisting of three independent optical systems for nadir, forward and backward (±23.8°) looking for along-track stereoscopy (Matsumoto et al. 2003). Each telescope consists of three mirrors and six or eight butted CCD lines. The nadir-looking radiometer can image a 70 km swath, while the forward and backward radiometers image a 35 km swath. Since each radiometer has ±1.5° across-track pointing to compensate for Earth's rotation, 35 km swath triplet images are obtained without mechanical scanning or yaw steering. An additional sensor, the Advanced Visible and Near Infrared Radiometer type 2 (AVNIR-2), provides 10 m resolution 70 km swath MB images (blue 420–500 nm, green 520–600 nm, red 610–690 nm, NIR 760–890 nm) using 7,000 pixels per CCD and with ±44° across-track pointing angle. Processing levels 1 and 2 are available in CEOS format. The mission objectives are topographic and land cover/use mapping, regional environment and disaster monitoring.

The Indian satellite Cartosat-2, launched 12 January 2007, has a single Pan sensor (450–850 nm, 0.8 m spatial resolution, 9.6 km swath) capable of providing imagery in spot, paint-brush and multi-view modes with 9.6–290 km strip lengths. The satellite can be steered up to ±45° along and across the track. Only Level-3 standard products (Precision and High Precision with 5–10 m and 3 m geometric accuracy, respectively) are provided in GeoTIFF, as well as thematic value-added products. However, Level-2A products can be provided over areas specified by users. The user-specified area may be covered by several multi-date image strips, not mosaicked but with sufficient side-lap. The data will be used for detailed mapping and other cartographic applications at cadastral level, urban and rural infrastructure development and management.

The satellite *WorldView-1*, the most agile commercial satellite, was launched on 18 September 2007, and belongs to the next generation commercial imaging satellite of DigitalGlobe Inc. The sensor is intended to provide imaging services to the National Geospatial-Intelligence Agency (NGA) as well as to the commercial customer base of DigitalGlobe. The camera has a single 0.45-m resolution Pan sensor (400–900 nm) using 35,000 pixels and TDI (six selectable levels from 8 to 64) for 17.6-km swath at nadir. The satellite can collect up to 750,000 square kilometres per day and with agility (up to ±45°) that allows rapid targeting and efficient in-track stereo collection. The same products as offered for QuickBird-2 will be provided to facilitate detailed 3D mapping and other cartographic applications at a cadastral level, as well as urban and rural infrastructure development and management.

The *Chinese–Brazilian satellite, CBERS-2B*, was launched on 21 November 2007. The instrument has one panchromatic High Resolution Camera (HRC) (500–800 nm) for along-track stereo (up to 26°) in addition to one moderate and one coarse resolution sensors. With four CCD arrays of 2592 elements and 2.1° FOV, the image width is 27 km with 2.7-m pixel spacing while the FOR can be up to 113 km. The operation mode is based on a 130 days revisit period. Among its applications are the generation of detailed national or state mosaics; thematic map production; product generation for local or county planning; DEM and 3D topographic mapping; urban and intelligence applications.

Digital frame cameras

Digital frame cameras were recently launched on 15 June 2006 for a three-year mission on the newest member of the Russia RESURS satellite fleet (http://www.sovinformsputnik.com/). This satellite was the first of an upgraded series of spacecraft with improved capabilities in imaging resolution and communications, e.g., extensive on-board memory, high-speed real-time

downlink system and GLObal Navigation Satellite System (GLONASS). Digital RESURS-DK system, apparently closely based on the Neman reconnaissance spacecraft used by the Russian military, will offer finer than 1 m spatial resolution, 28.3 km swath Pan images (580–800 nm) and 2–3-m V/NIR images (green 520–600 nm, red 600–700 nm, NIR 700–800 nm) of up to 105,000 km^2 daily (Table 8.2). The satellite has a non-sunsynchronous orbit at 360–690 km altitude and a body-pointing capability of ±30° across-track giving 448 km FOR. Highly specialized ground infrastructure includes not only receiving stations but also processing and fast-to-market hardware (Anshakov and Skirmunt 2000).

CONCLUDINGS REMARKS

Optical FSR/VFSR satellite sensory imagery raises operational issues related to coverage and cost, reliability and access to imagery, and national security. Each of these issues is discussed below.

The limited coverage and high cost do limit the value of VFSR imagery, especially for mapping on a regional scale. Because VFSR data (e.g., IKONOS, QuickBird, OrbView, Kompsat 2, EROS B, Cartosat-2, Worldview-1) are expensive (around US$20 per km^2 for the less processed level-1 image products, and up to five times for the most processed image levels 2B/3), cost efficiency for civilian applications can be difficult to achieve. As long as there is insufficient demand for VFSR space imagery from customers outside the defence/intelligence sector, price will likely remain high. Defence and government applications supported by different US programs (Clear View, NextView) are the largest market, and therefore tend to dominate the use of the Sensors (Gupta 1994), even though defence customers may though have access to 10 times better resolution data (reportedly, less than 0.1 m). More affordable for civilian applications are the FSR images (US$3–5 for level 1), which have a coarser resolution (2–5 m). The potential gains with coarser resolution are, however, not only in the reduced data cost, but also in the simplified processing (one SPOT-HRG image covers 36 IKONOS images) required to generate high-level image products. Unlike in the early years around 2000 when IKONOS was the only source of such data competition from multiple satellite data vendors should hopefully ensure that imagery prices are kept in check or even reduced while product quality should improve.

Reliability and access to imagery have been always important issues for users. One of the key concepts of the success of the Landsat and SPOT programs, which were supported by national governments, is continuity. This continuity is important not only to facilitate long-term monitoring, but also for maintaining consistency in operational applications, including the procedures, methods, hardware, software and generated products involved. The satellite business is risky and the FSR satellite business appears to be even riskier: four of the seven US VFSR/FSR commercial satellites so far have failed. It has become increasingly clear that there has been insufficient demand for VFSR imagery from customers outside the defence/intelligence sector to sustain a viable or profitable commercial industry based on this imagery (Petrie 2004). If it takes governments to keep major remote sensing programs afloat, what assurances are there that private corporations can remain viable over a long term in order that EO service companies develop a business plan centred on this new technology? So far, success has not been achieved, with limited demand for the very expensive image products and large financial losses for investors and operators, and staff cutbacks (Petrie 2004, 2005).

The issue of information access and dissemination within and between organizations should also be addressed. Internet technology facilitates simultaneous distribution of satellite-derived information to all levels within an organization or group of organizations. However, image licensing, which varies by vendor, can limit such sharing. Therefore before placing an order, purchasers should consider long-term use of data and degree of sharing needed. Ideally, organizations should have in place standard procedures to facilitate this decision making. Another issue relates to the timeliness of data. The time from image acquisition until the end product has been disseminated to users has generally been much too long to have practical use in time-critical situations, such as disaster response and humanitarian relief (Bjorgo 1999); further discussion of the use of remote sensing in disaster applications is given by Teeuw et al. (in this volume). It is important to realize that the technology to achieve rapid dissemination exists and is sometimes in use, but the institutional aspects need to be streamlined. Will we have near real time access to data? The experiment with the new launched micro-satellite Topsat could answer this question in the near future. In addition, the top-down information structure of many international organizations may also delay the dissemination of important satellite-derived information but this issue has to be resolved within each organization. Even though VFSR satellites in the defence/intelligence sector are 30 years in advance of civilian instruments (Li 2000), issues associated with accessibility and dissemination of these data raise many concerns, mainly related to national security (Gupta 1994).

National security can be a sensitive issue, generating seemingly irrational policies mainly related to the specific circumstances surrounding the application of the VFSR imagery. Policy decisions have been made to restrict some aspects of their commercialization (Gupta 1994), such as the 'shutter control'. The thorny issue of 'shutter control', tied to the licensing of the US commercial satellites, gives the government authority to determine what may and may not be imaged by US commercial-imaging satellites, but there are also examples with foreign companies and countries. During the first Gulf War the US military requested – and was granted – all available SPOT-HRV imagery of Kuwait and Iraq, the finest spatial resolution civilian available, with a total blackout for anyone imagery then else wanting these data. Yet, 'shutter control' provisions do not apply to foreign commercial-imaging satellites! What will happen now with the increasing number of internationally owned FSR/VFSR sensor systems? Further discussion on data policy issues is given in Harris (in this volume).

While the US administration has put considerable pressure on Germans to regulate their activities in an 'acceptable' way – and Germany has effectively implemented new security laws governing the way its data suppliers can do export business (Asian Surveying and Mapping 2006) – could a sovereign country control or 'turn-off' all imagery collected over or outside their territory, covering sensitive areas and situations where their interests are at sake? Several countries have launched satellites, thus making it harder for any one government to impose restrictions (Spyworld Actu 2006). While a 1986 United Nations resolution states broadly that data-gathering activities such as satellite photography 'shall not be conducted in a manner detrimental to the legitimate rights and interests of the sensed State', the times are changing, and with multiple eyes in the sky creating a transparent world, individual states may need to accept this transparency and to adapt to the advances in technology.

The outlook for the future is quite exciting: some 36 VSFR/FSR civilian satellites, with less than 3 m resolution, which include 26 optical sensors, are either in orbit, or planned for this decade: CBERS-3, GeoEye-1, OrbView-5, Worldview-2, Pleiades, Kompsat-3/4, Beijing-2 and higher, IKONOS Block-2, EKOSAT in the optical spectrum and RADARSAT-2, COSMO, Terra-SAR, RISAT and others in the microwave spectrum, as well as spy satellites. Such a large number of sensors will offer many choices, but it is not clear that this is an efficient way of developing Earth observation resources. Much money is being wasted because the need for land information is not even a consideration for many of these satellites. The main reason is often technical training or the development of national technical capability. Many of these satellite imaging systems are being built without any form of data distribution mechanism, and without any real input from potential users. Formosat-2 and Kompsat-2 provide examples of sensors that do appear to have a potential civilian role, as their data are distributed world-wide through the French SPOT-Image Company. Nevertheless, it seems in many cases it is more important to have the satellite in orbit with the national flag and/or to satisfy national security requirements, rather than to try to improve the availability of data.

REFERENCES

Altmaier, A. and C. Kany, 2002. Digital surface model generation from Corona satellite images. *ISPRS Journal of Photogrammetry and Remote Sensing*, 56(4): 221–235.

Anshakov, G. P. and V. K., Skirmunt, 2000. The Russian project of Resurs-DK 1 space complex development. Status, prospects, new opportunities for the consumers of space snapshots. *Acta Astronautica*, 47(2–9): 347–353.

Aplin, P., Atkinson, P. M. and P. J. Curran, 1997. Fine spatial resolution satellite sensors for the next decade. *International Journal of Remote Sensing*, 18(18): 3873–3881.

Asian Surveying and Mapping, 2006. Satellite data exports controlled. http://www.asmmag.com/ASM/content/2006/ASM_036/main_news_9.html. Last date accessed 28 December 2007.

Bakker, W., 2000. Satellite and Sensor Systems for Environmental Monitoring. In Meyers (ed.) *Encyclopedia of Analytical Chemistry: Applications, Theory and Instrumentation*. J. Wiley & Sons, Chichester, UK. Volume 10: 8693–8746.

Baudoin, A., 1999. The current and future SPOT program, In *Proceedings of the ISPRS Joint Workshop 'Sensors and Mapping from Space 1999'*, Sept. 27–30, Hannover, Germany, CD-ROM.

Bjorgo, E., 1999. Very high resolution satellites: a new source of information in humanitarian relief operations. *Bulletin of the American Society of Information Science and Technology*, 26(1) http://www.asis.org/Bulletin/Oct-99/bjorgo.html Last accessed 31 May 2007.

Bonneval, H., 1972. *Photogrammétrie générale, Tome 1: Enregistrement photographique des gerbes perspectives*. Editions Eyrolles, Paris, France. 232 p.

Centre National d'Études Spatiales (CNES), 1980. Le mouvement du véhicule spatial en orbite. In: *Cours de Technologie Spatiale*. CNES, Toulouse, France. 1031 p.

Cheng, P., and Th. Toutin, 1998. Unlocking the potential for IRS-1C data, *Earth Observation Magazine*, 6(3): 24–26.

Escobal, P. R., 1965. *Methods of Orbit Determination*. Krieger Publishing Company, Malabar, Florida, USA. 479 p.

GlobalSecurity.org, 2007. CORONA Summary. http://www.globalsecurity.org/space/systems/corona.htm. Last accessed 12/28/2007.

Grodecki, J. and G. Dial, 2003. Block adjustment of high-resolution satellite images described by rational functions. *Photogrammetric Engineering and Remote Sensing*, 69(1): 59–68.

Gupta, V., 1994. New satellite image for sale: The opportunities and risk ahead. http://www.llnl.gov/csts/publications/gupta/contents.html. Last accessed: 31 May 2007.

Harris, R., in this volume. Remote Sensing Policy. Chapter 2.

Kerekes, J. P., in this volume. Optical Sensor Technology. Chapter 7.

Kirillov, V., 2001. Russia on the market of high resolution images, *Moscow Defense Brief*, 2(6): 7p. http://mdb.cast.ru/mdb/6-2001/mas/rmhrsi/. Last accessed: 31 May 2007.

Lavrov, V., 1996. Space survey photocameras for cartographic applications. *International Archives of Photogrammetry and Remote Sensing*, Vienna, Austria, July, 31(B1): 105–109. http://www.sovinformsputnik.com/isprs96.pdf. Last accessed 31 May 2007.

Lavrov, V., 2000. Mapping with the use of Russian space high resolution images. *International Archives of Photogrammetry and Remote Sensing*, Amsterdam, Netherlands, 33. http://www.sovinformsputnik.com/lavrov2_eng.pdf. Last accessed 31 May 2007.

Li, Z., 2000. High-resolution satellite images: past, present and future. *Journal of Geospatial Engineering*, 2(2): 21–26. http://www.lsgi.polyu.edu.hk/sTAFF/zl.li/vol_2_2/03_li_zl_1.pdf. Last accessed: 31 May 2007.

Lukashevich, E. L., 1994. The space system 'Resurs-F' for the photographic survey of the Earth. *Space Bulletin*, I(4): 2–4.

Matsumoto, A., T. Hamazaki, Y. Osawa and D. Ichitsubo, 2003. Development Status of ALOS's Sensors. *International Archives of Photogrammetry, Remote Sensing and Spatial Information Science*, 34 (B7).

McDonald, R., 1995. CORONA: Success for space reconnaissance, a look into the Cold War, and a revolution for intelligence. *Photogrammetric Engineering and Remote Sensing*, 61(6): 689–720.

McDonald, R. (ed.), 1997. *CORONA Between the Sun and the Earth, The First NRO Reconnaissance Eye in Space*, American Society for Photogrammetry and Remote Sensing, Bethesda, Maryland, 440 p.

Morain, S. and A. M. Budge (eds.), 1998. *Earth Observing Platforms and Sensors*. John Wiley & Sons Inc., New York, CD-ROM.

Petrie, G., 2004. High-resolution imaging from space: a worldwide survey, part I North America and Part II Asia, *GEO Informatics*, 7(1&2): 22–27.

Petrie, G., 2005. Plenty of action on the high-res space imagery front. *GEO Informatics*, http://www.geoinformatics.com/asp/default.asp?t=article&newsid=1960. Last accessed: 31 May 2007.

Schade, O.H., 1948. Part IV: Correlation and evaluation of electro-optical characteristics of imaging systems. *RCA Review*, 9(4): 653–686.

Slama, C. C., C. Theurer and S. W. Henriksen (eds.), 1980. *Manual of Photogrammetry*, Fourth Edition, American Society of Photogrammetry, Falls Church, Virginia.

Smith F. D., 1997. The Design and Engineering of Corona's Optics. In R. A. McDonald (ed.), *Corona: Between the Sun and the Earth, The First NRO Reconnaissance Eye in Space*. American Society for Photogrammetry and Remote Sensing, Bethesda, Maryland, pp. 111–120.

Spyworld Actu, 2006. Le nombre de pays disposant d'images spatiales à un mètre de résolution se multiplie, surtout en Asie. http://www.spyworld-actu.com/spip.php?article2528. Last accessed 28 December 2007.

Teeuw, R., P. Aplin, N. McWilliam, T. Wicks, K. Matthieu and G. Ernst, in this volume. Hazard Assessment and Disaster Management using Remote Sensing. Chapter 32.

Toutin, Th., 2001. Elevation modelling from satellite visible and infrared (VIR) data: a review. *International Journal of Remote Sensing*, 22(6): 1097–1125.

Toutin, Th., 2004a. Photogrammétrie satellitale pour les capteurs de haute résolution : état de l'art, *Revue Française de Photogrammétrie et de Télédétection*, 175(3): 57–68.

Toutin, Th., 2004b. Comparison of stereo-extracted DTM from different high-resolution sensors: SPOT-5, EROS-A, IKONOS-II, and QuickBird. *IEEE Transactions on Geoscience and Remote Sensing*, 42(10): 2121–2129.

Moderate Spatial Resolution Optical Sensors

Samuel N. Goward, Terry Arvidson,
Darrel L. Williams, Richard Irish, and
James R. Irons

Keywords: moderate-resolution remote sensing, landsat, experimental, operational.

INTRODUCTION

Moderate spatial resolution, satellite-based, civilian land imaging systems originated from the Cold War between the United States (US) and the Union of Soviet Socialist Republics (USSR). In 1957, after Sputnik initiated the space race between the USSR and the US, the US defense intelligence community (Intel) placed into orbit the first land observatory, CORONA; this also addressed the intelligence vacuum created by the shooting down of Gary Powers' Utility-2 (U2) airplane over the USSR in 1960. In parallel, considerable efforts were underway in the US to develop civilian 'peaceful uses of outer space.' This included the convening of a National Research Council panel in 1967, focused on *Useful Applications of Earth-Oriented Satellites* (National Research Council 1969). This panel noted that, potentially, one of the most valuable applications would be the observation of the Earth's land areas. Thus began a conflict between the US Intel community and the civilian applied-sciences community concerning satellite land observations. One outcome was the placing of legal limits on civilian sensor resolution, which resulted in the current definition of moderate spatial resolution.

Today, the common characteristic of moderate-resolution land observatories is that they always are placed in near-polar, sun-synchronous, low Earth orbit (500–900 km). In fact, most of the finer spatial resolution land observation systems are also placed in similar orbits (Toutin, in this volume). Moderate-resolution systems vary considerably in pixel size, which is generally further related to sensor swath width and therefore observation repeat cycle (Figures 9.1 and 9.2). For systems with spatial resolutions finer than about 30 m, the nadir temporal repeat cycle is so low that agile, pointable platforms are typically employed to permit off-nadir imaging to increase temporal repeat for specific targets of interest. Off-nadir pointing sacrifices systematic nadir monitoring. Coarser, but still moderate spatial resolution (>30 m) satellite observatories provide local repeat cycles of 16 days or less, but not off-nadir pointing. These nadir-only observatories provide the opportunity to systematically monitor all the Earth's land areas.

Brief history of moderate-resolution land imaging

The first land observing satellite was the US Intel CORONA observatory placed into operation in 1960 (McDonald 1995, 2002, Toutin, in this volume). CORONA was a film-based camera system acquiring panchromatic 8 m-resolution

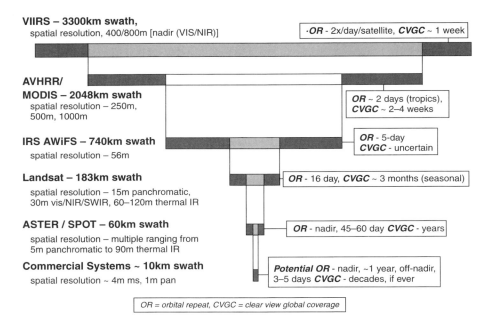

Figure 9.1   Comparison of image swaths for selected fine-, moderate-, and coarse-resolution sensors. This figure illustrates the continuum in which pixel resolution is in inverse proportion to the ability to provide global coverage. The operational strategy of 'point-and-shoot' is highly related to fine-resolution, while global surveying is facilitated by the coarser resolutions. (Figure created by Dr. Darrel Williams).

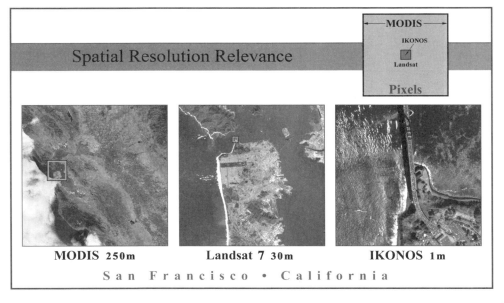

Figure 9.2   Comparative visual examples of coarse-, moderate-, and fine-resolution images of the central California, San Francisco region of the US. Note that the area of the region imaged is the squared product of the sample and line dimensions. (See the color plate section of this volume for a color version of this figure).
Source: Composition courtesy of Laura Rocchio, Landsat Project Science Office, NASA/GSFC.

and ultimately 2 m resolution imagery. Some of the earliest discussions concerning civilian space-acquired land observations suggested interest in using such film-based systems (Cloud 2001, Lowman 1998, Mack 1990). The Intel community strongly objected. They insisted that the civilian observatories have no better than an 80 m spatial resolution. In 1972, the first US civilian land observatory, the Earth Resources Technology Satellite (ERTS-1, later Landsat-1), was launched. ERTS-1 included two electronic sensors, the Return Beam Vidicon (RBV) and the Multispectral Scanner System (MSS) (USGS 2007) that operated at a nominal 80 m spatial resolution (Mika 1997).

Since the early 1970s, moderate-resolution land remote sensing has gone through three stages:

1 US-only phase (1970s),
2 commercial competitive phase (1980s), and
3 multiple observatory phase (1990s) (Stoney 2006).

We may be now entering a post-2000, fourth phase of government operations and international collaboration.

US-only phase
During the 1970s, the US operated the only civilian moderate-resolution land observatories – Landsats 1–3.

Commercial, competitive phase
In 1982 Landsat 4 was launched. Just a few years later, the first real international competitors – the French Système Pour l'Observation de la Terre (SPOT) system (launched in 1986) and the Indian Remote Sensing (IRS) system (launched in 1988) – began to challenge US leadership in this area (Stoney 2004, Green 2006). Further, there was considerable effort to move these government-funded satellite land observatories to a private or commercial basis (US Congress 1984). In the US this ultimately led to the founding of the Earth Observation Satellite (EOSAT) Corporation in 1985, to serve as a competitor to the French SPOT Image Corporation and help move the US assets into the commercial realm. In 1992, through the 1992 Land Remote Sensing Policy Act, EOSAT's role as the US agent for Landsat observations was repealed and the mission returned to the US Government (US Congress 1992).

Multiple observatory phase
From the 1990s to the present, more and more nations developed the interest and capability to launch and operate land observation satellites. As of 2006, 31 moderate spatial resolution satellites were in orbit and another 27 were planned, involving 23 nations of the world (Stoney 2006).

THE LANDSAT OBSERVATORY

Landsat, the keystone for moderate-resolution land remote sensing, continues to set the standards against which other moderate-resolution systems are judged. Landsat also remains, to this day, unique among moderate-resolution observatories in that it is the only observatory that is operated to systematically survey the Earth's land areas and is associated with a government-sanctioned archival system dedicated to preserving the historical record of these observations (Arvidson et al. 2006).

Technical evolution of the Landsat observatory

The Landsat observatory has been considered 'experimental' for over 36 years; during that period of time a considerable variety of sensors have been deployed.

Landsats 1–3: RBV and MSS
On ERTS-1, the Radio Corporation of America (RCA)-built RBV sensor – essentially a video equivalent to an 80 m spatial resolution film camera – was considered the primary observation instrument. The RBV system was designed to meet cartographic requirements for precision geometry. The secondary, more experimental sensor was the Hughes Santa Barbara MSS (Mika 1997) (Table 9.1). The MSS system was developed to test a new electro-optical scanner system concept and related multispectral analysis methods evolving from University of Michigan Willow Run Laboratories, Purdue University Laboratory for Applications of Remote Sensing, University of California Berkeley, and others (Freden and Gorden 1983). Some of the technical details included:

- Digital image transmission, compared to analog transmission of RBV images.
- Although MSS data were all digitized to 6 bits, the photomultiplier tubes used for bands 4–6[1] were highly nonlinear with radiance. The ground processing system at National Aeronautics and Space Administration's Goddard Space Flight Center (NASA/GSFC) 'linearized' the MSS band 4–6 signals through a logarithmic decompression to 7 bits.

The RBV system on ERTS-1 failed shortly after launch. By the launch of ERTS-2 (Landsat 2), MSS was regarded as the primary sensor; RBV images were infrequently requested. On Landsat 3, although the RBV sensor was redesigned as a 40 m panchromatic sensor, it was also rarely used.

Table 9.1   ERTS (Landsat 1) payload specifications (after Mika 1997)

Instrument	Spatial	Spectral	Radiometric

Return Beam Vidicon (RBV)
  3 co-aligned cameras, one
  for each band
	80 m IFOV
	185 km by 185 km framing camera
		B1 0.48 − 0.58 μm
		B2 0.58 − 0.68 μm
		B3 0.70 − 0.83 μm
			Analog video transmitted

Multispectral Scanner (MSS)	79 m IFOV 185 km swath scanning sensor (continuous strip image)	B4 0.5–0.6 μm B5 0.6–0.7 μm B6 0.7–0.8 μm B7 0.8–1.1 μm	6 bits per pixel, linear coding; logarithmic coding also available on bands 4, 5, 6

Thematic Mapper on Landsats 4–5

NASA proposed development of a new sensor for Landsats 4 and 5, called the Thematic Mapper (TM). The TM sensor was to be a six-spectral band instrument, with five bands in the solar reflective wavelengths and a sixth band in the thermal infrared region. Late in sensor development, a seventh band was added in the 2 μm wavelength region to mollify geological interests (Engel and Weinstein 1983). Plans called for Landsats 4 and 5 to be Space Shuttle retrievable – cancellation of the Vandenberg shuttle launch pad ended this possibility – and this, in combination with the optical-mechanical characteristics of the TM, resulted in Landsats 4 and 5 being placed in a lower 705 km orbit versus 909 km for Landsats 1–3. This produced a 16-day orbital repeat cycle (versus 18 days), which in turn required a new Worldwide Reference System (WRS-2) to designate satellite ground tracks (paths) and catalog data acquired along the ground tracks (path/row designations).

Pressure from high-volume users such as the US Department of Agriculture (USDA) prompted NASA to fly on Landsats 4 and 5 a companion MSS replicate, similar to that flown on Landsats 1–3. Use of a different platform for Landsats 4–5 forced a redesign of the MSS, changing the ground instantaneous field of view (IFOV) from 79 m to 82 m.

Enhanced Thematic Mapper on Landsat 6

Hughes Santa Barbara developed a new Enhanced Thematic Mapper (ETM) sensor for EOSAT, including a 15 m panchromatic band and dual gain capability. The MSS sensor was dropped. Landsat 6 was lost at launch in 1993 and the ETM was never operated in orbit.

Enhanced Thematic Mapper Plus on Landsat 7

Because of the Landsat 6 loss, Landsat 7's launch was targeted for 1998, earlier than originally planned. Replication of the ETM was deemed to be the quickest approach to replacement. The thermal IR (TIR) band resolution was improved from 120 m to 60 m, thus earning the 'Plus' designation (ETM+). The Air Force, initially a partner

in the Landsat-7 mission, proposed an additional sensor, the High Resolution Multispectral Stereo Imager (HRMSI), comparable to the SPOT High Resolution Visible (HRV) sensor (Mika 1997). HRMSI requirements continually increased and NASA declined to provide additional support for the resulting ground system upgrades. This, coupled with significant cuts in the Department of Defense budget after the fall of the Berlin Wall, caused the Air Force to withdraw from further participation in the Landsat mission. The National Oceanic and Atmospheric Administration (NOAA) and United States Geological Survey (USGS) then joined NASA to continue mission development with the single sensor, ETM+. NOAA ultimately also withdrew when their additional Landsat budget request was declined by Congress.

Landsat global coverage

The visionaries associated with the development of ERTS – William T. Pecora (USGS), Dr William Nordberg (NASA), and Dr Archibald Park (USDA) – had a clear sense of the immense value of global, systematic observations collected from this observatory (Pecora 1966 – Logsdon et al. 1998). The original Landsat sensor/platform design – near-polar orbit, ±5.78° swath, 80 m pixel – presented the opportunity to systematically acquire land surface images every 18 days. How often this was actually carried out depended upon recorder capacity and functionality, and commitment of the various operators (through time) to acquire systematic global coverage.

Early days

During the first 18 months of Landsat 1 observations, there was a valiant effort to acquire as much cloud-free imagery of the Earth as possible. Daily global cloud forecast maps were driven from NOAA's Suitland Maryland facility to NASA/GSFC, where mission operators would select cloud-free areas of the globe to image and uplink the commands to the spacecraft. Unfortunately, one of the first items to wear out on Landsats 1–3 was the on-board wideband video tape recorder. Once the recorder failed, it was only possible to downlink image data in real-time within line-of-sight to a receiving station. Although the US had established a number of international cooperator (IC) agreements during this early phase of Landsat, there were few arrangements for scenes transmitted to the ICs to be returned to the US archive. So the intensity of global coverage returned to the US fell off quickly after the first year or so of each of the Landsat-1 to -3 missions (Goward et al. 2006).

Landsats 4–5 and the Tracking and Data Relay Satellite (TDRS)

To avoid recorder limitations, Landsats 4–5 were designed with no recorder but rather a large boom antenna to interact with the TDRS system. Regrettably, Landsat 4 was launched one year before the first TDRS (TDRS-East); then the TDRS-West satellite was lost during the failed launch of the Challenger Space Shuttle in 1986. Despite these TDRS problems during early EOSAT–NOAA operations (1984–1988), the TDRS-East satellite was used aggressively to acquire nearly full coverage for North America, Europe, and Africa. Unfortunately, by the time TDRS-West was launched in 1988, EOSAT was being charged for the use of TDRS and therefore the acquisitions were driven by user demand rather than the more desirable goal of acquiring as much imagery as possible.

USGS EROS

The USGS contribution to the Landsat global vision was the establishment of an Earth Resources Observation System (EROS) Data Center (EDC)[2] in South Dakota to receive, process, and archive Landsat data. This location was selected for both political and technical reasons, the latter being that maximum real-time coverage of the continental US could be direct-downlinked to this location (or equivalent locations a little north or south at this longitude) (Johnson 1998). However, it was not until 1997–1998, in preparation for the Landsat 7 launch, that an antenna capable of receiving Landsat data was installed at EDC. From 1972–1984, the bulk of the data was downlinked to NASA/GSFC, with additional downlinks to receiving stations in Goldstone, California and near Fairbanks, Alaska, and then relayed to the EROS archive. Subsequently, EOSAT used a receiving facility in Norman, Oklahoma between 1984 and 1998. Data processing systems were developed at EROS, as well as a data distribution infrastructure and data archive.

One of the more significant outcomes of the 1992 Land Remote Sensing Policy Act was the creation of a federally-supported national archive of Landsat observations and the decision that the Landsat observatory should continue carrying out systematic surveys of the Earth's land areas (US Congress 1992). As with the innovation of the Landsat mission itself, the creation of a federally-mandated National Satellite Land Remote Sensing Data Archive (NSLRSDA) is unique, with no other nation involved in land remote sensing making a similar commitment. This decision to preserve and make readily accessible the growing collection of satellite observations ensured continued access to this history of the Earth's land areas (Goward et al. 2006). The NSLRSDA was implemented at EROS and its purview has been expanded to include other land observation datasets.

Landsat 7 Long-Term Acquisition Plan (LTAP)

The second important mandate from the 1992 law was a commitment to continue systematic monitoring of the Earth land areas, *ensuring that land remote sensing data of high priority locations will be acquired by the Landsat 7 system as required to meet the needs of the United States Global Change Research Program, as established in the Global Change Research Act of 1990, and to meet the needs of national security users* (US Congress 1992)[3]. To accommodate this need, the Landsat 7 platform included a large-capacity, solid-state recorder that permitted large volumes of imagery to be stored on the spacecraft until they could be downloaded at various US ground receiving stations. The Landsat *ad hoc* science team (1992–1995) determined that Landsat 7 should be tasked to systematically collect an annual, seasonally refreshed, essentially cloud-free set of imagery globally for inclusion in the NSLRSDA. This led the Landsat Project Science Office (LPSO) in 1995, to begin development of the LTAP (Arvidson et al. 2006). The LTAP restored the systematic global coverage dreamed of by the ERTS visionaries in the 1960s.

ETM+ scan line corrector failure

In late May 2003, the scan line corrector mechanism within the ETM+ failed, causing wedge-shaped gaps on the sides of each image. Compositing two or more clear images can fill these gaps (Storey et al. 2005). The LTAP acquisition strategy was therefore modified to enhance the probability of acquiring these pairs and the daily acquisition limit was increased from 250 to 300 scenes. A by-product of this is a slight decrease in the rate of global coverage acquisitions.

CURRENT MODERATE-RESOLUTION MISSIONS

As of late 2008, moderate-resolution land remote sensing is in a state of flux. In the US, most land missions are well past their design lifetimes. Development of the future Landsat Data Continuity Mission (LDCM) is just now being realized. Activities in this area by various nations around the world are equally in flux as we move away from the commercialization era toward an international, operational land-imaging era. Today there are many 'types' of moderate-resolution land imaging systems in orbit, producing quite a mix of imagery (Figure 9.3, Tables 9.2–9.4).

Status of US missions

The current US-operated moderate-resolution sensors include two Landsat sensors, the Earth Observing System (EOS) Terra satellite's Advanced Spaceborne Thermal Emission and Reflection Radiometer (ASTER), and the experimental Earth Observer (EO) -1 Advanced Land Imager (ALI) sensor. Details on these sensors are given in Table 9.2.

Landsat 5

Landsat 5 was launched in March 1984 by NASA and was then envisioned as the last Landsat satellite developed under direct federal government management. Its design life was 3 years. As of this writing, almost 25 years after its launch, the spacecraft and instrument are still miraculously functioning. Landsat 5 is sometimes referred to as the 'silent sentinel' because of its longevity and ceaseless watch over the Earth's land areas.

As of August 2008, EROS has received and archived over 760,000 Landsat-5 scenes with additional accumulations totaling ~5,000 scenes per month. This figure does not include the archive count for the international cooperator stations, which number 12 as of August 2008. Although Landsat-5 TM performance has degraded somewhat over time, cross-calibration with the Landsat-7 ETM+ instrument has been utilized to maintain the quality of Landsat 5 imagery (Teillet et al. 2001, Markham et al. 2004). In addition, these new calibration techniques are being retroactively

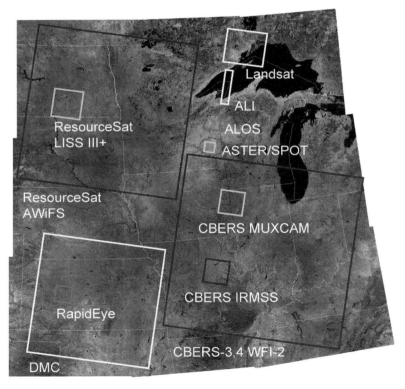

Figure 9.3   Comparative image footprint sizes for a variety of the moderate spatial resolution sensors discussed. Relative placement of the sensor footprints is for illustrative purposes only. The outer margins of the background image are approximately 2000 km across. (See the color plate section of this volume for a color version of this figure).
Source: Figure supplied by Jim Lacasse, USGS/EROS.

Table 9.2   Moderate-resolution sensors on current US satellites

Satellite	EO-1	LANDSAT 5	LANDSAT 7	Terra	Terra	Terra
Launch year	2000	1984	1999	1999	1999	1999
Sensor	ALI	TM	ETM+	ASTER-SWIR	ASTER-TIR	ASTER-VNIR[4]
Bands:[1]						
Blue	30 (2)	30	30			
Green	30	30	30			15
Red	30	30	30			15
NIR	30 (2)	30	30			15
SWIR	30 (3)	30 (2)	30 (2)	30 (6)		
MWIR						
TIR		120	60		90 (5)	
PAN	10		15			
Swath (km)	37/183[3]	185	183	60	60	60
Revisit (days)	16 nadir/7 off-nadir	16	16			5
Pointing?[2]	17.4	No	No	8.55	8.55	24

[1] Band resolutions are provided in meters, number in parentheses indicates the number of bands in that spectral range, if greater than 1.
[2] Pointing angle given in degrees, cross-track.
[3] 37 km is actual swath width; if focal plane was fully populated, swath width would be 183 km after removing band offsets.
[4] ASTER-VNIR has two telescopes, NIR backward pointing and MS nadir pointing.

Table 9.3   Moderate-resolution sensors on non-US satellites (as of late 2008)

Satellite	ALOS	CBERS-2	CBERS-2	CBERS-2	DMC (x5)	SPOT-2	SPOT-4	SPOT-5
Country	Japan	China/Brz	China/Brz	China/Brz	Multiple[4]	France	France	France
Launch year	2006	2003	2003	2003	2002–05	1990	1998	2002
Sensor	AVNIR-2	CCD[3]	IRMSS[3]	WFI[3]	SLIM6	HRV (2)	HRVIR (2)	HRG (2)
Bands:[1]								
Blue	10	20						
Green	10	20			26/32	20	20	10
Red	10	20		260	26/32	20	20	10
NIR	10	20		260	26/32	20	20	10
SWIR			80 (2)				20	20
MWIR								
TIR			160					
PAN		20	80		12/4	10	10	5
Swath (km)	70	113	120	890	600	60	60	60
						60	60	60
Revisit (days)	46 nadir/ 2 off-nadir	26 nadir/ 3 off-nadir	26	5	10 indiv/ daily with constellation	26 nadir/ 2-3 off-nadir	26 nadir/ 2-3 off-nadir	26 nadir/ 2-3 off-nadir
Pointing?[2]	44	32	No	No	No	31.06	31.06	31.06

[1] Band resolutions are provided in meters, number in parentheses indicates the number of bands in that spectral range, if greater than 1.
[2] Pointing angle given in degrees, cross-track.
[3] Only CCD is active, limited to 10 min per orbit. IRMSS and WFI are turned off.
[4] Algeria, Nigeria, Turkey (failed on orbit in 2006), UK, China.

applied to the existing archive of raw Landsat 5 TM imagery, such that improved products from these older image data have also become possible.

Landsat 7

The US government-owned Landsat 7 was successfully launched in April 1999. As of August 2008, EROS has received and archived over 840,000 Landsat 7 scenes with additional accumulations totaling ~9,000 scenes per month. It should be noted that the 840,000 ETM+ scenes in the NSLRSDA archive acquired during the first ten years of the Landsat 7 mission already surpasses the 760,000 TM scenes in the archive acquired during the 25 years of the Landsat 5 mission.

Table 9.4   Moderate-resolution sensors on non-US satellites (as of late 2008) (continued)

Satellite	IRS-1C	IRS-1C	IRS-1C	IRS-1D	IRS-1D	IRS-1D	IRS-P6	IRS-P6	IRS-P6
Country	India	India	India	India	India	India	India	India	India
Launch year	1995	1995	1995	1997	1997	1997	2003	2003	2003
Sensor	LISS-III	Pan	WiFS	LISS-III	Pan	WiFS	AWiFS	LISS-III	LISS-IV
Bands:[1]									
Blue									
Green	23.5			23.5			56	23.5	5.8
Red	23.5		188	23.5		188	56	23.5	5.8
NIR	23.5		188	23.5		188	56	23.5	5.8
SWIR	70.5			70.5			56	23.5	
MWIR									
TIR									
PAN		5.8			5.8				
Swath (km)	142/148 for SWIR	70	804	142/148 for SWIR	70	804	737	141	23 Mono/ 70 Pan
Revisit (days)	24	24 nadir/ 5 off-nadir	5	24	24 nadir/ 5 off-nadir	5	5	24	24 nadir/ 5 off-nadir
Pointing?[2]	No	26	No	No	26	No	25	No	26
Comments									

[1] Band resolutions are provided in meters, number in parentheses indicates the number of bands in that spectral range, if greater than 1.
[2] Pointing angle given in degrees, cross-track.

That big difference is directly attributable to a revised acquisition philosophy, the presence and reliability of the on-board solid-state recorder, and the use of the LTAP to drive the day-to-day acquisition strategy.

An additional innovation of the Landsat 7 program is the Image Assessment System or IAS, an element of the ground segment dedicated to the calibration, trending, and analysis of ETM+ data. Through routine on-board calibration acquisitions, imaging of calibration 'super sites' around the globe, daily inspection of a subset of acquired data, maintenance of a trending database, and analytical support from calibration scientists at EROS and NASA/GSFC, the ETM+ is arguably the best calibrated satellite land sensor ever placed in orbit. There are a few spacecraft problems with Landsat 7 (Goward et al. 2006). The Landsat 7 satellite has enough fuel reserves to achieve overlap of operations with the LDCM, thereby enabling cross-calibration to be performed with the new sensor.

EO-1

The EO-1 satellite was conceived as a one-year technology risk-reduction and demonstration mission in support of the LDCM. It carries the ALI, a hyperspectral imager (Hyperion), and a coarse-resolution hyperspectral atmospheric corrector – the Linear etalon imaging spectrometer array Atmospheric Corrector (LAC) (Beck 2003). At the end of the first year, the science and user communities were so pleased with the results of

the validation that they petitioned for continuance of the program and in some cases cooperated on funding continued operations. EO-1 continues to supply moderate-resolution multispectral data as well as hyperspectral data to this date. All this has been achieved despite the fact that the ALI focal plane is only partially populated with detectors, resulting in a swath width for ALI of 37 km versus the Landsat 7 standard width of 183 km.

Terra

The ASTER instrument package, carried on NASA's Terra satellite (launched in December 1999), consists of three separate sensors built by the Japanese Ministry of Economy, Trade, and Industry. One of the ASTER sensors collects data for five TIR bands and has contributed much to the science and applications that rely on thermal data. Few if any of the existing or planned missions carry sensors with TIR bands. Unfortunately, the duty cycle of the instrument is only 9.3%. That cycle coupled with the 60 km swath width limits the amount of global coverage that can be acquired. Terra has exceeded its design life, and it has enough fuel to last until between 2009 and 2015.

Status of international missions

Most of the current international moderate-resolution land remote sensing observatories include sensors that are in the finer spatial resolution range, around 10 m. None have the goal of global surveying but most have on-board recorders that enable acquisitions from anywhere on the globe. Most have a pointing capability that facilitates rapid revisits in the point-and-shoot mode of operation. Most have planned life spans of 5 years. Those beyond their mission life span are still active but could fail at any time – China/Brazil Earth Resources Satellite (CBERS)-2 is partially active. For some of these satellites, it is difficult to find information on the availability of the data, the current status, and system performance. Details on the sensors carried by the following international missions are found in Tables 9.3 and 9.4.

France – SPOT

There are currently three operational SPOT satellites in orbit – SPOT-2, -4, and -5 – launched in 1990, 1998, and 2002, respectively. SPOT-2 and -4 carry sensors with 20 m multispectral resolution and 10 m panchromatic resolution; SPOT-5 has improved resolutions of 10 m and 5 m, respectively. There is no attempt to systematically acquire a global archive; most acquisitions are based on user requests only. The platforms are agile and point-and-shoot is the nominal mode of operation. The data is marketed by SPOT Image, is sold under strict license conditions, and is relatively expensive. However, it has become a replacement dataset for many users who do not wish to work with ETM+ data gaps or gap-filled data products. Many IC stations have increased their SPOT downlinks to support projects previously based on Landsat data.

India – IRS

There are currently three operational IRS satellites in orbit – IRS-1C, -1D, and -P6 – launched in 1995, 1997, and 2003 respectively. The IRS-1C and -1D sensor configurations are identical: a Wide-Field Sensor (WiFS) with a coarse-resolution of 188 m, the multispectral Linear Imaging Self Scanner-3 (LISS-III), and a Panchromatic sensor (pan). The IRS-P6, also known as the ResourceSat-1, carries the Advanced Wide-Field Sensor (AWiFS) with a moderate resolution of 56 m, the LISS-III, and the next-generation LISS-IV which is operable in either a multispectral or panchromatic mode. The AWiFS is of great interest as a possible source of data in the event of a Landsat data gap prior to the launch of the LDCM. The wide swath of 737 km and the improved resolution of 56 m lend themselves to a systematic global surveying mission. However, it is unclear whether the on-board recorder capacity, the sensor duty cycle, and operational policies will allow this to actually happen. The USDA Foreign Agricultural Service, once the largest single user of Landsat data, is now using AWiFS data as a replacement for Landsat.

Japan – Advanced Land Observing Satellite (ALOS)

Launched by Japan in 2006, ALOS carries three instruments, of which one is moderate resolution; the other two are a fine-resolution sensor for stereomapping and a radar instrument. The Advanced Visible and Near Infrared Radiometer type 2 (AVNIR-2) provides 10 m in the visible/near-infrared (NIR) wavelengths and a swath of 70 km (37% of Landsat-7's swath). The off-nadir pointing capability of ±44° cross-track provides a revisit capability of two days; global surveying is not part of the mission. Limited ALOS data is downlinked to six stations outside of Japan. There is both an on-board calibration system for the AVNIR-2 and an on-board solid-state recorder. Improvements made to the AVNIR-2 over the original AVNIR sensor include improved resolution (10 m versus 16 m), a wider range of pointing angles (±44° versus ±40°), and an additional calibration lamp.

China and Brazil – CBERS-2

The CBERS-2, also known in China as the ZiYuan (ZY)-2, was launched by China and Brazil in 2003. CBERS-2 carries three instruments, two of which are moderate resolution and no longer operational; the third, a charge-coupled device (CCD) sensor, is an instrument with 260 m coarse-resolution that fulfills the global coverage requirement for the CBERS program. The CCD sensor is now limited to 10 min per orbit due to a battery anomaly. Only the CCD sensor data could be recorded on-board. As a result of the on-board problems and a delay in the launch of CBERS-3, a near-copy of the CBERS-2, called CBERS-2B, was launched in September 2007. The CCD and Wide Field Imager (WFI) sensors are the same as those on CBERS-2; there is a new pan instrument at fine-resolution. Unlike CBERS-2, both the CCD and WFI sensor data can be recorded on-board CBERS-2B. Science data is downlinked to China and Brazil. Brazil is working with several partners to establish CBERS receiving stations in at least four locations in Africa. CBERS products are provided to all users at no cost via Internet download. The average weekly distribution of scenes is >2000. CBERS data, central to the Brazil remote sensing program, is allowing Brazil to update their environmental database annually, rather than the previous 3–4 year basis with Landsat data. There was a thermal band on one of the inoperable CBERS-2 moderate resolution sensors; this sensor is not on the CBERS-2B, but a thermal capability is planned for CBERS-3/4. Like the IRS AWiFS, CBERS data has been studied by NASA and the USGS as a possible source of global data should there be a gap in Landsat-like data before the launch of the LDCM.

Various – Disaster Monitoring Constellation (DMC)

An innovative program to develop and deploy low-cost land observatories on small satellites (smallsats) was initiated at the University of Surrey in the United Kingdom (UK) around 1985. As of late 2008, this proof-of-concept constellation comprises five satellites (one of which failed in 2006):

- Alsat-1, launched by Algeria in 2002
- Nigeriasat-1, launched by Nigeria in 2003
- Bilsat-1, launched by Turkey in 2003 (battery failure)
- UK-DMC, launched by the UK in 2003
- Beijing-1, launched by China in 2005.

This is the first Earth-observation constellation of low-cost small satellites providing daily images for applications including global disaster monitoring. The sensors on these satellites are based on the Surrey Satellite Technology Ltd. Surrey Linear Imager Multispectral 6 Channel (SLIM6). An on-board recorder provides the capability to acquire data around the globe and record it for later downlink. The spectral bands are equivalent to ETM+ bands 2, 3, and 4. DMC data is marketed through DMC International Imaging on behalf of the Disaster Monitoring Constellation Consortium. Daily imaging capacity ranges from 10 to 80 scenes per satellite; revisit frequency is daily with five operating satellites. Up to five percent of this capacity is provided free by the member satellites for daily imaging of disaster areas through Reuters AlertNet and the International Charter 'Space and Major Disasters' (Teeuw et al., in this volume).

FUTURE OF MODERATE-RESOLUTION LAND IMAGING

The transition to the twenty-first Century, a half century after satellite Earth observations began, appears to be a critical turning point for moderate-resolution systematic, global land observations. Currently, the US is recognizing the need to transition from the 'experimental' exploration of new technologies to observe the Earth, to an 'operational' use of mature technologies to monitor the Earth's land cover dynamics. This comes at a time when we are increasingly aware that global change, particularly on the continents, is only satisfactorily documented with long-term land remote sensing systems (Interagency Working Group in Earth Observations 2005). Similar initiatives are beginning to take hold in Europe (Sentinel 2), India (AWiFS and LISS), China-Brazil (CBERS), and Japan (ALOS).

The increasing interest in moderate-resolution imagery coincident with the failed commercialization of these observing systems demonstrates the public role of these observatories in supporting societal decision-making. The primary goal in developing national land imaging programs is to support society's need to cope with increasing global change. In the US, the transition to an operational land imaging program is well underway (National Science and Technology Council 2006). Current international initiatives such as the Global Earth Observation System of Systems (GEOSS) and the Integrated Global Observations of Land (IGOL) are indicative of increased interest in international collaboration on moderate-resolution land remote sensing (Group on Earth Observations 2005, Integrated Global Observing Strategy Partnership 2004). However, it may well be that this operational transition will also stimulate the type of value-added marketplace originally sought under commercialization, such as we have previously experienced with systems such as the Global Positioning System (GPS) or the US interstate highways (Green 2006).

US land remote sensing – a turning point?

As noted previously, the 1992 Public Law, returning the Landsat mission to the federal government, created a substantial impact on US moderate-resolution land remote sensing as well as international activities in this area. This law not only determined the recent evolution of the Landsat mission (Landsat 7 and LDCM through the EO-1 ALI), it established standards including the LTAP and the IAS that will set the stage for the future of land imaging as discussed in the recent Office of Science and Technology Policy (OSTP) report *A Plan for a National Land Imaging Program* (National Science and Technology Council 2007).

Landsat Data Continuity Mission (LDCM)

The LDCM concept has evolved from a data purchase approach in 2001 (NASA Goddard Space Flight Center 2001, Berger 2003), a berth on the National Polar-orbiting Operational Environmental Satellite System (NPOESS) in 2004 (OSTP 2004), to an independent 'free-flyer' satellite in 2005 (Irons and Masek 2006, OSTP 2005). This last concept is now being implemented. The major program elements–spacecraft, sensor, launch vehicle, mission operations element–are all under contract. In 2012, the LDCM (also known as Landsat 8) will be launched (Table 9.7). Unlike previous Landsats, the LDCM will have the ability to point off-nadir by one ground track. The LDCM project plans to use this capability only for emergency response acquisitions, anticipated to be no more than a handful of times each year.

Building upon the lessons learned in designing and operating the EO-1 ALI sensor, the LDCM OLI represents the first major technical advance in Landsat measurement technologies in nearly a quarter century (Table 9.5). Not only will the sensor employ a solid-state, linear array, 12-bit radiometry, but it also includes substantially enhanced spectral coverage – two new spectral bands (termed coastal and cirrus) as well as substantially narrower band passes, primarily to avoid atmospheric water vapor absorption, following the EOS Moderate Resolution Imaging Spectroradiometer (MODIS) lead. Note also that the panchromatic band is now restricted to the visible spectral region, avoiding reduced contrast suffered from the previous broad, visible-near infrared band pass used on Landsat 7. Overall these OLI advances over the ETM+ should substantially enhance land observations in the LDCM era (Irons and Masek 2006).

US National Land Imaging Program (NLIP)

When the OSTP released an update to the August 2004 LDCM memo – thus removing Landsat from NPOESS and directing NASA and USGS to procure a free-flyer LDCM system – it also directed NASA and the Department of Interior (DOI)/USGS, in association with other relevant US agencies, to convene an effort to *develop a long-term plan to achieve technical, financial, and managerial stability for operational land imaging in accord with the goals and objectives of the U.S. Integrated Earth Observation System* (OSTP 2005). As important as the procurement of the LDCM is, of far more lasting importance was the formation of the OSTP Future of Land Imaging (FLI) committee (National Science and Technology Council 2006). This interagency committee worked actively from early 2006 to propose an approach to implement a US operational NLIP. In July 2006, the FLI committee held a public workshop during which the committee head,

Dr Gene Whitney, indicated that the committee was leaning toward recommendation of a single agency lead for the program, with that agency being DOI. Since that time the FLI published a report to the Office of the President providing a series of recommendations concerning the implementation of the NLIP (National Science and Technology Council 2007). Today the future of US moderate-resolution land remote sensing primarily awaits the implementation of NLIP, of which a critical first step is to establish a stable and reliable funding basis.

International plans for moderate-resolution sensors

Judging by the future launch plans of countries around the world, the importance of moderate-resolution, Landsat-like data has been fully recognized and incorporated into mission requirements (Tables 9.6 and 9.7). Three (HJ-1, RapidEye, Sentinel 2) of the eight[4] future systems briefly described below are constellation-based small-satellite programs, using the multiplicity of sensors to achieve both a global surveying capability and increased revisit frequency. An additional program – CBERS – may also have a constellation in orbit if the CBERS-2B operational life overlaps that of CBERS-3 and -4.

In spite of the forward-looking implementation of the constellation-based systems truly required to achieve routine global surveillance, the ability to depart from global surveying and point off-nadir is still deemed important, as demonstrated by the inclusion of such capability in seven of the eight future systems. Of the non-US programs, only the Sentinel 2 program has declared global surveying as a mission goal; one would assume that its off-nadir capability will only be used in exceptional circumstances, as is the case with the LDCM.

Another important constant of the Landsat program has been the availability of TIR data. Only three of the eight future systems have plans for

Table 9.5   LDCM OLI spectral band and ground sample distance (GSD) specifications

No.	Band	Minimum Lower Band Edge (nm)	Maximum Upper Band Edge (nm)	Center Wavelength (nm)	Maximum GSD
1	Coastal	433	453	443	30 m
2	Blue	450	515	482	30 m
3	Green	525	600	562	30 m
4	Red	630	680	655	30 m
5	NIR	845	885	865	30 m
6	SWIR 1	1560	1660	1610	30 m
7	SWIR 2	2100	2300	2200	30 m
8	Panchromatic	500	680	590	15 m
9	Cirrus	1360	1390	1375	30 m

Table 9.6   Moderate-resolution sensors launched in 2007/2008

Satellite	CBERS-2B	CBERS-2B	HJ-1	HJ-1	RapidEye (5)	THEOS
Country	China/Brzl	China/Brzl	China	China	Germany	Thailand
Launch year	2007	2007	2008	2008	2008	2008
Sensor	CCD	WFI	CCD	Infrared camera	REIS	MS camera
Bands:[1]						
Blue	20		30		6.5	15
Green	20		30		6.5	15
Red	20	258	30		6.5 (1/1 edge)	15
NIR	20	258	30	150	6.5	15
SWIR				150		
MWIR				150[3]		
TIR				300		
PAN	20					
Swath (km)	113	890	720	720	78	90
Revisit (days)	26	26	4 with 1 satellite; 2 with 2-sat constell; 1 with 4-sat constell	4 with 1 satellite; 2 with 2-sat constell; 1 with 4-sat constell	5.5 nadir/ 1 off-nadir	26 nadir/ 1–5 off-nadir
Pointing?[2]	32	No	No	No	25	50

[1] Band resolutions are provided in meters.
[2] Pointing angle given in degrees, cross-track.
[3] The MWIR spectral bandwidth for this sensor is 3.50–3.90 μm.

Table 9.7   Moderate-resolution sensors on satellites to be launched in the future

Satellite	CBERS-3/4	CBERS-3/4	CBERS-3/4	CBERS-3/4	INGENIO	LDCM	LDCM	Sentinel 2A/2B
Country	China/Brz	China/Brz	China/Brz	China/Brz	Spain	US	US	ESA
Launch year	2008/10[5]	2008/10[5]	2008/10[5]	2008/10[5]	2010	2012	2012	2012/13
Sensor	IRMSS	MuxCam	PanMux	WFI	Single sensor	OLI	TIRS[3]	Single sensor
Bands:[1]								
Blue		20			10	30(2)		60/10
Green		20	10	73	10	30		10
Red		20	10	73	10	30		20(2)/10
NIR	40	20	10	73	10	30		60/20(2)/10
SWIR	40(2)			73		30(2)		60/20(2)
MWIR								
TIR	80						120(2)	
PAN			5		2.5	15		
Swath (km)	120	120	60	866	60	185	185	289
Revisit (days)	26	26nadir/3off	26nadir/5off	5		16	16	10; 5 with 2 satellites
Pointing?[2]	No	32	32	No	35	15[4]	15[4]	20[4]

[1] Band resolutions are provided in meters, number in parentheses indicates the number of bands in that spectral range, if greater than 1.
[2] Pointing angle given in degrees, cross-track.
[3] Sensor inclusion is yet to be confirmed.
[4] Nominal operating mode is nadir-pointing.
[5] CBERS-3 was successfully launched 20 October 2008.

bands in the TIR range: CBERS-3/4 at 80 m resolution, HJ-1 at 300 m resolution, and possibly LDCM at 120 m resolution. (For reference, Landsat-5 TM has 120 m TIR resolution, Terra ASTER has 90 m resolution, and Landsat-7 ETM+ has 60 m resolution.) Inclusion of a TIR capability on the LDCM has been authorized by Congress but, as of late 2008, the required funding has not. Progress in

inclusion of TIR is the result, in no small part, of advocacy by the western US states that are taking advantage of the current Landsat TIR to monitor irrigated agriculture (Allen et al. 2005). The TIR data is also an important tool in determining cloud cover in acquired images and plays a crucial role in the current Landsat processing systems.

China and Brazil – CBERS-3/4

CBERS-3, launched in October 2008, and CBERS-4, scheduled for 2010, carry the same four sensors. Some of the improvements in the sensors include:

- The WFI resolution is 73 m (versus 258 m); the swath width is 866 km (versus 890 km).
- The InfraRed MultiSpectral Scanner (IRMSS) has dropped the pan band. The resolution of the legacy IR bands was improved from 80 m SWIR and 160 m TIR to 40 m SWIR and 80 m TIR. A near-IR band with 40 m resolution was also added. The swath width remains the same at 120 km.
- The CCD is replaced with a MuxCam sensor that covers the same multispectral bands at the same resolution.
- A new sensor, the PanMux, picks up the pan band from the CCD, and green/red/near-IR bands, with resolutions of 5 m pan and 10 m multispectral. The swath width is 60 km – half that of the CCD and IRMSS.

In addition to these sensor improvements, all sensor data will be recorded on board.

China – HJ-1

China successfully launched the first two satellites of Huan jing (HJ)-1, also called Environment-1, in September 2008. The program's first phase comprises two optical satellites (HJ-1A and HJ-1B) and one radar satellite (HJ-1C). HJ-1A carries multi-spectrum CCD cameras (two sets) and a hyperspectral imager. HJ-1B carries the CCD cameras and an infrared scanner. Both satellites have on-board recorder capability. The National Committee for Disaster Reduction is the source of requirements for the Environment-1 constellation, also known as the Small Satellite Constellation for Environment and Disaster Monitoring and Forecasting (SSCEDMF) program.

The constellation is designed to provide decision-making inputs, monitoring, and forecasting for disaster and environmental needs. The constellation will be implemented in two phases: (1) the 2+1 constellation comprising two small optical satellites and one small synthetic aperture radar satellite, to be launched no earlier than March 2008; (2) the final 4+4 constellation achieved with the addition of two more optical and three more radar satellites, to be launched in the 2010 time-frame. The revisit frequencies at the equator for the 2+1 constellation are 48 hours for the optical sensors and 96 hours for the radar sensor. For the 4+4 constellation, the revisit frequencies are 48 hours for the hyperspectral sensor and 24 hours for the other three sensors. The Chinese Government

is looking for international cooperation to make the second phase a reality.

Germany – RapidEye

This is the only fully commercial constellation, developed by RapidEye AG of Germany as a privately owned multisatellite system for rapid coverage of primarily agricultural areas. The constellation of five minisatellites was launched August 2008 on one launch vehicle. Each satellite carries a single sensor with near-fine-resolution of 6.5 m and an on-board recorder. The satellites are evenly spaced in orbit around the Earth and can provide daily revisits through off-nadir pointing.

European Space Agency (ESA) – Sentinel 2

This program is ESA's answer to Landsat-like and SPOT-like data continuity, with enhancements recommended by existing ESA programs. Two satellites are scheduled for launch, in 2012 and 2013, and will deliver systematic global coverage (less Antarctica) every five days with a resolution of 10 m. On-board sensor calibration and an on-board recorder will also be included. The spectral band heritage is the Environmental Satellite (ENVISAT) Medium Resolution Imaging Spectrometer (MERIS), SPOT, LDCM, Landsat, ALI, and MODIS. Off-track pointing is possible but planned for use in emergency situations only.

Spain – Spanish Earth Observation Satellite (INGENIO)

To be launched in 2010, INGENIO is Spain's first foray into Earth observing satellites. The satellite will be built by Spain under the management of ESA and will include a multispectral sensor. One of the sponsoring organizations within Spain is the Centre for the Development of Industrial Technology, under the Ministry of Industry, Commerce and Tourism. The Ministry of Education and Science is also involved. This is Spain's contribution to ESA's Global Monitoring for Environment and Security (GMES) program, while its primary mission is to "carpet map" Spain. There will be a second satellite, PAZ, which will carry a SAR payload.

Thailand – Thailand Earth Observation System (THEOS)

Thailand launched THEOS in October 2008. It carries a moderate-resolution SPOT-like multispectral camera having a 90 km swath width and a fine-resolution (2 m) panchromatic telescope, as well as an on-board recorder. Product distribution is handled through Thailand's Geo-Informatics and Space Technology Development Agency (GISTDA). Acquisitions are concentrated on Thailand. Pointing capability allows access to any part of Thailand in less than two days. Global repeat coverage is not possible as the sensor duty cycle is limited to 10 min per orbit. GISTDA is soliciting participation by international ground stations for data downlink receipt.

INTERNATIONAL COLLABORATION IN OPERATIONAL MODERATE-RESOLUTION LAND IMAGING

As noted previously, there are a number of emerging national and international efforts directed toward operational land imaging and coordination of resources – US NLIP, US Group on Earth Observations (GEO), GEOSS, and IGOL. It is becoming clear that there is a growing international consensus that we need to do a better job of monitoring and managing our home planet. Among existing and proposed systems, the US LDCM, ESA Sentinel 2, Indian IRS, China/Brazil CBERS, and the Japanese ALOS appear to have the capability of serving as the initial building blocks for an international constellation of moderate-resolution observatories that, working together, would provide a substantially refined monitoring capability. Already, there are even more far reaching concepts that employ smallsats and/or microsats to increase temporal frequency further and perhaps explore innovative orbits, as well as combinations of observations from a variety of sensor types, including passive optical, active optical (laser), and active microwave (radar). To foster these activities an international dialogue is already underway.

GEOSS

GEOSS dates from July 2003 when an Earth Observation Summit was held in Washington, DC. The stated goal of the summit was to *promote the development of a comprehensive, coordinated, and sustained Earth observation system or systems among governments and the international community to understand and address global environmental and economic challenges.* The summit formed the *ad hoc* GEO, which was charged with establishing a *10-year implementation plan for a coordinated, comprehensive, and sustained Earth observation system or systems.* In April 2004, then US Environmental Protection Agency administrator Michael Leavitt, along with other senior Cabinet officials, met in Japan with more than 50 environmental ministers from various nations to sanction a framework document centered on a 10-year implementation plan for GEOSS. 61 countries ratified the plan in February 2005, and in late 2008, membership

totaled 77 countries and 56 organizations. GEOSS is designed to:

- improve coordination of remote sensing collection systems and observation strategies;
- link all platforms including *in situ* measurements, aircraft, and satellite networks;
- identify gaps in global capacity;
- facilitate exchange of land, water, and atmospheric data and information; and
- improve decision-makers' abilities to address pressing policy issues (Group on Earth Observations 2005).

The ultimate vision for GEOSS is to merge existing software with new hardware and enhanced applications to supply Earth observation data and information to the world quickly and at no cost. The main data linking mechanism to be used within GEOSS is termed Clearinghouse. The outline for the mechanism was circulated for comment at the end of 2006; a pilot demonstration and pilot phase were started. GEOSS developers unveiled three Geo Portals in June 2008 with Clearinghouse search capability into multiple, interoperable, observation systems. The GEOSS program will also implement GEONetcast, an open and worldwide data and information exchange using the same technology as DirectTV. Until then, the web portals will be assessed, refined, and fully established in 2009.

IEOS

In April 2005, the OSTP released the final strategic plan for the US Integrated Earth Observation System (IEOS – the US contribution to GEOSS) (Interagency Working Group in Earth Observations 2005). The IEOS goal is to achieve coordinated and sustained observations of the Earth systems, thus producing societal benefits that include:

- improvement of weather forecasting.
- reduction of loss of life and property from disasters.
- protection and monitoring of the world's ocean resources.
- understanding, assessment, prediction, mitigation, and adaptation to climate variability and change.
- support of sustainable agriculture and combating against land degradation.
- understanding of the effect of environmental factors on human health.
- development of the capacity to make ecological forecasts.
- protection and monitoring of water resources.

- monitoring and management of energy resources.
- implementation of sustainable land use and management.

Integrated Global Observations of Land (IGOL)

Whereas GEOSS is directed toward all elements – atmosphere, oceans, and lands – of Earth observations, a more focused activity on integrated land observations has been developed under the auspices of the United Nations Food and Agriculture Organization – IGOL 2007. Under this activity, the aspects of our planet unique to land are given closer attention, as the specific character of land remote sensing observations is addressed. Guidance from this activity will most likely play a critical role in defining and encouraging the development of the moderate-resolution observatory constellation, made up of high-quality calibration satellites (calsats) and redundant smallsats providing substantial acquisitions of daily or more frequent repeat observations to track the seasonal dynamics, long-term growth trends, and disturbance patterns that define the Earth's land dynamics.

CONCLUSIONS

The early pioneers in moderate spatial resolution satellite land remote sensing had a clear vision of the potential of such observatories to enhance our understanding of our home planet. In the process of realizing that dream, we have managed to make, in retrospect, many unfortunate choices that have slowed adoption and application of these important twentieth century technological innovations. Only now – as we enter a new fourth phase of national government commitments and international cooperation to conduct global, systematic monitoring of the Earth, through deployment and operation of moderate-resolution land observatories – may we be seeing the beginning of achieving the mid-twentieth century goal of continued surveillance of our planet for the good of its occupants. We should know soon whether we are indeed moving forward toward achieving the promise so clearly articulated by the innovators of these satellite land observatories.

ACKNOWLEDGMENTS

The authors of this chapter were supported from funds provided by the Earth Science Program, Science Mission Directorate, NASA Headquarters and a US Geological Survey contract number 04CRFS0058, jointly funded by USGS and NASA.

NOTES

1. On Landsats 1–2 the RBV band designations were 1–3 and the MSS bands were 4–7. On Landsat 3, the 40 m RBV bands were 1–2, the MSS bands were left at 4–7 and a new MSS thermal infrared spectral band was designated band 8.
2. Renamed the Center for Earth Resources Observation and Science (EROS) in 2005.
3. The italicized portion is a direct quote from the Public Law 102-555, Title I, Sec. 101 (c) (1) (4).
4. In addition to the eight systems discussed here, it should be noted that the DMC has plans to launch three satellites in 2008–09 (for Nigeria, Spain, UK) and India is planning to launch ResourceSat-2 in 2009, with the same sensors as ResourceSat-1.

REFERENCES

Allen, R.G., M.Tasumi, A.T. Morse, and R. Trezza, 2005. A Landsat-based energy balance and evapotranspiration model in western US water rights regulation and planning. *Irrigation and Drainage Systems*, 19: 251–268.

Arvidson, T., S. N. Goward, J. Gasch, and D. L. Williams, 2006. Landsat 7 long-term acquisition plan: Development and validation. *Photogrammetric Engineering and Remote Sensing*, 70(10): 1137–1146.

Beck, R., 2003. *EO-1 Users Guide*, V2.3. University of Cincinnati, Cincinnati, Ohio.

Berger, B., 2003. NASA rejects only bid for landsat data buy contract. *Space News*. September 29, 2003: 1, 3.

Cloud, J., 2001. Re-viewing the Earth: remote sensing and Cold War clandestine knowledge production. *QUEST – The History of Spaceflight Quarterly*, 8(3): 5–16.

Engel, J. and O. Weinstein, 1983. The Thematic Mapper – An overview. *IEEE Transactions on Geoscience and Remote Sensing*, GE-21: 258–265.

Freden, S. C. and F. Gorden Jr., 1983. Landsat satellites. In: R. N. Colwell (ed.) *Manual of Remote Sensing*. American Society of Photogrammetry, Falls Church, Virginia. 1: 517–570.

Goward, S. N., T. J. Arvidson, D. L. Williams, J. Faundeen, J. Irons, and S. Franks, 2006. Historical record of Landsat global coverage: Mission operations, NSLRSDA, and international cooperator stations. *Photogrammetric Engineering and Remote Sensing* 70(10): 1155–1170.

Green, K., 2006. Landsat in context: The land remote sensing business model. *Photogrammetric Engineering and Remote Sensing*, 70(1): 1147–1154.

Group on Earth Observations, 2005. Global Earth Observations System of Systems (GEOSS) 10-year implementation plan reference document. European Space Agency, Noordwijk, The Netherlands.

Integrated Global Observing Strategy Partnership, 2004. Integrated global observations of the land – a proposed theme to the IGOS – partnership – Version 2. IGOS-P 11 Doc., Item 7.1, 11 pp (File code: ECV-T9-landcover-ref10-IGOS-land-theme).

Interagency Working Group in Earth Observations, 2005. Strategic plan for the U.S. integrated earth observation system. The National Science and Technology Council, The Committee on Environment and Natural Resources, Washington, DC.

Irons, J. and J. Masek, 2006. Requirements for a Landsat data continuity mission. *Photogrammetric Engineering and Remote Sensing*, 70(19): 1102–1110.

Johnson, R. L., 1998. *What it took: A history of the USGS EROS Data Center*. The Center for Western Studies, Augustana College, Sioux Falls, South Dakota.

Logsdon, J. H., R. G. Launius, D. H. Onkst, and S. J. Garber (eds), 1998. *Exploring the Unknown: Selected Documents in the History of the U.S. Civilian Space Program*. National Aeronautics and Space Administration, NASA History Division, Washington, DC Vol III – Using Space: 608.

Lowman, P. D., Jr., 1998. Landsat and Apollo: The forgotten legacy. *Photogrammetric Engineering and Remote Sensing*, 65(10): 1143–1147.

Mack, P. E., 1990. *Viewing the Earth: The Social Construction of the Landsat Satellite System*. MIT Press, Cambridge, Massachusetts.

Markham, B. L., J. C. Storey, M. M. Crawford, D. Goodenough, and J. Irons, 2004. Landsat Sensor Performance Characterization (Special Issue). *IEEE Transactions of Geoscience and Remote Sensing*, 42(12): 2687–2938.

McDonald, R. A., 1995. CORONA – Success for space reconnaissance, a look into the Cold War and a revolution for intelligence. *Photogrammetric Engineering and Remote Sensing* 61(6). 689–720.

McDonald, R. A. (ed.), 2002. *Beyond Expectations: Building an American National Reconnaissance Capability*. American Society of Photogrammetry and Remote Sensing, Bethesda, MD.

Mika, A. M., 1997. Three decades of Landsat instruments. *Photogrammetric Engineering and Remote Sensing*, 63(7): 839–852.

NASA Goddard Space Flight Center, 2001. Landsat Data Continuity Mission (LDCM) data specification. NASA Goddard Space Flight Center, Greenbelt, Maryland.

National Research Council, 1969. Useful applications of Earth-oriented satellites. National Academy of Sciences, Washington, DC.

National Science and Technology Council, 2006. Public workshop on the future of land imaging for the United States. National Science and Technology Council, Office of Science and Technology Policy, Washington, DC.

National Science and Technology Council, 2007. A plan for a national land imaging program (120 pp.). Washington, DC: Office of Science and Technology Policy, Future of Land Imaging Interagency Working Group.

OSTP, 2004. Landsat Data Continuity Strategy. Office of Science and Technology Policy, Executive Office of the President, Washington DC, 2 pp.

OSTP, 2005. Landsat Data Continuity Strategy Adjustment. Office of Science and Technology Policy, Executive Office of the President, Washington DC, 2 pp.

Pecora, W. T., 1966. Earth Resources Observation Satellite (EROS): A Department of Interior program to utilize space-acquired data for natural and human resource management. Unpublished internal memorandum, US Department of the Interior, US Geological Survey, Washington DC.

Stoney, W. E., 2004. Civil land observatory satellites. In H. Mark (ed.), *Encyclopedia of Space Science and Technology*. John Wiley & Sons, New York.

Stoney, W., 2006. *Guide to Land Imaging Satellites*. American Society of Photogrammetry and Remote Sensing, Bethesda, Maryland. http://www.asprs.org/news/satellites/satellites.html (last accessed: 31 December, 2008).

Storey, J., P. Scaramuzza, and G. Schmidt, 2005. Landsat 7 scan line corrector-off gap-filled product development. In: *Pecora 16 Conference Proceedings*, American Society of Photogrammetry and Remote Sensing, Bethesda, Maryland. Unpaginated CD-ROM.

Teeuw, R., P Aplin, N. McWilliam, T. Wicks, K. Matthieu, and G. Ernst, in this volume. Hazard Assessment and Disaster Management using Remote Sensing. Chapter 32.

Teillet, P. M., J. L. Barker, B. L. Markham, R. R. Irish, G. Fedosejevs, and J. C. Storey. 2001. Radiometric cross-calibration of Landsat 5 Thematic Mapper and Landsat 7 Enhanced Thematic Mapper Plus based on tandem data sets. *Remote Sensing of Environment*, 78(1–2): 39–54.

Toutin, T., in this volume. Fine Spatial Resolution Optical Sensors. Chapter 8.

US Congress, 1984. *Public Law 98–365: Land Remote-Sensing Commercialization Act of 1984*. 98th Congress, July 17, US Government Printing Office, Washington, DC, 17 pp.

US Congress, 1992. *Public Law 102–555: Land Remote Sensing Policy Act of 1992*. 102nd Congress, 28 October, US Government Printing Office, Washington, DC, 18 pp.

USGS, 2007. Landsat satellites. http://landsat.usgs.gov. (last accessed 31 December, 2008).

Coarse Spatial Resolution Optical Sensors

Christopher Owen Justice and
Compton James Tucker III

Keywords: coarse resolution remote sensing, AVHRR, MODIS, VIIRS.

INTRODUCTION

Previous chapters have discussed remote sensing imaging systems with a pixel dimension measured in meters. In this chapter, we move to coarser systems, where resolution is typically measured in hundreds of meters to kilometers. Coarse resolution sensors are therefore defined here as those with spatial resolutions between 250 m and 5 km. Until recently these were also referred to as 'moderate' resolution systems, but this latter term is now used to refer to the Landsat-class of sensors. This chapter focuses solely on land remote sensing, although it is recognized that coarse resolution sensors are also widely used for the study of oceans and atmosphere.

The origins of coarse resolution land remote sensing date to the early 1980s. The research community was faced with finding an alternative source of data to Landsat. In addition, there was a need to extrapolate globally ground-based remote sensing research that found spectral vegetation indices measured *in situ* through time to be highly related to net primary productivity and crop phenological stage (Tucker 1979).

Research by Tucker and colleagues at the Goddard Space Flight Center and the Beltsville Agricultural Research Center using handheld radiometry to study vegetation productivity and crop development showed the importance of high temporal frequency observations using spectral vegetation indices (Tucker et al. 1979). Combinations of visible and near-infrared data were used to form the normalized difference vegetation index (NDVI), which when integrated or summed over the growing season was highly related to total dry biomass production (Tucker et al. 1981). Subsequent research has shown a direct relationship between spectral vegetation indices and the absorbed fraction of photosynthetically-active radiation absorbed by plant canopies (Myneni et al. 1995; among others). An investigation of available satellite sensors with these visible and near infrared spectral bands and the requirement for frequent data acquisition suggested the applicability of the National Oceanic and Atmosphereic Administration (NOAA) Advanced Very High Resolution Radiometer (AVHRR) (Tucker et al. 1981).

Investigation of the possibility of using AVHRR data for land applications revealed a system designed for the weather community, with approximately 4 km pixels, daily global observations and coverage, spectral bands appropriate for vegetation monitoring, and thermal bands useful for cloud detection (Townshend and Tucker 1981). Despite skepticism concerning the utility of AVHRR imagery with such coarse spatial resolution, Tucker et al. (1983) successfully applied the vegetation index time-series approach to monitoring regional grassland productivity, and Justice et al. (1985)

demonstrated the ability to follow the phenology of vegetation from space with vegetation indices, that in turn set the stage for the application of these data to continental and global land cover mapping (Townshend et al. 1985).

THE AVHRR INSTRUMENTS

The AVHRR sensor was first flown on the TIROS-N polar-orbiting meteorological satellite in 1978. TIROS-N's AVHRR was configured originally with four channels (0.55–0.90 μm, 0.73–1.1 μm, 3.5–3.9 μm, and 10.5–11.5 μm) for meteorological applications that were expanded to five channels (by including a 11.5–12.5 μm channel) with the launch of NOAA-7 in June of 1981. A sixth channel (1.5–1.7 μm) was added with the launch of NOAA-15 in May 1998 (Table 10.1). In 1976–1977, before the launch of NOAA-6, Stanley Schneider and Dave McGinnis of the National Oceanic and Atmospheric Administration's National Earth Satellite Service succeeded in having future AVHRR's modified: the first channel was narrowed to 0.55–0.70 μm (Schneider and McGinnis, 1982). The principal reason for confining channel 1 to the visible part of the spectrum was to increase AVHRR effectiveness for snow mapping and vegetation monitoring. (S. Schneider 1977, personal communication.)

The NOAA-series of sun-synchronous polar-orbiting meteorological satellites orbit at an altitude of ~830 km. One half of these satellites in this series have a daytime overpass time suitable for obtaining AVHRR data for global vegetation studies (see Table 10.1) while the other satellites in the NOAA series have equatorial overpass times of 0730 and 1930 hours, that precludes global vegetation studies. The AVHRR sensor scans ~±55° from nadir and complete coverage of the earth is available at least twice daily with two resolutions at the satellite subpoint: 1.1 km, and a spatially-degraded resolution representing ~5.5 × 3.3 km, called Global Area Coverage (GAC). The GAC data are formed as a partial average of a 5 by 3 element block of 1.1 km pixels. The first four 1.1 km pixels in the first scan line of the block are averaged, and the fifth pixel is skipped, as well as the next two rows of five-pixels. Thus, the GAC data represent a 4/15 sample of the 5.5 by 3.3 km block, and a 15-fold reduction in data volume compared to the original 1.1 km data (Cracknell 1997).

NOAA National Environmental Satellite Data and Information Service (NESDIS) collected the AVHRR global area coverage (GAC) at a spatial resolution of approximately 4 km (600 megabytes/day) and selected local area coverage at a spatial resolution of 1.1 km. Because of constraints of limited on-board storage capability, 1.1 km AVHRR data were only collected by direct transmission to line-of-sight ground receiving stations or for a limited number of areas using the on-board tape recorder as requested.

Beginning with the launch of NOAA-6 in 1979, the calculation of vegetation indices from AVHRR data became possible, as channel 1 was confined to the upper portion of the visible spectrum, while channel 2 continued to cover the range 0.73–1.1 μm. Most of the vegetation indices derived from NOAA AVHRR data use the normalized difference vegetation index (NDVI), which is calculated from channels 1 (0.55–0.70 μm) and 2 (0.73–1.1 μm), often with

Table 10.1   A Summary of the AVHRR instruments, the satellites they have flown on, their periods of operation, and their daytime equatorial overpass times. Only those NOAA polar orbiting satellites where AVHRR data can provide global data suitable for optical remotely sensed studies are listed. For example, NOAA-6, NOAA-8, NOAA-10, NOAA-12, and NOAA-15 are not listed in this table because their daytime equatorial overpass times were 07:30 and 19:30 hours

Satellite	Period of operation	Daytime equatorial overpass times	AVHRR instrument type
NOAA-7	July 1981 to Dec. 1984	14:30	5-channel
NOAA-9	Jan. 1985 to Oct.1988	14:30	5-channel
NOAA-11	Oct. 1988 to July 1994	13:30	5-channel
NOAA-14	July 1994 to July 2000	13:30	5-channel
NOAA-16	July 2000 to July 2005	13:30	6-channel
NOAA-17	Aug. 2002 to present	10:00	6-channel
NOAA-18	July 2005 to present	13:30	6-channel
NOAA-19	Launch in 2009	10:00	6-channel
Metops-1	Aug. 2006 to present	09:30	6-channel
Metops-2	Launch in 2010	09:30	6-channel
Metops-3	Launch in 2014	09:30	6-channel

channel 5 (11.5–12.5 μm) used as a thermal cloud mask. The NDVI is calculated as (channel 2 − channel 1)/(channel 2 + channel 1). Beginning with NOAA-7, and with the exceptions of NOAA-8 and NOAA-10, the AVHRR's were all five-channel instruments. The fifth channel from 11.5–12.5 μm was added to enable sea-surface temperature determinations (Kidwell, 1988). For NOAA-14 through NOAA-19, a sixth channel was added from 1.5–1.7 μm. This channel alternates with the 3.5–3.7 μm channel, so only five channels transmit data at any one time. The 1.6 μm data are collected during the day for snow and ice studies, whereas the 3.6 μm is acquired at night, allowing night-time fire detection. This compromise has implications for the long-term record of data generated with the daytime 3.6 μm observations.

AVHRR studies

Several people interested in terrestrial or land applications, working independently, began to use NOAA-6 AVHRR data in 1980 and early 1981. Tucker et al. (1981) commented that the only satellite sensor system which might be used to make estimates of primary production using a vegetation index approach from space was the AVHRR on the NOAA series of satellites but noted the limitations of the wide field of view (±55°) and the wide bandwidths of channel 1 and channel 2. The daytime overpass time of NOAA-6 (0730 hours) was thought to be too early in the morning, while the daytime over pass time for NOAA-7 (1430 hours) was thought to possibly be too late in the afternoon because of cloud build-up.

One of the first AVHRR projects by Tucker was to look at AVHRR 1-km data from the Nile Delta. The U.S. Air Force was acquiring daily local area coverage (LAC) tape-recorded 1-km data over this part of the Mediterranean. The multiple scenes of data were inexpensive but the study pointed to the need to address cloud cover from data acquired from multiple images. Additional workers began evaluating these 1-km data for land vegetation purposes. Gray and McCrary (1981a, b), Townshend and Tucker (1981), Greegor and Norwine (1981), Schneider and McGinnis (1982), Ormsby (1982), Cicone and Metzler (1982), Yates and Tarpley (1982), Barnett and Thompson (1982), and Duggin et al. (1982) all reported on early attempts using NOAA-6 and NOAA-7 AVHRR data. Initially, almost all of the emphasis was on using selected cloud-free 1-km data and most of the topics investigated involved agricultural monitoring. Agricultural applications (crop yield forecasts, droughts, etc.) were viewed at that time as being a principal rationale for coarse resolution satellite-based land remote sensing (Macdonald and Hall 1980).

In late summer and early fall of 1981, Tucker and coworkers in Senegal attempted to compare normalized difference vegetation index 1-km data from NOAA-7's AVHRR instrument to ground-collected total above-ground herbaceous dry biomass measurements. A request was submitted to NOAA to tape-record 1-km LAC data for West Africa, centered over Senegal. All of the recorded overpasses, scheduled for about once every 9–10 days, were largely cloud-free for the area of interest (Tucker et al., 1983). However, subsequent acquisitions in 1982 were unsuccessful in obtaining sufficient cloud-free NOAA-7 AVHRR 1-km data over Senegal once every 9–10 days during the growing season. They complimented the 1 km data with AVHRR 4 km data from other overpasses, and started investigating different compositing approaches, atmospheric effects, cloud detection, and directional reflectance effects.

The work of Holben and Fraser (1984) and Holben (1986) established the scientific basis for forming maximum value normalized difference composite images; that is, selecting the highest normalized difference vegetation index value for a given location from several successive observations. In 1983, NOAA began producing the global vegetation index (Tarpley et al. 1984, Kidwell 1990) and several uses of the data for large-area study were subsequently reported (Townshend and Tucker 1981, Justice et al. 1985, Goward et al. 1987). In 1986 the Global Inventory Monitoring and Modeling Study (GIMMS) group at NASA/Goddard Space Flight Center published a special edition of the *International Journal of Remote Sensing* entitled 'Monitoring the grasslands of semi-arid Africa using NOAA-AVHRR data' (Justice 1986), which summarized several years of AVHRR 1-km and 4-km work from Sahelian Africa, with emphasis on rangeland monitoring. A subsequent special edition of the same journal entitled 'Coarse resolution remote sensing of the Sahelian environment' was published in 1991 (Prince and Justice 1991). These two special editions are excellent references on AVHRR land applications in the late 1980s, though the focus is upon African rangelands.

It became apparent by the mid-1980s that coarse resolution imagery formed into composites and analyzed over time could be used to address a wide range of terrestrial vegetation applications involving primary production, land cover mapping, famine early warning, and desert locust control, amongst others. In retrospect, the remote sensing community had not been able to exploit the potential of the temporal domain until AVHRR 1-km and 4-km data became available and were evaluated. Cracknell (2001) has summarized this unexpected development in an article entitled 'The exciting and totally unanticipated success of the AVHRR in applications for which it was never intended'.

Perhaps more importantly, the spectral vegetation indices from coarse resolution sensors such as the AVHRR are now the most widely used remote sensing products and more than 3,000 papers have been published in the scientific literature on this topic (as determined from the Web of Science in June 2007).

Satellite time-series analysis

The GIMMS Group at NASA/GSFC was formed in 1983 and developed the pioneering research on methods for coarse resolution remote sensing (Justice et al. 1985). The characteristics of the AVHRR sensing system created a number of obstacles to the development of useable data sets. The occurrence of clouds in the daily data meant that techniques had to be developed to optimize the time-series for cloud-free observations and a technique known as maximum value compositing was developed (Holben 1986). Although the Vegetation Index, as a spectral ratio, lent itself to minimizing multiplicative effects, the wide swath meant that viewing angle effects and atmospheric effects remained in the data (Holben and Fraser 1984, Holben 1986, Holben et al. 1986). The wide swath width of the AVHRR afforded by off-nadir viewing results in a spreading of the pixel size at extreme off-nadir views (55 degrees) of up to 1.5 times. The effects of water vapor contamination in the broad infrared channel were identified (Justice et al. 1991). Methods for atmospheric correction pioneered for the AVHRR by Tanre et al. (1992) and Vermote and Kaufman (1995) were further developed by Vermote et al. (1997) for Moderate Resolution Imaging Spectroradiometer (MODIS) data and subsequently applied to AVHRR data (El Saleous et al. 2000).

The absence of on-board calibration in the first two AVHRR bands meant that procedures needed to be developed for tracking sensor performance over the lifetime of the instrument and applying the appropriate calibration coefficients (Holben et al. 1990, Vermote and Kaufman 1995). With the removal of artifacts in the data, the NDVI data sets provided the opportunity to explore multi-temporal classification procedures, using phenology (Justice et al. 1985) as a means for discrimination of broad vegetation types (Tucker et al. 1985, Townshend et al. 1987, 1991). The multi-temporal data were also applied to a preliminary assessment of annual primary production at regional and global scales (Goward et al. 1985, Prince 1991a, b). Recognizing the potential of the AVHRR time-series for monitoring inter-annual variability, Tucker et al. (1991) explored year-to-year variability of vegetation in semi-arid regions to map desert spatial extent. With a longer time-series, Fung et al. (1987) and Tucker et al. (1986) were able to relate global land

photosynthesis to monthly fluctuations in atmospheric CO_2. Myneni et al. (1997) used the AVHRR record to detect changes in the length of growing season in boreal regions. AVHRR data have also been incorporated in ecological modeling activities (Potter et al. 1993, Running et al. 2004).

These early applications laid the foundation for a large amount of related research that has now been extended by more years of AVHRR data and by improved data from MODIS and SPOT Vegetation (Townshend 1994). The key to using NDVI time series data for longer time periods, now in excess of 25 years, has been the quantitative intercomparison and normalization among NDVIs from different sensor systems and/or the same sensor system flown on different satellites (Tucker et al. 2005, Brown et al. 2006).

Data processing and AVHRR products

Starting in the early 1980s, the GIMMS group, with limited funding, developed an AVHRR processing and analysis system on HP mini-computers to ingest data from 1600 bpi tapes from NOAA and different AVHRR receiving stations, to apply calibration coefficients, co-register the data, and create 10-day composites of the NDVI. These 10-day composites were assembled into monthly and annual data sets for analysis. The geometry of the AVHRR required careful visual inspection of images used in the compositing to ensure that the registration procedures were effective. As the utility of the data set was recognized, increasing pressure was put on the GIMMS research group to distribute the derived data products to the science community. In this way NASA and NOAA developed parallel processing systems for the AVHRR, with NASA developing products for the science community and NOAA developing products primarily for the National Weather Service (Gutman and Ignatov 1998).

The GIMMS NDVI data set was reprocessed several times and eventually NASA established the AVHRR Pathfinder project aimed at making a community consensus product broadly available (James and Kalluri 1994). The GIMMS group continued to develop enhancements to the data set and as a result the Pathfinder 2 processing code was developed, which forms the basis of the most recent data production and reprocessing (El Saleous et al. 2000).

Over the years a number of products have been developed using the AVHRR which can be considered standard. The NDVI has been described above. A land surface temperature product was developed by Pinheiro et al. (2006) improving on the split window method developed by Becker and Li (1990). Fire data sets were developed from the 1 km AVHRR data (e.g., Kendall et al. 1997,

Dwyer et al. 2000). A fractional cover data set was developed and distributed by NOAA (Gutman and Ignatov 1998, Gallo et al. 2001). A global land cover product was developed by Townshend (1994) and a percent tree cover product was developed by Hansen and DeFries (2004).

The requirement for products of improved spatial resolution was identified by scientists of the IGBP (Townshend et al. 1994). In response to this requirement, the IGBP-DIS developed an international effort to compile a global 1 km AVHRR data set from different ground stations for 1992 (Eidenshink and Faundeen 1994). This product was eventually used to generate global 1 km land cover and active fire products (Dwyer et al. 2000, Loveland et al. 2000). The in-depth research associated with utilizing these data sets led to a better characterization of the instruments and identification of the limitations of the products (Giglio et al. 1999, Csiszar and Sullivan 2002). The poor geolocation and resulting spatial registration problems of AVHRR data, the broad band widths and lack of calibration in the visible and near infrared channels, and the orbital drift during lifetime of each satellite, were early limitations to using AVHRR data for land studies that were overcome as the time span of the data increased.

Recent AVHRR developments: NOAA-19 and the METOP satellites

NOAA-19 will be the last NOAA polar-orbiting meteorological satellite before the NPOESS Preparatory Project (NPP) flies the next-generation operational instrument, currently planned for late 2009. However, this will not be the end of the AVHRR as a primary source of coarse-resolution earth science data: The European Meteorological Operational satellites, MetOp-1, MetOp-2, and MetOp-3, will fly AVHRR instruments in polar orbits possibly until 2015 (Table 10.1). MetOp-1's AVHRR is now operated by EUMETSAT, providing global 1-km data. It is remarkable that more than 20 AVHRR instruments have been flown since the launch of NOAA-6 in 1979. While the AVHRR instruments are today neither advanced nor very high resolution, they were 'mass-produced' in production runs of several instruments at a time, which enabled their operational success. However, although obvious and significant improvements to the AVHRR instruments have been recommended, such as improved spectral bandwidths for channels 1 and 2, on-board calibration of channels 1 and 2, routine global 1 km data acquisition (only 16 gigabits/day), and full-time operation of all six channels, changes to operational instruments have been extremely difficult for NOAA to implement for reasons not apparent to the user community. This is all the more surprising because the land

remote sensing community which would benefit greatly from these improvements is by far the largest user of data from the AVHRR (Cracknell 2001).

THE MODERATE RESOLUTION IMAGING SPECTRORADIOMETER

The MODIS instrument was designed as the coarse resolution imager for NASA's Earth Observing System (EOS), and was built by Santa Barbara Research to NASA specifications. The first MODIS was launched on the Terra (EOS AM-1) spacecraft on 18 December 1999 with a 10.30 am equatorial crossing time (descending node). The second MODIS was launched on the Aqua (EOS PM-1) spacecraft on 4 May 2002 with a 1.30 pm equatorial crossing time (ascending node). The MODIS instruments were designed with a six year life but it is hoped and expected that will both instruments will operate far beyond their life expectancy.

The MODIS instruments were designed to meet the needs of the land, ocean, and atmospheres communities and, in some cases, were a compromise among competing needs. The land requirements for the MODIS instrument grew from the experience with daily observations with the AVHRR and the spectral characteristics of the Landsat Thematic Mapper instrument. Additional bands were added in the middle infrared and thermal regions for land sensing and bands were included to enable atmospheric correction and cirrus cloud detection. MODIS has 36 spectral bands with a 12 bit quantization. Two bands are at 250 m resolution, five bands are at 500 m resolution and the remaining 29 are at 1 km. The higher spatial resolution bands were selected for land remote sensing to improve upon the AVHRR (Salomonson et al. 1989).

The EOS Terra and Aqua orbit is at ~705 km, and with the MODIS ±55 degree scan angle, this gives a swath width of 2,330 km and global near-daily coverage. The MODIS optical system consists of a two-mirror off-axis afocal telescope, which directs the incoming energy to four refractive objective assemblies, one for each of the visible, near infrared, shortwave/midwave infrared, and longwave infrared (VIS, NIR, SWIR/MWIR and LWIR) spectral regions to cover a total spectral range of 0.4 to 14.4 μm. A passive radiative cooler provides cooling to 83 K for the 20 infrared spectral bands on two mercury–cadmium–telluride focal plane assemblies. Photodiode-silicon readout technology for the visible and near infrared bands provides high quantum efficiency and low-noise readout with a good dynamic range. The MODIS system includes a space view and four on-board calibrators: a Solar Diffuser, a v-groove Blackbody, a Spectroradiometric Calibration Assembly and a Solar Diffuser Stability Monitor.

The instrument design is described in detail by Barnes et al. (1998), calibration and early performance is described by Guenther et al. (2002) and the instrument geolocation by Wolfe et al. (2002). Problems with early calibration adjustments render the MODIS Terra data prior to November 2000 largely unusable for land studies. As of the date of writing both instruments are performing nominally.

The MODIS data system and products

Data from the Terra and Aqua spacecraft are transmitted to a ground station at White Sands, New Mexico via the Tracking and Data Relay Satellite System (TDRSS). The data are then sent to the EOS Data and Operations System (EDOS) at the Goddard Space Flight Center. After Level 0 processing at EDOS, the MODIS Adaptive Processing System (MODAPS) produces the Level 1A, Level 1B, geolocation and cloud mask, and the higher level land products. The land products are distributed through the NASA Land Processes Distributed Active Archive Center (DAAC) at the U.S. Geological Survey EROS Data Center (EDC) (Justice et al. 2002). Cryosphere data products (snow and sea ice cover) are available from the National Snow and Ice Data Center (NSIDC) DAAC in Boulder, Colorado. The data are freely available with no restrictions on data sharing. Users with an appropriate x-band receiving system may capture regional data directly from the spacecraft using the MODIS Direct Broadcast signal.

The MODIS Land Products provide a major advance over those available from the AVHRR in both spatial resolution and quality, and include surface reflectance corrected for aerosols (Vermote et al. 2002), snow cover (Hall et al. 2002), land surface temperature (Wan et al. 2002), active fire, and burned area (Giglio et al 2003, Justice et al. 2002, Roy et al. 2005), leaf area index (Myeneni et al. 2002), albedo (Schaaf et al. 2002), land cover (Friedl et al. 2002), vegetation continuous fields (Hansen et al. 2002), and vegetation conversion (Zhan et al. 2002). An example of the MODIS surface reflectance product is shown in Figure 10.1. The NDVI for MODIS was augmented with the Enhanced Vegetation Index which built on a large body of research investigating indices designed to reduce the effect of soil background and atmospheric effects (Huete et al. 2002).

NASA placed a number of requirements on the MODIS Science Team members responsible for MODIS products. Firstly, a peer reviewed algorithm technical background document (ATBD) was to be generated for each product; secondly, a user guide was required; thirdly, the principle investigator was responsible for both assessing product quality (QA) and validating the product (i.e., determining the product accuracy); and fourthly, the code associated with the product was to be made openly available. These requirements have led to an improvement in the utility of the products. In addition two small groups were formed within the MODIS Land Group to coordinate the quality assessment (Roy et al. 2002) and product validation (Morisette et al. 2002, Nickeson et al. 2007). The activities of these two groups and the MODIS land team have created a paradigm shift for land remote sensing. It is now expected that products distributed to the science and applications communities be validated (i.e., have known and documented accuracies). To meet

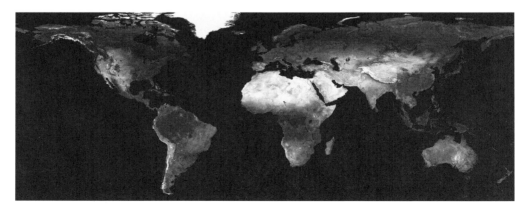

Figure 10.1   A MODIS surface reflectance, simulated true color image derived from the global 8 day surface reflectance products at 500 m (courtesy of E. Vermote and S. Kotchenova of the University of Maryland). (See the color plate section of this volume for a color version of this figure).

this expectation a Land Product Validation group has been established as part of the Committee on Earth Observation Satellites (CEOS). A large number of publications can now be found in the science and applications literature on the use of MODIS data. Through NASA funding, the MODIS instruments, data system, and the MODIS Land Science Team have provided an unprecedented and concerted contribution to coarse resolution land remote sensing, building on the success of the GIMMS group with the AVHRR. The various applications of coarse resolution data initiated with the AVHRR have been continued and refined using the MODIS system, with improved spatial resolution and the benefit of two satellites. New applications with higher accuracy products of surface reflectance, fire and burned area, albedo, land surface temperature have been developed. Rapid delivery of MODIS products within a few hours of acquisition has led to near real time applications of the data (Justice et al. 2002).

OTHER COARSE RESOLUTION SENSORS

The success with the AVHRR and developments with MODIS led to a number of other programs to provide global coarse resolution data. The most notable of these is the SPOT vegetation Program developed by CNES (Saint 1995). The vegetation sensor is a four-channel solid state wide-field of view instrument. It has spectral bands in the blue (0.43–0.47 μm), red (0.61–0.69 μm), near-infrared (0.79–0.89 μm), and short ware infrared (1.5–1.7 μm), with a swath-width of 2,250 km. There are no thermal bands on this sensor which limits cloud clearing approaches. Two vegetation instruments have been put into orbit, one on SPOT-4 in May 1998 and a second on SPOT-5 in May 2002. Since mid-2002 the vegetation sensor on SPOT-5 has been the primary instrument, collecting global daytime 1-km data. The vegetation sensor on SPOT-4 is now a 'back up' for the newer sensor.

Other coarse resolution sensors include the Along-Track Scanning Radiometer (ATSR) and the Advanced Along-Track Scanning Radiometer (AATSR) instruments launched by the European Space Agency. Presently the AATSR instrument is operating on the ENVISAT satellite and is the most recent in a series of instruments designed primarily to measure Sea Surface Temperature (SST), following on from ATSR-1 and ATSR-2 on board the ERS-1 and ERS-2 satellites, respectively. AATSR data have a resolution of 1 km at nadir, and are derived from measurements of reflected and emitted radiation taken at the following wavelengths: 0.55, 0.66, 0.87, 1.60, 3.70, 11.00, and 12:00 μm. Thus the spectral configuration of this instrument is very similar to the AVHRR. A primary difference is

the use of a conical scan to give a dual-view of the Earth's surface, on-board calibration targets, and use of mechanical coolers to maintain the thermal environment necessary for optimal operation of the infrared detectors. The swath-width is 500 km and the spatial resolution is 1 km at the satellite sub-point.

Also operating on the ENVISAT satellite launched in March 2002, is another coarse resolution instrument called the Medium Resolution Imaging Spectrometer (MERIS) (Rast et al. 1999). MERIS is a programmable, imaging spectrometer operating in the solar reflective spectral range. Fifteen spectral bands can be selected by ground command. The instrument scans the Earth's surface by the so called 'push-broom' method. Linear CCD arrays provide spatial sampling in the across-track direction, while the satellite's motion provides scanning in the along-track direction. MERIS is designed so that it can acquire data over the Earth whenever illumination conditions are suitable. The instrument's 68.5° (±34.25°) field of view around nadir covers a swath width of 1150 km. This wide field of view is shared between five identical optical modules arranged in a fan shape configuration. The spatial resolution of the instrument is 260 × 300 m at the satellite subpoint. Although designed primarily for ocean and coastal remote sensing, MERIS is increasingly used for land applications (Curran and Steele 2005).

The Japan Aerospace Exploration Agency (JAXA) launched a coarse resolution instrument called the Global Imager (GLI) on Japan's Advanced Earth Observing Satellite (ADEOS I) in 1996 and also on ADEOS II in 2002. Both missions were short-lived, although data were collected. The GLI instruments had 36 spectral bands with atmospheric correction channels necessary for high-precision observation of the ocean color and wide dynamic range channels for land and atmosphere observation (Murakami et al. 2005). Six 250-m resolution channels were selected for compatibility with Landsat TM.

In general there has been relatively little use of these sensors for land science and applications as compared to the AVHRR and MODIS. This is in part due to the paucity of information on the instrument and products, the difficulty in obtaining data and the lack of validation of the associated derived products.

NPP AND NPOESS VIIRS

In 1994, a U.S. Presidential Decision Directive was given to the Department of Commerce and the Department of Defense (DoD) to integrate the U.S. polar-orbiting systems into a single, converged, National Polar-orbiting Operational

Environmental Satellite System (NPOESS). To implement the project, the Departments of Commerce, Defense, and NASA were charged with creating an Integrated Program Office (IPO). To reduce development risk and to provide a bridging mission from the NASA EOS to NPOESS, the IPO and NASA jointly defined an NPOESS Preparatory Project (NPP), with the goal of providing continuing observations for studying global change after EOS Terra and Aqua and early user evaluation of NPOESS products. The Visible/Infrared Imager/Radiometer Suite (VIIRS) instrument was defined as the NPOESS equivalent of the MODIS instrument to be included on the NPP Mission (Murphy et al. 2001, Murphy 2006). The VIIRS instrument design review was completed in early 2002. It is currently planned for launch in 2010.

The VIIRS instrument design review was completed in early 2002. It is currently planned for launch in late 2009 and will be have a 13:30 (ascending node) overpass. The instrument has 21 bands from 0.412 µm to 11.500 µm, distributed on four focal plane assemblies, and a Day/Night Band centered at 0.700 µm on its own assembly. The VIIRS instrument design is described by Murphy et al. (2006).

The VIIRS instrument builds on the capabilities of the AVHRR, the MODIS and the Operational Linescanner, with spectral bands selected to satisfy the requirements for generating accurate operational and scientific products (16 moderate resolution bands at 750 m and five 'imagery' resolution bands at 375 m). The nominal altitude for NPOESS will be 833 km and the VIIRS scan will extend to ±56 degrees on either side of nadir giving a swath width of approximately 3600 km. The unique characteristic for land studies for VIIRS relative to MODIS is the controlled growth in spatial resolution across the scan, through use of segmented detectors and rotation of the ground instantaneous field of view. The aggregation of pixels across the scan results in only a doubling of pixel area at *c*. 55 degrees for VIIRS, compared to a five times growth for MODIS. Calibration will be performed onboard, using a solar diffuser for short wavelengths and a blackbody source and deep space view for thermal wavelengths. A solar diffuser stability monitor is included to track the performance of the solar diffuser.

A number of higher order products known as environmental data records (EDRs) are under development by the VIIRS contractor and managed by the IPO. For the community these are currently: albedo, land surface temperature, snow cover and depth, vegetation index, surface type, and active fire. Surface reflectance will be generated as an intermediate product. A number of these products, including the surface reflectance, build on the MODIS product algorithms.

The data system for NPOESS is being developed in the framework of NOAA's Comprehensive Large Array-data Stewardship System (CLASS). The challenges for this centralized system will be similar to those faced in the early development of the EOS Data and Information System (EOSDIS). In addition, as NPOESS is being designed for operational users, there is a strong requirement for near real-time data delivery and no requirement for reprocessing, which has proven to be essential for the science community. There is clearly a need for a data product generation system designed to meet the needs of the science community.

Although intended to reduce costs, the NPOESS program has already suffered serious cost overruns and as a result number of the climate instruments planned for the NPOESS platform have been cut. VIIRS remains a critical part of NPOESS and the instrument is being built. The leadership by DoD in the procurement resulted in an emphasis on operational rather than science products and set a different technical relationship with the contractor than NASA had with MODIS contractors. NASA did not play a lead role in the initial planning for NPOESS, VIIRS, and the required products, as at that time further EOS platforms were envisioned to meet the needs of the science community. This has left the science community to try and insert its requirements on the instrument and products with little overall influence on the system.

FUTURE DIRECTIONS FOR COARSE RESOLUTION REMOTE SENSING

Coarse resolution remote sensing has become an established approach for global to regional land studies. A programmatic challenge facing U.S. remote sensing community is now to ensure that the capabilities developed in the research domain for the AVHRR and MODIS are used in an operational status for NPP and NPOESS. This includes the products, their validation, and distribution systems (Townshend and Justice 2002).

Advances in the use of coarse resolution data continue to be developed and a number of areas of research and development are showing promising results. New standard products continue to be developed from MODIS to meet the needs of the global change science and applications communities. For example, a new vegetation moisture content product is currently being proposed to NASA (Zarco-Tejada et al. 2003).

Data assimilation methods, used extensively by the global modeling community are now being used experimentally by the land modeling community. The integration of data from different sensing systems remains a challenge for the research community although data fusion techniques are being developed for combining data from different systems and resolutions (Vermote et al. 2007). The interest of the global change community in studying monitoring long term trends requires a consistent long term data record. This has become increasingly important for a wide variety of climate studies.

Improvements in the spatial resolution of geostationary sensors have led to the investigation of these high temporal resolution coarse resolution systems for land applications. For example, high temporal resolution vegetation index data are being obtained from EuMetSat's Meteosat Second Generation satellite, Meteosat-8, with its Spinning Enhanced Visible and Infrared Imager (SEVIRI), which provides unprecedented multi-temporal data from a geostationary orbit (Fensholt et al. 2006b).

The two instruments of the Indian Advanced Wide Field Sensor (AWiFS), with their 350 km swath width, ~60 m pixel size, and a 4–5 day revisit frequency, result in multiple observations per month, allowing experimentation with coarse resolution time-series methods on moderate resolution data.

The focus of the Global Earth Observing System of Systems (GEOSS) is on facilitating international cooperation, with an emphasis on operational applications (Group on Earth Observations 2006). Coarse and moderate resolution data will be an integral part of the system. The availability of multiple coarse resolution systems provides desirable redundancy but raises the question as to the ease of their combined use. Comparisons between different sensors are starting to be undertaken, e.g., Fensholt et al. (2006a). As has been the case since the early beginnings of coarse resolution remote sensing with the investigation of agricultural applications of AVHHR data, new and improved systems are continuing to demonstrate their importance for regional monitoring. MODIS data are now being used to monitor agricultural conditions by the US Department of Agriculture (USDA) and new applications continue to be developed. It is hoped that the VIIRS instrument will continue the provision of high quality coarse resolution started by MODIS and transition both the data and applications into a truly operational domain.

The global coarse resolution satellite data, which started with the AVHRR of NOAA-7 in 1981 and has been improved and continued with MODIS and other sensors, forms the beginning of an uninterrupted long term record. Coarse resolution satellite data provide vital quantitative information for a wide variety of science and applications questions, including the study of climate change, and will continue to do so in the future. To ensure this, the land remote sensing community must continue to make the case, clearly and emphatically, for continued, well calibrated, coarse resolution instruments, producing science-quality observations and validated geophysical products.

REFERENCES

Barnes, W. L., T. S. Pagano, and V. V. Salomonson, 1998. Prelaunch characteristics of the moderate resolution imaging spectroradiometer (MODIS) on EOS AM1. *IEEE Transactions on Geoscience and Remote Sensing*, 36: 1088–1100.

Barnett, T. L. and D. R. Thompson, 1982. Large area relation of satellite spectral data to wheat yields. In *Proc. 8th Int. Symp. Machine Processing Remotely Sensed Data*, Purdue University, Indiana, USA, pp. 213–219.

Becker, F. and Z. L. Li, 1990. Towards a local split window method over land surfaces. *International Journal of Remote Sensing*, 11(3): 369–393.

Brown, M. E., J. E. Pinzon., K. Didan., J. T. Morisette, and C. J. Tucker, 2006. Evaluation of the consistency of long-term NDVI time series derived from AVHRR, SPOT-Vegetation, SeaWiFS, MODIS, and Landsat ETM+ sensors. *IEEE Transactions on Geoscience and Remote Sensing*, 44(7): 1787–1793.

Cicone, R. C. and S. D. Metzler, 1982. Comparisons of Landsat MSS, Nimbus-7 CZCS, and NOAA-6/7 AVHRR sensors for land use analysis. In *Proc. 8th Int. Symp. Machine Processing Remotely Sensed Data*, Purdue University, Indiana, USA, pp. 291–297.

Cracknell, A. P., 1997. *The Advanced Very High Resolution Radiometer*. Taylor and Francis: London.

Cracknell, A. P., 2001. The exciting and totally unanticipated success of the AVHRR in applications for which it was never intended. *Advances in Space Research*, 28: 233–240.

Csiszar, I. and J. Sullivan, 2002. Recalculated pre-launch saturation temperatures of the AVHRR 3.7 μm sensors on board the TIROS-N to NOAA–14 satellites. *International Journal of Remote Sensing*, 23: 5271–5276.

Curran, P. J. and C. M. Steele, 2005. MERIS: the re-branding of an ocean sensor. *International Journal of Remote Sensing*, 26(9): 1781–1798.

Duggin, M. L., D. Piwinski, V. Whitehead, and G. Tyland, 1982. Evaluation of NOAA AVHRR data for crop assessment. *Applied Optics*, 21(11): 1873–1875.

Dwyer, E., S. Pinnock, J. M. Gregoire, and J. M. C. Pereira, 2000. Global spatial and temporal distribution of vegetation fire as determined from satellite observations. *International Journal of Remote Sensing*, 21(6 and 7): 1289–1302.

Eidenshink, J. C. and J. L. Faundeen, 1994. The 1 km AVHRR global land data set – 1st stages in implementation. *International Journal of Remote Sensing*, 15(17): 3443–3462.

El Saleous, N. Z., E. F. Vermote, C. O. Justice, J. R. G. Townshend, C. J. Tucker, and S. N. Goward, 2000. Improvements in the global biospheric record from the Advanced Very High Resolution Radiometer (AVHRR), *International Journal of Remote Sensing*, 21(6): 1251–1277.

Fensholt, R., I. Sandholt, and S. Stisen, 2006a. Evaluating MODIS, MERIS, and VEGETATION Vegetation Indices Using

In Situ Measurements in a Semiarid Environment. *IEEE Transactions on Geoscience and Remote Sensing*, 44(7): 1774–1786.

Fensholt, R., I. Sandholt, S. Stisen, and C. J. Tucker, 2006b. Analyzing NDVI for the African continent using the geostationary meteosat second generation SEVIRI sensor. *Remote Sensing of Environment*, 101(2): 212–229.

Friedl, M. A., D. K. McIver, J. C. F. Hodges, X. Y. Zhang, D. Muchoney, A. H. Strahler, C. E. Woodcock, S. Gopal, A. Schneider, A. Cooper, A. Baccini, F. Gao, and C. Schaaf, 2002. Global land cover mapping from MODIS algorithms and early results. *Remote Sensing of Environment*, 83: 287–302.

Fung, I. Y., C. J. Tucker, and K. C. Prentice, 1987. Application of AVHRR Vegetation Index to Study Atmosphere-Biosphere Exchange of CO_2. *Journal of Geophysical Research*, 92: 2999–3015.

Gallo, K., D. Tarpley, K. Mitchell, I. Csiszar, T. Owen, and B. Reed, 2001. Monthly Fractional Green Vegetation Cover Associated with Land Cover Classes of the Conterminous USA. *Geophysical Research Letters*, 28: 2089–2092.

Giglio, L., J. D. Kendall, and C. O. Justice, 1999. Evaluation of global fire detection algorithms using simulated AVHRR infrared data. *International Journal of Remote Sensing*, 20: 1947–1985.

Giglio, L., J. Descloitres, C. O. Justice, and Y. Kaufman, 2003. An enhanced contextual fire detection algorithm for MODIS. *Remote Sensing of Environment*, 87: 273–282.

Goward, S. N., C. J. Tucker, and D. G. Dye, 1985. North-American vegetation patterns observed with the NOAA-7 advanced very high-resolution radiometer. *Vegetatio*, 64(1): 3–14.

Goward, S. N., A. Kerber, D. Dye, and V. Kalb, 1987. Comparison of North and South American biomes from AVHRR observations. *Geocarto International*, 2: 27–40.

Gray, T. I. and D. G. McCrary, 1981a. *Meteorological Satellite Data – A Tool to Describe the Health of the World's Agriculture*. AgRISTARS Report EW-NI-04042, Johnson Space Center, Houston, Texas, 7.

Gray, T. I. and D. G. McCrary, 1981b. The environmental vegetation index, a tool potentially useful for arid land management. In: *Proc. 15th Conf. on Agriculture and Forest Meteorology and 5th Conf. on Biometeorology*, American Meteorological Soc., pp. 205–207.

Greegor, D. H. and J. Norwine, 1981. *A Gradient Model of Vegetation and Climate Utilizing NOAA Satellite Imager – Phase I: Texas Transect*. AgRISTARS Report JSC-17435, FC-J1–04176, Johnson Space Center, Houston 58.

Group on Earth Observations, 2006, About GEO. http://www.earthobservations.org/about_geo.shtml (last accessed 29 December 2007).

Guenther B., X. Xiong, V. V. Salomonson, W. L. Barnes, and J. Young, 2002. On-orbit performance of Earth Observing System Moderate Resolution Imaging Spectroradiometer; first year of data. *Remote Sensing of Environment*, 83: 16–30.

Gutman, G., and A. Ignatov, 1998. The derivation of the green vegetation fraction from NOAA/AVHRR data for use in numerical weather prediction models. *International Journal of Remote Sensing*, 19(8): 1533–1543.

Hall, D. K., G. A. Riggs, V. V. Salomonson, N. E. Di Girolamo, and K. J. Bayr, 2002. MODIS Snow-Cover products. *Remote Sensing of the Environment*, 83: 181–194.

Hansen, M. C. and R. S. DeFries, 2004, Detecting long-term global forest change using continuous fields of tree-cover maps from 8-km advanced very high resolution radiometer (AVHRR) data for the years 1982–99. *Ecosystems*, 7(7): 695–716.

Hansen, M. C., R. S. DeFries, J. R. G. Townshend, L. Marufu, and R. Sohlberg, 2002. Towards an operational MODIS continuous field of percent tree cover algorithm examples using AVHRR and MODIS. *Remote Sensing of Environment*, 83: 303–319.

Huete, A., K. Didan, T. Miura, E. P. Rodríguez, X. Gao, and R. Ferreira, 2002. Overview of the radiometric and biophysical performance of the MODIS vegetation indices. *Remote Sensing of the Environment*, 83: 195–213.

Holben, B. N., 1986. Characteristics of maximum-value composite images from temporal AVHRR data. *International Journal of Remote Sensing*, 7: 1417–1434.

Holben, B. N., and R. S. Fraser, 1984. Red and near-infrared sensor response to off-nadir viewing. *International Journal of Remote Sensing*, 5: 145–160.

Holben, B. N., D. Kimes, and R. S. Fraser, 1986. Directional reflectance response in AVHRR red and near-ir bands for three cover types and varying atmospheric conditions. *Remote Sensing of Environment*, 19: 213–236.

Holben, B. N., Y. J. Kaufman, and J. Kendall, 1990. NOAA-11 AVHRR visible and near-infrared inflight calibration. *International Journal of Remote Sensing*, 11: 1511–1519.

James, M. E. and S. N. V. Kalluri, 1994. The Pathfinder AVHRR Land Data Set – an improved coarse resolution data set for terrestrial monitoring. *International Journal of Remote Sensing*, 15(17): 3347–3363.

Justice, C. O., 1986, Special edition entitled 'Monitoring the Grasslands of Semi-Arid Africa Using NOAA AVHRR Data'. *International Journal of Remote Sensing*, 7: 1383–1622.

Justice, C. O., J. R. G. Townshend, B. N. Holben, and C. J. Tucker, 1985. Analysis of the phenology of global vegetation using meteorological satellite data. *International Journal of Remote Sensing*, 6(8): 1272–1318.

Justice, C. O., T. Eck, B. N. Holben, and D. Tanre, 1991. The effect of water vapor on the NDVI derived for the Sahelian region from NOAA–AVHRR data. *International Journal of Remote Sensing*, 12(6): 1165–1188.

Justice, C. O., L. Giglio, S. Korontzi, J. Owens, J. T. Morisette, D. Roy, J. Descloitres, S. Alleaume, F. Petitcolin, and Y. Kaufman, 2002. The MODIS Fire Products. *Remote Sensing of Environment*, 83(1–2): 244–262.

Kendall, J. D., C. O. Justice, P. R. Dowty, C. D. Elvidge, and J. G. Goldammer, 1997. Remote Sensing of fires in southern Africa during the SAFARI 1992 campaign, In *Fire in Southern Africa*, edited by B. W. van Wilgen, M. O. Andreae, J. G. Goldammer, and J. A. Lindesay, pp. 89–133. Witwatersrand University Press, Johannesburg.

Kidwell, K. B., 1988. *NOAA Polar Orbital Data Users Guide* (NOAA National Climate Data Center, Washington, D.C.).

Kidwell, K. B., 1990. *Global Vegetation Index Users Guide* (NOAA National Climate Data Center, Washington, D.C.).

Loveland, T. R., B. C. Reed, J. F. Brown, D. O. Ohlen, Z. Zhu, L. Yang, and J. W. Merchant, 2000. Development of a global land cover characteristics database and IGBP Discover from 1 km AVHRR data. *International Journal of Remote Sensing*, 21(6–7): 1303–1330.

Macdonald, R. B. and F. G. Hall, 1980. Global crop forecasting. *Science*, 208(4445): 670–679.

Morisette, J. T., J. L. Privette, and C. O. Justice, 2002. A framework for the validation of MODIS land products. *Remote Sensing of Environment*, 83(1–2): 77–96.

Murakami, H., M. Yoshida, K. Tanaka, H. Fukushima, M. Toratani, A. Tanaka, and Y. Senga, 2005. Vicarious calibration of ADEOS-2 GLI visible to shortwave infrared bands using global datasets. *IEEE Transactions on Geoscience and Remote Sensing*, 43(7): 1571–1584.

Murphy, R. E., 2006 The NPOESS Preparatory Project. In J. J. Qu et al., Earth Science Satellite Remote Sensing Vol. 1: Science and Instruments, Chapter 10, pp. 183–198, Tsinghua University Press and Springer Verlag, Berlin.

Murphy, R. E., W. L. Barnes, A. I. Lyapustin, J. Privette, C. Welsch, F. De Luccia, H. Swenson, C. F. Schueler, P. E. Ardanuy, and P. S. M. Kealey, 2001. Using VIIRS to provide data continuity with MODIS. In *Proc., IEEE International Geoscience and Remote Sensing Symposium* (IGARSS 2001), vol. 3, 1212–1214.

Murphy, R. E., P. Ardanuy, F. J. DeLuccia, J. E. Clement, and C. F. Schueler, 2006. The Visible Infrared Imaging Radiometer Suite. In J. J. Qu et al., *Earth Science Satellite Remote Sensing Vol. 1: Science and Instruments*, Chapter 3, pp. 33–49, Tsinghua University Press and Springer Verlag.

Myneni, R. B., F. G. Hall, P. J. Sellers, and A. L. Marshak, 1995. The interpretation of spectral vegetation indexes. *IEEE Transactions on Geoscience and Remote Sensing*, 33: 481–486.

Myneni, R. B., C. D. Keeling, C. J. Tucker, G. Asrar, and R. R. Nemani, 1997. Increased plant growth in the northern high latitudes from 1981–1991. *Nature*, 386: 698–702.

Myneni, R. B., S. Hoffman, Y. Knyazikhin, J. L. Privette, J. Glassy, Y. Tian, Y. Wang, X. Song, Y. Zhang, G. R. Smith, A. Lotsch, M. Freidl, J. T. Morisette, P. Votova, R. R. Nemani, and S. W. Running, 2002. Global products of leaf area and fraction absorbed PAR from one year of MODIS data. *Remote Sensing of Environment*, 83: 214–231.

Nickeson, J. E., J. T. Morisette, J. L. Privette, C. O. Justice, and D. E. Wickland, 2007. Coordinating Earth Observation System Land Validation. *EOS Transactions*, 81–82 (Feb. 13).

Ormsby, J. P., 1982. Classification of simulated and actual NOAA-6 AVHRR data for hydrological land-surface feature definition. *IEEE Transactions on Geoscence and Remote Sensing*, GE-20: 262–268.

Pinheiro, A. C. T., R. Mahoney, J. L. Privette, and C. J. Tucker, 2006. Development of a daily long term record of NOAA-14 AVHRR land surface temperature over Africa. *Remote Sensing of Environment*, 103(2): 153–164.

Potter, C. S., J. T. Randerson, C. B. Field, P. A. Matson, P. M. Vitousek, H. A. Mooney, and S. A. Klooster, 1993. Terrestrial ecosystem production – a process model-based on global satellite and surface data. *Global Biogeochemical Cycles*, 7(4): 811–841.

Prince, S. D., 1991a. Satellite remote-sensing of primary production – comparison of results for Sahelian grasslands 1981–1988. *International Journal of Remote Sensing*, 12(6): 1301–1311.

Prince, S. D., 1991b. A model of regional primary production for use with coarse resolution satellite data. *International Journal of Remote Sensing*, 12(6): 1313–1330.

Prince, S. D. and C. O. Justice, 1991. Special edition entitled Coarse Resolution Remote Sensing of the Sahelian Environment. *International Journal of Remote Sensing*, 12: 1133–1421.

Rast, M., J. L. Bezy, and S. Bruzzi, 1999. The ESA Medium Resolution Imaging Spectrometer MERIS a review of the instrument and its mission. *International Journal of Remote Sensing*, 20(9): 1681–1702.

Roy, D. P., J. S. Borak, S. Devadiga, R. E. Wolfe, M. Zheng, and J. Descloitres, 2002. The MODIS land product quality assessment approach. *Remote Sensing of the Environment*, 83: 62–76.

Roy, D. P., Y. Jin, P. E. Lewis, and C. O. Justice, 2005. Prototyping a global algorithm for systematic fire-affected area mapping using MODIS time series data. *Remote Sensing of Environment*, 97: 137–162.

Running, S. W., R. R. Nemani, F. A. Heinsch, M. S. Zhao, M. Reeves, and H. Hashimoto, 2004. A continuous satellite-derived measure of global terrestrial primary production. *Bioscience*, 54(6): 547–560.

Saint, G. 1995. Spot-4 vegetation system – association with high-resolution data for multiscale studies. *Advances in Space Research*, 17: 107–110.

Salomonson, V., W. Barnes, P. Maymon, H. Montgomery, and Ostrow, H., 1989, MODIS: advanced facility instrument for studies of the Earth as a system. *IEEE Transactions on Geoscience and Remote Sensing*, 27(2): 145–153.

Schaaf, C. B, F. Gao, A. Strahler, W. Lucht, X. Li, T. Tsang, N. C. Strugnell, X. Zhang, Y. Jin, J. P. Muller, P. Lewis, M. Barnsley, P. Hobson, M. Disney, G. Roberts, M. Dunderdale, C. Doll, R. P. d'Entremont, B. Hu, S. Liang, J. L. Privette, and D. Roy, 2002. First operational BRDF, albedo, nadir reflectance products from MODIS. *Remote Sensing of Environment*, 83: 135–148.

Schneider, S. R. and D. F. McGinnis, 1982. The NOAA AVHRR: A new satellite sensor for monitoring crop growth. In: *Proc. 8th Int. Symp. Machine Processing Remotely Sensed Data*, Purdue University, Indiana, USA, pp. 281–290.

Tanre, D., B. N. Holben, and Y. J. Kaufman, 1992. Atmospheric correction algorithm for NOAA-AVHRR products – theory and application. *IEEE Transactions on Geoscience and Remote Sensing*, 30(2): 231–248.

Tarpley, J. P., S. R. Schneider, and R. L. Money, 1984. Global vegetation indices from NOAA-7 meteorological satellite. *Journal of Climate and Applied Meteorology*, 23, 491–494.

Townshend, J. R. G., 1994. Global data sets for land applications from the advanced very high-resolution radiometer – an Introduction. *International Journal of Remote Sensing*, 15(17): 3319–3332.

Townshend, J. R. G. and C. O. Justice, 2002. Towards operational monitoring of terrestrial systems by moderate-resolution remote sensing. *Remote Sensing of Environment*, 83(1–2), 351–359.

Townshend, J. R. G. and C. J. Tucker, 1981. Utility of AVHRR NOAA-6 and -7 for vegetation mapping. In: *Proc. Matching Remote Sensing Technologies and Their Applications Proceedings*, Remote Sensing Society, London, pp. 97–109.

Townshend, J. R. G., T. E. Goff, and C. J. Tucker, 1985. Multi-temporal dimensionality of images of normalized difference vegetation index at continental scales. *IEEE Transactions on Geoscience and Remote Sensing*, GE-23(6), 888–895.

Townshend, J. R. G., C. O. Justice, and V. Kalb, 1987. Characterization and classification of South American land cover types using satellite data. *International Journal of Remote Sensing*, 8: 1189–1207.

Townshend, J. R. G., C. O. Justice, W. Li, C. Gurney, and J. McManus, 1991. Global land classification by remote sensing: present capabilities and future prospects. *Remote Sensing of Environment*, 35: 243–256.

Townshend, J. R. G., C. O. Justice, D. Skole, J.-P. Malingreau, J. Cihlar, P. Teillet, F. Sadowski, and S. Ruttenbert, 1994. The 1 km resolution global data set – needs of the International Geosphere Biosphere Program. *International Journal of Remote Sensing*, 15(17): 3417–3441.

Tucker, C. J., 1979. Red and photographic infrared linear combinations for monitoring vegetation. *Remote Sensing of Environment*, 8(2): 127–150.

Tucker, C. J., J. H. Elgin, and J. E. McMurtrey, 1979. Temporal spectral measurements of corn and soybean crops. *Photogrammetric Engineering and Remote Sensing*, 45(5): 643–653.

Tucker, C. J., B. N. Holben, J. H. Elgin, and J. E. McMurtrey, 1981. Remote sensing of total dry matter accumulation in winter wheat. *Remote Sensing of Environment*, 11: 171–189.

Tucker, C. J., C. L. Vanpraet, E. Boerwinkel, and A. Gaston, 1983. Satellite remote sensing of total dry matter production in the Senegalese Sahel: 1980–1984. *Remote Sensing of Environment*, 13: 461–474.

Tucker, C. J., J. R. G. Townshend, and T. E. Goff, 1985. African land cover classification using satellite data. *Science*, 227: 369–375.

Tucker, C. J., I. Y. Fung, C. D. Keeling, and R. H. Gammon, 1986. Relationship between atmospheric CO_2 variations and a satellite-derived vegetation index. *Nature*, 319: 195–199.

Tucker, C. J., H. E. Dregne, and W. W. Newcomb, 1991. Expansion and contraction of the Sahara Desert from 1980 to 1990. *Science*, 253: 299–301.

Tucker C. J., J. E. Pinzon, M. E. Brown, D. A. Slayback, E. W. Pak, R. Mahoney, E. F. Vermote, and N. El Saleous, 2005. An extended AVHRR 8-km NDVI dataset compatible with MODIS and SPOT vegetation NDVI data. *International Journal of Remote Sensing*, 26(20): 4485–4498.

Vermote, E. F. and Y. J. Kaufman, 1995. Absolute calibration of AVHRR visible and near infrared channels using ocean and cloud views. *International Journal of Remote Sensing*, 16(13): 2317–2340.

Vermote, E. F., N. El Saleous, C. O. Justice, Y. K. Kaufman, J. L. Privette, L. Remer, J. C. Roger, and D. Tanre, 1997. Atmospheric correction of visible to middle infrared EOS MODIS data over land surfaces: background, operational algorithm and validation. *Journal of Geophysical Research*, 102(D14): 17131–17141.

Vermote, E. F., N. Z. El Saleous, and C. O. Justice, 2002. Atmospheric correction of MODIS data in the visible to middle-infrared first results. *Remote Sensing of Environment*, 83(1–2): 97–111.

Vermote, E. F., J. C. Roger, A. Sinyuk, N. Z. Saleous, and O. Dubovik, 2007. Fusion of MODIS-MISR aerosol inversion for estimation of aerosol absorption. *Remote Sensing of Environment*, 107(1–2): 81–89.

Wan, Z., Y. Zhang, Q. Zhang, and Z. Li, 2002. Validation of the land surface temperature products retrieved from Terra MODIS data. *Remote Sensing of Environment*, 83: 163–180.

Wolfe, R., M. Nishihama, A. J. Fleig, J. Kuyper, D. P. Roy, J. C. Storey, and F. Patt, 2002. Achieving sub-pixel geolocation accuracy in support of MODIS land science. *Remote Sensing of Environment*, 83: 31–49.

Yates, H. W. and J. D. Tarpley, 1982. The role of meteorological satellites in agricultural remote sensing. In: *Proc. 8th Int. Symp. Machine Processing Remotely Sensed Data*, Purdue University, Indiana, USA, pp. 23–32.

Zarco-Tejada, P. J., C. A. Rueda, and S. L. Ustin, 2003. Water content estimation in vegetation with MODIS reflectance data and model inversion methods. *Remote Sensing of Environment*, 85: 109–124.

Zhan, X., R. A. Sohlberg, J. R. G. Townshend, C. DiMiceli, M. L. Carroll, J. C. Eastman, M. C. Hansen, and R. S. DeFries, 2002. Detection of land cover changes using MODIS 250 m data. *Remote Sensing of Environment*, 83: 336–350.

Airborne Digital Multispectral Imaging

Douglas A. Stow, Lloyd L. Coulter, and
Cody A. Benkelman

Keywords: airborne, digital, multispectral, fine resolution, platforms, navigation systems, photography, videography, large format, line array, frame array, image processing, mapping.

INTRODUCTION

From a practical standpoint, digital image data from airborne multispectral sensors may be the most utilized type of remotely sensed data. The ASPRS 10-Year Industry Forecast (ASPRS 2004a) indicated that aerial imagery represents 66% of total US imagery sales and that this market share is expected to continue. The fine spatial resolution and color and/or color infrared (CIR) viewing attributes make this type of image data amenable to a wide range of applications and readily available at a reasonable cost. Geometric processing techniques are well advanced for generating orthoimages and image mosaics from airborne digital multispectral image data (ADMID). Such orthoimages (also known as image maps) are readily exploited by geographic information system (GIS) users as a backdrop for viewing, or as a primary or secondary data source for generating and/or updating thematic GIS layers. Scientists, managers, planners, business people, educators, and students alike, commonly utilize digital images from airborne multispectral systems for reconnaissance, inventory, mapping, and monitoring applications.

The main components of an airborne digital multispectral imaging system (ADMIS) are: (1) an aircraft platform, (2) an imaging sensor, (3) imaging support tools, and (4) software for improving the geometric and radiometric fidelity of the raw image data. Typically, fixed-wing aircraft platforms are utilized, though rotary-wing (i.e., helicopters) can also be employed. Sensor technology employed to capture digital multispectral image data varies markedly, with the two major categories being: (1) film-based aerial cameras combined with laboratory scanners and (2) direct digital imaging systems. The latter can be achieved through line scanner, line array, video, and single or multiple digital camera (i.e., framing array) sensors.

The advantages of ADMIS relative to satellite imaging systems are the potential for finer spatial resolution (typically in the 0.02–2 m range), lower cost for small area coverage, and greater flexibility in timing and spatial extent of coverage. Relative to film-based aerial photography, the key advantages are the greater ability to quantify and automate image interpretation, and greater flexibility in image enhancement and restoration. The key disadvantages of ADMIS relative to comparable satellite systems are greater platform instability, need to mobilize aircraft platforms for each mission, more limited spatial coverage, greater range of view angles and therefore anisotropic reflection effects, and typically, more limited spectral and radiometric range.

This chapter is structured into four sections: (1) Airborne Platforms, (2) Legacy Sensor Systems, (3) State-of-the-Art Sensor Systems, and (4) Image Processing and Analysis. The emphasis is on ADMIS that are or will soon be utilized most commonly to provide very fine spatial resolution, broad-band visible and near infrared (VNIR) image data for earth resource applications. Related sensor systems that are not covered or emphasized are true scanner, hyperspectral imaging, panchromatic-only, oblique-viewing camera, or discrete (i.e., operating in the ultraviolet, middle infrared, or thermal infrared portions of the EMR spectrum) sensor systems.

AIRBORNE PLATFORMS

Aerial platforms for digital multispectral imaging include unmanned aerial vehicles (UAVs), microlights, helicopters, single or multi-engine fixed-wing aircraft, and jet propelled aircraft. The selection of the airborne platform depends upon the anticipated imaging mission parameters and the imaging system to be used, as well as a clear business model for the aircraft/sensor operation (ASPRS 2004b). Mission parameters to consider include flying altitude, minimum velocity (for large-scale imaging), extent of areas to be imaged, and transit distances. Airborne digital multispectral imaging systems range from single-chip cameras to multiple sensor/array systems with associated information technology racks, each having varying size, weight, and power requirements.

Unmanned aerial vehicles (UAVs) are produced world wide and vary substantially in terms of size, weight, payload weight carrying capacity, space for sensors, and operating parameters (altitude, range, and duration) (NASA Research Park 2007). The use of UAVs for aerial imaging has largely been limited to the government and scientific community, but these platforms will eventually have a place in the photogrammetric community (ASPRS 2004b). With that said, commercial applications for UAV-based imaging are growing (Herwitz et al. 2004), and groups like the UAV Collaborative are working to further the use of UAVs for commercial very fine resolution imaging (NASA Research Park 2007). UAVs offer the potential for long duration flights and reduced operating costs. Physical space for instruments is often limited and remote operation or monitoring is required.

Microlight aircraft are suitable for small area, large-scale imaging with small format cameras. Microlight benefits include low purchase and/or operational costs and flexible take-off and landing capability near the imaging site, to take advantage of clear weather windows. Disadvantages generally

include limited (low) altitude, velocity, transit range, and stability.

Helicopters enable low altitude flying at reduced velocities for large-scale imaging with high percentage overlap. These rotary aircraft are less stable and operational costs are generally higher than those of fixed-wing aircraft. Helicopters also have relatively low altitude, velocity, and transit ranges.

Single engine fixed-wing aircraft are suitable for local area, large-scale imaging at altitudes less than 5,500 m (18,000 ft) (Falkner and Morgan 2001). Multi-engine fixed-wing aircraft are ideal for large area imaging and have high velocity, altitude, and transit ranges, as well as suitable space for equipment. Multi-engine aircraft commonly have cruising speeds greater than 200 knots and altitude ceilings between 7,600 m (25,000 ft) and 10,600 m (35,000 ft), with the higher altitudes within this range requiring turbo-prop engines. Fixed-wing single and multi-engine aircraft provide a stable operating environment and have sufficient space to house large camera systems. An informal survey conducted by the authors found that operators of large format digital camera systems (DMC, ADS40, and UltraCam) within the U.S. most commonly used multi-engine Piper PA-31 Navajo & Chieftain, Cessna 402, 404, 414, 441, T310R, Piper Aztecs, and Commander aircraft. Jet propelled aircraft are generally better suited for broad area coverage sensors such as synthetic aperture radar (SAR), but have also been used to collect ADMID using the Leica ADS40 sensor (ASPRS 2004b).

Every aspect of operating an aircraft becomes more expensive as the aircraft size increases. The costs include daily maintenance, mandatory periodic inspections, fuel costs, hanger-space, landing fees, insurance, etc. Warner et al. (1996) describe how both fixed annual costs and direct operating costs can be derived and used to calculate an hourly aircraft cost, based on an estimate of flight hours per year.

System mounting and platform stabilization

Camera mounts consist of three types: (1) internal (hole or window in floor), (2) external (attached externally to aircraft), and (3) extending (structure is attached to fuselage, but camera is outside) (Warner et al. 1996). Basic functional requirements for camera mounts include being able to: (1) level the camera and correct for aircraft pitch and roll, (2) correct for horizontal rotation (crabbing) of the aircraft, (3) dampen aircraft vibration and random aircraft movements, and (4) control the camera and mount corrections during flight. Advanced systems such as gyro-stabilized suspension mounts actively compensate for these angular movements and continuously measure the angular velocities of pitch, roll, and drift movements of the aircraft (ASPRS 2004b).

Forward motion compensation

Forward motion compensation (FMC) technology compensates for apparent image motion (AIM) and resulting image blur that is caused by movement of the ground (along the flight line axis) relative to the focal plane during the period of image exposure. AIM is a function of aircraft velocity, camera focal length, exposure time, and altitude of aircraft, and is calculated as:

$$\text{AIM} = \left(V_g \times f \times t\right)/H_g \qquad (1)$$

where V_g is the aircraft ground velocity (m/s), f is the focal length of the lens (mm), H_g is the aircraft height (m) above ground, and t is the camera shutter speed (s) (Warner et al. 1996). It follows that AIM is reduced with decreased velocity, focal length, or exposure time and with increased aircraft altitude. Most large format camera systems employ FMC technology to mitigate the effects of AIM, either mechanically through physical motion of the sensor or electronically by time delayed integration (TDI).

Navigation systems

GPS is a satellite-based system operated by the United States Department of Defense which enables users with GPS receivers to determine their position (in X, Y, and Z), velocity, and heading in real-time. The error of GPS positions can vary from 10 m with uncorrected pseudo-random code GPS data (with Selective Availability off), to 0.5–3 m with differentially corrected pseudo-random code data, to 2–5 cm with carrier phase differential positioning (ASPRS 2004b).

Avionics and auxiliary navigation systems incorporating GPS data are important tools for airborne imaging. Avionics receivers are normally used by pilots to navigate the aircraft during transit from one location to another. Aerial survey flight management systems (ASFMS) utilize hardware and software for accurate positioning of aerial imagery from the flight planning stages, to flight line navigation, to determination of actual camera exposure stations (referred to as airborne GPS or AGPS). Graphic displays assist the pilot with navigation and the operator with mission quality control. More complex ASFMS systems trigger camera exposure at pre-planned positions (ASPRS 2004b).

Inertial navigation systems (INS) have been used to provide position, velocity, attitude (roll, pitch, heading), accelerations, and angular rates of airborne vehicles since 1949 (Mostafa 2001). INSs use an inertial measurement unit (IMU) and navigation processor (NP) to measure the movements of the aircraft and to compute the current position and attitude (ASPRS 2004b). However, an INS is a dead reckoning system which navigates from an initial known point and is subject to smooth but unbounded (limitless) error and drift. GPS-derived position and velocity errors are noisy, but bounded (i.e., have limited magnitude). GPS and INS complement each other due to their opposing error characteristics, and GPS-aided INS (GPS-AINS) has become the preferred method for collecting accurate position (X, Y, and Z) and exterior orientation parameters (Ω, ϕ, and κ) data during flight. These exterior orientation data may be used to supplement or replace aero-triangulation through direct georeferencing (Mostafa 2001).

LEGACY SENSOR SYSTEMS

The focus of this section is on the legacy sensor systems that have used analog (film-based) or direct digital approaches to generate ADMID over the past few decades. The basic designs, advantages, and disadvantages of scanned aerial photography, airborne video, and single and multi-sensor digital camera systems are reviewed.

Aerial photography and film scanning

The first ADMID were generated with optical mechanical multispectral scanners flown on large fixed-wing aircraft, but the difficulty in geometrically correcting and georeferencing airborne scanner data led to the production of digital orthophotographs derived by scanning and geometrically processing aerial photographs. For multispectral digital imagery to be generated from aerial photography, color, or CIR aerial photographs are scanned by an electro-optical imaging device. Large format (23 × 23 cm) metric aerial photographs are normally utilized. Film scanners have evolved over the years, with most modern scanners being based on line- or area-array technologies that are similar to those used for direct digital airborne imaging (Jensen 2007). The main requirements for laboratory scanners when generating ADMID are high geometric and radiometric fidelity, and the ability to quantify image brightness in three colors (red, green, and blue, or RGB). Digital orthophotographs, such as the CIR Digital Orthophoto Quadrangles (DOQs) produced by the US Geological Survey, are a source of ADMID that have been generated by scanning of high-altitude CIR aerial photographs and subsequently orthorectifying the image scans with the aid of a digital elevation model (DEM).

Advantages and disadvantages

The primary advantage of scanning film is the relatively low cost of production, primarily due to the large format and extensive coverage associated with aerial photographs. An additional advantage to be considered is the potential to access vast archives of film that could be processed using modern techniques to extract valuable historical data.

There are several disadvantages that are primarily pertinent to quantitative analyses of film-based ADMID. First, quantitative analysis of scanned color or CIR aerial photography requires compensation for the fundamentally nonlinear nature of the film response. Also, varitions in the spectral sensitivity of the film negative and inconsistencies between different rolls of film, film handling, film development chemistry, and image scanning make it difficult to normalize spectral-radiometric response between frames and dates of imagery. Another disadvantage is the bulky and inefficient nature of processing and archiving analog aerial photographs.

Airborne video systems

ADMIS have been constructed using separate video cameras as the sensors (Everitt and Nixon 1985, Neale and Crowther 1994), and as an integrated color infrared video camera such as the BioVision sensor (Stutte 1991). Past airborne video systems required use of a 'frame grabber' to electronically capture individual still frames from the video stream. Such airborne video sensors have largely been replaced by single-frame digital cameras with far superior resolution and optics. However, for applications with a need for full motion video and/or near-real time data, a video system hosted on an airborne platform can provide significant value.

For most video applications, a single charge-coupled device (CCD) color camera is adequate, although some recent military applications also utilize non-CCD sensors (e.g., thermal infrared) and employ image fusion to merge visible imagery with data from other imaging modes. As a result, both single-camera and multi-camera video systems have been constructed, and they share many of the design considerations as CCD camera systems described in the following sections. Key challenges include the video standard, hardware ruggedness, and videotape, as discussed below.

Limitations of the video standard

For video cameras that meet the Electronics Industries Association (EIA) video standards (e.g., RS-170, RS-170A, or RS-330), the effective video image is comprised of 485 lines. This results in a resolution limit that, from today's perspective, seems very poor, with images typically captured at 480 lines by 640 pixels per line. Digitizing airborne video is further complicated by the refresh rate of 30 frames/second, with each image composed of two 'interlaced fields' showing only the odd or even lines, refreshed alternatively in 1/60 second. When this hardware is employed in an ADMIS, aircraft motion between (and during!) the individual 1/60 second video fields presents a challenge for both post-flight data processing, as well as effective coverage of a target site on the ground.

Recording on videotape

A past alternative to digitizing single video frames in flight with a frame grabber was to simply store all of the airborne data on videotape and then digitize with a laboratory frame grabber. This approach had the advantage of capturing all data, thus avoiding the risk of missing a critical image due to poor timing of the capture signal and a generally simpler design for the airborne equipment. Disadvantages of videotape include the incremental quality lost with analog recording and playback (presuming a frame grabber was used on the ground to digitize individual frames), as well as the inefficiency of sequential access to data on the videotape, rather than random access to images stored on a hard disk.

Multiple-sensor camera systems

Multiple-sensor systems are comprised of several imaging sensors (e.g., digital video or CCD array cameras), with filters used in the optical path to create a unique waveband image from each sensor. In multi-camera systems such as the DMSV/DMSC (Anderson et al. 1997), DSS (Applanix 2007), the Spectra-View (Airborne Data Systems 2006), or the ADAR sensor systems (Stow et al. 1996) that have separate optics for each sensor, spectral filtering is typically achieved using broadband absorption filters, although narrowband interference filters can also be employed. When multiple sensors are integrated into a single camera head, dichroic mirrors are normally used to provide spectral filtering (Geospatial Systems 2006).

Multi-sensor systems can utilize unique detector compounds (e.g., GaAs, InSb, etc.) to allow acquisition of imagery in spectral bands outside of the VNIR (e.g., ultraviolet, shortwave, or long-wave infrared, etc.). This discussion is limited to the VNIR sensors, most typically acquired using CCD detectors composed of silicon. For multi-sensor systems, the primary design challenges are sensor co-registration, spectral filtering, physical packaging, and metric optics, as discussed below.

Sensor co-registration

Alignment of the optics and sensors to allow accurate co-registration of all sensors (ideally to a fraction of a pixel) is very demanding, especially if the cameras and optics are commercial off-the-shelf (COTS) hardware. In a fully customized system (e.g., multiple sensors integrated into a single camera body, perhaps using a common optical path), the hardware problem associated with band-to-band registration is manageable, but with a corresponding increased cost in both mechanical design and manufacturing. If not solved in hardware, image co-registration must be addressed in software as a post-capture process, but software solutions remain approximate (Brown 1992, Goshtasby 2005).

Spectral filtering

Another design challenge is achieving both high throughput and sharp cut-on and cut-off for the spectral band-passes. Presuming relatively broad wavebands are desired, Schott glass absorption filters are often employed in multi-camera systems. Although these filters are rugged and stable across a broad temperature range, they do not provide sharp transitions from in-band throughput to out-of-band rejection. In addition, the number of choices of different spectral filters is very limited.

If narrow spectral bands are required, thin film interference filters (also called dichroic filters) can be employed, providing a much larger selection of distinct spectral band positions and bandwidths, and very high transmissions. Filters are available off-the-shelf with bandwidths as narrow as 3 nm, and custom-made filters are available. However, since the spectral bandpass is based on constructive interference, the cut-on and cut-off wavelengths depend on the angle of incidence of incoming light. Use of interference filters typically demands a sophisticated, custom optical system to ensure the spectral response is consistent across the full field-of-view. Figure 11.1 shows the configuration of the Geospatial Systems MS-3100 sensor, a three-band system based on dichroic interference filters, and a single collection and focusing optical assembly.

Physical packaging

Typical ADMIS are designed to fit within the 15′ or 18′ aerial film camera port, a requirement that can significantly limit the choice of sensors and optics for a multi-camera system. In the case of a custom-built camera, this design constraint can be overcome. In addition, elimination of aircraft vibration and thermal effects (temperature changes which can be relatively rapid and large in magnitude, from the ground to flight altitude) place additional demands on the physical packaging of the imaging sensors.

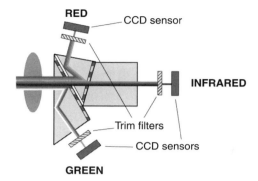

Figure 11.1   Block diagram showing custom multi-chip MS-3100 camera designed to acquire three band imagery in the green, red, and near infrared wavebands.
Source: Reprinted with permission from Geospatial Systems 2006.

Metric optical system

To enable generation of true orthorectified imagery, the optics of the imaging system must be metrically calibrated. Successful metric calibration generally requires that the optics must be rigidly mounted to the sensor, and should ideally have very low distortion characteristics. Metric calibration coefficients that specify the principal point (location where the optical axis reaches the sensor) and radial and tangential distortion coefficients are required for post-processing by photogrammetric software. As with other design considerations, including metric optics in an ADMIS is achievable, but can add a significant cost, especially in a multi-camera system. In addition, the desire for photogrammetric processing will increase other requirements of the ADMIS such as INS/IMUs.

Advantages and disadvantages

ADMIS comprised of individual sensors for each spectral band provide the best radiometric performance for quantitative remote sensing from an airborne platform. However, the higher costs associated with multiple sensors, as well as the technical challenges and costs of integrating multiple sensors into a common package have encouraged development of alternative, lower cost configurations, as described below.

Single-chip multispectral cameras

Given today's low cost digital cameras, a less expensive ADMIS can be built around a single camera, although with some significant trade-offs. CCD sensors are sensitive to incoming radiation from blue (approximately 400 nm) through near infrared (NIR, approximately 1000 nm).

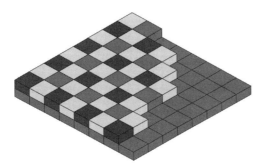

Figure 11.2 Bayer color filter pattern utilized for single-chip color and color infrared (CIR) digital cameras. Pixels shown in blue are sensitive to near infrared radiance in CIR mode. (See the color plate section of this volume for a color version of this figure).
Source: Image courtesy of Colin M. L. Burnett (http://en.wikipedia.org/wiki/Image:Bayer_pattern_on_sensor.svg).

Single-chip color cameras employ an alternating pattern of color filters bonded onto the front surface of the CCD to limit each pixel to a single color. The 'Bayer' color filter array alternates blue and green pixels on every other row, then green and red pixels on the intermediate rows, as shown in Figure 11.2 (Bayer 1976). The result is that 50% of the pixels record energy in the green portion of the spectrum, and 25% each are dedicated to blue and red. The three-color, full resolution image may then be created in software as a post-process by interpolating the values of neighboring pixels to fill in the missing information in each of the B (or NIR), G, and R color planes.

The Bayer array is the most common color configuration, but alternative approaches exist. In the 1990s, Kodak offered hand-held digital cameras (DCS CIR) that provided color infrared imagery. The CIR cameras included a Bayer array, but a NIR blocking filter in front of the array was removed and a blue-blocking filter was added, allowing NIR energy to reach the sensor. A modification to the post-capture interpolation software allowed extraction of the NIR signal and creation of CIR images. Kodak has recently announced new filtering configurations which provide red, green, blue, and unfiltered (panchromatic) pixels, with increased sensitivity in the panchromatic pixels reportedly resulting in an overall improvement in image sharpness and faster exposures (Tomkins 2007).

Advantages and disadvantages
The primary advantages of an imaging system based on a single-chip sensor are low cost and weight. Also, the image data can be stored in its raw form while on board the aircraft, thus keeping data storage requirements to a minimum.

A disadvantage of a single-chip system is that the interpolation process used to fill in spatially sub-sampled spectral values can create undesirable artifacts, such as aliasing, in the final imagery. Furthermore, since two-thirds of the image data are generated via software rather than direct measurement, their validity for quantitative analysis is questionable.

STATE-OF-THE-ART SENSOR SYSTEMS

Large format film (23 cm × 23 cm) cameras provide most of the aerial imagery for photogrammetry and GIS applications; however, large format ADMIS are now competing with film-based cameras, and the market share for large format ADMIS is expected to grow. Cramer (2005) projects that 25% or more of large format camera systems in operation in 2010 will be digital systems. Large format film cameras provide large area coverage with high detail at a relatively low cost. However, the radiometric range is generally limited to 7- or 8-bit quantization and the spectral quality is lower due to film characteristics and scanning.

State-of-the-art ADMIS are now providing large format imagery with the benefits of digital image acquisition. These benefits include: (1) finer radiometric resolution, (2) reduced operating costs per frame/image since film purchase, processing, and scanning are not required, (3) greater spectral coverage with B, G, R, and NIR wavebands acquired simultaneously, and (4) direct digital processing flows which may include the use of GPS-AINS and direct georeferencing. The availability of larger format CCD arrays, innovative image array technologies, larger hard drive capacities, and faster digital processing have made large format digital imaging more attractive.

Petrie (2003) defines large format digital sensors to be those capable of acquiring an image frame containing at least 36 megapixels. Current large format ADMIS and their specifications are listed in Table 11.1. These systems employ either push-broom line or frame arrays. The ADS40, DMC, and UltraCam$_X$ have been on the market longest and are the most commonly available systems.

Line array ADMIS

Line array ADMIS listed in Table 11.1 employ a three-line scanning technique, in which multiple linear CCDs are situated on the focal plane within

Table 11.1 State-of-the-art airborne line and frame array camera systems

Digital camera system	ADS40-SH52	HRSC-AX150	3-DAS-1	JAS 150	DMC	UltraCamₓ
Manufacturer	Leica Geosystems	DLR	Wehrli & Associates	Jena-Optronik	Intergraph Z/I Imaging®	Vexcel
Web page	http://gi.leica-geosystems.com	www.dlr.de	www.wehrliassoc.com	www.jena-optronik.com	www.intergraph.com	www.vexcel.com
First Delivery	SH40: 2001 SH52: 2007	2001	2007	2007	2003	UltraCam_D: 2004 UltraCamₓ: 2007
Number of systems delivered	46 (50 sold)[1]	3	Available in 2007	Available in 2007	46 (*51 sold*)[2]	51 (71 sold)[3]
Image size (pixels)	12,000 × any	12,000 × any	8,002 × any	12,000 × any	13,824 × 7,680	14,430 × 9,420
Number of sensor heads	1	1	3	1	8	8
Number of CCD sensors	4 pan 8 MS	5 pan 4 MS	3 blue, 3 green, 3 red	5 pan 4 MS	4 pan 4 MS	9 pan 4 MS
Size of CCD sensors (pixels)	12,000 × 1	12,000 × 1	8,002 × 1	12,000 × 1	7,000 × 4,000 (pan) 3,000 × 2,000 (MS)	4,992 × 3,328
Pixel size	6.5 μm	6.5 μm	9 μm	6.5 μm	12 μm	7.2 μm
Effective dynamic range	12 bit	12 bit	14 bit	12 bit	12 bit	13 bit
Spectral range (nm) 'Pan'	465–680	520–760	–	520–760	400–900	410–690
Spectral range (nm) 'Blue'	428–492	450–510	400–540	440–510	400–580	410–530
Spectral range (nm) 'Green'	533–587	530–570	500–610	520–590	500–650	470–660
Spectral range (nm) 'Red'	608–662	635–685	590–685	620–680	590–675	570–690
Spectral range (nm) 'NIR'	833–887	770–810	–	780–850	675–850 or 740–850 (alternate)	670–940
Maximum imaging Rate	1250 lines/s	1640 lines/s	745 lines/s	1250 lines/s	0.5 frames/s	0.75 frame/s
Field-of-view	64°	29.1°	36°	29.1°	69.3° × 42°	55° × 37°
Focal length (mm)	62.77	150	110	150	120 pan, 25 MS	100 pan, 33 MS
GPS/IMU	Mandatory Integrated	Mandatory Integrated	Mandatory Integrated	Mandatory Integrated	Optional Integrated	Optional Integrated
Pan-sharpened Imagery	No	No	No	No	Yes	Yes

[1] ADS40 information current as of 09 March 2007. Number of systems includes all ADS40 sensor models/configurations.

[2] DMC information current as of 14 February 2007. Number of systems includes all DMC sensor models/configurations.

[3] UltraCam information current as of 28 February 2007. Delivered systems include 47 UltraCam_D and 4 UltraCamₓ. Twenty additional sold but not delivered are UltraCamₓ.

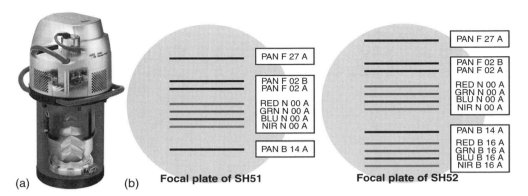

(a)   (b)   **Focal plate of SH51**   **Focal plate of SH52**

Figure 11.3   Leica ADS40 (a) digital sensor and (b) its latest sensor head focal plate configurations. Waveband, forward/nadir/backward/viewing direction, view angle, and sensor (A or B) is indicated on the focal plate diagram. (See the color plate section of this volume for a color version of this figure).
Source: Courtesy of Leica Geosystems Geospatial Imaging, LLC.

a single sensor head and view in the forward, nadir, and backward directions. This design yields 100% stereo viewing with consistent stereo angles, and multiple images are generated from a single pass of the aircraft (Chen et al. 2003). Depending upon the system and its configuration, stereo imagery may be panchromatic, color (RGB), and/or multispectral (VNIR). Further, configurations with stereo panchromatic and multispectral imagery from only one view direction are common.

Three-line scanners (TLS) use long CCD lines with 8,000–12,000 pixels that yield wide area (cross-track) image coverage per flight line. TLS systems have higher signal-to-noise ratios than digital frame cameras (Chen et al. 2003), and data readout rates from the TLS CCD lines are as fast as 1.25 ms (significantly shorter than frame camera shutter speeds). These factors enable faster acquisition and negate the need for forward motion compensation, since the apparent image motion is less then the ground resolution element of the pixel (Pacey and Fricker 2005). TLS systems are able to image large areas rapidly.

As each line of imagery is acquired over a unique time period by TLS systems, each line is associated with a unique aircraft position and orientation. Therefore, these line array systems differ from frame-based systems because they require integrated GPS/IMU systems for direct orientation support (ASPRS 2004b).

ADS40 airborne digital sensor
The Leica ADS40 is the most widely available commercial three-line scanner. Several ADS40 focal plate configurations have been produced for ADS40 operation. The most recently implemented ADS40 focal plate configurations are the SH51 and

SH52 (Figure 11.3). The SH51 and SH52 focal plates each have one forward, two nadir, and one backward viewing panchromatic line array. The two nadir viewing line arrays are staggered to enable panchromatic imaging with greater detail; images generated using the staggered arrays are known as high resolution (ASPRS 2004b). The SH51 has one set of nadir viewing line arrays for multispectral imaging, while the SH52 has one nadir viewing set and one backward viewing set (14° from nadir) for multispectral imaging. A tetrachroid optical system is used to split the B, G, R, and NIR light and record each waveband separately on individual linear CCD arrays. The tetrachroid uses cascaded dichroic beam splitters to minimize energy loss (ASPRS 2004b). The image layers generated during aerial acquisition with the SH52 focal plate are illustrated in Figure 11.4, which also illustrates the concept of a continuous 'carpet' of imagery generated from linear array pushbroom systems.

Frame array ADMIS

Large format ADMIS based on frame (area) arrays are more analogous to large format film cameras, in that images are captured from specific points and not continuously as with the line array systems. As with film cameras, relief displacement is two-dimensional and stereo angles vary as a function of the principal distance and the airbase. However, individual frame array CCDs are not, as of yet, large enough to replicate the detail and spatial coverage obtained from scanned aerial photographs. Therefore, digital frame cameras use multiple CCD arrays to create a single image. The DMC and UltraCam$_X$ cameras are the only large format frame

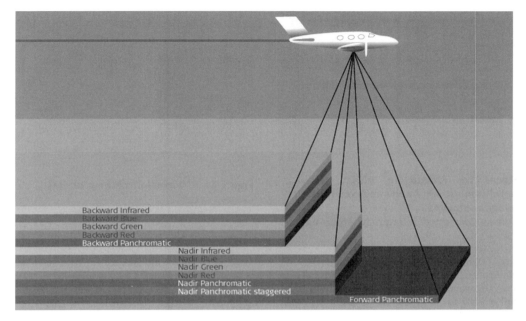

Figure 11.4   Image strips acquired using the Leica ADS40 with SH52 sensor head. (See the color plate section of this volume for a color version of this figure).
Source: Courtesy of Leica Geosystems Geospatial Imaging, LLC.

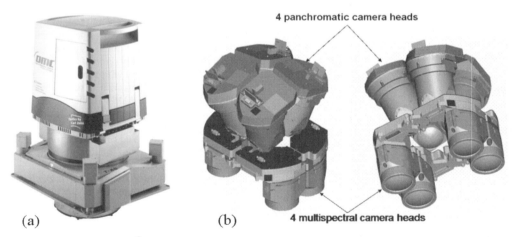

(a)   (b)

Figure 11.5   Z/I Imaging® DMC (a) digital camera system with mount and (b) camera head configuration.
Source: Courtesy of Intergraph Corporation.

array cameras on the market, and each system operates in a different manner.

DMC

Intergraph's Z/I Imaging® DMC is a modular digital camera system with four high resolution panchromatic camera heads (7000 × 4000 pixel CCDs, 120 mm focal length, f/4 lens) and four multispectral camera heads (3000 × 2000 pixel CCDs, 25 mm focal length, f/4 lens) (Figure 11.5). The four panchromatic camera heads are tilted off-nadir (approximately 10° along track and 20° across track) so that each image covers a distinct but slightly overlapping ground area (Figure 11.6) (Tang et al. 2000). These four panchromatic sub-images are processed to create a composite image with the properties of a single central perspective

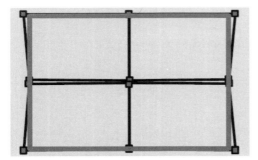

Figure 11.6 Z/I Imaging® DMC panchromatic sub-image and composite image footprints.
Source: Courtesy of Intergraph Corporation.

Figure 11.7 Vexcel UltraCam$_X$ digital camera system.
Source: Courtesy of Vexcel Corporation.

image (Tang et al. 2000). The four nadir-viewing multispectral images are acquired in B, G, R, and NIR wavebands with lower spatial resolution and cover the area of the panchromatic composite image.

Pan-sharpening is applied to merge the multispectral data with the panchromatic composite imagery to create very fine spatial resolution multispectral images (Heier et al. 2001). DMC forward motion compensation is handled electronically by operating the CCD array sensors in a time delayed integration mode. The number of TDI shifts (up to 50) occurring during camera exposure is factored into the image reconstruction.

UltraCam$_X$

The Vexcel UltraCam$_X$ has eight optical cones and thirteen CCD arrays (Figure 11.7). Unlike the DMC, the focals planes of all CCD arrays are on the same plane and all optical cones have the same field-of-view (ASPRS 2004b). Data collected with nine 4,992 × 3,328 pixel CCD arrays are combined to create a very fine spatial resolution panchromatic image, while the remaining four CCD arrays (also 4,992 × 3,328 pixels) acquire VNIR multispectral imagery at a lower spatial resolution than the panchromatic. The UltraCam$_X$ design is such that four panchromatic CCD arrays in a single master cone image the corners of the scene, while five panchromatic CCD arrays residing in three slave cones fill-in the remaining areas to create a final panchromatic composite image (Figure 11.8). The master cone defines the coordinate system for the composite image. The timing of panchromatic image acquisition is staggered by a few msec, so that each panchromatic image (captured from optical cones oriented along track) is essentially taken from the same position, and the images can be considered to be from a single perspective center. As with the DMC, pan-sharpening is applied to create

very fine resolution multispectral images, and FMC is accomplished using TDI.

Sensor comparison

From a quantitative remote sensing perspective, the key differences between the systems discussed above are the spectral-radiometric characteristics. The DMC and UltraCam$_X$ apply pan-sharpening to the final very fine resolution multispectral images. Therefore, these systems are not sampling spectrally pure data. The ADS40 (when not used in high resolution mode with the staggered panchromatic arrays) is the only large format airborne digital imaging system for which multispectral data are sampled directly for each pixel and in each waveband. Further, the ADS40 collects long strips of image data with consistent across-track view geometry. Therefore, less radiometric correction or tonal balancing is required to produce mosaics for a given scene compared to the digital frame array cameras.

From a photogrammetric perspective, the frame array sensors do not require GPS/IMU data for aero-triangulation, and any digital photogrammetric work station (DPWS) may be used to process the data (Tang et al. 2000). Line array ADMIS such as the ADS40 require specialized software to properly position the data. Frame array sensors inherently have more stable geometry (since images from individual CCD arrays are captured at the same instant) and are more appropriate for large-scale photogrammetry applications (Craig 2005).

PROCESSING AND ANALYSIS

Many aspects of digital image processing and analysis are unique to ADMID or require special considerations. These include some types of

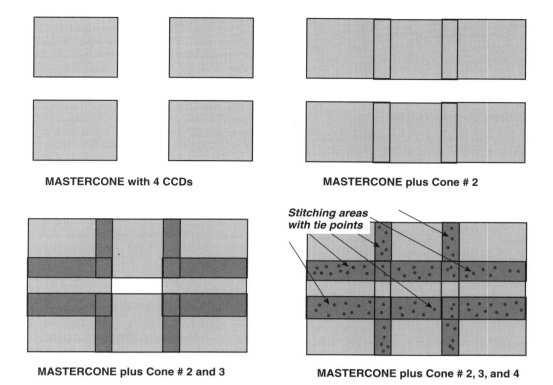

Figure 11.8   Vexcel UltraCam$_X$ panchromatic sub-image and composite image footprints.
Source: Courtesy of Vexcel Corporation.

geometric, radiometric, and image enhancement processing routines, as well as manual and semi-automated image analysis approaches.

Geometric processing

Sensor-specific rectification, georeferencing, registration, and mosaicking are geometric processing techniques required to generate image maps from raw ADMID. Rectification approaches depend on the nature of the two types of imaging strategies utilized by ADMIS, linear- and area-based sensing. Line-by-line sensing instruments (e.g., true scanners or pushbroom radiometers) generally rely on GPS and INS to track changes in platform position and sensor attitude, respectively (Schwarz et al. 1993). Each line of data must be corrected independently, even though systematic distortions are generally known and consistent for every line. Conversely, frame array systems (e.g., digital video and cameras) may or may not rely on data from synchronized GPS and INS for rectification because of the internal consistency of pixel locations within each frame. Systematic distortions of frame sensors can be characterized through calibration based on laboratory or ground targets, with the calibration data providing 'internal orientation' parameters (Toutin 2004).

'External orientation' information, utilized in both georeferencing and mosaicking procedures, can also be supplied with GPS and INS data, and/or with ground control points (GCPs). GCPs may be painted or temporary markers at known geographic locations, or inherent features in the scene for which geographic coordinates are known (from orthophotography or GPS surveys). GCPs may be used directly to calibrate polynomial warping or rubber sheeting transformations used to rectify and georeference frame array images, or to register one frame of imagery to another. The base or reference image may be a georeferenced orthoimage, a corresponding frame for a different waveband (i.e., band-to-band registration), or a corresponding frame from a different date of imagery (i.e., date-to-date registration). True orthoimages require correction of terrain-related relief displacement, based on DEMs (Toutin 1995).

Mosaicking is required to stitch together multiple line-sensed swaths or area-sensed frames, in order to cover an area larger than a single swath or frame. Secondary geometric adjustments are normally required to account for incomplete rectification in the overlap zone (normally manifested as offsets at cut lines).

Radiometric processing

Radiometric processing of ADMID particularly entails spatially normalizing pixel brightness values across an image swath, frame, or mosaic. The factors that cause inconsistencies in the image brightness to surface reflectance relationship within ADMID are anisotropic reflectance resulting from bidirectional reflectance effects and the wide field-of-view of many ADMIS, vignetting caused by unequal transmittance across the sensor optics, and variations in camera exposure settings for framing systems utilizing auto-exposure controls (Stow et al. 1996). Anisotropic reflectance adjustments are normally corrected through line, frame, or mosaic-based detrending of across image brightness patterns, since information on bi-directional reflectance distribution functions (BRDF) within a scene is not generally available (Barnsley et al. 1994). Vignetting is typically corrected by applying a digital anti-vignetting filter derived from laboratory imaging of a uniform brightness field to characterize the off-axis brightness fall-off of a given lens system (Stow et al. 1996). When such radiometric corrections are not or are unsuccessfully applied prior to mosaicking, brightness variations are noticeable and can affect subsequent visual or computer-assisted image analyses. Detrending routines may be applied to the entire image mosaic, or radiometric matching routines implemented at the seams of adjoining swaths or frames, with the latter termed 'color balancing.'

Ensuring that brightness values are consistent between different spectral band images (spectral normalization) and image dates (temporal normalization) can also be important. Radiometric normalization approaches can range between the purely deterministic, where brightness values are converted to spectral reflectances based on radiometric calibration and atmospheric correction through radiative transfer modeling, to empirical approaches that convert brightness values of one image to match another, based on radiometric control points. An empirical approach to deriving spectral reflectance values is possible with very fine spatial resolution ADMID, through the use of large calibration panels of varying and known reflectance placed in the scene at the time of imaging (Tarnavsky et al. 2004). Radiometric calibration of ADMID is often limited by the lack of availability of calibration coefficients, which tend to be more commonly available for satellite imaging systems.

Enhancement processing

Enhancement of remotely sensed image data can occur through application of contrast (brightness expansion), color, spatial, and spectral routines or transformations. Of these, two specific types of enhancements that are commonly utilized for ADMID are 'pan sharpening' and creation of pseudo-true color images from CIR images. The fusion entails synthetically disaggregating the multispectral image data to match the fine resolution grid of the panchromatic data, and then allocating the finer-resolution brightness components to each multispectral band. This process can also include spatial filtering. Pseudo-true color images can be generated from CIR data for natural scene visualization purposes, by estimating the blue waveband brightness based on the relative relationships between green, red, and NIR brightness values.

Image backdrops for GIS overlay

Image maps or orthomosaics derived from ADMID are commonly used as an image 'backdrop' for displaying and analyzing GIS layers. Fine spatial resolution images provide geographic context and a 'sense of place' when examining thematic GIS layers. Vector overlays of point, line, or polygon features are most effectively viewed in conjunction with image backdrops, since only a small subset of pixels are obscured by the vector features. Raster overlays can be achieved by blending the colors of a GIS theme with the high frequency details of a monochrome image, with the analyst controlling the degree of opacity-transparency of the overlay.

Image-based reconnaissance

Digital video and single-chip cameras provide a relatively inexpensive means for capturing reconnaissance imagery. The geometric and radiometric fidelity requirements are generally less stringent for reconnaissance purposes, where the objective is to extract qualitative information and document scene conditions in an efficient and timely fashion. Voice documentation can be conveniently integrated into an audio track by an airborne observer viewing the scene during image acquisition. Rapid image assessments can be made in-flight or shortly after landing, when timely, strategic information is required (e.g., in response to a natural disaster or emergency). Military developments in UAVs and telemetry technologies are beginning to make real-time reconnaissance a reality for civilian applications (e.g., wildfire suppression activities) (Ambrosia et al. 2003).

Image-based inventorying

A common practice in resource management, urban planning, and business is to use fine spatial resolution imagery as a primary data source for inventorying resources or infrastructural features. Inventorying normally involves counts and/or linear or areal measurements of scene features.

This may be achieved through a spatial sampling strategy or by a comprehensive, 'wall-to-wall' survey. Sampling strategies can reduce time and personnel costs, but entail statistically-based estimates that have a confidence range. ADMID are well suited for spatial sampling within a large area to be inventoried, as frames or strips of imagery can be acquired according to a pre-planned sample design (e.g., systematic grid, random, or stratified random). ADMID are often nested as an intermediate scale between the complete coverage of satellite image data and the sample locations of ground-based observations; this approach is known as multistage sampling (Hallum 1993).

Accounting or mensuration of scene objects with ADMID can be achieved with manual or automated approaches, with optimal strategies normally involving a combination of the two. Visual interpretation can be supported by interactive image enhancements. Counting tools such as clicking on objects, displaying a graphical symbol over counted objects (to minimize duplicative counting or omissions), and automatic counters support interactive enumeration (Chen et al. 2001). Digital planimeters are commonly incorporated in image analysis systems, enabling linear or areal measurements of objects by interactive ('heads-up') digitizing. Recent developments in automated object recognition enable certain types of scene objects such as buildings, automobiles, trees, and water bodies to be detected and counted (Navulur 2007). Most of the human intervention occurs in the form of developing algorithms, setting parameters, and manually editing errors and artifacts.

Image-based mapping

Mapping of land surface features is the most common task associated with ADMID. Map themes may be categorized as point, line, or polygonal features, or as continuous fields of surface properties. Continuous field maps may be further categorized as terrain, biophysical, or socio-demographic themes.

Unlike inventorying, mapping generally requires that ADMID have been rectified, georeferenced, and terrain corrected to generate an image map or orthomosaic. The viability of large format ADMIS for deriving fine spatial resolution stereo imagery has increased with the advent of soft-copy photogrammetric techniques for semi-automated generation of DEMs. However, as the benefits of ADMIS relative to scanning of analog stereo aerial photographs are being realized, other remote sensing technologies for generating DEMs such as LIDAR and InSAR are providing alternative sources for DEMs (Gamba and Houshmand 2000).

In many cases, the same ADMID used for deriving DEMs are also used for mapping non-terrain features, once a DEM is used to orthorectify the original ADMID. On-screen or 'heads-up' digitizing is a manual, interactive approach that is commonly implemented to generate vector maps from ADMID. To date, human interpreters are better able to exploit the fine spatial resolution characteristics of ADMID. Image analysts can interactively enhance images, digitize vector objects, and encode attributes about these objects. As with image-based inventorying, advances in automated object detection enable analysts to generate vector maps in a semi-automatic manner (i.e., manual editing is still required). This includes generation of land use and land cover maps (or special types such as vegetation maps) through image segmentation and hierarchical classification approaches (Benz et al. 2004, Navulur 2007). Per-pixel image classification approaches tend not to be as effective because of the high degree of image brightness heterogeneity associated with fine spatial resolution ADMID.

Deriving maps of continuous biophysical and socio-demographic properties with ADMID normally requires statistical-empirical (e.g., regression) models that are calibrated with a sample of field measurements of the property of interest. Given that ADMID are normally limited to VNIR wavelengths, the biophysical properties most commonly mapped are fractions of general surface cover types (e.g., vegetation, soil, impervious, and water) (Ridd 1995) or vegetation properties (e.g., above ground biomass or leaf area index) (Franklin et al. 1997). Socio-demographic properties (e.g., population or wealth) are typically modeled through dacymetric mapping approaches that integrate image and ground-based observations (Sutton et al. 2001).

Image-based monitoring and updating

Detailed monitoring of land surface changes and reliable updating of GIS layers can be readily achieved using multitemporal (i.e., time sequential) ADMID. Monitoring can be performed by repetitive reconnaissance, inventorying, or mapping. GIS updating can be achieved by interactively or semi-automatically comparing a dated GIS layer with recently acquired ADMID; if the GIS layer had originally been derived using ADMID, then the original image may also be incorporated in the updating approach (Cao et al. 2007).

While the fine spatial resolution characteristic of ADMID enables detailed land cover changes to be detected and identified, it also makes precise, pixel-level detection challenging because of mis-registration effects. Achieving accurate co-location or relative registration between two fine spatial resolution images can be difficult. A cost-effective approach to solving this challenge is to duplicate ADMIS camera stations when capturing subsequent dates of imagery. For example, a GPS-based triggering system can be used to capture image frames that nearly match the vertical and horizontal position of the camera station (and therefore, the sensor-view geometry) of the previous (baseline) image frames, which means that warping functions used for image registration are relatively simple and misregistration errors are minimal (Coulter et al. 2003).

Monitoring or GIS updating can be performed through heads-up interpretation and digitizing, or by using semi-automated change detection routines. Image processing and display functions such as swiping, flickering, or overlaying multi-temporal images are effective for manual, interactive change analyses. More automated procedures attempt to classify land cover transitions or quantify changes in surface properties on a per-pixel or per-object basis, by applying routines similar to those used for mapping single date images (Singh 1989).

REFERENCES

Airborne Data Systems, 2006. http://www.airbornedatasystems.com/spectra_view.htm (last accessed: 28 February 2007).

Ambrosia, V. G., S. S. Wegener, D. V. Sullivan, S. W. Buechel, S. E. Dunagan, J. A. Brass, J. Stoneburner, and S. M. Schoenung, 2003. Demonstrating UAV-acquired real-time thermal data over fires. *Photogrammetric Engineering and Remote Sensing*, 69(4): 391–402.

Anderson, J. E., G. B. Desmond, G. P. Lemeshewsky, and D. R. Morgan, 1997. Reflectance calibrated digital multi-spectral video: a test-bed for high spectral and spatial resolution remote sensing. *Photogrammetric Engineering and Remote Sensing*, 63(3): 224–229.

Applanix, 2007. http://www.applanix.com/products/dss_index.php (last accessed: 21 February 2007).

ASPRS, 2004a. 10-year industry forecast, phase I–III – study documentation. *Photogrammetric Engineering and Remote Sensing*, 70(1): 5–58.

ASPRS, 2004b. *Manual of Photogrammetry*, fifth edition. American Society for Photogrammetry and Remote Sensing, Bethesda, Maryland.

Barnsley, M. J., 1994. Environmental monitoring using multiple-view-angle (MVA) remotely-sensed data. In Foody, G. and Curran, P. (eds.), *Environmental Remote Sensing from Regional to Global Scales*. Wiley, Chichester, UK, pp. 181–201.

Bayer, B. E., 1976. Color imaging array. U.S. Patent 3,971,065.

Benz, U. C., P. Hoffman, G. Willhauck, I. Lingenfelder, and M. Heynen, 2004. Multi-resolution, object-oriented fuzzy analysis of remote sensing data for GIS-ready information. *ISPRS Journal of Photogrammetry and Remote Sensing*, 58: 239–258.

Brown, L. G., 1992. A survey of image registration techniques, *ACM Computing Surveys*, 24(4): 325–376.

Cao, L., D. Stow, J. Kaiser, and L. Coulter, 2007. Monitoring cross-border trails using airborne digital multispectral imagery and interactive image analysis techniques. *Geocarto International*, 22(2): 107–125.

Chen, D., D. Stow, J. Kaiser, and S. Daeschner, 2001. Detecting and enumerating new building structures utilizing very-high resolution image data and image processing. *Geocarto International*, 16: 69–81.

Chen, T., R. Shibasaki, and M. Shunji, 2003. Development and calibration of the airborne three-line scanner (TLS) imaging system. *Photogrammetric Engineering and Remote Sensing*, 69(1): 71–78.

Coulter, L., D. Stow, and S. Baer, 2003. A frame center matching technique for precise registration of multitemporal airborne frame imagery: methods and software approaches. *IEEE Transactions of Geoscience and Remote Sensing*, 41: 2436–2444.

Craig, J. C., 2005. Comparison of Leica ADS40 and Z/I Imaging DMC high-resolution airborne sensors. In: *Proceedings of SPIE*, Volume 5655: *Multispectral and Hyperspectral Remote Sensing Instruments and Applications II*, January 2005, Bellingham, Washington.

Cramer, M., 2005. Digital airborne cameras – status and future. In: *Proceedings of the ISPRS Workshop*: High Resolution Earth Imaging for Geospatial Information, May 17–20, University of Hannover, Hannover, Germany.

Everitt J. H. and P. R. Nixon, 1985. False color video imagery: a potential remote sensing tool for range management. *Photogrammetric Engineering and Remote Sensing*, 51(6): 675–679.

Falkner, E. and D. Morgan, 2001. *Aerial Mapping: Methods and Applications*, second edition. Lewis Publishers, Boca Raton, Florida.

Franklin, S. E., M. B. Lavigne, M. J. Deuling, M. A. Wulder, and E. R. Hunt, Jr., 1997. Estimation of forest leaf area index using remote sensing and GIS data for modeling net primary production. *International Journal of Remote Sensing*, 18(16): 3459–3471.

Gamba, P. and B. Houshmand, 2000. Digital surface models and building extraction: a comparison of IFSAR and LIDAR data. *IEEE Transaction on Geoscience and Remote Sensing*, 38(4): 1959–1968.

Geospatial Systems, 2006. http://www.geospatialsystems.com/wp-content/uploads/Multispectral-Cameras.pdf (last accessed: 21 February 2007).

Goshtasby, 2005. *2-D and 3-D Image Registration for Medical, Remote Sensing, and Industrial Applications*, John Wiley & Sons, New York.

Hallum, C., 1993. A change detection strategy for monitoring vegetative and land-use cover types using remotely-sensed, satellite-based data. *Remote Sensing of Environment*, 43(2): 171–177.

Heier, H., C. Dörstel, and A. Hinz, 2001. The new digital modular camera embedded into the Z/I Imaging workflow. In: *Proceedings of the ASPRS 2001 Conference: From Imagery to Geodata*, 26 April 2005, St. Louis, Missouri.

Herwitz, S. R., L. F. Johnson, S. E. Dunagan, R. G. Higgins, D. V. Sullivan, J. Zheng, B. M. Lobitz, J. G. Leung, B. Gallmeyer, M. Aoyagi, R. E. Slye, and J. Brass, 2004. Imaging from an unmanned aerial vehicle: agriculture surveillance and decision support. *Computers and Electronics in Agriculture*, 44: 49–61.

Jensen, J.R., 2007. *Remote Sensing of the Environment: An Earth Resource Perspective* (2nd edition), Prentice Hall, Upper Saddle River, NJ, USA.

Mostafa, M. M. R., 2001. History of inertial navigation systems in survey applications. *Photogrammetric Engineering and Remote Sensing*, 67(11): 1225–1228.

NASA Research Park, 2007. UAV collaborative. http://www.uav-applications.org (last accessed: 24 January 2007).

Navulur, K., 2007. *Multispectral Image Analysis Using the Object Oriented Paradigm*, CRC Press – Taylor Francis Group, Boca Raton, 163 pp.

Neale, C. M. U. and B. G. Crowther, 1994. An airborne multispectral video/radiometer remote sensing system: development and calibration. *Remote Sensing of Environment*, 49(3): 187–194.

Pacey, R. and P. Fricker, 2005. Forward motion compensation (FMC) – Is it the same in the digital imaging world? *Photogrammetric Engineering and Remote Sensing*, 71(11): 1241–1242.

Petrie, G., 2003. Airborne digital frame cameras: the technology is really improving! *Geoinformatics*, 6: 18–27.

Ridd, M. K., 1995. Exploring a V-I-S model for urban ecosystem analysis through remote sensing: comparative analysis for cities. *International Journal of Remote Sensing*, 16: 2165–2185.

Schwarz, K. P., M. A. Chapman, M. E. Cannon, and P. Gong, 1993. An integrated INS/GPS approach to the georeferencing of remotely sensed data. *Photogrammetric Engineering and Remote Sensing*, 59(11): 1667–1674.

Singh, Ashbindu, 1989. Digital chance detection techniques using remotely-sensed data. *International Journal of Remote Sensing*, 10: 989–1003.

Stow, D., A. Hope, A. Nguyen, S. Phinn, and C. Benkelman, 1996. Monitoring detailed land surface changes from an airborne multispectral digital camera system. *IEEE Transactions on Geoscience and Remote Sensing*, 34: 1191–1202.

Stutte, G. W., 1991. Videography: A management tool for sustainable agriculture. *Journal of Sustainable Agriculture*, 1(3): 1044–1046.

Sutton, P., D. Roberts, C. Elvidge, and K. Baugh, 2001. Census from heaven: an estimate of the global human population using night-time satellite imagery. *International Journal of Remote Sensing*, 22(16): 3061–3076.

Tang, L., C. Dörstel, K. Jacobsen, C. Heipke, and A. Hinz, 2000. Geometric accuracy potential of the digital modular camera. *International Archives of Photogrammetry and Remote Sensing*, XXXIII: 1051–1057.

Tarnavsky, E., D. Stow, L. Coulter, and A. Hope, 2004. Spatial and radiometric fidelity of airborne multispectral imagery in the context of land-cover change analyses. *GIScience and Remote Sensing*, 41: 62–80.

Tomkins, M. R., 2007. New sensor tech promises improved sensitivity. *The Imaging Resource*. http://www.imaging-resource.com/NEWS/1181811769.html (last accessed: 26 June 2007).

Toutin, T., 1995. Multi-source data integration with an integrated and unified geometric modelling. *EARSeL Journal of Advances in Remote Sensing*, 4: 118–129.

Toutin, T., 2004. Review article: Geometric processing of remote sensing images: models, algorithms and methods. *International Journal of Remote Sensing*, 25: 1893–1924.

Warner, W. S., R. W. Graham, and R. E. Read, 1996. *Small format aerial photography*. American Society for Photogrammetry and Remote Sensing, Bethesda, Maryland.

Imaging Spectrometers

Michael E. Schaepman

Keywords: imaging spectroscopy, imaging spectrometry, hyperspectral, airbone, spaceborne.

INTRODUCTION

Imaging spectrometers have significantly improved the understanding of interactions of photons with the surface and atmosphere. Spectroscopy has existed since the eighteenth century; the imaging part of this term became technically possible in the early 1980s. The first part of this chapter is devoted to a short historical background of this evolution. In subsequent sections, imaging spectroscopy is defined and the main acquisition principles are discussed. The main imaging spectrometers used for Earth observation are presented, as well as emerging concepts which give an insight in a broad range of air to spaceborne associated technology. Imaging spectroscopy has expanded to many other disciplines, and the approach is used in medicine, extraterrestrial research, process, and manufacturing industries, just to name a few areas. In addition, much development is currently seen in other wavelength domains such as the ultraviolet and the thermal. However, this chapter focuses on Earth observation based imaging spectrometers in the solar reflective wavelength range.

✓ HISTORICAL BACKGROUND

Three centuries ago Sir Isaac Newton published the concept of dispersion of light in his 'Treatise of Light,' and the concept of a spectrometer was born (Figure 12.1).

The corpuscular theory by Newton was gradually succeeded over time by the wave theory, resulting in Maxwell's equations of electromagnetic waves (Maxwell 1873). But it was only in the early nineteenth century that quantitative measurement of dispersed light was recognized and standardized by Joseph von Fraunhofer's discovery of the dark lines in the solar spectrum (Fraunhofer 1817) and their interpretation as absorption lines on the basis of experiments by Bunsen and Kirchhoff (1863). The term spectroscopy was first used in the late nineteenth century and provides the empirical foundations for atomic and molecular physics (Born and Wolf 1999). Following this, astronomers began to use spectroscopy for determining radial velocities of stars, clusters, and galaxies and stellar compositions (Hearnshaw 1986). A historical example of an astronomical spectrometer is George E. Hale's spectroheliograph (Figure 12.2) of the early twentieth century. The spectroheliograph was designed by this American astronomer to collect spectral images of the sun by simultaneously scanning the sun's image across the entrance slit and a film plate past the exit slit of a two-prism monochromator.

Advances in technology and increased awareness of the potential of spectroscopy in the 1960s to 1980s led to the development of the first analytical methods used in remote sensing (Arcybashev and Belov 1958, Lyon 1962), the inclusion of 'additional' spectral bands in multispectral imagers (e.g., the 2.09–2.35 μm band in Landsat for the detection of hydrothermal alteration minerals), as well as first airborne and later spaceborne imaging spectrometer concepts and instruments (Collins et al. 1982, Goetz et al. 1982, Vane et al. 1984, Vane 1986). Significant recent progress was achieved when in particular airborne imaging spectrometers became available on a wider basis (Goetz et al. 1985, Gower et al. 1987, Kruse et al. 1990, Rickard et al. 1993, Birk and McCord 1994, Rowlands et al. 1994, Green et al. 1998) helping to prepare for spaceborne imaging spectrometer activities (Goetz and Herring 1989). This initial phase of development lasted until the late 1990s, when the first imaging spectrometers were launched in space (e.g., MODIS (Salomonson et al. 1989), MERIS (Rast et al. 1999)). Nevertheless, true imaging spectrometers in space, satisfying a strict definition of a contiguity criterion, are still sparse (CHRIS/PROBA (Barnsley et al. 2004, Cutter 2006), Hyperion/EO-1 (Pearlman et al. 2003)).

Technological advances in the domain of focal plane (detector) development (Chorier and Tribolet 2001), readout electronics, storage devices, and optical design (Mouroulis and Green 2003) are leading to a significantly better sensing of the Earth's surface. Improvements in optical design (Mouroulis et al. 2000) signal-to-noise, finer and better defined bandwidths as well as contiguous spectral sampling combined with the goal of better

understanding of the modeled interaction of photons with matter (Schläpfer and Schaepman 2002) will allow for more quantitative, direct and indirect identification of surface materials, and atmospheric transmittance based on spectral properties from ground, air, and space.

DEFINITIONS OF IMAGING SPECTROSCOPY TERMS

Spectroscopy is defined as the study of light as a function of wavelength that has been emitted, reflected, or scattered from a solid, liquid, or gas. In remote sensing, the quantity most used is (surface) reflectance (expressed as a percentage). *Spectroradiometry* is the technology for measuring the power of optical radiation in narrow, contiguous wavelength intervals. The quantities measured are usually expressed as spectral irradiance (commonly measured in $W\ m^{-2}\ nm^{-1}$) and spectral radiance (commonly measured in $W\ sr^{-1}\ m^{-2}\ nm^{-1}$).

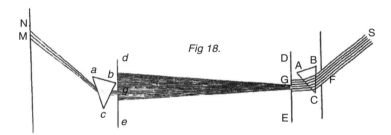

Figure 12.1   Sir Isaac Newton's 'Treatise of Light' discusses the concept of dispersion of light in 1704. He demonstrated that white light could be split up into component colors using prisms, and found that each pure color is characterized by a specific refrangibility (Newton 1704).

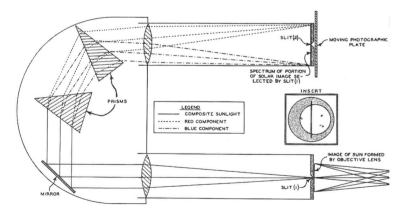

Figure 12.2   Schematic drawing of Hale's spectroheliograph, which was used to image the sun (Wright et al. 1972).
Source: Sky & Telescope.

Spectroradiometric measurements remain one of the least reliable of all physical measurements due to the multidimensionality of the problem, the instability of the measuring instruments and standards used, and sparse dissemination of the principles and techniques used for eliminating or reducing the measurement errors (Kostkowski 1997).

The term hyperspectral (alternatively also ultraspectral) is used most often for spectroscopy and spectrometry interchangeably and denotes usually the presence of a wealth of spectral bands without further specification. The variable use of the above terms expresses a variation in flavors, but usually not a fundamental physical difference. Hyperspectral denotes many spectral bands, which potentially can be used to solve an $n-1$ dimensional problem, where n represents the number of spectral bands. An imaging spectrometer with 200 spectral bands (i.e., dimensions $= 200$) can theoretically solve a spectral unmixing based problem with 199 end members, or can be used in a model inversion approach with 199 unknowns. Practically, there are instrument performance limitations (e.g., signal-to-noise ratio (SNR)), or strong correlations between adjacent bands, as well as ill-posed problems in model inversion, which reduce this dimensionality significantly.

The original definition for imaging spectrometry was coined by Goetz et al. (1985) as being 'the acquisition of images in hundreds of contiguous, registered, spectral bands such that for each pixel a radiance spectrum can be derived' (Figure 12.3). A more detailed definition is that imaging spectrometry (imaging spectroscopy, or also hyperspectral imaging) is a passive remote sensing technology for the *simultaneous* acquisition of spatially *coregistered images*, in many, *spectrally contiguous bands*, measured in *calibrated radiance units*, from a *remotely operated platform* (Schaepman et al. 2006).

In the specific case of imaging spectrometry, the focus of the refined definition is on many, spectrally *contiguous* bands, de-emphasizing the need of 'hundreds of contiguous bands' (Goetz, 2007). The contiguity criteria or the proximity requirement of spectral bands is usually poorly defined, in particular since all imaging spectrometers in remote sensing undersample the Earth. The Nyquist–Shannon theorem requires that a perfect reconstruction of the signal is possible when the sampling frequency is greater than twice the maximum frequency of the signal being sampled, which is not the case in space based imaging spectrometers. The rate of undersampling requires compromises to be made in the resolution-acquisition-time domains, which in turn has fostered the development of deconvolution theories. Initially, instruments having at least 10 adjacent spectral bands with a spectral resolution (or full width at half the maximum (FWHM)) of 10 nm were considered as imaging spectrometers, however, nowadays the understanding is that imaging spectrometers must be able to sample individual relevant features (absorption, reflectance, transmittance, and emittance) with at least three or more contiguous spectral bands at a spectral resolution smaller than the spectral width of the feature itself.

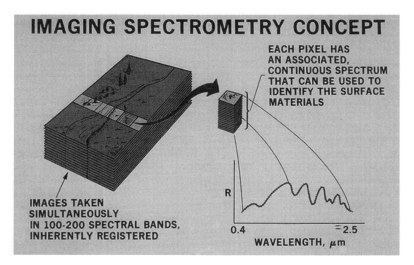

Figure 12.3   Original imaging spectrometry concept drawing as used by G. Vane and A. Goetz (courtesy of NASA JPL).

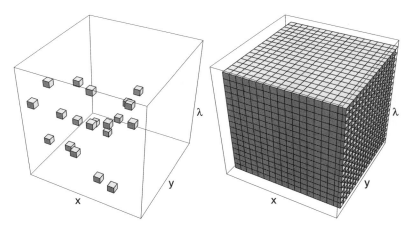

Figure 12.4   Conceptual imaging spectrometer data cube with two spatial domains (*x* and *y*) and the spectral domain (λ). Randomly distributed voxels each represent individual 'radiometers' (left) and a fully acquired data cube (right).

IMAGING SPECTROMETER PRINCIPLES

In imaging spectrometry a generalized data concept, called the *data cube*, is used to visualize the relation between the spatial and the spectral domains present. The spatial data is acquired by imaging a scene using techniques such as staring filter wheels, pushbroom, or whiskbroom scanner, amongst others. When acquiring data in only one spectral band (monochromatic acquisition), each individual element may be referred to as a pixel with a spatial extent and a single wavelength. By adding many spectral bands all pixels can be represented conceptually as voxels. A voxel ('volume element') is a three-dimensional equivalent of a pixel, representing individual radiometers having 3D units of length (x, y) and wavelength (z). All these individual radiometers represent the data cube as a 3D discrete regular grid (Figure 12.4).

Imaging spectrometer data is often visualized as a data cube formed by a series of image layers, each layer of which is an individual wavelength interval. The sides of this cube are color-coded spectra by intensity, and the top is a three-band color composite (Figure 12.5).

IMAGING SPECTROMETER DATA CUBE ACQUISITION

The acquisition of the data cube is performed differently by different imaging spectrometer technologies. In general, whiskbroom imaging spectrometers collect series of pixels, pushbroom

Figure 12.5   Imaging spectrometer data cube acquired by an airborne HyMap system on 26 August 1998 in Switzerland (Limpbach Valley). The sides are color-coded spectra by intensity. (See the color plate section of this volume for a color version of this figure).

scanners series of lines, and staring systems, filter wheel systems, or snap shot cameras series of monochromatic images at different spectral wavelengths (Figure 12.6). The data acquisition process is usually performed until a complete data cube is filled with voxels. In the following sections, each of the major acquisition approaches is discussed in more detail.

Whiskbroom scanners

Whiskbroom scanners are usually opto- or electromechanical sensors that cover the field-of-view (FOV) by a mechanized angular movement using a scanning mirror sweeping from one edge of the swath to the other, or by the mechanical rotation of the sensor system. The inherent sensor instantaneous field-of-view (IFOV) is therefore a single spatial pixel and its associated spectral component. The image lines are collected using an across-track scanning mechanism and the image is acquired by the forward movement of the platform used (Figure 12.7). The particular advantages of the whiskbroom scanning principle for imaging spectrometers include a higher spectral uniformity since all pixels are recorded using the same detector line array and allowing the optical design to accommodate a larger detector pixel size.

Because the whiskbroom system can rely on single detectors, the calibration effort is usually much simpler than with other systems. In addition, this technology supports in-flight calibration with scanning reference sources located at the end or the beginning of each scan line. The disadvantages of this design include the presence of a mechanical scanning system, the shorter integration time that is available than in pushbroom based systems, and the image forming geometry which is dependent on the scanning speed, scan mirror arrangement and the orthogonal platform movement. Imaging spectrometers based on the whiskbroom scanning principle include the airborne AVIRIS (Green et al. 1998), DAIS (Chang et al. 1993), and HyMap (Cocks et al. 1998) instruments, as well as the spaceborne MODIS instrument (Salomonson et al. 1989).

Pushbroom scanners

A pushbroom scanner is a sensor typically without mechanical scanning components for the data acquisition. The image formation is solely based on the (forward) movement of the sensor. A pushbroom sensor is an imaging system which acquires a series of one-dimensional samples orthogonal to the platform line of flight with the along-track spatial dimension constructed by the forward motion of the platform. The spectral component is acquired by dispersing the incoming radiation onto an area array. Translated to the concept of the data cube, a pushbroom scanner records the across track dimension x and the spectral dimension λ, representing lines, simultaneously (Figure 12.8) and the along track (y) component is acquired with the platform movement.

Pushbroom scanners have the advantage that they allow a longer integration time for each individual detector element in comparison with whiskbroom based instruments (e.g., the inverse of the line frequency is equal to the pixel dwell time).

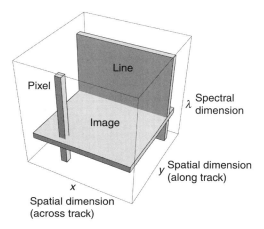

Figure 12.6   Data cube schematic depicting the three major acquisition principles of imaging spectrometers: Pixels are acquired by whiskbroom systems, lines by pushbroom systems, and images by filter wheel systems (or staring cameras).

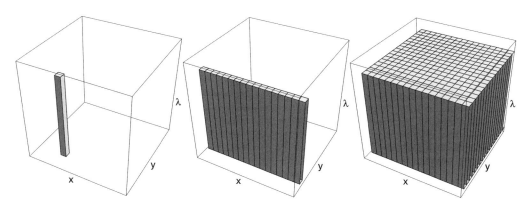

Figure 12.7   Whiskbroom scanning and its representation in the data cube (left single spectrum (one pixel), middle one scan line, right full data cube).

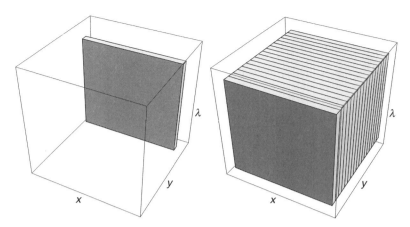

Figure 12.8   Pushbroom scanning and its representation in the data cube (a single scan line is composed out of the across track pixels *x* and the spectral bands λ (left), resulting in a full data cube (right)).

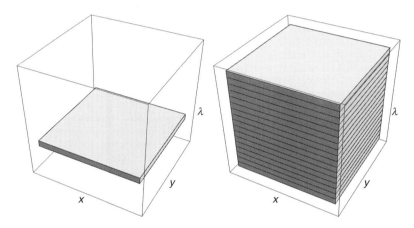

Figure 12.9   Filter wheel acquisition and its representation in the data cube (left single monochromatic image, right full data cube).

In addition, there are distinct and fixed geometric relations between the pixels within a scan line.

Since area arrays are used as focal planes in these systems, the uniform calibration of the detector response is critical. In a combined analysis of SNR, uniformity, and stability, pushbroom scanners might not necessarily outperform whiskbroom systems even though they have a longer integration time.

Examples of pushbroom based imaging spectrometers include the airborne CASI (Babey and Anger 1989) and ROSIS (Kunkel et al. 1991) instruments and the spaceborne MERIS (Rast et al. 1999) and Hyperion (Pearlman et al. 2003).

Filter wheel cameras

The filter wheel camera is an opto-mechanical sensor that changes the spectral sensitivity of various channels using a turnable filter wheel in the optical path. The field-of-view (FOV) therefore represents a full monochromatic frame, represented in the data cube by the *x* and *y* axis. The spectral component is collected by rotating the filter wheel to different band pass filters, which have different transmissions for different spectral regions. The data cube is filled by 'stacking' individual (monochromatic) images on top of each other (Figure 12.9).

The particular advantages of the filter wheel camera consist in the coherent spatial coregistration if not used on a moving platform. This makes this technology very suitable for staring telescope applications in astronomy. Most filter wheel cameras use area arrays for the simultaneous coverage of the spatial extent. If operated from a moving platform, mosaicing the individual frames is the most important feature. The calibration of the filter wheel camera can be performed by using a calibrated spectrometer or band pass filter/detector to test the sensitivity and non-uniformity of the detector elements. The nonuniformity calibration of the detector array is the most challenging issue with this technology. The disadvantages of this design include the presence of the mechanical turning filter wheel, which necessitates a fast change of the individual filters on moving platforms. Even so, individual spectral images may not be aligned satisfactorily. A major advantage is that it is easy to change the spectral bands by replacing the filter wheel for different applications. In general, very few airborne or spaceborne imaging spectrometers are based on the filter wheel camera approach, mostly due to its limitation in simultaneously collecting many spectral bands. However, the concept has been demonstrated in airborne instruments (e.g., Airborne POLDER (Leroy and Bréon 1996)), spaceborne (e.g., STRV-2 MWIR imager (Cawley 2003) and in astronomy staring telescopes using a filter wheel approach.

Other, less frequently used imaging spectrometer concepts

Wedge spectrometers (Figure 12.10) are based on a linear spectral wedge filter, which can be mated directly to an area array, avoiding the use of often bulky and complex optics required for imaging spectrometers that use gratings or prisms. This approach acquires a 2D FOV, consisting of $x \cdot y$ lines (corresponding to the number of $x \cdot y$ pixels of the area array). The difference compared to a filter wheel is that here the y pixels (in the along-track direction) record y different spectral channels but for different adjacent spatial swaths. With the movement of the platform along-track, the across-track ground images are sequentially sampled at the range of wavelengths supported by the wedge filter.

The major advantage of a wedge spectrometer is the compact design because many optical elements can be avoided. The major disadvantage is accommodating the Earth rotation (or the platform movement), which can generate spectral smearing (or spectral mis-registration).

The calibration of wedge spectrometers is comparable to pushbroom imagers in static environments, although the wider FOV in the along-track direction may introduce different challenges. An example of a wedge spectrometer flow in space is the LAC instrument onboard of EO-1 (Reuter et al. 2001), others are in planning (Puschell et al. 2001), but the concept has not yet seen significant data distribution and use.

Other interesting imaging spectrometer concepts include the computed tomography imaging spectrometer (CTIS) (Descour et al. 1997). CTIS is a non-scanning instrument capable of simultaneously acquiring full spectral information from every position within its FOV. The raw image collected by the CTIS consists of 49 diffraction orders. The 0th diffraction order is located at the center of the image. This order represents a direct view of the spatial radiance distribution in the field stop

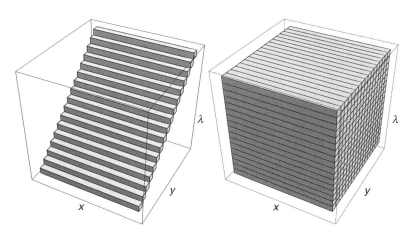

Figure 12.10   Wedge spectrometer and its representation in the data cube (left single scan, right optimal filled cube).

and exhibits no dispersion. The remaining diffraction orders exhibit dispersion increasing with order number. Reconstruction of the data cube from the raw data requires knowledge of how individual voxels map to the imaging array. Each voxel corresponds to an object volume, measuring $\Delta x \Delta y \Delta \lambda$, where Δx, Δy, and $\Delta \lambda$ are the spatial and spectral sampling intervals, respectively.

Another emerging technology for imaging spectrometers is the use of acousto-optical tunable filters (AOTF) allowing a rapid change of spectral bands (Calpe-Maravilla et al. 2004). Conceptually AOTF based systems are similar to filter wheel instruments.

EVOLUTION FROM AIRBORNE TO SPACEBORNE IMAGING SPECTROMETERS

This section presents an overview of selected instruments which have had an impact on the evolution and development of imaging spectrometers. Comprehensive and detailed overviews are difficult to generate; however, Kramer (2002) presents a very complete list of existing and planned instruments.

Early concepts of acquiring spectral (and directional) information from natural targets were discussed already in 1958 in the former Soviet Union (Arcybashev and Belov, 1958). The idea was to acquire a scene – a forest in this case – under various view angles by using a complex flight pattern (Figure 12.11). In addition, the camera – a

spectrograph at this time named Spectrovisor – was tilted to different view directions to increase the angle of observation.

Inherent limitations of computer I/O performance and – at these times – tape or band recorder capacity, resulted in first designs of spectrometers that were not capable of imaging the full swath width continuously. Consequently they were called profiling systems, having across-track gaps in the spatial coverage. One of the first profiling instruments that was deployed on an aircraft was the GER (Geophysical and Environmental Research Corp., of Millbrook, NY, USA, a company that is no longer in business) MARK II Airborne Infrared Spectroradiometer (Chiu and Collins 1978, Figure 12.12).

Much of the technology development for imaging spectrometers took place in the 1970s and 1980s at the NASA Jet Propulsion Laboratory (JPL) in Pasadena (USA). At that time, Alex Goetz and Gregg Vane proposed successfully to use NASA internal funds to use a hybrid focal plane array with 32×32 elements, allowing the construction of an imaging spectrometer that covered the spectral region beyond the 1100 nm cutoff of silicon arrays (Goetz 2007). These efforts resulted in the Airborne Imaging Spectrometer (AIS) (Vane et al. 1984) (Figure 12.12).

Following the successful deployment of AIS, the spectroscopists at JPL proposed a fully fledged imaging spectrometer program that would range from the airborne AIS1 and AIS2, the Airborne Visible/Infrared Imaging Spectrometer (AVIRIS), as well as two orbiting sensors, the Shuttle Imaging

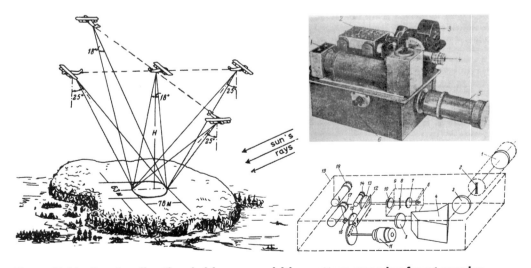

Figure 12.11 **Spectro-directional airborne acquisition pattern assessing forest angular spectral reflectance (left) and the Spectrovisor imaging spectrograph (right) (Arcybashev and Belov 1958, Kol'cov 1959) (Reprinted with permission of Juris Druck + Verlag AG).**

Figure 12.12 Airborne imaging spectrometers. From *left, top*: GER MARK II Airborne Infrared Spectroradiometer, Airborne Imaging Spectrometer (AIS) instrument assembly, Airborne Visible/Infrared Imaging Spectrometer (AVIRIS); *Middle*: Fluorescence Line Imager/ Programmable Multispectral Imager (FLI/PMI), Digital Airborne Imaging Spectrometer (DAIS7915), Reflective Optics Imaging Spectrometer (ROSIS); *Bottom*: Shortwave Infrared Full Spectrum Imager (SFSI), Hyperspectral Digital Imagery Collection Experiment (HYDICE) detector assembly, and Hyperspectral Mapper (HyMap). (See the color plate section of this volume for a color version of this figure).
Source: Photos courtesy of: S.-H. Chang, G. Vane, R. Green, R. Baxton, A. Müller, H. van der Piepen, B. Neville, M. Landers, and M. E. Schaepman.

Spectrometer Experiment (SISEX) and a satellite-borne instrument, the High Resolution Imaging Spectrometer (HIRIS).

The development of AVIRIS started in 1984 and the imager first flew aboard a NASA ER-2 aircraft at 20 km altitude in 1987 (Vane 1987). Since then it has gone through major upgrades as technology changed in detectors, electronics, and computing. AVIRIS can be seen as the major driver for the development of imaging spectrometry.

The AVIRIS instrument (Figure 12.12) contains 224 different detectors, each with a wavelength sensitivity range (also known as spectral bandwidth) of approximately 10 nanometers (nm), allowing it to cover the entire range between 380 nm and 2,500 nm. AVIRIS uses a scanning mirror to sweep back and forth (whiskbroom fashion), producing 614 pixels for the 224 detectors each scan. The pixel size and swath width of the AVIRIS data depend on the altitude from which the data is collected. When collected by the ER-2 from 20 km above the ground, the so-called 'high altitude option,' each pixel produced by the instrument covers an area approximately 20 × 20 m on the ground (with some overlap between pixels), thus yielding a ground swath about 11 km wide. When collected by the lower flying Twin Otter at a 4 km altitude, the 'low altitude option,' each ground pixel is 4 × 4 m, and the swath is 2 km wide.

In Canada, the development of imaging spectrometers has been intensive and G. Borstad proposed the Fluorescence Line Imager/Programmable Multispectral Imager (FLI/PMI) instrument with 288 spectral bands ('spectral mode') and 512 pixels swath width ('spatial mode'). The first data were acquired with this instrument around 1985 (Figure 12.12) (Borstad et al. 1985).

An interesting further development of the FLI/PMI instrument is considered to be the CASI (Compact Airborne Spectrographic Imager), developed as a combination of a spectral or spatial imager similar to the FLI/PMI but subsequently enhanced to an operational and commercially flown instrument for many years (Babey and Anger 1989, Gower et al. 1992). ITRES Corp., the manufacturer of the CASI line of instruments, can be seen as an offspring of Moniteq, who produced the FLI/PMI instrument.

In Europe, airborne imaging spectrometers were first primarily flown by leasing instruments from the US or Canada. However, thanks to the efforts of the German Aerospace Centre (DLR, Oberpfaffenhofen (GER)) two instruments became available on a broader basis for the user community. First, a European Commission supported purchase of a GER imaging spectrometer (Collins and Chang, 1990) with particular features such as the inclusion of a thermal range (Chang et al. 1993) prompted a European proposal for an Airborne Remote Sensing Capability (EARSEC) (Carrère et al. 1995). The instrument was named GER DAIS 7915 (Digital Airborne Imaging Spectrometer) and incorporated 72 solar reflective and 7 mid-infrared/thermal bands (79 in total). Its operation was eventually discontinued in 2005, after having served 10 years in Europe fostering the use of imaging spectrometers (Figure 12.12).

The second development by DLR can be considered an airborne forerunner instrument for the spaceborne MERIS on ENVISAT and was named ROSIS (Kunkel et al. 1991). The Reflective Optics System Imaging Spectrometer was tested first in 1989 (van der Piepen et al. 1989) and featured a CCD–based pushbroom design. The instrument included a choice of selectable bands (32 out of 128) covering the wavelength range significant for coastal zones and oceans (400–1100 nm). A totally revised version of ROSIS was presented under the name ROSIS-03 in 1998 (Gege et al. 1998) (Figure 12.12).

Two other interesting airborne instruments were also developed. One was the SWIR Full Spectrum Imager (SFSI) (Neville and Powell 1992). SFSI employs a two-dimensional platinum silicide Schottky barrier CCD array with 488 rows of 512 detector elements. In operation, a region of 480 lines by 496 columns is used; two adjacent lines are summed together on the detector array to yield an effective array of 240 by 496

detector elements. This gives 240 spectral bands for each of 496 pixels in the across-track dimension for each integration period, which is hardware selectable by various clock speeds (40–67 ms). The data are digitized to 13 bits and recorded as 16 bits (Figure 12.12).

The HYDICE instrument is also worth mentioning. HYDICE (Rickard et al. 1993) was a program to develop a state-of-the-art imaging spectrometer to support utility studies of high spectral resolution measurements in the 400–2500 nm range. The program was initiated by the U.S. Congress to investigate the application of hyperspectral data to the needs of federal agencies (forest assessment for the U.S. Department of Agriculture, mineral exploration for the U.S. Geological Survey, and so forth). The sensor was built by Hughes Danbury Optical Systems, Inc., and integrated into a Convair 580 aircraft operated by the Environmental Research Institute of Michigan (ERIM). The HYDICE sensor made its first data collection flight on 26 January 1995. The HYDICE sensor is a pushbroom imaging spectrometer that uses a biprism dispersing element and a two-dimensional focal plane array to enable a single optical path design. The array is a 320×210 element InSb array fabricated by Hughes Santa Barbara Research Center, with multiple gain regions to support operation over the full 400–2500 nm spectral range. The array is electronically shuttered with a fixed read time of 7.3 ms. The frame rate can be adjusted from 8.3 to 50 ms, allowing one to use nearly the full range of velocity to height (V/H) ratios within the flight envelope of the CV 580. In particular, the altitude range from 5,000 to 25,000 feet can be used to achieve spatial resolutions from 0.8 to 4 meters (Figure 12.12).

A widely available instrument is the HyMap. Manufactured by Integrated Spectronics of Australia, and operated by HyVista Corp., also of Australia, HyMap (and its predecessor called Probe1) became operational in 1996 (Cocks et al. 1998). The HyMap series of airborne hyperspectral scanners have been deployed in a large number of countries, undertaking hyperspectral remote sensing surveys in support of a wide variety of applications ranging from mineral exploration to defense research to satellite simulation. The evolution of the HyMap series continues with the development of a system providing hyperspectral coverage across the solar wavelengths (0.4–2.5 μm) and 32 bands in the thermal infrared (8–12 μm) (Figure 12.12).

Spaceborne imaging spectrometers are currently still only sparsely available. Following a strict interpretation of the definition of an imaging spectrometer, only Hyperion on EO1 (Pearlman et al. 2003) and CHRIS on PROBA (Barnsley et al. 2004) can be considered true imaging spectrometers. MERIS on ENVISAT (Rast et al. 1999) is designed like a true imaging spectrometer using a pushbroom concept with continuous dispersion on the CCD array, but does not read out all contiguous spectral bands for the use in the ground processing and archiving facility. MODIS (Salomonson et al. 1989) on Terra and Aqua satisfies the classical 10 spectral bands and 10 nm spectral resolution definitions. However, MODIS does not satisfy the continuity criterion and is technologically built as an advanced multiband spectrometer using whiskbroom acquisition.

EMERGING INSTRUMENTS AND CONCEPTS

Earth observation based on imaging spectroscopy has been transformed in less than 30 years from a sparsely available research tool into a commodity product available to a broad user community. Currently, imaging spectrometer data are widespread and they prove, for example, that distributed models of biosphere processes can assimilate these observations to improve estimates of Net Primary Production, and that in combination with data assimilation methods, can estimate complex variables such as soil respiration, at various spatial scales (Schaepman 2007). However, a lack of data continuity of airborne and spaceborne imaging spectrometer missions remain a continuing challenge to the user community.

In addition, imaging spectrometers do not only cover the solar reflective part of the electromagnetic spectrum of land surfaces, they increasingly also cover atmospheric sounding (e.g., SCIA-MACHY (Bovensmann et al. 1999), and GIFTS (Elwell et al. 2006)) and the thermal region (SEBASS (Hackwell et al. 1996)).

In any case, there is an emerging need to converge from exploratory mission concepts (e.g., former ESA's Earth Explorer Core Mission proposal SPECTRA (Rast et al. 2004)) and technology demonstrators (e.g., NASA's Hyperion on EO-1), and operational precursor missions (e.g., SSTL's CHRIS on the ESA PROBA mission), toward systematic measurement and operational missions (e.g., ESA's MERIS on ENVISAT, NASA's MODIS on Terra/Aqua).

Several initiatives proposing space operated Earth Observation imaging spectrometers in the above categories have been submitted for evaluation and approval (e.g., EnMAP (Hofer et al. 2006)) or Flora (Asner et al. 2005)). However for the time being, existing and future airborne imaging spectrometer initiatives (e.g., ARES (Richter et al. 2005), APEX (Schaepman et al. 2004)) will continue to provide regular access to imaging spectrometer data, before routine collection at regional and global scales will be available.

REFERENCES

Arcybashev, E. S. and S. V. Belov, 1958. The reflectance of tree species [orig. Russ.]. In D. Steiner and T. Guterman (eds.), *Russian Data on Spectral Reflectance of Vegetation, Soil, and Rock Types*. Juris Druck & Verlag, Zurich.

Asner, G., R. Knox, R. Green, and S. Ungar, 2005. *The FLORA Mission for Ecosystem Composition, Disturbance and Productivity*. Mission Concept for the National Academy of Sciences Decadal Study, 14.

Babey, S. K. and C. D. Anger, 1989. A compact airborne spectrographic imager. *IGARSS '89, 12th Canadian Symposium on Remote Sensing*, Vancouver, B. C., July 10–14.

Barnsley, M. J., J. J. Settle, M. A. Cutter, D. R. Lobb, and F. Teston, 2004. The PROBA/CHRIS mission: A low-cost smallsat for hyperspectral multiangle observations of the earth surface and atmosphere. *IEEE Transactions on Geoscience and Remote Sensing*, 42(7): 1512–1520.

Birk, R. J. and T. B. McCord, 1994. Airborne hyperspectral sensor systems, *IEEE AES Systems Magazine*, October 1994, 26–33.

Born, M. and E. Wolf, 1999. *Principles of Optics*, 7th edition. Cambridge University Press, Cambridge.

Borstad, G. A., H. R. Edel, J. F. R. Gower, and A. B. Hollinger, 1985. Analysis and test flight data from the fluorescence line imager. *Canadian Special Publication of Fisheries and Aquatic Sciences*, 83, Dept. of Fisheries and Oceans, Ottawa, Ontario.

Bovensmann, H., J. P. Burrows, M. Buchwitz, J. Frerick, S. Noel, V. V. Rozanov, K. V. Chance, and A. P. H. Goede, 1999. SCIAMACHY: Mission objectives and measurement modes. *Journal of the Atmospheric Sciences*, 56(2): 127–150.

Bunsen, R. and G. Kirchhoff, 1863. Untersuchungen über das Sonnenspektrum und die Spektren der Chemischen Elemente. *Abh. kgl. Akad. Wiss.*

Calpe-Maravilla, J., J. Vila-Frances, E. Ribes-Gomez, V. Duran-Bosch, J. Munoz-Mari, J. Amoros-Lopez, L. Gomez-Chova, and E. Tajahuerce-Romera, 2004. 400–1000 nm imaging spectrometer based on acousto-optic tunable filters. *Proceedings of SPIE*, 5570: 460–471.

Carrère, V., D. Oertel, J. Verdebout, G. Maracci, G. Schmuck, and G. Sieber, 1995. The optical component of the European Airborne Remote Sensing Capabilities (EARSEC). *Proceedings of SPIE*, 2480: 186–194.

Cawley, S. J., 2003. The Space Technology Research Vehicle-2 Medium Wave Infrared Imager. *Acta Astronautica*, 52: 717–726.

Chang, S., M. J. Westfield, F. Lehmann, D. Oertel, and R. Richter, 1993. A 79-channel airborne imaging spectrometer. *Proceedings of SPIE*, 1937: 164–172.

Chiu, H.-Y. and W. Collins, 1978. A spectroradiometer for airborne remote sensing. *Photogrammetric Engineering and Remote Sensing*, 44: 507–517.

Chorier, P. and P. Tribolet, 2001. High performance HgCdTe SWIR detectors for hyperspectral instruments. *Proceedings of SPIE*, 4540: 328–341.

Cocks, T., R. Jenssen, A. Steward, I. Wilson, and T. Shields, 1998. The HyMap Airborne Hyperspectral Sensor: The system, calibration and performance, *Proc. 1st EARSeL*

Workshop on Imaging Spectroscopy (M. Schaepman, D. Schläpfer, and K. I. Itten, Eds.), 6–8 October 1998, Zurich, EARSeL, Paris.

Collins, W. and S. Chang, 1990. The Geophysical Environmental Research Corp. 63 Channel Airborne Imaging Spectrometer and 12 Band Thermal Scanner, *Proceedings of SPIE*, 1298: 62–71.

Collins, W., S. H. Chang, and G. L. Raines, 1982. Mineralogical mapping of sites near Death Valley, California and Cross-man Peak, Arizona, using airborne near–infrared spectral measurements. In *Proc. Intl. Symp. on Remote Sens. of Environ.*, 2nd edn. Thematic Conference on Remote Sensing for Exploration Geology, ERIM, Fort Worth, TX, 26–27.

Cutter, M. A., 2006. The PROBA-1/CHRIS Hyperspectral Mission – five years since launch. *European Space Agency Special Publication*, 625: 6.

Descour, M. R., C. E. Volin, E. L. Dereniak, K. J. Thome, A. B. Schumacher, D. W. Wilson, and P. D. Maker, 1997. Demonstration of a high-speed nonscanning imaging spectrometer. *Optics Letters*, 22(16): 1271–1273.

Elwell, J. D., G. W. Cantwell, D. K. Scott, R. W. Esplin, G. B. Hansen, S. M. Jensen, M. D. Jensen, S. B. Brown, L. J. Zollinger, V. A. Thurgood, M. P. Esplin, R. J. Huppi, G. E. Bingham, H. E. Revercomb, F. A. Best, D. C. Tobin, J. K. Taylor, R. O. Knuteson, W. L. Smith, R. A. Reisse, and Hooker, R., 2006. A Geosynchronous Imaging Fourier Transform Spectrometer (GIFTS) for hyperspectral atmospheric remote sensing, instrument overview and preliminary performance results. *Proceedings of SPIE*, 6297: 62970S.

Fraunhofer, J., 1817. Bestimmung des Brechungs- und Farbenzerstreuungs-Vermoegens verschiedener Glasarten, in Bezug auf die Vervollkommnung achromatischer Fernroehre. *Gilberts Annalen der Physik*, 56: 264–313.

Gege, P., D. Beran, W. Mooshuber, J. Schulz, and H. van der Piepen, 1998. System Analysis and Performance of the New Version of the Imaging Spectrometer ROSIS, *Proc. 1st EARSeL Workshop on Imaging Spectroscopy* (M. Schaepman, D. Schläpfer, and K. I. Itten, Eds.), 6–8 October 1998, Zurich, EARSeL, Paris.

Goetz, A. F. H., 2007. Twenty-seven years of hyperspectral remote sensing of the Earth: A personal view. *Remote Sensing of Environment* (in press.)

Goetz, A. F. H. and M. Herring, 1989. The High-Resolution Imaging Spectrometer (Hiris) for Eos. *IEEE Transactions on Geoscience and Remote Sensing*, 27: 136–144.

Goetz, A. F. H., L. C. Rowan, and M. J. Kingston, 1982. Mineral identification from orbit – initial results from the shuttle multispectral infrared radiometer, *Science*, 218: 1020–1024.

Goetz, A. F. H., G. Vane, J. Solomon, and B. N. Rock, 1985. Imaging spectrometry for Earth remote sensing. *Science*, 228: 1147–1153.

Gower, J. F. R., G. A. Borstad, and H. R. Edel, 1987. Fluoresence line imager: First results from PAssive Imaging of Chlorophyll Fluoresence. In *International Geoscience and Remote Sensing Symposium (IGARSS)*, IEEE, Michigan, 1605.

Gower, J. F. R., G. A. Borstad, G. D. Anger, and H. R. Edel, 1992. CCD–based Imaging Spectroscopy for Remote Sensing: The FLI and CASI Programs, *Canadian Journal of Remote Sensing*, 18: 4.

Green, R. O., M. L. Eastwood, C. M. Sarture, T. G. Chrien, M. Aronsson, B. J. Chippendale, J. A. Faust, B. E. Pavri, C. J. Chovit, M. Solis, M. R. Olah, and O. Williams, 1998. Imaging spectroscopy and the Airborne Visible/Infrared Imaging Spectrometer (AVIRIS). *Remote Sensing of Environment*, 65: 227.

Hackwell, J. A., D. W. Warren, R. P. Bongiovi, S. J. Hansel, T. L. Hayhurst, D. J. Mabry, M. G. Sivjee, and J. W. Skinner, 1996. LWIR/MWIR imaging hyperspectral sensor for airborne and ground-based remote sensing. *Proceedings of SPIE*, 2819: 102–107.

Hearnshaw, J. B., 1986. *The Analysis of Starlight. One Hundred and Fifty Years of Astronomical Spectroscopy*. Cambridge University Press.

Hofer, S., H. J. Kaufmann, T. Stuffler, B. Penné, G. Schreier, A. Müller, A. Eckardt, H. Bach, U. C. Benz, and R. Haydn, 2006. EnMAP hyperspectral imager: An advanced optical payload for future applications in Earth observation programs. *Proceedings of SPIE*, 6366: 63660E.

Kol'cov, V. V., 1959. The application of the Spectrovisor to the investigation of small objects from an aircraft [in Russian]. Shikh Nazemnykh Ob, Ektov S Samoleta, Trudy Lab. Aeromet., Akad. Nauk. SSSR, Moscow/Leningrad, 7, 58–69.

Kostkowski, H. J., 1997. *Reliable Spectroradiometry*. 1st edn. Spectroradiometry Consulting, La Plata.

Kramer, H., 2002. *Observation of the Earth and its Environment – Survey of Missions and Sensors*. 4th edn. Springer, Berlin.

Kruse, F. A., K. S. Kierein-Young, and J. W. Boardman, 1990. Mineral mapping at Cuprite, Nevada with a 63-channel imaging spectrometer. *Photogrammetric Engineering and Remote Sensing*, 56: 83–92.

Kunkel, B., F. Blechinger, D. Viehmann, H. van der Piepen, and R. Doerfer, 1991. ROSIS Imaging Spectrometer and its potential for ocean parameter measurements (airborne and spaceborne). *International Journal of Remote Sensing*, 12(4): 753–761.

Leroy, M. and F.-M. Bréon, 1996. Angular signatures of surface reflectances from airborne POLDER data. *Remote Sensing of Environment*, 57(2): 97–107.

Lyon, R. J. P., 1962. *Evaluation of Infrared Spectroscopy for Compositional Analysis of Lunar and planetary oils*. Stanford. Res. Inst. Final Rep. Contract NASA, 49(04).

Maxwell, J. C., 1873. *A Treatise on Electricity and Magnetism*. Clarendon Press, Oxford.

Mouroulis, P. and R. O. Green, 2003. Optical design for imaging spectroscopy. *Proceedings of SPIE – The International Society for Optical Engineering*, 5173: 18–25.

Mouroulis, P., R. O. Green, and T. G. Chrien, 2000. Design of pushbroom imaging spectrometers for optimum recovery of spectroscopic and spatial information. *Applied Optics*, 39(13): 2210–2220.

Neville, R. A. and I. Powell 1992. Design of SFSI: An Imaging Spectrometer in the SWIR, *Canadian Journal of Remote Sensing*, 18(4): 210–222.

Newton, I., 1704. Opticks: Or, a Treatise of the Reflexions, Refractions, Inflexions and Colours of Light. Sam Smith and Benj. Walford, London. Book I, Plate IV, Part I, Fig. 18.

Pearlman, J. S., P. S. Barry, C. C. Segal, J. Shepanski, D. Beiso, and S. L. Carman, 2003. Hyperion, a space-based imaging spectrometer. *IEEE Transactions on Geoscience and Remote Sensing*, 41(6, Part I): 1160–1173.

Puschell, J. J., H.-L. Huang, and H. M. Woolf, 2001. Geostationary wedge-filter imager-sounder, *Proceedings of SPIE*, 4151: 68–76.

Rast, M., J. L. Bézy, and S. Bruzzi, 1999. The ESA Medium Resolution Imaging Spectrometer MERIS – A review of the instrument and its mission. *International Journal of Remote Sensing*, 20(9): 1681–1702.

Rast, M., F. Baret, B. van de Hurk, W. Knorr, W. Mauser, M. Menenti, J. Miller, J. Moreno, M. E. Schaepman, and M. M. Verstraete, 2004. *SPECTRA – Surface Processes and Ecosystem Changes Through Response Analysis*. ESA SP-1279(2), Noordwijk.

Reuter D. C., G. H. McCabe, R. Dimitrov, S. M. Graham, D. E. Jennings, M. M. Matsumura, D. A. Rapchun, and J. W. Travis, 2001. The LEISA/atmospheric corrector (LAC) on EO-1. *International Geoscience and Remote Sensing Symposium (IGARSS)*, 46–48.

Richter, R., A. Mueller, M. Habermeyer, S. Dech, K. Segl, and H. Kaufmann, 2005. Spectral and radiometric requirements for the airborne thermal imaging spectrometer ARES. *International Journal of Remote Sensing*, 26: 3149–3158.

Rickard, L. J., R. W. Basedow, E. F. Zalewski, P. R. Silverglate, and M. Landers, 1993. HYDICE: an airborne system for hyperspectral imaging. *Proceedings of SPIE*, 1937: 173–179.

Rowlands, N., R. A. Neville, and I. P. Powell, 1994. Short-wave infrared (SWIR) imaging spectrometer for remote sensing. *Proceedings of SPIE*, 2269: 237–242.

Salomonson, V. V., W. L. Barnes, P. W. Maymon, H. E. Montgomery, and H. Ostrow, 1989. MODIS: advanced facility instrument for studies of the earth as a system. *IEEE Transactions on Geoscience and Remote Sensing*, 27(2): 145–153.

Schaepman, M. E., 2007. Spectrodirectional remote sensing: from pixels to processes. *International Journal of Applied Earth Observation and Geoinformation*, 9: 204–223.

Schaepman, M. E., K. I. Itten, D. Schläpfer, J. W. Kaiser, J. Brazile, W. Debruyn, A. Neukom, H. Feusi, P. Adolph, R. Moser, T. Schilliger, L. de Vos, G. M. Brandt, P. Kohler, M. Meng, J. Piesbergen, P. Strobl, J. Gavira, G. J. Ulbrich, and R. Meynart, 2004. APEX: current status of the airborne dispersive pushbroom imaging spectrometer. In *Proceedings of SPIE: Sensors, Systems, and Next-Generation Satellites VII* (R. Meynart, ed.), 5234: 202–210.

Schaepman, M. E., R. O. Green, S. Ungar, B. Curtiss, J. Boardman, A. J. Plaza, B.-C. Gao, S. Ustin, R. Kokaly, J. Miller, S. Jacquemoud, E. Ben-Dor, R. Clark, C. Davis, J. Dozier, D. Goodenough, D. Roberts, G. Swayze, E. J. Milton, and A. F. H. Goetz, 2006. The future of imaging spectroscopy – prospective technologies and applications. In *Proc. of the Geoscience and Remote Sensing Symposium (IGARSS)*. IEEE, Denver (USA), 2005–2009.

Schläpfer, D. and M. Schaepman, 2002. Modeling the noise equivalent radiance requirements of imaging spectrometers based on scientific applications. *Applied Optics*, 41(27): 5691–5701.

van Der Piepen, H., R. Doerffer, and B. Kunkel, 1989. ROSIS – ein abbildendes Spektrometer für die Umweltforschung, *DLR-Nachrichten*, 58: 21–24.

Vane, G., 1986. Introduction Airborne Imaging Spectrometer (AIS-1, AIS-2). In G. Vane and A. F. H. Goetz, eds. *Proc. Second Airborne Imaging Spectrometer Data Analysis Workshop*, NASA/Jet Propulsion Laboratory Publ. 86–35, Pasadena, CA, 1–16.

Vane, G. (ed.), 1987. Airborne Visible/Infrared Imaging Spectrometer (AVIRIS) – A description of the sensor, ground data, Processing Facility, Laboratory Calibration and First Results, *JPL Publ. 87–38*, November 15, Jet Propulsion Laboratory, Pasadena, CA.

Vane, G., A. F. H. Goetz, and J. B. Wellman, 1984. Airborne imaging spectrometer – a new tool for remote-sensing. *IEEE Transactions on Geoscience and Remote Sensing*, 22: 546-549.

Wright, H., J. Warnow-Blewett, and C. Weiner, 1972. *The Legacy of George Ellery Hale*, MIT Press, Cambridge, Massachusetts.

13

Active and Passive Microwave Systems[1]

Josef Martin Kellndorfer and
Kyle C. McDonald

Keywords: Imaging radar (SAR), Synthetic Aperture radar (SAR), polarimetry, interferometry, InSAR, PolInSAR, DInSAR, altimeters, scatterometers, radiometers.

MICROWAVE SENSORS OVERVIEW

Remote sensors operating in the microwave region of the electromagnetic (EM) spectrum can be grouped into two basic categories: (1) Sensors that *transmit* microwaves and measure the energy of the returned signal, and (2) sensors that measure energy from microwaves *emitted* from observed targets. The former systems are referred to as *active*, the latter as *passive* microwave sensors. Active sensors are based on *ra*dio *d*etection *a*nd *r*anging (*RADAR*[2]) technology and can be distinguished into *imaging radars*, *altimeters*, and *scatterometers* (Elachi 1988). Passive sensors work conceptually similarly to optical thermal sensors and are generally referred to as *radiometers* or *microwave scanners*.

A variety of text books cover the aspects of microwave remote sensing in depth, e.g., *Introduction to Microwave Remote Sensing* (Woodhouse 2006), Fundamentals of *Microwave Remote Sensing* (Ulaby et al. 1981, 1982, 1986), *Application and Techniques of Spaceborne Radar Remote Sensing* (Elachi 1988), *Radar Polarimetry for Geoscience Applications* (Ulaby and Elachi 1990), *Radargrammetric Image Processing* (Leberl 1990), *Synthetic Aperture Radar: Systems and Signal Processing* (Curlander and McDonough

1991), *SAR Geocoding* (Schreier 1993), *Principles and Applications of Imaging Radar* (Henderson and Lewis 1998), *Understanding Synthetic Aperture Radar Images* (Oliver and Quegan 1998), *Synthetic Aperture Radar Processing* (Franceschetti and Lanari 1999), and *Digital Processing of Synthetic Aperture Radar Data* (Cumming and Wong 2005).

This chapter gives an overview of the various microwave remote sensing systems and their associated typical data products and processing techniques.

ACTIVE MICROWAVE SENSORS (RADARS)

Active microwave sensors (i.e., RADARs) are characterized by a system design which encompasses the transmission and subsequent reception of EM energy in the form of microwaves through *antennas*. If a transmitter and receiver are using the same antenna, a radar sensor is referred to as a *monostatic* system. Such a system typically consists of six components: (1) a *pulse* generator which discharges timed pulses of energy, (2) a *transmitter*, (3) a *duplexer* to switch between transmission and reception, (4) an *antenna*, (5) a *receiver*, and

(6) a *data recorder*. Due to the vast amounts of data generated by active microwave sensors, on-board processing is generally not feasible and raw microwave signals are processed in ground-segment based *signal processors*. Most spaceborne and airborne radar systems operate as monostatic systems. A *bistatic* radar system contains two antennas, one used for transmission, and the other one used for reception. Bistatic systems can also be configured where one or both of the two antennas act as transmitter and receiver. An example of spaceborne deployment of a bistatic radar system with one transmit/receive and one receive only antenna was the sensor configuration flown during the *Shuttle Radar Topography Mission*. Bistatic systems where both antennas transmit and receive are generally referred to as operating in *ping-pong* mode.

Ulaby et al. (1981) define an *antenna* as the region of transition between an EM wave propagating in free space and a guided wave propagating in a transmission line, or vice versa. It acts as a coupler between the two media, performing the same function that a lens does in the optical sensors. A key characteristic of each antenna is its *antenna pattern* (or radiation pattern, beam pattern, lobe pattern) which describes how an antenna beam is focused in the direction of the target. Transmitted microwave energy is strongest in the center of the main *lobe* of an antenna. The region of the beam, radiated from an antenna, where the peak energy drops by about 50% is referred to as the half-power or 3 dB[3] *beamwidth* of an antenna. The beamwidth of an antenna, together with platform altitude and beam direction, i.e., pointing angle toward a target in nadir or off-nadir, determines in principle the ground area 'illuminated'[4] by the radar sensor at any given time. This area is also referred to as the *footprint* of the radar main beam on the surface. Microwave energy focused through the antenna is generated in the transmitter and released in the form of short pulses (bursts). The reflected energy returning (backscattered) to the sensor, i.e., the radar *echo*, is detected by the same (monostatic case) or a different (bistatic case) antenna which is connected to the receiver module. The *gain* of an antenna determines the ability to separate target information in a return signal from noise. While altimeters and scatterometers are generally based on conical beam antenna designs, modern antenna designs for imaging radars provide for a variety of imaging modes (see below).

An important distinction in the operation of active microwave sensors is whether the transmitter and receiver of the energy pulses retain information on the *phase* of the signal in addition to its *amplitude*. If phase information is retained, an active microwave sensor is operating as a *coherent* system, if not, the system is *incoherent*. Coherence is the basis for high-resolution imaging radars

and interferometers, by using the *Doppler-effect*[5] of signal phase shifts to produce high resolution imagery with the signal processor.

Imaging radars

Imaging radars are used to generate high- to moderate-resolution (1 to ~100 m) images of Earth's or another celestial body's surface or subsurface. Each image pixel contains information on the microwave energy detected by the receiver of a radar sensor. As described in Elachi and van Zyl (2006) imaging radars generate images that look very similar to passive visible and infrared images, but the principle behind image generation is fundamentally different in the two cases. Imaging radars use the echoes of the transmitted pulses that are backscattered to the receiving antenna(s) from different surface elements. The backscattered echoes are then separated in the *range* (*cross-track*) dimension and the angular size (in the case of real-aperture radar) of the antenna pattern, or the Doppler history (in the case of synthetic aperture radar) to separate surface pixels in the *azimuth* (*along-track*) dimension.

Principles of imaging radar – real and synthetic aperture radar

In principle, imaging radars operate as side-looking sensors which transmit and receive EM waves through the antenna interface. Most spaceborne sensors operate as monostatic sensors, although future mission scenarios discuss bi- and multi-static concepts (Sarabandi et al. 2003).

In contrast to optical sensors based on photo-electronic scanning elements with fixed spatial resolution, the spatial (pixel) resolution of an imaging radar system has two components, range and azimuth resolution, which are variable across the imaged ground swath. For the pulsed radar system, the range resolution is determined by the bandwidth of a transmitted pulse. A larger bandwidth results in better range resolution. A wide bandwidth can be generated by a short pulse duration. However, a short pulse results in lower energy transmitted, which has a negative effect on the signal-to-noise ratio, and thus decreased radiometric resolution. To address these limitations, modern SAR sensors use a technique to generate longer pulses with linear frequency modulation, termed *chirping*, where the chirp-bandwidth is based on the desired range resolution. With this technique range resolution can be optimized without compromising radiometric resolution.

The azimuth resolution of a *real aperture radar* (RAR) system is a direct function of the antenna length, where a longer antenna results in better azimuth resolution. Since reasonable azimuth resolution from spaceborne sensors cannot be achieved with real aperture systems, the required antenna length is achieved by synthesizing a large aperture through signal processing based on the Doppler-effect. The Doppler-shift of target objects results from the change in relative velocity between the object and the sensor. As long as the measured radar-echoes sustain high-quality Doppler processing the azimuth resolution possible is one half of the physical length of the antenna. The imaging technique based on Doppler processing and thus synthesizing a long antenna is termed *synthetic aperture radar (SAR)* and is the most common sensor design for imaging radars (Figure 13.1).

Frequency

EM waves in the microwave range are often expressed by the *frequency band* in which a radar sensor operates. During World War II a letter system was designed for secrecy to code the various bands. The most commonly used bands for radar remote sensing and their corresponding wavelengths[6] are listed in Table 13.1.

Polarization

Polarization of an EM wave is related to the position of the tip of the electric field vector as it evolves with time. EM waves propagate along an invisible plane where the electric field vector traces a line

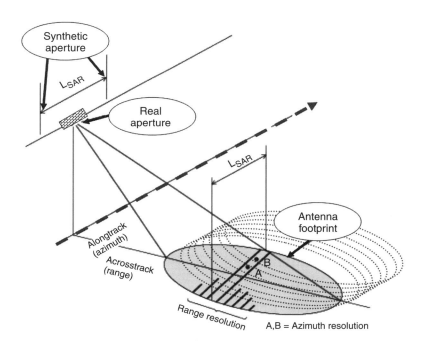

Figure 13.1 Concept of side-looking imaging radar. Resolution in range is related to the pulse width of the transmitted signal, resolution in azimuth is dependent on (a) the beamwidth for real aperture radar, and (b) the length L of the synthesized antenna for synthetic aperture radar.

Table 13.1 Commonly used frequency bands for radar remote sensing

Band name	Frequency	Wavelength	Platform examples
Ka Band	40,000–26,000 MHz	0.8–1.1 cm	
K Band	26,500–18,500 MHz	1.1–1.7 cm	
X Band	12,500–8,000 MHz	2.4–3.8 cm	SIR-C, SRTM, TerraSAR-X, COSMO-SkyMed
C Band	8,000–4,000 MHz	3.8–7.5 cm	Airsar, SIR-C, ERS-1/2, Radarsat 1/2, Envisat, SRTM, Sentinel-1
L Band	2,000–1,000 MHz	15.0–30.0 cm	Airsar, Seasat, SIR-A/B/C, JERS-1, ALOS, DESDynl
P Band	1,000–300 MHz	30.0–100.0 cm	Airsar, BIOMASS

(*linear polarization*), a circle, or ellipse (*circular or elliptical, right- or left rotating polarization*). Radar remote sensors can be designed to transmit and receive a single state of polarization of a wave predominately in the *horizontal* (parallel to the surface of the Earth) or *vertical* (perpendicular to the surface of the Earth) polarization (Figures 13.2 and 13.3). If sensors are designed to transmit and receive microwaves in the same polarization they are referred to as *like-polarized* or *co-polarized* and designated as hh or vv polarized systems.[7,8] *Cross-polarized* transmit/receive modes are designated as hv or vh modes. Sensors can also be designed to allow transmission of one polarization with reception of two polarizations (*dual-polarization* modes), or transmission and reception of two polarizations (*quad-polarization* mode). *Fully polarimetric* sensors allow the measurement of the entire polarization state of a circularly or elliptically transmitted wave. In addition to the detection of the backscatter values, additional *polarimetric signatures* can be derived when SAR images are acquired in multi- or full polarization sensor modes. Radar *polarimetry* is an active field of research with more and more airborne (e.g., NASA JPL Airsar, DLR

E-SAR, CCRS CONVAIR, JAXA PI-SAR) and spaceborne (e.g., SIR-C/X-SAR, Envisat/ASAR, ALOS/PALSAR, Radarsat, TerraSAR-X) sensors coming online which allow for measurement of the polarimetric signatures of microwaves (Ulaby and Elachi 1990, Zebker and van Zyl 1991, Boerner 2003, Hajnsek et al. 2003). Given tight power budgets inherent to spaceborne remote sensing missions, developing radar imaging sensors capable of more efficient polarimetric measurements using *compact (partial) polarimetry* technology is an important research topic (Souyris et al. 2005, Inglada et al. 2006).

Look/incidence angle

Since imaging radars operate as side-looking, off-nadir instruments, the sensor characteristic related to the geometry of the observation is an important factor during radar image acquisition and processing. At the sensor side the *look angle* or *depression angle*[9] are considered. The latter is defined as the angle between the slant range of the radar beam at any given point on the ground and an imaginary plane which contains the radar antenna

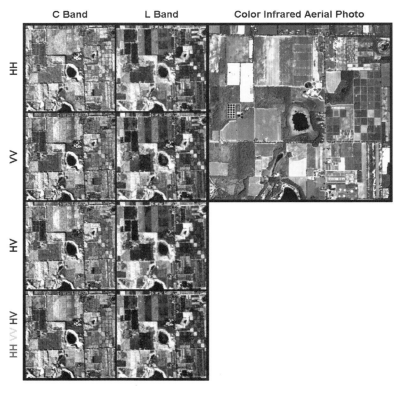

Figure 13.2   Back scatter images of JPL AIRSAR C/L-band channels in hh, vv, and hv polarizations. Kellogg Biological Station in Michigan, USA. Image size is 2.8 by 2.6 sq km. (See the color plate section of this volume for a color version of this figure).

ALOS/PALSAR
Xingu Basin,
Mato Grosso, Brazil
June/July 2007
L-HH/HV/HH-HV

Figure 13.3   JAXA ALOS PALSAR dual-polarimetric (hh/hv) L-Band data depicting Amazon deforestation. Xingu river basin, Brazil. Zoomed image is 70 by 50 sq km. (See the color plate section of this volume for a color version of this figure).

and is parallel to the Earth surface. The look angle is defined as 90° minus the depression angle and describes the off-nadir direction. At the target side, the *incidence angle* is defined as the angle between the vector of the surface normal (pointing skywards) of the Earth's ellipsoid and the vector of the back scattered wave direction (pointing towards the sensor). Thus, without considering Earth's curvature, the magnitude of the incidence angle would equal that of the look angle. Due to the curvature, the incidence angle is always slightly larger than the look angle, assuming no terrain. When true surfaces of terrain are considered, the angle between the surface normal and the slant range are termed the *local incidence angle*. For radiometric calibration of imaging radar data it is imperative to determine the local incidence angle and correct for its effects on the radar signal (Bayer et al. 1991). As a rule of thumb, if slopes are tilted toward the sensor an increase in the backscattered energy results (brightening of targets), whereas slopes tilted away from the sensor result in reduced backscatter (darkening of targets). Due to greater altitude, look angles vary less across an image swath acquired from satellite platforms compared to acquisitions from airborne platforms. Thus more uniform illumination is seen in satellite compared to airborne radar images. The viewing geometry between sensor and terrain also determines the effects of radar *foreshortening*, *layover* and *shadow* (Figure 13.4) (Jensen 2000).

These effects are experienced in steep terrain and are more prominent in airborne acquisitions. Foreshortening and layover effects occur at smaller look angles while radar shadow effects are more pronounced at larger look angles. *Radargrammetry* is an application which utilizes the effects of imaging radar's geometric characteristics for topographic mapping (Leberl 1990).

SAR interferometry

SAR interferometry (InSAR[10]) is based on the principle of measuring phase differences in the EM wave in the range to a target from different antenna locations (Figure 13.5) which allows for topographic mapping, motion detection, vegetation height, and volume measurements, and the detection of very subtle changes (centimeters) in surface elevation as caused for example by subsidence and earthquakes (Massonnet et al. 1993, Zebker et al. 1994, Madsen and Zebker 1998, Hanssen 2001, Kellndorfer et al. 2004, Sarabandi and Lin 2000, Oveisgharan and Zebker 2007, Walker et al. 2007). The interferometrially based elevation change techniques are referred to as *differential SAR interferometry* (DInSAR). Note that the term InSAR often refers to *across-track interferometry*[11], which means that the antennas are displaced in the across-track direction. InSAR data are generated from two or more complex SAR

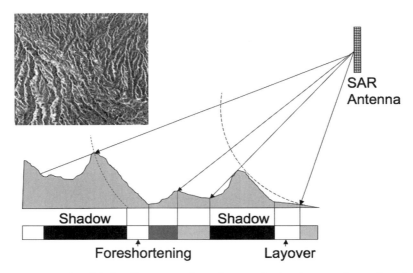

Figure 13.4   Due to the side-looking operation of SARs, the backscatter intensity of the return signal is influenced by the observed terrain characteristics. Common effects are foreshortening, layover, and shadow in imaging radars.

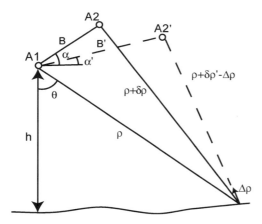

Figure 13.5   Principle of SAR interferometry. B/B' denote the baseline lengths between two (or more) antenna locations (A1/A2/A2'). InSAR measures the phase difference ($\delta\rho$, $\delta\rho'$) to compute elevation and elevation changes (Courtesy NASA/JPL-Caltech).

images by forming *interferograms*. Since the phase of the radar echo can only be measured modulo 2π, the interferogram is essentially an image depicting regions of equal phases, called *fringes*. Post-processing in the form of *phase-unwrapping* of the interferograms is then necessary to extract topographic information (Figure 13.6) (Goldstein et al.

1988, Zebker and Lu 1998). The key challenge for phase-unwrapping is to resolve ambiguities in the phase changes, i.e., whether the change from one fringe to the next is due to an increase or a decrease in surface elevation. Optimization of phase-unwrapping algorithms is an active field of InSAR research and a detailed review on InSAR techniques and applications is given by Gens and van Genderen (1996) and Massonnet and Feigl (1998). The processing of SAR data with a focus on interferometry is described by Hein (2004).

From a sensor configuration point of view, InSAR systems are specified by additional parameters besides the basic configurations inherent to all SAR systems. An important factor for the successful generation of an interferogram is the distance between the InSAR antenna locations, referred to as the *baseline* of the SAR interferometric measurements. For example, usable baselines for spaceborne InSAR measurements of topography at C-band with an altitude of approximately 700 km range to about 200 m. Longer baselines would lead to geometric decorrelation due to the differences of lines of sight between the InSAR image pairs (Zebker and Villasenor 1992). The requirement for baseline length scales linearly with the wavelength of the InSAR system, i.e., longer wavelengths allow for longer baselines. Furthermore, a distinction is made as to how the InSAR image pairs are acquired. *Single-pass* interferometry refers to the acquisition of InSAR data during one overpass where two receiving antennas are mounted on one platform, for example as was

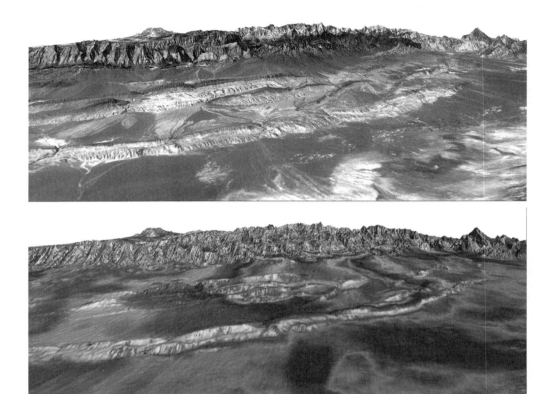

Figure 13.6   Digital elevation data derived from interferometric radar data. (See the color plate section of this volume for a color version of this figure).
Source: Courtesy NASA/JPL-Caltech.

configured during the Shuttle Radar Topography Mission (SRTM), or when satellite or airborne sensors are flying in formation to provide synchronous reception of a radar echo, e.g., as planned for the German TerraSAR-X Tandem Mission (Moreira et al. 2004). During *repeat-pass interferometry*, data are collected from SAR sensors imaging a target area during repeat overpasses with the same SAR sensor or an InSAR compatible constellation of two or more sensors as demonstrated by the ERS-1/2 mission (Sansosti 1999, Santoro et al. 2007). The ideal baseline for surface deformation detection with repeat-pass interferometry is zero. If one image pair is acquired, a *single-baseline* InSAR dataset can be generated, while several repeat overpasses with slightly varying orbital tracks allow for the generation of *multi-baseline* InSAR datasets which provide an enhanced feature space for InSAR data analysis (Stebler et al. 2002). Since *interferometric coherence* between two or more SAR images is key for the formation of fringes in an interferogram, an attempt is usually made to minimize effects of temporal and volume decorrelation. Those effects are avoided

during single-pass interferometry. However, coherence changes measured from repeat-pass interferometry have been demonstrated to contain valuable signals, e.g., for vegetation studies (Treuhaft and Siqueira 2000, Santoro et al. 2007).

Polarimetric SAR interferometry (PolInSAR) is a relatively new and growing field in SAR interferometry and is gaining increasing attention since technological advances for future SAR systems will provide space borne PolInSAR capabilities (Treuhaft and Cloude 1999, Ferretti et al. 2001, Papathanassiou and Cloude 2001, Mette et al. 2002, Stebler et al. 2002, Boerner 2003, Cloude and Papathanassiou 2003, Yamada et al. 2003). Selected current topics researched by the PolInSAR community include SAR tomography, persistent scatterers, combination of multiple interferograms, interferometry with different frequencies, coherent/non-coherent stitching of interferometric phases, and wide areas SAR interferometry. An assessment of advances and the state-of-art in SAR interferometry is available from the PolInSAR workshop series (http://earth.esa.int/polinsar/).

Special SAR imaging modes

The development of phased array antennas[12] which enable electronic beam steering, i.e., the changing of the direction and width of the main lobe of the antenna, provides for special imaging modes of recent and future spaceborne imaging SAR systems (Mailloux 1994, Hansen 1998). The first spaceborne imaging SAR sensor which had a host of imaging modes was Canada's Radarsat-1. Today, Europe's Envisat/ASAR and Japan's ALOS/PALSAR sensors are also based on phased array antennas and provide different imaging modes as will Germany's TerraSAR-X sensor. The special SAR sensor modes are generally a trade-off between swath width, spatial resolution and polarimetric imaging capability (Figure 13.7). The characteristics of the various SAR imaging modes are listed in Table 13.2.

The repeat imaging of a specific target area is another advantage of the beam steering capability and provides for high radiometric enhancement, high coverage rate, high temporal repeat rates, multi-angle interferometry capability, and repeated coverage at different viewing angles.

SAR image data characteristics

Since a major component of imaging radar is the signal processor, radar image data come in many different forms and processed states. The *level* of processing is application-driven and major advances have been made in recent years to automate SAR image processing (Hein 2004). This section discusses some of the main characteristics of SAR image data which are important for radar image analysis and value extraction.

Foremost, each image pixel value is some form of expression of the power ratio of transmitted versus received microwave energy which is often expressed in the dB scale. The conversion formula to dB is $10 \log_{10}$ (power ratio), or $20 \log_{10}$ (sqrt(power)) ratio. Without further processing this ratio is termed β^0, also known as the *radar brightness*. If the local incidence angle θ_i is known, the normalized radar backscatter coefficient σ^0 can be

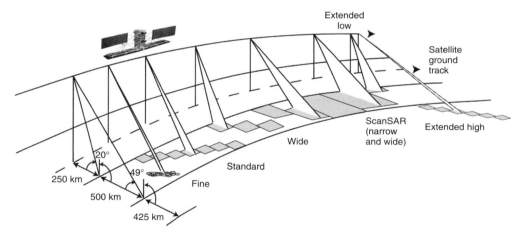

Figure 13.7   SAR beam modes. New antenna designs provide the capability to focus, widen, or steer the imaging radar beam, providing for many imaging beam modes, e.g., for Radarsat-1.

Table 13.2   SAR imaging modes

Standard Beam (Stripmap)	Standard imaging mode of a SAR sensor defined by the mission plan given power availability and sensor design. For example, ALOS/PALSAR features a base mission with medium resolution (15 m) and dual-polarimetric (hh/hv) imaging. ALOS swath width at the standard beam modes is ca. 80 km.
ScanSAR	In this mode spatial and polarization resolution is sacrificed for achieving a larger swath width, e.g., 350 km at 100 m resolution for ALOS/PALSAR. Note that the effects of incidence angle are larger for ScanSAR images. The principle of ScanSAR modes is the division of the SAR antenna in several portions imaging a number of subswaths.
Fine Beam	Swath width is compromised to achieve higher spatial resolution and/or polarimetric imaging capability.
Spotlight	Very high resolution in an SAR image is achieved by steering the radar beam such that a target is kept within the beam for a longer time which results in the formation of longer synthetic aperture and thus better resolution.

computed as $\sigma^0 = \beta^0/\cos\theta_i$. σ^0 is of interest to compare radar measurements between sensors of different resolution and from different incidence angles, and can also be expressed as the radar cross section σ normalized by the surface area (A), i.e., $\sigma^0 = \sigma/A$.

Due to the coherent nature of SAR image generation, interference between randomly positioned scatterers within a resolution cell causes effects of *random fading noise*, most commonly known as image *speckle*. This noise is expressed in a SAR image as a 'salt-and-pepper effect', i.e., random brightening and darkening of neighboring image pixels. In a single-look image, i.e., an image processed to full resolution of the azimuth and range resolution for each image cell, the speckle effect is most pronounced and follows statistically a Raleigh distribution for *amplitude images* or exponential distribution for *power images*. Whether image data are presented as amplitude or power data is processor dependent. Important for the processing of SAR image data is that many image operations, e.g., the averaging of neighboring pixels, need to be performed on the data in power units to be mathematically correct. A commonly used procedure in SAR image processing to reduce the effect of speckle and thus gain radiometric stability in the imagery is *multi-look* processing where several neighboring image cells are integrated (averaged), usually in the azimuth direction since azimuth resolution is often several times better than range resolution. This effect of resolution differences often causes single-look processed SAR images to appear stretched in the azimuth direction.

Another distinction in processing raw image data is whether data are retained in a *complex* format, i.e., amplitude and phase information is preserved, or whether data are processed to provide amplitude/power values directly, i.e., power values are *detected*. In the complex case, data of a single polarization are provided in two channels, an in-phase (I) and a quadrature (Q) component, i.e., the amplitude of the received wave at 0 phase and $\pi/2$ offset. With these two values power P can be computed as $P = I^2 + Q^2$. Complex polarimetric data are stored in a special compressed data

format which optimizes the storage requirements. The most common forms are the *Stokes matrix* and complex *Mueller matrix* format to store polarimetric radar image data (Held et al. 1988, Dubois et al. 1989, Ulaby and Elachi 1990). Preservation of all SAR image information in the complex format is necessary for interferometric processing.

Common image data formats after initial processing of the raw radar imagery, i.e., azimuth and range *compression*, are listed in Table 13.3. The Committee on Earth Observing Systems (CEOS) has standard definitions for these formats (http://www.ceos.org/).

Depending on the SAR image sensor, image data also can vary with respect to the polarizations. SAR image data can have single polarized, multi-polarized, or fully polarimetric origin and only selected polarizations, e.g., like-polarized or cross-polarized data are extracted. Furthermore, data are distinguished into slant range or ground range geometry.

SAR data processing

Depending on the application and (In)SAR image data source, processing of SAR data can be accomplished in many different ways. For example, the detection of point scatterers will involve different image filtering techniques than those typically used with smoothing area targets like forests or agricultural fields.

Nonetheless, some basic SAR data processing steps are almost always performed. All further SAR data analysis requires processing of the raw (level 0) data in a SAR processor to level 1 data. Various flavors of SAR processors exist (Bamler 1992), but most common are processors based on the *Range-Doppler* algorithm. This algorithm is computationally efficient and phase-preserving. Level 1 data are generally generated in the SAR processor to slant range geometry. The native azimuth resolution is often several times better than the native range resolution which generally makes SAR image data appear compressed in the range direction. Thus further processing is often performed which produces somewhat squared image pixels by *multi-look processing* in the form of

Table 13.3 Typical processing levels of SAR image data according to CEOS standards

Level 0	Raw signal	Needs to be processed in SAR processor to visualize data
Level 1	Single-look complex (SLC)	Primarily used for interferometry as it retains amplitude and phase information
Level 1	Multi-look complex	Data are multi-look processed
Level 2	Multi-look detected	Data are converted to amplitude (or power)[1] and multi-look processed
Level 2	Ellipsoid corrected	Data are geocoded with respect to a reference ellipsoid
Level 2	Terrain corrected	Data are geocoded with a digital elevation model. Ideally local incidence angle, layover, and shadow layers are provided and radiometric slope corrections are performed for full radiometric calibration

[1] Magnitude = sqrt(Power)

averaging several pixels in the azimuth direction. For flat terrain, *slant-to-ground range conversion* generally yields adequate pixel geometries. However, in even moderate terrain, full *terrain correction* must be performed in order for SAR images to conform to a map projection. The effects of topography on SAR geometric and radiometric calibration are well understood and can be corrected for with most SAR post-processors (Bayer et al. 1991, Schreier 1993, van Zyl et al. 1993, Kellndorfer et al. 1996, Ulander 1996). Advances are also made in calibration of slope effects on the radar cross section as a function of polarization states (Lee et al. 2000). If radiometric distortion effects in the range direction which result from an uncompensated antenna pattern are visible in a SAR image, an *antenna pattern correction* needs to be performed. An overview of general SAR calibration is described by Freeman (1992). Backscatter analysis for SAR data classification and bio-geophysical parameter retrieval (like biomass, soil moisture, snow depth) is often preceded by signal upgrading through speckle noise reduction. A host of research has been performed on the best *speckle filtering* (Figure 13.8) methods for various SAR imagery Lopes et al. 1990, 1993, Lee et al. 1991, 1999, Baraldi and Panniggiani 1995, Touzi 2002, Xie et al. 2002, Lopez-Martinez et al. 2005). For interferometric processing the challenge lies in the formulation of interferograms and the subsequent process of phase unwrapping, i.e., the generation of height information from fringes in the interferogram. In recent years many approaches to phase unwrapping have been developed (Goldstein et al. 1988, Fornaro et al. 1996, Ghiglia and Pritt 1998, Zebker and Lu 1998, Ferretti et al. 1999, Xu et al. 1999).

(In)SAR applications

Today, microwave imaging is used in a range of applications to map and monitor Earth and other celestial bodies in the solar system. Earth observation of both ocean and land surfaces makes increasing use of SAR imaging technology (Henderson and Lewis 1998).

Environmental monitoring has greatly improved with the widespread operational availability of SAR data from the European (ERS-1/2), Japanese (JERS-1, ALOS), and Canadian (Radarsat) satellites. Today, SAR is used for the operational monitoring of shipping traffic on the world's oceans which includes the detection of oil spills and real-time monitoring of ice flows in shipping lanes (Rey et al. 1990, Fahnestock et al. 1993, Gade and Ufermann 1998). With the availability of microwaves to penetrate clouds, SAR imaging plays an increasingly important role in ecological monitoring of tropical vegetation. The improvement of annual deforestation monitoring with SAR (Figure 13.3) is an invaluable contribution for a post-Kyoto climate treaty where credits for avoided deforestation to developing countries are discussed. SAR's sensitivity to vegetation structure has been demonstrated to allow for the retrieval of vegetation biophysical parameters like vegetation height and biomass (Dobson et al. 1992, 1995, Foody et al. 1997, Kellndorfer et al. 1998, 2003, 2004, Simard et al. 2000, Treuhaft et al. 2003). SAR's sensitivity to moisture and flooding conditions make it an invaluable tool for observations of wetlands and flooding events (Pope et al. 1994, Hess et al. 1995, Alsdorf et al. 2000). Typical InSAR applications include digital elevation model generation (Madsen et al. 1993, Gens and van Genderen 1996, van Zyl 2001, Smith 2002), vegetation structure mapping (Treuhaft et al.

Figure 13.8   Application of a speckle filter (right) to an unfiltered SAR image (left).

1996, Wegmuller and Werner 1997, Treuhaft and Siqueira 2000, Kellndorfer et al. 2004), the observations of ice dynamics (Goldstein et al. 1993, Mohr et al. 1998), and detection of subtle changes in the earth surface elevation as a tool to measure subsidence and to predict earthquakes and volcanic activity (Rosen et al. 1996, Massonnet and Feigl 1998, Ferretti et al. 2000).

Altimeters

Principles of radar altimeters

Radar altimeters use the ranging capability of radar to measure surface topography (Elachi et al. 1990, Elachi and van Zyl 2006). These sensors transmit a short radar pulse and precisely measure the time delay between pulse transmission and the reception of the signal returned to the sensor from the Earth's surface. The distance between the sensor and the observed target is determined by multiplying the time delay by the speed of light and dividing by 2 to account for the difference between round-trip and one-way distance. The resolution of the range measurement is given by the product of the speed of light and half the effective pulse length.

The radar echo represents a reflection from an extended area that corresponds to the footprint of the radar antenna beam. When the antenna beam is narrow and the corresponding footprint is small relative to the width of the transmitted pulse, the altimeter is beam-limited. Such narrow beam systems are called for when operating over regions where surface elevation may vary significantly over short distances, and are commonly used to characterize land surface topography. Pulse-limited systems correspond to cases when the antenna beam is wide and the antenna footprint large relative to the pulse width. Such systems are commonly used to characterize topography of smooth surfaces where the surface topography does not change significantly over the antenna footprint, such as the ocean. In such cases, range resolution may be improved by fitting a model-derived echo function to the measured radar return. Efficient mapping of large regions over a short period of time may be achieved through imaging altimetry, by mechanically scanning the altimeter back and forth across a wide swath or by spinning the antenna in azimuth. Scanning can also be implemented through electronically steering a phased array antenna. For real aperture radar, high spatial resolution requires a large antenna area. Alternatively, synthetic aperture approaches may be used to enable high resolution mapping with a reduced antenna size. Imaging altimeters typically have lower spatial resolution than traditional SARs, but require less mass and volume.

Data characteristics and processing

Spaceborne altimeters require careful calibration for provision of accurate scientific data. Radar altimeters are each unique, requiring careful characterization of the instrument's response to estimation errors. The shape of the waveform returned signal provides the precise measure of the surface elevation. Key factors for system calibration include characterization of the shape of the transmitted pulse, which affects the shape of the return pulse, and characterization of the receiver's measure of the returned pulse. Waveform fitting may be employed to assess the accuracy of the height retrievals (Hayne et al. 1994). These calibration issues are generally addressed in data processing prior to distribution of the data products.

Altimeter applications

An example of ocean surface altimetry is provided by the joint U.S./France Jason-1 mission program. Launched in 2001 as a follow-up to the TOPEX/Poseidon mission, the Jason satellite utilizes a dual-frequency (13.6 and 5.3 GHz) radar altimeter to study ocean height. Figure 13.9 shows an example of a 10-day composite ocean surface height product, illustrating ocean surface height distribution relative to the mean surface height over the seasonal cycle. Jason data have provided measurements of ocean surface height to a 3.3 cm accuracy. Such data allow understanding of ocean circulation and associated impacts to global climate.

Figure 13.9   Jason radar altimeter derived ocean surface height. (See the color plate section of this volume for a color version of this figure).
Source: Courtesy NASA/JPL-Caltech.

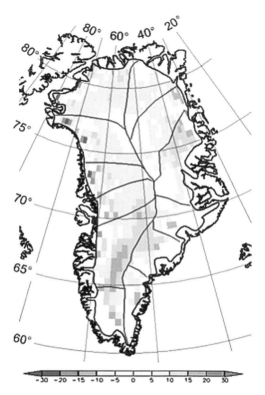

Figure 13.10 Greenland ice thickness derived from ERS-1/2/ radar altimeters (Johannessen et al. 2005). (See the color plate section of this volume for a color version of this figure).

Multi-year ERS-1 and ERS-2 altimeter data have been used to examine ice sheet height over Greenland (Johannessen et al. 2005). The continuous data set, collected from 1992 to 2003, provides a capability for examining the eleven year trend in ice sheet elevation change (Figure 13.10). Analysis shows an increase of +6.4 cm/year over the high elevation interior, or 5.4 cm/year over the entire region. This is equivalent to 60 cm of increased ice elevation over the 11 year period, or 54 cm when corrected for uplift of the underlying bedrock.

Scatterometers

Principles of scatterometers

Scatterometers are non-imaging radars designed to measure the radar backscattering cross section of the target illuminated by a radar beam (Elachi and van Zyl 2006). Scatterometers may utilize any of a variety of observation geometries. Geometries and instrument characteristics of some spaceborne scatterometers are shown in Figure 13.11. Side-looking, tilted fan-beam

configurations illuminate a region perpendicular to the flight-track direction, with incidence angles varying widely from near to far range. Forward-looking fan-beam scatterometers allow observation only along a narrow swath along the flight direction, but allow observation of a given region at a variety of incidence angles as the instrument travels along the flight track. Squint-looking scatterometers observe the surface with a fan-beam tilted between the forward-looking and side-looking directions. Utilizing more than one antenna in fore and aft-looking squint modes allows observation of a given ground point at two azimuth view angles. Because of range ambiguities, fan-beam radars view off-nadir, such that a gap in radar coverage exists directly under the instrument flight path. Utilizing a scanning pencil beam configuration eliminates the nadir gap, allowing wide, regional coverage over a large scanning area, at a constant incidence angle, with measurements acquired for each region at two azimuth look directions. Large swath width or scanning geometries allow spaceborne systems to observe a high percentage of the global surface with high temporal repetition, often better than 3 days.

Sounding or profiling radars are scatterometers that measure backscatter as a function of distance from the radar (Im et al. 2005). Design of these systems requires careful attention to medium absorption and scattering properties, as well as radar system sensitivity, resolution, and other radar parameters to optimize the radar's capability for characterizing the target medium.

Data characteristics and processing

Spaceborne scatterometers commonly have relatively coarse spatial resolution. Native resolution of these real aperture sensors is commonly 25–50 km (see Figure 13.11). In response to these limitations, enhanced resolution imaging techniques have been developed to improve the utility of the spaceborne scatterometer data (Long et al. 1993, Early and Long 2001). Utilizing the spatial overlap of multi-pass scatterometer measurements, this innovative processing technique has enabled wider use of these data for many climate studies. This capability facilitates the use of scatterometer data originally developed for non-imaging purposes.

More recently, concepts have been considered which combine scanning antenna architectures with synthetic aperture processing techniques (Entekhabi et al. 2004). As illustrated in Figure 13.12, such systems utilize constant Doppler (iso-Doppler) contours in synthetic aperture processing to improve the spatial resolution of the radar system. This architecture capitalizes on the wide coverage allowed by the scanning pencil beam geometry, thus providing frequent repeat global observations at significantly improved spatial resolution.

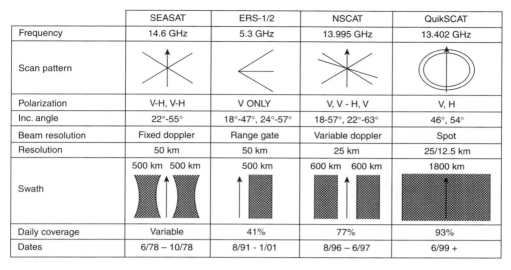

	SEASAT	ERS-1/2	NSCAT	QuikSCAT
Frequency	14.6 GHz	5.3 GHz	13.995 GHz	13.402 GHz
Scan pattern				
Polarization	V-H, V-H	V ONLY	V, V - H, V	V, H
Inc. angle	22°-55°	18°-47°, 24°-57°	18-57°, 22°-63°	46°, 54°
Beam resolution	Fixed doppler	Range gate	Variable doppler	Spot
Resolution	50 km	50 km	25 km	25/12.5 km
Swath	500 km 500 km	500 km	600 km 600 km	1800 km
Daily coverage	Variable	41%	77%	93%
Dates	6/78 – 10/78	8/91 - 1/01	8/96 – 6/97	6/99 +

Figure 13.11   Observation geometries and instrument characteristics of typical spaceborne scatterometers (reprinted from Liu and Xie 2006, with permission from the American Society of Photogrammetry and Remote Sensing). SEASAT, ERS-1/2 and NSCAT are examples of side-looking, fan-beam scatterometers. The QuikSCAT scatterometer is an example of a scanning pencil beam instrument.

Scatterometer applications

Scatterometers have proven particularly useful in the measurement of wind speed over the ocean surface (Elachi and van Zyl 2006, Liu and Xie 2006). Such ocean viewing scatterometers utilize multiple azimuth viewing angles to measure backscatter from the ocean surface and derive the corresponding near-surface wind vector. This technique has its origin in the relationship of the ocean capillary waves, and hence backscatter amplitude, to the strength of the near-surface wind field. Measurement of the backscatter at different azimuth view angles allows derivation of the wind direction as well as wind speed.

Application of spaceborne scatterometers to terrestrial ecosystem studies have capitalized on the high temporal coverage afforded by these systems for characterizing seasonality of land surface freeze–thaw state for study of high latitude carbon and water cycles (McDonald and Kimball 2005). Resolution enhanced data are particularly well-suited for studies of vegetation such as in the Amazon basin (Long and Hardin 1994) and for studies of vegetation phenology (Frolking et al. 2005, 2006). Enhanced resolution scatterometer data are also widely applied in the study of sea ice and cryosphere (Long and Drinkwater 1999, Remund et al. 2000). Figure 13.13 illustrates the powerful utility of spaceborne scatterometers for acquiring data for global-scale studies.

NASA's CloudSat mission utilizes a 94-GHz nadir-looking Cloud Profiling Radar (CPR) which

measures the power backscattered by clouds as a function of distance from the radar (Figure 13.14) (Im et al. 2005). The scatterometer is designed to collect information about the vertical structure of clouds and aerosols unavailable from other Earth observing satellites. The choice of a 94-GHz radar was driven by the need to achieve sufficient cloud detection sensitivity. A lower frequency radar would require a larger antenna and higher transmit power. At frequencies much greater than 100 GHz, a large antenna and high transmit power would also be needed to account for rapid signal attenuation through cloud absorption. The 94-GHz frequency chosen by CPR offers the best compromise, for both performance and radar system implementation.

PASSIVE MICROWAVE SENSORS (RADIOMETERS AND SCANNERS)

Principles of passive microwave sensors

Microwave radiometers are passive microwave sensor systems that observe emission (emissivity) in the microwave portion of the EM spectrum. Radiometers measure a medium's brightness temperature, T_b, also referred to as radiobrightness. A material's microwave brightness temperature is characterized by the product of its emissivity and its physical temperature (in Kelvin). The effective

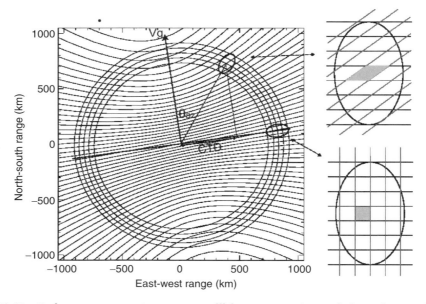

Figure 13.12   Radar measurement geometry utilizing a processing technique that combines the scanning scatterometer measurement geometry with synthetic aperture processing. The velocity vector v_g shows the instrument flight direction. Iso-Doppler contours are shown overlain on the circular constant range contours traced out 360° by the scanning azimuth look angle θ_{az}. The ellipsoids shown at right represent the coverage of the antenna pencil beam at two different look directions along the scanning azimuth; the relative position of the iso-Doppler and constant range contours are shown. The spatial resolution is optimized where the iso-Doppler contours are perpendicular to the range contours. Hence, spatial resolution varies with azimuth look angle. (© 2004 IEEE, reprinted with permission from Entekhabi, D., E. Njoku, P. Hauser, M. Spencer, T. Doiron, J. Smith, R. Girard, S. Belair, W. Crow, T. Jackson, Y. Kerr, J. Kimball, R. Koster, K. McDonald, P. ONeill, T. Pultz, S. Running, J. C. Shi, E. Wood, and J. van Zyl, 2004. The Hydrosphere State (HYDROS) mission concept: An earth system pathfinder for global mapping of soil moisture and land freeze/thaw. *IEEE Transactions on Geoscience and Remote Sensing*, 42(10): 2184–2195).

brightness temperature observed by a downward-looking sensor is the sum of the brightness temperature of the medium and the brightness temperature of the sky, T_{sky}, reflected from that medium toward the sensor. Earth's atmospheric absorption is relatively small for frequencies less than 10 GHz. At higher measurement frequencies Earth's atmosphere can have a significant effect on the measured effective brightness temperature through emission (T_{sky}) and through atmospheric absorption and scattering. For radiometers designed to observe Earth's surface, lower frequencies or frequencies chosen to correspond to atmospheric windows are commonly used.

Emissivity is a unitless variable ranging from 0 for a perfectly non-emitting material, to 1 for a perfect emitter (blackbody) (Ulaby et al. 1986). For a gray surface, which is neither perfectly non-emitting nor perfectly emitting, brightness

temperature is equivalent to the product of the surface temperature and a factor that depends on the surface emissivity (Elachi and van Zyl 2006). Emissivity is a function of surface composition and roughness. The sensitivity of brightness temperature to variations in emissivity driven by surface composition or roughness is generally much larger than the effect of surface temperature. As in the active (radar) case, emissivity and hence brightness temperature are also functions of polarization and the radiometer's observation angle.

Spaceborne and airborne radiometer systems commonly employ scanning antenna geometries, as in the case of the scanning pencil-beam scatterometers, with real aperture radiometers having been deployed on satellites since the 1970s. More recently, aperture synthesis techniques have been developed and continue to mature for application

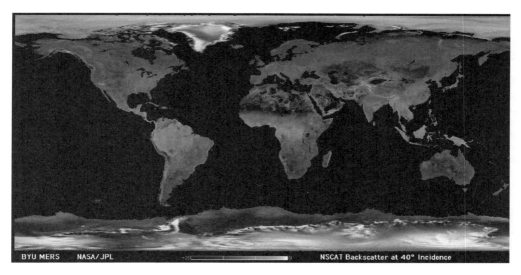

Figure 13.13   The global land and cryosphere as observed by the NASA Scatterometer (NSCAT). Data acquired by the fan-beam scatterometer for incidence angles ranging from approximately 20–60° and normalized to a 40° effective incidence angle. Resolution enhancement has been applied to improve the spatial resolution of the scatterometter data to better than 10 km. The brightest regions are glacial ice sheets in Greenland and Antarctica. For other regions, the image brightness is related to the vegetation cover and surface moisture. Tropical rainforests are relatively bright while desert regions are dark. Very dry, sandy deserts show up as black in this image. (Courtesy of NASA/JPL-Caltech and the Scatterometer Climate Record Pathfinder project, Microwave Earth Remote Sensing Lab, Brigham Young University).

Figure 13.14   Cross-section of tropical clouds and thunderstorm cells from CloudSat profiling radar. (Courtesy of NASA/JPL-Caltech and the Cooperative Institute for Research in the Atmosphere (CIRA), Colorado State University). (See the color plate section of this volume for a color version of this figure).

from aircraft and satellite platforms (Le Vine et al. 1994, Le Vine 1999, Kerr et al. 2001).

Data characteristics and processing

Global coverage utilizing microwave radiometry from space has been on-going since the launch of the Nimbus-7 Scanning Multichannel Microwave Radiometer (SMMR) in 1978, which continued

operation until 1987. Global coverage continues with the Defense Meteorological Satellite Program (DMSP) series of satellites which fly the Special Sensor Microwave/Imager (SSM/I) with operations beginning in 1987. The Advanced Microwave Scanning Radiometer on-board NASA's Earth Observing System (AMSR-E) was launched in 2002, providing additional global monitoring capability. Gridded global data sets and derived products are available from SMMR, SSM/I, and AMSR-E from the National Snow and Ice Data Center Distributed Active Archive Center (NSIDC-DAAC).

As is the case with spaceborne scatterometers, real aperture spaceborne radiometers generally have relatively coarse spatial resolution, with native resolution of these real aperture sensors commonly 25 km or more. Similar to scatterometers, resolution enhancement techniques have been applied to expand the utility of spaceborne radiometry (Long and Daum 1998). This capability is illustrated with data from AMSR-E (Figure 13.15).

Radiometer applications

Radiometers have been utilized in a multitude of applications involving studies of Earth's ocean, land, and atmosphere. Multi-decade monitoring of the global cryosphere has been particularly noteworthy owing to the importance of these regions in global climate. Satellites have made continual observations of Arctic sea ice since 1978, recording a general decline in extent throughout that period (Parkinson 2000). Passive microwave sensors have been used extensively in monitoring snow covered area, snow water equivalent (SWE) and snow

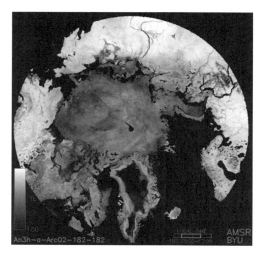

Figure 13.15   North polar view with the AMSR-E radiometer. Data shown are from the H-polarized 19 GHz channel, and were acquired on July 2, 2002. (Courtesy of NASA/JPL-Caltech and the Scatterometer Climate Record Pathfinder project, Microwave Earth Remote Sensing Lab, Brigham Young University). (See the color plate section of this volume for a color version of this figure).

melt processes (Kelly et al. 2003), crucial to the understanding of global climate and water, carbon, and energy cycles.

The estimation of surface soil moisture with microwave radiometry has been investigated extensively. With the launch of AMSR-E, standard soil moisture products have become available from spaceborne radiometers (Njoku et al. 2003). Planned and proposed radiometry missions have soil moisture estimation as a central theme (Kerr et al. 2001, Entekhabi et al. 2004). Ocean studies with microwave radiometry have included retrieval of near-surface ocean winds (Atlas et al. 1996, Wentz 1997), ocean temperature, and more recently, studies of ocean salinity. Objectives of ESA's SMOS mission include measurement of ocean salinity (Kerr et al. 2001). NASA's Aquarius mission, planned for launch in 2009, includes an L-band radiometer with a primary focus of measuring sea surface salinity globally (Koblinsky et al. 2003).

NOTES

1. Portions of the work were carried out at the Jet Propulsion Laboratory, California Institute of Technology, under a contract with the National Aeronautics and Space Administration.

2. In general, classifications of the electromagnetic spectrum distinguish microwaves from the longer waves in the 'radio frequency' spectrum. Although *radar* is an acronym for *radio detection and ranging*, historically all active remote sensors operating in the microwave portion of the electromagnetic spectrum are referred to as radar systems.

3. dB = decibel. This is an important 'unit' in microwave remote sensing to describe power ratios. It is defined as 10 times the logarithm with a base of 10 ($10 \log_{10}$). Examples for multiplication factors expressed in dB are: 3 dB = 2, 10 dB = 10, 20 dB = 100, 30 dB = 1000, −3 dB = 0.5, −10 dB = 0.1, −20 dB = 0.01, −30 dB = 0.001.

4. Analogous to an area illuminated by a spotlight shining in the dark.

5. Named after Christian Doppler who published in 1842 (at age 39) the effects of shifting frequencies when observing moving targets. Active microwave sensors use the effect by moving sensors which observe stationary (as well as moving) targets.

6. Frequency (f) and wavelength (w) are inverse proportional given the speed of light (c) with $f = c/w$. Thus, the shorter the wavelength the larger the frequency and vice versa.

7. The now accepted convention associates the first letter with the transmitting, the second letter with the receiving polarization.

8. To describe a radar sensor's frequency and polarization characteristic, frequency band letter

codes are generally followed by the transmit/receive polarization state, e.g., L-hh or C_w.

9. A good mnemonic for depression angle is considering an equivalent to looking down with a lowered head (being depressed) from the straight ahead look direction.

10. *InSAR* is now more commonly accepted; the synonymous acronym *IfSAR* is still found in the surveying and geomatics community.

11. As distinguished from *along-track interferometry* where antennas are displaced in flight direction of the sensor. TerraSAR-X has an antenna configuration where the antenna can be spilt in two sections allowing along-track interferometry. Discussion here is limited to across-track interferometry.

12. Arrangement of dipoles on a radar antenna where each dipole's phase can be controlled so that the beam can scan very rapidly.

REFERENCES

Alsdorf, D. E., J. M. Melack, T. Dunne, L. A. Mertes, L. L. Hess, and L. C. Smith, 2000. Interferometric radar measurements of water level changes on the Amazon flood plain. *Nature* 404(6774): 174–177.

Atlas, R., R. N. Hoffman, S. C. Bloom, J. C. Jusem and J. Ardizzone, 1996. A multiyear global surface wind velocity dataset using SSM/I wind observations. *Bulletin of the American Meteorological Society*, 77(5): 869–882.

Bamler, R., 1992. A comparison of range-Doppler and wavenumber domain SAR focusing algorithms. *IEEE Transactions on Geoscience and Remote Sensing*, 30(4): 706–713.

Baraldi, A. and F. Panniggiani, 1995. A refined gamma MAP SAR speckle filter with improved geometrical adaptivity. *IEEE Transactions on Geoscience and Remote Sensing*, 33(5): 1245–1257.

Bayer, T., R. Winter, and G. Schreier, 1991. Terrain influences in SAR backscatter and attempts to their correction. *IEEE Transactions on Geoscience and Remote Sensing*, 29(3): 451–462.

Boerner, W.-M., 2003. Recent advances in extra-wide-band polarimetry, interferometry and polarimetric interferometry in synthetic aperture remote sensing and its applications. *IEE Proceedings on Radar and Sonar Navigation*, 150(3): 113–124.

Cloude, S. R. and K. P. Papathanassiou, 2003. Three-stage inversion process for polarimetric SAR interferometry. *IEE Proceedings – Radar, Sonar and Navigation*, 150(3): 125–134.

Cumming, I. G. and F. H.-c. Wong, 2005. *Digital Processing of Synthetic Aperture Radar Data: Algorithms and Implementation.* Artech House, Boston.

Curlander, J. C. and R. N. McDonough, 1991. *Synthetic Aperture Radar: Systems and Signal Processing.* Wiley, New York.

Dobson, M. C., F. T. Ulaby, T. LeToan, A. Beaudoin, E. S. Kasischke, and N. Christensen, 1992. Dependence of radar backscatter on coniferous forest biomass. *IEEE Transactions on Geoscience and Remote Sensing*, 30(2): 412–415.

Dobson, M. C., F. T. Ulaby, L. E. Pierce, T. L. Sharik, K. M. Bergen, J. Kellndorfer, J. R. Kendra, E. Li, Y. C. Lin, A. Nashashibi, K. Sarabandi, and P. Siqueira, 1995. Estimation of forest biophysical characteristics in Northern Michigan with SIR-C/X-SAR. *IEEE Transactions on Geoscience and Remote Sensing*, 33(4): 877–895.

Dubois, P., L. Norikane, J. van Zyl, and H. Zebker, 1989. Data volume reduction for imaging radar polarimeter. *Antennas and Propagation Society International Symposium, 1989. AP-S. Digest*, 1354–1357.

Early D. S. and D. G. Long, 2001. Image reconstruction and enhanced resolution imaging from irregular samples, *IEEE Transactions on Geoscience and Remote Sensing*, 39(2): 291–302

Elachi, C., 1988. *Spaceborne Radar Remote Sensing: Applications and Techniques.* IEEE Press, New York.

Elachi, C. and J. J. van Zyl, 2006. *Introduction to the Physics and Techniques of Remote Sensing.* Wiley, Hoboken.

Elachi, C., K. E. Im, and E. Rodriguez, 1990. Global digital topography mapping with a synthetic aperture scanning radar altimeter. *International Journal of Remote Sensing*, 11(4): 585–601.

Entekhabi, D., E. Njoku, P. Hauser, M. Spencer, T. Doiron, J. Smith, R. Girard, S. Belair, W. Crow, T. Jackson, Y. Kerr, J. Kimball, R. Koster, K. McDonald, P. ONeill, T. Pultz, S. Running, J. C. Shi, E. Wood, and J. van Zyl, 2004. The Hydrosphere State (HYDROS) mission concept: An earth system pathfinder for global mapping of soil moisture and land freeze/thaw. *IEEE Transactions on Geoscience and Remote Sensing*, 42(10): 2184–2195.

Fahnestock, M., R. Bindschadler, R. Kwok, and K. Jezek, 1993. Greenland ice sheet surface properties and ice dynamics from ERS-1 SAR imagery. *Science*, 262(5139): 1530.

Ferretti, A., C. Prati, and F. Rocca, 1999. Multibaseline InSAR DEM reconstruction: the wavelet approach. *IEEE Transactions on Geoscience and Remote Sensing*, 37(2 Part 1): 705–715.

Ferretti, A., C. Prati, and F. Rocca, 2000. Nonlinear subsidence rate estimation using permanent scatterers indifferential SAR interferometry. *IEEE Transactions on Geoscience and Remote Sensing*, 38(5 Part 1): 2202–2212.

Ferretti, A., C. Prati, and F. Rocca, 2001. Permanent scatterers in SAR interferometry. *IEEE Transactions on Geoscience and Remote Sensing*, 39(1): 8–20.

Foody, G. M., R. M. Green, R. M. Lucas, P. J. Curran, M. Honzak, and I. D. Amaral, 1997. Observations on the relationship between SIR-C radar backscatter and the biomass of regenerating tropical forests. *International Journal of Remote Sensing*, 18(4): 687–694.

Fornaro, G., G. Franceschetti, and R. Lanari, 1996. Interferometric SAR phase unwrapping using Green's formulation. *IEEE Transactions on Geoscience and Remote Sensing*, 34(3): 720–727.

Franceschetti, G. and R. Lanari, 1999. *Synthetic Aperture Radar Processing.* CRC Press Boca Raton.

Freeman, A., 1992. SAR calibration – An overview. *IEEE Transactions on Geoscience and Remote Sensing*, 30(6): 1107–1121.

Frolking, S., M. Fahnestock, T. Milliman, K. McDonald, and J. Kimball, 2005. Interannual variability in North American grassland biomass/productivity detected by SeaWinds scatterometer backscatter. *Geophysical Research Letters*, 32(21): 409.

Frolking, S., T. Milliman, K. McDonald, J. Kimball, M. Zhao and M. Fahnestock, 2006. Evaluation of the Sea-Winds scatterometer for regional monitoring of vegetation phenology. *Journal of Geophysical Research*, 111(D17): 17302.

Gade, M. and S. Ufermann, 1998. Using ERS-2 SAR images for routine observation of marine pollutionin European coastal waters. *IEEE Geoscience and Remote Sensing Symposium, IGARSS'98*, 2: 757–759.

Gens, R. and J. L. van Genderen, 1996. SAR interferometry: issues, techniques, applications. *International Journal of Remote Sensing*, 17(10): 1803–1836.

Ghiglia, D. C. and M. D. Pritt, 1998. *Two-dimensional Phase Unwrapping: Theory, Algorithms, and Software*. Wiley, Hoboken.

Goldstein, R. M., H. Engelhardt, B. Kamb, and R. M. Frolich, 1993. Satellite radar interferometry for monitoring ice-sheet motion: Application to an Antarctic ice stream. *Science*, 262(5139): 1525–1530.

Goldstein, R. M., H. A. Zebker, and C. L. Werner, 1988. Satellite radar interferometry – Two-dimensional phase unwrapping. *Radio Science*, 23(8): 713–720.

Hajnsek, I., E. Pottier, and S. R. Cloude, 2003. Inversion of surface parameters from polarimetric SAR. *IEEE Transactions on Geoscience and Remote Sensing*, 41(4): 727–744.

Hansen, R. C., 1998. *Phased Array Antennas*. Wiley-Interscience, New York.

Hanssen, R., 2001. *Radar Interferometry: Data Interpretation and Error Analysis*. Kluwer Academic Publishers, The Netherlands.

Hayne, G. S., D. W. Hancock, C. L. Purdy, and P. S. Callahan, 1994. The corrections for significant wave height and attitude effect in the TOPEX radar altimeter: TOPEX/POSEIDON: geophysical evaluation. *Journal of Geophysical Research*, 99(C12): 24941–24955.

Hein, A., 2004. *Processing of SAR Data: Fundamentals, Signal Processing, Interferometry*. Springer, Berlin, New York.

Held, D. N., W. E. Brown, A. Freeman, J. D. Klein, H. Zebker, T. Sate, T. Miller, Q. Nguyen, and Y. Lou, 1988. The NASA/JPL multifrequency, multipolarisation airborne SAR System. *IEEE Geoscience and Remote Sensing Symposium, IGARSS'88*, 1: 345–349.

Henderson, F. M. and A. J. Lewis, 1998. *Principles and Applications of Imaging Radar*. Wiley, New York.

Hess, L. L., J. M. Melack, and S. Filoso, 1995. Delineation of inundated area and vegetation along the Amazon floodplain with the SIR-C synthetic aperture radar. *IEEE Transactions on Geoscience and Remote Sensing*, 33(4): 896–904.

Im, E., C. Wu, and S. L. Durden, 2005. Cloud profiling radar for the CloudSat mission. *Aerospace and Electronic Systems Magazine, IEEE*, 20(10): 15–18.

Inglada, J., J. C. Souyris, C. Henry, C. Tison, F. S. Agency, and F. Toulouse, 2006. Incoherent SAR polarimetric analysis over point targets. *IEEE Geoscience and Remote Sensing Letters*, 3(2): 246–249.

Jensen, J. R., 2000. *Remote Sensing of the Environment: An Earth Resource Perspective*. Prentice Hall, New York.

Johannessen, O. M., K. Khvorostovsky, M. W. Miles, and L. P. Bobylev, 2005. Recent ice-sheet growth in the interior of Greenland. *Science*, 310(5750): 1013–1016.

Kellndorfer, J. M., M. C. Dobson, and F. T. Ulaby, 1996. Geocoding for classification of ERS/JERS-1 SAR composites. *IEEE International Geoscience and Remote Sensing Symposium, IGARSS'96*, 4: 2335–2337.

Kellndorfer, J. M., L. E. Pierce, M. C. Dobson, and F. T. Ulaby, 1998. Toward consistent regional-to-global-scale vegetation characterization using orbital SAR systems. *IEEE Transactions on Geoscience and Remote Sensing*, 36(5 Part 1): 1396–1411.

Kellndorfer, J. M., M. C. Dobson, J. D. Vona, and M. Clutter, 2003. Toward precision forestry: plot-level parameter retrieval for slash pine plantations with JPL AIRSAR. *IEEE Transactions on Geoscience and Remote Sensing*, 41: 1571–1582.

Kellndorfer, J. M., W. S. Walker, L. E. Pierce, M. C. Dobson, J. A. Fites, C. T. Hunsaker, J. D. Vona, and M. Clutter, 2004. Vegetation height estimation from Shuttle Radar Topography Mission and National Elevation Datasets. *Remote Sensing of Environment*, 93: 339–358.

Kelly, R. E., A. T. Chang, L. Tsang, and J. L. Foster, 2003. A prototype AMSR-E global snow area and snow depth algorithm. *IEEE Transactions on Geoscience and Remote Sensing*, 41(2): 230–242.

Kerr, Y. H., P. Waldteufel, J. P. Wigneron, J. Martinuzzi, J. Font, and M. Berger, 2001. Soil moisture retrieval from space: the Soil Moisture and Ocean Salinity (SMOS) mission. *IEEE Transactions on Geoscience and Remote Sensing*, 39(8): 1729–1735.

Koblinsky, C. J., P. Hildebrand, D. LeVine, F. Pellerano, Y. Chao, W. Wilson, S. Yueh, and G. Lagerloef, 2003. Sea surface salinity from space: Science goals and measurement approach. *Radio Science*, 38(4): 8064.

Le Vine, D. M., 1999. Synthetic aperture radiometer systems. *IEEE Transactions on Microwave Theory and Techniques*, 47(12): 2228–2236.

Le Vine, D. M., A. J. Griffis, C. Swift, and T. J. Jackson, 1994. A synthetic aperture microwave radiometer for remote sensing applications. *Proceedings of the IEEE*, 82(12): 1787–1801.

Leberl, F. W., 1990. *Radargrammetric Image Processing*. Artech House, Norwood, MA.

Lee, J. S., M. R. Grunes, and S. A. Mango, 1991. Speckle reduction in multipolarization, multifrequency SAR imagery. *IEEE Transactions on Geoscience and Remote Sensing*, 29(4): 535–544.

Lee, J. S., M. R. Grunes, and G. de Grandi, 1999. Polarimetric SAR speckle filtering and its implication forclassification. *IEEE Transactions on Geoscience and Remote Sensing*, 37(5): 2363–2373.

Lee, J. S., D. L. Schuler, and T. L. Ainsworth, 2000. Polarimetric SAR data compensation for terrain azimuth slope variation. *IEEE Transactions on Geoscience and Remote Sensing*, 38(5): 2153–2163.

Liu, W. T. and X. Xie, 2006. Measuring ocean surface wind from space. In: J. Gower (ed.), *Manual of Remote Sensing: Remote Sensing of the Marine Environment*. American

Society for Photogrammetry and Remote Sensing, Annapolis Junction.

Long, D. G. and D. L. Daum, 1998. Spatial resolution enhancement of SSM/I data. *IEEE Transactions on Geoscience and Remote Sensing*, 36(2): 407–417.

Long, D. G. and M. R. Drinkwater, 1999. Cryosphere applications of NSCAT data. *IEEE Transactions on Geoscience and Remote Sensing*, 37(3): 1671–1684.

Long, D. G. and P. J. Hardin, 1994. Vegetation studies of the Amazon basin using enhanced resolution Seasat scatterometer data. *IEEE Transactions on Geoscience and Remote Sensing*, 32(2): 449–460.

Long D. G., P. J.Hardin, and P. T. Whiting, 1993. Resolution enhancement of spaceborne scatterometer data, *IEEE Transactions on Geoscience and Remote Sensing*, 31(3): 700–715

Lopes, A., R. Touzi, and E. Nezry, 1990. Adaptive speckle filters and scene heterogeneity. *IEEE Transactions on Geoscience and Remote Sensing*, 28(6): 992–1000.

Lopes, A., E. Nezry, R. Touzi, and H. Laur, 1993. Structure detection and statistical adaptive speckle filtering in SAR images. *International Journal of Remote Sensing* 14(9): 1735–1758.

Lopez-Martinez, C., X. Fabregas, and E. Pottier, 2005. Multidimensional speckle noise model. *EURASIP Journal on Applied Signal Processing*, 2005(20): 3259–3271.

Madsen, S. N., H. A. Zebker, and J. Martin, 1993. Topographic mapping using radar interferometry: processing techniques. *IEEE Transactions on Geoscience and Remote Sensing*, 31(1): 246–256.

Madsen, S. N. and H. A. Zebker, 1998. Imaging radar interferometry. In: F. M. Henderson and A. J. Lewis (eds.), *Manual of Remote Sensing: Principals and Applications of Imaging Radar*. Wiley, New York.

Mailloux, R. J., 1994. *Phased Array Antenna Handbook*. Artech House Boston,

Massonnet, D. and K. L. Feigl, 1998. Radar interferometry and its application to changes in the Earth's surface. *Rev. Geophys*, 36(4): 441–500.

Massonnet, D., M. Rossi, C. Carmona, F. Adragna, G. Peltzer, K. Feigl and T. Rabaute, 1993. The displacement field of the Landers earthquake mapped by radar interferometry. *Nature*, 364(6433): 138–142.

McDonald, K. C. and J. S. Kimball, 2005. Hydrological application of remote sensing: Freeze–thaw states using both active and passive microwave sensors. *Encyclopedia of Hydrological Sciences*, 2(part 5): 783–798.

Mette, T., K. P. Papathanassiou, I. Hajnsek, and R. Zimmerman, 2002. Forest biomass estimation using polarimetric SAR interferometry. *IEEE Geoscience and Remote Sensing Symposium, IGARSS'02*, 2: 817–819.

Mohr, J. J., N. Reeh, and S. N. Madsen, 1998. Three Dimensional Glacial Flow and Surface Elevation Measured with Radar Interferometry. *Nature*, 391: 273–276.

Moreira, A., G. Krieger, I. Hajnsek, D. Hounam, M. Werner, S. Riegger, and E. Settelmeyer, 2004. TanDEM-X: a TerraSAR-X add-on satellite for single-pass SAR interferometry. *IEEE International Geoscience and Remote Sensing Symposium, IGARSS'04*, 2: 1000–1003.

Njoku, E. G., T. J. Jackson, V. Lakshmi, T. K. Chan, and S. V. Nghiem, 2003. Soil moisture retrieval from AMSR-E.

IEEE Transactions on Geoscience and Remote Sensing, 41(2): 215–229.

Oliver, C. and S. Quegan, 1998. *Understanding Synthetic Aperture Radar Images*. Artech House, Boston.

Oveisgharan, S. and H.A. Zebker, 2007. Estimating snow accumulation from InSAR correlation observations. *IEEE Transactions on Geoscience and Remote Sensing*, 45(1): 10–20.

Papathanassiou, K. P. and S. R. Cloude, 2001. Single-baseline polarimetric SAR interferometry. *IEEE Transactions on Geoscience and Remote Sensing*, 39(11): 2352–2363.

Parkinson, C. L., 2000. Variability of Arctic sea ice: The view from space, An 18-year record. *Arctic*, 53(4): 341–358.

Pope, K. O., J. M. Rey-Benayas, and J. F. Paris, 1994. Radar remote sensing of forest and wetland ecosystems in the Central American tropics. *Remote Sensing of Environment*, 48(2): 205–219.

Remund, Q. P., D. G. Long, and M. R. Drinkwater, 2000. An iterative approach to multisensor sea ice classification. *IEEE Transactions on Geoscience and Remote Sensing*, 38(4): 1843–1856.

Rey, M. T., J. K. Tunaley, J. T. Folinsbee, P. A. Jahans, J. A. Dixon, and M. R. Vant, 1990. Application of radon transform techniques To wake detection in Seasat-A SAR images. *IEEE Transactions on Geoscience and Remote Sensing*, 28(4): 553–560.

Rosen, P. A., S. Hensley, H. A. Zebker, F. H. Webb, and E. J. Fielding, 1996. Surface deformation and coherence measurements of Kilauea Volcano, Hawaii, from SIR C radar interferometry. *Journal of Geophysical Research*, 101(E10): 23109–23126.

Sansosti, E., 1999. Digital elevation model generation using ascending and descending ERS-1/ERS-2 tandem data. *International Journal of Remote Sensing*, 20(8): 1527–1547.

Santoro, M., A. Shvidenko, I. McCallum, J. Askne, and C. Schmullius, 2007. Properties of ERS-1/2 coherence in the Siberian boreal forest and implications for stem volume retrieval. *Remote Sensing of Environment*, 106(2): 154–172.

Sarabandi, K. and Y. C. Lin, 2000 Simulation of interperometric SAR response for characterizing the scattering phase center statistics of forest canopies. *IEEE Transactions on Geoscience and Remote Sensing*, 38(1): 115–125.

Sarabandi, K., J. Kellndorfer, and L. Pierce, 2003. GLORIA: Geostationary/Low-Earth Orbiting Radar Image Acquisition System: a multi-static GEO/LEO synthetic aperture radar satellite constellation for Earth observation. *IEEE International Geoscience and Remote Sensing Symposium, IGARSS'03*, 2: 773–775.

Schreier, G., 1993. *SAR Geocoding: Data and Systems*. Wichmann, Karlsruhe.

Simard, M., S. S. Saatchi, and G. De Grandi, 2000. The use of decision tree and multiscale texture for classification of JERS-1 SAR data over tropical forest. *IEEE Transactions on Geoscience and Remote Sensing*, 38(5): 2310–2321.

Smith, L. C., 2002. Emerging applications of interferometric synthetic aperture radar (InSAR) in geomorphology and hydrology. *Annals of the Association of American Geographers*, 92(3): 385–398.

Souyris, J. C., P. Imbo, R. Fjortoft, S. Mingot, and J.-S. Lee, 2005. Compact polarimetry based on symmetry properties of geophysical media: the /spl pi//4 mode. *IEEE Transactions on Geoscience and Remote Sensing*, 43(3): 634–646.

Stebler, O., E. Meier, and D. Nuesch, 2002. Multi-baseline polarimetric SAR interferometry – first experimental space-borne and airborne results. *ISPRS Journal of Photogrammetry and Remote Sensing*, 56: 149–166.

Touzi, R., 2002. A review of speckle filtering in the context of estimation theory. *IEEE Transactions on Geoscience and Remote Sensing*, 40(11): 2392–2404.

Treuhaft, R. N. and S. R. Cloude, 1999. The structure of oriented vegetation from polarimetric interferometry. *IEEE Transactions on Geoscience and Remote Sensing*, 37(5): 2620–2624.

Treuhaft, R. N. and P. R. Siqueira, 2000. Vertical structure of vegetated land surfaces from interferometric and polarimetric radar. *Radio Science*, 35(1): 141–177.

Treuhaft, R. N., S. N. Madsen, M. Moghaddam, and J. J. van Zyl, 1996. Vegetation characteristics and underlying topography from interferometric radar. *Radio Science*, 31(6): 1449–1486.

Treuhaft, R. N., G. P. Asner, and B. E. Law, 2003. Structure-based forest biomass from fusion of radar and hyperspectral observations. *Geophysical Research Letters*, 30(9): 1472–1475.

Ulaby, F. T. and C. Elachi, 1990. *Radar Polarimetry for Geoscience Applications*. Artech House, Norwood, MA.

Ulaby, F. T., R. K. Moore, and A. K. Fung, 1981. *Microwave Remote Sensing: Active and Passive. Microwave Remote Sensing Fundamentals and Radiometry*. Addison-Wesley, Reading, MA.

Ulaby, F. T., R. K. Moore, and A. K. Fung, 1982. *Microwave Remote Sensing: Active and Passive. Radar Remote Sensing and Surface Scattering and Emission Theory*. Addison-Wesley, Reading, MA.

Ulaby, F. T., R. K. Moore, and A. K. Fung, 1986. *Microwave Remote Sensing: Active and Passive. From Theory to Applications*. Addison-Wesley, Reading, MA.

Ulander, L. M. H., 1996. Radiometric slope correction of synthetic-aperture radar images. *IEEE Transactions on Geoscience and Remote Sensing*, 34(5): 1115–1122.

van Zyl, J. J., 2001. The Shuttle Radar Topography Mission (SRTM): A breakthrough in remote sensing of topography. *Acta Astronautica*, 48(5–12): 559–565.

van Zyl, J. J., B. D. Chapman, P. Dubois, and J. Shi, 1993. The effect of topography on SAR calibration. *IEEE Transactions on Geoscience and Remote Sensing*, 31(5): 1036–1043.

Walker, W. S., J. Kellndorfer, E. LaPointe, M. Hoppus, and J. Westfall, 2007. An Empirical SRTM-based Approach to Mapping Vegetation Canopy Height for the Conterminous United States. *Remote Sensing of Environment* (In Press)

Wegmuller, U. and C. Werner, 1997. Retrieval of vegetation parameters with SAR interferometry. *IEEE Transactions on Geoscience and Remote Sensing*, 35(1): 18–24.

Wentz, F., 1997. A well-calibrated ocean algorithm for special sensor microwave/imager. *Journal of Geophysical Research – Oceans*, 102(C4): 8703–8718.

Woodhouse, I., 2006. *Introduction to Microwave Remote Sensing*. Taylor & Francis, Boca Raton, London, New York.

Xie, H., L. E. Pierce, and F. T. Ulaby, 2002. SAR speckle reduction using wavelet denoising and Markov random field modeling. *IEEE Transactions on Geoscience and Remote Sensing*, 40(10): 2196–2212.

Xu, W., I. Cumming, D. MacDonald, A. Ltd, and B. C. Richmond, 1999. A region-growing algorithm for InSAR phase unwrapping. *IEEE Transactions on Geoscience and Remote Sensing*, 37(1): 124–134.

Yamada, H., Y. Yamaguchi, and W. M. Boerner, 2003. Forest height feature extraction in polarimetric SAR interferometry by using rotational invariance property. *IEEE International Geoscience and Remote Sensing Symposium, IGARSS'03*, 3: 1426–1428.

Zebker, H. A. and Y. Lu, 1998. Phase unwrapping algorithms for radar interferometry: residue-cut, least-squares, and synthesis algorithms. *Journal of the Optical Society of America*, 15(3): 586–598.

Zebker, H. A. and J. J. van Zyl, 1991. Imaging radar polarimetry: a review. *Proceedings of the IEEE*, 79(11): 1583–1606.

Zebker, H. A. and J. Villasenor, 1992. Decorrelation in interferometric radar echoes. *IEEE Transactions on Geoscience and Remote Sensing*, 30(5): 950–959.

Zebker, H. A., T. G. Farr, R. P. Salazar, and T. H. Dixon, 1994. Mapping the world's topography using radar interferometry: the TOPSAT mission. *Proceedings of the IEEE* 82(12): 1774–1786.

Airborne Laser Scanning

Juha Hyyppä, Wolfgang Wagner,
Markus Hollaus, and Hannu Hyyppä

Keywords: airborne laser scanning, point clouds, small-footprint, intensity, laser ranging, opto-mechanical scanning, PRF, beam, speckle, first pulse, last pulse, full-waveform, strip adjustment, point classification, filtering, DTM, DSM, forest inventory, 3D city models.

INTRODUCTION TO AIRBORNE LASER SCANNING

A short history of airborne laser scanning

Airborne laser scanning (ALS) is a remote sensing technique used for mapping of topography, vegetation, urban areas, ice, and infrastructure. It is often referred to as light detection and ranging (lidar) because it uses a laser to illuminate the Earth's surface and a photodiode to register the backscatter radiation. Lasers were developed by the 1964 Nobel Prize winners Townes, Basov, and Prokhorov in 1958. Since 1960, when Theodore Maiman demonstrated that light amplification by stimulated emission of radiation (laser) is possible in the infrared and optical part of the electromagnetic spectrum, lasers have been widely used for military intelligence and civil surveying. As early as the 1960s NASA used lasers for tracking a corner cube reflector on the moon. In 1975, NASA and other organizations developed an airborne oceanographic lidar for measuring chlorophyll concentration and other biological and chemical substances. Non-scanning lidars were used for bathymetry, forestry, and other applications in the 1970s and 1980s (e.g., Nelson et al. 1984, Schreier et al. 1985, Guenther 2007), which established the basic principles of using lasers for remote sensing purposes. The first experiments with modern laser scanner instruments were conducted in the early 1990s, and in 1993 the first prototype of a commercial ALS system dedicated to topographic mapping was introduced. For the development of ALS, the reader is referred to Bufton (1989), Lohr and Eibert (1995), Flood and Gutelius (1997), and Wever and Lindenberger (1999).

Definition of airborne laser scanning

Airborne laser scanning is a lidar-based method for acquiring range measurements from an airborne platform and the precise orientation of these measurements (Figure 14.1). Short laser pluses (4–10 ns) are emitted with a high frequency (e.g., 50–200 kHz) and are continuously deflected in across-flight direction using various scanning methods. The position and rotation of the sensor is continuously recorded along the flight path using a Global Positioning System (GPS) and an Inertial Measurement Unit (IMU). The recorded measurements of sensor position and orientation, beam deflection, and range can be converted to a georeferenced three-dimensional (3D) point cloud representing the surface targets that reflected the laser pulses. Since it is *a priori* not known which targets generated the echoes, models are required for converting the 3D point cloud into geospatial data products such as Digital Terrain Models (DTM), Digital Surface Models (DSM) (Figure 14.2), Canopy Height Model (CHM),

Figure 14.1 Principle of airborne laser scanning. Airborne laser scanning is a method for acquiring lidar range measurements from an airborne platform and the precise orientation of these measurements. Short laser pluses (4–10 ns) are emitted with a high frequency (e.g., 50–200 kHz) and are continuously deflected in across-flight direction using various scanning methods. The position and rotation of the sensor is continuously recorded along the flight path using a Global Positioning System (GPS) and an Inertial Measurement Unit (IMU). The recorded measurements of sensor position and orientation, beam deflection and range can be converted to a georeferenced three-dimensional (3D) point cloud representing the surface targets that reflected the laser pulses. Courtesy of Hannu Hyyppä. (See the color plate section of this volume for a color version of this figure).

or normalized Digital Surface Model (nDSM, CHM = nDSM = DSM-DTM). These geospatial basic products are increasingly used in a broad range of applications, e.g., in hydrology, construction engineering, infrastructure monitoring, and forestry.

Because many of the commercially attractive applications require spatially detailed models with a high geometric accuracy, ALS sensor and flight specifications are typically chosen such that the

size of the laser footprint on the ground is less than 1 m and the sampling density is several points per square meter. Overviews of laser scanning can be found in Baltsavias (1999a), Wehr and Lohr (1999), and Fowler et al. (2006). Probably the most extensive review of commercially available resources can be found Baltsavias (1999b). As ALS is a rapidly growing technology, several technical improvements have occurred since this review was published. Specifically, the sampling frequency, positioning accuracy, and number of recorded echoes have been improved during the last seven years. Also, small-footprint waveform-digitizing ALS systems have become available (Wagner et al. 2006).

In addition to these small-footprint ALS systems, several large-footprint systems have been developed (Blair et al. 1994, Rabine et al. 1996, Garvin et al. 1996). These systems have been mainly used to test spaceborne lidar techniques. Using lidar measurements acquired by the ICESat satellite (Zwally et al. 2002) over forested areas, Lefsky et al. (2005) were able to demonstrate the potential capability of this technique for global mapping of canopy height.

Advantages of airborne laser scanning

There are several features that distinguish ALS from traditional imaging techniques. They are active systems, generating their own energy to survey the target. Thus, ALS systems are not dependent on the sun as a source of illumination. This means that subject to flight regulations and adverse weather conditions surveys can be performed at any time of the day or night in any season. Also, the interpretation of laser scanner data is not hampered by shadows caused by clouds or neighboring objects such as trees or buildings. In forested areas the laser pulses may travel unimpeded back and forth along the same path through small openings in a forest canopy, providing ground echoes even in closed canopy forests (Kraus and Pfeifer 1998). As summarized in Ackermann (1999), in European-type coniferous and deciduous forests the percentage of laser point penetrating to the forest ground is about 20–40%, with up to 70% in deciduous forests in wintertime. Finally, ALS also delivers spatially dense range measurements over homogeneous targets, where stereoscopic image matching techniques may encounter problems in reconstructing the 3D surface due to lack of image texture. Disadvantages of ALS compared to traditional photogrammetric techniques are the lower spatial resolution and the lack of spectral information. However, new ALS systems are capable of recording the intensity and sometimes the full-waveform of the target, which improves the semi-automatic classification of features from laser data.

Figure 14.2   Example of lidar-derived digital surface models DSM obtained for forested area. Trees or tree groups are visible. DSMs are typically coded by colors red referring to the highest and blue to the lowest elevations. Courtesy of Hannu Hyyppä. (See the color plate section of this volume for a color version of this figure).

Data collection costs

During 1996–2005, approximately 150 ALS systems were manufactured (Fowler et al. 2006). Today, at least 20–30 systems are sold or put on the market annually. Commercial systems are becoming more flexible in their operation, also allowing for a more application-specific service. However, there is still a market for proprietary systems with an application-specific design, where commercial systems cannot compete. An example for such a proprietary system is NASA's Experimental Advanced Airborne Research Lidar (EAARL) designed to map near-shore bathymetry, topography and vegetation structure simultaneously (Nayegandhi et al. 2006). Data acquisition costs have dropped steadily in the last few years, reflecting the increasing competition between service providers. Nowadays, prices as low as 100–200 €/km^2 are sometimes charged for acquiring data sets for large areas (>1,000 km^2 or even >10,000 km^2) with a relatively high point density (4–10 pts/m^2) and even lower prices (50–100 €/km^2) are possible for lower point densities (0.5–1 pts/m^2). The costs are strongly dependant on the surveyed area. Thus, ALS has become an economic source of data even for regional or country-wide surveys, such as what has happened for example in the Finland, Netherlands, Switzerland, Germany, Austria, and US.

To aid georeferencing, interpretation of the laser point cloud and error analysis, optical images are often acquired simultaneously using small-format or medium-format digital cameras. Since optimum flight conditions for laser scanning differ from optimum imaging conditions, these images are typically not of photogrammetric quality and they are often just used for visual interpretation.

AIRBORNE LASER SCANNING SYSTEMS

Nowadays, a large variety of different ALS instruments are used in different platforms (e.g., plane, helicopter), operating at different altitudes above ground. The flying altitude (e.g., 100–6,000 m) depends on the specifications of the laser used and the required footprint size and point density on the ground. The following gives a short description of the main ALS components and their functions. A typical ALS system consists of (1) a laser ranging unit, (2) an opto-mechanical scanner, (3) a position and orientation unit, and (4) a control, processing, and storage unit (Figure 14.3). The data measured from all the units are time stamped and saved in the storage unit. The control and processing unit provides the system set up, monitoring, and control of all the ALS system components.

Laser ranging unit

The laser ranging unit can be subdivided into a transmitter, a receiver, and the optics for both units. In current ALS, the transmitter is usually a semiconductor diode laser or an Nd:YAG (neodymium:yttrium-aluminium-garnet) solid-state laser pumped by semiconductor lasers, covering the optical band between 800 and

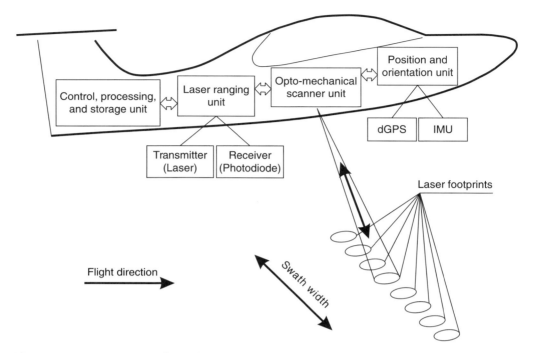

Figure 14.3 Components of a typical airbone laser scanner system. Courtesy of Markus Hollaus.

1,600 nm (Wehr and Lohr 1999). As summarized by Weitkamp (2005), the transmitter optics of ALS systems serve the dual purpose of expanding the laser-beam diameter in order to reduce the area density of the laser pulse energy and of reducing the divergence of the laser beam. The receiver is a photodiode, which converts the incident power received by the detector to a current at the output. The receiver optic collects the backscattered light and focuses it onto the detector, which converts the photons to electrical impulses. For ideal optical detection, one incident photon releases one electron from the detector. In real detectors a fraction of the photons goes undetected. Since in ALS the backscattered echo often has extremely low power, it is important that the fraction of recorded photons is high. This fraction is referred to as the quantum efficiency. Types of photodiodes used in ALS include avalanche photodiodes or PIN (positive-intrinsic-negative) photodiodes (Wehr and Lohr 1999).

The whole laser ranging unit is monostatic, which means that the transmitting and receiving apertures are mounted on the aircraft so that the transmitting and receiving paths share the same optical path (Wehr and Lohr 1999). For practical realization, the transmitter and the receiver are mounted coaxially, with the axes of the transmitted beam and receiver field of view (FOV)

either coinciding, side by side or biaxial, with the two axes parallel or near-parallel, but not identical (Weitkamp 2005). An illuminated surface point is therefore ensured of being in the FOV of the optical receiver. For the range measurements, two different principles are applied, namely pulse ranging and phase difference measurement. The pulse ranging principle measures the traveling time of short pulses transmitted from the laser, backscattered from an object, and detected from a receiver. This measurement principle is commonly used for airborne systems. The phase difference measurement is used for what are called continuous wave (CW) lasers, which continuously emit laser light (Wehr and Lohr 1999). Currently, these laser systems are seldom used, so they are not described any further here.

One of the most crucial factors for an exact range measurement is the echo detection algorithm used (Wagner at al. 2004, Figure 14.4). As the length of the laser pulse is longer than the accuracy needed (a few meters versus a few centimeters), a specific timing in the return pulse needs to be defined. In a non-waveform ranging system, analogue detectors are used to derive discrete, time-stamped trigger pulses from the received signal in real time during the acquisition process. The timing event should not change when the level of signal varies, which is an important requirement in the design

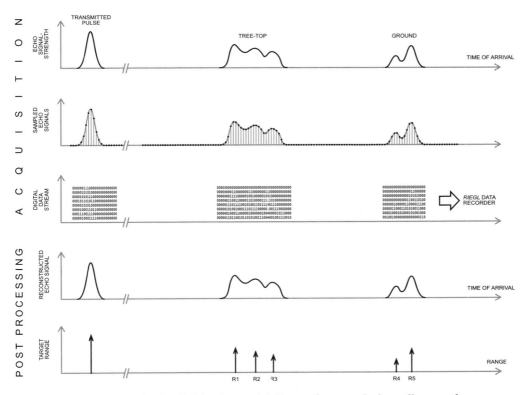

Figure 14.4   Principle of echo digitization and full-waveform analysis outlines as the process of sampling. Courtesy of Wolfgang Wagner.

of analog detections as discussed by Palojärvi (2003). Unfortunately, in the case of commercial ALS systems detailed information concerning the analog detection method is normally lacking, even though different detection methods may yield quite different range estimates. For full-waveform digitizing ALS systems several algorithms can be used in the post-processing stage (e.g., leading edge discriminator/threshold, center of gravity, maximum, zero crossing of the second derivative, and constant fraction). The most basic technique for pulse detection is to trigger a pulse whenever the rising edge of the signal exceeds a given threshold (leading edge discriminator). Although conceptually simple and easy to implement, this approach suffers from a serious drawback: the timing of the triggered pulse (and thus the accuracy of any distance measurements derived from it) is rather sensitive to the amplitude and width of the signal. If the amplitude of the pulse changes then the timing point also changes. The same holds for the center of gravity when computed over all points above a fixed threshold. More sophisticated schemes are based on finite differences concerning numerical derivatives (e.g., the detection of local maxima or the zero crossings of the second

derivative) or, more generally, the zero-crossings of a linear combination of time-shifted versions of the signal. An example of the latter approach is the constant fraction discriminator, which determines the zero crossings of the difference between an attenuated and a time-delayed version of the signal (Gedcke and McDonald 1968). Maximum, zero crossing, and constant fraction are invariant with respect to amplitude variations and also, to some degree, changes in pulse width. In practice, these detectors should only trigger for signal amplitudes above a given threshold (this 'internal' threshold is not to be confused with the threshold detector) in order to suppress false positives, i.e., spurious trigger-pulses due to noise. This is especially true for operators based on higher order derivatives such as zero crossing, which are known to be rather noise-sensitive.

Opto-mechanical scanning unit

The ALS collects data in a strip fashion; the motion of the aircraft in combination with the deflection of the laser beam across the flight path generates strip-wise measurements of the illuminated surface.

The opto-mechanical scanning unit is responsible for the deflection of the transmitted laser beams across the flight path. The design of the deflection unit (e.g., oscillating mirror/zig-zag scanning, rotating mirror/line scanning, pushbroom/fiber scanning, Palmer/conical scanning) defines the scan pattern on the ground. In an oscillating mirror, as used by Optech in ALTM and Leica in ALS, the mirror rotates back and forth within limited extents, producing a zig-zag line on the ground. Due to the necessary changes in mirror velocity, a greater point density is generated at the edges, decreasing at the nadir. In this configuration, the last points on a scan which are near the edges are typically ignored due to errors caused by the accelerating mirror (Katzenbeisser 2003). In the rotating mirror scanning used by Riegl, the mirror is rotated 360 degrees at a constant velocity, producing points along parallel lines on the ground. Using this system, no measurements can be recorded when the mirror is pointed away from the target. The fiber scanners used by Toposys apply pushbroom technology by providing a large number of identical fiber arrays mounted in the focal plane of the receiving and transmitting lenses. The fibers are arranged in a circle at the transmitter/receiver end and in a fixed linear array at the other end producing a pushbroom type pattern on the ground. With this type of scanner is used, the pulse can be sent and received from each fiber separately. In the Palmer scanner used by TopEye in MK-II and ScaLARS, an elliptical pattern is produced on the ground. Higher point density is obtained near the edges. The advantage of these scanning techniques include the redundant data (forward and backward looking) that can be used in calibration of the data, as they produce a relatively large number of edge hits from vertical objects (houses, trees), improving edge detection, and since the scanning angle is almost constant, there is no variation in intensity due to incidence angle, and thus calibrating the intensity information is easier.

Position and orientation unit

In addition to the distances between the laser ranging unit and the objects measured on the ground, the positions and orientations of the ALS system must be measured. A differential global positioning system (dGPS) consisting of an onboard GPS receiver and one or more ground stations provides the position of the laser ranging unit. Their orientation is described by the pitch, roll, and heading of the aircraft, which are measured by an inertial navigation system (INS), and are also referred to as inertial measurement unit (IMU). Furthermore, the beam direction relative to the ALS system is recorded for each laser measurement.

CHARACTERISTICS OF ALS

Main ALS parameters

Important technical parameters to characterize the physical properties of laser scanner systems are the laser wavelength (μm), pulse duration (ns), pulse energy (mJ), pulse repetition frequency (PRF) (kHz), beamwidth (mrad), scan angle (deg), scan rate (Hz), flying height (m), and size (m) of laser footprint on the ground. Some of the technical parameters related to the data acquisition that can be modified by the user are as follows:

Flight altitude and scan angle
For a selected laser scanner, the main user-selected parameters are flight altitude and maximum scan angle (FOV/2, Field of View). Flight altitudes can vary between 100 to 6,000 m, depending on the systems and flying platform (UAV, helicopter, aircraft). The flight altitude also determines the PRF of systems, which allow typically only one pulse in the air. The type of scanner used (scanning mechanism) and, for a fixed PRF and scanner rate, the maximum scan angle and flying height, determine the area coverage and the distribution and density of points for a given pulse ALS system. The density of the pulses on the target is the main parameter affecting the quality of the surveyed data. Increase in flight height increases planimetric errors (errors in plane direction (X–Y) are mainly due to IMU). In addition, the echo intensity decreases with increasing flight altitude, which has a negative effect on the pulse detection. For example, the published accuracies for Leica ALS-50 II for hard target are 6 cm in elevation and 7 cm in plane from the altitude of 500 m (FOV 40°), and 23 cm in elevation and 64 cm in plane from the altitude of 6,000 m (FOV 40°).

Pulse repetition frequency
In 1993, the fastest pulse rate (pulse repetition frequency, PRF) was 2 kHz, while the best systems in 2007 provided samples with a rate of 150–200 kHz. Two years ago, the systems have had only one pulse in the air, which has limited the maximum PRF from higher altitudes (the speed of light acts as a limitation). For example, in the Leica ALS-50 II, which has a 20 degree maximum scan angle (FOV 40°), the maximum PRF is 150, 99.3, 58.4, 41.1, 31.8, 26.0, and 21.9 kHz from flight altitudes of 500, 1,000, 2,000, 3,000, 4,000, 5,000, and 6,000 m, respectively. At the moment several manufacturers have announced systems that allow several transmitted pulses in the air simultaneously.

Flight planning
As a sophisticated data collection system, ALS is subject to systematic and random errors caused by

its components, i.e., its range detection, scanning mechanism, GPS, and especially IMU. In order to minimize these errors, calibration is performed during each flight. When ALS records the test area on flights in two opposite directions and typically with perpendicular cross-strips, the ALS results can then be compared with the results obtained using other flight directions. The process of correcting orientation with multiple strips is called strip adjustment. The use of a Palmer scanner reduces the amount of flying in corridor mapping, since use of forward and backward looking modes already collects redundant data.

Beam size

The effect of beam size on the quality of the model obtained has not been well studied and findings are contradictory. The ability to detect either the ground or tree crowns depends not only on the beam size but also on the signal strength, the pulse detection algorithm and the proper sampling (footprint size versus sampling density).

In laser scanning there is a tendency toward the design of laser beams with smaller and smaller beam divergence (tendency to 'single mode' signals). With these sensors the number of multiple returns per emitted pulse will decrease, as a smaller surface patch is illuminated. Since the information acquired per beam decreases, classification of the data is only possible in relation to neighboring echoes. An interesting aspect for future system design may eventually be the combination of narrow (only a single return with high quality range information) and wide (recording the full-waveform information) beams in order to use the advantages of both techniques and to optimize both the derivation of the 3D object and DTM.

Intensity

As stated earlier, ALS records the intensity and sometimes the full-waveform, in addition to the point cloud data. In ALS data processing the intensity of each point recorded has been mainly used only as a predictor for classification (e.g., Holmgren and Persson 2004), for matching aerial imagery and laser scanner data, and for lidargrammetry (Fowler et al. 2006). More effective use of intensity values has been lacking, partly because there have been no techniques to calibrate them.

Speckle

The laser pulse illuminates a given surface area that consists of several scattering points. Thus, the returned echo comprises a coherent combination of individual echoes from a large number of points

(as with radars, see Elachi 1987). The result is a single vector representing the amplitude V and phase $f(I \sim V^2)$ of the total echo, which is a vector sum of the individual echoes. This means that as the sensor moves, the successive beam intensities (I) will result in different values of I. This variation is called fading. Thus, an image of a homogeneous surface with constant reflectivity will result in intensity variation from one resolution element to the next. Speckle gives images recorded with laser light a grainy texture.

Correcting intensity as a function of incidence angle

The term surface 'roughness' is always understood as measured in wavelengths. The Fraunhofer criterion states that the surface is smooth if the phase difference $\Delta\phi < \pi/8$ (Schanda 1986). This criterion evidently dictates that for a surface to be effectively smooth at normal incidence, irregularities must be less than about $\lambda/32$ in height. Thus, for a surface to give specular reflection at ALS wavelengths (say $\lambda = 1 \, \mu m$) surface height variations must be less than about 30 nm. This is a condition of smoothness likely to be met with only on certain man-made surfaces such as sheets of glass or metal. In ALS it is therefore a sensible *a priori* assumption that most targets will behave similar to Lambertian targets while the intensity variation as a function of incidence angle is low and the incidence angle correction of intensity can be based on relatively simple models.

Backscattered intensity and calibration of intensity

The recorded intensity value is affected by the emitted energy, the strength and shape of the outgoing pulse, the range, the signature of the reflecting surface, and atmospheric effects (attenuation). In general, the intensity can be considered to be related to the power received, which can be given in the form (Wagner et al. 2006):

$$P_r = \frac{P_t D_r^2}{R^4 \beta_t^2 \Omega} \rho A_s \qquad (1)$$

where P_r and P_t are the transmitted and receiver power, D_r is the receiver aperture size, R is the range, β_t is the beam divergence, Ω corresponds to the bidirectional properties of the scattering, ρ is the reflectivity of the target surface, and A_s is the receiving area of the scatterer. Thus, the recorded intensity is inversely related to R^2 for homogenous targets filling the full footprint, to R^3 for linear objects (e.g., wire) and to R^4 for individual large scatterers. The intensity can be calibrated assuming that the recorded laser scanner intensity is a function of target reflectivity, range (including incidence angle), and PRF (see Ahokas et al. 2006 for details).

First and last pulse

The first commercially available airborne laser scanners were pure ranging systems, which only recorded the time of one backscattered pulse to determine the range from the sensor to the target. The recording of only one pulse is sufficient if there is only one target within the laser footprint. In this case the shape of the reflected pulse is 'single mode' and straightforward to interpret. However, even for small laser footprints there may be several objects within the travel path of the laser pulse that generate individual backscatter pulses. More advanced laser scanners have therefore been built which are capable of recording more than one pulse. State-of-the-art commercial laser scanners typically measure the first and last pulse; some are able to measure up to five pulses. Still, the problem is that it is not always clear how to interpret these measurements for different targets. Pragmatically, one may for example assume over forested terrain that the first pulse is associated with the top of the canopy and the last pulse, with some probability, with the forest floor.

Full-waveform

Digitizing and recording the complete backscattered waveform during acquisition for later post-processing has the advantages that algorithms can be adjusted to tasks, intermediate results are respected, and the neighborhood relations of pulses can be considered (Jutzi and Stilla 2003). In addition to determining the 3D position of the targets, the waveform measurements can also be used to characterize their scattering behavior in terms of the backscatter cross section and the width of the echoes (Wagner et al. 2006). This is an important advantage of full-waveform ALS systems which is not properly understood yet. The main appeal of using the backscatter cross section rather than the intensity as measured by conventional ALS systems is that it is a physical measure of the electromagnetic energy intercepted and reradiated by objects which allows for linking measurements to electromagnetic scattering theory. Also, the information provided by the echo width is very useful. As shown by Doneus and Briese (2006), it is possible to detect low vegetation and other objects close to the terrain surface using the echo width. Thus such objects can be removed before filtering the ALS point cloud to derive a digital terrain model. Further, the echo width potentially provides information on the surface roughness, the slope of the

target, or the depth of a volumetric target, e.g., a layer of fog.

PROCESSING OF ALS DATA

Georeferencing, system calibration, and strip adjustment

The three-dimensional (3D) positions of the targets are calculated in a post-processing mode, whereas all the different data sources are linked using the time stamp. The dGPS and IMU are used to measure the time-dependent position and orientation of the airborne platform and thus their flight trajectory can be computed. Consequently, the accuracies of the trajectories are significantly influenced by the measurement frequencies of dGPS and IMU. Finally, the ranges are converted to 3D elevation data using the instrument mounting parameters and the aircraft trajectories. The results of these calculations are strip-wise 3D points in a global coordinate system, e.g., WGS84. Typical format for ALS data is LAS (e.g., Samberg 2007)

There are systematic and dynamic biases in the raw laser data that can be reduced in a system calibration (after set-up and after changes in the system) and in the strip adjustment process (for each campaign). System calibration is typically used for correcting errors due to range, encoder offset, torsion, encoder latency, encoder scale factor, misalignment angles, and intensity based range correction. The parameters typically solved in the strip adjustment are the mirror scale parameter and the misalignment between the laser scanner and the inertial measurement unit expressed as heading, roll and pitch angular correction. The resulting angular offsets must then be applied to the data. The optimum site for calibration flights comprises both flat and sloped surfaces with no low vegetation. The typical flight pattern for calibration is four flight passes over the same area in a cross-type pattern, with sloped surfaces located at the center of the cross. Additionally, time-stamped trajectory information and laser points linked to the trajectory position are needed. The systematic shift in elevation and roll offset can be corrected even using flat surfaces. Correction of pitch offset requires slopes in the flight direction. Heading correction requires slopes on both sides of the flight line. These orientation errors can also be corrected with a simultaneous 3D adjustment using least squares as proposed by Kager (2004). The process (Figure 14.5) is similar to block adjustment, with the modification that homogeneous planes replace homogeneous points as tying features, and strips replace images. Also, time-dependent parameters need to be corrected.

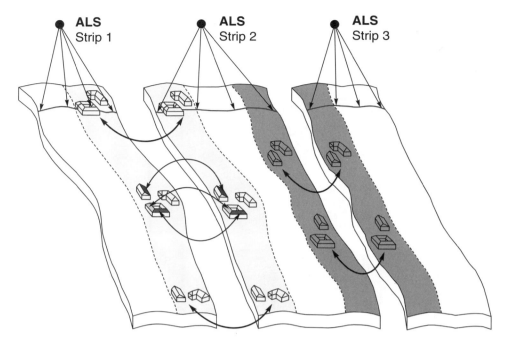

Figure 14.5　Principle of strip adjustment. Image adapted from Kager (2004). (See the color plate section of this volume for a color version of this figure).

Well-distributed ground control points can finally be used to assess the quality of the corrected point cloud and DSM data. Any remaining systematic errors should be removed. The random errors should meet the requirements of the user.

Classification of points

Laser points have coordinate, echo, reflectance, and time information to be used in the classification. Statistical classification methods are commonly used. Points can be divided into various classes using intensity values, echoes (first, intermediate or last pulse), or height information or time. Height can be relative to the ground or absolute values. Various filtering techniques have been developed for low point detection, power line, building and ground classification purposes (Ahokas 2007).

Classification of low points
Removal of low points is an important pre-processing part of data analysis and is usually done before ground classification. Low points are below the ground surface and their origin may be manholes on the street or multiple reflections from trees or buildings. A single or group of points may cause an anomaly in the correct ground surface. A point can be defined as being low if all of its neighboring points in a search window are

more than a predefined value higher than the point (Ahokas 2007).

Classification of ground points
Photogrammetrists have developed various methods for obtaining DTMs from laser scanning point clouds. Kraus and Pfeifer (1998) developed a DTM algorithm for which laser points between terrain points and non-terrain points were distinguished using an iterative prediction of the DTM and weights attached to each laser point, depending on the vertical distance between the expected DTM level and the corresponding laser point. The method is implemented in SCOP++ (Kraus and Otepka 2005). Axelsson (2000) developed a progressive TIN densification method which is implemented into Terrascan software (Figure 14.6). Laser point clouds are first classified to separate ground points from all other points. The program selects local low points on the ground and makes an initial triangulated model. New laser points are then added to the model iteratively and the actual ground surface is then described more and more precisely. Maximum building size, iteration angle, and distance parameters determine which points are accepted. A comparison of the filtering techniques used for DTM extraction can be found in a report on ISPRS comparison of filters (Sithole and Vosselman 2004). Selection of the filtering strategy is not a simple process. In practice, the amount

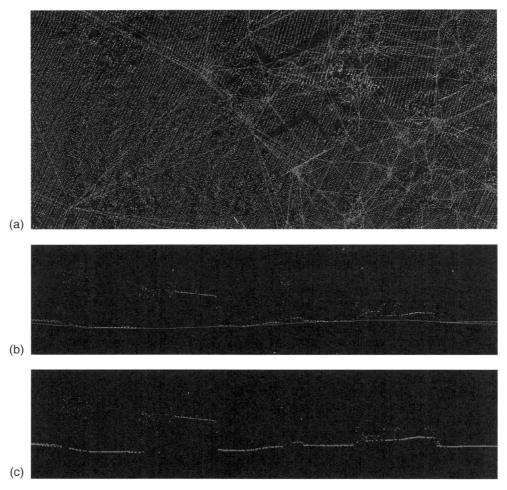

(a)

(b)

(c)

Figure 14.6   The creation of triangulated terrain model in progress in TerraScan using TIN densification method (top). Ground points are partly found and searching *for more* ground points continues (middle). All the ground points are found. Houses and trees are excluded (bottom). Courtesy of Arttu Soininen. (See the color plate section of this volume for a color version of this figure).

of interactive work determines the final quality of the product. Examples of commercial software that include DTM generation are REALM, TerraScan, and SCOP++.

Classification of building points

Previous research has shown that automatic detection of buildings from laser scanner data is possible with relatively good accuracy (e.g., Hug 1997, Matikainen et al. 2003, Vögtle and Steinle 2003, Rottensteiner et al. 2005a, b, Forlani et al. 2006, Zhang et al. 2006). Different types of information have been used to separate buildings and vegetation, including local co-planarity, height texture or surface roughness, reflectance

information from images or laser scanning, height differences between first pulse and last pulse laser scanner data, and shape and size of objects.

Classification of power line points

Melzer and Briese (2004) have developed a fully automated method for the 3D reconstruction of power lines based on ALS data. After common preprocessing steps, they use a 2D Hough transformation to locate groups of parallel power lines (corridors) in the projection of the point cloud onto the x–y-plane. In a final step, a 3D fit is computed locally within its corresponding corridor for each power line. TerraScan also has a semi-automatic technique for power line detection.

SOME APPLICATIONS

Elevation models

A detailed comparison of the filtering techniques used for DTM extraction was made within an ISPRS comparison of filters (Sithole and Vosselman 2004). Reutebuch et al. (2003) reported random errors of 14 cm for clearcut, 14 cm for heavily thinned forest, 18 cm for lightly thinned forest and 29 cm for uncut forest, using TopEye data with 4 pulses per m^2. The variation in ALS-derived DTM quality with respect to date, flight altitude, pulse mode, terrain slope, forest cover, and within-plot variation was reported in Hyyppä et al. (2005).

Forest inventory

The first scientific studies of ALS for forestry purposes included standwise mean height and volume estimation (e.g., Næsset 1997a, b), individual-tree-based height determination and volume estimation (e.g., Hyyppä and Inkinen 1999, Hyyppä et al. 2001a, b), tree-species classification (e.g., Brandtberg et al. 2003, Holmgren and Persson 2004), and measurement of forest growth and detection of harvested trees (e.g., Yu et al. 2004, 2006). A Scandinavian summary of laser scanning in forestry can be found in Næsset et al. (2004). Methods applicable to forest measurements are reviewed in Hyyppä et al. (2008). Distribution-based techniques based on Næsset (2002) are operative in Norway. Individual-tree-based methods are semi-operative in several countries. An ISPRS and EuroSDR comparison of the laser-based and photogrammetric methods for tree extraction has recently been finished (Kaartinen and Hyyppä 2008).

3D city models

A short summary of state-of-the-art building extraction methods can be obtained from Baltsavias (2004), Maas and Vosselman (1999), and Brenner (2005). Brenner (2005) provides a good up-to-date review of various approaches to and aspects of building extraction using carefully selected examples. By combining the good height assessment accuracy of laser scanner and good planimetric accuracy of aerial images, both high accuracy and higher automation can be obtained. Automated processing of integrated laser scanning data and digital images is at a very early stage of research (Brenner, 2005). Whereas automatic building extraction methods based on aerial imagery typically concentrate on finding edges, methods based on laser scanner data typically concentrate on finding planes, and edges are detected either as plane intersections or height changes in DSM, or given as additional data (ground plans). A comparison of various laser and photogrammetry based methods for building extraction can be found in an EuroSDR comparison of building extraction (Kaartinen and Hyyppä 2006).

REFERENCES

Ackermann, F., 1999. Airborne laser scanning – Present status and future expectations. *ISPRS Journal of Photogrammetry and Remote Sensing*, 54(2–3): 64–67.

Ahokas, E., S. Kaasalainen, J. Hyyppä, and J. Suomalainen, 2006. Calibration of the Optech ALTM 3100 laser scanner intensity data using brightness targets. ISPRS Commission I Symposium, 3–6 July 2006, Marne-la-Vallee, France, In *International Archives of Photogrammetry, Remote Sensing and Spatial Information Sciences*, 36(A1), CD-ROM.

Ahokas, E., 2007. On the quality of airborne laser scanning data. Lic. Sc. thesis, Department of Surveying, Helsinki University of Technology Espoo, 72p.

Axelsson P., 2000. DEM generation from laser scanner data using adaptive TIN models. In *International Archives of Photogrammetry and Remote Sensing*. 33(B4): pp. 110–117.

Baltsavias, E. P., 1999a. Airborne laser scanning: basic relations and formulas. *ISPRS Journal of Photogrammetry and Remote Sensing*. 54(2–3): 199–214.

Baltsavias, E. P., 1999b. Airborne laser scanning: existing systems and firms and other resources. *ISPRS Journal of Photogrammetry and Remote Sensing*, 54(2–3): 164–198.

Baltsavias, E. P., 2004. Object extraction and revision by image analysis using existing geodata and knowledge: current status and steps towards operational systems. *ISPRS Journal of Photogrammetry and Remote Sensing*, 58(3–4): 129–151.

Blair, J. B., D. B. Coyle, J. L. Bufton, and D. J. Harding, 1994. Optimization of an airborne laser altimeter for remote sensing of vegetation and tree canopies. *IEEE Geoscience and Remote Sensing Symposium (IGARSS '94)*, Vol. 2, 939–941.

Brandtberg, T., T. Warner, R. Landenberger, and J. McGraw. 2003. Detection and analysis of individual leaf-off tree crowns in small footprint, high sampling density lidar data from the eastern deciduous forest in North America. *Remote Sensing of Environment*, 85: 290–303.

Brenner, C., 2005. Building reconstruction from images and laser scanning. *International Journal of Applied Earth Observation and Geoinformation*, 6: 186–198.

Bufton, J. L., 1989. Laser altimetry measurements from aircraft and spacecraft. *Proc. IEEE*, 77(3): 463–477.

Doneus, M. and C. Briese, 2006. Digital terrain modelling for archaeological interpretation within forested areas using full-waveform laserscanning. In M. Ioannides, D. Arnold, F. Niccolucci and K. Mania (eds), *The 7th International Symposium on Virtual Reality, Archaeology and Cultural Heritage*, VAST (2006).

Elachi, C., 1987. *Spaceborne Radar Remote Sensing: Applications and Techniques*. IEEE Press, New York, 255 p.

Flood, M. and B. Gutelius, 1997. Commercial implications of topographic terrain mapping using scanning airborne laser radar. *Photogrammetric Engineering and Remote Sensing*, 63(4): 327–329, 363–366.

Forlani, G., C. Nardinocchi, M. Scaioni, and P. Zingaretti, 2006. Complete classification of raw LIDAR data and 3D reconstruction of buildings. *Pattern Analysis and Applications*, 8(4): 357–374.

Fowler, R., A. Samberg, M. Flood, and T. Greaves, 2006. Topographic and terrestrial lidar, In: D. Maune (ed.), *Digital Elevation Model Technologies and Applications*, 2nd Edn, ASPRS. Chapter 7: 199–252.

Garvin, J., J. Blair, J. Bufton, and D. Harding, 1996. The Shuttle Laser Altimeter (SLA-01) Experiment: Topographic Remote Sensing of Planet Earth. *EOS Trans., AGU*, 77(7): 239 p.

Gedcke, D. and McDonald, 1968. Design of the constant fraction of pulse height trigger for optimum time resolution. *Nuclear Instruments and Methods*, 58: 253–260.

Guenther, G. C., 2007. Digital Elevation Model Technologies and Applications: The DEM Users Manual, 2nd Edn, D. Maune (ed.), American Society for Photogrammetry and Remote Sensing, Chapter 8: Airborne lidar bathymetry, 253–320.

Holmgren, J. and Å. Persson, 2004. Identifying species of individual trees using airborne laser scanning. *Remote Sensing Environment*, 90: 415–423.

Hug, C., 1997. Extracting artificial surface objects from airborne laser scanner data. In Gruen et al. (eds.): *Automatic Extraction of Man-Made Objects from Aerial and Space Images* (II), Birkhäuser Verlag, Basel, pp. 203–212.

Hyyppä, H., X. Yu, J. Hyyppä, H. Kaartinen, E. Honkavaara, and P. Rönnholm, 2005. Factors affecting the quality of DTM generation in forested areas. In *Proceedings of ISPRS Workshop Laser Scanning 2005*, September 12–14, 2005, Enschede, Netherlands, XXXVI(Part 3/W19): pp. 85–90.

Hyyppä, J. and M. Inkinen, 1999. Detecting and estimating attributes for single trees using laser scanner. *The Photogrammetric Journal of Finland*, 16: 27–42.

Hyyppä, J., M. Schardt, H. Haggrén, B. Koch, U. Lohr, H. U. Scherrer, R. Paananen, H. Luukkonen, M. Ziegler, H. Hyyppä, U. Pyysalo, H. Friedländer, J. Uuttera, S. Wagner, M. Inkinen, A. Wimmer, A. Kukko, E. Ahokas, and M. Karjalainen, 2001a. HIGH-SCAN: The first European-wide attempt to derive single-tree information from laser-scanner data. *The Photogrammetric Journal of Finland*, 17: 58–68

Hyyppä, J., O. Kelle, M. Lehikoinen and M. Inkinen, 2001b. A segmentation-based method to retrieve stem volume estimates from 3-dimensional tree height models produced by laser scanner. *IEEE Transactions of Geoscience and Remote Sensing*, 39: 969–975.

Hyyppä, J., H. Hyyppä, F. Leckie, F. Gougeon, X. Yu, and M. Maltamo, 2008. Review of methods of small-footprint airborne laser scanning for extracting forest inventory data in boreal forests. *International Journal of Remote Sensing*, 29: 1339–1366.

Jutzi, B. and U. Stilla, 2003. Laser pulse analysis for reconstructing and classification of urban objects. *International Archives of Photogrammetry and Remote Sensing*, XXXIV, Part 3/W8: 151–156.

Kaartinen, H. and J. Hyyppä, 2006. Evaluation of building extraction. *EuroSDR Official Publication*, 50: 9–110.

Kaartinen, H., and Hyyppä, J., 2008. Tree Extraction, *EuroSDR Official Publication*, 53.

Kager, H., 2004. Discrepancies between overlapping laser scanner strips – simultaneous fitting of aerial laser scanner strips. *International Archives of Photogrammetry, Remote Sensing and Spatial Information Sciences*, XXXV (Part B/1): 555–560.

Katzenbeisser, R., 2003. Technical Note on: Echo Detection. http://toposys.de/pdf-ext/Engl/echo-detec3.pdf, 12 p. (Last accessed 2 February 2008.)

Kraus, K. and J. Otepka, 2005. DTM Modeling and Visualization – The SCOP Approach. In *Photogrammetric Week '05*, D. Fritsch (ed.); Herbert Wichmann Verlag, Heidelberg. pp. 241–252.

Kraus, K. and N. Pfeifer, 1998. Determination of terrain models in wooded areas with airborne laser scanner data. *ISPRS Journal Photogrametry and Remote Sensing*, 53: 193–203.

Lefsky, M. A., D. J. Harding, M. Keller, W. B. Cohen, C. C. Carabajal F. Del Bom Espirito-Santo, M. O. Hunter, and R. de Oliveira, 2005. Estimates of forest canopy height and aboveground biomass using ICESat. *Geophysical Research Letters*, 32(22): L22S02: 1–4.

Lohr, U. and M. Eibert, 1995. The TopoSys Laser Scanner-System. *Photogrammetrische Woche 1995*, Fritsch/Hobbie (eds.), pp. 263–267.

Maas, H.-G. and G. Vosselman, 1999. Two algorithms for extracting building models from raw laser altimetry data. *ISPRS Journal of Photogrammetry and Remote Sensing*, 54(2–3): 153–163.

Matikainen, L., J. Hyyppä, and H. Hyyppä, 2003. Automatic detection of buildings from laser scanner data for map updating. *International Archives of Photogrammetry, Remote Sensing and Spatial Information Sciences*, Dresden, Germany, Vol. XXXIV, Part 3/W13: pp. 218–224.

Melzer, T. and C. Briese, 2004. Extraction and modeling of power lines from ALS point clouds. *Austrian Association for Pattern Recognition (ÖAGM)*, Vol., 47–54.

Nayegandhi, A., J. C. Brock, C. W. Wright, and M. O'Connell, 2006. Evaluating a small footprint, waveform-resolving lidar over coastal vegetation communities. *Photogrammetric Engineering and Remote Sensing*, 72(12): 1407–1417.

Næsset, E., 1997a. Determination of mean tree height of forest stands using airborne laser scanner data. *ISPRS Journal of Photogrammetry and Remote Sensing*, 52: 49–56.

Næsset, E., 1997b. Estimating timber volume of forest stands using airborne laser scanner data. *Remote Sensing of Environment*, 61: 246–253.

Næsset, E., 2002. Predicting forest stand characteristics with airborne scanning laser using a practical two-stage procedure and field data. *Remote Sensing of Enviroment*, 80: 88–99.

Næsset, E. T. Gobakken, J. Holmgren, H. Hyyppä, J. Hyyppä, M. Maltamo, M. Nilsson, H. Olsson, Å. Persson, and U. Söderman, 2004. Laser scanning of forest resources: the Nordic experience. *Scandinavian Journal of Forest Research*, 19(6): 482–499.

Nelson, R., W. Krabill and G. Maclean, 1984. Determining forest canopy characteristics using airborne laser data. *Remote Sensing of Environment*, 15: 201–212.

Palojärvi, P., 2003. Integrated electronic and optoelectronic circuits and devices for pulsed time-of-flight laser rangefinding. PhD thesis, Department of Electrical and Information Engineering and Infotech Oulu, University of Oulu, 54 p.

Rabine, D. L., J. L. Bufton, and C. R. Vaughn, 1996. Development and test of a raster scanning laser altimeter for high resolution airborne measurements of topography. *International Geoscience and Remote Sensing Symposium, IGARSS '96. 'Remote Sensing for a Sustainable Future'*. 27–31 May 1996, vol. 1: 423–426.

Reutebuch, S., R. McGaughey, H. Andersen, and W. Carson, 2003. Accuracy of a high-resolution lidar terrain model under a conifer forest canopy. *Canadian Journal of Remote Sensing*, 29: 527–535.

Rottensteiner, F., J. Trinder, S. Clode, and K. Kubik, 2005a. Using the Dempster–Shafer method for the fusion of LIDAR data and multi-spectral images for building detection. *Information Fusion*, 6: 283–300.

Rottensteiner, F., G. Summer, J. Trinder, S. Clode, and K. Kubik, 2005b. Evaluation of a method for fusing lidar data and multispectral images for building detection. *International Archives of Photogrammetry, Remote Sensing and Spatial Information Sciences*, Vienna, Austria, Vol. XXXVI: Part 3/W24. http://www.commission3.isprs.org/cmrt05/papers/CMRT05_Rottensteiner_et_al.pdf (accessed 15 March 2006)

Samberg, A., 2007, An implementation of ASPRS LAS format, In *International Archives of Photogrammetry, Remote Sensing and Spatial Information Sciences*, XXXVI(3/W52): pp. 363–372.

Schanda, E., 1986. *Physical Fundamentals of Remote Sensing*, Springer Verlag, Berlin.

Schreier, H., J. Lougheed, C. Tucker, and D. Leckie, 1985, Automated measurements of terrain reflection and height variations using airborne infrared laser system. *International Journal of Remote Sensing*, 6(1): 101–113.

Sithole, G. and G. Vosselman, 2004. Experimental comparison of filter algorithms for bare-Earth extraction from airborne laser scanning point clouds. *ISPRS Journal of Photogrammetry and Remote Sensing*, 59: 85–101.

Vögtle, T. and E. Steinle, 2003. On the quality of object classification and automated building modelling based on laserscanning data. *International Archives of Photogrammetry, Remote Sensing and Spatial Information Sciences*, Dresden, Germany, Vol. XXXIV, Part 3/W13: pp. 149–155.

Wagner, W., A. Ullrich, T. Melzer, C. Briese, and K. Kraus, 2004. From single-pulse to full-waveform airborne laser scanners: potential and practical challenges. *International Archives of Photogrammetry, Remote Sensing and Spatial Information Sciences*, 35(B3): 201–206.

Wagner, W., A. Ullrich, V. Ducic, T. Melzer, and N. Studnicka, 2006. Gaussian decomposition and calibration of a novel small-footprint full-waveform digitizing airborne laser scanner. *ISPRS Journal of Photogrammetry and Remote Sensing*, 60: 100–112.

Wehr, A. and U. Lohr, 1999. Airborne laser scanning – an introduction and overview. *ISPRS Journal of Photogrammetry and Remote Sensing*, 54(2–3): 68–82.

Weitkamp, C., 2005. Lidar: Introduction. In *Laser Remote Sensing*. T.Fujii, and T. Fukuchi (eds), Taylor & Francis, Boca-Raton: pp. 1–36.

Wever, C. and J. Lindenberger, 1999. Experiences of 10 years laser scanning. *Proceedings of the Photogrammetry Week '99*, pp. 125–132.

Yu, X., J. Hyyppä, H. Kaartinen, and M. Maltamo, 2004. Automatic detection of harvested trees and determination of forest growth using airborne laser scanning. *Remote Sensing of Environment*, 90: 451–462.

Yu, X., J. Hyyppä, A. Kukko, M. Maltamo, and H. Kaartinen, 2006. Change detection techniques for canopy height growth measurements using airborne laser scanner data. *Photogrammetric Engineering and Remote Sensing* 72(12): 1339–1348.

Zhang, K., J. Yan, and S.-C. Chen, 2006. Automatic construction of building footprints from airborne LIDAR data. *IEEE Transactions on Geoscience and Remote Sensing*, 44 (99): 2523–2533.

Zwally, H. J., B. Schutz, W. Abdalati, J. Abshire, C. Bentley, A. Brenner, J. Bufton, J. Dezio, D. Hancock, D. Harding, T. Herring, B. Minster, K. Quinn, S. Palm, J. Spinhirne, and R. Thomas, 2002. ICESat's laser measurements of polar ice, atmosphere, ocean, and land. *Journal of Geodynamics*, 34(3–4): 405–445.

Remote Sensing Analysis: Design and Implementation

Radiometry and Reflectance: From Terminology Concepts to Measured Quantities

Gabriela Schaepman-Strub,
Michael E. Schaepman, John Martonchik,
Thomas Painter, and Stefan Dangel

Keywords: radiometry, reflectance, terminology, reflectance measurements, reflectance quantities, albedo, reflectance factor, bidirectional reflectance, BRDF.

INTRODUCTION

The remote sensing community devotes major efforts to calibrate sensors, improve measurement setups, and validate derived products to quantify and reduce measurement uncertainties. Given recent advances in instrument design, radiometric calibration, atmospheric correction, algorithm and product development, validation, and delivery, the lack of standardization of reflectance terminology and products has emerged as a source of considerable error.

Schaepman-Strub et al. (2006) highlighted the fact that the current use of reflectance terminology in scientific studies, applications, and publications often does not comply with physical standards. Biases introduced by using an inappropriate reflectance quantity can exceed minimum sensitivity levels of climate models (i.e., ±0.02 reflectance units (Sellers et al. 1995)). Further, they may introduce systematic, wavelength dependent errors in reflectance and higher level product validation efforts, in data fusion approaches based on different sensors, and in applications.

These differences are especially important in long term, large area trend studies, as the latter are mostly based on multiple sensors with different spectral and angular sampling, modeling, as well as atmospheric correction schemes.

Optical remote sensing is based on the measurement of reflected and emitted electromagnetic radiation. This chapter will deal with the reflected portion of optical remote sensing. Given the inherent anisotropy of natural surfaces and the atmosphere, the observed reflected radiance depends on the actual solar zenith angle, the ratio of direct to diffuse irradiance (including its angular distribution), the observational geometry, including the swath width (field of view, FOV) and the opening angle of the remote sensing instrument (i.e., the instantaneous field of view, IFOV). Current atmospheric correction schemes compensate for the part of the observed signal which is contributed by the atmosphere. However, these schemes mostly rely on the assumption of Lambertian surfaces, thus neglecting their anisotropy and corresponding geometrical-optical effects introduced by the variation of illumination as well as differing IFOVs. Resulting at-surface reflectance products may differ considerably by physical definition further contributing to a numerical bias of products.

The aim of this chapter is to explain geometric-optical differences in remote sensing observations and reflectance quantities, in order to give users the background to choose the appropriate product for their applications, to design experiments with field instrumentation, and process the measurements accordingly.

This chapter presents a systematic and consistent definition of radiometric units, and a conceptual model for the description of reflectance quantities. Reflectance terms such as BRDF, HDRF, BRF, BHR, DHR, black-sky albedo, white-sky albedo, and blue-sky albedo are defined, explained, and exemplified, while separating conceptual from measurable quantities. The reflectance conceptual model is used to specify the measured quantities of current laboratory, field, airborne, and satellite sensors. Finally, the derivation of higher-level reflectance products is explained, followed by examples of operational products. All symbols and main abbreviations used in this chapter are listed in Table 15.1.

RADIOMETRY AND GEOMETRICAL-OPTICAL CONCEPTS

Radiometry is the measurement of optical radiation, which is electromagnetic radiation within the wavelength range 0.01–1000 micrometers (μm). Photometry follows the same definition as radiometry, except that the measured radiation is weighted by the spectral response of the human eye. Photometry is thus restricted to the wavelength range from about 360 to 830 nanometers (nm; 1000 nm = 1μm), typical units used in photometry include lumen, lux, and candela. Remote sensing detectors are usually not adapted to the response function of the human eye; therefore this chapter concentrates on radiometry. The following two sections on radiometry are primarily based on an extended discussion of Palmer (2003).

Basic quantities and units - energy, power, projected area, solid angle

Radiometric units are based on two conceptual approaches, namely those based on (a) power or energy, or (b) geometry.

(a) Power and energy
- **Energy** is an SI derived unit, measured in Joules, with the recommended symbol Q.

Table 15.1   Symbols used in this chapter

S	distribution of direction of radiation
A	surface area [m^2]
A_p	projected area [m^2]
E	irradiance, incident flux density; $\equiv d\Phi/dA$[W m^{-2}]
I	radiant intensity; $\equiv d\Phi/d\omega$[W sr^{-1}]
L	radiance; $\equiv d^2\Phi/(dA\cos\theta d\omega)$[W m^{-2}sr^{-1}]
M	radiant exitance, extent flux density; $\equiv d\Phi/dA$[W m^{-2}]
Q	energy [J]
ρ	reflectance; $\equiv d\Phi_r/d\Phi_i$[dimensionless]
R	reflectance factor; $\equiv d\Phi_r/d\Phi_r^{id}$[dimensionless]
t	time
β	plane angle [rad]
Φ	radiant flux, power [W]
θ	zenith angle, in a spherical coordinate system [rad]
ϕ	azimuth angle, in a spherical coordinate system [rad]
ω	solid angle; $\equiv \int d\omega \equiv \iint \sin\theta d\theta d\phi$[sr]
Ω	projected solid angle; $\equiv \int \cos\theta d\omega \equiv \iint \cos\theta \sin\theta d\theta d\phi$[sr]
λ	wavelength of the radiation [nm]

Sub- and superscripts

i	*i*ncident
r	*r*eflected
id	*i*deal (lossless) and *d*iffuse (isotropic or Lambertian)
atm	*atm*ospheric
dir	*dir*ect
$diff$	*diff*use

Terms

BHR	BiHemispherical Reflectance
BRDF	Bidirectional Reflectance Distribution Function
BRF	Bidirectional Reflectance Factor
DHR	Directional – Hemispherical Reflectance
HDRF	Hemispherical – Directional Reflectance Factor

- **Power**, also known as *radiant flux*, is another SI derived unit. It is the derivative of energy with respect to time, dQ/dt, and the unit is the watt (W). The recommended symbol for power is Φ. Energy is the integral over time of power, and is used for integrating detectors and pulsed sources, whereas power is used for non-integrating detectors and continuous sources.

(b) Geometry
- The **projected area**, A_p, is defined as the rectilinear projection of a surface of any shape onto a plane normal to the unit vector (Figure 15.1, top). The differential form is $dA_p = \cos(\theta)\,dA$ where θ is the angle between the local surface normal and the line of sight. The integration over the surface area leads to $A_p = \int_A \cos(\theta)dA$.

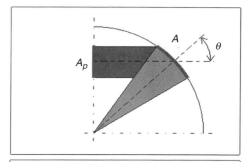

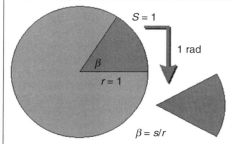

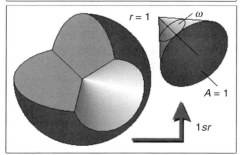

Figure 15.1 Top: Projected area A_p. Middle: Plane angle β. Bottom: Solid angle ω. (Reproduced with permission from International Light Technologies Inc, Peabody, MA).

- The *plane angle* is defined as the length of an arc (s) divided by its radius (r), $\beta = s/r$ (Figure 15.1, middle). If the arc that is subtended by the angle is exactly as long as the radius of the circle, then the angle spans 1 radian. This is equivalent to about $57.2958°$. A full circle covers an angle of 2π radians or $360°$, therefore the conversion between degrees and radians is 1 rad $= (180/\pi)$ degrees. In SI-terminology the above reads as follows: The radian is the plane angle between two radii of a circle that cuts off on the circumference an arc equal in length to the radius (Taylor 1995).

- The *solid angle,* ω, extends the concept of the plane angle to three dimensions, and equals the ratio of the spherical area to the square of the radius (Figure 15.1, bottom). As an example we consider a sphere with a radius of 1 metre. A cone that covers an area of 1 m^2 on the surface of the sphere encloses a solid angle of 1 steradian (sr). A full sphere has a solid angle of 4π steradian. A round object that appears under an angle of $57°$ subtends a solid angle of 1 sr. In comparison, the sun covers a solid angle of only 0.00006 sr when viewed from Earth, corresponding to a plane angle of $0.5°$. The *projected solid angle* is defined as $\Omega = \int \cos(\theta)d\omega$.

Radiometric units - irradiance, radiance, reflectance, reflectance factor, and wavelength dependence

Referring to the above concepts, we can now approach the basic radiometric units as used in remote sensing (Figure 15.2).

- *Irradiance, E* (also know as incident flux density), is measured in $W\,m^{-2}$. Irradiance is power per unit area incident from all directions in a hemisphere onto a surface that coincides with the base of that hemisphere ($d\Phi/dA$). A similar quantity is *radiant exitance, M*, which is power per unit area leaving a surface into a hemisphere whose base is that surface.

- *Reflectance,* ρ, is the ratio of the radiant exitance (M [$W\,m^{-2}$]) with the irradiance (E [$W\,m^{-2}$]), and as such dimensionless. Following the law of energy conservation, the value of the reflectance is in the inclusive interval 0 to 1.

- *Radiant intensity, I,* is measured in $W\,sr^{-1}$, and is power per unit solid angle ($d\Phi/d\omega$). Note that the atmospheric radiation community mostly uses the terms intensity and flux as they were defined in Chandrasekhar's classic work (Chandrasekhar 1950). More recent textbooks on atmospheric radiation still propose the use of intensity, with units $W\,m^{-2}\,sr^{-1}$, along with a footnote saying that intensity is equivalent to radiance. An extensive discussion on the (mis-)usage of the term intensity and corresponding units is given in Palmer (1993). He concludes that following the SI system definition of the base unit candela, the SI derived unit for radiant intensity is $W\,sr^{-1}$, and for radiance $W\,m^{-2}\,sr^{-1}$. In this chapter, we follow the SI definitions.

- *Radiance, L,* is expressed in the unit $W\,m^{-2}\,sr^{-1}$, and is power per unit projected area per unit solid

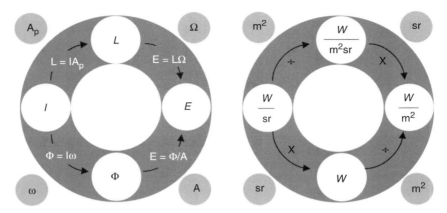

Figure 15.2   Overview of radiometric quantities (left) and corresponding units (right) as used in remote sensing, namely radiant intensity *I*, radiance *L*, irradiance *E*, and radiant flux (or power) Φ. (See the color plate section of this volume for a color version of this figure.)

angle ($d^2\Phi/d\omega\, dA\, cos(\theta)$), where θ is the angle between the surface normal and the specified direction).

- The ***reflectance factor, R***, is the ratio of the radiant flux reflected by a surface to that reflected into the same reflected-beam geometry and wavelength range by an ideal (lossless) and diffuse (Lambertian) standard surface, irradiated under the same conditions. Reflectance factors can reach values beyond 1, especially for strongly forward reflecting surfaces such as snow (Painter and Dozier 2004). For measurement purposes, a Spectralon panel commonly approximates the ideal diffuse standard surface. This is a manufactured standard having a high and stable reflectance throughout the optical region, approximates a Lambertian surface, and is traceable to the U.S. National Institute of Standards and Technology (NIST). Its use in field spectroscopy and additional references are described in Milton et al. (in press). We assume further that an isotropic behavior implies a spherical source that radiates the same in all directions, i.e., the intensity [W sr^{-1}] is the same in all directions. The Lambertian behavior refers to a flat reflective surface. The intensity of light reflected from a Lambertian surface falls off as the cosine of the observation angle with respect to the surface normal (Lambert's cosine law), whereas the radiance L [W m^{-2} sr^{-1}] is independent of observation angle. The reflected radiant flux from a given area is reduced by the cosine of the observation angle, but the observed area has increased by the cosine of the angle, and therefore the observed radiance is the same independent of observation angle. Note that

Lambertian always refers to a flat surface with the *reflected intensity* falling off as the cosine of the observation angle with respect to the surface normal (Lambert's cosine law). Isotropic on the other hand means 'having the same properties in all directions', and does not refer to a specific physical quantity. Therefore, a perfectly diffuse or Lambertian surface element dA is one for which the *reflected radiance* is isotropic, with the same value for all directions into the full hemisphere above the element dA of the reflecting surface.

Geometrical-optical concepts – directional, conical, and hemispherical

The anisotropic reflectance properties of a surface (Figure 15.3) can mathematically be described by the bidirectional reflectance distribution function (BRDF). The term bidirectional implies single directions for the incident and reflected radiances (entering and emanating from differential solid angles, respectively). This mathematical concept can only be approximated by measurements, since infinitesimal elements of solid angle do not include measurable amounts of radiant flux (Nicodemus et al. 1977), and unlimited small light sources, as well as an unlimited small sensor instantaneous field of view (IFOV) do not exist. Consequently, all measurable quantities of reflectance are performed either in the *conical* or *hemispherical* domain. From a physical point of view, we therefore differentiate between conceptual (directional) and measurable quantities (involving conical and hemispherical solid angles of observation and illumination). According to Nicodemus et al. (1977),

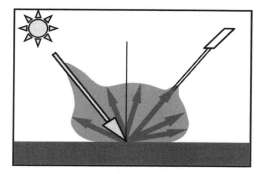

Figure 15.3 Reflectance anisotropy of a vegetation canopy showing the dependence of the observed reflectance on the viewing direction and IFOV of the remote sensing instrument (black). The shape of the reflectance distribution changes with solar angle and ratio of direct solar radiation and diffuse (radiation scattered by the atmosphere) illumination. It can also be seen that vegetation is mainly a backward scattering object, with a so-called reflectance hot spot toward the main illumination direction (as opposed to snow, which is a forward scatterer), while waxy leaves may introduce a forward scattering component.

the angular characteristics of the incoming radiance are named first in the reflectance term, followed by the angular characteristics of the reflected radiance. This results in nine different cases of reflectance quantities, illustrated in Figure 15.4. The mathematical derivations for the different cases are given below, followed by sections elaborating the geometrical configurations of most common measurement setups and on the resulting recommended terminology.

REFLECTANCE QUANTITIES IN REMOTE SENSING – BRDF, BRF, HDRF, DHR, BHR

Based on the above concepts, we can develop the corresponding mathematical formulations for the most relevant quantities used in remote sensing, namely the BRDF (Bidirectional Reflectance Distribution Function), BRF (Bidirectional Reflectance Factor), HDRF (Hemispherical-Directional Reflectance Factor), DHR (Directional-Hemispherical Reflectance), and BHR (Bihemispherical Reflectance). The same concepts may be extended to the transmittance behavior (BTDF, Bidirectional Transmittance

Distribution Function), for example when measuring and modeling leaf optical properties.

We symbolize reflectance and reflectance factor as

$$\rho(S_i, S_r, \lambda) = \text{reflectance, and}$$
$$R(S_i, S_r, \lambda) = \text{reflectance factor}$$

where S_i and S_r describe the angular distribution of all incoming and reflected radiation observed by the sensor, respectively. S_i and S_r only describe a set of angles occurring with the incoming and reflected radiation and not their intensity distributions. S_r represents a cone with a given solid angle corresponding to a sensor's instantaneous field of view (IFOV), but no sensor weight functions are included here. This becomes only necessary if the sensitivity of the sensor depends on the location within the rim of the cone. When a sensor has a different IFOV for different wavelength ranges, then S_r depends on the wavelength.

The terms S_i and S_r can be expanded into a more explicit angular notation to address the remote sensing problem:

$$\rho(\theta_i, \phi_i, \omega_i; \theta_r, \phi_r, \omega_r; \lambda), \text{ and}$$
$$R(\theta_i, \phi_i, \omega_i; \theta_r, \phi_r, \omega_r; \lambda)$$

where the directions (θ and ϕ are the zenith and azimuth angle, respectively) of the incoming (subscript i) and the reflected (subscript r) radiation, and the associated solid angles of the cones (ω) are indicated. This notation follows the definition of a general cone.

For surface radiation measurements made from space, aircraft or on the ground, under ambient sky conditions, the cone of the incident radiation is of hemispherical extent ($\omega = 2\pi [\text{sr}]$). The incident radiation may be divided into a direct sunlight component and a second component, namely sunlight which has been scattered by the atmosphere, the terrain, and surrounding objects, resulting in an anisotropic, diffuse illumination, sometimes called 'skylight'.

The above reflectance and reflectance factor definitions lead to the following special cases:

- ω_i or ω_r are omitted when either is zero (directional quantities).
- If $0 < (\omega_i \text{ or } \omega_r) < 2\pi$, then θ, ϕ describe the direction of the center axis of the cone (e.g., the line from a sensor to the center of its ground field of view – conical quantities).
- If $\omega_i = 2\pi$, the angles θ_i, ϕ_i indicate the direction of the incoming direct radiation (e.g., the position of the sun). For remote sensing applications, it is often useful to separate the natural incoming radiation into a direct (neglecting the sun's size)

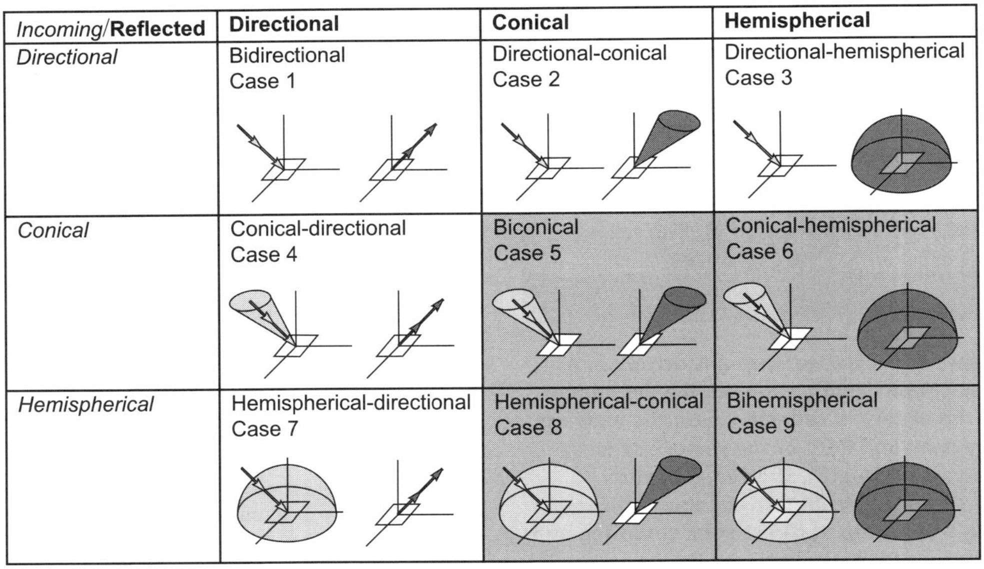

Incoming/**Reflected**	**Directional**	**Conical**	**Hemispherical**
Directional	Bidirectional Case 1	Directional-conical Case 2	Directional-hemispherical Case 3
Conical	Conical-directional Case 4	Biconical Case 5	Conical-hemispherical Case 6
Hemispherical	Hemispherical-directional Case 7	Hemispherical-conical Case 8	Bihemispherical Case 9

Figure 15.4   Geometrical-optical concepts for the terminology of at-surface reflectance quantities. All quantities including a directional component (i.e., Cases 1–4, 7) are conceptual quantities, whereas measurable quantities (Cases 5, 6, 8, 9) are shaded in gray. (Reprinted from G., Schaepman-Strub, M. E. Schaepman, T. H. Painter, S. Dangel, and J. V. Martonchik, 2006. Reflectance quantities in optical remote sensing–definitions and case studies. *Remote Sensing of Environment*, 103: 27–42). (See the color plate section of this volume for a color version of this figure.)

and hemispherical diffuse part. One may also include a terrain reflected diffuse component that is calculated with a topographic radiation model such as TOPORAD (Dozier 1980). Consequently, the preferred notation for the geometry of the incoming radiation under ambient illumination conditions is $\theta_i, \phi_i, 2\pi$. Note that in this case, θ_i, ϕ_i describe the position of the sun and not the center of the cone (2π). In the case of an isotropic diffuse irradiance field, without any direct irradiance component (closest approximated in the case of an optically thick cloud deck), θ_i, ϕ_i are omitted. Isotropic behavior implies that the intensity [W sr^{-1}] is the same in all directions.

• If $\omega_r = 2\pi$, θ_r and ϕ_r are omitted.

It should be noted that the nine standard reflectance terms defined by (Nicodemus et al. 1977) 'are applicable only to situations with uniform and isotropic radiation throughout the incident beam of radiation'. They then state that, 'If this is not true, then one must refer to the more general expressions'. This implies that any significant change to the nine reflectance concepts when the incident radiance is anisotropic lies in the mathematical expression used in their definition.

Based on this implication, Martonchik et al., 2000, adapted the terminology to the remote sensing case, which involves direct and diffuse sky illumination. In the following, we give the mathematical description of the most commonly used quantities in remote sensing, thus the general expressions for *non-isotropic* incident radiation. When applicable, we simplify the expression for the special case of isotropic incident radiation. Further, the particular wavelength dependency is omitted as well in most cases to improve readability of the equations. However, it must be understood that all interaction of light with matter is wavelength dependent, and may not simply be ignored.

The bidirectional reflectance distribution function (BRDF) – Case 1

The *bidirectional reflectance distribution function (BRDF)* describes the scattering of a parallel beam of incident light from one direction in the hemisphere into another direction in the hemisphere. The term BRDF was first used in the literature in the early 1960s (Nicodemus 1965). Being expressed as the ratio of infinitesimal quantities, it cannot be directly measured (Nicodemus et al. 1977). The

BRDF describes the intrinsic reflectance properties of a surface and thus facilitates the derivation of many other relevant quantities, e.g., conical and hemispherical quantities, by integration over corresponding finite solid angles.

The spectral BRDF, $f_r(\theta_i, \phi_i; \theta_r, \phi_r; \lambda)$ can be expressed as:

$$BRDF_\lambda = f_r(\theta_i, \phi_i; \theta_r, \phi_r; \lambda)$$

$$= \frac{dL_r(\theta_i, \phi_i; \theta_r, \phi_r; \lambda)}{dE_i(\theta_i, \phi_i; \lambda)} [sr^{-1}]. \quad (1)$$

For reasons of clarity, we will omit the spectral dependence in the following. We therefore write for the BRDF:

$$BRDF = f_r(\theta_i, \phi_i; \theta_r, \phi_r)$$

$$= \frac{dL_r(\theta_i, \phi_i; \theta_r, \phi_r)}{dE_i(\theta_i, \phi_i)} [sr^{-1}]. \quad (2)$$

Reflectance factors – Definition of Cases 1, 5, 7 and 8

When reflectance properties of a surface are measured, the procedure usually follows the definition of a reflectance factor. The reflectance factor is the ratio of the radiant flux reflected by a sample surface to the radiant flux reflected into the identical beam geometry by an ideal (lossless) and diffuse (Lambertian) standard surface, irradiated under the same conditions as the sample surface.

The *bidirectional reflectance factor (BRF; Case 1)* is given by the ratio of the reflected radiant flux from the surface area dA to the reflected radiant flux from an ideal and diffuse surface of the same area dA under identical view geometry and single direction illumination:

$$BRF = R(\theta_i, \phi_i; \theta_r, \phi_r) = \frac{d\Phi_r(\theta_i, \phi_i; \theta_r, \phi_r)}{d\Phi_r^{id}(\theta_i, \phi_i)} \quad (3)$$

$$= \frac{\cos\theta_r \sin\theta_r dL_r(\theta_i, \phi_i; \theta_r, \phi_r)d\theta_r d\phi_r dA}{\cos\theta_r \sin\theta_r dL_r^{id}(\theta_i, \phi_i)d\theta_r d\phi_r dA} \quad (4)$$

$$= \frac{dE_i(\theta_i, \phi_i)}{dL_r^{id}(\theta_i, \phi_i)} \cdot \frac{dL_r(\theta_i, \phi_i; \theta_r, \phi_r)}{dE_i(\theta_i, \phi_i)} \quad (5)$$

$$= \frac{f_r(\theta_i, \phi_i, \theta_r, \phi_r)}{f_r^{id}(\theta_i, \phi_i)} = \pi f_r(\theta_i, \phi_i; \theta_r, \phi_r). \quad (6)$$

An ideal Lambertian surface reflects the same radiance in all view directions, and its BRDF is $1/\pi$. Thus, the BRF [unitless] of any surface can be expressed as its BRDF $[sr^{-1}]$ times π (Equation (6)). For Φ_r^{id} and L_r^{id}, we omit the view zenith and azimuth angles, because there is no angular dependence for the ideal Lambertian surface.

The concept of the *hemispherical-directional reflectance factor (HDRF; Case 7)* is similar to

the definition of the BRF, but includes irradiance from the entire hemisphere. This makes the quantity dependent on the actual, simulated or assumed atmospheric conditions and the reflectance of the surrounding terrain. This includes spectral effects introduced by the variation of the diffuse to direct irradiance ratio with wavelength (e.g., Strub et al. 2003).

HDRF

$$= R(\theta_i, \phi_i, 2\pi; \theta_r, \phi_r) = \frac{d\Phi_r(\theta_i, \phi_i, 2\pi; \theta_r, \phi_r)}{d\Phi_r^{id}(\theta_i, \phi_i, 2\pi)} \quad (7)$$

$$= \frac{\cos\theta_r \sin\theta_r L_r(\theta_i, \phi_i, 2\pi; \theta_r, \phi_r)d\theta_r d\phi_r dA}{\cos\theta_r \sin\theta_r L_r^{id}(\theta_i, \phi_i, 2\pi)d\theta_r d\phi_r dA} \quad (8)$$

$$= \frac{L_r(\theta_i, \phi_i, 2\pi; \theta_r, \phi_r)}{L_r^{id}(\theta_i, \phi_i, 2\pi)}$$

$$= \frac{\int_{2\pi} f_r(\theta_i, \phi_i; \theta_r, \phi_r)d\Phi_i(\theta_i, \phi_i)}{\int_{2\pi} (1/\pi)d\Phi_i(\theta_i, \phi_i)} \quad (9)$$

$$= \frac{\int_0^{2\pi}\int_0^{\pi/2} f_r(\theta_i, \phi_i; \theta_r, \phi_r)\cos\theta_i \sin\theta_i L_i(\theta_i, \phi_i)d\theta_i d\phi_i}{(1/\pi)\int_0^{2\pi}\int_0^{\pi/2}\cos\theta_i \sin\theta_i L_i(\theta_i, \phi_i)d\theta_i d\phi_i}. \quad (10)$$

If we divide L_i into a direct (E_{dir} with angles θ_0, ϕ_0) and diffuse part, we may continue:

$$= \frac{\begin{pmatrix} f_r(\theta_0, \phi_0; \theta_r, \phi_r)E_{dir}(\theta_0, \phi_0) \\ + \int_0^{2\pi}\int_0^{\pi/2} f_r(\theta_i, \phi_i; \theta_r, \phi_r)\cos\theta_i \sin\theta_i L_i^{diff}(\theta_i, \phi_i)d\theta_i d\phi_i \end{pmatrix}}{\begin{pmatrix} (1/\pi)(E_{dir}(\theta_0, \phi_0) \\ + \int_0^{2\pi}\int_0^{\pi/2}\cos\theta_i \sin\theta_i L_i^{diff}(\theta_i, \phi_i)d\theta_i d\phi_i) \end{pmatrix}} \quad (11)$$

then, if and only if L_i^{diff} is isotropic (i.e., independent of the angles), we may continue:

$$= \frac{\begin{pmatrix} f_r(\theta_0, \phi_0; \theta_r, \phi_r)E_{dir}(\theta_0, \phi_0) \\ + L_i^{diff}\int_0^{2\pi}\int_0^{\pi/2} f_r(\theta_i, \phi_i; \theta_r, \phi_r)\cos\theta_i \sin\theta_i d\theta_i d\phi_i \end{pmatrix}}{\left((1/\pi)\left[E_{dir}(\theta_0, \phi_0) + L_i^{diff}\int_0^{2\pi}\int_0^{\pi/2}\cos\theta_i \sin\theta_i d\theta_i d\phi_i\right]\right)} \quad (12)$$

$$= \pi f_r(\theta_0, \phi_0; \theta_r, \phi_r)\frac{(1/\pi)E_{dir}(\theta_0, \phi_0)}{(1/\pi)E_{dir}(\theta_0, \phi_0) + L_i^{diff}}$$

$$+ \int_0^{2\pi}\int_0^{\pi/2} f_r(\theta_i, \phi_i; \theta_r, \phi_r)\cos\theta_i \sin\theta_i d\theta_i d\phi_i$$

$$\times \frac{L_i^{diff}}{(1/\pi)E_{dir}(\theta_0, \phi_0) + L_i^{diff}} \quad (13)$$

$$= R(\theta_0, \phi_0; \theta_r, \phi_r)d + R(2\pi; \theta_r, \phi_r)(1-d) \quad (14)$$

where d corresponds to the fractional amount of direct radiant flux (i.e., $d\in[0, 1]$).

The *biconical reflectance factor (conical–conical reflectance factor, CCRF; Case 5)*, is defined as:

$$CCRF$$
$$= R(\theta_i, \phi_i, \omega_i; \theta_r, \phi_r, \omega_r)$$
$$= \frac{\underset{\omega_r}{\int} \underset{\omega_i}{\int} f_r(\theta_i, \phi_i; \theta_r, \phi_r) L_i(\theta_i, \phi_i) d\Omega_i d\Omega_r}{(\Omega_r/\pi) \underset{\omega_i}{\int} L_i(\theta_i, \phi_i) d\Omega_i} \quad (15)$$

where $\Omega = \int d\Omega = \int \cos\theta d\omega = \int \int \cos\theta \sin\theta d\theta d\phi$ is the projected solid angle of the cone.

Formally, the CCRF can be seen as the most general quantity, because its expression contains all other cases as special ones: for $\omega = 0$ the integral collapses and we obtain the directional case, and for $\omega = 2\pi$ we obtain the hemispherical case. However, the BRF and BRDF remain the most fundamental and desired quantities because they are the only quantities not integrated over a range of angles.

For large IFOV measurements performed under ambient sky illumination, the assumption of a zero interval of the solid angle for the measured reflected radiance beam does not hold true. The resulting quantity most precisely could be described as *hemispherical-conical reflectance factor (HCRF; Case 8)*, obtained from Equation (15) by setting $\omega_i = 2\pi$:

$$HCRF = R(\theta_i, \phi_i, 2\pi; \theta_r, \phi_r, \omega_r)$$
$$= \frac{\underset{\omega_r}{\int} \underset{2\pi}{\int} f_r(\theta_i, \phi_i; \theta_r, \phi_r) L_i(\theta_i, \phi_i) d\Omega_i d\Omega_r}{(\Omega_r/\pi) \underset{2\pi}{\int} L_i(\theta_i, \phi_i) d\Omega_i}. \quad (16)$$

Reflectance – Definition of Cases 3 and 9

When applying remote sensing observations to surface energy budget studies, for example, the total energy reflected from a surface is of interest, rather than a reflectance quantity directed into a small solid angle. In the following, we describe the hemispherical reflectance as a function of different irradiance scenarios including (i) the special condition of pure direct irradiance, (ii) common Earth irradiance, composed of diffuse and direct components, and (iii) pure diffuse irradiance.

The *directional-hemispherical reflectance (DHR; Case 3)* corresponds to pure direct illumination (reported as black-sky albedo in the MODIS (Moderate Resolution Imaging Spectroradiometer) product suite (Lucht et al. 2000)). It is the ratio of the radiant flux for light reflected by a unit surface area into the view hemisphere to the illumination radiant flux, when the surface is illuminated

with a parallel beam of light from a single direction.

$$DHR$$
$$= \rho(\theta_i, \phi_i; 2\pi) = \frac{d\Phi_r(\theta_i, \phi_i; 2\pi)}{d\Phi_i(\theta_i, \phi_i)}$$
$$= \frac{dA \int_0^{2\pi} \int_0^{\pi/2} dL_r(\theta_i, \phi_i; \theta_r, \phi_r) \cos\theta_r \sin\theta_r d\theta_r d\phi_r}{d\Phi_i(\theta_i, \phi_i)} \quad (17)$$
$$= \frac{d\Phi_i(\theta_i, \phi_i) \int_0^{2\pi} \int_0^{\pi/2} f_r(\theta_i, \phi_i; \theta_r, \phi_r) \cos\theta_r \sin\theta_r d\theta_r d\phi_r}{d\Phi_i(\theta_i, \phi_i)} \quad (18)$$
$$= \int_0^{2\pi} \int_0^{\pi/2} f_r(\theta_i, \phi_i; \theta_r, \phi_r) \cos\theta_r \sin\theta_r d\theta_r d\phi_r. \quad (19)$$

The *bihemispherical reflectance (BHR; Case 9)*, generally called albedo, is the ratio of the radiant flux reflected from a unit surface area into the whole hemisphere to the incident radiant flux of hemispherical angular extent:

$$BHR$$
$$= \rho(\theta_i, \phi_i, 2\pi; 2\pi) = \frac{d\Phi_r(\theta_i, \phi_i, 2\pi; 2\pi)}{d\Phi_i(\theta_i, \phi_i, 2\pi)} \quad (20)$$
$$= \frac{dA \int_0^{2\pi} \int_0^{\pi/2} dL_r(\theta_i, \phi_i, 2\pi; \theta_r, \phi_r) \cos\theta_r \sin\theta_r d\theta_r d\phi_r}{dA \int_0^{2\pi} \int_0^{\pi/2} dL_i(\theta_i, \phi_i) \cos\theta_i \sin\theta_i d\theta_i d\phi_i} \quad (21)$$
$$= \frac{\left(\begin{array}{c} \int_0^{2\pi} \int_0^{\pi/2} \int_0^{2\pi} \int_0^{\pi/2} f_r(\theta_i, \phi_i; \theta_r, \phi_r) \cos\theta_r \sin\theta_r d\theta_r d\phi_r \\ L_i(\theta_i, \phi_i) \cos\theta_i \sin\theta_i d\theta_i d\phi_i \end{array} \right)}{\int_0^{2\pi} \int_0^{\pi/2} L_i(\theta_i, \phi_i) \cos\theta_i \sin\theta_i d\theta_i d\phi_i} \quad (22)$$
$$= \frac{\int_0^{2\pi} \int_0^{\pi/2} \rho(\theta_i, \phi_i; 2\pi) L_i(\theta_i, \phi_i) \cos\theta_i \sin\theta_i d\theta_i d\phi_i}{\int_0^{2\pi} \int_0^{\pi/2} L_i(\theta_i, \phi_i) \cos\theta_i \sin\theta_i d\theta_i d\phi_i}. \quad (23)$$

If as before we divide L_i into a direct (E_{dir} with angles θ_0, ϕ_0) and diffuse part, and assume that L_i^{diff} is isotropic we can write:

$$= \frac{\left(\begin{array}{c} \rho(\theta_0, \phi_0; 2\pi) E_{dir}(\theta_0, \phi_0) \\ + \pi L_i^{diff} (1/\pi) \int_0^{2\pi} \int_0^{\pi/2} \rho(\theta_i, \phi_i; 2\pi) \cos\theta_i \sin\theta_i d\theta_i d\phi_i \end{array} \right)}{E_{dir}(\theta_0, \phi_0) + \pi L_i^{diff}} \quad (24)$$
$$= \rho(\theta_0, \phi_0; 2\pi) d + \rho(2\pi; 2\pi)(1-d) \quad (25)$$

where d again corresponds to the fractional amount of direct radiant flux.

For the special case of pure diffuse isotropic incident radiation, a situation that may be most closely approximated in the field by a thick cloud or aerosol layer, the resulting BHR (reported as white-sky albedo in the MODIS product suite (Lucht et al. 2000), and sometimes also referred to as BHR$_{iso}$)

can be described as follows:

$$BHR = \rho(2\pi; 2\pi)$$

$$= \frac{1}{\pi} \int_0^{2\pi} \int_0^{\pi/2} \rho(\theta_i, \phi_i; 2\pi) \cos\theta_i \sin\theta_i \mathrm{d}\theta_i \mathrm{d}\phi_i.$$

$$(26)$$

Under ambient illumination conditions, the albedo is influenced by the combined diffuse and direct irradiance. To obtain an approximation of the albedo for ambient illumination conditions (also reported as blue-sky albedo in the MODIS product suite), it is suggested that the BHR for isotropic diffuse illumination conditions and the DHR be combined linearly (see Equation (25)), corresponding to the actual ratio of diffuse to direct illumination (Lewis and Barnsley 1994, Lucht et al. 2000). The diffuse component then can be expressed as a function of wavelength, optical depth, aerosol type, and terrain contribution. The underlying assumption of an isotropic diffuse illumination may lead to significant uncertainties due to ignoring the actual distribution of the incoming diffuse radiation (e.g., Pinty et al. 2005).

All the above-mentioned albedo values, with the exception of the BHR for pure diffuse illumination conditions, depend on the actual illumination angle of the direct component. Thus it is highly recommended to include the illumination geometry in the metadata of albedo quantities.

OBSERVATIONAL GEOMETRY OF REMOTE SENSING INSTRUMENTS

This section discusses the geometric configuration of selected operational sensors, including laboratory and field instruments, as well as airborne and spaceborne sensors, using the basic conceptual model as presented in Figure 15.4. From a strict physical point of view, the most common measurement setup of satellites, airborne, and field instruments corresponds to the hemispherical-conical configuration (Case 8), while laboratory conditions are mostly biconical (Case 5).

Laboratory instruments

Laboratory conditions provide the ability to measure reflectance properties under controlled environmental conditions, being independent of irradiance variations due to a changing atmosphere, time of the day, or season. This is desirable when inherent reflectance properties (i.e., the BRDF) of a surface are investigated. Laboratory measurements involve an artificial light source, which is usually

non-parallel (due to internal beam divergence and collimating limitations), whereas solar direct illumination can be approximated as being parallel (within 0.5°). The diffuse illumination component in the laboratory can be minimal when the reflections are minimized (e.g., walls are painted black and black textiles cover reflecting objects). For the Laboratory Goniometer System (LAGOS), the diffuse-to-total illumination ratio was shown to be lower than 0.5% in the spectral range of 400–1000 nm (Dangel et al. 2005). The instantaneous field of view (IFOV) of a few degrees of the non-imaging spectroradiometer employed corresponds to a conical opening angle. Given the above conditions, the typical measurement setup of laboratory spectrometer measurements corresponds to the biconical configuration, resulting in conical–conical reflectance factors (CCRF – Case 5). For a perfectly collimated light source and a small IFOV, measurements may approximate the bidirectional quantity (e.g., the SpectroPhotoGoniometer (SPG) to measure leaf optical properties (Combes et al. 2007)).

Ground based field instruments

In the field, ambient illumination always includes a diffuse fraction. Its magnitude and angular distribution depend on the actual atmospheric conditions, surrounding terrain and objects, and wavelength. Thus, outdoor measurements always include hemispherical illumination, which can be described as a composition of a direct and an anisotropic diffuse component. Shading experiments are discussed in Schaepman-Strub et al. (2006), where it is concluded that they are only suitable to separate direct and diffuse illumination if the shading object exactly covers the solar disc (0.00006 sr). The reason is that a significant fraction of diffuse illumination is located within a small cone in the direction of the direct illumination of the sun.

The partitioning into direct and diffuse illumination influences the radiation regime within vegetation canopies. Based on a modeling study, Alton (2007) showed that the light use efficiency (LUE) of three forest canopies increases by 6–33% when the irradiance is dominated by diffuse rather than direct sunlight. This demonstrates the importance of accompanying field spectrometer campaigns with sun photometer measurements to assess the contribution of direct and diffuse irradiance. For field instruments with an IFOV full cone angle of about 4–5 degrees (e.g., PARABOLA (Portable Apparatus for Rapid Acquisition of Bidirectional Observation of the Land and Atmosphere, Abdou et al. 2001) or ASG (Automated Spectro-Goniometer, Painter et al. 2003)), the surface directional reflectance variability across the opening angle needs to be investigated. As long as this variability is unknown or not neglectable and corrected for, the measurements should be reported as HCRF (Case 8). This is especially true for sensors with larger IFOV, such as the ASD (Analytical Spectral Devices) FieldSpec series (25°, while fore-optics allow a restriction to 8° or less), and in cases where the sensitivity of the sensor outside of the cone only gradually falls off across several degrees outside of the half power point. More details concerning field spectrometer measurements can be found in Milton et al. (in press).

Albedometers are designed to cover the full down- and upward hemisphere (two pyranometers with an IFOV of 180° each) and approximate the bihemispherical configuration (Case 9) (e.g., Kipp and Zonen 2000).

Airborne sensors

The surface illumination conditions for airborne sensor observations are the same as for field measurements, thus of hemispherical extent (see above). The IFOV of airborne sensors is usually very small, e.g., 0.021° for the Airborne Multiangle Imaging SpectroRadiometer (AirMISR), 0.057° for the Airborne Visible/InfraRed Imaging Spectrometer (AVIRIS), 0.189° for the Digital Airborne Imaging Spectrometer 7915 (DAIS 7915), and 0.129° for the HyMap airborne hyperspectral scanner. In a strict physical sense, airborne observations therefore correspond to the hemispherical-conical configuration (HCRF – Case 8), while numerically approaching the hemispherical-directional configuration (HDRF – Case 7). Most correction schemes for airborne data do not correct for the hemispherical irradiance and thus the resulting at-surface reflectances approximately correspond to HDRFs (Schaepman-Strub et al. 2006).

Satellite sensors

Ambient illumination conditions of hemispherical extent are also present at the Earth surface when observed from spaceborne sensors. Generally, space-based instruments with a spatial resolution of about 1 km have an IFOV with a full cone angle of approximately 0.1° (e.g., Multiangle Imaging SpectroRadiometer (MISR), MODerate resolution Imaging Spectroradiometer (MODIS), Advanced Very High Resolution Radiometer (AVHRR)). If the HDRF is constant over the full cone angle of the instrument IFOV, then the HCRF numerically equals the HDRF. This approximation is mostly used when processing satellite sensor data.

Multi-angular sampling principles

All approaches correcting for the bias introduced by varying sun and view angles rely on multi-angular information to infer the BRDF as intrinsic reflectance property of the surface. The BRDF is a function of the solar *and* observational angles, thus measurements are performed under changing illumination or viewing geometries or a combination of both. Instantaneous multi-angular sampling is very rare, as most sampling schemes rely on tilting sensors and thus changing their viewing geometry. The MISR satellite has nine cameras with fixed viewing angles, and approximately 7 min lapses between the first and the last camera overpass for a selected area. For non-instantaneous multi-angular measuring concepts, assumptions on the temporal stability of the surface or the atmospheric composition (e.g., aerosol optical depth) are often made. This is a disadvantage for highly variable surfaces such as vegetation canopies which change their physiological state throughout a day, or snowmelt events lasting several days. A selection of the most common multi-angular sampling principles of laboratory, field, airborne, and satellite sensors is given in Figure 15.5.

PROCESSING OF REFLECTANCE PRODUCTS

While the preceding section explained the observation geometry of operational sensors and the multi-angular sampling principles, this section will focus on the derivation of various reflectance quantities from the observations.

Most state of the art atmospheric correction schemes convert top of atmosphere radiance to one singular view angle at-surface reflectance, while preserving the influence of the diffuse illumination on the surface reflectance, thus representing HDRF data. However, this reflectance quantity does not exactly represent what is required for many applications, such as energy budget studies, multi-temporal investigations, and studies relying on multiple sensor data. The main product pathways to obtain higher level reflectance products are discussed below, namely (a) removing the effect of the diffuse hemispherical illumination in single view angle observations to obtain inherent reflectance properties of the surface (i.e., the BRDF), (b) interpolating and extrapolating the single-angle observations to the entire reflected hemisphere to obtain albedo quantities, and (c) normalizing the single-angle observations to a standardized viewing geometry (i.e., to compute Nadir BRDF Adjusted Reflectance, NBAR). All three approaches are based on the derivation of the BRDF, thus on the extraction of the intrinsic

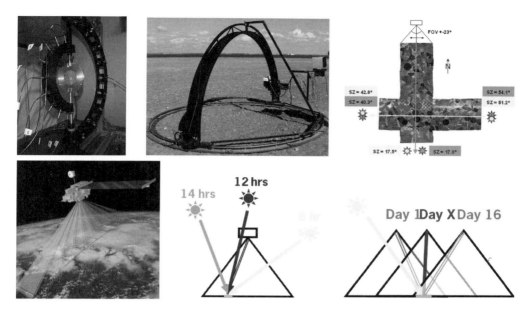

Figure 15.5   Examples of multi-angular sampling principles: (a) laboratory facility to measure leaf optical properties (SPG) (photo courtesy of Stéphane Jacquemoud), (b) field goniometer system (FIGOS), (c) airborne multi-angular sampling during DAISEX'99 using the HyMap sensor (SZ = solar zenith angle (Berger et al. 2001)), (c) spaceborne near-instantaneous multi-angular sampling (MISR with nine cameras), (e) daily composites of geostationary satellite sensors (e.g., Meteosat), (f) multiple-day compositing of polar orbiting satellite sensors (e.g., MODIS 16 days). (See the color plate section of this volume for a color version of this figure.)

reflectance properties of the surface, using multi-angular sampling of the observations through variation of sun and/or viewing angles. The derivation of the BRDF based on laboratory measurements requires a correction for conicity and inhomogeneity of the artificial illumination source (for details see Dangel et al. 2005). The derivation of different at-surface reflectance quantities from measurements usually requires a sophisticated processing scheme (Figure 15.6), as for example implemented for the MISR surface reflectance products (Martonchik et al. 1998). Unfortunately, this issue does not always receive sufficient attention in remote sensing when implementing processing schemes for 'reflectance' products. Below, the main required processing steps to infer the whole suite of reflectance quantities are described, including some examples as implemented in operational algorithms, and the assumptions used. For each heading, the input reflectance quantity of the processing scheme is specified on the left hand side of the arrow, and the resulting reflectance quantity on the right.

1   *HCRF (Case 8)* → *HDRF (Case 7)*
   The basic retrieval scheme starts with hemispherical-conical observations (Case 8).

Currently, most of the existing processing approaches assume that the HDRF is constant over the full cone angle of the instrument IFOV, thus the HCRF numerically equals the HDRF (Case 7) without further correction. Given a sufficient number of viewing angles (e.g., MISR), the BHR is directly derived through interpolation. The algorithm for retrieving the HDRF and BHR from MISR top-of-atmosphere (TOA) radiances is virtually independent of any particular kind of surface BRF model and its accuracy mainly depends on the accuracy of the atmospheric information used (Martonchik et al. 1998).

2   *HDRF (Case 7)* → *BRDF and BRF (Case 1)*
   BRF data are derived using a parameterized BRDF model to eliminate the diffuse illumination effects present in the HDRFs (e.g., Modified Rahman Pinty Verstraete (MRPV) for MISR (Martonchik et al. 1998) and ground based measurements (Lyapustin and Privette 1999)). An alternative approach is used for MODIS, where the atmospheric correction is performed under the assumption of a Lambertian surface. The resulting surface reflectances, collected during a

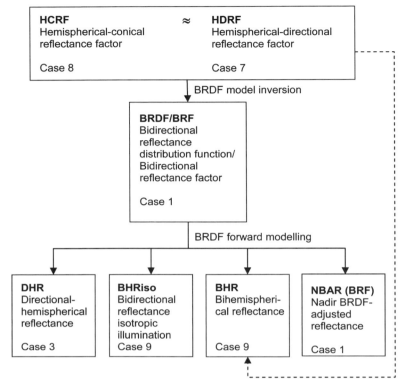

Figure 15.6   Recommended processing pathway of reflectance products. The pathway of direct BHR retrieval from HDRF data can be performed through interpolation, thus not relying on a BRDF model (e.g., MISR, FIGOS).

period of 16 days, are subsequently used to fit a BRDF model (i.e., RossThickLiSparseReciprocal for MODIS (Lucht et al. 2000)). In the case of dense angular sampling (sun and viewing geometry) and potentially measured irradiance, such as in ground level experiments, the BRDF alternatively can be retrieved using radiative transfer solutions, and thus without relying on a BRDF model (Martonchik 1994). Note that inverting a BRDF model using HDRF data without previous correction of the diffuse illumination may result in a distortion of the BRDF shape in the visible and near-infrared, even for low aerosol content in the atmosphere (Lyapustin and Privette 1999).

3 **BRDF (Case 1) → DHR (Case 3), BHR, and BHR$_{iso}$ (Case 9)**
The angular integration of the BRDF under different illumination conditions results in hemispherically integrated reflectance quantities. DHR (also referred to as black-sky albedo) corresponds to a direct illumination beam only, BHR (named blue-sky albedo occasionally) to a combination of direct and diffuse illumination, whereas BHR$_{iso}$ (known also as white-sky albedo) involves an isotropic diffuse illumination only. The calculations are based on forward modeling using the BRDF model parameters as previously obtained by the HDRF–BRDF retrieval. Only a limited selection of hemispherical products are usually delivered for a particular sensor (e.g., DHR (solar angle corresponding to mean solar noon within 16 days) and BHR$_{iso}$ for MODIS, while a routine is provided to calculate the BHR as a linear combination of DHR and BHR$_{iso}$). For MISR, only the DHR product relies on BRDF forward modeling, representing the solar geometry at the time of observation. The MISR BHR product is directly inferred based on HDRF data – both products do not involve BRDF forward modeling and rely on non-isotropic illumination conditions, thus on actual diffuse and direct illumination corresponding to atmospheric conditions and sun geometry at the time of observation. Thus, only MISR delivers BHR products which include direct and anisotropic diffuse illumination, most

closely representing actual conditions. However, a modeling study showed that the assumption of an isotropic diffuse illumination as compared to anisotropic diffuse illumination leads to relative albedo biases within bounds of 10% (Pinty et al. 2005).

4 BRDF → NBAR BRF

The MODIS product suite additionally contains a Nadir BRDF Adjusted Reflectance (NBAR) product that is a BRF modeled for the nadir view at the mean solar zenith angle of the 16-day period. This means that angular effects introduced by the large swath width of MODIS are corrected.

Several studies showed that directional and hemispherical illumination reflectance products from current operational instruments are highly correlated and that the differences are generally small. Schaepman-Strub et al. (2006) calculated a relative difference in single-view angle products (HDRF versus BRF) of up to 14%. The relative bias of hemispherically integrated reflectance quantities, i.e., BHR versus DHR, was smaller, and reached a maximum of 5.1% in the blue spectral band, under a relatively thick atmosphere (with an Aerosol Optical Depth (AOD) of 0.36 at 558 nm). The differences generally increase with increasing aerosol optical depth, and decrease with increasing wavelength. On the other hand, a systematic numerical study of surface albedo based on the radiative transfer equation for 12 land cover types investigated the dependence on atmospheric conditions and solar zenith angle (Lyapustin 1999). For a large number of vegetation and soil surfaces, the range of relative variation of surface albedo with atmospheric optical depth did not exceed 10–15% at a solar zenith angle smaller than 50° and 20–30% at solar zenith angles larger than 70°. At 52–57° solar zenith angle the albedo is almost insensitive to the atmospheric optical depth, resulting in a DHR that is equal to the BHR. The above biases may thus introduce a systematic error when neglected (e.g., in vegetation indices).

The above data processing pathway illustrates that the at-surface reflectance quantities inferred from satellites are not directly observed quantities, but are based on sophisticated algorithms addressing the atmospheric correction and the BRDF retrieval as well as forward modeling. The differing sampling schemes of the observations, the applied BRDF models, as well as the atmospheric correction schemes (see, e.g., the aerosol optical depth comparison between MODIS and MISR (Kahn et al. 2007)) introduce a bias between the reflectance products of different satellite sensors, which has not yet been assessed. It is therefore highly recommended to select the appropriate reflectance product carefully, and to pay attention to sensor-specific assumptions and restrictions when integrating multiple sensor data.

CONCLUSIONS AND RECOMMENDATIONS

This chapter presents a basic conceptual model of reflectance terminology, complemented by examples of sensors and products in order to help the user to critically review the products of his/her choice, select the appropriate reflectance quantity and name, and process measurements according to their physical meaning.

The variety in physical quantities resulting from different sensor sampling schemes, preprocessing, atmospheric correction, and angular modeling, requires a rigorous documentation standard for remotely sensed reflectance data. Beyond the algorithm theoretical basis document with a detailed description of the data processing steps performed, a short and standardized description on the physical character of the delivered reflectance product must be accessible as well. This necessarily includes the accurate listing of opening angles and directions of illumination and observation, revealing whether the product represents inherent reflectance properties of the surface or contains a diffuse illumination component corresponding to the atmospheric and terrain conditions of the observations. Relying on this standardized reflectance description, users can choose the appropriate reflectance products and evaluate whether approximations will introduce relevant biases to their applications. Numerically, differences between hemispherical, conical, and directional quantities depend on various factors, including the anisotropy of the surface, the sensitivity distribution within the sensor IFOV, and its fall off outside the cone, the viewing and sun geometry, atmospheric conditions, and the scattering properties of the area surrounding the observed surface. This implies that numerical differences are wavelength dependent according to the involved absorption and scattering processes of the atmosphere and the observed surface. The diffuse illumination component generally decreases with increasing wavelength, resulting in decreasing numerical differences from the blue toward longer wavelengths.

ACKNOWLEDGMENTS

This chapter is dedicated to the memory of Jim Palmer, who passed away on January 4 2007.

The authors would like to thank Stephen Warren, University of Washington, for comments on usage of the term intensity.

REFERENCES

Abdou, W. A., M. C. Helmlinger, J. E. Conel, C. J. Bruegge, S. H. Pilorz., J. V. Martonchik, and B. J. Gaitley, 2001. Ground measurements of surface BRF and HDRF using PARABOLA III. *Journal of Geophysical Research – Atmospheres,* 106: 11967–11976.

Alton, P. B., 2007. The impact of diffuse sunlight on canopy light-use efficiency, gross photosynthetic product and net ecosystem exchange in three forest biomes. *Global Change Biology,* 13: 776–787.

Berger, M., M. Rast, P. Wursteisen, E. Attema, J. Moreno, A. Muller, U. Beisl, R. Richter, M. Schaepman, G. Strub, M. P. Stoll, F. Nerry, and M. Leroy, 2001. The DAISEX campaigns in support of a future land-surface-processes mission. *ESA Bulletin,* European Space Agency, 105: 101–111.

Chandrasekhar, S., 1950. *Radiative Transfer,* Claredon Press, Oxford (Reprinted in 1960 by Dover Publications).

Combes, D., L. Bousquet, S. Jacquemoud, H. Sinoquet, C. Varlet-Grancher, and I. Moya, 2007. A new spectrogoniophotometer to measure leaf spectral and directional optical properties. *Remote Sensing of Environment,* 109: 107–117.

Dangel, S., M. M. Verstraete, J. Schopfer, M. Kneubuhler, M. Schaepman, and K. I. Itten, 2005. Toward a direct comparison of field and laboratory goniometer measurements. *IEEE Transactions on Geoscience and Remote Sensing,* 43: 2666–2675.

Dozier, J., 1980. A clear-sky spectral solar-radiation model for snow-covered mountainous terrain. *Water Resources Research,* 16: 709–718.

Kahn, R. A., M. J. Garay, D. L. Nelson, K. K. Yau, M. A. Bull, B. J. Gaitley, J. V. Martonchik, and R. C. Levy, 2007. Satellite-derived aerosol optical depth over dark water from MISR and MODIS: Comparisons with AERONET and implications for climatological studies. *Journal of Geophysical Research – Atmospheres,* 112: D18205, doi:10.1029/2006JD008175.

Kipp and Zonen, 2000. *Instruction Manual CM11 Pyranometer/ CM14 Albedometer.* Delft, The Netherlands.

Lewis, P. and M. J. Barnsley, 1994. Influence of the sky radiance distribution on various formulations of the earth surface albedo. *Proceedings of 6th International Symposium on Physical Measurements and Signatures in Remote Sensing, ISPRS*: 707–715. Val d'Isere, France.

Lucht, W., C. B. Schaaf, and A. H. Strahler, 2000. An algorithm for the retrieval of albedo from space using semiempirical BRDF models. *IEEE Transactions on Geoscience and Remote Sensing,* 38: 977–998.

Lyapustin, A. I., 1999. Atmospheric and geometrical effects on land surface albedo. *Journal of Geophysical Research – Atmospheres,* 104: 4127–4143.

Lyapustin, A. I. and J. L. Privette, 1999. A new method of retrieving surface bidirectional reflectance from ground measurements: Atmospheric sensitivity study. *Journal of Geophysical Research – Atmospheres,* 104: 6257–6268.

Martonchik, J. V., 1994. Retrieval of surface directional reflectance properties using ground-level multiangle measurements. *Remote Sensing of Environment,* 50: 303–316.

Martonchik, J. V., D. J. Diner, B. Pinty, M. M. Verstraete, R. B. Myneni, Y. Knyazikhin, and H. R. Gordon, 1998. Determination of land and ocean reflective, radiative, and biophysical properties using multiangle imaging. *IEEE Transactions on Geoscience and Remote Sensing,* 36: 1266–1281.

Martonchik, J. V., C. J. Bruegge, and A. Strahler, 2000. A review of reflectance nomenclature used in remote sensing. *Remote Sensing Reviews,* 19: 9–20.

Milton, E. J., M. E. Schaepman, K. Anderson, M. Kneubuhler, and N. Fox, in press. Progress in field spectroscopy. *Remote Sensing of Environment.* Doi: 10.1016/j.rse.2007.08.001.

Nicodemus, F. E., 1965. Directional Reflectance and Emissivity of an Opaque Surface. *Applied Optics,* 4: 767–773.

Nicodemus, F. E., J. C. Richmond, J. J. Hsia, I. W. Ginsberg, and T. Limperis, 1977. *Geometrical Considerations and Nomenclature for Reflectance.* NBS Monograph 160. National Bureau of Standards, US Department of Commerce, Washington, D.C. http://physics.nist.gov/Divisions/Div844/ facilities/specphoto/pdf/geoConsid.pdf (last accessed: January 11 2008).

Painter, T. H. and J. Dozier, 2004. Measurements of the hemispherical-directional reflectance of snow at fine spectral and angular resolution. *Journal of Geophysical Research – Atmospheres,* 109, D18115, doi: 10.1029/2003JD004458.

Painter, T. H., B. Paden, and J. Dozier, 2003. Automated spectro-goniometer: A spherical robot for the field measurement of the directional reflectance of snow. *Review of Scientific Instruments,* 74: 5179–5188.

Palmer, J. M., 1993. Getting intense on intensity. *Metrologia,* 30: 371.

Palmer, J. M., 2003. *Radiometry FAQ.* Version 1.1: Oct. 2003. http://www.optics.arizona.edu/Palmer/rpfaq/rpfaq.pdf (last accessed: January 11 2008).

Pinty, B., A. Lattanzio, J. V. Martonchik, M. M. Verstraete, N. Gobron, M. Taberner, J. L. Widlowski, R. E. Dickinson, and Y. Govaerts, 2005. Coupling diffuse sky radiation and surface albedo. *Journal of the Atmospheric Sciences,* 62: 2580–2591.

Schaepman-Strub, G., M. E. Schaepman, T. H. Painter, S. Dangel, and J. V. Martonchik, 2006. Reflectance quantities in optical remote sensing–definitions and case studies. *Remote Sensing of Environment,* 103: 27–42.

Sellers, P. J., B. W. Meeson, F. G. Hall, G. Asrar, R. E. Murphy, R. A. Schiffer, F. P. Bretherton, R. E. Dickinson, R. G.Ellingson, C. B. Field, K. F. Huemmrich, C. O. Justice, J. M. Melack, N. T. Roulet, D. S. Schimel, and P. D. Try, 1995. Remote-sensing of the land-surface for studies of global change – models, algorithms, experiments. *Remote Sensing of Environment,* 51: 3–26.

Strub, G., M. E. Schaepman, Y. Knyazikhin, and K. I. Itten, 2003. Evaluation of spectrodirectional Alfalfa canopy data acquired during DAISEX '99. *IEEE Transactions on Geoscience and Remote Sensing,* 41: 1034–1042.

Taylor, B. N., 1995. *Guide for the Use of the International System of Units (SI).* Special Publication 811. NIST (Ed.). Washington: National Institue of Standards and Technology (NIST), US Department of Commerce, Washington, D.C.

Pre-Processing of Optical Imagery

Freek van der Meer, Harald van der Werff, and Steven M. de Jong

Keywords: spectrometer, pre-processing, calibration, radiometric, atmospheric, geometric, SNR, NER, noise.

INTRODUCTION

The pre-processing chain from raw image data to a fully georeferenced and parametrized (calibrated and in units that have a physical meaning) data that an end user typically would deploy for environmental studies is the scope of this chapter. Many calibration and processing steps are similar for multi-band, multispectral and high-spectral resolution imagery. Hence, the processing chain of hyperspectral remote sensing imagery is chosen as a proxy for the broader range of optical, visible, and infrared, data sets. Active sensor systems are excluded in this review as these require a significantly different pre-processing. Thermal infrared imagery is, however, included as it shows similar characteristics to data acquired in the visible and shortwave infrared regions, despite the fact that the physics and engineering behind this type of imagery is different.

The objective of hyperspectral remote sensing (e.g., imaging spectrometry/spectroscopy) is typically to measure quantitatively the components of the Earth system, such as radiance, upwelling radiance, emissivity, temperature, and reflectance, from calibrated spectra acquired as images for scientific research and applications (see also Schaepman, in this volume). These measurements facilitate the estimation of biophysical parameters

such as carbon balance, yield/volume, nitrogen, cellulose, chlorophyll, soil constituents, inherent optical properties of water and water quality, optical properties, and atmospheric absorption characteristics of ozone, oxygen, water vapor, and other trace gases.

In the following sections, the pre-processing chain (based on van der Meer and De Jong 2001, and De Jong and van der Meer 2004) and individual pre-processing steps are outlined. Radiometric calibration, geometric aspects, and atmospheric influences will be discussed concluding with an evaluation of inherent sensor and image noise and image defects. Table 16.1 provides an overview of the full processing chain. The intermediate steps highlighted in the table are discussed below, in the same order.

THE PRE-PROCESSING CHAIN

Introduction

Pre-processing operations are intended to correct for sensor- and platform-specific radiometric and geometric distortions of data. Radiometric corrections may be necessary due to variations in image illumination and viewing geometry, atmospheric conditions, and sensor noise and

Table 16.1 Overview of pre-processing chain for spectrometer data (after Peter Strobl Personal Communication)

Influence source	Factors	Correction method	Result
Sensor	Sensitivity Intrinsic signal Off-set Response Degradation	Sensor model System correction	At-sensor radiance
Atmosphere	Transmission Path radiance	Atmosphere model Ground truth data	Surface reflectance
Carrier	Internal orientation (roll, pitch, yaw) External orientation (long., lat., altitude)	Model carrier movement Sensor gyros Navigation data Geo-correction	Georeferenced reflectance
Image	Dropped lines Salt and pepper Resonance	MNF cleaning Fourier filtering Smoothing	Improved georeferenced reflectance

response. Each of these may vary depending on sensor and platform characteristics and conditions during data acquisition. For comparative purposes and for quantification of environmental variables, it is desirable to convert data to known (absolute) radiance, L, in $W\ m^{-2}\ sr^{-1}$, which can be defined as $d\Phi = L\ dA\ d\Omega\ \cos\theta$ for an area dA, arriving at a direction θ to the normal of dA in the range of directions forming a solid angle of $d\Omega$ steradians or scaled reflectance units. Radiance can be integrated over a specific range of wavelengths, spectral radiance, and expressed as the photon flux power per unit area per solid angle per unit of wavelength interval. Martonchik et al. (2000) and Schaepman-Strub et al. (in this volume) provide reviews of nomenclature of units used in remote sensing. From an engineering perspective, quantifying radiance requires knowledge of the following sensor characteristics:

- the aperture response function; defining the sensitivity of the entrance slit to the radiation reaching the detector
- the geometric response function; often confused with the pixel size, defining the sensitivity of the instrument to radiation from a certain ground resolution cell
- the spectral response function; defining the sensitivity of the detector to radiation of different wavelengths
- the temporal response function; defining the temporal variability in the sensor sensitivity to radiation.

Pre-processing is largely done by the manufacturers, institutes, agencies, or commercial companies that maintain and operate a sensor.

In most, if not all, cases, the user is provided with at-sensor radiance data and the necessary instrument characteristics (e.g., spectral response functions, band passes, etc.) needed for processing the data to higher level products. The calibration of a spectrometer consists of a sequence of procedures providing the calibration data files and ensuring the radiometric, spectrometric, and geometric stability of the instrument. These include:

- measurement of noise characteristics of the sensor channels
- measurement of the dark current of the channels
- measurement of the relative spectral response function
- derivation of the effective spectral bandwidth of the channels
- definition of the spectral separation of the channels
- measurement of the absolute radiometric calibration coefficients (the 'transfer functions' between absolute spectral radiance at the entrance of the aperture of the sensor and the measured radiation dependent part of the output signal of each channel)
- definition of the Noise Equivalent Radiance (NER) and the Noise Equivalent Temperature Difference (NE DeltaT) of the channels
- measurement of the Instantaneous Field of View (IFOV), the spectral resolution and the deviations in the band-to-band registration.

The pre-processing tasks for optical imagery are conducted in successive steps, starting from sensor calibration, which involves correcting for sensitivity response and signal degradation, and results in at-sensor radiance (i.e., translating digital count values into physical units of measurements). Next, atmospheric calibration, compensating for transmission through the atmosphere, path radiance, and scattering using either empirical approaches or atmosphere models, results in surface reflectance. Finally, carrier calibration, which involves correction for factors such as roll, pitch, yaw, altitude, and relief, is undertaken based on modeling carrier movement and the topography, resulting in georeferenced surface reflectance. Typically, this product is delivered to the user. In addition, the user may decide to perform sensor and image noise calibration involving factors such as dropped lines and salt and pepper noise removal by convolution filtering which results in enhanced separation of the signal and noise components in the image data.

Khurshid et al. (2006) provide an overview of preprocessing of EO-1 satellite hyperspectral HYPERION data, and their paper is a good example of the typical pre-processing chain.

Sensor calibration

Sensor or radiometric calibration is typically done under laboratory conditions; while regular in-flight calibration monitors sensor degradation (see Gianinetto and Lechi (2006) for a recent example of in-flight calibration of Multispectral Infrared and Visible Imaging Spectrometer (MIVIS) data) and requires estimating the aperture, temporal, geometric, and spectral response functions. Sensor calibration requires the development of a linear transform function, the radiometric response function, between photon-counts (converted to digital numbers (DN)) and radiance, for each channel. Thus each image channel needs to be calibrated to derive its specific radiometric response function, which translates raw signal (photon counts, $N_{photons}$) into (at-sensor) spectral radiance through a linear relationship. $N_{photons}$ is not only a function of the incident spectral radiance but is also related to sensor characteristics of which the spectral response function and the point spread function (PSF) are the two most important. The PSF describes the spatial decline of the measured signal. The PSF 'describes the two-dimensional normalized distribution of intensity measured on the focal plane from an ideal point source located in the scene' (Kerekes, in this volume). One detector catering for a certain channel or band theoretically should sense (be sensitive) only to photons of one particular wavelength (energy intensity); however, in practice each band measures radiance in a wavelength range that stretches from a few nanometers lower and a few nanometers higher wavelength than the center wavelength of the channel. The curve describing the decline of the

destector response levels around the central channel wavelength is the spectral response function (SRF), which is found using monochromator measurements. Similarly the PSF describes the spatial decline of signal strength, which is found using a monochromator covering the sensor designated wavelength coverage and different pinhole targets. In line with the definition of the PSF defined above, the SRF is defined as 'the normalized response of the detector response as a function of wavelength'. Brazile et al. (2006) present a review of methods for estimating SRF's for spectral imagers.

Schott et al. (1988) describes radiometric calibration procedures. Guanter et al. (2007) discuss sensor calibration with specific reference to Compact Airborne Spectrographic Imager (CASI) data while Hook et al. (2007) compare in-flight and laboratory calibration results for the Advanced Spaceborne Thermal Emission and Reflection Radiometer (ASTER) and Moderate Resolution Imaging Spectroradiometer (MODIS) systems. See also Liang (in this volume).

Spectrometers take indirect measurements of physical features in the sense that the digitized signal recorded is directly proportional to the incoming photon energy but not measured in any physically meaningful unit. The relation between the raw digitized signal and a physical meaningful variable, such as the photon flux power per unit solid angle per wavelength interval, is established after radiometric correction. The radiometric response function is the relation between the signal caused by the bombardment of $N_{photons}$ on the detector and the incoming spectral radiance, L_λ. This function allows for converting each channel from photon counts, represented by DN, to radiance and subsequently spectral radiance. The radiometric function is measured by mounting the sensor onto an integrating sphere. The set of lamps in the sphere produce light, of known spectral radiance, which can be compared with the measurements of the sensor, which in turn are cross calibrated using a field spectrophotometer with a known standard. The radiometric response function corrected for the spectral response and geometric response as well for the temporal response (i.e., the temporal drift not further elaborated here) gives the at-sensor spectral radiance. The radiometric response function is a linear relationship, which for each channel is determined by two coefficients; the gain and offset, c_1 and c_0. Calculating at-sensor radiance, $L(\text{mW cm}^{-2}\,\text{sr}^{-1}\,\mu\text{m}^{-1})$, from raw recorded, DN, using given gain and offset values is done by:

$$L = c_0 + c_1(\text{DN}). \qquad (1)$$

Similarly, for thermal sensors, two black bodies (*bb*), one with a low temperature, T_{bb1} and a second with a high temperature T_{bb2}, give a linear response to derive at-sensor radiance L_{bb} as:

$$L_{bb} = L_1(T_{bb1})$$
$$+ \frac{L_2(T_{bb2}) - L_1(T_{bb1})}{DN_{bb2} - DN_{bb1}}(DN - DN_{bb1})$$
$$(2)$$

where DN_{bb1} and DN_{bb2} are the digital numbers for black body 1 and 2 and $L_1(T_{bb1})$ and $L_2(T_{bb2})$ are the corresponding spectral radiances for black body 1 and 2. Figure 16.1 schematically shows the various components of atmospheric attenuation and the signal that finally reaches the sensor, which goes through a chain of conversion from analogue to digital numbers. After radiometric calibration, at-sensor radiance results from a linear scaling of the photon counts to radiation levels.

Atmosphere calibration

All materials with a temperature above 0 K emit electromagnetic energy. Objects on the Earth's surface are able to reflect or scatter incident electromagnetic radiation. In the visible, near-infrared, middle-infrared and shortwave infrared parts of the electromagnetic spectrum, solar radiation reflected by objects at the Earth's surface is measured. In the thermal-infrared part, particularly in the atmospheric window at approximately 10 μm, emitted radiation by objects at the Earth's surface is measured. In all cases, the atmosphere influences reflected solar radiation or emitted radiation by the Earth's surface before it reaches an airborne or spaceborne sensor. The atmosphere consists mainly of molecular nitrogen and oxygen (clean dry air), water vapor, and particles (aerosols) such as dust, soot, water droplets, and ice crystals. The atmospheric effects on the signal,

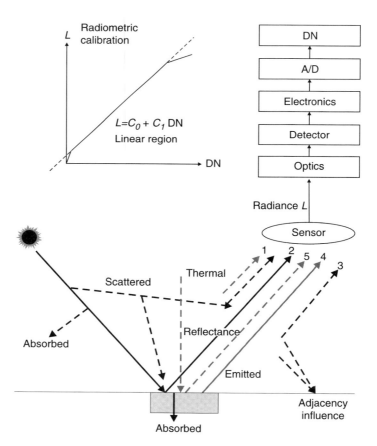

Figure 16.1   Schematic diagram showing the various components contributing to loss of signal in the atmosphere, the various signal conversion from analogue to digital numbers and the radiometric calibration using linear empirical models to estimate at-sensor radiance (after Richter 1996).

scattering (see Herman et al. 1993 for a review) and absorption (see LaRocca 1993 for a review), of these constituents on radiation can vary with wavelength, condition of the atmosphere and the solar zenith angle. Scattering by aerosols can be divided into Rayleigh, Mie, and non-selective scattering. These processes lead to the formation of diffuse radiation. A portion of the diffuse radiation goes back to space and a portion reaches the ground. Radiation reaching the sensor can be split into four components: path radiance, reflected diffuse radiance, reflected direct radiance, and reflected radiance from a neighborhood (Figure 16.1). The radiation, which has not been scattered, is called direct radiation. Absorption of energy is caused by the presence of gases and water vapor in the atmosphere. Raw calibrated spectrometer data have the general appearance of the solar irradiance curve, with radiance decreasing toward longer wavelengths. In addition, major atmospheric water vapor (H_2O) bands are found at 0.94, 1.14, 1.38, and 1.88 μm, an oxygen (O_2) band at 0.76 μm, carbon dioxide (CO_2) bands near 2.01 and 2.08 μm, and trace gases including ozone (O_3), carbon monoxide (CO), nitrous oxide (N_2O), and methane (CH_4), producing noticeable absorption features in the 0.4–2.5 μm wavelength region (Figure 16.2).

The aim of atmospheric calibration algorithms is to re-scale the raw radiance data provided by spectrometers to reflectance by reducing the atmospheric influence. On land, the surface reflected signal is of interest to the scientist, while for water bodies the surface reflected signal is considered noise, and is composed of the reflected component of diffuse skylight and of the direct sunlight impinging on the water surface. Water bodies in general reflect (as subsurface irradiance reflectance) in the range of 1–15% of downwelling irradiance (Morel and Gentili 1993). The majority of waters reflect between 2 and 6% of downwelling irradiance. Thus, to obtain, for example, in the order of 40 levels of irradiance reflectance in the range of 2–6% reflectance, a minimal accuracy of atmospheric correction to 0.1% reflectance is needed for water studies.

Atmosphere correction can be done using radiative transfer (RT) models which aim at modeling atmosphere transmission given knowledge on the composition of the atmosphere in terms of gases and aerosols. Alternatively, scene dependent empirical models can be used. The first set of methods provides absolute reflectance values, whereas the second set of methods provides relative (to the target or standard used) reflectance values. Richter et al. (2002, 2006) present an automatic atmospheric correction algorithm for hyperspectral imagery. These methods are discussed in more detail below. A review of the performance of atmospheric correction methods is found in Ben-Dor and Levin (2000).

Relative atmosphere correction methods

In relative reflectance data, reflectivity is measured relative to a standard target from the image. Correction methods currently available for this purpose include flat-field correction, internal average relative reflectance correction, and empirical line correction.

The purpose of the flat-field correction is to reduce the atmospheric influence in the raw imaging spectrometer data and eliminate the solar irradiance drop-off, as well as any residual instrument effects. This is achieved by dividing the whole data set by the mean value (per spectral band) of an area within the image which is spectrally and morphologically flat, and spectrally homogeneous. The flat-field chosen should have a high albedo to avoid decreasing the signal to noise ratio (SNR). If applied properly, the flat-field method removes the solar irradiance curve and major gaseous absorption features.

The Internal Average Relative Reflectance correction method uses a reference spectrum which is calculated as the average pixel spectrum of an entire image. This spectrum is divided into each image radiance spectrum for each pixel to produce a relative reflectance spectrum. Care should be taken when cover types with strong absorption features are present in the image. In these cases, absorption features are ratioed into the reflectance spectra and appear as spikes on the derived spectra.

The Empirical Line method (Figure 16.3) uses two calibration targets, a bright and a dark reference target. These targets are characterized using radiance pixel spectra as well as field reflectance spectra, which allows for the derivation of a linear relationship between radiance and reflectance for each channel. Applying this linear relationship to each radiance value for each pixel in each channel results in a best fit between sets of field spectra and image spectra characterizing the same ground areas thus removing atmospheric effects, residual instrument artefacts, and viewing geometry effects.

Absolute atmosphere correction methods

Absolute reflectance data without *a priori* knowledge of surface characteristics can be obtained using atmosphere models. These models correct for scattering and absorption in the atmosphere due to water vapor and mixed gases as well as for topographic effects and differences in image illumination. The 0.94 and 1.1 μm water absorption bands are used to calculate water vapor in the atmosphere (e.g., Frouin et al. 1990, Carrere and Conel 1993) while transmission spectra of the mixed gases in the 0.4–2.5 μm wavelength region are simulated on basis of the water vapor values

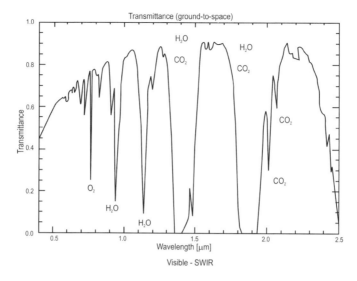

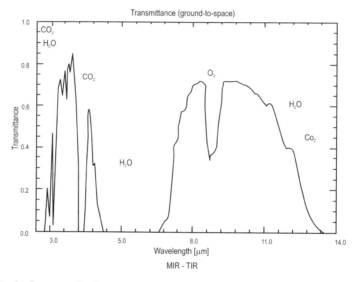

Figure 16.2 Typical atmospheric transmittance curves for the visible-shortwave infrared region (top) and the mid-infrared-thermal infrared region (bottom). Areas of absorption are indicated as well as major gases and water vapor responsible for the absorption.

found and the solar and observational geometry. Scattering effects in the atmosphere are modeled using RT codes (such as LOWTRAN, MODTRAN, 5S and 6S, see for example Van den Bosch and Alley 1990) using standard atmosphere types such as a rural or urban atmosphere under tropical or arctic conditions.

A RT code (see also Liang, in this volume) models the atmosphere's behavior on incident radiation through deriving transmission spectra of the main atmospheric gases which are integrated with

effects of atmospheric scattering from aerosols and molecules. Examples of studies dealing with scaled surface reflectance estimates from hyperspectral data using atmosphere models include work with sensors such as the Airborne Visible Infrared Imaging Spectrometer (AVIRIS; Gao et al. 1993, Gaddis et al. 1996), Digital Airborne Imaging Spectrometer (DAIS; Richter 1996), CASI (Olbert 1998) and Medium Resolution Imaging Spectrometer (MERIS; Moore et al. 1999) as well as studies deploying multispectral sensors (Liang et al. 1997).

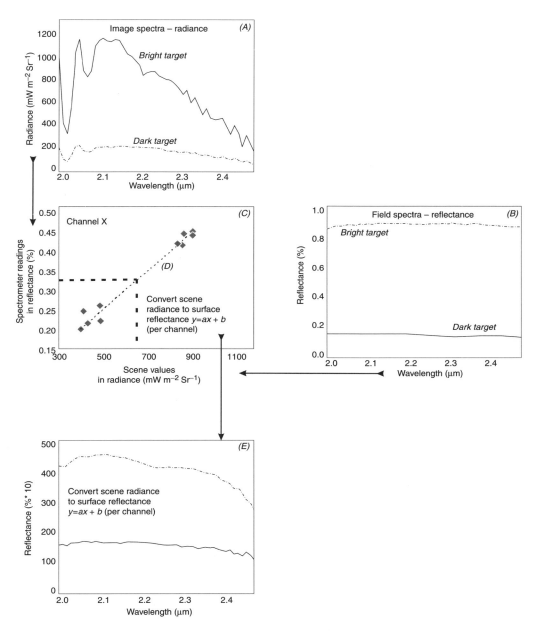

Figure 16.3   Process of atmospheric correction using the empirical line method. Two targets, a bright and a dark target, are characterized in the image (A; in radiance, raw at sensor radiance image data) and in field (B; in reflectance with a spectrometer). For each channel, a linear relationship (C) can be found by plotting the radiance and reflectance values of the two targets in the wavelength region of the specific band. This relationship allows to transfer (D) any given image radiance value into a surface reflectance value (E).

Albert and Gege (2006) illustrate the use of inversion of irradiance to reflectance for shallow waters using water optical properties and bottom properties.

User inputs of atmosphere condition usually required include the date, time and location of data acquisition, ozone depth, aerosol type, visibility, and elevation. These variables can be derived partly from radiosonde data or meteorological stations in the area. Readily used atmosphere correction models deploying radiative transfer codes include ATmosphere REMoval program (ATREM), AT-mos-pheric COR-rection (ATCOR) and Atmospheric COrrection Now (ACORN). Atmospheric correction in the reflective visible (VIS), shortwave infrared (SWIR), and thermal (TIR) spectral regions are treated separately since the influence of the Sun dominates the solar reflective region while this can be largely neglected in the thermal region. The basic relation defining spectral radiance in the VIS-SWIR region is given by:

$$L_\lambda = L_0(\lambda) + \frac{E_g(\lambda)}{\pi}[\tau_{dir}(\lambda) + \tau_{dif}(\lambda)]\rho(\lambda) \quad (3)$$

where $L_0(\lambda)$ is the path radiance for a black body ($\rho = 0$), $E_g(\lambda)$ is the global irradiance on the ground, $\tau_{dir}(\lambda)$ is the direct atmospheric transmittance (ground to sensor), $\tau_{dif}(\lambda)$ is the diffuse atmospheric transmittance (ground to sensor), and $\rho(\lambda)$ is the reflectance of a Lambertian surface. The basic relation defining spectral radiance in the thermal region is given by:

$$L_\lambda = [\varepsilon_\lambda L_{bb\lambda}(T) + (1 - \varepsilon_\lambda)L_{sky\lambda}]\tau_\lambda + L_{atm\lambda} \quad (4)$$

where L_λ is the spectral radiance for the wavelength λ, ε_λ is the surface emissivity at wavelength λ, $L_{bb\lambda}(T)$ is spectral radiance from a black body at surface temperature T, $L_{sky\lambda}$ is spectral radiance incident upon the surface from the atmosphere, τ_λ is spectral atmospheric transmission, and $L_{atm\lambda}$ is spectral radiance from atmospheric emission and scattering that reaches the sensor. Atmospheric values in these equations are typically simulated using RT codes in conjunction with radiosonde data to derive wavelength dependent surface emissivity (TIR) or reflectance (VIS-SWIR).

Calibration of thermal infrared sensor data (see Sobrino et al. (2006) for an example of land surface temperature derivation from thermal data) differs from optical data acquired in the visible-shortwave infrared windows in that the data exhibits both a temperature and an emissivity component that need to be separated. Gangopadhyay et al. (2005) describe techniques to separate temperature and emissivity. Validation of the TIR atmospheric corrected data is not generally possible as this would require measurements of surface temperature and surface spectral emissivity and atmospheric properties at the time of overpass (acquisition). As the measured radiance in the TIR window is reworked into both emissivity as well as temperature of a target or pixel the interplay of the two factors is needed to be considered to assess the quality of the outcome. A way around this is to make use of the fact that the spectral emissivity of water is known and constant, thus over water bodies only temperature changes. For ASTER TIR data in this way it was shown that the instrument produces an error in surface radiance of about 5% which results in an error of 3 °C in surface temperature.

Figure 16.4 shows a DAIS hyperspectral image of an area in south east France before and after atmospheric correction. The original at-sensor radiance spectra show a strong resemblance to the solar radiation curve with a marked decline of the signal toward longer wavelength and a peak in the visible part of the spectrum. The reflectance spectra show distinct absorption features that can be attributed to various surface materials present in the area.

From bidirectional surface reflectance to albedo

Albedo is the unitless fraction of incoming solar radiation reflected by the land surface integrated over all viewing directions. Directional or black-sky albedo is calculated for a given solar zenith angle, whereas hemispheric or white-sky albedo is integrated over all illumination directions. For most (non-Lambertian) surfaces, the albedo is direction dependent and hence the bidirectional reflectance distribution function (BRDF) characterizing the directional scattering properties of a material, is used to derive albedo. More specifically, the albedo at a particular solar zenith angle is often approximated by summing the directional-hemispherical reflectance at that solar zenith angle and the bi-hemispherical reflectance. Schaepman-Strub et al. (2006) and Schaepman-Strub et al. (in this volume) provide a review of BRDF and albedo estimation as well as an overview of nomenclature used to describe reflectance-based products.

The albedo of the Earth is around 0.39 and this is in equilibrium with the overall temperature of the Earth. The greenhouse effect, by trapping infrared radiation and thus reducing the area of highly reflective snow and ice, can lower the albedo of the Earth and exacerbate global warming. For this reason, land surface albedo is one of the most important variables characterizing the Earth's radiative regime and accurate measurements of global albedo are of utmost importance in climatic models. Hence surface albedo is a prime component of Earth observation satellite missions and instruments, such

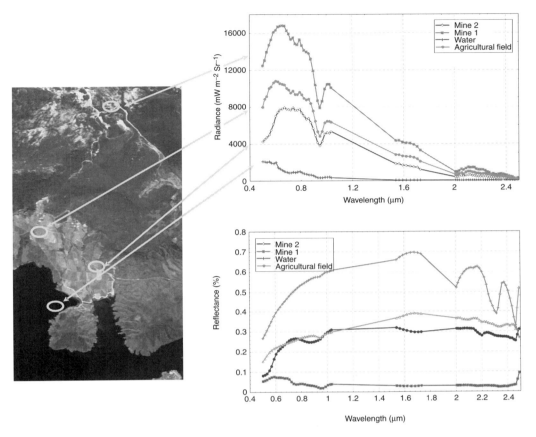

Figure 16.4   Example of image radiance spectra (top graph) and image reflectance spectra (bottom graph). The radiance spectra have the overall appearance of a solar curve, while the reflectance spectra show distinct areas of absorption. Mine 2 is a dolomite mine showing strong absorption at 2.3 μm attributed to Ca-Mg OH, while mine 1 is a bauxite mine showing absorption features in the visible part of the spectrum attributed to iron-oxide. (See the color plate section of this volume for a color version of this figure).

as MODIS onboard the Terra and Aqua satellites, provide regular estimates of global albedo. As the total amount of reflected radiation cannot be directly measured by the sensor, an approximation is found by modeling the BRDF allowing the conversion of satellite reflectance measurements into estimates of directional-hemispherical reflectance and bi-hemispherical reflectance and hence albedo.

Carrier calibration

Viewing geometry and swath coverage are a function of altitude and topography. Aircraft carrying sensors fly at altitudes varying from less than 1,000 to 22,000 m. Although AVIRIS typically operates at 22 km when flying in the high altitude option, most airborne hyperspectral sensors are flown at around

3 km altitude for operational reasons. An airborne sensor must image a wide range of incidence angles (up to 70 degrees) in order to achieve relatively wide swaths. Image characteristics such as pixel displacement, reflectance, and resolution will be subject to wide variations across large incidence angle range. Space borne sensors are able to avoid some of these imaging geometry problems since they operate at high altitudes and can produce comparable swath widths, but over a much narrower range of incidence angles (typically ranging from 5 to 15 degrees). Gu and Anderson (2003) provide an overview of geometric calibration of image data.

Airborne sensor data suffer from spatial distortions related to the carrier movements and also from topographic relief. To be able to correct for these problems, modern systems are equipped with an onboard Global Positioning System (GPS)

for absolute location of the aircraft at acquisition time and with onboard gyroscope in an inertial navigation system to record tilt of the aircraft in terms of roll, pitch, and yaw. In geometric correction, for each pixel the original observation geometry is reconstructed based on the flight line, aircraft altitude, surface topography and aircraft navigational information. The result of the correction is georeferenced at-sensor radiance data. Geometric distortion of airborne image data resulting from data recording can be characterized as being associated with four effects:

1 *Panoramic effect*: Due to the scanning with constant angular velocity, pixels cover greater ground area at the edges of the scan line.
2 *Over- and undersampling*: Due to non-perfect synchrony between air-speed, altitude, and scan speed in the flight direction, redundant information is scanned or data holes occur.
3 *Geometric distortions due to projection*: Due to movements of the aircraft, roll pitch and/or yaw distortions may occur.
4 *Drift effect*: Due to side wind or other effects the airplane may drift from its course and as a result distortions occur in the flight path.

In addition, topographic effects may shift pixel locations (georeferencing errors) from their true position and cause deviations across the image of the pixel size. The onboard gyroscopic measurements, recording the attitude of the aircraft in terms of roll, pitch, and yaw, and differential GPS measurements recording the flight path in x, y, and z absolute coordinates, allows geometric correction of the image data. Three approaches can be applied (Richter 1997, 1998):

1 georeferencing using control points and registration to a map base
2 georeferencing using pixel transformations through gyroscope data
3 parametric georeferencing using both gyroscopic data, flight line information, and a digital terrain model.

Schläpfer and Richter (2002) provide an overview of a combined geometric and atmospheric processing of airborne imaging spectrometry data resulting in parametric orthorectification. Richter and Schläpfer (2002) provide a methodology for geometric corrections including topographic information.

In addition, images may suffer from illumination differences that systematically change over an image. This is mainly caused by the orientation of an image with respect to the Sun (solar elevation and azimuth angle) and sensor characteristics (viewing angle). Light incident from the same direction as the sensor view causes a so-called hotspot, an area on the ground from which most radiation is reflected, the effect of which decreases parabolically over the image. In addition, bi-directional effects that are inherent in optical properties of specific targets in the image (including both vegetation and water bodies) may be another source of illumination differences in an image. Correction for combined cross-track illumination effects typically include computing cross-track illumination differences by assuming an equal average illumination over the entire image (average radiance values computed in columns). A linear or polynomial function is fit for each channel through the radiance or reflectance values to simulate cross-track illumination differences. Applying this function to the data results in a cross-track correction that standardizes the reflectance values to one common image level.

Sensor and image noise calibration

In imaging spectrometry, signal is considered to be the signal strength measured by an imaging spectrometer, whereas noise describes the random variability of that signal. The quantification of the level of noise alone is not a very useful measure for the quality of a spectrometer since the effect of noise is higher when the signal strength is low. Therefore, a SNR expressed as the ratio of the signal's mean to its standard deviation is frequently used. Spectrometers produce both periodic (coherent) sensor noise that can be removed, and random noise that cannot be removed. The SNR is calculated on data sets with periodic noise removed. The remaining random noise can be additive noise, which is independent of the signal, and multiplicative noise which is proportional to signal strength. The major part of noise in spectrometer data is additive and decreases sharply with both an increase in wavelength and atmospheric absorption. The random noise component consists of random sensor noise (which is image independent), intra-pixel variability (resulting from spatially heterogeneous pixel contents), and inter-pixel variability.

Images can suffer from spatially coherent noise and/or spectrally coherent noise (Schowengerdt 1997). Spatial noise in images hampers visual interpretation, while spectral noise in images blurs absorption features (depth, width, symmetry) and hampers statistical analyses used to estimate Earth surface properties from spectra.

Noise in images can be attributed to the detector itself, to the system's electronics and to resonance caused by moving parts (mirrors) in the instrument. Sensor calibration noise (e.g. striping) includes:

- sensor radiometric resolution
- sensor dynamic range (saturation)
- sensor drift (change of sensitivity of the sensor over time)
- signal conversion (AC into DC, then voltage into DN).

Electronic vibrations and moving parts in sensor that may cause noise include:

- platform vibrations, especially aircraft engines
- irradiance variation (solar energy, aerosols, water vapor)
- atmospheric attenuation (scattering, absorption)
- atmospheric path radiance (multiple scattering).

Figure 16.5 shows different types of noise than can be seen in image data, including (1) local noise such as striping, dropped, or degraded (bad) lines due to transmission loss, sensor saturation, or electronic problems, (2) periodic/coherent noise showing repetitive patterns that may be caused by electronic interference or sensor calibration differences, (3) global noise and random variation in DN that result in a grainy image, and (4) white/local noise are salt and pepper noises.

Quantification of noise levels is done by sensor engineers and the image user community. Both groups have developed their own approach and jargon. While sensor manufacturers typically use noise equivalent radiance or reflectance (NER), the user community works with SNR. There are no easy ways to convert noise quantified using engineering concepts to noise quantified using operational use. Schläpfer and Schaepman (2002) tried to find a link between NER and noise from a user perspective.

Noise estimation; the engineering perspective

Sensor manufacturers work with NER. NER is measured as the input radiance to an optical sensor that produces an output SNR of 1. Alternatively a unit known as the Noise Equivalent Reflectance Difference is used which is the differential measure that indicates the ground reflectance difference producing an image contrast of 1. From a user's perspective, the disadvantages of using these units are that only noise that is generated by the instrument is considered (i.e., the noise is instrument specific) and that this is measured under controlled conditions in the laboratory (often in an integrating sphere) which is far from comparable to a field situation.

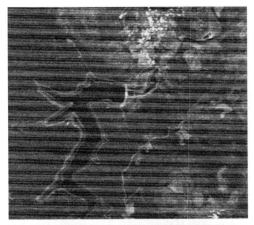
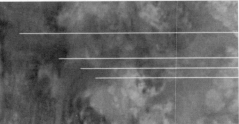
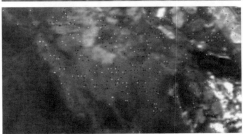

Figure 16.5   Three images suffering from noise. Local noise such as striping due to resonance is shown in the topmost image, the middle image has bad lines while the lower image shows distinct salt and pepper noises.

A concept that is also considered by engineers is referred to as sensor cross-talk or cross band leakage, which is an adjacency effect between channels. Cross-talk can occur if two adjacent bands have similar frequencies such that the channel will 'lock on' to the other's frequency or a harmonic thereof and energy is leaked from one channel into the other. When channels partly overlap, their spectral response function overlaps, energy may be recorded in both channels and in such a way cross-talk may occur. Finally, cross-talk also may occur between adjacent CCDs in an array.

Noise estimation; the user perspective

There is no consensus in the literature about the best method to assess signal versus noise. SNR estimates should be image specific, not sensor specific and typically are expressed as the ratio of average reflectance and standard deviation of dark and bright targets.

Signal improvement

Several techniques have been developed to increase the SNR or to improve the image signal including:

- removal of bad bands
- low pass filtering
- separation of signal and noise using principal component transforms minimizing noise components.

Removing bad lines is done by replacing the pixel values in these lines by the average or median of the adjacent pixels from adjacent scan lines. Low pass filters have proven to be quick, robust methods for noise reduction (Xu et al. 1994). Due to fact that the majority of these filters are based on smoothing, loss of information is inevitable).

A readily used alternate technique for noise reduction based on principal component analysis is the minimum noise fraction transform (MNF; Switzer 1980, Green et al. 1988) also referred to as noise-adjusted principal component (NAPC). This method aims at the isolation of noise, assuming that noise has no spatial coherence. Two assumptions prevail, namely (1) image noise is weakly autocorrelated compared to the autocorrelation of the image signal, and (2) correlation between pixels yield information for MNF rotation referred to as MAF: min/max autocorrelation factors (noise would yield a flat variogram: no spatial correlation). Similar transforms are also used for dimensionality reduction of hyperspectral data (Bruce et al. 2002).

The MNF transform (Lee et al. 1990) is essentially a two-way cascaded principal component transform. The forward rotation results in a new image cube in which ordering of signal and noise is done in the transformed bands. The user can interactively decide which factors are used for reversed transformation, thus how much noise (components/bands) is removed and how much is retained. The noise for the MNF transform can be estimated from a dark current image (image taken with the shutter of the sensor closed), dark parts of the image (e.g., lakes, shaded terrain), or the entire image itself.

To remove noise from image data, the MNF transformation can be used. The common approach is to apply low-pass filtering to the low-order MNF images which contain most of the low spatially

correlated noise and a degraded signal component and a removal of the high MNF factor images that contain nearly only noise. After filtering and removal, the cleaned MNF factor images are back-transformed to the original data space using the original transformation coefficients.

IN-ORBIT CALIBRATION AND TIME SERIES

Often it is observed that the channel response of a sensor degrades in orbit. Since most sensors have no onboard calibration device, the earlier described pre-launch calibration routines cannot be performed. Thus it is necessary to develop post-launch calibration procedures leading to post-launch coefficients that can be applied to correct for the in-orbit degradation of the sensor's signal. This is particularly of importance when dealing with time series of data products (e.g., BRDF, albedo, fPAR, etc.) that are input to climate and vegetation models. These often require estimation of physical variables to an accuracy of the order of 2–3%. Thus, in-orbit instrument calibration is essential to ensure time series of measurements.

Post-launch calibration typically is done through so-called vicarious calibration techniques based on radiometric calibration of a satellite sensor by a method independent of that used to perform the initial laboratory calibration. Two approaches are commonly taken when making vicarious measurements (Secker et al. 2001): data from the sensor is compared with radiance measured by a sensor mounted on an aircraft or radiative transfer modeling is used to estimate the top-of-atmosphere radiance from a ground target of known reflectance. This requires a ground calibration target that is stable meaning that the long-term mean value of the top-of-the-atmosphere albedo remains uniform in time and variations over a year are reproduced from one year to the next.

Some sensors (e.g., MERIS and the Sea-viewing Wide-Field-of-view Sensor (SeaWiFS)) are equipped with onboard diffuser panels that measure solar irradiance. These measurements can be used to monitor sensor signal degradation, but since the characteristics of the diffuser panel also changes over time, in-orbit vicarious calibration is still needed. An additional means of instrument in-orbit calibration is lunar calibration (e.g., SeaWiFS and MODIS), which is based on the assumption that the moon is an invariant diffuse reflector. By measuring radiance reflected from the moon, the temporal degradation and sensitivity of a sensor can be assessed.

Pre-launch and in-orbit vicarious calibration forms the basis of Earth observation and monitoring by deriving time series of validated geophysical products. Current Earth observation missions deliver not only physical products (e.g., Top-Of-Atmosphere, TOA, radiance and scaled surface reflectance) to the user, but in addition also provide numerous derived geophysical products. MODIS for example provides over 140 products including: surface reflectance, surface temperature, land cover, snow and ice estimates, vegetation indices, fire maps, photosynthetically active radiation (fPAR), Leaf Area Index (LAI), BRDF, albedo, etc. The ESA MERIS instrument provides Level 2 geophysical products based on TOA radiance including: normalized reflectance at surface, chlorophyll and other water constituent concentration (ocean), reflectance at surface, vegetation indices (land), aerosol type and optical thickness, water-vapor column content, cloud top height, optical thickness, and albedo (atmosphere).

Kokhanovsky et al. (2007) provide an overview of MERIS products while Xiong and Barnes (2006) discuss MODIS products.

CONCLUDING REMARKS

Pre-processing of Earth observation data is important as it is the basis for deriving calibrated and validated physical measurements of the Earth's surface. Current practices in pre-processing of optical sensor data are to some extent standardized, but nevertheless this is not a static field. Each new sensor that is built and each new instrument that is launched requires a modified approach to derive calibrated at-sensor radiance or scaled surface reflectance data. There is also a clear difference in user needs depending on the type of application the data should cater for. In water quality estimation, researchers deal with a low and diffuse signal. In order to be able to predict water quality to a reasonable accuracy, high precision reflectance measurements are required. Also in the area of atmospheric chemistry, soil chemistry, and biophysics, high accuracy is required in order to link spectral properties to surface or atmosphere constituents. On the other hand, when dealing with surface mineralogical and geological mapping one can often rely on relative reflectance measurements, for example, reflectance scaled against a calibration target rather than absolute reflectance from radiative transfer models, as the products are typically derived from statistical matching of absorption features. The level of accuracy reached often also depends on the ancillary data that can be obtained. Absolute modeling of atmospheric transmission requires detailed knowledge on atmosphere composition which is not always at hand.

Trial and error approaches using atmosphere models linked with ground data may yield satisfactory results in the absence of proper calibration data sets although these approaches do not build on proper scientific probing. Proper georeferencing requires high precision aircraft positioning and attitudinal response and a sound topographic model. Regrettably despite all the calibration efforts we are often faced with low SNR data sets. This is partly because we wish to achieve higher spectral resolution and narrower sampling which inevitably is achieved at the cost of the measured and available signal. In addition, low albedo targets and static and dynamic sources of resonance contribute to the inherent noise in data sets. Sophisticated techniques for improving the SNR by means of noise filtering and isolating noise components may be helpful, but starting with high quality data, thus acquiring data under ideal illumination conditions, can often help in reducing the need for all these subsequent efforts.

REFERENCES

Albert, A. and P. Gege, 2006. Inversion of irradiance and remote sensing reflectance in shallow water between 400 and 800 nm for calculations of water and bottom properties. *Applied Optics*, 45(10): 2331–2343.

Ben-Dor, E. and N. Levin, 2000. Determination of surface reflectance from raw hyperspectral data without simultaneous ground data measurements: a case study of the GER 63-channel sensor data acquired over Naan, Israel. *International Journal of Remote Sensing*, 21(10): 2053–2074.

Brazile, J., R. A. Neville, K. Staenz, D. Schläpfer, L. X. Sun, and K. I. Itten, 2006. Scene-based spectral response function shape discernibility for the APEX imaging spectrometer. *IEEE Geosciencs and Remote Sensing Letters*, 3(3): 414–418.

Bruce, L. M., C. H. Koger, and J. Li, 2002. Dimensionality reduction of hyperspectral data using discrete wavelet transform feature extraction. *IEEE Transactions on Geoscience and Remote Sensing*, 40(10): 2331–2338.

Carrere, V. and J. E. Conel, 1993. Recovery of atmospheric water vapour total column abundance from imaging spectrometer data around 940 nm – sensitivity analysis and application to Airborne Visible/Infrared Imaging Spectrometer (AVIRIS) data. *Remote Sensing of Environment*, 44: 179–204.

De Jong, S. and F. Van der Meer, 2004. *Remote Sensing Image Analysis: Including the Spatial Domain*. Springer-Kluwer Academic Publishers, Dordrecht, the Netherlands.

Frouin, R., P. Y. Deschamps, and P. Lecomte, 1990. Determination from space of atmospheric total water amounts by differential absorption near 940 nm: Theory and airborne verification. *Journal of Applied Meteorology*, 29: 448–460.

Gaddis, L. R., L. A. Soderblom, H. H. Kieffer, K. J. Becker, J. Torson, and K. Mullins, 1996. Decomposition of AVIRIS spectra: extraction of surface-reflectance, atmospheric and instrumental components. *IEEE Transactions on Geosciences and Remote Sensing*, 34: 163–178.

Gangopadhyay, P. K., B. Maathuis, and P. Van Dijk, 2005. ASTER-derived emissivity and coal-fire related surface temperature anomaly: a case study in Wuda, north China. *International Journal of Remote Sensing*, 26(24): 5555–5571.

Gao, B. C., K. B. Heidebrecht, and A. F. H. Goetz, 1993. Derivation of scaled surface reflectance from AVIRIS data. *Remote Sensing of Environment*, 44: 165–178.

Gianinetto, M. and G. Lechi, 2006. A new methodology for in-flight radiometric calibration of the MIVIS imaging sensor. *Annals of Geophysics*, 49(1): 65–70.

Green, A. A., M. Berman, P. Switzer, and M. D. Craig, 1988. A transformation for ordering multispectral data in terms of image quality with implications for noise removal. *IEEE Transactions on Geoscience and Remote Sensing*, 26: 65–74.

Gu, Y. and J. M. Anderson, 2003. Geometric processing of hyperspectral image data acquired by VIFIS on board light aircraft. *International Journal of Remote Sensing*, 24(23): 4681–4698.

Guanter, L., V. Estelles, and J. Moreno, 2007. Spectral calibration and atmospheric correction of ultra-fine spectral and spatial resolution remote sensing data. Application to CASI-1500 data. *Remote Sensing of Environment*, 109(1): 54–65.

Herman, B., A. J. LaRocca, and R .E. Turner, 1993. Atmospheric scattering. In: W. L. Wolfe and G. J. Zissis (eds.), *The Infrared Handbook*. Environmental Research Institute of Michigan (ERIM), Ann Arbor, Michigan, pp. 4–1 to 4–76.

Hook, S. J., R. G. Vaughan, H. Tonooka, and S. G. Schladow, 2007. Absolute radiometric in-flight validation of mid infrared and thermal infrared data from ASTER and MODIS on the terra spacecraft using the Lake Tahoe, CA/NV, USA, automated validation site. *IEEE Transactions on Geosciences and Remote Sensing*, 45(6): 1798–1807.

Kerekes, J. P., in this volume. Optical Sensor Technology. Chapter 7.

Khurshid, K. S., K. Staenz, L. X. Sun, R. Neville, H. P. White, A. Bannari, C. M. Champagne, and R. Hitchcock, 2006. Preprocessing of EO-1 hyperion data. *Canadian Journal of Remote Sensing*, 32(2): 84–97.

Kokhanovsky, A. A., K. Bramstedt, von W. Hoyningen-Huene, and J. P. Burrows, 2007. The intercomparison of top-of-atmosphere reflectivity measured by MERIS and SCIAMACHY in the spectral range of 443–865 nm. *IEEE Geoscience and Remote Sensing Letters*, 4(2): 293–296.

LaRocca, A. J., 1993. Atmospheric absorption. In: W. L. Wolfe and G. J. Zissis (eds), *The Infrared Handbook*. Environmental Research Institute of Michigan (ERIM), Ann Arbor, Michigan, pp. 5–1 to 5–132.

Lee, J. B., S. Woodhyatt, and M. Berman, 1990. Enhancement of high spectral resolution remote-sensing data by a noise-adjusted principal components transform. *IEEE Transactions on Geoscience and Remote Sensing*, 28: 295–304.

Liang, S., in this volume. Quantitative Models and Inversion in Optical Remote Sensing. Chapter 20.

Liang, S., H. FallahAdl, S. Kalluri, J. JaJa, Y. J. Kaufman, and J. R. G. Townshend, 1997. An operational atmospheric correction algorithm for Landsat Thematic Mapper imagery over the land. *Journal of Geophysical Research*, 102: 17173–17186.

Martonchik, J. V., C. J. Bruegge, and A. Strahler, 2000. A review of reflectance nomenclature used in remote sensing. *Remote Sensing Reviews*, 19: 9–20.

Moore, G. F., J. Aiken, and S. J. Lavender, 1999. The atmospheric correction of water colour and the quantitative retrieval of suspended particulate matter in Case II waters: application to MERIS. *International Journal of Remote Sensing*, 20: 1713–1733.

Morel, A. and B. Gentili, 1993. Diffuse reflectance of oceanic waters. II. Bidirectional aspects. *Applied Optics*, 32: 6864–6879.

Olbert, C., 1998. Atmospheric correction for CASI data using an atmospheric radiative transfer model. *Canadian Journal of Remote Sensing*, 24: 114–127.

Richter, R., 1996. Atmopsheric correction of DAIS hyperspectral image data. *Computers and Geosciences*, 22: 785–793.

Richter, R., 1997. Correction of atmospheric and topographic effects for high spatial resolution satellite imagery. *International Journal of Remote Sensing*, 18: 1099–1111.

Richter, R., 1998. Correction of satellite imagery over mountainous terrain. *Applied Optics*, 37: 4004–4015.

Richter, R. and D. Schläpfer, 2002. Geo-atmospheric processing of airborne imaging spectrometry data. Part 2: atmospheric/topographic correction. *International Journal of Remote Sensing*, 23(13): 2631–2649.

Richter, R., A. Muller, and U. Heiden, 2002. Aspects of operational atmospheric correction of hyperspectral imagery. *International Journal of Remote Sensing*, 23(1): 145–157.

Richter, R., D. Schläpfer, and A. Muller, 2006. An automatic atmospheric correction algorithm for visible/NIR imagery. *International Journal of Remote Sensing*, 27(9–10): 2077–2085.

Schaepman, M. E., in this volume. Imaging Spectrometers. Chapter 12.

Schaepman-Strub, G., M. E. Schaepman, J. V. Martonchik, T. H. Painter, and S. Dangel, in this volume. Radiometry and Reflectance: From Terminology Concepts to Measured Quantities. Chapter 15.

Schaepman-Strub. G., M. E. Schaepman, T. H. Painter, S. Dangel, and J. V. Martonchik, 2006. Reflectance quantities in optical remote sensing-definitions and case studies. *Remote Sensing of Environment*, 103(1): 27–42.

Schläpfer, D. and R. Richter, 2002. Geo-atmospheric processing of airborne imaging spectrometry data. Part 1: parametric orthorectification. *International Journal of Remote Sensing*, 23(13): 2609–2630.

Schläpfer, D. and M. Schaepman, 2002. Modeling the noise equivalent radiance requirements of imaging spectrometers based on scientific applications. *Applied Optics*, 41(27): 5691–5701.

Schott, J. R., C. Salvaggio., and W. J. Volchok, 1988. Radiometric scene normalization using pseudoinvariant features. *Remote Sensing of Environment*, 26(1): 1–16.

Schowengerdt, R. A., 1997. *Remote Sensing. Models and Methods for Image Processing*. Second edition, Academic Press, Chestnut Hill, MA.

Secker, J., K. Staenz, R. P. Gauthier, and P. Budkewitsch, 2001. Vicarious calibration of airborne hyperspectral sensors in operational environments. *Remote Sensing of Environment*, 76(1): 81–92.

Sobrino, J. A., J. C. Jimenez-Munoz, P. J. Zarco-Tejada, G. Sepulcre-Canto, and E. De Miguel, 2006. Land surface temperature derived from airborne hyperspectral scanner thermal infrared data. *Remote Sensing of Environment*, 102(1–2): 99–115.

Switzer, P., 1980. Extension of discriminant analysis for statistical classification of remotely sensed satellite imagery. *Journal of the International Association for Mathematical Geology*, 12: 367–376.

Van den Bosch, J. and R. Alley, 1990. Application of LOWTRAN 7 as an atmospheric correction to Airborne Visible/InfraRed Imaging Spectrometer (AVIRIS) data. In: R. O. Green (ed.), *Proceedings Second AVIRIS Workshop.*

JPL Publication 90–54, Jet Propulsion Lab., Pasadena, California, US., pp. 78–81.

Van der Meer, F. and S. De Jong, 2001. *Imaging Spectrometry: Basic Principles and Prospective Applications.* Springer-Kluwer Academic Publishers, Dordrecht, the Netherlands.

Xiong, X. X. and W. Barnes, 2006. An overview of MODIS radiometric calibration and characterization. *Advances in Atmospheric Sciences*, 23(1): 69–79.

Xu, Y., J. Weaver, D. Healy, and J. Lu, 1994. Wavelet transform domain filters: a spatially selective noise filtration technique. *IEEE Transactions on Image Processing*, 3(6): 747–758.

Surface Reference Data Collection

Chris J. Johannsen and
Craig S. T. Daughtry

Keywords: ground truth, ancillary data, *in situ* measurements, calibration, sampling design, Global Positioning Systems (GPS), surface references, ground reference.

INTRODUCTION

Applications of remote sensing must define relationships between image data and conditions at corresponding points on the ground. Unfortunately, many published remote sensing reports focus on image processing techniques with little detail regarding the methods used for collecting ground truth data (McCoy 2005). In this chapter, we summarize components that should be considered for planning and acquiring surface reference data for a remote sensing project. The specific surface characteristics that may be measured are quite diverse and vary by application and by analysis methodology. Our discussion will focus on example applications in agriculture, but the general qualities of the reference data are extensible to most endeavors that involve remotely sensed observations.

Before a researcher can apply any of these methods, he must determine the number of samples required to be reasonably confident of detecting specific differences in the surface features measured. Therefore, statistical determinations of the number of needed samples are also briefly discussed.

Definition of surface reference data

'Ground truth data', which can be more accurately described as surface reference data, consist of observations collected at or near the surface to support analysis of remote sensing data. Accurate reference data permit analysts to match points or areas in imagery to corresponding regions on the earth's surface and to establish relationships between the image and conditions on the surface (Campbell 2007).

Through the years 'ground truth' has meant many things depending on the method used to collect the data, the experience of the person collecting the data, the purpose for the collection of the remote sensing data, and the time interval between when the remote sensing data were collected and when the ground truth data were collected. As McCoy (2005) explained ground truth has been replaced by 'reference information' to be more inclusive than 'ground' and less absolute than 'truth'. Therefore the term 'surface reference data' is intended to mean any data or information collected to support the analysis and interpretation of remote sensing data obtained for studying air, land, and water resources.

Surface reference data can be as simple as general soils maps and meteorological data for interpreting AVHRR classification results (Justice and Townshend 1981). Surface reference data can also be as complex as diurnal plant temperatures which are a function of many factors including the evaporative demands of the atmosphere, the water holding capacity of the soil, the rooting depth, and available soil moisture (Pinter et al. 2003). An important consideration in planning is that surface reference data may be either time-critical or time stable (Lillesand et al. 2004). Thus, a decades-old soil map may be adequate for assessing the AVHRR classification results, but plant temperatures should coincide within minutes of the remotely sensed thermal imagery.

Purposes of surface reference data

Reference data collection should be an integral part of the analytical plan. In one of the most thorough publications devoted to procedures for gathering surface reference data, Joyce (1978) outlined procedures for systematic collections of data for land cover classification using multispectral satellite imagery. Surface reference data must be fully compatible with the resolution of the sensor, the spatial scale of the landscape, and the kinds of digital analyses to be employed. Criteria for determining size, number, uniformity, and predominant land cover of sample sites were outlined.

Surface reference data generally serve one of three purposes (Campbell 2007):

- Guide the analysis process by providing training sites for supervised classifications.
- Assess and evaluate the accuracy of the results of the remote sensing analysis.
- Characterize and model the spectral behavior of radiation within the scene.

Surface reference data minimally includes attributes or measurements that describe surface conditions at a specific location and time. Additional information for reference data archives often include weather records, instruments used and their calibrations, descriptions of sampling and analysis methods, description of experiment, and people involved (Biehl and Robinson 1983, McCoy 2005, Carter et al. 2008).

If possible, it is almost always advantageous for the analyst to be involved in the collection of the reference data. This provides the analyst with additional information that may not be recorded such as a grass waterway, an old fence row or an eroded spot through the center of the training sample site. In other words, the analyst becomes more aware of anomalies within the study area.

Carter et al. (2008) provided a classification system for identification for anomalies in agricultural crops that included details for recognizing anomalies caused by weather, nutrients, drainage, weeds, insects, disease, crop residue, chemicals, mechanical, animals, and management conditions. Thus familiarity with the land condition whether in agriculture, forestry, urban, or other land uses is extremely important for selecting representative sample sites.

A basic understanding of the relationship between biophysical variables and spectral responses of materials in the study area (e.g., vegetation, soil, rock, water, asphalt, and concrete) is essential for selecting which variables to observe and measure in the field. In addition, knowledge of what factors cause variations in the basic spectral responses will provide crucial insights. For example, how do variations in surface soil moisture affect the reflectance of vegetation throughout the growing season?

Clear, detailed objectives are the foundation for the project and should determine appropriate methods for each stage of the work. The expected final product should drive the field data acquisition (McCoy 2005). Numerous methods of acquiring surface reference data have been developed that vary greatly in their precision, accuracy, and difficulty of performance. The method of choice depends largely on the feature to be measured, the accuracy required, the amount of material to be sampled, and amount of time and equipment available. Researchers must balance competing demands to acquire the appropriate data with the realities of the time and budget constraints of the project.

TYPES OF SURFACE REFERENCE DATA

Surface reference data may be broadly classified as discrete or continuous. Discrete data are qualitative descriptions or nominal designations that convey basic land-use and land-cover differences among regions in the imagery, examples include water, urban, forest, rangeland, cropland, and wetland. Accurate and precise nominal labels concisely communicate significant information. Continuous data or quantitative measurements of the physical and biological characteristics of surface features complement nominal labels with more specific data to document the precise meaning of the labels. Biophysical data may include slope, soil organic matter content, vegetation biomass, leaf area index, leaf chlorophyll content, and leaf angles.

All field observations and measurements must include a means of determining reliable locations for each sample site. Before GPS, land surveying

Figure 17.1   Researchers use GPS, maps, and remotely sensed images on a notebook computer to locate and annotate conditions of crops.

techniques using maps, compasses, and land features were used to locate sample sites. Today, with a relatively inexpensive GPS receiver, a researcher can determine coordinates of sample sites in the field or navigate to a point on the ground with coordinates derived from a map or geo-referenced image (Figure 17.1). Although all GPS coordinates contain errors, a researcher can often mitigate common GPS errors resulting from multi-path reflections, electrical interference, poor satellite geometry, and obstructions (McCoy 2005). For more information on GPS, numerous excellent texts (e.g., French 1999) and on-line tutorials (e.g., www.trimble.com/gps/; www8.garmin.com/aboutGPS/) are available.

In this section, we will focus on discrete observations and quantitative measurements of the biophysical characteristics of soils and vegetation that affect the spectral reflectance of terrestrial scenes. The optical properties and the geometric arrangement of each component in the scene determine its contribution to the overall scene reflectance. Other major factors influencing scene reflectance are discussed in the Field Spectrometry section.

Discrete observations

Other researchers/workers more readily accept established classification systems that provide precise class descriptions and define the relationships between classes than *ad hoc* classifications. The USGS Land-Use and Land-Cover Classification System is a hierarchical design that lends itself to use images of varied scales and resolutions (Anderson et al. 1976). Level I categories are designed for broad-scale, coarse-resolution imagery and include urban land, agricultural land, range land, forest land, and water. For Level II, more detailed classes within each Level I category are identified such as croplands and pastures, orchards, confined feeding operations, within the category of agricultural land. More detailed classes at Level III and below may be defined by the analyst.

Other efforts to standardize the collection of surface reference data include the multi-agency Federal Geographic Data Committee (FGDC) document, 'Existing Vegetation Classification and Mapping Technical Guide' which describes a protocol to produce a consistent classification of vegetation across the National Forest System. This document can be located at the US Forest Service website: http://www.fs.fed.us/emc/rig/. McCoy (2005) also provides a listing of metadata resources related to standards.

Descriptive reports, usually collected for a specific purpose, are available from government agencies, extension services, professional organization, and private companies. For example, USDA National Agricultural Statistics Service provides reports that include the area planted and production estimates of major crops (http://www.nass.usda.gov/).

Many types of maps, e.g., topographic, hydrologic, highway, and land ownership, can be used as reference data for interpreting imagery.

Although all maps contain errors, some maps such as USGS topographic maps conform to a set of national mapping standards. The scale of a map is not a good indication of the pixel size for all features found on the map because some features are added for improving the map appearance such as a stream or river that may have its width or flow path change depending on the season or year (Monmonier 1991).

Important reference information can be interpreted from other sources. For example, slope and aspect, which impact the vegetation type and quantity, can be interpreted from digital topographic maps. Sanchez-Flores and Yool (2004) demonstrated that key vegetation parameters of a fire management system, such as fine woody fuels, could be predicted by vegetation type and slope while coarse woody fuels could be predicted by elevations.

Nominal data are often easy to collect for small areas, but can be challenging for larger areas. It is often convenient to annotate maps or aerial photos to relate point observations to areal units. Sketches, notes, photographs, and GPS coordinates are also valuable for documenting the inherent variability within nominal classes. Because seasonal changes may affect physical characteristics of some classes, the timing of surface reference data collections must be coordinated with the acquisition of the remote sensing imagery. Observations should be collected using standardized procedures tailored for each project to ensure that uniform data are collected at each site. McCoy (2005) included a compilation of suggested forms for taking field notes. Much of data collection involves good visual skills by the person who is collecting the data. These skills can be obtained by observing others during their data collection but also by practice. Training is crucial for consistency, particularly if more than one observer is involved.

Unfortunately, one often realizes that additional observations are needed during the analysis process. This may be too late for time-critical reference data, but not necessarily for time-stable data. Global Position Systems (GPS) have greatly improved our abilities to collect data at specific locations and to return to the location as needed (Parsons et al. 1995). With a tablet PC and GPS receiver, we can find specific location on a digital image with great accuracy and record updated information (Figure 17.1).

Biogeophysical properties of soils

The spectral response of soils is significantly influenced by water, organic matter, iron oxides, surface roughness, soil texture, and types of clays (Stoner and Baumgardner 1981, Irons et al. 1989, Campbell, in this volume, and van Leeuwen, in

this volume). Most of these soil properties are relatively static and change slowly. Detailed soil maps contain much information about soil properties throughout the soil profile and are useful for selecting sampling sites. However, soil moisture and surface roughness are dynamic properties and are easily altered by weather and tillage. The presence of crop residues further complicates the remote sensing of soil characteristics as viewed from above.

Soil color is a differentiating characteristic for many classes in all modern soil classification systems and is an essential part of the definitions for both surface and subsurface diagnostic horizons (Baumgardner et al. 1985). Soil color is most commonly determined by a human observer making a visual comparison between a soil sample and color chips arranged according to hue, value, and chroma in the Munsell Soil Color Charts (ASC Scientific 2007) The Munsell soil color notations are widely used throughout the world and provide the standard by which accurate information is communicated about soil color.

Soil moisture is an important factor affecting many environmental processes including plant growth, soil erosion, soil biogeochemistry, and the exchange of heat and water between the land and the atmosphere. Accurate and timely estimates of soil moisture are required to understand and model these processes. However, soil water content is highly variable in time and space as a result of heterogeneous soils and variable climate conditions. Landscape position, slope, soil properties, and surface roughness affect surface and subsurface water movement from up-slope sites and to down-slope sites. Subsurface restricting layers also preferentially funnel water flow processes.

Water strongly affects soil dielectric properties and numerous remote sensing approaches have focused on microwave wavelengths for estimating soil moisture (Ulaby et al. 1996). Water also greatly influences the reflectance of soils in the 400 to 2500 nm wavelength region (Hoffer and Johannsen 1969, Baumgardner et al. 1985). Reflectance spectra of moist soils also include prominent absorption bands centered near 1450 and 1940 nm and weaker absorptions bands near 970, 1200, and 1770 nm (Irons et al. 1989). The overall decrease in soil reflectance upon wetting is a result of water filling the pore spaces in the soil and changing the index of refraction. However, quantification of soil moisture using these wavelengths is difficult because of variability from other chemical and physical properties (Lobell and Asner 2002).

Water in the soil is measured as a function of a volume of the bulk soil and may be expressed on mass basis (g H_2O/g soil) or on a volume basis (cm^3 H_2O/cm^3 soil). The frequently used terms, field capacity, permanent wilting point, and plant available soil water, are conceptual and the actual soil moisture values vary as functions of soil texture and structure. Thus, soil water content must be defined in specific terms. A variety of quantitative and qualitative methods are available for measuring soil moisture. The gravimetric method which involves collecting, weighing, drying, and reweighing soil samples, is the direct, absolute technique to estimating total water content of soils. Other techniques estimate soil water indirectly by measuring neutron scattering, dielectric properties, or electrical resistance (Gardner 1986). The method and equipment selected will depend on applicability to local soil conditions, ease of use, cost of equipment, and desire to monitor real-time changes in soil moisture. Soil moisture measurements should be taken at a minimum of three representative sample locations at each monitoring site. Sample locations should have similar soil type, slope, aspect, elevation, and vegetation. For regional watershed scales, synthetic aperture radar (SAR) measurements can provide surface soil moisture information with reasonable accuracy, but variations in soil surface roughness and vegetation biomass contributed significant errors at local scales (Moran et al. 2004).

Although there are many methods for measuring chemical, physical, and biological properties of soils (Klute et al. 1986, Sparks et al. 1996), most soil properties are time-consuming and complex to characterize especially for large areas. Robust surrogates are required which can be unambiguously interpreted. A range of such surrogates and their applications at different spatial scales was discussed by Hamblin (1991) who recommended that researchers use surrogates with caution. Campbell (in this volume) and van Leeuwen (in this volume) discuss the spectral reflectance of soils and their biophysical properties.

Soil survey maps provide a wealth of information on soil properties. Map units are influenced by the properties of the soil profile that may not be evident in reflectance images and may have inclusions of other minor soil types that may have very different soil properties. Because inherent fertility, drainage, and moisture holding capacity differences among soils influence vegetation growth, phenology, and morphology, soil patterns can often be discerned even when soils are totally or partially obscured by green or senesced vegetation (Baumgardner et al. 1985). The accurate identification of vegetation cover is especially important for mapping soils in non-cultivated vegetation areas.

Biophysical properties of vegetation

Classical growth analysis studies and radiation use efficiency by plant canopies require frequent measurements of the amount and organization of above ground vegetation. The location and orientation of each piece of foliage is also a key input to models of radiation scattering characteristics of canopies at both optical and microwave wavelengths. Numerous methods of measuring vegetation characteristics have been developed that vary greatly in their precision, accuracy, and difficulty of performance. The method of choice depends largely on the (a) morphological features of the vegetation to be measured, (b) accuracy required, (c) amount of vegetation to be sampled, and (d) amount of time and equipment available (Daughtry 1990).

Frequency, density, biomass, cover, and leaf area index are broad categories of biophysical measurements typically used to describe the quantity of terrestrial vegetation. For remote sensing studies, the condition (health and vigor) of the vegetation and the orientation of the foliage are also important descriptors. In this section, we will briefly describe each term and cite sources for more information. Because many of these biophysical characteristics vary diurnally and seasonally, the time of their measurements must be coordinated with the image acquisitions.

Frequency

Frequency is the probability of finding a certain species when a particular size of sampling frame is randomly located with study area. It is easiest of all for quantitative vegetation measurements, but it is also the most difficult characteristic to interpret. While frequency is very efficient for detecting change in vegetation, additional measurements are needed to provide a complete analysis of the nature of the change. Bonham (1989) provides guidance for selecting appropriate sample sizes based on vegetation characteristics. For example, the presence or absence of leafy spurge (*Euphorbia esula* L.) was used to assess the accuracy of detecting the noxious weed with remotely sensed images (Parker-Williams and Hunt 2004).

Density

Total number of individual plants per unit area is density. Although simple in concept, the identification of an individual plant is often one of the greatest difficulties in determining plant density. This is easy for trees and other single stemmed species, but challenging for perennial species that spread vegetatively and may have multiple stems with interconnected root systems. Clearly, an understanding of the morphology and ecology of the species of interest is necessary for accurate estimates of plant density. The contribution of a species to total biomass will vary from year to year as a result of differing environmental conditions. Estimates of density are useful for monitoring plant responses to natural and man-made perturbations of the environment.

The primary techniques to estimate density may be classified as either plot or distance techniques (Bonham 1989). In the plot technique, the number of individuals within a specific area (plot) is counted to estimate density. Important considerations are size and shape of the plot and the number of observations needed to estimate density adequately. The distance or plotless technique is based on the concept that the number of plants per unit area can be estimated from the average distance between two plants.

Plant density for agricultural row crops is often expressed as plant population (plants/m^2 or plants/hectare). Row spacing and mean distance between plants in a row determine plant population. Farmers plant crops at populations that will maximize their yields. For non-tillering crops (e.g., corn), yields are reduced when plant populations are outside a relatively narrow optimum range. For crops that can compensate by branching or tillering (e.g., soybean, wheat, and cotton), yields are less sensitive to plant populations. Row spacing significantly affects when the crop effectively covers the soil and intercepts maximum sunlight for photosynthesis. Plant populations are often higher in irrigated areas than in dryland or non-irrigated areas.

Biomass

Above-ground biomass or primary production is one of the best indicators of species importance within a plant community. Direct measurement of vegetation biomass typically involves clipping and weighing plants. Measurements of below-ground biomass are challenging and will not be discussed. Although mass may be determined on fresh or dried plants, dry mass is more reliable. Accurate measurements of fresh mass are difficult to obtain because water loss may occur rapidly when plants are harvested. For large plants, it may be advantageous to double sample, i.e., harvest and determine the fresh mass of a large volume of plant material and then take small subsamples to determine the proportion of dry matter (Daughtry 1990). Dry mass for the whole sample can be calculated from the fresh mass of the large sample and the proportion of dry matter in the subsamples. Double sampling is useful for trees and shrubs or when oven space for drying the samples is limited. Plant samples should be dried at 60–80°C to constant mass rather than for a fixed time. High temperatures may cause loss of volatile compounds and transformations of non-structural carbohydrates. Representative sampling is critical to minimize errors. A brush shredder or forage chopper may be used to chop and mix plant parts prior to subsampling.

Indirect or nondestructive measurements of biomass eliminate or reduce clipping and can be used for repeated measurements of the same area. These allometric techniques are well-developed in forestry and ecological sciences. Typically they involve the use of empirical relationships between a variable that is difficult to measure (e.g., total tree biomass) and a more easily measured variable (e.g., diameter at breast height, dbh) that is highly correlated to tree biomass (Bonham 1989). Other major components of tree canopies can also be estimated with simple allometry. The underlying concept is that a given amount of xylem or sapwood can physiologically and mechanically support only a certain amount of foliage. Thus foliage area and mass can be estimated if one measures the cross-sectional area of the sapwood (Long et al. 1981). Although these equations must be calibrated for each species, there seems to be consistency in the coefficients (Whittaker and Marks 1975, Causton 1985, Bonham 1989).

Crop yield monitors and GPS have provided farmers with unprecedented information about the spatial variability within their fields. Because yield integrates the cumulative effects of weather conditions and management practices over the growing season, attempts to link yield monitor maps directly with soil properties and remotely sensed images have had mixed success (Taylor et al. 1997, Basso et al. 2001, Dobermann and Ping 2004). Approaches that incorporate multi-temporal remotely sensed data into agrometeorological crop growth models are generally robust (Moran et al. 1997). Within-field variability of yields is a function of complex interactions involving local weather, plant genetics, nutrient status, and soil water dynamics. Horizontal subsurface water flow pathways also exist in some soils that redistribute water and have beneficial impacts on corn grain yields during drought years (Gish et al. 2005). In some cases, timely remotely sensed images can identify these subsurface flow pathways indirectly through their effects on crop growth and yields.

Cover

There are many subtle variations in the definition of cover including: (a) vegetation cover – total cover of vegetation on an area, (b) ground cover – cover by plants, litter, and rocks within a vegetation type, (c) range plant cover – cover of all plants available to livestock, (d) habitat cover – cover of vegetation to protect wildlife, (e) crown cover – the canopy of trees including leaves and branches, and (f) crop residue cover – cover by plant residues remaining on the soil after harvest For each definition, cover is the vertical projection of vegetation onto the ground. Cover is less than the total leaf area because many leaves overlap each other.

Methods of measuring terrestrial vegetation cover may be classified into nine basic categories (Bonham 1989). In general, the more reliable the method, the more impractical its use in assessing many plots and fields. Only techniques in the intercept and photographic categories are appropriate for direct measurements of cover in agricultural fields, rangelands, and forests.

The intercept techniques typically determine the presence or absence of vegetation at a finite number of points that may be regularly-spaced, randomly-located, or clustered. Although many significant variations of the intercept technique have been described (Bonham 1989, Morrison et al. 1995), all rely on visual interpretation by a human observer. The standard technique used by USDA Natural Resource Conservation Service (NRCS) to quantify crop residue cover is the line-point transect (Morrison et al. 1993). Accuracy of the line-point transect depends on the length of the line, the number of points used, and the skill of the observers.

Photographic methods consist of taking single vertical photographs or vertical stereographic pairs of photographs and estimating the fraction of the soil covered by vegetation from the photographs. Digital images can be classified into soil and vegetation classes using objective procedures. However, significant classification errors occur when the spectral differences between classes (vegetation and soil or litter) are not sufficient for discrimination.

Leaf area

Leaf area index (LAI) is the area of one side of green leaves per unit area of ground (Ross 1981). This definition implies that a leaf receives light mainly from one direction and is appropriate for broadleaf plants and grasses; however for conifers and plants with spirally twisted or cylindrical leaves total area of the assimilatory surface may be appropriate. In any case, LAI is a fundamental parameter of vegetation canopies that influences both photosynthesis and evapotranspiration.

Direct methods of measuring leaf area may be classified into five categories (Daughtry 1990). Scanning optical planimeters that measure the area of leaves as they move through the instrument on transparent conveyor belts have greatly improved the speed and accuracy of measuring leaf area compared to other methods. As with biomass sampling, double sampling may be advantageous. A large sample of leaves is used to estimate total leaf mass per unit of area of soil, which has a relatively high coefficient of variation (CV), and a subsample of leaves is used to estimate specific leaf area (SLA = leaf area/leaf mass), which has a low coefficient of variation (Daughtry and Hollinger 1984). Thus resources are focused on the source

of largest variance. The question that a researcher must address is whether the gain in efficiency sufficiently offsets the increase in overall error associated with double sampling.

Indirect methods of estimating LAI generally involve measuring radiation inside and outside the canopy and inverting gap fraction data for estimating canopy structure (Figure 17.2). The gap fraction of a canopy is fraction of view in some direction from beneath the canopy that is not blocked by foliage. The fractional area of sun flecks is equivalent to gap fraction in the solar direction. Gap fraction information for a range of angles contains much structural information for continuous canopies and can be adapted for row crops and individual trees. Direct and indirect methods of measuring LAI are generally equivalent within experimental errors. Welles (1990) reviewed commercially-available instruments for estimating LAI using gap fraction data, identified their strengths and limitations, and concluded that the best method for a particular application should be based on constraints imposed by need

Figure 17.2   Researchers use a Plant Canopy Analyzer (LAI-2000, LI-COR, Inc, Lincoln, NE) to measure incoming radiation above the plant canopy. When these measurements are combined with measurements beneath the canopy, leaf area index can be calculated using the gap fraction method. (Photo by Scott Bauer, USDA-ARS, K9534-1, available at http://www.ars.usda.gov/is/graphics/photos/aug01/k9534-1.htm)

for reference readings and the frequency of suitable sky conditions.

Canopy architecture

The radiation scattering properties of vegetation depend not only on the quantity of vegetation but also on the location, inclination, and orientation of the foliage elements. Six variables are required to describe the location (x, y, z) and direction (slope and aspect) with time (t) of each piece of foliage in the canopy. This is a formidable task when one considers that the most reliable method of measuring foliage location and directions uses a protractor, a compass, and a ruler (Ross 1981). Alternative methods of measuring canopy structure generally involve gap fraction data (Welles 1990) or specialized mechanical devices (Lang 1990).

The orientation of crop rows has an impact on the proportion of sunlit and shaded leaves and sunlit and shaded soil throughout the day. The proportion of sunlit soil is greatest when the sun shines down (parallel to) the rows and is least when the sun shines across (perpendicular to) the rows. The interaction of sun angle and crop row direction caused much greater changes in scene reflectance for visible wavelengths than for near infrared wavelengths (Kollenkark et al. 1982b). Thus, information on row spacing and row direction should be recorded and may affect the selection of training samples (Hoffer and Johannsen 1969).

Condition

Development stage, nutrient availability, water status, and weed, disease, and insect infestations determine the health and vigor of vegetation which affect the optical and thermal properties of plant canopies (Pinter et al. 2003). Knipling (1970) and Bauer (1975) provided seminal reviews of physical and physiological basis of remote sensing of vegetation. The underlying premise for assessing vegetation condition is that the spectral reflectance and temperature of vegetation are related to photosynthesis and evapotranspiration, two key physiological processes of all vegetation.

The development of annual plants is divided into vegetative and reproductive phases with subdivisions that represent the progress toward physiological maturity. Management decisions are often based on crop development stage. Illustrated development stage scales are available for most major crops including corn (Ritchie 2005), soybeans (Pedersen 2004), and wheat (Simmons et al. 1995). Crop development stages also influence the spectral reflectance of crops (Kollenkark et al. 1982a, Pinter et al. 2003).

Leaf pigments, including chlorophyll, carotene, and xanthophyll, dominate the spectral properties of leaves in the visible (400–700 nm) wavelengths.

Chlorophyll absorbs strongly in the blue and red regions and generally masks the other pigments. As leaves senesce or experience environmental stresses (e.g., nutrients or water), chlorophyll content will decrease and reflectance in blue and red wavelengths will increase. Because of the close link between leaf chlorophyll and leaf N concentration, chlorophyll meters (e.g., SPAD-502, Minolta Osaka Co., Japan) have been used to monitor relative plant N status. In-field reference areas, where N is not limiting, are required to normalize differences among cultivars, growth stages, and environmental conditions (Blackmer and Schepers 1995). Remote sensing techniques, based on measuring the reflected radiation from plant canopies, have the potential of evaluating the spatial variability of plant N status in large fields. However the relatively subtle differences in canopy reflectance associated with changes in leaf chlorophyll are often confounded with major changes in plant growth and development due to N availability (Daughtry et al. 2000). Other nutrient deficiencies also cause changes in leaf and canopy reflectance, but can be difficult to detect against variations plant growth across fields (Pinter et al. 2003).

Field spectrometery

Field spectroscopy is the *in situ* measurement of the spectral properties of surface features in the environment (McCoy 2005). It provides a bridge between the macro-scale observations made by aircraft and satellite sensors and the micro-scale observations made in the laboratory (Bauer et al. 1986). Often the relationships between spectral properties and biophysical characteristics can best be studied in controlled experiments where complete data describing the condition of the targets are attainable and where frequent, timely spectral measurements can be made. In other cases, the natural variability of surface features must be sampled.

Knowledge and understanding of the interactions of solar radiation with various surface features has led to detailed radiative transfer models that can often be inverted to predict key biophysical characteristics from reflectance measurements. For example, canopy reflectance can be simulated with the SAIL (Scattering by Arbitrarily Inclined Leaves) model as a function of leaf area index, leaf angle distribution, leaf optical properties, soil background reflectance, view angles, and solar angles. The inverse process can estimate biophysical features from the reflectance data (Goel 1989). A version of the SAIL model, WinSAIL, can be downloaded at http://www.ars.usda.gov/Services/Services.htm?modecode=12-65-06-00.

The spectral properties of surface features are usually expressed as bidirectional reflectance factor (BRF) which is the ratio of radiant flux reflected by a sample to that reflected by an ideal, perfectly diffuse reference surface irradiated in exactly the same way as the sample (Robinson and Biehl 1979, Schaepman-Strub et al. in this volume). Corrections for the actual reflectance of the reference surface as a function of illumination and view angles must be applied (Jackson et al. 1992). The most commonly used reference surface is Spectralon (Labsphere, North Sutton, NH), a sintered polytetrafluoroethylene-based material that is durable, washable, and well suited for reference panels. The two directions required for BRF are the source incident direction (solar zenith θ_s and azimuth angles φ_s) and the sensor view direction (view zenith θ_v and azimuth angles φ_v).

Key assumptions and protocols for accurate measurements of BRF in the field are provided by Robinson and Biehl (1979), Milton (1987), and McCoy (2005). Measurement of BRF requires observations of the reference panel and observations of the target in the same illumination conditions and view angles. The most accurate method is to use the same instrument for both observations. When this is not possible, one instrument can observe the reference panel while the other instrument observes the targets; however, the instruments must be cross-correlated over the reference panel (Walter-Shea and Biehl 1990).

Scene reflectance has been measured from handheld, pole-mounted, tripod-mounted, truck-mounted, or helicopter-mounted multi-band radiometers or spectroradiometers (Bauer et al. 1986, Deering 1989). For example in Figure 17.3, the 18° fore optic of a spectroradiometer (FieldSpec Pro, ASD Inc, Boulder, CO) and a digital camera were aligned and mounted on a pole at 2.3 m above the soil at a 0° view zenith angle to measure the reflectance of crop residues in agricultural fields (Daughtry and Hunt 2008). The diameter of the field of view of spectroradiometer at the soil surface was 0.7 m and was large enough to include a representative portion of the scene. Multiple observations were acquired along transects through numerous fields. Because the camera and the fore optics were aligned, the fractions of green vegetation, crop residue, and soil in the field of view of the spectroradiometer could be determined. The digital images also serve for quality-checking the reflectance data for shadows and unwanted objects in the field of view.

Some scenes are relatively uniform (agricultural crops) while other scenes are much more variable and random (rangelands and forests). The area of the sensor's field of view must be large enough to include a representative portion of the scene. The variance of the BRF measurements decreased as the area viewed increased (Daughtry et al. 1982). Extreme caution must be exercised in analyzing and interpreting BRF data acquired at distances

Figure 17.3   Researchers use a spectroradiometer (FieldSpec Pro, Analytical Spectral Devices, Boulder, CO) to measure the reflectance of crop residues and soils. The fore optic of the spectroradiometer and a digital camera are mounted at the top a telescoping pole with a bubble level mounted about waist high. The fields of view of the spectro-radiometer and the camera are highlighted on the photograph.

where the diameter of the sensor's field of view is smaller than the inherent variability within the scene. Ideally, the diameter of the field of view should be several multiples of the row spacing. When this is not possible, many observations of different areas of the scene are required to obtain an accurate composite of scene reflectance.

Spectral measurements of uniform bright and dark target within a study site can be used to correct for atmospheric effects and empirically calibrate remotely sensed imagery (Moran et al. 2003). These target areas should be relatively flat, homogeneous, and exceed the pixel size plus registration errors of the image, preferably greater than 5×5 pixels. Examples of possible targets include asphalt, concrete, sand, bare soil, and rock outcrops. Chemically-treated canvas tarps or panels can also be deployed as reference targets, but are difficult to keep clean under normal field conditions. Targets with vegetation must be used with caution because plant leaves and canopies are non-Lambertian and preferentially reflect radiation in certain directions (Deering 1989). It is best to measure the reflectance of these targets as close to the acquisition time of the image as possible.

Aerial images

Although it may seem incongruous to consider aerial imagery as a form of surface reference data, such imagery can provide detailed data for interpretation of other imagery at coarser resolution (Campbell 2007). Aerial photography and high resolution satellite images are commonly used for assisting in locating sample sites, interpreting land cover/land use classes, and verifying results obtained from analysis of remotely sensed data. Their use requires knowledge of how and when the photography/images were obtained, knowledge of the changing cover types within a scene, and photo-interpretation skills (Lillesand et al. 2004). Traditionally, manned, fixed-wing aircraft and helicopters have been used extensively for acquiring high spatial resolution imagery. Remotely piloted light aircraft or unmanned airborne vehicles (UAV) have also been used for reconnaissance by the military for decades. Recent advances in control systems, airframes, and sensors have produced a host of new civilian applications for UAVs (Figure 17.4) as platforms for monitoring surface conditions, particularly in agriculture and forestry (Hunt et al. 2005).

SAMPLING DESIGNS

The number and arrangement of sampling locations is crucial for adequately representing large

Figure 17.4   An unmanned aerial vehicle (UAV) taking-off from a grass runway. The UAV, developed by Intellitech Microsystems, can carry a variety of remote sensing instruments and can be programmed to acquire images over targets of interest.

study areas. Inadequate sampling diminishes the accuracy of the product. Commonly used sampling patterns include simple random, stratified random, systematic (grid), and stratified systematic non-aligned patterns (Campbell 2007). Many modifications of these basic sampling patterns exist (Bonham 1989, McCoy 2005). The sampling pattern of choice varies with the complexity of the landscape and the feature observed. Landscapes with large agricultural fields can often be represented with widely spaced observations, while landscapes with complex topography require more intensive sampling patterns. Closely spaced observations may underestimate the true variability of the feature of interest. Spatial autocorrelation, the degree to which samples provide independent information, and the experience of other investigators can provide valuable guidance in designing an efficient sampling strategy.

Discrete classes

The most important guide for selecting samples sites of training data is to account for all the variability within the classes. Homogeneous classes require only a few sites while classes with high variability require many sites. Jensen (1996) offered a rule-of-thumb that one should select a number of pixels that is at least 10 times the number of bands used in developing the training statistics for each class for valid variance-covariance matrices. For example, data with six bands would require at least 60 pixels per class. Furthermore, distributing these 60 pixels among six sites will represent the class better than one site with 60 pixels (McCoy 2005).

Accuracy assessment requires that an adequate number of samples per land cover class be acquired for statistical analysis. Sample sizes calculated using the binomial distribution are appropriate for overall accuracy of the classification. Error matrices show the classification accuracies for individual classes as well as the errors of omission and commission. The procedure for calculating appropriate sample size from the multinomial distribution is summarized by Congalton and Green (1999) and Stehman and Foody (in this volume). They recommended a minimum of 50 samples for each land cover category in the error matrix as a general guideline for planning purposes. For areas >250,000 hectares in size or analyses with >12 categories, the minimum number of samples should be increased to 75–100 samples per category.

Continuous data

Many of the methods discussed earlier in this chapter give sufficiently accurate measurements of biophysical properties of surface features. However, the inherent variability among plant and soil samples is an additional source of experimental error that must be considered. Even in uniform crop canopies variability in leaf area per plant may exceed 10% (Daughtry and Hollinger 1984). In natural stands the variability is often much larger.

A researcher needs to know the number of measurements of each biophysical property that must be acquired to detect differences with the desired confidence. If the researcher acquires too few samples per experimental unit, his estimates of the true value of the biophysical property of interest will be too inaccurate to be useful. On the other hand, taking too many samples limits the scope of the experiment. The power of an experiment is a function of (1) the probability of a Type I error (i.e., probability of finding a difference that does not truly exist), (2) the true alternative hypothesis, and (3) the sample size. A good approximation of the true power of experiment is ably described by Howell (1987).

A first step for any researcher is to decide how small a difference among treatments must be detected, or conversely how large an error can be tolerated. This demands careful thinking about the consequences of a sizeable measurement error. Efficient and creative multistage sampling schemes can minimize experimental errors and costs. A preliminary sampling, followed by a statistical analysis, is very helpful in designing effective experiments for measuring biophysical properties.

FUTURE CHALLENGES

Surface reference data will continue to be an integral part of effective plans for analyzing and interpreting remotely sensed images. Overall objectives and the realities of time and budget constraints determine the appropriate methods for acquiring surface reference data.

Technology will drive surface reference data collection in the future. With GPS, we are now able to locate the specific vegetation by pixel, which was not possible 10 years ago. With the further development of hardware and software, we will be able to merge datasets such as climate, soils, geology, topographic, vegetation, and similar parameters at the pixel site locations to better understand the observations that are being collected. Therefore, the slope and aspect along with specific measurements of nitrogen or other elements can tell us why a specific group of plants are different from its neighbors. This also provides the opportunity for on-site management changes in addition to understanding remote sensing measurements.

We are confident that understanding the limitations of specific measurements, such as soil moisture, will lead to measurements that can help model this important parameter. Measurements of vegetation that provide the researcher/user with specifics on nutrient levels, disease detection, insect damage, and other factors that have not been possible will be better understood in the future so that corrective management practices can be undertaken. The importance of understanding observations and measurements made at the surface will become more important but the technology to provide the interpretations will also be there for the average user. Therefore, the importance of taking and understanding surface observations will not diminish in the future.

REFERENCES

Anderson, J. R., E. E. Hardy, J. T. Roach, and R. E. Witmer, 1976. A Land Use and Land Cover Classification for Use with Remote Sensor Data (U.S. Geological Survey Professional Paper 964), U.S. Government Printing Office, Washington, DC.

ASC Scientific, 2007. *Munsell Soil Color Chart*. Carlsbad, California, USA, http://www.ascscientific.com/books.html (Last accessed January 3 2008).

Basso, B., J. T. Ritchie, F. J. Pierce, R. P. Braga, and J. W. Jones, 2001. Spatial validation of crop models for precision agriculture. *Agricultural Systems*, 68: 97–112.

Bauer, M. E., 1975. The role of remote sensing in determining the distribution and yield of crops. *Advances in Agronomy*, 27: 271–304.

Bauer, M. E., C. S. T. Daughtry, L. L. Biehl, E. T. Kanemasu, and F. G. Hall, 1986. Field spectroscopy of agricultural crops. *IEEE Transactions of Geoscience and Remote Sensing*, GE-24: 65–75.

Baumgardner, M. F., L. F. Silva, L. L. Biehl, and E. R. Stoner, 1985. Reflectance properties of soils. *Advances in Agronomy*, 38: 1–44.

Biehl, L. L. and B. R. Robinson, 1983. Data acquisition and preprocessing techniques for remote sensing field research. *Proceedings of Society of Photo-optical Instrumentation Engineers*, 356: 143–149.

Blackmer, T. M. and J. S. Schepers, 1995. Use of a chlorophyll meter to monitor nitrogen status and schedule fertilization for corn. *Journal of Production Agriculture*, 56–60.

Bonham, C. D., 1989. *Measurements for Terrestrial Vegetation*. John Wiley, New York.

Campbell, J. B., 2007. *Introduction to Remote Sensing*. 4th edn. Guilford Press, New York.

Campbell, J. B., in this volume. Remote Sensing of Soils. Chapter 24.

Carter, P. G., C. J. Johannsen, and B. A. Engel, 2008. Recognizing patterns within cropland vegetation: A crop anomaly classification system. *Journal of Terrestrial Observation*, 1: in press.

Causton, D. R., 1985. Biometrical, structural, and physiological relationships among tree parts. In M. G. R. Cannell and J. E. Jackson (eds), *Attributes of Trees as Crop Plants*. Institute of Terrestrial Ecology, Natural Environment Research Council, Huntington, UK pp. 137–159.

Congalton, R. G. and K. Green, 1999. *Assessing the Accuracy of Remotely Sensed Data: Principles and Practices*. Lewis, Boca Raton, FL.

Daughtry, C. S. T., 1990. Direct measurements of canopy structure. *Remote Sensing Reviews,* 5: 45–60.

Daughtry, C. S. T. and S. E. Hollinger, 1984. Costs of measuring leaf area index in corn. *Agronomy Journal,* 76: 836–841.

Daughtry, C. S. T. and E. R. Hunt, Jr., 2008. Mitigating the effects of soil and residue water contents on remotely sensed estimates of crop residue cover. *Remote Sensing of Environment,* in press.

Daughtry, C. S. T., V. C. Vanderbilt, and V. J. Pollara, 1982. Variability of reflectance measurements with sensor altitude and canopy type. *Agronomy Journal,* 74: 744–751.

Daughtry, C. S. T., C. L. Walthall, M. S. Kim, E. Brown de Colstoun, and J. E. McMurtrey III, 2000. Estimating corn leaf chlorophyll concentration from leaf and canopy reflectance. *Remote Sensing of Environment,* 74: 229–239.

Deering, D. W., 1989. Field measurements of bidirectional reflectance, In G. Asrar (ed.), *Theory and Applications of Optical Remote Sensing*. John Wiley, New York pp. 14–65.

Dobermann, A. and J. L. Ping, 2004. Geostatistical integration of yield monitor data and remote sensing improves yield maps. *Agronomy Journal,* 96: 285–297.

French, G. T., 1999. *Understanding the GPS*. Delmar Learning, Albany, NY.

Gardner, W. H., 1986. Water content. In *Methods of Soil Analysis Part 1. Physical and Mineral Methods*. Agronomy Monograph 9 (2nd Edn), American Society of Agronomy, Soil Science Society of America, Madison, WI pp. 493–544.

Gish, T. J., C. L. Walthall, C. S. T. Daughtry, and K.-J. S. Kung, 2005. Using soil moisture and spatial yield patterns to identify subsurface flow pathways. *Journal of Environmental Quality,* 34: 274–286.

Goel, N. S., 1989. Inversion of canopy reflectance models for estimation of biophysical parameters from reflectance data. In G. Asrar (ed.), *Theory and Applications of Optical Remote sensing*. John Wiley, New York pp. 205–251.

Hamblin, A., 1991. Sustainable agricultural systems – What are the appropriate measures for soil structure. *Australian Journal of Soil Research,* 29: 709–715.

Hoffer, R. M. and C. J. Johannsen, 1969. Ecological potentials in spectral signature analysis. In P. Johnson (ed.), *Remote Sensing in Ecology*. University of Georgia Press, Athens, GA pp. 1–16.

Howell, D. C., 1987. *Statistical Methods for Psychology*, 2nd edn. PWS-Kent, Boston, MA.

Hunt, E. R., Jr., M. Cavigelli, C. S. T. Daughtry, J. E. McMurtrey III, and C. L. Walthall, 2005. Evaluation of digital photography from model aircraft for remote sensing of crop biomass and nitrogen status. *Precision Agriculture,* 6: 359–378.

Irons, J. R., R. A. Weismiller, and G. W. Petersen, 1989. Soil reflectance. In G. Asrar (ed.), *Theory and Applications of Optical Remote Sensing*. John Wiley, New York pp. 66–106.

Jackson, R. D., T. R. Clarke, and M. S. Moran, 1992. Bidirectional calibration results for 11 Spectralon and 16 $BaSO_4$ reference reflectance panels. *Remote Sensing of Environment,* 40: 231–239.

Jensen, J. R., 1996. *Introductory Digital Image Processing: A Remote Sensing Perspective*. 2nd edn. Prentice-Hall, Upper Saddle, NJ.

Joyce, A. T., 1978. Procedures for gathering ground truth information for a supervised approach to a computer-implemented land cover classification of Landsat-acquired multispectral scanner data, US Government Printing Office, Washington, DC pp. 43.

Justice, C. O. and J. R. G. Townshend, 1981. Integrating ground truth data with remote sensing. In J. R. G. Townshend (ed.), *Terrain Analysis and Remote Sensing*. Allen and Unwin, London pp. 38–58.

Klute, A., C. S. Campbell, R. D. Jackson, M. M. Mortland, and D. R. Nielsen, 1986. *Methods of Soil Analysis: Part 1. Physical and Mineralogical Methods*. Soil Science Society of America, Madison, Wisconsin USA.

Knipling, E. B., 1970. Physical and physiological basis for reflectance of visible and near infrared radiation from vegetation. *Remote Sensing of Environment,* 1: 155–159.

Kollenkark, J. C., C. S. T. Daughtry, M. E. Bauer, and T. L. Housley, 1982a. Effects of cultural practices on agronomic and reflectance characteristics of soybean canopies. *Agronomy Journal,* 74: 751–758.

Kollenkark, J. C., V. C. Vanderbilt, C. S. T. Daughtry, and M. E. Bauer. 1982b. Soybean canopy reflectance as influenced by solar illumination angle. *Applied Optics,* 21: 1179–1184.

Lang, R. G., 1990. An instrument for measuring canopy structure. *Remote Sensing Reviews,* 5: 61–72.

Lillesand, T. M., R. W. Kiefer, and J. W. Chipman, 2004. *Remote Sensing and Image Interpretation*, 5th edn. John Wiley and Sons, New York.

Lobell, D. B. and G. P. Asner, 2002. Moisture effects on soil reflectance. *Soil Science Society of America Journal,* 66: 722–727.

Long, J. N., F. W. Smith, and D. R. M. Scott, 1981. The role of Douglas-fir stem sapwood and heartwood in the mechanical and physiological support pf crowns and development of stem form. *Canadian Journal of Forest Research,* 11: 459–464.

McCoy, R. M., 2005. *Field Methods in Remote Sensing*. Guilford Press, New York.

Milton, E. J., 1987. Principles of field spectroscopy. *International Journal of Remote Sensing,* 8: 1807–1827.

Monmonier, M., 1991. *How to Lie with Maps*. University of Chicago Press, Chicago, IL.

Moran, M. S., Y. Inoue, and E. M. Barnes, 1997. Opportunities and limitations for image-based remote sensing in precision crop management. *Remote Sensing of Environment,* 61: 319–346.

Moran, M. S., C. D. Peters-Lidard, J. M. Watts, and S. McElroy, 2004. Estimating soil moisture at the watershed scale with satellite-based radar and land surface models. *Canadian Journal of Remote Sensing,* 30: 805–826.

Moran, S., G. Fitzgerald, A. Rango, C. Walthall, E. Barnes, W. Bausch, T. Clarke, C. Daughtry, J. Everitt, J. Hatfield,

K. Havstad, T. Jackson, N. Kitchen, W. Kustas, M. McGuire, P. Pinter, K. Sudduth, P. Schepers, T. Schmugge, P. Starks, and D. Upchurch, 2003. Sensor development and radiometric correction for agricultural applications. *Photogrammetric Engineering and Remote Sensing*, 69: 705–718.

Morrison, J. E., C. Huang, D. T. Lightle, and C. S. T. Daughtry, 1993. Residue cover measurement techniques. *Journal of Soil and Water Conservation*, 48: 479–483.

Morrison, J. E., J. Lemunyon, and H. C. Bogusch Jr., 1995. Sources of variation and performance of nine devices when measuring crop residue cover. *Transactions of the ASAE*, 38: 521–529.

Parker-Williams, A. E. and E. R. Hunt, Jr., 2004. Accuracy assessment for detection of leafy spurge with hyperspectral imagery. *Journal of Range Management*, 57: 106–112.

Parsons, S. D., D. R. Ess, and C. J. Johannsen, 1995. How GPS works and why differential correction (DC) is needed. *Precision Decisions '95*. Foundation for Ag Research, Champaign, IL pp. 28–32.

Pedersen, P., 2004. *Soybean Growth and Development*. Iowa State University PM 1945.

Pinter, P. J., J. L. Hatfield, J. S. Schepers, E. M. Barnes, M. S. Moran, C. S. T. Daughtry, and D. R. Upchurch, 2003. Remote sensing for crop management. *Photogrammetric Engineering and Remote Sensing*, 69: 647–664.

Ritchie, S. W., 2005. *How a Corn Plant Develops*. Iowa State University Special Report 48.

Robinson, B. R. and L. L. Biehl, 1979. Calibration procedures for measurement of reflectance factor in remote sensing field research. *Proceedings of Society of Photo-Optical Instrumentation Engineers*, 196: 16–20.

Ross, J., 1981. *The Radiation Regime and Architecture of Plant Stands*. Dr W. Junk, The Hague, The Netherlands.

Sanchez-Flores, E. and S. R. Yool, 2004. Site environment characterization of downed woody fuels in the Rincon Mountains, Arizona: regression tree approach. *International Journal of Wildland Fire*, 13: 467–477.

Schaepman-Strub, G., M. E. Schaepman, J. V. Martonchik, T. H. Painter, and S. Dangel, in this volume. Radiometry and Reflectance: From Terminology Concepts to Measured Quantities. Chapter 15.

Simmons, S. R., E. A. Oelke, and P. M. Anderson, 1995. *Growth and Development Guide for Spring Wheat*. University of Minnesota Extension FO-02547.

Sparks, D. L., A. L. Page, P. A. Helmke, R. H. Loeppert, P. N. Soltanpour, M. A. Tabatabai, C. T. Johnson, and M. E. Summer, 1996. *Methods of Soil Analysis: Part 3. Chemical Methods*. Soil Science Society of America, Madison, Wisconsin, USA.

Stehman, S. V. and G. M. Foody, in this volume. Accuracy Assessment. Chapter 21.

Stoner, E. R. and M. F. Baumgardner, 1981. Characteristic variations in reflectance of surface soils. *Soil Science Society of America Journal*, 45: 1161–1165.

Taylor, J. C., G. A. Wood, and G. Thomas, 1997. Mapping yield potential with remote sensing. *Precision Agriculture*, 2: 713–720.

Ulaby, F. T., P. C. Dubois, and J. V. Zyl, 1996. Radar mapping of surface soil moisture. *Journal of Hydrology*, 184: 57–84.

van Leeuwen, W. J. D., in this volume. Visible, Near-IR and Shortwave IR Spectral Characteristics of Terrestrial Surfaces. Chapter 3.

Walter-Shea, E. A. and L. L. Biehl, 1990. Measuring vegetation spectral properties. *Remote Sensing Reviews*, 5: 179–206.

Welles, J. M., 1990. Some indirect methods of estimating canopy structure. *Remote Sensing Reviews*, 5: 31–55.

Whittaker, R. H. and P. L. Marks, 1975. Methods of assesing terrestrial productivity. In H. Leith and R. H. Whittaker (eds), *Primary Productivity of the Biosphere*. Springer-Verlag, New York pp. 55–118.

Integrating Remote Sensing and Geographic Information Systems

James W. Merchant and Sunil Narumalani

Keywords: data integration, geospatial data fusion, image understanding, image processing.

INTRODUCTION

Remote sensing and geographic information systems (GIS) comprise the two major components of geographic information science (GISci), an overarching field of endeavor that also encompasses global positioning systems (GPS) technology, geodesy and traditional cartography (Goodchild 1992, Estes and Star 1993, Hepner et al. 2005). Although remote sensing and GIS developed quasi-independently, the synergism between them has become increasingly apparent (Aronoff 2005). Today, GIS software almost always includes tools for display and analysis of images, and image processing software commonly contains options for analyzing 'ancillary' geospatial data (Faust 1998). The significant progress made in 'integration' of remote sensing and GIS has been well-summarized in several reviews (Ehlers 1990, Mace 1991, Hinton 1996, Wilkinson 1996). Nevertheless, advances are so rapid that periodic reassessment of the state-of-the-art is clearly warranted.

In this chapter, we focus on integration of remote sensing and GIS. Our definition of integration includes the use of each technology to benefit the other, as well as the application of both technologies, in concert, for modeling and decision-support. The discussion will range from consideration of simple visualization to data extraction and database development, and analyses based on 'multisource' data. We will, throughout, emphasize that, although progress in remote sensing-GIS integration has clearly benefited from advances in computing (hardware and software) and global positioning system (GPS) technology, equally important have been developments in theory and analytical methods, including innovations in use of decision trees, neural networks, and evidential reasoning. Finally, we will provide a look to the future and issues yet to be adequately addressed.

Historical perspective

Although the literature on remote sensing-GIS integration is relatively recent, the precursors of integration extend back to the early twentieth century. Aerial stereoscopic photography and analogue photogrammetric analysis methods have been routinely used since the 1930s to produce topographic maps, soils surveys, and land use maps. The digitized versions of such maps, of course, comprise key databases in today's geographic information systems. Modern digital aerial and satellite-borne remote sensing systems, and data analysis procedures such as digital photogrammetry and image classification, continue to be important means by which data for GIS are acquired, updated, and enhanced.

Likewise, persons engaged in interpreting aerial photography have recognized for at least 80 years that data from other sources, collateral or 'ancillary data' such as maps portraying topography, are critical in such work (see Estes et al. 1983, Jensen 2005). Photo interpretation demands the application of reasoning and logic, based on use of multiple data sources and formalized using aids such as interpretation keys (Campbell 1978, 2007, Estes et al. 1983). In recent decades, there have been many efforts to implement, and improve upon, strategies and methods for image analysis developed during the pre-digital era. Advances in digital remote sensing systems and data conversion (e.g., digitizing maps) have, for example, dramatically increased both the number and variety of geospatial datasets available, and strides in computing have greatly enhanced capabilities for processing and analyzing such data.

Contemporary approaches to integration of remote sensing and GIS

Ehlers (1990) provided a three-level taxonomy within which to consider the current state-of-the-art of remote sensing-GIS integration (Table 18.1). Most of the work we discuss in this chapter can be categorized as *first-level* or *second-level integration*, but progress toward the *third-level* is being made.

Gao (2002) points out that Ehlers' three-level classification must today be augmented to include GPS. The linear, interactive, hierarchical, and complex models of GPS–GIS–remote sensing integration he presents emphasize the directionality and sophistication of data flow between the three technologies, and in some respects parallel Ehlers' classification (i.e., trending toward 'seamless' integration that facilitates complex analyses). Although we will infrequently discuss GPS in the following pages, it should be understood that GPS is an essential component of contemporary GIS-remote sensing integration and is often employed in studies cited. Indeed, all of the geographic information science technologies constitute important components of an increasingly unified geospatial information analysis infrastructure trending toward Ehlers' third-level.

An example of first-level integration of remote sensing and GIS is the overlay of a digital image (e.g., a digital orthophoto) with a cartographic dataset derived from a GIS (e.g., roads selected from a digital line graph), producing a merged product that allows an analyst to visualize information derived from both. It is assumed that the two datasets complement one another in some fashion. For instance, if the image is more recent than the GIS data, the analyst might commonly use information extracted from the image, through photo interpretation and 'heads-up' digitizing, to update the cartographic (e.g., roads) dataset. Similarly, a 'perspective view' of a landscape can be created by 'draping' an image over a three dimensional rendering of topography derived from a digital elevation model (DEM) manipulated in a GIS, thus enabling one to visualize associations between land cover (depicted on the image) and terrain configuration. Such products may be animated to produce virtual 'fly-overs' (see NASA's Scientific Visualization Studio for good examples – http://svs.gsfc.nasa.gov).

Table 18.1   Levels of remote sensing-GIS integration (modified from Ehlers 1990)

Level of integration	Principal characteristics	Examples of analyses supported
First-level integration	Achieved via data exchange between separate GIS and image analysis systems	(a) Simultaneous display of GIS (usually vector) data and remotely-sensed (raster) images; (b) Ability to move the results of low-level image processing to the GIS and the results of GIS-based analyses to image-analysis software.
Second-level integration	Permits 'seamless' tandem or combined raster–vector processing facilitated by a common user interface	(a) Capability to incorporate GIS data directly into image processing; (b) Ability to accommodate heterogeneous data input in a coherent manner; (c) Ability to generate simulations combining a GIS and image data with temporal evolution.
Third-level integration	Remote sensing and GIS operate as a single, integrated system – 'telegeoprocessing'	(a) Accommodate raster and vector data in a unified data structure; (b) Facilitate real time analysis using sensor networks and linkages to other technologies.

It is noteworthy that even relatively simple integration of remote sensing and GIS, such as the examples noted above, can be quite powerful. Visualization of integrated datasets may, for example, lead one to new insights and hypotheses regarding inter-relationships between geospatial variables. Today, however, it is common to encounter instances of remote sensing-GIS integration that are far more complex than visualization. In the next sections of this chapter, we emphasize *second-level* and *third-level* integration, first summarizing, respectively, ways in which data acquired via remote sensing are commonly used in GIS, and the importance of GIS (geospatial) data and analytical methods in remote sensing. In the course of this discussion we illuminate more sophisticated modes of integration encapsulated by current terminology such as 'multisource analysis,' 'data fusion,' and 'evidential reasoning.' Attention is also given to recent developments in sensor networks and 'telegeoprocessing.'

INTEGRATION OF DATA DERIVED FROM REMOTE SENSING IN GIS

As noted above, aerial photography has long been used to generate analogue geospatial products that, now, in a digital form, often constitute important components of GIS databases. With the advent of digital remote sensing systems and image processing software, the importance of remote sensing in GIS has expanded considerably. Applications of remote sensing range from the use of orthoimagery as a GIS base layer, to the development of thematic data on land use and the generation of unique geospatial datasets via extraction of cartographic features such as buildings and roads from imagery.

Orthoimagery as base data for GIS

In a GIS database, all features must be positioned as accurately as possible. Additionally, all data layers must be registered to one another and should be geo-referenced to a specific map projection and coordinate system. Base maps provide the frame-of-reference for positioning, registration and geo-referencing. In recent years, orthoimages generated from aerial photography (i.e., digital orthophotos) or fine spatial resolution satellite data have been used as base layers with increasing frequency (Davis and Wang 2003). Such images have been corrected to remove spatial displacements arising from sensor perspective and topographic relief. Orthoimages offer several advantages over products such as digitized USGS 7.5 quadrangles ('digital raster graphics') traditionally used as base maps. They are, for example, usually quite recent and they provide a 'realistic' view of the landscape with recognizable features being easily discerned (e.g., road intersections, buildings).

Developing thematic data for GIS

Remote sensing is the primary source for many kinds of thematic data critical to GIS analyses, including data on land use and land cover characteristics and surface elevation. Aerial and satellite imagery are also often used to assess landscape change and to update existing geospatial databases (e.g., roads, hydrography). Details on methods used to create geospatial data from remote sensing are found throughout this volume. Here we highlight just a few examples of thematic geospatial data commonly derived via remote sensing.

Surface elevation

DEMs (Deng, in this volume) are widely used in GIS. As noted above, photogrammetry has long been the principal means by which surface elevation is mapped. Although aerial photography is still often used in such work, satellite imagery such as that acquired by IKONOS and ASTER is increasingly competitive in terms of resolution and accuracy (Hirano et al. 2003). Increasingly, active remote sensing methods that use lasers (airborne laser scanning (ALS) or LIDAR (Hyyppä et al., in this volume)) or microwave energy (IFSAR or InSAR (Kellndorfer and McDonald, in this volume)) are also being employed for producing DEMs (Hodgson et al. 2003). LIDAR data, for example, have been used to prepare DEMs with less than 30-cm resolution, especially important for GIS-based flood risk assessment and transportation planning (Post et al. 2000). IFSAR provides unique cloud penetration, day/night operation and wide-area coverage. Shuttle Radar Topography Mission (SRTM) IFSAR data are being used to produce GTOPO30, a DEM having a 30 arc-second resolution that covers most of the earth's land areas (Gesch et al. 2001; http://www2.jpl.nasa.gov/srtm/).

Land use and land cover mapping

Campbell (2007) and Jensen (2005) provide excellent summaries of the state-of-the-art in mapping the type and condition of land use and land cover via multispectral image classification. As an example, the U.S. National Land Cover Dataset (NLCD), widely used in GIS analyses, was based primarily on classification of Landsat imagery (Homer et al. 2004; http://www.mrlc.gov/mrlc2k_nlcd.asp). Many advances in land use and land cover mapping are stemming from new methods of analysis founded on integration of remote sensing and ancillary GIS data, as described later in this chapter.

Biophysical phenomena

Remote sensing can also be used to provide unique thematic data regarding a variety of biophysical characteristics, including surface temperature, imperviousness, water clarity, evapotranspiration, vegetation pigments, biomass, canopy structure and height, leaf area index (LAI), and soil moisture (see Wulder 1998, Yang et al. 2003, Courault et al. 2005, Jensen 2005). Such data are required for GIS-based hydrologic, meteorological, wildfire risk, and crop simulation modeling (e.g., Keane et al. 2001, Nemani et al. 2002). Biophysical data derived from sensors such as MODIS are increasingly key inputs to global scale biophysical models focused on carbon dynamics, ecosystem processes, vegetation productivity, and oceanic phytoplankton distribution (Huete 2005).

Feature extraction

Feature extraction procedures are used to automate identification and mapping of physical objects (e.g., buildings, roads) from imagery collected by remote sensing. These procedures have been used to update, or increase accuracy of, existing GIS databases, or to create new data layers (Gruen and Li 1997). Mena (2003) reviewed current capabilities to automatically extract roads from aerial and satellite imagery. Shan and Lee (2005) reported on procedures to develop databases of buildings with IKONOS imagery. Hu et al. (2003) surveyed a suite of integrated remote sensing techniques that are being used for 3-D feature extraction in urban areas.

Landscape change

Remote sensing provides many opportunities to identify and map changes in geospatial features (Lu et al. 2004, Jensen 2005). Today, aerial and satellite remote sensing are often used for updating GIS databases on land use, hydrography and transportation networks (e.g., Jensen et al. 1993, Laliberte et al. 2001). Impacts of natural hazards such as hurricanes, and progression of dynamic processes such as soil erosion, can also be assessed with remote sensing (e.g., Lupo et al. 2001). Multitemporal data are used to monitor seasonal and interannual landscape events at local, regional, and global scales. For example, Zhang et al. (2006) described use of MODIS data for assessing global patterns of vegetation greenness.

USING GEOSPATIAL DATA AND GIS IN REMOTE SENSING

Geospatial datasets depicting phenomena such as surface elevation, soils, transportation, hydrography, and land use are common components of geographic information systems. Although, as noted above, such datasets are frequently developed, at least in part, through analyses of aerial and satellite imagery, from a remote sensing point-of-view they are usually considered 'ancillary data.' The value of ancillary data in photo interpretation and digital image analysis is widely recognized (Tso and Mather 2001, Jensen 2005, Campbell 2007). Digital elevation models have been used particularly often in such work (see Florinsky 1998). Jensen (2005) notes that ancillary data may also include outputs of GIS analysis such as topographic slope and aspect (derived from a DEM), soils recoded into hydric and non-hydric classes, or polygons defining proximity to roads or streams (see Lunetta et al. 2003). Here we examine three ways in which ancillary (GIS) data are, today, frequently used in remote sensing: (1) geometric correction and orthorectification of imagery, (2) radiometric correction, and (3) image classification. All examples represent cases of Ehlers' (1990) 'second-level' integration.

Geometric correction and orthorectification

Images acquired via remote sensing exhibit spatial displacements of objects and scale variations that stem principally from sensor orientation and topographic relief (Aronoff 2005, Jensen 2007). Such displacements must be removed in order to create a planimetrically-correct ('orthorectified') image, i.e., an image that has properties similar to a map, such as consistent scale. Software for creating digital orthoimages is a component of many image analysis systems, but data required for rectification, including geodetic control (often obtained via GPS) and digital elevation data, must typically be imported from a GIS. The output of rectification, an orthoimage, is then often exported back to the GIS to support subsequent analyses. Orthoimages, as noted above, are now frequently used as base maps for registering and geo-referencing other layers in a GIS (Aronoff 2005). Geometric correction and registration are prerequisites to 'fusion' of multi-source data required for creation of products such as perspective views (Toutin 2004).

Orthoimages created from fine spatial resolution (~1 m) sensors on satellites such as IKONOS and Quickbird have become more widely available in recent years. The quality of orthoimage products is closely related to the availability of accurate digital elevation models having horizontal and vertical resolution appropriate to the sensor (Toutin 2004). As discussed earlier, much current effort focuses on development of improved DEMs. LIDAR, for example, promises to provide a means of dealing with problems related to tall structures (e.g., buildings in urban areas), the presence of which can introduce errors in orthoimages generated from DEMs that ordinarily portray only 'bare earth' topography (Zhou et al. 2004).

Radiometric correction

Digital elevation models are also often used in radiometric correction of digital images (Tso and Mather 2001, Aronoff 2005, Jensen 2005). Spectral reflectance from the earth's surface is a complex function of land cover, angle of solar illumination, atmospheric condition, and topographic position (slope and aspect). In a given image, the brightness values for pixels of the same land cover type can vary substantially depending on whether they are situated in full sunlight (e.g., south facing slopes in the northern hemisphere) or shadow. These effects are especially evident in areas of high relief. If such artifacts are not removed they can significantly degrade results of multispectral classification.

Image processing designed to remove (or reduce) brightness variation stemming from topographic position is called 'topographic normalization' or 'terrain correction.' Topographic normalization is sometimes accomplished using simple band ratios that tend to compensate somewhat for reflectance differences, but more robust methods require the use of a DEM and GIS-derived slope and aspect (Gu et al. 1999, Riaño et al. 2003). As with geometric correction, the success of terrain correction depends a great deal on the availability of sufficiently fine resolution and accurate DEMs.

Image classification

Although both the generation of orthoimagery and topographic normalization demonstrate a form of remote sensing-GIS integration, we currently see the most sophisticated types of integration manifested in digital image classification and modeling. The principal goal of such work is to enhance information extraction and thematic mapping by means of 'data fusion' and 'multisource analysis.' Enhancement usually entails improving the accuracy and/or categorical resolution of products traditionally generated solely from multispectral analysis of digital aerial or satellite imagery. Hutchinson (1982) provided an early, but still quite relevant, assessment of the use of multisource data in image classification. Excellent, more current,

overviews are found in Campbell (2007), Jensen (2005), Jensen et al. (in this volume), Tso and Mather (2001), and Richards and Jia (1999).

A wide variety of methods for employing multi-source data in digital image classification have been developed. All may be viewed as attempts to exploit the long-recognized value of bringing to bear multiple sources of data (ancillary data and/or multiple types of imagery) on extraction of information from images (e.g., through image interpretation) and to invoke, to some extent, the reasoning and logic employed for most of a century in visual interpretation of images (see Jensen 2005, 2007, Campbell 2007). Hallmarks of image interpretation logic include (1) use of a systematic strategy that proceeds from 'knowns' to 'unknowns,' and (2) use of inference and 'convergence of evidence' exploiting observed relationships between multiple data types (image and ancillary data). The process of image interpretation often involves use of heuristics and/or 'rules' based on expert knowledge and observation (e.g., rules about biogeographic relationships between vegetation zonation, elevation, slope, and aspect in mountainous areas) (Campbell 1978, 2007, Estes et al. 1983). Here, we examine several approaches to integrating GIS with remote sensing to improve extraction of information from digital images. These include image stratification, classification modification, postclassification sorting, and advanced methods for multisource data analysis.

Image stratification

Stratification is a procedure for subdividing an image into regions ('strata') that are each considered 'internally homogeneous.' In digital image analysis, one approach is to use geospatial data for stratification and to subsequently classify each stratum separately. For example, Homer et al. (1997) used a GIS ecoregions dataset to stratify Landsat TM data for classification of land cover in Utah. Taking a different tack, Smith and Fuller (2001) used a digital map of agricultural field parcel boundaries to stratify Landsat TM, SPOT HRV and IRS imagery (see also Mason et al. 1988). Classification of land cover (e.g., crops) was subsequently carried out using a per-field or 'object-based' (as opposed to per-pixel) approach. The resulting maps were both more accurate and cleaner (exhibiting less visual noise) than a conventionally-produced per-pixel classification. GIS-assisted stratification has also been used as a guide for selecting training data for use in supervised classification (e.g., see Hutchinson 1982, Ortiz et al. 1997, Mesev 1998).

Classification modification

Most image classification relies on use of parametric supervised or unsupervised statistical techniques (Jensen 2005). A number of investigators have developed methods by which ancillary data can be integrated with imagery and used to improve multispectral classification. Ancillary geospatial data have, for example, sometimes been used to modify prior probabilities ('weights,' commonly assumed to be equal) in digital image classification using Bayesian Maximum Likelihood decision rules. GIS-derived data can be used to adjust prior probabilities to reflect frequencies with which informational classes are expected to occur in the image. Mesev (1998), for instance, showed that SPOT HRV classification of urban land cover could be improved by adjusting prior probabilities using GIS-derived census data (e.g., dwelling density).

Another approach has been to create, for each pixel, a 'stacked vector' comprised of both (multi)spectral image data and digital GIS data (a 'logical channel' such as elevation), in effect increasing the number of channels available for analysis and providing a means to incorporate non-spectral features (Maselli et al. 2000, Jensen 2005). For example, Wulder et al. (2004) used slope and aspect variables, derived in GIS from a DEM, and Landsat TM data to classify land cover in a mountainous area of British Columbia. They achieved the highest mapping accuracies using unsupervised classification of combined elevation and image data, after first stratifying the image into areas of shadow and nonshadow. Ricchetti (2000) found that slope, used as a logical channel, significantly improved Landsat TM classification of geological units.

Integrating GIS-derived data into image analysis can also be a means to improve class characterization and labeling of spectral classes. Ma et al. (2001), for example, used DEM-derived slope and aspect data in a two-stage 33-scene Landsat TM classification of Montana's land cover. They found that the approach improved both the efficiency of multiscene classification and robustness of class labeling. Focusing on agricultural land cover mapping, Ortiz et al. (1997) employed historical cropping pattern data in a GIS to assist in describing spectral classes developed through classification of multidate Landsat TM imagery. Hall et al. (2001) found that census data (e.g., population density and housing condition) integrated with Radarsat and Landsat TM imagery could be used to assess urban poverty in *radios* (census tracts) of Rosario, Argentina.

Other researchers have selectively used ancillary geospatial data to add information to a traditional multispectral classification when conventional classification methods fail to adequately identify important 'information classes.' For instance, Vogelmann et al. (1998) merged National Wetlands Inventory (NWI) data into a Landsat TM classification of the Mid-Atlantic region to compensate for inability of a spectrally-based classifier to satisfactorily identify wetlands (see also Lunetta et al. 2003). Working with ERS-1 synthetic aperture radar imagery, Brivio et al. (2002) demonstrated that a DEM-derived 'least accumulated cost-distance' dataset could be used to enhance maps of flooded areas when timing of image acquisition was suboptimal for the event.

Postclassification class sorting

It has long been observed that important land cover classes and conditions may be spectrally inseparable using conventional classification algorithms, and, conversely, that some spectral classes may need to be combined for cartographic purposes (Jensen 2005). For example, in areas of high relief, pixels representing water and coniferous forest in shadow may, inadvertently, be placed in the same spectral class because they exhibit similar reflectance. To a user of the product generated, this type of confusion is clearly undesirable. On the other hand, it is also common to find that pixels of one land cover type may be placed in several spectral classes because they represent subtle subclasses of the same cover type (e.g., Ponderosa Pine forest on different slopes and aspects). In this case, an analyst may wish to assemble pixels into a new combined informational class (e.g., Ponderosa Pine) for the final product.

Geospatial data can be used to help resolve such problems. Commonly this is accomplished with decision rules implemented through Boolean operations (IF–THEN–ELSE) to sort pixels from initial spectral classes into new 'informational classes' (Hutchinson 1982). For example, Cibula and Nyquist (1987) employed a DEM and climate data to increase categorical resolution of a Landsat MSS classification of Olympic National Park from 9 to 21 classes, while maintaining an accuracy greater than 85% for most classes. Vogelmann et al. (1998) assembled elevation data (and derived slope, aspect, and shaded relief), Defense Meteorological Satellite Program city lights data, prior land use/land cover maps, digital line graphs, NWI data, and population census data to enhance an unsupervised Landsat TM-based land cover classification of the Mid-Atlantic region. Brown et al. (1993) used digitized ecoregions, major land resource areas, climate, and elevation data to enhance a land cover classification of the conterminous U.S. derived initially from multitemporal AVHRR NDVI imagery. An iterative spectral class splitting, merging and refinement procedure was implemented to produce 189 final informational classes from 70 initial spectral-temporal classes. Other good examples are reported by Harris and Ventura (1995), Mesev (1998), and Driese et al. (2001).

Advanced methods for multisource data analysis

In recent years, a number of 'advanced methods' for image classification have been developed to augment, and potentially improve on, traditional statistically-based image classification procedures. These methods include rule-based or 'knowledge-based' classification ('expert systems'), artificial neural networks (ANN), decision tree classifiers and evidential reasoning (Lawrence and Wright 2001, Mertikas and Zervakis 2001, Jensen 2005, Jensen et al., in this volume;). Such approaches to data analysis lend themselves well to the integration of imagery with non-image (geospatial) data because, unlike traditional classification algorithms, they do not require assumptions regarding the normality of the data distribution, and they can readily accommodate heterogeneous data (discrete and continuous, numerical and categorical) that may have varying degrees of accuracy. In addition, these methods often facilitate structured, hierarchical strategies that mimic some aspects of human reasoning. Jensen (2005), Tso and Mather (2001) and Richards and Jia (1999) provide excellent introductions to advanced methods for multisource data analysis.

Rule-based classification methods employ a 'knowledge base' that stores facts, understandings, and heuristics specific to a particular domain (e.g., land cover classification based on IKONOS imagery and ancillary data). Rules and conditions, that draw on the knowledge base, guide classification (Jensen 2005). Huang and Jensen (1997) showed that a rule-based approach, using SPOT HRV data combined with terrain, soils and wind fetch data to classify land cover in the vicinity of a wetland, was superior to both conventional maximum likelihood and unsupervised classification. Their study employed decision tree classification, a procedure that involves systematic binary splitting of explanatory variables, such as spectral responses and ancillary data, to achieve specified classification goals. The decision tree method 'automatically' identifies critical variables, separating them from those less useful in classification, without extensive *a priori* expert knowledge (Campbell 2007). Homer et al. (2004) employed decision tree classification for development of the 2001 NLCD from multitemporal Landsat TM data and a variety of ancillary data (see http://www.mrlc.gov/mrlc2k_nlcd.asp). Lawrence and Wright (2001) and Rogan et al. (2003) provide other good examples of the use of decision trees in integrated GIS-remote sensing analysis.

Rule-based classification can also be implemented through 'evidential reasoning' (Wilkinson 1996, Mertikas and Zervakis 2001). This procedure has many of the advantages of the decision tree approach, but also enables use of subjective judgment and provides the user with a measure of the risk (confidence) of the decision made for each class. Franklin et al. (2002), for instance, used evidential reasoning to map grizzly bear habitat in Alberta. Thirty-seven variables derived from Landsat TM imagery, a DEM, digital vegetation maps, and ecological unit maps were analyzed. Resulting maps were substantially better than those achieved by traditional maximum likelihood classification.

Artificial neural networks (ANN) constitute yet another method by which expert knowledge and ancillary data can be integrated with image analysis. An ANN is designed to simulate human reasoning and decision-making processes, including rule development through learning and adaptation based on training (Civco 1993, Foody 1995, Jensen 2005). Neural networks have sometimes been shown to perform well in comparison to traditional classification methods (see Bruzzone et al. 1997 and Mas 2004), but other investigators have found them less useful and more difficult to implement than other methods (Skidmore et al. 1997). Moreover, even when successfully used, it is sometimes difficult to explain how specific classification outcomes were achieved.

INTEGRATION OF GPS, GIS, AND REMOTE SENSING

The importance of GPS in contemporary GIS and remote sensing analyses can hardly be overstated (see Gao 2002). GPS may be employed at many different steps in analysis including image rectification, georeferencing thematic data in a GIS, collection of field data to support image analysis (e.g., for ground truth, calibration, or accuracy assessment), and development, or updating, of GIS databases portraying features such as roads and utilities. Although most instances of GPS–GIS–remote sensing integration continue to fall within Ehlers' (1990) first- and second-levels, the number of applications that demand the use of all three technologies in concert continues to expand. Examples include work in precision farming, wildlife management, emergency response, and mobile mapping (see Gao 2002, Hong et al. 2006, Sampson and Delgiudice 2006).

FUTURE PROSPECTS AND RESEARCH ISSUES

We have reviewed a wide variety of ways in which GIS and remote sensing are integrated. The synergism between the two technologies has been clear for many years, and our appraisal follows in a long line of periodic assessments of the state-of-the-art and research challenges (e.g., Ehlers 1990, Mace 1991, Estes and Star 1993, Goodchild 1994, Hinton 1996, Wilkinson 1996, McMaster and Usery 2005). In recent decades theoretical and technical advances have fostered increasingly 'seamless' data analysis; nevertheless, in order for the full potential of remote sensing–GIS integration to be realized, significant, often inter-related, issues remain to be addressed.

Advances in technology

Geographic information science is a dynamic arena in which there are continual innovations. In remote sensing, for example, increasing use of hyperspectral and microwave sensors, LIDAR, very high resolution imagery, and time-series data present both opportunities (e.g., improved characterization of land cover) and challenges (e.g., handling large data volumes, image understanding) for data fusion and integrated data analysis (Hepner et al. 2005). Enhancements in internet, wireless and satellite communications, and innovations in *in-situ* sensors, are paving the way for increasingly robust 'real time' applications of remote sensing and GIS, a process some have termed 'telegeoprocessing' (Xue et al. 2002, Aksoy and Aksoy 2004). Web-based tools, such as Google Earth (http://earth.google.com/) and Internet Map Service (IMS) applications, now provide an increasingly larger audience with ready access to geospatial data, and allow elementary integration of imagery and graphics, but more sophisticated implementation of web-based integrated geospatial analysis will require resolution of issues related to metadata standards, data transmission formats, client/server computation and communication protocols (Xue et al. 2002, Tsou 2004).

Data availability and characteristics

Geospatial datasets are more widely available, and less costly, than ever before. Web-based search tools and portals (e.g., http://gos2.geodata.gov/wps/portal/gos and http://nationalmap.gov/) provide increasingly efficient means to locate and access data. Yet, issues of identifying and characterizing varying, and often uncertain, data quality (e.g., spatial and categorical resolution, positional and attribute accuracy) remain to be resolved (Davis et al. 1991, Lunetta et al. 1991, Florinsky 1998, Shi et al. 2005; Zhu 2005). Moreover, traditional differences in the ways of representing geospatial data continue to present issues. Images in raster format must often be merged with GIS data in vector format, and outputs frequently are desired in vector format (for use in GIS-based

analyses). Although strides are being taken toward development of new (e.g., object-oriented) geospatial data models, the long-standing dichotomy between raster and vector data structures (and 'field' and 'object' models of the world), and difficulties in integrating data represented in these disparate modes, remains problematic (Goodchild 1994, Blaschke et al. 2000). Recently, it has been recognized that refinement of ontologies that facilitate data integration may be critical to improving multisource data analysis (Fonseca et al. 2002).

Analytical methods

We have described, above, some 'advanced methods' of integrated data analysis that show substantial promise. However, building knowledge bases and establishing rules required to implement most such methods are difficult. Several investigators have explored use of data mining and machine learning procedures, although much remains to be accomplished (Huang and Jensen 1997, Yuan et al. 2005). Underpinning such efforts must be research designed to improve understanding of the human dimensions of geospatial/image data analysis, including, for example, articulation of the ways humans make decisions in image interpretation (Argialas and Harlow 1990). Additional progress is also needed in the use of fuzzy concepts to, for example, reduce artifacts of integrating data having differing formats (e.g., to better represent indeterminate boundaries such as gradients) and to improve multisource classification (Goodchild 1994, Jensen 2005). It is clear, too, that we require much better methods for tracking and characterizing errors generated when multisource data having different inherent scales, resolutions and accuracies are fused for analysis (Ehlers and Wenzhong 1996, Wilkinson 1996, Shi et al. 2005).

Environmental modeling in a fully-integrated data processing environment

The synergism between remote sensing and GIS is enhanced when these technologies are used in concert with ancillary technologies such as GPS and *in-situ* sensor networks and advanced telecommuncations. 'Telegeoprocessing' in concept embodies the third-level integration envisioned by Ehlers (1990). Although, as noted above, many challenges remain, significant progress has been made toward this goal. Now we see increasing efforts to take the next step – full integration of telegeoprocessing with models designed to address specific issues and support decision-making. Keane et al. (2001), for example, review progress toward integration of remote sensing, GIS, and biophysical models for wildland fire risk assessment and management. Nemani et al. (2002) describe the Terrestrial Observation and Prediction System , a prototype endeavor that integrates satellite-derived data (e.g., MODIS LAI), ancillary geospatial data (e.g., DEM, soils), surface weather observations, and a terrestrial ecosystem model to forecast biophysical conditions such as soil moisture in near-real time. In applications of this type, integration of GIS and remote sensing is important in virtually the entire process from data capture and assimilation to database development, data analysis and delivery of information to users. As we approach realization of the potential of telegeoprocessing, the need to develop and test real-world applications will be a continuing priority.

Education

Finally, but certainly not least important, we believe that advances in integration of remote sensing and GIS call for periodic re-examination of how these technologies are taught (Lauer et al. 1991). At introductory and intermediate levels, instruction in remote sensing and GIS will likely continue to be offered in separate courses as has been traditional. However, the synergism between the two technologies is now quite evident, and technical means to integrate data are rapidly expanding. Next-generation geospatial scientists need to be conversant with both the intellectual foundations and the technical methods by which integrated data analysis can be conducted. We suggest that this might best be accomplished through capstone courses, practicums or seminars in which students are compelled to consider remote sensing–GIS integration and telegeoprocessing from the perspective of problem-solving rather than of the individual technologies.

REFERENCES

Aksoy, D., and A. Aksoy, 2004. Satellite-linked sensor networks for planetary scale monitoring. http://ieeexplore.ieee.org/iel5/9623/30415/01404836.pdf (last accessed July 25 2007).

Argialas, D. P. and C. A. Harlow, 1990. Computational image interpretation models: An overview and a perspective. *Photogrammetric Engineering and Remote Sensing*, 56(6): 871–886.

Aronoff, S., 2005. *Remote Sensing for GIS Managers*. ESRI Press, Redlands, CA.

Blaschke, T., S. Lang, E. Loup, J. Strobl, and P. Zeil, 2000. Object-oriented image processing in an integrated GIS/remote sensing environment and perspectives for environmental applications. In A. Cremers and K. Greve (eds),

Environmental Information for Planning, Politics and the Public. Metropolis, Marlburg, Germany, pp. 555–570.

Brivio, P. A., R. Colombo, M. Maggi, and R. Tomasoni, 2002. Integration of remote sensing data and GIS for accurate mapping of flooded areas. *International Journal of Remote Sensing*, 23(3): 429–441.

Brown, J. F., T. R. Loveland, J. W. Merchant, B. C. Reed, and D. O. Ohlen, 1993. Using multisource data in global land cover characterization: concepts, requirements and methods. *Photogrammetric Engineering and Remote Sensing*, 59(6): 977–987.

Bruzzone, L., C. Conese, F. Maselli, and F. Roli, 1997. Multisource classification of complex rural areas by statistical and neural-network approaches. *Photogrammetric Engineering and Remote Sensing*, 63(5): 523–533.

Campbell, J. B., Jr., 1978. A geographical analysis of image interpretation methods. *The Professional Geographer*, 30(3): 264–269.

Campbell, J. B., 2007. *Introduction to Remote Sensing* (4th edn). The Guilford Press, New York, NY.

Cibula, W. G. and M. O. Nyquist, 1987. Use of topographic and climatological models in a geographical data base to improve Landsat MSS classification for Olympic National Park. *Photogrammetric Engineering and Remote Sensing*, 53(1): 67–75.

Civco, D. L., 1993. Artificial neural networks for land-cover classification and mapping. *International Journal of Geographical Information Systems*, 7(2): 173–186.

Courault, D., B. Seguin, and A. Olioso, 2005. Review on estimation of evapotranspiration from remote sensing data: From empirical to numerical modeling approaches. *Irrigation and Drainage Systems*, 19: 223–249.

Davis, C. H. and X. Wang, 2003. Planimetric accuracy of IKONOS 1 m panchromatic orthoimage products and their utility for local government GIS basemap applications. *International Journal of Remote Sensing*, 24(22): 4267–4288.

Davis, F. W., D. A. Quattrochi, M. K. Ridd, N. S-N. Lam, S. J. Walsh, J. C. Michaelsen, J. Franklin, D. A. Stow, C. J. Johannsen, and C. A. Johnston, 1991. Environmental analysis using integrated GIS and remotely sensed data: some research needs and priorities. *Photogrammetric Engineering and Remote Sensing*, 57(6): 689–697.

Deng, Y., in this volume. Making Sense of the Third Dimension Through Topographic Analysis. Chapter 22.

Driese, K. L., W. A. Reiners, and R. C. Thurston, 2001. Rule-based integration of remotely-sensed data and GIS for land cover mapping in NE Costa Rica. *Geocarto International*, 16(1): 35–44.

Ehlers, M., 1990. Remote sensing and geographic information systems: Towards integrated spatial information processing. *IEEE Transactions on Geoscience and Remote Sensing*, 28(4): 763–766.

Ehlers, M. and S. Wenzhong, 1996. Error modelling for integrated GIS. *Cartographica* 33(1): 11–21.

Estes J. E. and J. L. Star, 1993. *Remote Sensing and GIS Integration: Towards a Prioritized Research Agenda*, Technical Report 93–4. National Center for Geographic Information and Analysis, Santa Barbara, CA. http://www.ncgia.ucsb.edu/Publications/Tech_Reports/93/93–4.PDF (last accessed: March 21 2007).

Estes, J. E., E. J. Hajic, and L. R. Tinney, 1983. Fundamentals of image analysis: Analysis of visible and thermal infrared data. In: D. Simonett (ed.), *Manual of Remote Sensing*. American Society for Photogrammetry and Remote Sensing, Bethesda, MD. Chapter 24: 987–1124.

Faust, N., 1998. Raster GIS. In: T. W. Foresman (ed.), *The History of Geographic Information Systems: Perspectives from the Pioneers*. Prentice-Hall, Upper Saddle River, NJ. Chapter 5: 59–72.

Florinsky, I. V., 1998. Combined analysis of digital terrain models and remotely sensed data in landscape investigations. *Progress in Physical Geography*, 22(1): 33–60.

Fonseca, F. T., M. J. Egenhofer, P. Agouris, and G. Camara, 2002. Using ontologies for integrated geographic information systems. *Transactions in GIS*, 6(3): 231–257.

Foody, G. M., 1995. Land cover classification by an artificial neural network with ancillary information. *International Journal of Geographical Information Systems*, 9(5): 527–542.

Franklin, S. E., D. R. Peddle, J. A. Dechka, and G. B. Stenhouse, 2002. Evidential reasoning with Landsat TM, Dem and GIS data for landcover classification in support of grizzly bear habitat mapping. *International Journal of Remote Sensing*, 23(21): 4633–4652.

Gao, J., 2002. Integration of GPS with remote sensing and GIS: Reality and prospect. *Photogrammetric Engineering and Remote Sensing*, 68(5): 447–453.

Gesch, D., J. Williams, and W. Miller, 2001. A comparison of US Geological Survey seamless elevation models with Shuttle Radar Topography Mission data. *Geoscience and Remote Sensing Symposium IGARSS '01*, 2: 754–756. http://ieeexplore.ieee.org/iel5/7695/21046/00976625.pdf (last accessed: July 10 2007).

Goodchild, M. F., 1992. Geographical information science. *International Journal of Geographical Information Systems*, 6(1): 31–45.

Goodchild, M. F., 1994. Integrating GIS and remote sensing for vegetation analysis and modeling: methodological issues. *Journal of Vegetation Science*, 5(5): 615–626.

Gruen, A. and H. Li, 1997. Semi-automatic linear feature extraction by dynamic programming and LSB-Snakes. *Photogrammetric Engineering and Remote Sensing*, 63(8): 985–995.

Gu, D., A. R. Gillespie, J. B. Adams, and R. Weeks, 1999. A statistical approach for topographic correction of satellite images by using spatial context information. *IEEE Transactions on Geoscience and Remote Sensing*, 37(1): 236–246.

Hall, G. B., N. W. Malcolm, and J. M. Piwowar, 2001. Integration of remote sensing and GIS to detect pockets of urban poverty: the case of Rosario, Argentina. *Transactions in GIS*, 5(3): 235–253.

Harris, P. M. and S. J. Ventura, 1995. The integration of geographic data with remotely-sensed imagery to improve classification in an urban area. *Photogrammetric Engineering and Remote Sensing*, 61(8): 993–998.

Hepner, G. F., D. J. Wright, C. J. Merry, S. J. Anderson, and S. D. DeGloria, 2005. Remotely acquired data and information in GIScience. In: R. B. McMaster and E. L. Usery (eds),

A Research Agenda for Geographic Information Science. CRC Press, Boca Raton, FL. Chapter 13: 351–364.

Hinton, J. C., 1996. GIS and remote sensing integration for environmental applications. *International Journal of Geographical Information Systems*, 10(7): 877–890.

Hirano A., R. Welch, and H. Lang, 2003. Mapping from ASTER stereo image data: DEM validation and accuracy assessment. *ISPRS Journal of Photogrammetry and Remote Sensing*, 57(5): 356–370.

Hodgson, M., J. R. Jensen, L. Schmidt, S. Schill, and B. Davis, 2003. An evaluation of LIDAR- and IFSAR-derived digital elevation models in leaf-on conditions with USGS Level 1 and Level 2 DEMs. *Remote Sensing of Environment*, 84: 295–308

Homer, C., R. D. Ramsey, T. C. Edwards, Jr., and A. Falconer, 1997. Landscape cover-type modeling using a multiscene Thematic Mapper mosaic. *Photogrammetric Engineering and Remote Sensing*, 63(1): 59–67.

Homer C, C. Huang, L. Yang, B. Wylie, and M. Coan, 2004. Development of a 2001 national land-cover database for the United States. *Photogrammetric Engineering and Remote Sensing*, 70: 829–840.

Hong, N., J. G. White, R. Weisz, C. R. Crozier, M. L. Gumpertz, and D. K. Cassel, 2006. Remote sensing-informed variable-rate nitrogen management of wheat and corn: Agronomic and groundwater outcomes. *Agronomy Journal*, 98: 327–338.

Hu, J., S. You, and U. Neumann, 2003. Approaches to large-scale urban modelling. *IEEE Computer Graphics and Applications*, 23(6): 62–69.

Huang, X. and J. R. Jensen, 1997. A machine-learning approach to automated knowledge-base building for remote sensing image analysis with GIS data. *Photogrammetric Engineering and Remote Sensing*, 63(10): 1185–1194.

Huete, A. R., 2005. Global variability of terrestrial surface properties derived from MODIS visible to thermal-infrared measurements. IEEE 0–7803–9050–4–4/05. http://ieeexplore.ieee.org/iel5/10226/32601/01526782.pdf. (last accessed: March 19 2007).

Hutchinson, C. F., 1982. Techniques for combining Landsat and ancillary data for digital classification improvement. *Photogrammetric Engineering and Remote Sensing*, 48(1): 123–130.

Hyyppä, J., W. Wagner, M. Hollaus, and H. Hyyppä, in this volume. Airborne Laser Scanning. Chapter 14.

Jensen, J. R., 2005. *Introductory Digital Image Processing: A Remote Sensing Perspective* (3rd edn). Prentice-Hall, Upper Saddle River, NJ.

Jensen, J. R., 2007. *Remote Sensing of the Environment: An Earth Resource Perspective* (2nd edn). Prentice-Hall, Upper Saddle River, NJ.

Jensen, J. R., D. J. Cowen, J. D. Althausen, S. Narumalani, and O. Weatherbee, 1993. An evaluation of the Coast-Watch change detection protocol in South Carolina. *Photogrammetric Engineering and Remote Sensing*, 59(6): 1039–1046.

Jensen, J. R., J. Im, P. Hardin, and R. R. Jensen, in this volume. Image Classification. Chapter 19.

Keane, R. E., R. Burgan, and J. van Wagtendonk, 2001. Mapping wildland fuels for fire management across multiple scales: Integrating remote sensing, GIS, and bio-physical modeling. *International Journal of Wildland Fire*, 10: 301–319.

Kellndorfer, J. and K. McDonald, in this volume. Active and Passive Microwave Systems. Chapter 13.

Laliberte, A. S., D. E. Johnson, N. R. Harris, and G. M. Casady, 2001. Stream change analysis using remote sensing and Geographic Information Systems (GIS). *Journal of Range Management*, 54: A22–A50.

Lauer, D. T., J. E. Estes, J. R. Jensen, and D. D. Greenlee, 1991. Institutional issues affecting the integration and use of remotely sensed data and geographic information systems. *Photogrammetric Engineering and Remote Sensing*, 57(6): 647–654.

Lawrence, R. L. and A. Wright, 2001. Rule-based classification systems using classification and regression tree (CART) analysis. *Photogrammetric Engineering and Remote Sensing*, 67(10): 1137–1142.

Lu, D., P. Mausel, E. Brondízio, and E. Moran, 2004. Change detection techniques. *International Journal of Remote Sensing*, 25(12): 2365–2407.

Lunetta, R. S., R. G. Congalton, L. K. Fenstermaker, J. R. Jensen, K. C. McGwire, and L. R. Tinney, 1991. Remote sensing and geographic information system data integration: error sources and research issues. *Photogrammetric Engineering and Remote Sensing*, 57(6): 677–687.

Lunetta, R. S., J. Ediriwickrema, J. Iiames, D. M. Johnson, J. G. Lyon, A. McKerrow, and A. Pilant, 2003. A quantitative assessment of a combined spectral and rule-based land-cover classification in the Neuse River Basin of North Carolina. *Photogrammetric Engineering and Remote Sensing*, 69(3): 299–310.

Lupo, F., I. Reginster, and E. F. Lambin, 2001. Monitoring land-cover changes in West Africa with SPOT Vegetation: impact of natural disasters in 1998–1999. *International Journal of Remote Sensing*, 22(13): 2633–2639.

Ma, Z., M. M. Hart, and R. L. Redmond, 2001. Mapping vegetation across large geographic areas: Integration of remote sensing and GIS to classify multisource data, *Photogrammetric Engineering and Remote Sensing*, 67(3): 295–307.

Mace, T. H. (Guest Editor), 1991. Special Issue: Integration of Remote Sensing and GIS. *Photogrammetric Engineering and Remote Sensing*, 57(6).

Mas, J. F., 2004. Mapping land use/cover in a tropical coastal area using satellite sensor data, GIS and artificial neural networks. *Estuarine, Coastal and Shelf Science*, 59(2): 219–230.

Maselli, F., A. Rodolfi, L. Bottai, S. Romanelli, and C. Conese, 2000. Classification of Mediterranean vegetation by TM and ancillary data for the evaluation of fire risk. *International Journal of Remote Sensing*, 21(17): 3303–3313.

Mason, D. C., D. G. Corr, A. Cross, D. C. Hogg, D. H. Lawrence, M. Petrou, and A. M. Tailor, 1988. The use of digital map data in the segmentation and classification of remotely-sensed images. *International Journal of Geographical Information Systems*, 2(3): 192–215.

McMaster, R. B. and E. L. Usery (eds), 2005. *A Research Agenda for Geographic Information Science*. CRC Press, Boca Raton, FL.

Mena, J. B., 2003. State of the art on automatic road extraction for GIS update: a novel classification. *Pattern Recognition Letters*, 24: 3037–3058.

Mertikas P. and M. E. Zervakis, 2001. Exemplifying the theory of evidence in remote sensing image classification. *International Journal of Remote Sensing*, 22(6): 1081–1095.

Mesev, V., 1998. The use of census data in urban image classification. *Photogrammetric Engineering and Remote Sensing*, 64(5): 431–438.

Nemani, R., P. Votava, J. Roads, M. White, S. Running, and J. Coughlan, 2002. Terrestrial Observation and Prediction System: integration of satellite and surface weather observations with ecosystem models. *Geoscience and Remote Sensing Symposium Proceedings, IGARSS '02*, 4: 2394–2396.

Ortiz, M. J., A. R. Formaggio, and J. C. N. Epiphanio, 1997. Classification of croplands through integration of remote sensing, GIS and historical database. *International Journal of Remote Sensing*, 18(1): 95–105.

Post J. L., L. Borer, G. Priestnall, J. Jaafar, and A. Duncan, 2000. Extracting urban features from LIDAR digital surface models. *Computers, Environment and Urban Systems*, 24(2): 65–78.

Riaño, D., E. Chuvieco, J. Salas, and I. Aguado, 2003. Assessment of different topographic corrections in Landsat-TM data for mapping vegetation types. *IEEE Transactions on Geoscience and Remote Sensing*, 41(5): 1056–1061.

Ricchetti, E., 2000. Multispectral satellite image and ancillary data integration for geological classification. *Photogrammetric Engineering and Remote Sensing*, 66(4): 429–435.

Richards, J. A. and X. Jia, 1999. *Remote Sensing Digital Image Analysis: An Introduction* (3rd edn). Springer, Berlin, Germany.

Rogan, J., J. Miller, D. Stow, J. Franklin, L. Levien, and C. Fischer, 2003. Land-cover change monitoring with classification trees using Landsat TM and ancillary data. *Photogrammetric Engineering and Remote Sensing*, 69(7): 793–804.

Sampson, B. A. and G. D. Delgiudice, 2006. Tracking the rapid pace of GIS-related capabilities and their accessibility. *Wildlife Society Bulletin*, 34: 1446–1454.

Shan, J. and S. D. Lee, 2005. Quality of Building Extraction from IKONOS Imagery. *Journal of Surveying Engineering*, 131(1): 27–32.

Shi, W. Z., M. Ehlers, and M. Molenaar, 2005. Uncertainties in integrated remote sensing and GIS. *International Journal of Remote Sensing*, 26: 2911–2915.

Skidmore, A. K., B. J. Turner, W. Brinkhof, and E. Knowles, 1997. Performance of a neural network: Mapping forests using GIS and remotely sensed data. *Photogrammetric Engineering and Remote Sensing*, 63(5): 501–514.

Smith, G. M. and R. M. Fuller, 2001. An integrated approach to land cover classification: an example in the Island of Jersey. *International Journal of Remote Sensing*, 22(16): 3123–3142.

Toutin, T., 2004. Geometric processing of remote sensing images: Models, algorithms and methods. *International Journal of Geographical Information Systems*, 25(10): 1893–1924.

Tso, B. and P. M. Mather, 2001. *Classification Methods for Remotely Sensed Data*. Taylor and Francis, London, U.K.

Tsou, M.-H., 2004. Integrating web-based GIS and image processing tools for environmental monitoring and natural resource management. *Journal of Geographical Systems*, 6: 155–174.

Vogelmann J. E., T. Sohl, S. M. Howard, and D. M. Shaw, 1998. Regional land cover characterization using Landsat Thematic Mapper data and ancillary data sources. *Environmental Monitoring and Assessment*, 51: 415–428.

Wilkinson, G. G., 1996. A review of current issues in the integration of GIS and remote sensing data. *International Journal of Remote Sensing*, 10(1): 85–101.

Wulder, M., 1998. Optical remote-sensing techniques for the assessment of forest inventory and biophysical parameters. *Progress in Physical Geography*, 22(4): 449–476.

Wulder, M. A., S. E. Franklin, J. C. White, M. M. Cranny, and J. A. Dechka, 2004. Inclusion of topographic variables in an unsupervised classification of satellite imagery. *Canadian Journal of Remote Sensing*, 30(2): 137–149.

Xue, Y., A. P. Cracknell, and H. D. Guo, 2002. Telegeoprocessing: the integration of remote sensing, geographic information system (GIS), global positioning system (GPS) and telecommunication. *International Journal of Remote Sensing*, 23(9): 1851–1893.

Yang L., C. Huang, C. Homer, B. Wylie, and M. Coan, 2003. An approach for mapping large-area impervious surfaces: synergistic use of Landsat 7 ETM+ and high spatial resolution imagery. *Canadian Journal of Remote Sensing*, 29: 230–240.

Yuan, M., B. P. Buttenfield, M. N. Gahegan, and H. Miller, 2005. Geospatial data mining and knowledge discovery. In: R. B. McMaster and E. L. Usery (eds), *A Research Agenda for Geographic Information Science*. CRC Press, Boca Raton, FL. Chapter 14: 365–388.

Zhang, X., M. A. Friedl, and C. B. Schaaf, 2006. Global vegetation phenology from Moderate Resolution Imaging Spectroradiometer (MODIS): Evaluation of global patterns. *Journal of Geophysical Research*, 111: G04017, doi:10.1029/2006JG000217.

Zhou, G., W. Schickler, A. Thorpe, P. Song, W. Chen, and C. Song, 2004. True orthoimage generation in urban areas with very tall buildings. *International Journal of Geographical Information Systems*, 25(22): 5163–5180.

Zhu, A-X., 2005. Research issues on uncertainty in geographic data and GIS-based analysis. In: R. B. McMaster and E. L. Usery (eds), *A Research Agenda for Geographic Information Science*. CRC Press, Boca Raton, FL. Chapter 7: 197–223.

19

Image Classification

John R. Jensen, Jungho Im, Perry Hardin,
and Ryan R. Jensen

Keywords: artificial neural networks, decision trees, machine learning,
maximum likelihood, object oriented classification, support
vector machines, spectral mixture analysis.

INTRODUCTION

Humans have been able to visually identify different types of land-use and land-cover in aerial photography ever since they were first acquired in the mid-1800s. Human visual image classification is based on the use of the fundamental *elements of image interpretation* such as size, shape, shadow, tone, color, texture, pattern, site, association, etc. Visual image classification is still very important and is performed by millions of people every day as they browse remote sensing images on the internet (e.g., using Google Earth or Virtual Earth) to identify features of interest.

Since the mid-1960s, humans have been able to extract land-use/land-cover and biophysical information directly from remote sensor data using digital computers and special-purpose image classification algorithms. Digital image classification algorithms may be based on *parametric* statistics (which assume normally distributed data), *nonparametric* statistics (which do not require normally distributed data), and *non-metric* [those that can operate on both real-valued data (e.g., spectral reflectance data with values from 0 to 100%) and nominal scaled data (e.g., category 1 $= \leq 5\%$ slope, category 2 $= 6–10\%$ slope)]. Digital image classification algorithms make use of *supervised* or *unsupervised* classification logic. Furthermore, the classification algorithms can process individual pixels (referred to as *per-pixel classifiers*) or groups of contiguous pixels (referred to as *object-oriented classifiers*). This chapter introduces some of the fundamental types of supervised and unsupervised classification algorithms based on per-pixel and object-oriented classification logic.

Land-use and/or land-cover information derived from remote sensor data should be categorized using a documented classification system such as the U.S. Geological Survey's *Land Use and Land Cover Classification System for Use with Remote Sensor Data* or the American Planning Association's *Land Based Classification System* so that the information extracted may be shared. Jensen (2005) summarizes several of the most important image classification schemes.

SUPERVISED CLASSIFICATION

The objective of a supervised classification is to categorize every image pixel into one of several predefined landtype classes. The five phases of supervised classification are class definition, preprocessing, training, automated pixel assignment, and accuracy assessment. These phases are seldom performed in isolation. For example, if the accuracy assessment phase reveals that certain landtype classes are poorly distinguished, the pixel assignment phase might be revisited using a different classifier.

The problem of supervised landtype classification is one of *pattern recognition*. Pattern recognition is sometimes called *machine learning*. A *classifier* is an algorithm that learns patterns associated with a set of landtype categories and then uses that knowledge to place new patterns into one of the learned categories. The *pattern* or *pattern vector* (**x**) is a set of characteristics used to classify the pixel. The individual measurements in the pattern vector are called *features*. The typical feature pattern for a landtype category is often referred to as its *signature*. Features typically include reflectance values from the several image bands (e.g., green, red, and near-infrared reflectance) or new transformed spectral information created based on principal component scores, vegetation index values or texture measures. Sometimes features include *ancillary data* such as pixel slope and elevation measured from a digital terrain models.

Simple landtype classification with multispectral satellite imagery is usually done on a per-pixel basis. This means that each pixel in the study area is presented to the classifier in-turn for categorization without regard to the information in the pixels surrounding it. In supervised classification, this presentation is called the *pixel assignment step*. More innovative approaches do not treat pixels in isolation but consider neighborhoods. *Texture metrics* are used to quantify the distribution of spectral reflectance in a pixel's immediate vicinity. *Contextual classifiers* look at the landtypes within a pixel's neighborhood to detect and fix illogical class assignments. *Object oriented approaches* categorize polygons rather than individual pixels.

Maximum likelihood classification

Even after 40 years of use, the *maximum likelihood classifier* is still the workhorse of supervised image classification. When conducting a maximum likelihood classification, one must know quite a bit about the land-cover present in the study area. This knowledge is usually obtained on the ground or through exposure to many types of ancillary data such as elevation and slope, etc. This allows the analyst to stipulate the relative landtype proportions within the study area before the supervised classification begins. These proportions can be expressed as *prior probabilities* (or *priors*). In image classification, priors are the probability that an unclassified pixel belongs to a certain class before its pattern vector is examined. For example, if the landtype schema has c classes, prior probability for a class can be represented by P_i with $i \in \{1, 2 \ldots c\}$. In the absence of any other information, a simple rule for classification would be to assign an unlabeled pixel to the class which has the highest prior probability:

Assign pixel to class i if $P_i > P_j$ for all $i \neq j$.
$$(1)$$

Although this rule is easy to understand, it isn't helpful since it would produce a map entirely of one landtype class, the landtype class with the highest prior. A better rule would utilize the information available in the pattern vectors. This rule would use conditional probabilities, i.e., the probability that a pixel is a member of class i conditional on its pattern vector **x**. If we represent these conditional probabilities as $P_{i|\mathbf{x}}$, the assignment rule is:

Assign pixel to class i if $P_{i|\mathbf{x}} > P_{j|\mathbf{x}}$ for all $i \neq j$.
$$(2)$$

This is called *Bayes rule*. By assigning a pixel to the class with the highest conditional probability, Bayes rule minimizes the total error of classification. For remote sensing problems where hundreds of spectral patterns are possible, the difficulty of determining $P_{i|\mathbf{x}}$ for each pattern precludes the rule's use. In contrast, it is much easier to estimate $P_{\mathbf{x}|i}$. The value of $P_{\mathbf{x}|i}$ is the probability of observing **x** within each class i. The relationship between $P_{\mathbf{x}|i}$ and $P_{i|\mathbf{x}}$ is stated by *Bayes' theorem* as (Duda et al. 2001):

$$P_{i|\mathbf{x}} = \frac{(P_{\mathbf{x}|i})(P_i)}{\sum_{k=1}^{c}(P_{\mathbf{x}|k})(P_k)} \qquad (3)$$

Substituting Bayes' theorem into Bayes' rule and reducing the expression produces a workable classification rule (James 1985):

Assign pixel to class i if $(P_{\mathbf{x}|i})(p_i) > (P_{\mathbf{x}|j})(p_j)$
for all $i \neq j$.
$$(4)$$

All of the elements required to serve this rule can be found by sampling the remote sensor data. Once the landtype schema is specified, a simple geographic random sample and visual inspection of satellite imagery can estimate P_i for each class. The probability density function of $P_{\mathbf{x}|i}$ for each class is found by locating a set of representative training pixels for each class, extracting the **x** for each, summarizing them, and modeling the $P_{\mathbf{x}|i}$ distribution. The Gaussian maximum likelihood classifier uses Bayes' rule to classify unlabeled pixels. It estimates $P_{\mathbf{x}|i}$ as a multivariate normal distribution:

$$E_{\mathbf{x}|i} = \frac{1}{(2\pi)^{b/2}\,|\mathbf{\Sigma}_i|^{1/2}}$$
$$\times \exp\left[-\frac{1}{2}(\mathbf{x}-\mathbf{\mu}_i)^T \mathbf{\Sigma}_i^{-1}(\mathbf{x}-\mathbf{\mu}_i)\right] \quad (5)$$

where $E_{x|i}$ is the estimate of $P_{x|i}$ for an unlabeled pixel with pattern x. The symbol Σ_i represents the covariance matrix for group i whereas $|\Sigma_i|$ and Σ_i^{-1} represent its determinant and inverse. The value of $(x - u_i)$ is the difference between the pattern vector to be classified (x) and the vector of band means for the class (μ_i). For any pixel needing classification, Equation (5) is used to calculate $E_{x|i}$ for each class. Following Equation (4), the pixel would then be assigned to the class producing the highest product $(E_{x|i})(P_i)$.

For a hyperspectral image, the use of Equation (5) can create a significant computing burden. By algebraic manipulation, a simpler formula will produce identical calculation results when a slightly different decision rule is used (James 1985).

$$\text{Let } d_i(x) = \ln|\Sigma_i| + (x - \mu_i)^T \, \Sigma_i^{-1}(x - \mu_i)$$
$$- \ln(P_i), \quad \text{and} \qquad (6)$$

assign pixel to class i if $d_i(x) < d_j(x)$ for all $i \neq j$.
$$(7)$$

Equation (6) is called a *discriminant function*. Discriminants used in remote sensing include the linear and quadratic functions (Hardin 1989). Each requires different assumptions about the class covariance matrices and both can use decision rules weighted according to prior probabilities (Duda et al. 2001).

Supervised maximum-likelihood classification has consistently been the first choice of remote sensing practitioners but it has drawbacks. Although maximum-likelihood is a popular classifier, it can produce mediocre classification results under certain conditions:

- *Condition 1. When different landtypes have similar spectral signatures.*
- *Condition 2. When a single landtype has multiple signatures.* For example, assume that a single 'grassland' landtype has been specified in a rangeland classification schema. Depending on grazing pressure, soil moisture, species mix, etc., several grassland signatures may be manifest. In cases like this, it would seem best to treat all spectral variants of a landtype separately in the training and classification phase and combine them into a single map category afterwards.
- *Condition 3. When many parameters need estimation.* The maximum likelihood classifier and its related discriminant functions require the estimation of several parameters. If we represent the number of features in each pattern vector as b, each group mean vector requires b estimates. Each covariance matrix requires $(b^2/2 + b/2)$ parameters. For a problem involving c classes, the

parameter total would equal $cb + c(b^2/2 + b/2)$ estimates. Sufficient training data must be collected to accurately estimate these parameters. Richards (1986) recommends 10 to 100 training pixels per class, but we recommend 30 to 100 pixels for *each mean* in the problem. For a simple problem with four bands and three classes, this would equal 360 to 1,200 training pixels divided evenly between the categories. In practice, although the requisite total number of pixels may be collected, they are seldom distributed uniformly between the classes in the schema. As a result, some training classes are parameterized adequately while others are poorly characterized. The problem of parameter estimation with limited training data is addressed by using the *minimum distance to means* rule and the *parallelpiped classifier* rule (Jensen 2005: 371–373). The number of parameters required for the two rules are cb and $2cb$, respectively. Although these rules require the estimation of only a few parameters per class, they seldom produce higher accuracy than the maximum-likelihood classifier.

- *Condition 4. When spectral distributions are not multivariate normal.* When maximum likelihood classification is performed, the spectral distribution of brightness values for each class is assumed multivariate normal. Although many practitioners accept the precondition, Richards (1986) does not consider multivariate normality 'a demonstrable property of natural spectra or information classes.' Severe normality problems can arise when a practitioner appends an ordinal or nominal ancillary dataset to a satellite image stack and treats it like any other spectral band. Vegetation indexes can similarly cause problems.
- *Condition 5. When prior probabilities are unknown.* Mather (1985) suggests that the accuracy of maximum likelihood classification degenerates when prior probabilities for each landtype class are unknown. Under these conditions, analysts usually assume the priors are equal. This is understandable, but results in a high misclassification rate among pixels where no class produces a high $E_{x|i}$ i.e., those pixels on the periphery of the class distributions.

Many innovations in supervised classification are designed to mitigate the problems created by these conditions.

Nearest-neighbor classification

Because multivariate normality in spectral data cannot be assured, significant research has focused on the use of distribution-free (i.e., nonparametric) classifiers. For example, the first nearest neighbor classifier utilizes no assumptions about group shape in the classification process to minimize the overall error rate (James 1985). Given a pixel to be classified (z) and a training pixel (t), the equation for the nearest neighbor distance between them is:

$$d_{nn} = \sum_{i=1}^{b} (x_{zi} - x_{ti})^2 \qquad (8)$$

where x_{zi} is the brightness value for band i for unlabeled pixel z, and x_{ti} is the brightness value for same band for a training pixel. When doing pixel assignment, the pixel is given the label of the group that its nearest neighbor belongs to. The first nearest neighbor is the pixel that generates the smallest value of d_{nn} among all the training pixels. Hardin (1994) reported a variety of extensions to the first nearest neighbor rule where a label is assigned by taking a majority vote of the training pixels in the unlabeled pixel's spectral neighborhood.

Because d_{nn} is the square of the Euclidean distance[1] between two points, the nearest neighbor classifier is sensitive to differences in the scaling of the bands and the relative magnitude of the digital numbers within the bands. This deficiency can be problematic for image data sets where rescaled ratios and texture bands are used in conjunction with brightness values. Under these conditions data should be standardized before using a nearest neighbor classifier. Although it is possible to weight the distance values according to prior probability, it is unnecessary if the proportion of training pixels in the various classes approximate their priors in the target scene (James 1985). Because the first nearest neighbor classifier is nonparametric, inefficiency is expected when (1) classes meet the parametric multivariate qualifications; or (2) classes are relatively distinct and normality deviations are inconsequential.

When the nearest-neighbor classifier was first introduced to remote sensing practitioners, its slow execution speed precluded its adoption. This issue was later solved by storing the training pixels in a search tree rather than a linear array (Hardin and Thomson 1992).

Artificial Neural Network classification

Artificial Neural Networks (ANNs) have found wide use in image classification. Research on ANNs has been motivated by the recognition that human brains are very efficient processing vast quantities of information from different sources. The neuron is the fundamental processing unit of an ANN and is analogous to the human brain's biological neurons. ANN behavior resembles the brain in two respects: (1) knowledge is acquired through a repetitive training process, and (2) inter-neuron connection strengths known as synaptic weights are used to store the knowledge.

Neural network classifiers use an interconnected assembly of neurons, whose processing capability is stored in weights learned from a set of training patterns (Gurney 1997). In a simple classification task, an ANN contains an input layer, hidden layers, and an output layer. The input layer is given feature patterns to learn or classify whereas the output layer represents the nominal landtype categories. The hidden layers adaptively model the output based on the input patterns by adjusting weighted synapses at the network nodes (Jensen 2005).

ANNs are nonparametric. They do not require that any statistical assumption be satisfied. They adaptively estimate continuous functions from data without mathematically describing how outputs depend on inputs (e.g., adaptive model-free function estimation using a nonalgorithmic strategy; Gopal and Woodcock 1996). ANNs have been used in remote sensing to classify images (Hardin 2000), extract biophysical characteristics (Jensen et al. 2000), and incorporate multisource data. ANNs can accommodate non-normally distributed numerical and categorical GIS data and image spatial information. ANN classifiers usually outperform statistically-based remote sensing classifiers because they are distribution free, easily incorporate multidimensional datasets, and capture nonlinearity in the data. Numerous studies have demonstrated the utility of employing ANNs with remote sensing data.

- Gopal and Woodcock (1996) found that ANNs were able to monitor California conifer mortality more accurately than traditional change detection algorithms.
- Weiss and Baret (1999) used ANNs to measure photosynthetically active radiation and leaf area index (LAI) for the International Satellite Land Surface Climatology Project. They found that ANNs predicted biophysical variables more accurately than NDVI-based methods.
- Hardin (2000) found that a probabilistic ANN significantly outperformed a parametric maximum likelihood classifier in six different image scenes.
- Jensen et al. (2000) used an ANN to discriminate conifer stand age in Brazil and concluded that ANNs were able to model the complex nonlinearity of biophysical processes and were better at extracting conifer stand age than traditional image processing techniques.
- Jensen and Binford (2004) found that ANNs were more accurate than multiple regression techniques for estimating LAI in Florida. Likewise,

Menzies et al. (2007) discovered that ANNs were better estimators of LAI near Santarem, Brazil. Jensen and Hardin (2005) determined that ANNs were more accurate than statistical techniques for estimating urban LAI.

Different factors affect ANN performance and accuracy, including: number of hidden layers and neurons, and learning/momentum rate. ANNs with few neurons and only one hidden layer usually generalize training patterns more effectively. Conversely, complex networks can more precisely characterize training sample patterns but have a lower capacity to generalize and correctly classify pixel patterns dissimilar to training set patterns. If a high learning rate is selected, the network will learn training patterns quickly. Unfortunately, this network may be unstable. If different training data are applied to the network, a different solution may result (e.g., Menzies et al. 2007). This instability can sometimes be mitigated by modifying the momentum rate to allow for relatively fast learning while not allowing quick changes in learning direction (i.e., momentum). The primary problem with slow learning is that the user has to wait until a network trains and cannot conveniently experiment with alternative network configurations in an interactive manner.

There are a large number of network types available and each type has several parameters that require specification. Given this flexibility, it is difficult to select the most accurate architecture without time-consuming experiments. Current research focuses on making ANN classification more automated and providing user interfaces that are more easily understood. To this end, new image processing software packages contain ANN classification algorithms with more default parameters and architectures. Despite these improvements the 'black box' nature of ANNs has been difficult for some researchers to overcome. ANNs are being demystified as books and articles are published that describe ANNs in language more understandable to the nonspecialist.

Decision tree classification

Decision trees have tree-like structures where leaves in the tree represent classes and branches represent conjunctions of features that lead to the classes (Jensen 2005). Since decision trees make no assumptions about data distributions and independency, they have been used in remote sensing classification as alternatives to traditional parametric classifiers such as *maximum-likelihood* classification.

A decision tree takes a set of attributes as input, which can be discrete or continuous, and returns an output (i.e., decision) through a sequence of tests (Russell and Norvig 2003). The output value can be discrete or continuous. For example in image classification, analysts are interested in extracting discrete class information (e.g., forest, agriculture). In this case the process is based on *discrete-valued functions* and is called *classification learning*. However, many applications require the extraction of biophysical information about a pixel. Such classification is based on the use of *continuous functions* and is called *regression learning* (Lawrence and Wright 2001).

Decision trees predict class membership by recursively partitioning a given dataset into more homogeneous sub-groups (DeFries and Chan 2000). Binary (i.e., Boolean) decision trees, wherein each node including input is evaluated to determine if the condition is true (positive) or false (negative), have been widely adopted for image classification. More sophisticated (e.g., trinary) decision trees can be used to solve more complicated classification problems.

Remote sensing scientists can create decision trees using their knowledge base about a subject. But, this human domain process may result in a well-known problem, which is commonly referred to as the *knowledge acquisition bottleneck*. There are two reasons for the bottleneck, including, (1) the amount of time required to develop the knowledge base; and (2) the human experts' inability to express their knowledge explicitly in a systematic, correct, and complete format that can be used in a computer application. To overcome these difficulties, *machine learning* has been introduced to automate the building of knowledge bases for the creation of decision trees. Either inductive or deduction inference strategies can be used in machine learning. For remote sensing applications, *machine-learning decision trees* based on inductive inference strategy have gained significant momentum (e.g., Huang and Jensen 1997, Hodgson et al. 2003, Tullis and Jensen 2003). The reason for this popularity can be found in their simplicity and the speed of learning from training data. Decision trees based on the inductive machine learning process are considered to be *nonmetric* rather than parametric or non-parametric due to their use of heuristic search techniques from example data.

Numerous machine learning decision tree programs have been developed. Quinlan's C5.0 inductive machine learning decision tree has been widely used for remote sensing image classification (e.g., Im and Jensen 2005). Jensen (2005) summarized the characteristics of C5.0-derived decision trees: (1) it generates a decision tree from an example dataset by using a recursive 'divide and conquer' strategy; (2) based on the entropy concept from communications theory, the most efficient attributes are selected at each node of the decision tree; (3) while decision trees are often difficult to read and interpret, C5.0 converts a generated decision tree to production rules, which consist of a series of *if–then* statements; and (3) when converting the decision tree to production rules, rule confidences and a default class are utilized in a voting procedure to avoid non-classified data. Numerous studies have demonstrated the utility of decision trees with remotely sensed data, including:

- Chan et al. (2001) detected change using different machine learning algorithms including multilayer perceptrons, learning vector quantization, decision tree classifiers, and maximum likelihood classifiers.
- Stefanov et al. (2001) used a decision tree to classify land-cover to monitor urban change. They used Landsat TM data and ancillary data such as digital elevation models, texture, and land-use data to improve the classification accuracy.
- Pal and Mather (2003) assessed the effectiveness of decision trees for land-cover classification versus ANN and/or maximum likelihood methods. Decision trees were successful classifying multispectral data, but did not perform as well as other methods using hyperspectral data.
- Hodgson et al. (2003) mapped imperviousness using high spatial resolution digitized color orthophotography and lidar data. They used three different approaches including maximum likelihood classification, spectral clustering and decision trees.
- Im and Jensen (2005) introduced local correlation image analysis for change detection and created *neighborhood correlation images* (NCIs) associated with bi-temporal high spatial resolution datasets. They used a machine learning decision tree classifier (C5.0) to extract change information from the bi-temporal data including the newly created NCI feature layers.

Decision trees are often compared to ANNs because both techniques use artificial intelligence to extract information. Jensen (2005) summarized differences and similarities associated with decision trees and ANNs:

- Both techniques can be used in classification and/or regression. No assumptions about data distribution and independency are required. Both deal effectively with nonlinear relationships associated with the input variables and can be applied to hyperspectral imagery.
- Categorical data can be easily handled in decision trees while they must be converted to numerical values before use in ANNs.

- Decision trees are simple and rapid and have fewer parameters to adjust than ANNs. However, adjusting the parameters in both techniques can yield dramatically different results.
- Only a few passes through the dataset are needed to generate a decision tree from example data, whereas many cycles (i.e., epochs) are needed to construct neural networks depending on the learning algorithm adopted.
- Decision trees can be converted into a series of 'if – then' rules, which are easier to understand in the decision-making process. It is often difficult to understand and interpret the weights and biases formed during the creation of an ANN.

A decision tree may be used to isolate areas where a neural network might focus. A decision tree approach could also be used to identify the optimum architecture of a neural network.

Support vector machine classification

Support vector machines (SVMs) have been recently used for remote sensing classification and target detection. Comparative studies of SVMs versus ANNs and machine learning decision trees have demonstrated the robustness of SVMs, especially for small training sets (e.g., Huang et al. 2002, Foody and Mathur 2004). SVMs find an optimal separating hyperplane (OSH) between classes using the training samples, which are represented as the support vectors (Foody and Mathur 2006). SVMs focus on solving the optimization problem with training samples, e.g., $(x_1, y_1), \ldots, (x_\lambda, y_\lambda), x \in R^n, y \in \{-1, +1\}$ using the following equations:

$$\min_{w,b,\zeta} \left(\frac{w^T \cdot w}{2} + C \sum_{i=1}^{\lambda} \zeta_i \right) \qquad (9)$$

$$y_i(w^T \cdot \Phi(x_i) + b) \geq 1 - \zeta_i, \quad \zeta_i \geq 0$$

where ζ_i are positive slack variables used to allow some of the samples to fall on the wrong side of the hyperplane, $C \sum_{i=1}^{\lambda} \zeta_i$ is a penalty term to penalize solutions for which ζ_i are very large, and $w^T \cdot \Phi(x_i) + b$ is a hyperplane in a highly dimensional feature space (Su et al. 2006). The basic SVM approach may be extended to optimize nonlinear surfaces using the following decision function:

$$f(x) = \sum_{i=1}^{\lambda} \alpha_i y_i K(x, x_i) + b \qquad (10)$$

where α_i are non-negative Lagrange multipliers used to search the OSH, and $K(x, x_i)$ is a kernel function, which replaces the inner product $(x \cdot x_i)$ in order to solve computational problems in a higher dimensional space (Wang et al. 2005). The selection of the kernel function k is critical for producing successful results. Several widely used SVM types of kernels include linear, polynomial, Gaussian, sigmoid, and spectral angle mapper kernels. One advantage of SVMs is that small training samples, selected on the boundaries of classes, where mixed pixels are common, can lead to accurate classification. This is because only pixels close to the decision boundary constitute support vectors, and therefore contribute to the fitting of OSH (Foody and Mathur 2006).

Object-oriented classification

The availability of high-spatial resolution multispectral imagery from satellite sensors (e.g., QuickBird 61 × 61 cm panchromatic data from *DigitalGlobe, Inc.*) requires new approaches to classify remote sensor data. Traditional classification algorithms based on *per-pixel* analysis may not be ideal to extract the information desired from the high spatial resolution remote sensor data (Visual Learning Systems 2002). The high spatial resolution remote sensor data typically exhibits high frequency components with high contrast and horizontal layover of objects such as buildings and trees due to off-nadir look angles, which are not easily handled in tradition classification algorithms (Jensen 2007, Im and Jensen 2005). Furthermore, Townshend et al. (2000) point out that a proportion of the signal apparently coming from a given pixel may actually be coming from the surrounding terrain pixels.

New *object-oriented* image classification algorithms based on *image segmentation* techniques have been developed to meet this need. *Object-oriented image segmentation* analysis has demonstrated significant advantages for the analysis of high-spatial resolution imagery (e.g., Frohn et al. 2005, Jensen et al. 2006, Im et al. 2008). *Object-oriented* analysis typically incorporates both spectral and spatial information in the *image segmentation* phase by subdividing the image into meaningful homogeneous regions based on shape, texture, size, and other features as well as spectral characteristics, and organizing them hierarchically as image objects (commonly referred to as image segments) (Benz et al. 2004, Blaschke 2005). Image objects created using image segmentation techniques have advantages for image classification, including (1) numerous meaningful feature layers can be derived; and (2) the objects can be extracted from any spatially distributed variables such as elevation and population density. Once homogeneous image objects are created, any classification algorithm such as nearest neighbor, maximum likelihood, decision trees, and neural networks can be used to classify the objects (Civco et al. 2002).

Numerous image segmentation algorithms have been developed and can be roughly grouped into two categories: edge-based algorithms and area-based algorithms. However, only a few algorithms integrate both spectral and spatial characteristics to produce image objects. One of the algorithms widely used in remote sensing digital image classification was developed by Baatz et al. (2001). Their algorithm incorporates individual pixel values and their surrounding pixels to produce image objects based on the following two criteria:

- color criterion (h_{color})
- shape criterion (h_{shape}).

The *color criterion* (h_{color}) is computed as the sum of the standard deviations (σ_k) of spectral values of band k multiplied by its weight (w_k) (Definiens 2003):

$$h_{color} = \sum_{k=1}^{n} w_k \times \sigma_k. \qquad (11)$$

The *shape criterion* (h_{shape}) uses two landscape ecology metrics: compactness and smoothness. It is computed as the sum between compactness ($h_{compact}$) multiplied by its weight ($w_{compact}$) and smoothness (h_{smooth}) multiplied by the weight ($1 - w_{compact}$) (Definiens 2003):

$$h_{shape} = w_{compact} \times h_{compact} + (1 - w_{compact}) \times h_{smooth}. \qquad (12)$$

The user-defined compactness weight ($w_{compact}$) ranges between 0 and 1. Both compactness and smoothness represent shape heterogeneity (or homogeneity): a higher compactness weight helps separate objects with different shapes but not necessarily a lot of spectral contrast. Conversely, a higher smoothness weight helps identify objects that have greater variability between features (Baatz et al. 2004).

Based on these two criteria (i.e., h_{color} and h_{shape}), image objects consisting of relatively homogeneous pixels are created using the following general segmentation function ($I_{segment}$) (Baatz et al. 2001, Definiens 2003):

$$I_{segment} = w_{color} \times h_{color} + (1 - w_{color}) \times h_{shape}. \qquad (13)$$

The user-defined color weight (w_{color}) also ranges between 0 and 1. If the user wants to put greater emphasis on the spectral characteristics of the pixels in the dataset when creating homogeneous objects, w_{color} is weighted more heavily. If the user believes that spatial characteristics of the dataset are more important, then shape (h_{shape}) should be weighted more heavily (Jensen 2005).

The size of resultant image objects is also one of the important factors in the image segmentation process. A *pixel neighborhood* function may be used to determine whether an image object must continue to be grown or whether a new image object should be generated (Definiens 2003). Smaller objects are merged into larger objects based on the user-specified parameters including the spectral and spatial shape parameters (compactness and smoothness) criteria and the neighborhood function logic. Once the smallest growth of an object exceeds a user-specified parameter which specifies the maximum possible change of heterogeneity when several objects are merged, the segmentation process is stopped.

An example of object-oriented image segmentation and classification of an area in Las Vegas, NV is found in Figure 19.1. Pan-sharpened QuickBird high spatial resolution (0.61 × 0.61 cm) multispectral data obtained on 18 May 2003 was used. Two levels of image segmentation are shown in Figure 19.1(a and b) (i.e., scales 100, 150). Image segmentation was performed using color and shape criteria of 0.80 and 0.20, respectively. The spatial parameter was more heavily weighted to smoothness (0.80) than compactness (0.20). Simple nearest neighbor classifiers were applied to each set of image segmentation scale data. Two classification maps based on each segmentation level data are shown in Figure 19.1(c and d).

Object-oriented classification using image objects is substantially different than traditional *per-pixel* classification and has several advantages, including:

- **Flexibility and robustness** Image objects can have diverse attributes of not only spectral information such as *mean reflectance* but also of shape measures associated with each object such as *area*. However, when using some of the attributes, users must define directly their relationships (e.g., membership function), which are sometimes difficult to determine.
- **Rapid processing** The classification process is usually fast since the classification algorithms use much smaller number of processing units (i.e., image objects) than the number of pixels in the dataset. In many studies, the number of image objects for classification is <1% of the number of the pixels in the dataset (e.g., Jensen 2005, Im et al. 2007).
- **Scale** Users may want to use fine-scale image objects to classify detailed heterogeneous

information such as building rooftops in the dataset. Sometimes, larger image objects are of interest when trying to identify large homogeneous areas (e.g., water and forest). Users must decide what scale (level) of image segmentation is most appropriate for their projects.

Image segmentation techniques allow the use of geographical and landscape ecology concepts involving neighborhood, distance, and location to extract useful information from remotely sensed data. Merging or fusion of multiple types of remote sensor data can be conducted through *image segmentation* for a particular application. *Object-oriented image segmentation* and classification is an emerging paradigm when compared with *per-pixel* classification. It will become increasingly important not only for classification but also for use in change detection. Recent studies based on *object-oriented image segmentation* and classification include:

- Frohn et al. (2005) mapped thaw lakes and lake basins on the North Slope of Alaska using Landsat ETM+ imagery. They classified the data using texture analysis, spectral transformations, and object-oriented analysis based on image segmentation.
- Jensen et al. (2006) monitored agricultural land-use with multiple-date Landsat TM/ETM+ and SPOT data in South Africa. They evaluated four classification methods including maximum likelihood, ISODATA, object-oriented image segmentation, and mixture-tuned matched filtering techniques. The object-oriented approach outperformed all other algorithms yielding classification accuracies around 85%.
- Im et al. (2008) performed object-based change detection using correlation image analysis and image segmentation using bi-temporal QuickBird datasets. They created *object/neighborhood correlation images* to extract change information from the composite imagery based on the single set of objects showing bi-temporal topology using two different classification methods (i.e., machine learning decision trees and nearest neighbor classifiers).

Spectral mixture analysis

A spectral *endmember* is the spectral reflectance curve for a pure land-cover in the landscape, such as water and bare soil. *Spectral mixture analysis* identifies the fractions of each spectral endmember within each mixed pixel (Jensen 2005). For example, consider the single mixed pixel consisting

Image Segmentation and Object-based Classification

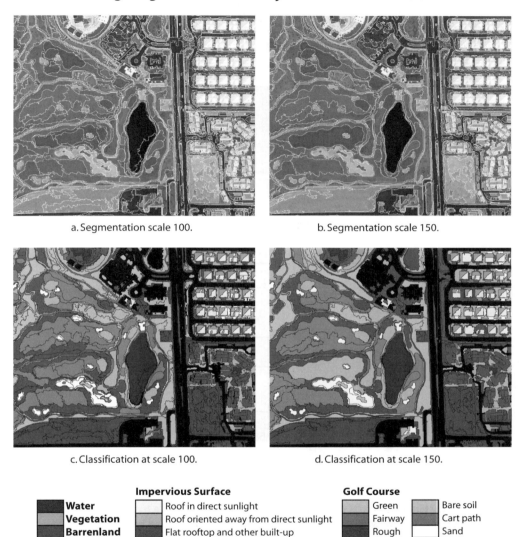

a. Segmentation scale 100.

b. Segmentation scale 150.

c. Classification at scale 100.

d. Classification at scale 150.

■ **Water**	**Impervious Surface**	**Golf Course**	
■ **Vegetation**	☐ Roof in direct sunlight	■ Green	■ Bare soil
■ **Barrenland**	■ Roof oriented away from direct sunlight	■ Fairway	■ Cart path
	■ Flat rooftop and other built-up	■ Rough	☐ Sand
	■ Asphalt		

Figure 19.1 Example of object-based classification at two different image segmentation scales. The pan-sharpened QuickBird high spatial resolution (61 × 61 cm) multispectral imagery of Las Vegas, NV was acquired on 18 May 2003. (See the color plate section of this volume for a color version of this figure).

of 50% vegetation, 30% bare soil, and 20% water (Figure 19.2a). Assuming linear mixing, vegetation spectral contribution to the pixel is 50%. Similarly, the linear mixing associated with any pixel in the scene can be described using the following rules:

$$BV_{ij} = \sum_{k=1}^{n} F_k \cdot R_k + E_{ij} \qquad (14)$$

where BV_{ij} is the apparent surface reflectance (energy) of the pixel at location i, j under investigation, F_k is the fraction of endmember k, R_k is the reflectance (energy) of endmember k, n is the number of spectral endmembers found in the scene, and E_{ij} represents the residual error. The linear combination of all three endmembers (if the spectral abundance of the three endmembers comprise the spectral response of the pixel) must

a. Linear Mixing Model for a Single Pixel

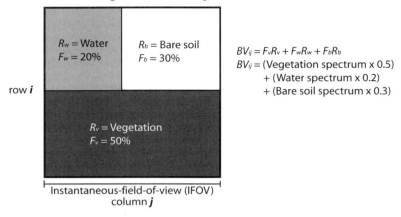

$$BV_{ij} = F_vR_v + F_wR_w + F_bR_b$$
$$BV_{ij} = (\text{Vegetation spectrum} \times 0.5)$$
$$+ (\text{Water spectrum} \times 0.2)$$
$$+ (\text{Bare soil spectrum} \times 0.3)$$

b. Endmembers in Two-dimensional Feature Space

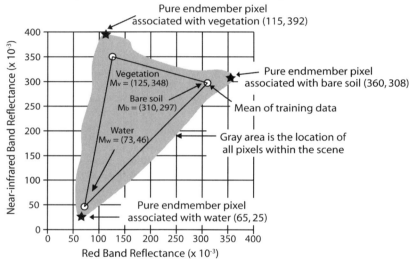

Figure 19.2 (a) Linear spectral mixture modeling associated with a hypothetical pixel consisting of vegetation, water, and bare soil. (b) Hypothetical mean vectors and endmembers in two-dimensional feature space using red and near-infrared bands (modified from Jensen 2005).

sum to 1 as follows:

$$F_v + F_w + F_b = 1 \qquad (15)$$

where F_v, F_w, and F_b are spectral vegetation, water, and bare soil fractions.

Figure 19.2(b) depicts the location of hypothetical mean vectors and endmembers in two-dimensional feature space. In order to understand how linear spectral mixture analysis works, think about a two-band (red and near-infrared) image dataset. Let us assume that there are only three types of pure endmembers (i.e., vegetation, water,

and bare soil) in the data and that we have a sample pixel that has the scaled reflectance of (150, 300). To calculate the proportion (abundance) of vegetation, water, and bare soil endmembers, Formula 14 and the scaled reflectance of the endmembers in Figure 19.1(b) can be used assuming that the data contain no noise (i.e., $E_{ij} = 0$):

$$150 = 115 \times F_v + 65 \times F_w + 360 \times F_b$$
$$300 = 392 \times F_v + 25 \times F_w + 308 \times F_b. \qquad (16)$$

These two equations are not sufficient since we have three unknown parameters (i.e., there are fewer bands than endmembers). Fortunately, because we have Formula 15 (i.e., $F_v + F_w + F_b = 1$), we can calculate the proportion (abundance) of three endmembers. The proportion of vegetation, water, and bare soil found within this sample pixel would be 61%, 21%, and 18%, respectively. This logic can be extended to hyperspectral imagery, where there are generally more bands than endmembers.

Although spectral mixture analysis is usually applied to hyperspectral data to produce a suite of abundance (fraction) images for each endmember, it can be also used with multispectral data. Each fraction image represents a subpixel estimate of an endmember's relative abundance as well as the spatial distribution (i.e., area) of the endmember. Although the spectral mixture analysis is intuitively appealing, many remote sensing scientists report that it is difficult to identify all of the pure endmembers from a dataset. Examples of the use of spectral mixture analysis include:

- McGwire et al. (2000) found that endmembers (average green leaf, soil, and shadow) derived from hyperspectral data were more highly correlated with vegetation percent cover than when the hyperspectral data were processed using traditional narrow-band and broadband vegetation indices (e.g., NDVI, SAVI).
- Lu and Weng (2006) explored thermal features and their relationship with biophysical descriptors in an urban environment using ASTER data. They used linear spectral mixture analysis to extract hot-object and cold-object fraction images from the thermal bands and to extract impervious surface, green vegetation, and soil fractions from visible to shortwave infrared bands.

UNSUPERVISED CLASSIFICATION

In an *unsupervised classification*, the identity of the land-cover types to be specified as classes within a scene are not generally known *a priori* (i.e., before the fact). Consequently, the analyst instructs the computer to use a series of statistically-based heuristic rules to group (identify) pixels with similar spectral characteristics (often referred to as clustering). The analyst must then relabel and combine the spectral clusters (e.g., cluster 1, cluster 2) into traditional information classes (e.g., cluster 1 = water, cluster 2 = forest).

Clustering

Clustering does not require the supervised collection of training data. Rather, the analyst

(1) selects an appropriate clustering algorithm; and (2) requests the computer to identify n mutually exclusive clusters (groups of pixels) in the scene that share common spectral reflectance characteristics (Duda and Canty 2002). Two of the most common clustering methods are the chain method (sometimes called sequential clustering) and the *Iterative Self-Organizing Data Analysis* (ISODATA). It is useful to review the characteristics of these two clustering algorithms.

When using the chain method of clustering, the spectral data are evaluated using two sequential passes through the dataset. During the first pass, a user-specified number of clusters are identified based on the distance to spectral class means. In this pass, the first pixel in the image (row, column 1, 1) becomes the first class mean. The algorithm then moves to the next pixel (1, 2). If this pixel is further away from the first class mean than a user-specified spectral distance, then this pixel becomes a new class mean. If the pixel falls within the distance, then this pixel becomes part of the first class mean. This process occurs sequentially for each pixel in the image. As pixels are assigned to different classes, the class means shift to reflect the new pixel values. After the class means are defined, the spectral data go through a second pass where each pixel is assigned to a spectral class based on Euclidean distance (Jensen 2005).

ISODATA uses a different approach. First, the analyst specifies how many clusters should be identified. Initial mean vectors for the n clusters to be derived are distributed throughout the spectral space based on input band mean and standard deviation characteristics. The algorithm then makes a first pass through the data where each pixel is compared to each cluster mean and assigned to the cluster whose mean is closest. ISODATA is iterative because it makes many passes through the dataset, and in the second pass (and succeeding passes) the cluster means are modified based on actual spectral values of pixels assigned to each cluster. Each pixel is compared to every class mean and assigned to the nearest cluster, and clusters may be merged (if two cluster means are within a pre-defined distance) or separated (if a cluster mean has a large standard deviation). Clusters with less than a specified number of pixels are deleted. This iterative process continues until there is little change in class assignments between iterations or the number user-defined iterations is reached (Jensen 2005).

Clustering can be performed with little or no *a priori* knowledge of the study area. However, after clustering the analyst must assign information classes to the spectral clusters that were generated. In some cases, the spectral class means may not correlate at all with any information class that would be of use to the analyst. Also, land-cover types with very subtle spectral differences may not 'break-out' during the clustering processes. In these cases, analysts usually request many more clusters than they really need and then collapse many spectral clusters into fewer information classes.

Ancillary data as features for use in supervised or unsupervised classification

Ancillary data such as elevation, soil type, socioeconomic variables, etc. are often used to improve the accuracy of multispectral classifications. Generally, a classification based on both spectral and ancillary data should be an improvement over a classification using either data source alone. Ancillary data can also be used *a priori* to regionalize the image into strata that can then be processed independently (i.e., geographical stratification; Jensen 2005). Per-pixel logical channel classification incorporates ancillary data as one of the channels (or bands) used by the classification algorithm.

Ancillary data must be used carefully. The majority of the ancillary data are produced for specific purposes that may be inconsistent with remote sensing image classification. Limitations in ancillary data such as logical consistency, inaccurate topology (e.g., sliver polygons), and small map scale may also preclude their use with remotely sensed data. Ancillary data may also have distribution characteristics that make them unsuitable for parametric classification. Nevertheless, numerous types of ancillary data are often used in the classification process.

THE FUTURE

Great strides will be made in image classification in the future. Instead of down-linking just *data* from a satellite, much of the data will be processed onboard and transmitted as thematic *information* (e.g., vegetation type, LAI). Greater emphasis will be placed on the use of object-oriented image segmentation algorithms that incorporate spatial and well as spectral information in the classification process. Improved landscape ecology metrics will make the spatial component more robust. There will be numerous advancements in feature selection and data dimensionality reduction, especially with regard to the use of hyperspectral data for image classification. Major advances will be made in the development of vegetation and urban transformation indices that will enable more useful information to be extracted from agriculture areas and urban landscapes.

NOTES

1. Although Euclidean distance measurement is common, there are alternative distance measurement that can be used. (Jensen 2005: 374).

REFERENCES

Baatz, M., U. Benz, S. Dehghani, M. Heymen, A. Holtje, P. Hofmann, I. Ligenfelder, M. Mimler, M. Sohlbach, M. Weber, and G. Willhauck, 2001. *eCognition User Guide*, Definiens Imaging GmbH, Munich, 310 p.

Baatz, M., U. Benz, S. Dehghani, and M. Heynen, 2004. *eCognition User Guide 4*. Definiens Imagine GmbH, Munchen, Germany.

Benz, U., P. Hofmann, G. Willhauck, I. Lingenfelder, and M. Heynen, 2004. Multi-resolution, object-oriented fuzzy analysis of remote sensing data for GIS-ready information, *ISPRS Journal of Photogrammetry and Remote Sensing*, 58: 23.

Blaschke, T., 2005. A framework for change detection based on image objects. In S. Erasmi, B. Cyffka, and M. Kappas, (eds.), *Göttinger Geographische Abhandlungen, 113*, Göttingen, pp. 1–9.

Chan, J. C., K. Chan, and A. G. Yeh, 2001. Detecting the nature of change in an urban environment: A comparison of machine learning algorithms. *Photogrammetric Engineering and Remote Sensing*, 67(2): 213–225.

Civco, D. L., J. D. Hurd, E. H. Wilson, M. Song, and Z. Zhang, 2002. A comparison of land use and land cover change detection algorithms, *Proceedings, ASPRS-ACSM Annual Conference and FIG XXII Congress*, ASP&RS, Bethesda, 12 p.

Definiens, 2003. *eCognition Professional*, Version 3.0 Munich: Definiens Imaging GmbH, www.definiens-imaging.com (last date accessed September 7 2007).

DeFries, R. S. and J. C. Chan, 2000. Multiple criteria for evaluating machine learning algorithms for land cover classification. *Remote Sensing of Environment*, 74: 503–515.

Duda, T. and M. Canty, 2002. Unsupervised classification of satellite imagery: choosing a good algorithm. *International Journal of Remote Sensing*, 23: 2193–212.

Duda, R. O., P. E. Hart, and D. G. Stork, 2001. *Pattern Classification*. New York: John Wiley & Sons, 654 p.

Foody, G. M. and A. Mathur, 2004. A relative evaluation of multiclass image classification by support vector machines. *IEEE Transactions on Geoscience and Remote Sensing* 42: 1335–1343.

Foody, G. M. and A. Mathur, 2006. The use of small training sets containing mixed pixels for accurate hard image classification: Training on mixed spectral responses for classification by a SVM. *Remote Sensing of Environment*, 103: 179–189.

Frohn, R. C., K. M. Hinkel, and W. R. Eisner, 2005. Satellite remote sensing classification of thaw lakes and drained thaw lake basins on the North Slope of Alaska. *Remote Sensing of Environment*, 97: 116–126.

Gopal, S. and C. E. Woodcock, 1996. Remote sensing of forest change using artificial neural networks. *IEEE*

Transactions on Geoscience and Remote Sensing, 34: 398–404.

Gurney, K., 1997. *An Introduction to Neural Networks*. UCL Press Ltd., London.

Hardin, P. J., 1989. A Logic for the Pixel Assignment Step of Imagery Classification Predicated on Class Structure, Ph.D. Dissertation, Department of Geography, University of Utah.

Hardin, P. J., 1994. Parametric and nearest-neighbor methods for hybrid classification. *Photogrammetric Engineering and Remote Sensing*, 60: 1439–1448.

Hardin, P. J., 2000. Neural networks versus nonparametric neighbor-based classifiers for semisupervised classification of Landsat Thematic Mapper imagery. *Optical Engineering*, 39: 1898–1908.

Hardin, P. J. and C. Thomson, 1992. Fast nearest-neighbor classification methods for multispectral imagery. *The Professional Geographer*, 44: 191–201.

Hodgson, M. E., J. R. Jensen, J. A. Tullis, K. D. Riordan, and C. M. Archer, 2003. Synergistic use of LIDAR and color aerial photography for mapping urban parcel imperviousness. *Photogrammetric Engineering and Remote Sensing*, 69(9): 973–980.

Huang, X. and J. R. Jensen, 1997. A machine learning approach to automated construction of knowledge bases for image analysis expert systems that incorporate GIS data. *Photogrammetric Engineering and Remote Sensing*, 63(10): 1185–1194.

Huang, C., L. S. Davis, and J. R. Townshend, 2002. An assessment of support vector machines for land cover classification. *International Journal of Remote Sensing*, 23(3): 725–749.

Im, J. and J. R. Jensen, 2005. A change detection model based on neighborhood correlation image analysis and decision tree classification, *Remote Sensing of Environment*, 99: 326–340.

Im, J., J. R. Jensen, and J. A. Tullis, 2008. Object-based change detection using correlation image analysis and image segmentation techniques. *International Journal of Remote Sensing*, 29(2): 399–423.

James, M., 1985. *Classification Algorithms*. Wiley-Interscience, New York.

Jensen, J. R., 2005. *Introductory Digital Image Processing: A Remote Sensing Perspective 3rd*. Prentice Hall, Upper Saddle River, NJ, 525 p.

Jensen, J. R., 2007. Remote *Sensing of the Environment: An Earth Resource Perspective*. Prentice Hall, Upper Saddle River, NJ. 592 p.

Jensen, R. R. and M. W. Binford, 2004. Measurement and comparison of leaf area index estimators derived from satellite remote sensing techniques. *International Journal of Remote Sensing*, 25: 4251–4265.

Jensen, R. R. and P. H. Hardin, 2005. Estimating urban leaf area using field measurements and satellite remote sensing data. *Journal of Arboriculture*, 31(1): 21–27.

Jensen, J. R., Q. Fang, and J. Minhe, 2000. Predictive modelling of coniferous forest age using statistical and artificial neural network approaches applied to remote sensor data. *International Journal of Remote Sensing*, 20: 2805–2822.

Jensen, J. R., M. Garcia-Quijano, B. Hadley, J. Im, Z. Wang, A. L. Nel, E. Teixeira, and B. A. Davis, 2006. Remote sensing agricultural crop type for sustainable development in South Africa. *Geocarto International*, 21(2): 5–18.

Lawrence, R. L. and A. Wright, 2001. Rule-based classification systems using classification and regression tree (CART) analysis. *Photogrammetric Engineering and Remote Sensing*, 67(10): 1137–1142.

Lu, D. and Q. Weng, 2006. Spectral mixture analysis of ASTER images for examining the relationship between urban thermal features and biophysical descriptors in Indianapolis, Indiana, USA. *Remote Sensing of Environment*, 104: 157–167.

Mather, P. M., 1985. A computationally-efficient maximum-likelihood classifier employing prior probabilities for remotely sensed data. *International Journal of Remote Sensing*, 6: 369–376.

McGwire, K., T. Minor, and L. Fenstermaker, 2000. Hyperspectral mixture modeling for quantifying sparse vegetation cover in arid environments. *Remote Sensing of Environment*, 72: 360–374.

Menzies, J., R. R. Jensen, E. Brondizio, E. Moran, and P. Mausel, 2007. The accuracy of neural network and regression leaf area estimators in the Amazon Basin. *GIScience and Remote Sensing*, in press.

Pal, M. and P. M. Mather, 2003. An assessment of the effectiveness of decision tree methods for land cover classification. *Remote Sensing of Environment*, 86: 554–565.

Richards, J. A., 1986. *Remote Sensing Digital Image Analysis: An Introduction*. Springer-Verlag, Berlin.

Russell, S. J. and P. Norvig, 2003. *Artificial Intelligence: A Modern Approach*, 2nd edn, Prentice-Hall, Upper Saddle River, NJ, 1080 p.

Stefanov, W. L., M. S. Ramsey, and P. R. Christensen, 2001. Monitoring urban land cover change: An expert system approach to land cover classification of semiarid to arid urban centers. *Remote Sensing of Environment*, 77: 173–185.

Su, L., M. J. Chopping, A. Rango, H. V. Martonchik, and D. P. Peter, 2006. Support vector machines for recognition of semi-arid vegetation types using MISR multi-angle imagery. *Remote Sensing of Environment*, 107: 299–311.

Townshend, J. R. G., C. Huang, S. Kalluri, R. DeFries, S. Liang, and K. Yang, 2000. Beware of per-pixel characterization of land cover. *International Journal of Remote Sensing*, 21(4): 839–843.

Tullis, J. A. and J. R. Jensen, 2003. Expert system house detection in high spatial resolution imagery using size, shape, and context. *Geocarto International*, 18(1): 5–15.

Visual Learning Systems, 2002. *User Manual: Feature Analyst Extension for ArcView/ArcGIS*, Visual Learning Systems, Missoula, MT.

Wang, J. G., P. Neskovic, and L. N. Cooper, 2005. Training data selection for support vector machines. *Lecture Notes in Computer Science*, 3600: 554–564.

Weiss, M. and F. Baret, 1999. Evaluation of canopy biophysical variable retrieval performances from the accumulation of large swath satellite data. *Remote Sensing of Environment*, 70: 293–306.

Quantitative Models and Inversion in Optical Remote Sensing

Shunlin Liang

Keywords: radiative transfer, land, modeling, inversion, biogeophysical parameters.

INTRODUCTION

In the past several decades, vast amounts of remotely sensed Earth observations have been acquired and accumulated. With a series of new satellite programs planned, the data volume will continue to increase. Automating the procedures for processing and analyzing these data is critical. Furthermore, various numerical models characterizing Earth's environments, each associated with a different decision support system, have to be calibrated and run with spatially and temporally explicit data sets produced only from remotely sensed data at the appropriate scales.

Quantitative estimation of land surface variables from satellite observations is challenging. It requires not only the understanding of the remotely sensed signals through the physical modeling approaches, but also effective inversion algorithms. This chapter focuses on models and inversion algorithms for estimating land surface variables with the emphasis on the visible and near-infrared (IR) spectrum. The next section provides the overview of physical modeling of remote sensing signals. It starts with atmospheric radiative transfer, followed by surface radiation modeling and the sensor models. The following section presents the state-of-the-art inversion

algorithms and summarizes some of the key products generated using these algorithms, followed by a section on validation with ground measurements.

MODELING TECHNIQUES

To understand remote sensing signals, land surface variables must be linked to at-sensor radiance recorded at the top (or middle) of the atmosphere using various physically based models (Liang 2004). These models can be used for simulating remote sensing signals by changing surface and atmospheric conditions, and for developing practical inversion algorithms to estimate land surface variables from satellite observations. Physically based models in three areas are discussed below: atmosphere, land surface, and sensor.

Atmospheric radiative transfer modeling

The signals recorded by a sensor include both atmosphere and surface information. Atmospheric gases, aerosols, and clouds scatter and absorb incoming solar radiation and the reflected and/or emitted radiation from the surface. The atmosphere greatly modulates the spectral dependence and spatial distribution of the surface radiation. In order to estimate land surface variables, it is very important to understand the radiative transfer within the atmosphere so that atmospheric effects can be effectively removed from remotely sensed data.

If the atmospheric particles can be assumed to be isotropic, the one-dimensional (1D) radiative transfer equation for radiance $I(\mu, \phi)$ at any direction $(\Omega: [\mu, \phi])$ in the solar spectrum can be written as:

$$
\mu \frac{dI(\tau, \mu, \phi)}{d\tau} = I(\tau, \mu, \phi)
$$
$$
- \frac{\omega}{2\pi} \int_0^{2\pi} \int_{-1}^1 I(\tau, \mu_i, \phi_i)
$$
$$
\times P(\mu, \phi, \mu_i, \phi_i) \, d\mu_i \, d\phi_i \quad (1)
$$

where μ is the consine of the viewing zenith angle, ϕ is the viewing azimuth angle, τ is called *optical depth* or *optical thickness*, depending on the geometric height z and the extinction coefficient $(\sigma_e)\tau = \int_0^z \sigma_e(z)dz$, ω is the single scattering albedo, and P is the phase function. These optical parameters $(\sigma_e, \omega,$ and $P)$ can be determined from the aerosol models that are characterized by the particle size distribution, refractive index and so on.

Eq. (1) cannot be solved unless the boundary conditions are specified. For a 1D atmosphere, two boundary conditions are needed. At the top of the atmosphere, there is only direct solar radiation. The lower boundary condition at the Earth surface is characterized by the surface bidirectional reflectance factor (BRF).

Numerous algorithms have been developed to solve Equation (1) and some of them have been incorporated into software packages to calculate the atmospheric radiation fields, such as MODTRAN (http://www.ontar.com/Software/ProductDetails.aspx?item=MODTRAN4) and 6S (ftp://kratmos.gsfc.nasa.gov/pub/eric/6S/).

Surface radiation modeling

All land surface radiation models can be classified into three groups: radiative transfer (RT), geometric-optical (GO), and computer simulations. For vegetation canopies, RT models are more suitable for dense vegetation, while GO models are more suitable for sparse vegetation. The distinction between RT and GO models has become fuzzy lately because hybrid models that integrate RT and GO models have been developed. Computer simulation models can characterize canopy radiation field accurately by taking into account three-dimensional (3D) structures explicitly but require

extensive computer resources most appropriate for simulations.

RT modeling

RT models deal with radiation transport within a turbid medium that is composed of small particles. The average optical properties of the medium or its sub-region are used to calculate the radiation quantities based on the RT equation. The 1D RT equation of a flat homogeneous canopy for radiance I at the direction (Ω) is simplified as:

$$
- \mu \frac{\partial I(\tau, \Omega)}{\partial \tau} + h(\tau, \Omega)G(\Omega)I(\tau, \Omega)
$$
$$
= \frac{1}{\pi} \int_0^{2\pi} \int_{-1}^1 \Gamma(\Omega' \to \Omega)I(\tau, \Omega') \, d\Omega' \quad (2)
$$

where $G(\Omega)$ is a geometric function usually defined to represent the mean projection of a unit foliage area in the direction Ω and largely determined by leaf angle distribution (LAD), $\Gamma(\Omega', \Omega)$ is the area scattering phase function which is mainly dependent on leaf optics (reflectance and transmittance) and LAD, $h(\tau, \Omega)$ is an empirical correction function to account for the variation of the extinction coefficient. The detailed formulae for these functions are available elsewhere (e.g., Liang 2004) The optical depth τ can be defined by the leaf area density $u_l(z)$ as $\tau(z) \equiv \int_0^z u_l(z) \, dz$, and the total optical depth of the canopy layer is called leaf area index (LAI).

It is easy to see that Equations (1) and (2) are very similar, and the similar algorithms for solving atmospheric RT equation can be equally applied to land surfaces RT equations. The RT equations for soil and snow are even much more similar to the atmospheric RT (Liang 2004).

There are also 3D RT models developed for canopies (e.g., Myneni et al. 1990). Gastellu-Etchegorry et al. (1996) developed a 3D canopy model by representing the canopy as voxels with assumed turbid-medium scattering within each cell.

Developing new radiative transfer models has slowed significantly in recent years. Very few examples can be found in the recent literature. Pitman et al. (2005) applied the numerical RT algorithm to calculate quartz emissivity. Kokhanovsky et al. (2005) developed an approximate snow reflectance model based on the asymptotic solution to the RT equation. Li and Zhou (2004) simulated the snow-surface BRF and hemispherical directional reflectance factor (HDRF) of snow-covered sea ice through a multilayered azimuth- and zenith-dependent plane-parallel RT model.

For canopies, recent efforts have been mainly focused on accounting for 3D structure of the canopy field using 1D models (Smolander and Stenberg 2003, Pinty et al. 2004a, Rautiainen and

Stenberg 2005, Widlowski et al. 2005) or stochastic radiative transfer models (Shabanov et al. 2000, Kotchenova et al. 2003, Shabanov et al. 2005). Liangrocapart and Petrou (2002) developed a two-layer model of the bidirectional reflectance of homogeneous vegetation canopies, taking into account the anisotropic scattering of both the vegetation canopy and the background, such as bare soil or leaf litter. A new modeling approach with the concept of spectrally invariants is to develop simple algebraic combinations of leaf and canopy spectral transmittance and reflectance using a small set of wavelength independent structure specific variables, and provide a simple and accurate parameterization of canopy reflectance at any wavelength in the solar spectrum (Huang et al. 2007). Community efforts to compare some vegetation radiative transfer models are ongoing (Pinty et al. 2004b, Widlowski et al. 2007).

GO modeling

GO modeling assumes light traveling along a direct line, interacting with surfaces that are modeled as geometric objects with specific shapes. A GO model explicitly calculates the radiance of each pixel as an area-weighted sum of radiance from subpixel components within a given field of view, for example, sunlit crown, shadowed crown, and sunlit background. In the Li and Strahler (1985) GO model, a tree is represented simply as an ellipsoid on a stick; tree counts vary from pixel to pixel as a Poisson function. The reflectance of a pixel (S) is an area-weighted sum of the signatures for four components: sunlit crown (c), sunlit background (g), shadowed crown (t), and shadowed background (z):

$$S = K_g G + (1 - K_g)(K_c C + K_z Z + K_t T) \quad (3)$$

In this expression, K_c, K_z, and K_t are the areal proportions of sunlit crown, shadowed background, and shadowed crown. They are normalized so that $K_c + K_z + K_t = 1$. K_g is simply the proportion of the pixel remaining uncovered by trees or shadows.

The classic GO models essentially characterize the interaction of direct solar radiation with land surfaces. Including the diffuse radiation field into the GO model leads to a hybrid RT/GO models (Nilson and Peterson 1991, Ni et al. 1999, Chen and Leblanc 2001, Huemmrich 2001, Peddle et al. 2004). The basic principles and methods of GO modeling are reviewed by Chen et al. (2000).

Computer simulation modeling

Computer simulation modeling in optical remote sensing includes typically ray-tracing Monte Carlo modeling and radiosity modeling. These models can explicitly represent 3D structures of the canopy and provide accurate and complete descriptions of the canopy radiation field. Ray-tracing methods are based on a sampling of photon trajectories within the vegetation canopies and other media (Disney et al. 2000). Modeling surface reflectance (Disney et al. 2006), light detection, and ranging (lidar) (Govaerts et al. 1996, Lewis 1999) and synthetic aperture radar (SAR) interference effects (Lin and Sarabandi 1999b) are examples of this application.

Radiosity methods are widely used in computer graphics for realistic scene-rendering and have also found application in canopy reflectance modeling (Borel et al. 1991, Goel et al. 1991). Qin et al. (2002) apply a radiosity model to calculate the canopy radiation regime, including canopy directional reflectance. A major advantage of the method is that once a solution is found for radiative transport, canopy reflectance can be simulated at any view angle. Although various acceleration methods can be applied, a major limitation of the method is the initial computational load in forming the view factor matrix and solving for radiative transport.

Sensor modeling

Since common detector materials do not respond across the entire optical spectrum, most sensors have separate focal planes and noise mechanisms for each spectral region. The sensor model can describe the effects of an imaging spectrometer on the spectral radiance mean and covariance statistics of a land surface. The input radiance statistics of every spectral channel are modified by electronic gain, radiometric noise sources, and relative calibration error to produce radiance signal statistics that represent the scene as imaged by the spectrometer.

Sensor models have been developed for both a dispersive spectrometer and a Fourier transform spectrometer that is most commonly used for the thermal infrared spectral region (Kerekes and Landgrebe 1991, Kerekes and Baum 2002, 2005). The generic dispersive spectrometer model covers separately the visible and near IR, shortwave IR, middle IR, and longwave IR spectral regions. The sensor model includes approximations for the spectral response functions and radiometric noise sources.

Besides the sensor spectral response function and radiometric noise, the sensor model can also include spatial effects. In the sensor ground instantaneous field of view (IFOV), surface elements do not contribute to the pixel value equally but rather the central part contributes most to the pixel value. This kind of spatial effect is usually specified by the sensor point spread function (PSF) in the spatial domain (Mobasseri et al. 1980). The Fourier transform of PSF is called modulation transfer function (MTF), a precise measurement of details and contrast made in the frequency domain.

The MTF of a sensor system can be measured in a laboratory using different techniques, such as the impulse input method, sinusoidal input method, knife-edge input method, and pulse input method. However, only some of these methods can be used for estimating the sensor MTF from remote sensing imagery. For example, the pulse input method was used to measure the Landsat TM MTF in which the source for the pulse is the bridge over San Francisco Bay (Schowengerdt et al. 1985). Ruiz and Lopez (2002) derived the normalized PSF matrix of SPOT imagery using the edge method. These measured PSF or MTF can be used for sensor modeling. Huang et al. (2002) demonstrated that incorporating the sensor PSF can improve land cover classification.

INVERSION ALGORITHMS

This section begins with an introduction to the typical radiometric pre-processing techniques, such as calibration, atmospheric correction, and topographic correction. The overview of inversion algorithms in four different categories is provided. Different land surface products using these inversion methods are introduced separately.

Radiometric preprocessing algorithms

Three types of preprocessing are presented: radiometric calibration, atmospheric correction, and topographic correction.

Radiometric calibration

Radiometric calibration is a process that converts recorded sensor voltages or digitized counts to an absolute scale of radiance that is independent of the image-forming characteristics of the sensor. This process can be a relative or an absolute calibration. *Absolute calibration* is performed using a ratio of the digital numbers acquired by the sensor to the value of an accurately known, uniform-radiance field at the sensor's entrance pupil. *Relative calibration* is determined by normalizing the outputs of the detectors to a given, often average, output from all the detectors in the band.

Calibration measurements can be conducted in three stages. *Pre-flight calibration* measures a sensor's radiometric properties before that sensor is sent into space. *In-flight calibration* is usually performed on a routine basis with onboard calibration systems. *Post-launch calibration* data have to be obtained from vicarious calibration techniques that

typically make use of selected natural or artificial sites on the surface of the Earth.

There are two common post-launch calibration methods. The first is to fly an aircraft with a calibrated radiometer that measures the spectral radiance of the target, under the same illumination and view directions as those that existed for the satellite sensor when the data of interest were collected. This approach often incorporates simultaneous radiometric measurements of spatially and spectrally homogeneous Earth targets. This method is usually called *radiance-based calibration method*. The *reflectance-based method* requires an accurate measurement of the spectral reflectance of the ground target and measurement of spectral extinction depths and other meteorological variables.

Atmospheric correction

Since the observed radiance recorded by a spaceborne or airborne sensor contains both atmospheric and surface information, atmospheric effects must be removed to estimate land surface biogeophysical variables. Such atmospheric correction consists of two major steps: atmospheric parameter estimation and surface reflectance retrieval. If all atmospheric parameters are known, atmospheric correction is made easier. The most challenging aspect of atmospheric correction is to estimate atmospheric properties (particularly water vapor content and aerosol optical depths) from the imagery itself.

The differential absorption technique has been widely used to estimate the total water vapor content of the atmosphere directly from multispectral or hyperspectral imaging systems (Green et al. 1998). The general idea is to utilize one spectral band at the water absorption region (e.g., 0.94 μm) and one or more bands outside of the absorption region. The difference among these bands indicates the amount the water vapor in the atmosphere. In a recent study, Liang and Fang (2004) applied a neural network to estimate water vapor from hyperspectral data. Miesch et al. (2005) developed a water vapor correction algorithm for hyperspectral data using Monte Carlo simulations. Two thermal bands have been used to estimate the precipitable water content by means of the split-window algorithm or simple ratio (Li et al. 2003, Jimenez-Munoz and Sobrino 2005), although it has been warned that these formulae may not be stable under different conditions (Barton and Prata 1999, Ottle and Francois 1999).

A relatively long history exists for estimating aerosol loadings from remotely sensed imagery. Besides various statistical methods, the physically-based methods include those using *spatial signatures*, such as spatial contrast of reflectance (Tanre and Legrand 1991, Guanter et al. 2007), matching histograms of hazy and clear regions (Richter 1997), matching individual clusters in both hazy and clear regions (Liang et al. 2001, 2002); *spectral signatures*, such as the 'Dark-object' methods (e.g., Liang et al. 1997, Kaufman et al. 2000) and the hyperspectral algorithm (Liang and Fang 2004); *angular signatures* (Martonchik et al. 2002, North 2002); *polarization signatures* (Deuze et al. 2001); and *temporal signatures* (Tang et al. 2005, Liang et al. 2006b, Zhong et al. 2007).

Topographic correction

Topographic correction of remotely sensed imagery over mountainous regions is at least as important as atmospheric correction. Topographic shading and shadowing modulates any remote sensing signals and affects the inversion of land surface parameters. Topographic correction requires high-resolution and accurate digital elevation model (DEM) data.

The simplest and earliest approaches are ratio algorithms. If it is assumed that illumination effects caused by the topography are proportional at different bands, the ratio of two bands can eliminate topographic effects. However, these ratio algorithms generally cannot produce satisfactory results since the radiometric variations caused by topography are wavelength dependent and their differences at different bands are not simply increased or decreased by the same proportion.

The cosine correction method is empirical, but easy to implement and thus has been widely used. A much more complex method was developed by Dozier et al. (Dozier 1989, Dozier and Frew 1990). It formulates the total incoming shortwave irradiance to be the sum of three components: direct solar radiation (F_I), diffuse solar radiation (F_D), and radiation reflected from the neighboring pixels. DEM data are used to calculate these components. Zheng et al. (2007) applied this formulation to calculation of incident PAR from GOES data. Schaaf et al. (1994) corrected the topographic effect in using a GO model and Wang et al. (2005) corrected the topographic effect for estimating incident solar radiation.

Inversion algorithms

Model simulation and statistical analysis

Statistical models have proven to be very useful in various remote sensing applications. They are usually created using ground measurements (Johannsen and Daughtry, in this volume). Because it is very expensive to collect extensive ground measurements under various conditions, the major weakness of models based on ground measurements is limited representation.

An alternative solution is to simulate remotely sensed data using a physically-based radiation model that may have been calibrated and validated by field measurements. Different statistical methods can be used to relate inputs and outputs of the model simulations. Besides the conventional multivariate regression analysis, different machine learning methods and other advanced statistical analysis techniques have been used, such as *artificial neural network (ANN) methods* for estimating various biophysical variables (Smith 1993, Baret and Fourty 1997, Kimes et al. 1998, Fang and Liang 2003, 2005); *genetic algorithms* for estimating land surface roughness and soil moisture (Jin and Wang 1999, Wang and Jin 2000), biophysical parameters (such as tree density, tree height, trunk diameter, and soil moisture) of a forest stand from microwave data (Lin and Sarabandi 1999a) and LAI (Fang et al. 2003), *regression tree methods* for estimating fractional vegetation coverage (Hansen et al. 2002); and *Bayesian networks* for estimating LAI (Kalacska et al. 2005).

An important drawback of most machine learning techniques (e.g., ANN) has been their lack of explanation capability. It is increasingly apparent that, without some form of explanation capability, the full potential of trained ANNs may not be realized. Many studies have focused on mechanisms, procedures, and algorithms designed to insert knowledge into ANNs (knowledge initialization), extract rules from trained ANNs (rule extraction), and utilize ANNs to refine existing rule bases (rule refinement).

Optimization algorithms

The optimization algorithms estimate the parameters (Ψ) of the surface radiation model by minimizing the cost function defined as follows:

$$F^2 = \sum_{i=1}^{n} \omega_i [R_i - f_i(\psi)]^2 \qquad (4)$$

where R_i, $i = 1,2,...n$ denote the remotely sensed signals (radiance, reflectance or brightness temperature), ω_i are the weighting coefficients, and $f_i(\psi)$ are the predicted values of the surface radiation model. After giving the initial values, a searching algorithm can be used to determine the parameter set Ψ iteratively.

This approach has been widely used by the land remote sensing community (Liang and Strahler 1993, 1994, Kimes et al. 2000, Liang 2001a, b). There are several recent works available. For example, Gascon et al. (2004) estimated LAI, crown coverage, and leaf chlorophyll concentration from SPOT and IKONOS imagery, using a 3D canopy radiative transfer model. To reduce the computational requirements, some parametric functions were fit using look-up tables created by the 3D reflectance model. Meroni et al. (2004) applied this algorithm to invert LAI from hyperspectral data. Schaepman et al. (2005) inverted biophysical and biochemical variables from multiangular and hyperspectral remote sensing data using a coupled leaf–canopy–atmosphere radiative-transfer model. The MISR science team is also producing land surface products using the four-band multiangular MISR data (Diner et al. 2005). Qin et al. (2007) incorporated the adjoint algorithm of the canopy RT model in the optimization process to speed up the computation.

The high computational demands of optimization approaches has led to use of more simplified surface reflectance models, rather than forcing optimization algorithm efficiencies. One of the general trends in optical remote sensing is to use simpler empirical or semi-empirical models (Lucht and Roujean 2000). The optimization algorithms are used to estimate the parameters in these simple models. These parameters are then related to surface properties. For example, Widlowski et al. (2004) fitted a simple bidirectional reflectance distribution model (BRDF) model to multiangular observations and then linked the surface structural properties to one of the parameters. Chen et al. (2005) mapped the global clumping index from multiangular observations.

Look-up table algorithms

Optimization algorithms are computationally expensive and very slow conducting the inversion process with a huge amount of remotely sensed data. The look-up table (LUT) approach has been used extensively to speed up the inversion process. It pre-computes the model reflectance for a large range of combinations of parameter values. In this manner, the most computationally expensive aspect can be completed before the inversion is attempted, and the problem is reduced to searching a LUT for the modeled reflectance set which most resembles the measured set (Kimes et al. 2000).

This method has been used for a variety of remote sensing inversion issues, such as atmospheric correction (Liang et al. 1997) and estimating LAI (Knyazikhin et al. 1998a, b, Weiss et al. 2000) and incident solar radiation (Liang et al. 2006a). In an ordinary LUT approach, the dimensions of the table have to be large enough in order to achieve a high accuracy, which leads to a much slower on-line searching. Moreover, many parameters have to be fixed in the LUT method. To reduce the dimension of the LUTs for rapid table searching, Gastellu-Etchegorry et al. (2003) developed empirical functions to fit the LUT values so that a table searching procedure becomes a simple calculation of the local functions. Alternatively, Liang et al. (2005b) developed a simple linear regression instead of table searching for each angular bin in the solar illumination and sensor viewing geometry.

Data assimilation

In the previous sections, various methods for estimating land surface variables have been discussed. The values of these variables, estimated from different sources, may not be physically consistent. Most techniques do not take advantage of observations acquired at different times and cannot handle observations with different spatial resolutions. In particular, these techniques estimate only variables that significantly affect radiance received by the sensors. In many cases, the estimation of some variables not directly related to radiance is desirable.

Given the ill-posed nature of remote sensing inversion (the number of unknowns is far greater than the number of observations) and vast expansion of the amount of observation data, a challenge emerges: how best can observations derived from many different sources be combined and integrated? How shall observations specific to location, time, and setting be connected to an understanding that comes from a diverse body of nonspecific theory? Data fusion techniques that simply register and combine data sets together from multiple sources may not be adequate. The data assimilation (DA) method allows use of all available information within a time window to estimate various unknowns of land surface models (Liang and Qin 2007). The information that can be incorporated includes observational data, existing pertinent *a priori* information, and, importantly, a dynamic model that describes the system of interest and encapsulates theoretical understanding. Data assimilation has been widely used in meteorology and oceanography, but more efforts are needed in the land remote sensing community to explore its potential for characterizing land surface environments.

Various land DA schemes have different characteristics, but they may have the following features in common: (1) a forward land dynamic model that describes the time evolution of state variables such as surface temperature, soil moisture and carbon stocks; (2) an observation model that relates the model estimates of state variables to satellite observations and vice versa; (3) an objective function that combines model estimates and observations along with any associated prior information and error structure; (4) an optimization scheme that adjusts forward model parameters or state variables to minimize the discrepancy between model estimates and satellite observations; and (5) error matrices that specify the uncertainty of the observations, model and any background information (these are usually included in the objective function).

The meteorology and oceanography communities have been at the forefront in developing and using data assimilation methods. In recent years, meteorologists and oceanographers have tended to view data assimilation as a model state estimation problem. The land community has aggressively tried to catch up and has applied data assimilation methods in recent years. Land surfaces are much more complex and the dynamic models for characterizing the land surface environment is still at the early stage. Moreover, there is no real spatial component to much land modeling, but spatial dynamics are the core to ocean and atmosphere models. Thus, the DA methods in meteorology and oceanography may need to be changed accordingly for land applications.

The examples in hydrology and the water cycle include the global land data assimilation system (Rodell et al. 2004) and the North American Land data assimilation system (Mitchell et al. 2004). Williams et al. (2005) developed a data assimilation approach that combines stock and flux observations with a dynamic model to improve estimates of, and provide insights into, ecosystem carbon exchanges. Their approach has been further extended lately (Quaife et al. 2007). Rayner et al. (2005) developed a terrestrial carbon cycle data assimilation system for determining the space-time distribution of terrestrial carbon fluxes for the period 1979–1999. Hazarika et al. (2005) integrated the MODIS LAI product with an ecosystem model for accurate estimation of net primary productivity (NPP). Their research demonstrates the utility of combining satellite observation with an ecosystem process model to achieve improved accuracy in estimates and monitoring global NPP. Fang et al. (2007) assimilated MODIS LAI product into a crop growth model for estimation of crop yield by determining some critical parameters of the crop model.

Mapping LAI and FPAR

LAI characterizes vegetation canopy functioning and energy absorption capacity, and is a variable that is used in most land surface process models. There are some differences in defining LAI between remote sensing and user communities in terms of total LAI, green LAI, or effective LAI. A common procedure to estimate LAI is to establish an empirical relationship between vegetation indices (VI) and LAI by statistically fitting observed LAI values to the corresponding VI. The nonlinear functions optimized with RT simulations for different sensors have been demonstrated to be very effective (Gobron et al. 2000).

Besides the VI-based statistical model, all other inversion methods mentioned in the previous section have been used to map LAI. Several instrument science teams are providing global LAI maps. MODIS, MISR, AVHRR, MERIS, POLDER, and the SPOT VEGETATION are among some other notable satellite instruments that provide a global LAI product with various spatial and temporal resolutions, which are summarized in Table 20.1. The detailed validation and inter-comparisons are needed. Integrating these products to provide a more accurate and unified product to the user community is highly desirable.

The fraction of the absorbed photosynthetically active radiation (PAR) by green vegetation (FPAR) has been recognized as one of the fundamental terrestrial variables in the context of global change science. Many production efficiency models for calculating gross primary productivity (GPP) and NPP are based on the following formulation: GPP/NPP $\propto$ FPAR $\cdot$ PAR. Different FPAR products are being generated from multiple satellite sensors using various methods, as summarized in Table 20.2.

Mapping broadband albedo

Land surface broadband albedo is a critical variable affecting the Earth's climate and is still among the main radiative uncertainties of current climate modeling (Wang et al. 2006, Dickinson 2007). A typical albedo mapping algorithm, such as that used for generating the MODIS albedo product, may include three components (Schaaf et al. 2002): (1) atmospheric correction that converts top-of-atmosphere (TOA) radiance to surface directional reflectance; (2) BRDF modeling that converts directional reflectance to spectral albedos;

Table 20.1 LAI products

Product	Spatial resolution (km)	Temporal resolution (day)	Time coverage	Reference
MOD15	1	8	2000–	Knyazikhin et al. (1999)
MISR	1.1	9	2000–	Knyazikhin et al. (1998a)
CYCLOPES	1	10	1997–2003	Bacour et al. (2006)
VGT	1.12	10	2003	Deng et al. (2006)
GLOBCARBON	10	14–30	1998–2003	Plummer and Fierens (2006)

Table 20.2 FPAR products

Product	Spatial resolution (km)	Temporal resolution (day)	Time coverage	Reference
MOD15	1	8	2000–	Knyazikhin et al. (1999)
MISR	1.1	9	2000–	Knyazikhin et al. (1998a)
CYCLOPES	1	10	1997–2003	Bacour et al. (2006)
MERIS	1.2	Daily	2002–	Gobron et al. (1999)
GLOBCARBON	10	14–30	1998–2003	Plummer and Fierens (2006)
SeaWiFS	1.5	1/10/30	1997–2004	Gobron et al. (2006)

Table 20.3 Albedo products

Product	Spatial resolution (km)	Temporal resolution (day)	Time coverage	Reference
MOD43	0.5	8	2000–	Schaaf et al. (2002)
MISR	1.1	9	2000–	Diner et al. (2005)
CYCLOPES	1	10	1997–2003	Bacour et al. (2006)
MERIS	0.05°	16	2002–2006	Muller et al. (2006)
VIIRS	0.5	1	2009–	Liang et al. (2005a)

and (3) narrowband to broadband conversion that converts spectral albedos to broadband albedos. Atmospheric correction algorithms and surface BRDF modeling have been discussed earlier. Different statistical formulae have been developed to convert narrowband albedo to broadband albedo (Liang 2001a, Liang et al. 2003, 2005a). Using such a physical approach, albedo will depend on the performance of all the procedures that characterize the known processes, such as performance of the atmospheric correction and the accuracy of the angular model used to describe the directional distribution of the reflectance. It is unknown whether errors associated with each procedure cancel or enhance each other.

Instead of retrieving most of the variables explicitly from remote sensing data, an alternative method to a physically-based retrieval is to combine all procedures together in one step through regression analysis aiming only to make a best-estimate broadband albedo. The direct retrieval method primarily consists of two steps (Liang et al. 1999, Liang 2003, Liang et al. 2005b). The first step is to produce a large database of TOA directional reflectance and surface albedo for a variety of surface and atmospheric conditions using radiative transfer model simulations. The second step is to link the simulated TOA reflectance with surface broadband albedo using nonparametric regression algorithms (e.g., neural networks and projection pursuit regression) or linear regression analysis. This method will be used for producing the albedo product from VIIRS in the future.

Land surface albedo can also be mapped from multiangular sensors, such as MISR. Some of the albedo products are summarized in Table 20.3. Inter-comparisons, extensive validation, and integration are also all greatly needed.

Mapping incident solar radiation

Incident solar radiation, either PAR in the visible spectrum (400–700 nm) or insolation in the total shortwave (300–4000 nm), is a key variable required by almost all land surface models. Many ecosystem models calculate biomass accumulation as linearly proportional to incident PAR. Information on the spatial and temporal distribution of PAR, by control of the evapotranspiration process, is required for modeling the hydrological cycle and for estimating global oceanic and terrestrial NPP.

The only practical means of obtaining incident PAR at spatial and temporal resolutions appropriate for most modeling applications is through remote sensing. For calculating incident solar radiation, the methods fall into roughly two types of algorithms. The first approach is to use the retrieved cloud and atmosphere parameters from other sources, with the measured TOA radiance/flux acting as a constraint. The Clouds and Earth's Radiant Energy System (CERES) algorithm (Wielicki et al. 1998) employs the cloud and aerosol information from MODIS, and TOA broadband fluxes as a constraint, to produce both insolation and PAR at the spatial resolution of 25 km with the instantaneous sensor footprint.

The second approach is to establish the relationship between the TOA radiance and surface incident insolation or PAR based on extensive radiative transfer simulations. This method was first applied to analyze Earth Radiation Budget Experiment (ERBE) data. Liang et al. (2006a) and Wang et al. (2007a) generated the PAR and insolation products at 1 km from MODIS data directly using similar approaches. This method has been revised to map PAR over China (Liu et al. 2007). This algorithm has also been extended to GOES data (Wang et al. 2007b, Zheng et al. 2007) and is being revised for other satellite data (e.g., AVHRR, SeaWiFS) as well.

Mapping emissivity and skin temperature

Upwelling thermal radiation mainly depends on the land surface temperature (LST, T) and emissivity (ε):

$$F_u = (1 - \varepsilon)F_d + \varepsilon\sigma T^4 \qquad (5)$$

where F_d is downward thermal radiation and σ is a constant. For dense vegetation and water surfaces, broadband emissivity is almost one (0.96–1). For non-vegetated surfaces, it is far below one. Unfortunately, most GCMs and land surface models have assumed constant emissivity values, which may lead to errors in net radiation and other radiative fluxes (Rowntree 1991, Jin and Liang 2006).

Estimation of both emissivity and land surface temperature simultaneously from thermal infrared remotely sensed data is a very challenging issue. Radiance received by the sensor contains information about the atmosphere (e.g., temperature and water vapor profiles) and surface properties (emissivity and LST). Therefore, the first step for retrieving surface emissivity and LST is to perform atmospheric correction. The second step is to separate emissivity and temperature from the retrieved surface leaving radiance.

For sensors that have two thermal bands, such as AVHRR and GOES, a known emissivity is assumed (or inferred from land cover maps or vegetation indices) in order to estimate LST using the so-called split-window algorithms. Based on two thermal bands in the 10.5–12.5 μm range, the split-window algorithms have been widely applied for estimating LST, given surface emissivity.

Some new generation sensors, such as ASTER and MODIS, have multiple thermal bands, which allow estimations of spectral emissivities and LST simultaneously. The MODIS team has developed two approaches for retrieving LST and emissivity (MOD17). The first approach for generating 1 km LST product is based on the generalized split-window algorithm with emissivity values assigned from land cover classes (Wan and Dozier 1996). The second approach for producing 5 km LST is based on the 'day/night' algorithm using both day and evening observations (Wan and Li 1997). The MODIS atmospheric temperature profile product (MOD11) also includes LST. In a recent validation study (Wang et al. 2007c), it was found that the MOD17 product has comparable accuracy to MOD11. These two approaches differ significantly from the ASTER algorithm (Gillespie et al. 1998), which is based on atmospheric correction and a temperature/emissivity separation algorithm.

Table 20.4 lists the LST products from different sensors. LST can also be estimated from other sensors, such as TM (Sobrino et al. 2004a), SEVIRI (Sobrino and Romaguera 2004), AVHRR (Sobrino et al. 1999, 2001), and ATSR (Sobrino et al. 1996).

FIELD MEASUREMENTS AND VALIDATION

With various land products available, users need quantitative information on product uncertainties in order to select the most suitable product, or combination of products, for their specific needs. As remote sensing observations are generally merged with other sources of information or assimilated within process models, evaluation of product accuracy is required. Making quantified accuracy information available to the user can ultimately provide developers the necessary feedback for improving the products, and can possibly provide methods for their fusion to construct a consistent long-term series of surface status (Morisette et al. 2006a).

The Committee on Earth Observation Satellites (CEOS), through the work and consensus of Land Product Validation (LPV), has defined the following validation hierarchy for global land products.

Table 20.4 LST products

Product	Spatial resolution (km)	Temporal resolution (day)	Time coverage	Reference
MOD11	1	1	2000–	Wan and Dozier (1996); Wan and Li (1997)
MOD17	1	1	2000–	Seemann et al. (2006)
AATSR	1.1	3	2002–	Sobrino et al. (2004b)
ASTER	0.09	Discrete	2000–	Gillespie et al. (1998)
VIIRS	1	6 hours	2009–	Yu et al. (2005)

Stage 1 Validation

Product accuracy has been estimated using a small number of independent measurements obtained from selected locations, time periods, and ground-truth/field program efforts.

Stage 2 Validation

Product accuracy has been assessed over a widely distributed set of locations and time periods via several ground-truth and validation efforts.

Stage 3 Validation

Product accuracy has been assessed, and the uncertainties in the product well-established, via independent measurements made in a systematic and statistically robust way that represents global conditions.

Land product validation has to rely on ground measurements, which may be time consuming and very expensive. Because of its importance, such product validation must involve the efforts of the entire community. Sharing the validation methodologies, instruments, measured data, and results pilots the way to success and progress. The global land product validation results are recently published in a journal special issue (Morisette et al. 2006b) and more publications should follow.

FUTURE CHALLENGES

Immense amounts of data available from satellite observations offer promise yet also present great challenges. Considerable investments have been put into developing physical models to understand surface radiation regimes. State-of-the-art remote sensing modeling and inversion is well advanced. Some of these models have been incorporated into useful algorithms for estimating land surface variables from satellite observations.

Developing realistic and computationally simplified surface radiation models mostly suitable for inversion of land surface variables from satellite data is urgently required. Inversion of land surface parameters is generally a nonlinear ill-posed problem, and use of regularization methods by incorporating *a priori* knowledge and integrating multiple-source data from different spectra and instruments deserves further research. We are also faced with the computational challenges of massive data volumes, and exploring and using grid technologies should be encouraged. Validating, inter-comparing, and eventually integrating multiple products of the same biogeophysical variable are urgently needed. Effective communications between remote sensing scientists and the user communities are also necessary.

ACKNOWLEDGMENTS

The author greatly appreciates the funding supports from NASA in the last decade for conducting research in this area, and thanks two reviewers and Dr. Tim Warner for their valuable comments and suggestions.

REFERENCES

Bacour, C., F. Baret, D. Beal, M. Weiss, and K. Pavageau, 2006. Neural network estimation of LAI, fAPAR, fCover and LAIxC(ab), from top of canopy MERIS reflectance data: Principles and validation. *Remote Sensing of Environment*, 105: 313–325.

Baret, F. and T. Fourty, 1997. Estimation of leaf water content and specific leaf weight from reflectance and transmittance measurements. *Agronomie*, 17: 455–464.

Barton, I. J. and A. J. Prata, 1999. Difficulties associated with the application of covariance-variance techniques to retrieval of atmospheric water vapor from satellite imagery. *Remote Sensing of Environment*, 69: 76–83.

Borel, C. C., S. A. W. Gerstl, and B. J. Powers, 1991. The radiosity method in optical remote sensing of structured 3-D surfaces. *Remote Sensing of Environment*, 36: 13–44.

Chen, J., X. Li, T. Nilson, and A. Strahler, 2000. Recent advances in geometrical optical modelling and its applications. *Remote Sensing Reviews*, 18: 227–262.

Chen, J. M. and S. G. Leblanc, 2001. Multiple-scattering scheme useful for geometric optical modeling. *IEEE Transactions on Geoscience and Remote Sensing*, 39: 1061–1071.

Chen, J. M., C. H. Menges, and S. G. Leblanc, 2005. Global mapping of foliage clumping index using multi-angular satellite data. *Remote Sensing of Environment*, 97: 447–457.

Deng, F., J. M. Chen, S. Plummer, M. Z. Chen, and J. Pisek, 2006. Algorithm for global leaf area index retrieval using satellite imagery. *IEEE Transactions on Geoscience and Remote Sensing*, 44: 2219–2229.

Deuze, J. L., F. M. Breon, C. Devaux, P. Goloub, M. Herman, B. Lafrance, F. Maignan, A. Marchand, F. Nadal, G. Perry, and D. Tanre, 2001. Remote sensing of aerosols over land surfaces from POLDER-ADEOS-1 polarized measurements. *Journal of Geophysical Research – Atmospheres*, 106: 4913–4926.

Dickinson, R., 2007. Applications of terrestrial remote sensing to climate modeling. In S. Liang (ed.), *Advances in Land Remote Sensing: System, Modeling, Inversion and Application*. Springer, New York.

Diner, D. J., J. V. Martonchik, R. A. Kahn, B. Pinty, N. Gobron, D. L. Nelson, and B. N. Holben, 2005. Using angular and spectral shape similarity constraints to improve MISR aerosol and surface retrievals over land. *Remote Sensing of Environment*, 94: 155–171.

Disney, M., P. Lewis, and P. Saich, 2006. 3D modelling of forest canopy structure for remote sensing simulations in the optical and microwave domains. *Remote Sensing of Environment*, 100: 114–132.

Disney, M. I., P. Lewis, and P. R. J. North, 2000. Monte Carlo ray tracing in optical canopy reflectance modeling. *Remote Sensing Reviews*, 18: 163–196.

Dozier, J., 1989. Spectral signature of alpine snow cover from the Landsat Thematic Mapper. *Remote Sensing of Environment*, 28: 9–22.

Dozier, J. and J. Frew, 1990. Rapid calculation of terrain parameters for radiation modeling from digital elevation data. *IEEE Transctions on Geoscience and Remote Sensing*, 28: 963–969.

Fang, H. and S. Liang, 2003. Retrieve LAI from Landsat 7 ETM+ data with a neural network method: simulation and validation study. *IEEE Transactions on Geoscience and Remote Sensing*, 41: 2052–2062.

Fang, H. and S. Liang, 2005. A hybrid inversion method for mapping leaf area index from MODIS data: experiments and application to broadleaf and needleleaf canopies. *Remote Sensing of Environment*, 94: 405–424.

Fang, H., S. Liang, and A. Kuusk, 2003. Retrieving Leaf Area Index (LAI) using a genetic algorithm with a canopy radiative transfer model. *Remote Sensing of Environment*, 85: 257–270.

Fang, H., S. Liang, G. Hoogenboom, J. Teasdale, and M. Cavigelli, 2007. Crop yield estimation through assimilation of remotely sensed data into DSSAT-CERES. *International Journal of Remote Sensing*, in press.

Gascon, F., J. P. Gastellu-Etchegorry, M. J. Lefevre-Fonollosa, and E. Dufrene, 2004. Retrieval of forest biophysical variables by inverting a 3-D radiative transfer model and using high and very high resolution imagery. *International Journal of Remote Sensing*, 25: 5601–5616.

Gastellu-Etchegorry, J. P., V. Demarez, and F. Zagolski, 1996. Modeling radiative transfer in heterogeneous 3D vegetation canopies. *Remote Sensing of Environment*, 58: 131–156.

Gastellu-Etchegorry, J. P., F. Gascon, and P. Esteve, 2003. An interpolation procedure for generalizing a look-up table inversion method. *Remote Sensing of Environment*, 87: 55–71.

Gillespie, A. R., S. Rokugawa, T. Matsunaga, J. Cothern, S. Hook, and A. Kahle, 1998. A temperature and emissivity separation algorithm for Advanced Spaceborne Thermal Emission and Reflection Radiometer (ASTER) images. *IEEE Transactions on Geosciences and Remote Sensing*, 36: 1113–1126.

Gobron, N., B. Pinty, M. Verstraete, and Y. Govaerts, 1999. The MERIS Global Vegetation Index (MGVI): description and preliminary application. *International Journal of Remote Sensing*, 20: 1917–1927.

Gobron, N., B. Pinty, M. M. Verstraete, and J.-L. Widlowski, 2000. Advanced vegetation indices optimized for upcoming sensors: design, performance, and applications. *IEEE Transactions on Geoscience and Remote Sensing*, 38: 2489–2504.

Gobron, N., B. Pinty, O. Aussedat, J. M. Chen, W. B. Cohen, R. Fensholt, V. Gond, K. F. Huemmrich, T. Lavergne, F. Melin, J. L. Privette, I. Sandholt, M. Taberner, D. P. Turner, M. M. Verstraete, and J. L. Widlowski, 2006. Evaluation of fraction of absorbed photosynthetically active radiation products for different canopy radiation transfer regimes: Methodology and results using Joint Research Center

products derived from SeaWiFS against ground-based estimations. *Journal of Geophysical Research – Atmospheres*, 111, Art. No. D13110, ISI: 000239217300005.

Goel, N. S., I. Rozehnal, and R. I. Thompson, 1991. A computer graphics based model for scattering from objects of arbitrary shapes in the optical region. *Remote Sensing of Environment*, 36: 73–104.

Govaerts, Y. M., S. Jacquemoud, M. M. Verstraete, and S. L. Ustin, 1996. Three-dimensional radiation transfer modeling in a dicotyledon leaf. *Applied Optics*, 35: 6585–6598.

Green, R. O., M. L. Eastwood, C. M. Sarture, T. G. Chrien, M. Aronsson, B. J. Chippendale, J. A. Faust, B. E. Pavri, C. J. Chovit, M. Solis, M. R. Olah, and O. Williams, 1998. Imaging spectroscopy and the airborne visible/infrared imaging spectrometer (AVIRIS). *Remote Sensing of Environment*, 65: 227–248.

Guanter, L., M. D. Gonzalez-Sanpedro, and J. Moreno, 2007. A method for the atmospheric correction of ENVISAT/MERIS data over land targets. *International Journal of Remote Sensing*, 28: 709–728.

Hansen, M., R. S. DeFries, J. R. G. Townshend, R. Sohlberg, C. Dimiceli, and M. Carroll, 2002. Towards an operational MODIS continuous field of percent tree cover algorithm: examples using AVHRR and MODIS data. *Remote Sensing of Environment*, 83: 303–319.

Hazarika, M. K., Y. Yasuoka, A. Ito, and D. Dye, 2005. Estimation of net primary productivity by integrating remote sensing data with an ecosystem model. *Remote Sensing of Environment*, 94: 298–310.

Huang, C., J. R. G. Townshend, S. Liang, S. N. V. Kalluri, and R. S. DeFries, 2002. Impact of sensor's point spread function on land cover characterization: assessment and deconvolution. *Remote Sensing of Environment*, 80: 203–212.

Huang, D., Y. Knyazikhin, R. E. Dickinson, M. Rautiainen, P. Stenberg, M. Disney, P. Lewis, A. Cescatti, Y. Tian, W. Verhoef, and R. B. Myneni, 2007. Canopy spectral invariants for remote sensing and model applications. *Remote Sensing of Environment*, 106: 106–122.

Huemmrich, K. F. 2001. The GeoSail model: a simple addition to the SAIL model to describe discontinuous canopy reflectance. *Remote Sensing of Environment*, 75: 423–431.

Jimenez-Munoz, J. C. and J. A. Sobrino, 2005. Atmospheric water vapour content retrieval from visible and thermal data in the framework of the DAISEX campaigns. *International Journal of Remote Sensing*, 26: 3163–3180.

Jin, M. and S. Liang, 2006. Improve land surface emissivity parameter for land surface models using global remote sensing observations. *Journal of Climate*, 19: 2867–2881.

Jin, Y. and Y. Wang, 1999. A genetic algorithm to simultaneously retrieve land surface roughness and soil moisture. In, *25th Annual Conference and Exhibition of the Remote Sensing Society Earth Observation: From Data to Information*. University of Wales at Cardiff and Swansea.

Johannsen, C. J. and C. S. T. Daughtry, in this volume. Surface Reference Data Collection. Chapter 17.

Kalacska, M., A. Sanchez-Azofeifa, T. Caelli, B. Rivard, and B. Boerlage, 2005. Estimating leaf area index from satellite imagery using Bayesian networks, *IEEE Transactions on Geoscience and Remote Sensing*, 43: 1866–1873.

Kaufman, Y. J., A. Karnieli, and D. Tanre, 2000. Detection of dust over deserts using satellite data in the solar wavelengths. *IEEE Transactions on Geoscience and Remote Sensing*, 38: 525–531.

Kerekes, J. P. and J. E. Baum, 2002. Spectral imaging system analytical model for subpixel object detection. *IEEE Transactions on Geoscience and Remote Sensing*, 40: 1088–1101.

Kerekes, J. P. and J. E. Baum, 2005. Full-spectrum spectral imaging system analytical model. *IEEE Transactions on Geoscience and Remote Sensing*, 43: 571–580.

Kerekes, J. P. and D. A. Landgrebe, 1991. An analytical model of Earth-observational remote sensing systems. *IEEE Transactions on Systems, Man, and Cybernetics*, 21: 125–133.

Kimes, D., R. Nelson, M. Manry, and A. Fung, 1998. Attributes of neural networks for extracting continuous vegetation variables from optical and radar measurements. *International Journal Remote Sensing*, 19: 2639–2663.

Kimes, D. S., Y. Knyazikhin, J. L. Privette, A. A. Abuelgasim, and F. Gao, 2000. Inversion Methods for Physically-Based Models. *Remote Sensing Review*, 18: 381–440.

Knyazikhin, Y., J. V. Martonchik, D. J. Diner, R. B. Myneni, M. M. Verstraete, B. Pinty, and N. Gobron, 1998a. Estimation of vegetation canopy leaf area index and fraction of absorbed photosynthetically active radiation from atmosphere-corrected MISR data. *Journal of Geophysical Research*, 103: 32239–32256.

Knyazikhin, Y., J. V. Martonchik, R. B. Myneni, D. J. Diner, and S. W. Running, 1998b. Synergistic algorithm for estimating vegetation canopy leaf area index and fraction of absorbed photosynthetically active radiation from MODIS and MISR data. *Journal of Geophysical Research*, 103: 32257–32275.

Knyazikhin, Y., J. Glassy, J. L. Privette, Y. Tian, A. Lotsch, Y. Zhang, Y. Wang, J. T. Morisette, P. Votava, R. B. Myneni, R. R. Nemani, and S. W. Running, 1999. MODIS Leaf Area Index (LAI) and Fraction of Photosynthetically Active Radiation Asborved by Vegetation (FPAR) Product (MOD15) Algorithm Theoretical Basis Document. NASA.

Kokhanovsky, A. A., T. Aoki, A. Hachikubo, M. Hori, and E. P. Zege, 2005. Reflective properties of natural snow: Approximate asymptotic theory versus *in situ* measurements. *IEEE Transactions on Geoscience and Remote Sensing*, 43: 1529–1535.

Kotchenova, S. Y., N. V. Shabanov, Y. Knyazikhin, A. B. Davis, R. Dubayah, and R. B. Myneni, 2003. Modeling lidar waveforms with time-dependent stochastic radiative transfer theory for remote estimations of forest structure. *Journal of Geophysical Research – Atmospheres*, 108, Art. No. 4484

Lewis, P., (1999). Three-dimensional plant modelling for remote sensing simulation studies using the Botanical Plant Modelling System. *Agronomie*, 19: 185–210.

Li, S. S. and X. B. Zhou, 2004. Modelling and measuring the spectral bidirectional reflectance factor of snow-covered sea ice: an intercomparison study. *Hydrological Processes*, 18: 3559–3581.

Li, X. and A. Strahler, 1985. Geometric-optical modeling of a coniferous forest canopy. *IEEE Transactions on Geoscience and Remote Sensing*, 23: 705–721.

Li, Z. L., L. Jia, Z. B. Su, Z. M. Wan, and R. H. Zhang, 2003. A new approach for retrieving precipitable water from ATSR2 split-window channel data over land area. *International Journal of Remote Sensing*, 24: 5095–5117.

Liang, S., 2001a. Narrowband to broadband conversions of land surface albedo. *Remote Sensing of Environment*, 76: 213–238.

Liang, S., 2001b. An optimization algorithm for separating land surface temperature and emissivity from multispectral thermal infrared imagery. *IEEE Transactions on Geosciences and Remote Sensing*, 39: 264–274.

Liang, S., 2003. A direct algorithm for estimating land surface broadband albedos from MODIS imagery. *IEEE Transactions on Geosciences and Remote Sensing*, 41: 136–145.

Liang, S., 2004. *Quantitative Remote Sensing of Land Surfaces*. John Wiley and Sons Inc., New York.

Liang, S. and H. Fang, 2004. An improved atmospheric correction algorithm for hyperspectral remotely sensed imagery. *IEEE Geoscience and Remote Sensing Letters*, 1: 112–117.

Liang, S. and J. Qin, 2007. Data assimilation methods for land surface variable estimation. In S. Liang (ed.), *Advances in Land Remote Sensing: System, Modeling, Inversion and Applications*, Chapter 11, Springer.

Liang, S. and A. H. Strahler, 1993. An analytic BRDF model of canopy radiative transfer and its inversion. *IEEE Transaction on Geoscience and Remote Sensing*, 31: 1081–1092.

Liang, S. and A. H. Strahler, 1994. Retrieval of surface BRDF from multiangle remotely sensed data. *Remote Sensing of Environment*, 50: 18–30.

Liang, S., H. Fallah-Adl, S. Kalluri, J. JáJá, Y. J. Kaufman, and J. R. G. Townshend, 1997. An operational atmospheric correction algorithm for Landsat Thematic Mapper Imagery over the land. *Journal of Geophysical Research – Atmospheres*, 102: 17173–17186.

Liang, S., A. Strahler, and C. Walthall, 1999. Retrieval of land surface albedo from satellite observations: A simulation study. *Journal of Applied Meteorology*, 38: 712–725.

Liang, S., H. Fang, and M. Chen, 2001. Atmospheric correction of Landsat ETM+ Land Surface Imagery: I. Methods. *IEEE Transactions on Geosciences and Remote Sensing*, 39, 2490–2498.

Liang, S., H. Fang, M. Chen, C. Shuey, C. Walthall, and C. Daughtry, 2002. Atmospheric Correction of Landsat ETM+ Land Surface Imagery: II. Validation and applications. *IEEE Transactions on Geoscience and Remote Sensing*, 40: 2736–2746.

Liang, S., C. Shuey, H. Fang, A. Russ, M. Chen, C. Walthall, C. Daughtry, and R. Hunt, 2003. Narrowband to broadband conversions of land surface albedo: II. Validation. *Remote Sensing of Environment*, 84: 25–41.

Liang, S., Y. Yu, and T. P. Defelice, 2005a. VIIRS narrowband to broadband land surface albedo conversion: formula and validation. *International Journal of Remote Sensing*, 26, 1019–1025.

Liang, S. L., J. Stroeve, and J. E. Box, 2005b. Mapping daily snow/ice shortwave broadband albedo from Moderate Resolution Imaging Spectroradiometer (MODIS): The improved direct retrieval algorithm and validation with Greenland in situ measurement. *Journal of Geophysical Research – Atmospheres*, 110, Art. No. D10109.

Liang, S., T. Zheng, R. Liu, H. Fang, S. C. Tsay, and S. Running, 2006a. Mapping incident photosynthetically active radiation (PAR) from MODIS data. *Journal of Geophysical Research – Atmospheres*, 111, Art. No. D15208, doi: 15210.11029/12005JD006730.

Liang, S., B. Zhong, and H. Fang, 2006b. Improved estimation of aerosol optical depth from MODIS imagery over land surfaces. *Remote Sen. Environ.*, 104: 416–425.

Liangrocapart, S. and M. Petrou, 2002. A two-layer model of the bidirectional reflectance of homogeneous vegetation canopies. *Remote Sensing of Environment*, 80: 17–35.

Lin, Y. C. and K. Sarabandi, 1999a. Retrieval of forest parameters using a fractal-based coherent scattering model and a genetic algorithm. *IEEE Transactions on Geoscience and Remote Sensing*, 37: 1415–1424.

Lin, Y. C. and K. Sarabandi, 1999b. A Monte Carlo coherent scattering model for forest canopies using fractal-generated trees. *IEEE Transactions on Geoscience and Remote Sensing*, 37: 440–451.

Liu, R., S. Liang, H. He, J. Liu, and T. Zheng, 2007. Mapping photosynthetically active radiation from MODIS data in China. *Remote Sen. Environ.*, in press.

Lucht, W. and J. L. Roujean, 2000. Considerations in the parametric modeling of BRDF and albedo from multiangular satellite sensor observations. *Remote Sensing of Reviews*, 18: 343–380.

Martonchik, J. V., D. J. Diner, K. A. Crean, and M. A. Bull, 2002. Regional aerosol retrieval results from MISR. *IEEE Transactions on Geoscience and Remote Sensing*, 40: 1520–1531.

Meroni, M., R. Colombo, and C. Panigada, 2004. Inversion of a radiative transfer model with hyperspectral observations for LAI mapping in poplar plantations. *Remote Sensing of Environment*, 92: 195–206.

Miesch, C., L. Poutier, W. Achard, X. Briottet, X. Lenot, and Y. Boucher, 2005. Direct and inverse radiative transfer solutions for visible and near-infrared hyperspectral imagery. *IEEE Transactions on Geoscience and Remote Sensing*, 43, 1552–1562.

Mitchell, K. E., D. Lohmann, P. R. Houser, E. F. Wood, J. C. Schaake, A. Robock, B. A. Cosgrove, J. Sheffield, Q. Y. Duan, L. F. Luo, R. W. Higgins, R. T. Pinker, J. D. Tarpley, D. P. Lettenmaier, C. H. Marshall, J. K. Entin, M. Pan, W. Shi, V. Koren, J. Meng, B. H. Ramsay, and A. A. Bailey, 2004. The multi-institution North American Land Data Assimilation System (NLDAS): Utilizing multiple GCIP products and partners in a continental distributed hydrological modeling system. *Journal of Geophysical Research – Atmospheres,* 109, Art. No. D07S90.

Mobasseri, B. G., P. E. Anuta, and C. D. McGillem, 1980. A parametric model for multispectral scanners. *IEEE Transactions on Geoscience and Remote Sensing*, 18: 175–179.

Morisette, J., F. Baret, and S. Liang, 2006a. Special issue on global land product validation. *IEEE Transactions on Geoscience and Remote Sensing*, 44: 1695–1697.

Morisette, J. T., F. Baret, and S. Liang, 2006b. Global land product validation, Special issue of *IEEE Transactions on Geoscience and Remote Sensing*, 44: 1695–1937.

Muller, J. P., B. Brockmann, N. Fomferra, J. Fischer, R. Preusker, and P. Regner, 2006. Algorithm for MERIS land surface BRDF/albedo retrieval and its validation using contemporaneous EO data products. In *The 3rd Geoland Forum*, Feb. 9th, 2006. Vienna.

Myneni, R. B., G. Asrar, and S. A. W. Gerstl, 1990. Radiative transfer in three-dimensional leaf canopies. *Transport Theory and Statistical Physics*, 19: 205–250.

Ni, W., X. Li, C. E. Woodcock, R. Caetano, and A. H. Strahler, 1999. An analytical model of bidirectional reflectance over discontinuous plant canopies. *IEEE Transactions on Geoscience and Remote Sensing*, 37: 1–13.

Nilson, T. and U. Peterson, 1991. A forest canopy reflectance model and a test case. *Remote Sensing of Environment*, 37: 131–142.

North, P. R. J. 2002. Estimation of aerosol opacity and land surface bidirectional reflectance from ATSR-2 dual-angle imagery: Operational method and validation. *Journal of Geophysical Research – Atmospheres,* 107, Aricle no. 4149.

Ottle, C. and C. Francois, 1999. Further insights into the use of the split-window covariance technique for precipitable water retrieval. *Remote Sensing of Environment*, 69: 84–86.

Peddle, D. R., R. L. Johnson, J. Cihlar, and R. Latifovic, 2004. Large area forest classification and biophysical parameter estimation using the 5-Scale canopy reflectance model in Multiple Forward-Mode. *Remote Sensing of Environment*, 89: 252–263.

Pinty, B., N. Gobron, J. L. Widlowski, T. Lavergne, and M. M. Verstraete, 2004a. Synergy between 1-D and 3-D radiation transfer models to retrieve vegetation canopy properties from remote sensing data. *Journal of Geophysical Research – Atmospheres*, 109, Art. No. D21205.

Pinty, B., J. L. Widlowski, M. Taberner, N. Gobron, M. M. Verstraete, M. Disney, F. Gascon, J. P. Gastellu, L. Jiang, A. Kuusk, P. Lewis, X. Li, W. Ni-Meister, T. Nilson, P. North, W. Qin, L. Su, S. Tang, R. Thompson, W. Verhoef, H. Wang, J. Wang, G. Yan, and H. Zang, 2004b. Radiation Transfer Model Intercomparison (RAMI) exercise: Results from the second phase. *Journal of Geophysical Research – Atmospheres,* 109, Art. No. D06210, ISI: 000220622400004.

Pitman, K. M., M. J. Wolff, and G. C. Clayton, 2005. Application of modern radiative transfer tools to model laboratory quartz emissivity. *Journal of Geophysical Research – Planets,* 110, Art. No. E08003.

Plummer, S. and F. Fierens, 2006. THE GLOBCARBON INITIATIVE: Multi-sensor estimation of global biophysical products for global terrestrial carbon studies. In, *3rd Geoland FORUM,* Feb 9th, 2006. Vienna.

Qin, J., S. Liang, X. Li, and J. Wang, 2007. Development of the adjoint model of a canopy radiative transfer model for sensivity study and inversion of leaf area index. *IEEE Transactions on Geoscience and Remote Sensing*, revised.

Qin, W. H., S. A. W. Gerstl, D. W. Deering, and N. S. Goel, 2002. Characterizing leaf geometry for grass and crop canopies from hotspot observations: A simulation study. *Remote Sensing of Environment*, 80: 100–113.

Quaife, T., P. Lewis, M. De Kauwe, M. Williams, B. E. Law, M. Disney, and P. Bowyer, 2007. Assimilating canopy reflectance data into an ecosystem model with an Ensemble Kalman Filter. *Remote Sensing of Environment*, in press.

Rautiainen, M. and P. Stenberg, 2005. Application of photon recollision probability in coniferous canopy reflectance simulations. *Remote Sensing of Environment*, 96: 98–107.

Rayner, P. J., M. Scholze, W. Knorr, T. Kaminski, R. Giering, and H. Widmann, 2005. Two decades of terrestrial carbon fluxes from a carbon cycle data assimilation system (CCDAS). *Global Biogeochem. Cycles*, 19, Art. No. GB2026.

Richter, R. 1997. Correction of atmospheric and topographic effects for high spatial resolution satellite imagery. *International Journal of Remote Sensing*, 18: 1099–1111.

Rodell, M., P. R. Houser, U. Jambor, J. Gottschalck, K. Mitchell, C. J. Meng, K. Arsenault, B. Cosgrove, J. Radakovich, M. Bosilovich, J. K. Entin, J. P. Walker, D. Lohmann, and D. Toll, 2004. The global land data assimilation system. *Bulletin of the American Meteorological Society*, 85: 381–394.

Rowntree, P. 1991. Atmospheric parameterization schemes for evaporation over land: Basic concepts and climate modeling aspects. In: T. Schmugge and J. Abdre (eds), *Land Surface Evaporation: Measurement and Parameterization*. Springer-Verlag, New York (pp. 5–34).

Ruiz, C. P. and F. J. A. Lopez, 2002. Restoring SPOT images using PSF-derived deconvolution filters. *International Journal of Remote Sensing*, 23, 2379–2391.

Schaaf, C. B., X. Li, and A. Strahler, 1994. Topographic effects on bidirectional and hemispherical reflectancescalculated with a geometric-optical canopy model. *IEEE Transactions on Geoscience and Remote Sensing*, 32: 1186–1193.

Schaaf, C., F. Gao, A. Strahler, W. Lucht, X. LI, T. Tsung, N. Strugll, X. Zhang, Y. Jin, P. Muller, P. Lewis, M. Barnsley, P. Hobson, M. Disney, G. Roberts, M. Dunderdale, C. Doll, R. d'Entremont, B. Hu, S. Liang, J. Privette, and D. Roy, 2002. First operational BRDF, albedo nadir reflectance products from MODIS. *Remote Sensing of Environment*, 83: 135–148.

Schaepman, M. E., B. Koetz, G. Schaepman-Strub, and K. I. Itten, 2005. Spectrodirectional remote sensing for the improved estimation of biophysical and -chemical variables: two case studies. *International Journal of Applied Earth Observation and Geoinformation*, 6: 271–282.

Schowengerdt, R., C. Archwamety, and R. C. Wrigley, 1985. Landsat Thematic Mapper image derived MTF. *Photogrammetric Engineering and Remote Sensing*, 51: 1395–1406.

Seemann, S. W., E. E. Borbas, J. Li, W. P. Menzel, and L. E. Gumley, 2006. MODIS Atmospheric profile retrieval: Algorithm Theoretical Basis Document, Version 6, Oct. 25, 2006.

Shabanov, N. V., Y. Knyazikhin, F. Baret, and R. B. Myneni, 2000. Stochastic modeling of radiation regime in discontinuous vegetation canopies. *Remote Sensing of Environment*, 74: 125–144.

Shabanov, N. V., D. Huang, W. Yang, B. Tian, Y. Knyazikhin, R. B. Myneni, D. E. Ahl, S. T. Gower, A. R. Huete, L. E. O. C. Arogao, and Y. E. Shimabukuro, 2005. Analysis and optimization of the MODIS leaf area index algorithm retrievals over broadleaf forests. *IEEE Transactions on Geoscience and Remote Sensing*, 43: 1855–1865.

Smith, J. A. 1993. LAI inversion using a backpropagation neural network trained with a multiple scattering model. *IEEE Transactions on Geoscience and Remote Sensing*, 31: 1102–1106.

Smolander, S. and P. Stenberg, 2003. A method to account for shoot scale clumping in coniferous canopy reflectance models. *Remote Sensing of Environment*, 88: 363–373.

Sobrino, J. A. and M. Romaguera, 2004. Land surface temperature retrieval from MSG1-SEVIRI data. *Remote Sensing of Environment*, 92: 247–254.

Sobrino, J. A., Z. L. Li, M. P. Stoll, and F. Becker, 1996. Multi-channel and multi-angle algorithms for estimating sea and land surface temeprature with ATSR data. *International Journal of Remote Sensing*, 17: 2089–2114.

Sobrino, J. A., J. C. Jimenez-Munoz, and L. Paolini, 2004a. Land surface temperature retrieval from LANDSAT TM 5. *Remote Sensing of Enviroment*, 90: 434–440.

Sobrino, J. A., N. Raissouni, J. Simarro, F. Nerry, and F. Petitcolin, 1999. Atmospheric water vapor content over land surfaces derived from the AVHRR data: application to the Iberian Peninsula. *IEEE Transactions on Geoscience and Remote Sensing*, 37: 1425–1434.

Sobrino, J. A., N. Raissouni, and Z. L. Li, 2001. A comparative study of land surface emissivity retrieval from NOAA data. *Remote Sensing of Environment*, 75: 256–266.

Sobrino, J. A., G. Soria, and A. J. Prata, 2004b. Surface temperature retrieval from Along Track Scanning Radiometer 2 data: Algorithms and validation. *Journal of Geophysical Research – Atmospheres,* 109, Art. No. D11101.

Tang, J., Y. Xue, T. Yuc, and Y. N. Guan, 2005. Aerosol optical thickness determination by exploiting the synergy of TERRA and AQUA MODIS. *Remote Sensing of Environment*, 94: 327–334.

Tanre, D. and M. Legrand, 1991. On the satellite retrieval of Saharan dust optical thickness over land: two different approaches. *Journal of Geophysical Research*, 96: 5221–5227.

Wan, Z. and J. Dozier, 1996. A generalized split-window algorithm for retrieving land-surface temperature measurement from space. *IEEE Transactions on Geoscience and Remote Sensing*, 34: 892–905.

Wan, Z. and Z. L. Li, 1997. A physics-based algorithm for retrieving land-surface emissivity and temperature from EOS/MODIS data. *IEEE Transactions on Geoscience and Remote Sensing*, 35: 980–996.

Wang, Y. and Y. Jin, 2000. A genetic algorithm to simultaneously retrieve land surface roughness and soil moisture. *Journal of Remote Sensing*, 4: 90–94.

Wang, D., S. Liang, and T. Zheng,.2007a. Estimation of daily-integrated PAR from MODIS data. *IEEE Geoscience and Remote Sensing Letters*, revised.

Wang, K., S. Liang, T. Zheng, and D. Wang, 2009b. Simultaneous estimation of surface photosynthetically active radiation and albedo from GOES. *Remote Sensing of Environment*, in press.

Wang, S., A. P. Trishchenko, K. V. Khlopenkov, and A. Davidson, 2006. Comparison of International Panel on Climate Change Fourth Assessment Report climate model simulations of surface albedo with satellite products over northern latitudes. *J. Geophys. Res.*, 111, Art. No. D21108. DOI: 10.1029/2005JD006728.

Wang, W., S. Liang, and T. Meyer, 2008. Validating MODIS land surface temperature products. *Remote Sensing of Environment*, 112(3): 623–635.

Wang, K., X. Zhou, J. Liu, and M. Sparrow, 2005. Estimating surface solar radiation over complex terrain using moderate-resolution satellite sensor data. *International Journal of Remote Sensing*, 26(1): 47–58.

Weiss, M., F. Baret, R. B. Myneni, A. Pragnere, and Y. Knyazikhin, 2000. Investigation of a model inversion technique to estimate canopy biophysical variables from spectral and directional reflectance data. *Agronomie*, 20: 3–22.

Widlowski, J. L., B. Pinty, N. Gobron, M. M. Verstraete, D. J. Diner, and A. B. Davis, 2004. Canopy structure parameters derived from multi-angular remote sensing data for terrestrial carbon studies. *Climatic Change*, 67: 403–415.

Widlowski, J. L., B. Pinty, T. Lavergne, M. M. Verstraete, and N. Gobron, 2005. Using 1-D models to interpret the reflectance anisotropy of 3-D canopy targets: Issues and caveats. *IEEE Transactions on Geoscience and Remote Sensing*, 43: 2008–2017.

Widlowski, J. L., M. Taberner, B. Pinty, V. Bruniquel-Pinel, M. Disney, R. Fernandes, J. P. Gastellu-Etchegorry, N. Gobron, A. Kuusk, T. Lavergne, S. Leblanc, P. E. Lewis, E. Martin, M. Mottus, P. R. J. North, W. Qin, M. Robustelli, N. Rochdi, R. Ruiloba, C. Soler, R. Thompson, W. Verhoef, M. M. Verstraete, and D. Xie, 2007. Third Radiation Transfer Model Intercomparison (RAMI) exercise: Documenting progress in canopy reflectance models.

Journal of Geophysical Research – Atmospheres, 112, Art. No. D09111.

Wielicki, B. A., B. R. Barkstrom, B. A. Baum, T. P. Charlock, R. N. Green, D. P. Kratz, R. B. I. Lee, P. Minnis, G. L. Smith, T. Wong, D. F. Young, R. D. Cess, J. A. J. Coakley, D. A. Crommelynck, L. Donner, R. Kandel, M. D. King, A. J. Miller, V. Ramanathan, D. A. Randall, L. L. Stowe, and R. M. Welch, 1998. Clouds and the Earth's Radiant Energy System (CERES): Algorithm overview. *IEEE Transactions on Geosciences and Remote Sensing*, 36: 1127–1141.

Williams, M., P. A. Schwarz, B. E. Law, J. Irvine, and M. R. Kurpius, 2005. An improved analysis of forest carbon dynamics using data assimilation. *Global Change Biology*, 11: 89–105.

Yu, Y. Y., J. L. Privette, and A. C. Pinheiro, 2005. Analysis of the NPOESS VIIRS land surface temperature algorithm using MODIS data. *IEEE Transactions on Geoscience and Remote Sensing*, 43: 2340–2350.

Zheng, T., S. Liang, and K. C. Wang, 2007. Estimation of incident PAR from GOES imagery. *Journal of Applied Meteorology and Climatology*, in press.

Zhong, B., S. Liang, and B. Holben, 2007. Validating a new algorithm for estimating aerosol optical depths from MODIS imagery. *International Journal of Remote Sensing*, 28: 4207–4214.

Accuracy Assessment

Stephen V. Stehman and Giles M. Foody

Keywords: cross validation, design-based inference, error matrix, land-cover change, probability sampling, reference data error.

INTRODUCTION

Remote sensing can be used to generate products representing a variety of features of the Earth's environment. These products may consist of categorical outputs such as maps of land-cover or continuous outputs such as maps of surface temperature or leaf area index. Assessing the accuracy of these products is critical to understanding their potential utility and the possible impact of error in the product on the intended application. Accuracy assessment is an important element of 'validation', where validation is defined as 'the process of assessing by independent means the quality of the data products derived from the system outputs' (Justice et al. 2000, p. 3383).

Accuracy is defined as the degree to which the map corresponds to what is on the ground. Therefore, the fundamental basis of accuracy assessment is a comparison of the derived product to ground condition. Typically, it is not possible to ascertain perfectly the ground condition, so in reality we assess 'agreement' with the best available information on ground condition rather than 'accuracy' based on the true ground condition. In recognition of the limitations of the data on ground condition it is also preferable to avoid the expression 'ground truth' and instead refer to ground or reference data.

An accuracy assessment consists of three components, a response design, sampling design, and analysis (Stehman and Czaplewski 1998). The response design is the protocol for collecting information to determine the ground condition

associated with a point location, a pixel, an object, or an areal unit, and translating that information into a label or quantity against which the map label or quantity is compared. Because it is usually impractical to determine the ground condition for the entire area mapped, sampling is a fundamental component of accuracy assessment. The sampling design specifies the locations at which the ground data will be collected. The analysis protocol specifies the measures used to describe accuracy and how to estimate these measures from the sample.

One of the most common applications of accuracy assessment related to terrestrial remote sensing is evaluating the accuracy of categorical products such as land-cover produced by a supervised classification analysis. Two types of classification are commonly employed. In a hard or crisp classification, each pixel is assigned to one and only one class, whereas for a soft classification, each pixel is assigned a grade of membership for each class. Applications in which both the map and ground data classifications are hard are the most prevalent and the focus of attention in this chapter. But recognition of the 'mixed pixel' problem (i.e., pixels representing an area of more than one class) has led to greater use of methods involving soft classification.

The accuracy of categorical products is typically evaluated using a 'site-specific' approach in which the map and ground data are compared on a location-by-location basis. In contrast, 'nonsite-specific' accuracy assessment compares the proportion of area classified or mapped for

Table 21.1   Population Error Matrix. The classes are the categories displayed by the map. For example, the classes might be Water, Developed, Barren, Forest, Agriculture, and Wetland for a land-cover map. The cell entry, p_{ij}, is the proportion of area mapped as class i and labeled class j in the ground condition. The row margin p_{i+} is the sum of all p_{ij} values in row i and represents the proportion of area classified as class i. The column margin p_{+j} is the sum of all p_{ij} values in column j, and represents the proportion of area that is truly class j. The main diagonal (p_{ii}) indicates correctly allocated pixels. Off-diagonal elements represent mis-classified pixels, those for which there has been an error in labeling. Accuracy may be expressed on a per-class basis by relating the relevant entry in the main diagonal to either the associated row (user's accuracy) or column marginal (producer's accuracy)

		Ground Condition or Reference Class					
		1	2	3	...	c	Total
	1	p_{11}	p_{12}	p_{13}	...	p_{1c}	p_{1+}
Map	2	p_{21}	p_{22}	p_{23}	...	p_{2c}	p_{2+}
Class	3	p_{31}	p_{32}	p_{33}	...	p_{3c}	p_{3+}
	.	.	.	.	...	.	.
	c	p_{c1}	p_{c2}	p_{c3}	...	p_{cc}	p_+^c
Total		p_{+1}	p_{+2}	p_{+3}	...	p_{+c}	p_{c+}

each class to the corresponding true proportion of area of each class, and thus evaluates an aggregate feature of the map. With a site-specific approach, each location is associated with two labels, the class predicted by the classification that is shown on the map and the class observed as the ground condition. Cross-tabulating the set of class labels provides a simple summary of the quality of a classification. The derived confusion or error matrix highlights those cases for which the labels agree and the classification is accurate (main diagonal of Table 21.1), and, where there is error (off-diagonal elements of Table 21.1), the nature of the error is indicated (e.g., omission and commission error). The matrix may be used to derive estimates of accuracy on a per-class as well as overall basis. Per-class accuracy is defined from two perspectives, producer's accuracy (the complement of omission error) and user's accuracy (the complement of commission error).

This problem is acute for heterogeneous regions where misregistration by 1 or 2 pixels will have a high impact, whereas misregistration within a homogeneous region will have little effect on the accuracy estimates. An additional concern with pixel-based assessments is that the minimum mapping unit of mapped data sets sometimes used to derive the reference class labels may be coarser than the pixel. In such circumstances, it is important to note that the map label may represent the dominant class contained by the unit but it is possible for the geographical area represented by the pixel to be of another class. Map units often contain such secondary class inclusions but their presence may not be apparent. The impurity of map units, however, can add to uncertainty in accuracy assessment and be a major source of error. A polygon or parcel based assessment evaluates accuracy of the objects mapped (e.g., crop fields).

RESPONSE DESIGN

A key element of the response design is deciding the spatial support on which the accuracy assessment will be based. Options include a point, pixel, land-cover polygon, or other spatial unit (e.g., a regular areal unit such as a 3 × 3 pixel block). Conventionally, the pixel has been used as the basic spatial unit for site-specific accuracy assessments. A pixel-based assessment evaluates accuracy for the smallest spatial unit of the classification. A major concern with pixel-based assessments is that spatial misregistration between the map and ground location confounds classification error with location error.

SAMPLING DESIGN

The sampling design is chosen depending on the project's objectives and the available budget for the assessment. Four sampling designs, simple random sampling (SRS), systematic sampling, stratified random sampling, and cluster sampling, provide basic options applicable to pixel-based accuracy assessment (Stehman 1999). More complex sampling designs may need to be considered if the assessment must meet multiple objectives within constraints imposed by a limited budget.

SRS and systematic sampling are the simplest designs to implement and analyze, with systematic sampling generally yielding the more precise estimates. The relative precision of systematic sampling to SRS involves a complex interaction between sample size and the spatial correlation of the classification errors, so for any given application, it can be difficult to predict which design will be most precise (Stehman 1992, Moisen et al. 1994). An appealing feature of systematic sampling is that the sample locations are spatially well distributed across the region of interest. Similar to SRS, systematic sampling allows for unbiased estimators of the target accuracy measures. But unbiased estimators of the variance of these accuracy estimators are not available for systematic sampling. A disadvantage of both SRS and systematic sampling is that the sample sizes for rare classes will be small unless the overall sample size is very large. Both SRS and systematic sampling are equal probability sampling designs, so the proportion each class represents in the sample will be approximately equal to the proportion the class represents in the region mapped. For example, a class that comprises 0.3% of the map area would require a sample size of 10,000 pixels to yield 30 pixels of the rare class in the sample.

Stratifying by map class may be employed either to increase the precision of the estimate of overall accuracy or to achieve the objective of precisely estimating class-specific accuracy. The strategy for allocating the sample to strata depends on the objective motivating stratification. If stratification is chosen to decrease the standard error of estimated overall accuracy, optimal allocation should be implemented in which the sample size allocated to stratum (map class) i is $n_i = nN_iz_i / \sum N_iz_i$, where n is the total sample size, N_i is the number of pixels in map class i, and $z_i = \sqrt{U_i(1 - U_i)/c_i}$, where U_i is the user's accuracy and c_i is the cost of sampling a pixel from class i, and the summation is over all strata (Cochran 1977, p. 109). Larger sample sizes are thus allocated to those strata representing a larger proportion of the area mapped, greater variability of classification error (i.e., U_i near 0.50), and smaller cost to sample. Optimal allocation results in different sample sizes for each stratum, and rare classes are typically allocated small sample sizes because they represent a small proportion of the area. However, if estimating class-specific accuracy is the reason for stratifying by map class, equal allocation per-stratum is commonly used to ensure that the estimated user's accuracy for each class is based on a pre-determined sample size n_i. A simple, practical approach to guide the sample size decision is to specify a range of sample sizes (n_i) and a best guess of user's accuracy (U_i) for each class i, and then calculate the approximate standard error for estimated user's accuracy, $\sqrt{U_i(1 - U_i)/n_i}$. The sample size decision can be made balancing the anticipated standard error with the cost of obtaining a sample of the chosen size.

Cluster sampling employs two or more sizes of sampling units. When two sizes are used, the larger unit or 'cluster' is called a primary sampling unit (PSU), and the smaller unit is called a secondary sampling unit (SSU). The PSU may be a block of pixels, a linear cluster of pixels (Edwards et al. 1998), an aerial photograph, or a satellite sensor image, and the SSU is typically a pixel. Cluster sampling is initiated by selecting a sample of PSUs. If all SSUs within each sampled PSU are selected, the design is one-stage cluster sampling, and if a subsample of SSUs is selected, the design is two-stage cluster sampling.

The primary motivation for cluster sampling is that it reduces the per-pixel cost of obtaining ground data. For example, if field visits are required to obtain reference data, travel costs are reduced because of the spatial proximity of many of the sample pixels. Similarly, if aerial photography is used to collect reference data, cluster sampling reduces the cost and handling time by grouping the sample pixels within fewer photographs. Mitigating the cost advantage of cluster sampling is the problem that classification error is typically spatially correlated leading to positive within cluster correlation and higher variance of the accuracy estimates (Moisen et al. 1994). Two-stage cluster sampling achieves some of the desired cost advantage while diminishing the detrimental effect on precision of high positive within cluster correlation. By selecting fewer sample pixels within each PSU and re-allocating resources to sample additional PSUs, two-stage cluster sampling increases precision relative to one-stage cluster sampling.

A second motivation for cluster sampling occurs when the objectives of the accuracy assessment require a spatial unit larger than a pixel. Assessing the accuracy of landscape features such as patch size and shape distributions (Dungan 2006) and some analyses incorporating reference data error (Hagen 2003) are two applications requiring reference data for blocks of contiguous pixels.

Probability sampling and design-based inference

Because accuracy assessment is so dependent on sampling, the statistical inferences generalizing from the sample to the full map must be rigorous. Design-based inference underlies much of the sampling theory and practice of accuracy assessment (Stehman 2000). Rigorous design-based inference requires a probability sampling design and statistically consistent estimators (see *Estimation* section). The sampling designs presented in this chapter are all probability sampling designs.

Implementing a probability sampling design can be inconvenient, as, for example, when access to ground locations is impractical or prevented (e.g., denied access to private property). Limited funding for accuracy assessment may lead to consideration of a less costly, non-probability, sample of conveniently accessible locations, but the cost of adopting such an approach is an inability to make rigorous generalization from the sample to the population. Even when limited resources for an accuracy assessment are available, it is often possible to implement a probability sampling design that is practical and cost effective (Stehman 2001). To do so, however, it may be necessary to sacrifice some of the objectives of the assessment or to accept higher standard errors for some estimates. Statistically credible inferences should be the norm in accuracy assessment, and rigorous inference begins with implementing a probability sampling design. Despite the inconvenience and cost this might entail, the advantage of rigorous inference based on a probability sample generally far outweighs any perceived benefit of a convenience or judgment non-probability sample.

ANALYSIS

The error matrix is central to the analysis component of most site-specific accuracy assessments. Table 21.1 represents a population error matrix for a hard classification of c classes derived from a census of the ground data for the entire area mapped. The columns of the error matrix represent the ground data and the rows represent the map class. It is not unusual to see the row and column identities reversed when error matrices are presented, so it is critical to clearly label the row and column identities of the error matrix. The proportion of area that is map class i and reference class j is denoted by p_{ij}. The diagonal cells of the error matrix (p_{ii}) represent the proportion of area correctly classified, and the off-diagonal entries identify classification errors. Errors can be viewed from two perspectives. For example, consider a pixel contributing to the entry in matrix element p_{13}. This pixel is actually a member of class 3 but has been allocated erroneously to class 1 by the classifier. This pixel can, therefore, be viewed as having been omitted from class 3 and committed to class 1.

The row and column marginal proportions of the error matrix contain important information. The row margin (p_{i+}) represents the proportion of area classified as class i, and the column margin (p_{+j}) represents the true proportion of area of class j. The column margins are typically unknown in practice, but the row margins are known because the complete classification of the region is available. The difference between p_{i+} and p_{+i} reflects the nonsite-specific or quantity error (Pontius 2000) of the classification.

Several site-specific accuracy measures can be derived from the population error matrix. Overall accuracy, the sum of the c diagonal entries (often expressed as a percentage of the total sample size), provides a single-value summary measure. To describe accuracy by class, the map classification or ground data may be used to specify a relevant subset of the region. If the subset is all locations mapped as class i (i.e., 'conditional' on the locations mapped as class i), we obtain user's accuracy, $U_i = p_{ii}/p_{i+}$, the proportion of area mapped as class i that is found to have class i in the ground data. For example, if class i is deciduous forest, user's accuracy is the proportion of area mapped as deciduous forest that is in reality deciduous forest. The commission error proportion for class i is $1 - U_i$. Conditioning on the subset of all locations in the ground data set that are class j leads to producer's accuracy, $P_j = p_{jj}/p_{+j}$, representing the proportion of area with class j in the ground data that is actually mapped as class j. The omission error proportion for class j is $1 - P_j$.

Overall, user's, and producer's accuracies have straightforward interpretations in terms of proportions of area of the map being evaluated, and these measures incorporate the appropriate marginal distributions of the error matrix. In this regard, it is useful to recognize that overall accuracy, written as a parameter derived from the population error matrix (Table 21.1), can be viewed as an area-weighted, average user's accuracy, $O = \sum_{i=1}^{c} p_{i+}U_i$, or as an area-weighted, average producer's accuracy, $O = \sum_{j=1}^{c} p_{+j}P_j$. Average user's accuracy, $\sum U_i/c$, is an example of a measure that does not incorporate appropriate marginal proportions and therefore does not represent a real feature of the target map. Average user's accuracy represents the overall accuracy of a hypothetical map in which all classes cover an equal proportion of area (i.e., p_{i+} is the same for all classes). Because it is very unlikely that a real-world classification would produce equal marginal proportions, average user's accuracy is not a practically meaningful measure of overall accuracy.

Measures combining user's and producer's accuracies (e.g., averaging the two for a particular class) are similarly difficult to justify because of their questionable interpretation. User's and producer's accuracies are conditional on different subsets of the region classified (i.e., user's accuracy conditions on the area mapped as class k, producer's accuracy conditions on the area that is truly class k). An accuracy measure combining the two would be conditional on the union of these two subsets, and the resulting subset, while well-defined, is unlikely to be a meaningful subset on which to condition an accuracy measure. Accuracy measures that fail to represent meaningful features of a real classification such as average user's accuracy and the average of user's and producer's accuracy for a class should be avoided or used only with great caution.

The analyses discussed thus far are applicable when both the map and ground data have been subject to a hard classification. Recently, considerable attention has focused on soft or fuzzy classifications, especially as a means of addressing uncertainty. For example, Gopal and Woodcock (1994) developed an assessment applicable when the map is a hard classification but there is uncertainty in the ground data labeling. For each sample pixel, a linguistic scale value is assigned to each class based on the estimated degree of membership of the pixel to that class. A variety of operators may then be used to assess accuracy. Mickelson et al. (1998), Muller et al. (1998), Franklin et al. (2001), and Laba et al. (2002) are examples implementing this approach. If both the map and ground data are soft classifications a variety of approaches may be used for accuracy assessment. One basic approach is to 'harden the data' by translating the soft classification into a single class label. Obviously this approach loses much of the information contained in the soft classification and alternative analyses have been developed. For example, Binaghi et al. (1999) propose employing a fuzzy set intersection operator to produce an error matrix from which accuracy measures may be derived. Pontius and Cheuk (2006) propose a modification of this approach that affects the off-diagonal entries of the confusion matrix.

The estimation procedure should be statistically consistent. The important practical implication of consistent estimation is that the estimation formulas (Table 21.2) depend on the sampling design implemented (Card 1982, Stehman 2001). For any equal probability sampling design (i.e., a design for which each pixel has the same probability of being included in the sample), the analysis of the sample error matrix mimics that of the population error matrix. Simple random, systematic, one-stage cluster, and stratified random sampling with proportional allocation are all equal probability designs. Stratified random sampling with equal allocation is a commonly used unequal probability sampling design. Czaplewski (1994), Stehman and Czaplewski (1998), and Stehman et al. (2003) derive accuracy estimators for more complex sampling designs.

Accuracy estimates should be accompanied by standard errors to quantify the uncertainty attributable to sampling variability. The standard error formula depends on the sampling design and the accuracy estimator. Standard error formulas for the four basic designs are provided in Table 21.3. Czaplewski (1994), Edwards et al. (1998), Nusser and Klaas (2003), and Stehman et al. (2003) document standard error formulas for more complex sampling designs. For large sample sizes, a 95% confidence interval for an accuracy parameter can be constructed as the parameter estimate plus or minus two standard errors.

Estimation

The accuracy measures defined for the population must be estimated from the sample data.

Chance-corrected accuracy

Several accuracy measures have been derived to incorporate an adjustment for random

Table 21.2 Formulas for Accuracy Measures Associated with the Error Matrix of a Pixel-based Assessment

Characteristic	Population	Sample-based estimates	
		Equal probability	Stratified random
Proportion of area in cell (i,j)	p_{ij}	n_{ij}/n	$(n_{ij}N_{i+})/(n_{i+}N)$
Overall accuracy	$\sum_{i=1}^{c} p_{ii}$	$\sum_{i=1}^{c} n_{ii}/n$	$(1/N)\sum_{i=1}^{c}(n_{ii}N_{i+}/n_{i+})$
User's accuracy for class i	p_{ii}/p_{i+}	n_{ii}/n_{i+}	n_{ii}/n_{i+}
Producer's accuracy for class j	p_{jj}/p_{+j}	n_{jj}/n_{+j}	$\dfrac{n_{jj}(N_{j+}/n_{j+})}{\sum_{i=1}^{c} n_{ij}(N_{i+}/n_{i+})}$

Notes:

p_{ij}, p_{i+}, p_{+j} defined as in Table 1

n_{ij} = number of sample pixels in cell (i, j) of the error matrix

n_{i+} = number of sample pixels in row (map class) i of the error matrix

n_{+j} = number of sample pixels in column (reference class) j of the error matrix

n = number of pixels in the sample

N_{i+} = number of pixels in the entire region classified (mapped) as class i

c = number of classes

Table 21.3 Variance Estimation Formulas for Overall, User's, and Producer's Accuracies. The standard error in each case is the square root of the variance estimator. The formulas do not include 'finite population correction factors' because it is assumed that the sample will represent such a small proportion of the area that these correction factors will have negligible impact

Simple Random Sampling (Cochran 1977)

Overall accuracy: $\hat{O}(1 - \hat{O})/(n - 1)$

User's accuracy for map class i: $\hat{U}_i(1 - \hat{U}_i)/(n_{i+} - 1)$

Producer's accuracy for reference class j: $\hat{P}_j(1 - \hat{P}_j)/(n_{+j} - 1)$

Stratified Random Sampling (assuming the strata are the c map classes)

Overall accuracy: $\displaystyle\sum_{i=1}^{c} \frac{N_{i+}^2}{N^2} \frac{\hat{U}_i(1 - \hat{U}_i)}{n_{i+} - 1}$

User's accuracy for map class i: $\hat{U}_i(1 - \hat{U}_i)/(n_{i+} - 1)$

Producer's accuracy for reference class j:

$$\frac{1}{\hat{N}_{+j}^2} \left[N_{j+}^2 (1 - \hat{P}_j)^2 \hat{U}_j(1 - \hat{U}_j)/(n_{j+} - 1) + \hat{P}_j^2 \sum_{i \neq j}^{c} N_{i+}^2 \frac{n_{ij}}{n_{i+}} \left(1 - \frac{n_{ij}}{n_{i+}}\right)/(n_{i+} - 1) \right]$$

where:

$\displaystyle\hat{N}_{+j} = \sum_{i=1}^{c} \frac{N_{i+}}{n_{i+}} n_{ij}$ is the estimated marginal total for reference class j

Simple Random Sampling of Clusters: One-stage cluster sampling, equal size clusters (Stehman 1997)

Overall accuracy: $\dfrac{\displaystyle\sum_{u=1}^{m} (p_u - \bar{p})^2}{m(m - 1)}$,

where:

p_u is the proportion of pixels classified correctly in cluster u,

$\bar{p} = \displaystyle\sum_{u=1}^{m} p_u/m$, and

m = number of clusters in the sample

User's accuracy for map class i: $\dfrac{\displaystyle\sum_{u=1}^{m} (a_{u,i} - \hat{U}_i r_{u,i})^2}{\bar{r}_i^2 m(m - 1)}$,

where:

$r_{u,i}$ is the number of pixels in sample cluster u with map class i,

$\bar{r}_i = \displaystyle\sum_{u=1}^{m} r_{u,i}/m$ is the sample mean of $r_{u,i}$, and

$a_{u,i}$ is the number of pixels in sample cluster u for which both the map and reference class are class i.

Producer's accuracy for reference class j: $\dfrac{\displaystyle\sum_{u=1}^{m} (a_{u,j} - \hat{P}_j c_{u,j})^2}{\bar{c}_j^2 m(m - 1)}$,

where:

$c_{u,j}$ is the number of pixels in sample cluster u with reference class j,

$\bar{c}_j = \displaystyle\sum_{u=1}^{m} c_{u,j}/m$ is the sample mean of $c_{u,j}$, and

$a_{u,j}$ is the number of pixels in sample cluster u for which both the map and reference class are class j.

Systematic Sampling

The variance estimator formulas shown for simple random sampling are commonly used to approximate the variances resulting from a systematic sampling design. Unbiased variance estimators do not exist for systematic sampling so an approximation is required (Cochran 1977, Section 8.11).

chance agreement. These measures are motivated by the premise that even a random assignment of class labels will result in a proportion of the study area being correctly classified. The kappa coefficient is the most commonly used chance-adjusted measure in accuracy assessment. Despite its widespread use, the underlying rationale and practical utility of kappa as a measure of map accuracy is questionable (similar concerns apply to conditional kappa, the class-specific version of kappa). If the objective is to describe accuracy of a particular classification, it is not relevant whether locations are classified by random chance or by 'good cause'. Subtracting chance agreement, as incorporated in kappa, unnecessarily and unfairly penalizes the map's accuracy and, therefore, does not reflect the true proportion of area actually correctly classified that a user would experience when using this map. Indeed, Liu et al. (2007) conducted an extensive study of published error matrices and found that kappa is highly correlated with overall accuracy, thereby demonstrating that kappa typically does not represent a dimension or component of accuracy different from overall accuracy, but instead is just a downward rescaling of overall accuracy.

Another conceptual flaw of kappa is that it incorporates the row marginal proportions of the classification (i.e., p_{i+}) in the adjustment for chance agreement. These marginal proportions are an outcome of the classification, not the result of random chance, and it is not sensible to define chance agreement using a property of the classification itself. Kappa's dependence on the row margins leads to the illogical property that two classifications of the same region based on the same legend will have different chance agreement if the row margins produced by the classifications are different. If kappa is truly adjusting only for chance agreement, the chance correction should be the same for two classifications applied to the same region and using the same legend. Foody (1992) suggested a chance corrected measure that is not dependent on the row marginal proportions produced by the classification. This measure is $(O - 1/c)/(1 - 1/c)$, where O is overall accuracy and c is the number of classes.

Because it is not possible to determine if agreement at a specific location is attributable to random chance or to the ability of the classifier (i.e., 'good cause'), random chance agreement is a hypothetical entity, not a real feature of a classification. Recognizing the hypothetical nature of random chance agreement and the problems arising from the dependence of kappa on the marginal distributions, Agresti (1996) suggested use of models to specify unambiguously the meaning of chance agreement, in contrast to the nebulous definition of chance agreement incorporated in some accuracy measures. Developing such models and incorporating them in analyses would lead to a better understanding of a map's accuracy than the current common practice of calculating chance-corrected accuracies that are based on unspecified underlying models of chance agreement.

Normalized error matrices

Normalizing an error matrix by iterative proportional fitting (Congalton and Green 1999) reconfigures the error matrix to force equality of all row and column margins (i.e., $p_{i+} = p_{+j} = 1/c$ for every i and j) while maintaining the interaction structure of the matrix. Accuracy estimates derived from a normalized error matrix do not represent a property of an actual map because the equal row and column marginal proportions produced by the normalized matrix represent a scenario that is unlikely to occur in reality. Consequently, the accuracy measures derived from the normalized matrix apply to a hypothetical population that possesses an unrealistic distribution of area of the classes. Normalizing an error matrix can lead to large bias in the accuracy estimates (Stehman 2004) and this procedure, dating from the early days of accuracy assessment, should not be used.

Information theory based measures

An example of an accuracy measure derived from an information theory perspective is the average mutual information index (Finn 1993). An advantage of information-based measures is that they are applicable when the map and ground data are based on different legends. A disadvantage of these measures is that high values can occur when the map consistently misclassifies one class as another (see Tables 1a and 1c of Finn 1993). Although this feature is a useful diagnostic tool to indicate when re-labeling the categories may resolve a consistent misclassification, it is an undesirable feature for a measure intended to describe accuracy. As Finn (1993) points out, the average mutual information index is a measure of consistency of the classification with the ground data, not a measure of the correctness of the classification. The distinction between consistency and correctness is important because ultimately the goal of accuracy assessment is to determine if the classification is correct.

Analyses accounting for error in reference data

The reference data label represents another classification and, therefore, may contain error attributable to factors such as interpreter mislabeling, temporal mismatch between image and reference data collection, and spatial misregistration of the image and reference data (Congalton and Green 1993, Foody 2002). Powell et al. (2004) provide a case study in which the effects of different sources of reference data error are quantified. Two approaches to account for reference data error in the analysis are to incorporate this error directly into an accuracy measure, or to report several accuracy measures that together provide insight into the potential impact of reference data error. Hagen (2003) provides an example of the first approach. He constructs an accuracy measure that incorporates a fuzzy intersection operator to reflect the seriousness of different categorical misclassifications and a distance-based weighting scheme to account for location error.

Stehman et al. (2003) and Wickham et al. (2004) illustrate the second approach using an ensemble of accuracy measures to deduce the potential effect of reference data error on the accuracy estimates. The reference data labeling protocol permits assigning both a primary and secondary class label to account for ambiguity in discerning a single hard reference class. The marginal increase in accuracy observed when agreement is defined as a match between the map label and either the primary or secondary reference label compared to accuracy observed when agreement is defined as a match between the map label and only the primary label quantifies the impact of 'categorical fuzziness' on accuracy. Another analytic option is to employ a 3 × 3 pixel window centered on each sample pixel and then to compare the modal class (as determined from the classification) of the 3 × 3 window centered on the sample pixel to the ground data label of the sample pixel. This analysis provides an indication of accuracy that accommodates the possibility of location error in the reference data. Lastly, an analysis contrasting accuracy derived from the subset of heterogeneous areas versus accuracy derived from the subset of homogeneous areas (as defined by the 3 × 3 pixel window) provides another indication of how much of the observed misclassification could be associated with location error. The rationale for this approach is that accuracy derived from the heterogeneous subset includes the confounding effect of location error, whereas accuracy of the homogeneous subset is mostly immune from location error effects.

None of these alternative analyses completely resolve the problems associated with reference data error. However, it is necessary to recognize that reference data error exists, to take all practical precautions to minimize these errors, and then to analyze the data in ways that yield insight into the potential impact of these errors on the accuracy results. It is worth re-stressing that accuracy assessment is, in reality, an evaluation of the degree of agreement between the predicted and reference class labels. Errors in labeling may occur in both data sets, so reference data quality is an important concern in the assessment and interpretation of classification accuracy statements.

Spatial variation in map quality

Conventional accuracy assessments yield typically an aspatial index of overall map quality but convey no information on the spatial distribution of error (McGwire and Fisher 2001). To-date, approaches to illustrate the spatial variation in map quality have focused on the mapping of uncertainty measures of the class labeling which can be output from some classification analyses. Thus, for example, the posterior probabilities of class membership calculated in the course of a maximum likelihood classification analysis may be used to provide a per-pixel indicator of map quality which could usefully accompany the conventional classification accuracy statement (Foody et al. 1992, Maselli et al. 1994). McIver and Friedl (2001) provide an example in which a per-pixel measure of classification equality is derived from a machine-learning classification. An alternative, albeit coarser representation of the spatial variation in map quality can be obtained by partitioning the accuracy assessment sample data into subregions and estimating accuracy for each of the subregions (Foody 2005).

ASSESSING ACCURACY OF CONTINUOUS PRODUCTS

Most of the accuracy assessment literature focuses on categorical products, but some analyses of remotely sensed data yield continuous products in which the output is a quantity such as elevation, percent impervious surface, temperature, net change, or percent composition of land-cover classes. The fundamental principles of response design, sampling design, and analysis remain applicable to the assessment of these continuous products. In the evaluation of a continuous product, however, the specific details of implementation of each of the three components of an accuracy assessment may differ from those typically implemented in the evaluation of a categorical product.

The response design for a continuous product may be more challenging than for a categorical product because it is generally more difficult to determine a quantity than it is to determine a category such as land-cover. For example, determining the percent urban imperviousness for a 30 m × 30 m pixel may require detailed interpretation of 1 m^2 resolution imagery. Obtaining ground measurements for other quantities, such as gross primary production estimated over areas measured in km^2, may require expensive instrumentation that is not readily portable. The difficulty of obtaining reference data for some map quantities may exert a strong influence on the sampling design. If the per-sample cost is very high such that only a small sample size is affordable, a judgment sample of representative locations may be warranted. However, the downside of adopting this approach is diminished rigor of the inferences.

If the quantity is such that a reasonable sample size is practical, deciding on a sampling design will require taking into consideration the usual issues of stratification and clustering. Is it necessary to create strata to ensure that the sample includes an adequate number of locations where the map quantity takes on values that are not common? Will clustering the sample locations result in substantial savings of time and cost? Relatively little investigation has focused on sampling designs for assessing accuracy of continuous products.

The measures describing the accuracy of continuous products will differ from those used with categorical products. For a given assessment unit i, denote the classified quantity as x_i and the true quantity as y_i (e.g., x_i is the classified percent impervious surface and y_i is the true percent impervious surface). If the region classified has N assessment units (e.g., N pixels), measures for quantifying accuracy include the Pearson or Spearman correlation between x_i and y_i, mean deviation $MD = (1/N) \sum_{i=1}^{N} (x_i - y_i)$, mean absolute deviation $MAD = (1/N) \sum_{i=1}^{N} |x_i - y_i|$, and root mean square error $RMSE = \sqrt{(1/N) \sum_{i=1}^{N} (x_i - y_i)^2}$. Ji and Gallo (2006) review many other measures proposed for describing accuracy of a continuous product. These accuracy measures must be estimated from the sample data, and the estimators used should be consistent following the methodology outlined for analyzing categorical outputs.

ASSESSING ACCURACY OF CHANGE

Because remote sensing is an effective and practical approach to monitor change in land-cover or other attributes, assessing the accuracy of change is an important problem. Accuracy of both gross change and net change may be of interest. Gross change quantifies the directional change from one class to another whereas net change quantifies the difference between the total area of a class at two dates. For example, the total area changed from forest to agriculture is gross change, whereas the difference between the area of forest at time 1 and the area of forest at time 2 is net change. Complementary gross changes negate each other in the derivation of net change. For example, a gross change of 5% of the area from forest to agriculture combined with a gross change of 5% from

agriculture to forest results in no net change for either forest or agriculture. Gross change and net change accuracy are assessed at different spatial supports. A pixel is typically the support for gross change assessments, while a region such as a block of pixels or even the entire mapped area is the support for net change assessments.

Assessing the accuracy of gross change follows the same protocols used for assessing a conventional (static) map (see Biging et al. (1998) for the basic sampling designs and analyses and Van Oort (2007) for additional considerations pertaining to analysis). An error matrix is used to describe gross change accuracy, where the rows and columns of the error matrix represent the different possible states of no change (e.g., forest remains forest, wetland remains wetland) and of gross change (e.g., forest to agriculture, wetland to urban). If the legend has c classes, this 'change/no change' error matrix has $c^2 \times c^2$ cells to account for all possible states of change and no change. Overall, user's and producer's accuracies can be calculated from the change/no change error matrix. Because it is generally not practical to estimate precisely user's and producer's accuracy for all possible change types, a more restricted set of objectives may need to be targeted. For example, instead of estimating accuracy of each type of gross change from forest to another class, it may be more practical to estimate accuracy of forest change to any other class (e.g., forest gross loss).

Assessing the accuracy of net change requires defining one or more levels of support at which the assessment will be made. The data for assessing the accuracy of net change are the quantities of net change for all classes derived from the classification (for each support unit) and the quantities of net change for all classes derived from the reference data. Net change accuracy is then summarized by measures such as mean absolute deviation, root mean square error, and correlation (Stehman and Wickham 2006) or by a confusion matrix and associated measures (Pontius and Cheuk 2006).

Because change in land-cover is often a relatively rare event, the sampling design for assessing gross change accuracy may require stratification to increase the sample size within change areas (Biging et al. 1998). The typically large number of possible change types makes it impractical to define every type of change as a stratum, but it may be realistic to define strata using the more common or important change types. Alternatively, strata could be defined using more general gross change categories, for example, forest loss of any type. Stratification is also strongly motivated for the objective of estimating the accuracy of net change. Because each support unit possesses different amounts of net change for the different classes, the stratification must take into account this multivariate feature (Stehman and Wickham 2006).

Assessing change accuracy presents some additional challenges not encountered with conventional (static) product assessments. A major difficulty is the requirement to know the class labels both before and after the change occurred. For example, if the response design specifies an on-the-ground visit to the sample locations, it would be impossible to stratify by mapped change simply because the change had not happened at the time the ground visits were needed for the initial date. This problem may be reduced if, for example, historical aerial photography is the source of reference data, but many problems remain (e.g., the photography may have been acquired at an inappropriate scale or time).

CROSS-VALIDATION

In cross-validation, a subset of the training data is set aside to form a validation set on which an initial evaluation of the classifier's ability to discriminate the classes may be based. A variety of approaches may be used, often involving the process being repeated for different sub-sets of the training sample. Cross-validation is particularly useful in helping to fine-tune the parameter settings of some classifiers including classification trees and neural networks, and as an in-process evaluation of the performance of the automated portion of a classification. If cross-validation accuracy is unacceptably low, remedial steps to enhance the classification may be taken.

Cross-validation requires only the training data used to develop the classification, and thus avoids the added expense of collecting reference data independent of the training data. However, cross-validation is not always appropriate to assess end-product accuracy. First, it may be difficult to implement if multiple classification algorithms are combined, or if significant manual interpretation is incorporated in the classification protocols. More critically, if the training data input into cross-validation do not originate from a probability sampling design, then the accuracy estimates are biased (Steele 2005) because the sample does not represent the population well. For example, training data are often obtained from convenient locations, selected from homogeneous areas, or chosen to represent exemplar cases of a class. The accuracy of a population of such special-case locations may differ substantially from accuracy of the entire region (Hammond and Verbyla 1996), and cross-validation estimates accuracy of the former population, whereas it is the accuracy of the entire region that is of interest. Steele et al. (2003) and Steele (2005) offer options to diminish the potential bias of cross-validation estimates derived from a non-probability sample. If cross-validation

accuracy is based on training data derived from a probability sample, the bias concern is eliminated.

ACCURACY OF LARGE-AREA MAPS

Assessing the accuracy of national, continental, or global maps may present additional challenges because these maps are typically subjected to diverse applications by multiple users. Mayaux et al. (2006), Scepan (1999), Stehman et al. (2003), Wickham et al. (2004), and Wulder et al. (2006) are examples of large-area accuracy assessments. Strahler et al. (2006) provide general recommendations for accuracy assessment of global products. One recommendation is to conduct a qualitative, systematic accuracy review designed to provide a timely assessment useful for improving and updating the map before it reaches the final stage. The contribution of probability sampling to rigorous inference and the utility of the error matrix for description and analysis are still relevant to global map assessments (Strahler et al. 2006). The sampling designs for large-area assessments typically incorporate both strata and clusters (Edwards et al. 1998, Nusser and Klaas 2003, Stehman et al. 2003, Wickham et al. 2004), with stratification chosen to increase the precision of the class-specific accuracy estimates, and cluster sampling chosen to reduce costs. Strahler et al. (2006) propose developing a global ground data set for accuracy assessment that would be appropriate for a broad variety of applications. While the cost of such an endeavor would be significant, the long term cost efficiency and benefits gained by a coordinated, integrated accuracy protocol applicable to different global land-cover products would easily justify the investment.

BEYOND SITE-SPECIFIC ASSESSMENTS

Historically, assessments have focused on site-specific accuracy and the question of how well the map represents the ground condition on a location-by-location basis. A common application of land-cover data is to aggregate the information to some spatial unit (e.g., 10 km × 10 km blocks or watersheds) and use the percent composition of the classes within each spatial unit as input into a model or statistical analysis. Assessing composition accuracy for a support larger than a pixel is a type of nonsite-specific accuracy assessment. To assess composition accuracy of aggregated map data, the ground data must be determined for a sample of the support units. The area proportions for each class derived from the classification and from the ground data are then compared and accuracy is quantified on a per class basis, for example, by mean absolute deviation, root mean square error or correlation, by a multivariate distance measure such as Mahalanobis distance, or by a confusion matrix approach (Pontius and Cheuk 2006).

The potential also exists to expand the set of accuracy measures to include similarity measures derived from a map comparison perspective (see the special issue of the *Journal of Geographical Systems* (2006, volume 8, issue 2)). To frame accuracy assessment as a map comparison problem, the target classification to be evaluated and a map representing the ground data of the same region are considered the two maps to be compared. The comparison can be made on the basis of features (i.e., spatially contiguous collections of pixels of constant attribute such as land-cover polygons) to evaluate how well attribute, location, size, and shape, as well as relationships among features are mapped (Dungan 2006). Other map comparison measures focus on similarity of location and quantity (Pontius 2000, 2002), account for fuzziness of location and fuzziness of category (Hagen 2003, Hagen-Zanker 2006), examine similarity in pattern based on areal intersection of land use polygons (White 2006), measure similarity using a polygon based fuzzy local matching technique (Power et al. 2001), and quantify goodness of fit of either map categories or individual raster patches (Hargrove et al. 2006).

Clearly, many potential options for accuracy measures become available if accuracy assessment is viewed as a map comparison problem. These map comparison measures have rarely been employed in practice, so questions remain about their ease of interpretation and utility. For the most part, the map comparison literature discusses only the situation where complete coverage information is available for both maps, whereas in accuracy assessment, ground data are usually available for only the sample locations. Before these map comparison methods can be applied for accuracy assessment, it must be demonstrated that they can be estimated from a sample. Because some of the map comparison measures require information for complete blocks of pixels, incorporating these measures in accuracy assessment would affect the sampling design. Not only would the size of the block required be a relevant sampling design concern, but the need to obtain reference data for adjacent pixels would rule out all but cluster sampling of the basic sampling designs typically implemented in a pixel-based assessment.

FUTURE DIRECTIONS

Several developments that are needed in accuracy assessment have been alluded to in the separate sections of this chapter. One of the more important needs in accuracy assessment is simply to improve basic practice using existing theory and methods. Non-probability sampling designs are too commonly implemented, and incorrect analyses are still conducted and reported (e.g., applying formulas that assume simple random sampling to data acquired by a stratified design). Another pressing need is to construct protocols to address multiple accuracy objectives within a comprehensive, coherent assessment framework. That is, in addition to the per-pixel assessments traditionally implemented, it is desirable to evaluate accuracy of patch size and shape distributions, accuracy of land-cover composition for data aggregated to a larger spatial support, and accuracy of other characteristics of a classification that are derived at a support larger than a pixel. Developing protocols that would allow for simultaneously assessing accuracy of static map products along with gross and net change products is another manifestation of the theme that accuracy assessments need to address multiple objectives within a common framework. To-date, accuracy assessment has largely addressed the problem of describing accuracy, with less emphasis on the interface between descriptive accuracy and the impact of error on the applications of the map products. Developing the theory and methods for evaluating how accuracy assessment can better integrate with error propagation analyses will further enhance the value of accuracy assessment.

REFERENCES

Agresti, A., 1996. *An Introduction to Categorical Data Analysis*. John Wiley and Sons, New York.

Biging, G. S., D. R. Colby, and R. G. Congalton, 1998. Sampling systems for change detection accuracy assessment. In R. S. Lunetta and C. D. Elvidge (eds), *Remote Sensing Change Detection: Environmental Monitoring Methods and Applications*. Ann Arbor Press, Chelsea, Michigan, pp. 281–308.

Binaghi, E., P. A. Brivio, P. Ghezzi, and A. Rampini, 1999. A fuzzy set-based accuracy assessment of soft classification. *Pattern Recognition Letters*, 20: 935–948.

Card, D. H., 1982. Using known map category marginal frequencies to improve estimates of thematic map accuracy. *Photogrammetric Engineering and Remote Sensing*, 48: 431–439.

Cochran, W. G., 1977. *Sampling Techniques* (3rd edn.). John Wiley and Sons, New York.

Congalton, R. G. and K. Green, 1993. A practical look at the sources of confusion in error matrix generation. *Photogrammetric Engineering and Remote Sensing*, 59: 641–644.

Congalton, R. G. and K. Green, 1999. *Assessing the Accuracy of Remotely Sensed Data: Principles and Practices*. CRC Press, Boca Raton, FL.

Czaplewski, R. L., 1994. *Variance Approximations for Assessments of Classification Accuracy*. Research Paper RM-316, U.S. Department of Agriculture, Forest Service, Rocky Mountain Forest and Range Experiment Station, Fort Collins, CO, 29 pp.

Dungan, J. L., 2006. Focusing on feature-based differences in map comparison. *Journal of Geographical Systems*, 8: 131–143.

Edwards, T. C., Jr., G. G. Moisen, and D. R. Cutler, 1998. Assessing map accuracy in an ecoregion-scale cover-map. *Remote Sensing of Environment*, 63: 73–83.

Finn, J. T., 1993. Use of the average mutual information index in evaluating classification error and consistency. *International Journal of Geographical Information Systems*, 7: 349–366.

Foody, G. M., 1992. On the compensation for chance agreement in image classification accuracy assessment. *Photogrammetric Engineering and Remote Sensing*, 58: 1459–1460.

Foody, G. M., 2002. Status of land cover classification accuracy assessment. *Remote Sensing of Environment*, 80: 185–201.

Foody, G. M., 2005. Local characterization of thematic classification accuracy through spatially constrained confusion matrices. *International Journal of Remote Sensing*, 26: 1217–1228.

Foody, G. M., N. A. Campbell, N. M. Trodd, and T. F. Wood, 1992. Derivation and applications of probabilistic measures of class membership from the maximum-likelihood classification. *Photogrammetric Engineering and Remote Sensing*, 58: 1335–1341.

Franklin, J., D. Beardsley, H. Gordon, D. K. Simons, and J. M. Rogan, 2001. Evaluating errors in a digital vegetation map with forest inventory data and accuracy assessment using fuzzy sets. *Transactions in GIS*, 5: 285–304.

Gopal, S. and C. Woodcock, 1994. Theory and methods for accuracy assessment of thematic maps using fuzzy sets. *Photogrammetric Engineering and Remote Sensing*, 60: 181–188.

Hagen, A., 2003. Fuzzy set approach to assessing similarity of categorical maps. *International Journal of Geographical Information Science*, 17: 235–249.

Hagen-Zanker, A., 2006. Map comparison methods that simultaneously address overlap and structure. *Journal of Geographical Systems*, 8: 165–185.

Hammond, T. O. and D. L. Verbyla, 1996. Optimistic bias in classification accuracy assessment. *International Journal of Remote Sensing*, 17: 1261–1266.

Hargrove, W. W., F. M. Hoffman, and P. F. Hessburg, 2006. Mapcurves: a quantitative method for comparing categorical maps. *Journal of Geographical Systems*, 8: 187–208.

Ji, L. and K. Gallo, 2006. A new agreement coefficient for comparing remotely sensed data acquired from different sensors. *Photogrammetric Engineering and Remote Sensing*, 72: 823–833.

Justice, C., A. Belward, J. Morisette, P. Lewis, J. Privette, and F. Baret, 2000. Developments in the 'validation' of satellite sensor products for the study of land surface. *International Journal of Remote Sensing*, 21: 3383–3390.

Laba, M., S. K. Gregory, J. Braden, D. Ogurcak, E. Hill, E. Fegraus, J. Fiore, and S. D. DeGloria, 2002. Conventional and fuzzy accuracy assessment of the New York Gap Analysis Project land cover maps. *Remote Sensing of Environment*, 81: 443–455.

Liu, C., P. Frazier, and L. Kumar, 2007. Comparative assessment of the measures of thematic classification accuracy. *Remote Sensing of Environment*, 107: 606–616.

Maselli, F., C. Conese, and L. Petkov, 1994. Use of probability entropy for the estimation and graphical representation of the accuracy of maximum-likelihood classifications. *ISPRS Journal of Photogrammetry and Remote Sensing*, 49(2): 13–20.

Mayaux, P., H. Eva, J. Gallego, A. H. Strahler, M. Herold, S. Agrawal, S. Naumov, E. E. De Miranda, C. M. Di Bella, C. Ordoyne, Y. Kopin, and P. S. Roy, 2006. Validation of the Global Land Cover 2000 map. *IEEE Transactions on Geoscience and Remote Sensing*, 44: 1728–1739.

McGwire, K. C. and P. Fisher, 2001. Spatially variable thematic accuracy: Beyond the confusion matrix. In C. T. Hunsaker, M. F. Goodchild, M. A. Friedl, and T. J. Case (eds), *Spatial Uncertainty in Ecology: Implications for Remote Sensing and GIS Application*. Springer, New York, pp. 308–329.

McIver, D. K. and M. A. Friedl, 2001. Estimating pixel-scale land cover classification confidence using non-parametric machine learning methods. *IEEE Transactions on Geoscience and Remote Sensing*, 39: 1959–1968.

Mickelson, Jr., J. G., D. L. Civco, and J. A. Silander, Jr., 1998. Delineating forest canopy species in the Northeastern United States using multi-temporal TM imagery. *Photogrammetric Engineering and Remote Sensing*, 64: 891–904.

Moisen, G. G., T. C. Edwards, Jr., and D. R. Cutler, 1994. Spatial sampling to assess classification accuracy of remotely sensed data. In W. K. Michener, J. W. Brunt, and S. G. Stafford (eds), *Environmental Information Management and Analysis: Ecosystem to Global Scales*. Taylor and Francis, New York, pp. 159–176.

Muller, S. V., D. A. Walker, F. E. Nelson, N. A. Auerbach, J. G. Bockheim, S. Guyer, and D. Sherba, 1998. Accuracy assessment of a land-cover map of the Kuparuk River Basin, Alaska: Considerations for remote regions. *Photogrammetric Engineering and Remote Sensing*, 64: 619–628.

Nusser, S. M. and E. E. Klaas, 2003. Survey methods for assessing land cover map accuracy. *Environmental and Ecological Statistics*, 10: 309–331.

Pontius, R. G., 2000. Quantification error versus location error in comparison of categorical maps. *Photogrammetric Engineering and Remote Sensing*, 66: 1011–1016.

Pontius, R. G., Jr., 2002. Statistical methods to partition effects of quantity and location during comparison of categorical maps at multiple resolutions. *Photogrammetric Engineering and Remote Sensing*, 68: 1041–1049.

Pontius, R. G., Jr., and M. L. Cheuk, 2006. A generalized cross-tabulation matrix to compare soft-classified maps at multiple spatial resolutions. *International Journal of Geographical Information Science*, 20: 1–30.

Powell, R. L., N. Matzke, C. de Souza, Jr., M. Clark, I. Numata, L. L. Hess, and D. A. Roberts, 2004. Sources of error in accuracy assessment of thematic land-cover maps in the Brazilian Amazon. *Remote Sensing of Environment*, 90: 221–234.

Power, C., A. Simms, and R. White, 2001. Hierarchical fuzzy pattern matching for the regional comparison of land use maps.

International Journal of Geographical Information Science, 15: 77–100.

Scepan, J., 1999. Thematic validation of high-resolution global land-cover data sets. *Photogrammetric Engineering and Remote Sensing*, 65: 1051–1060.

Steele, B. M., 2005. Maximum posterior probability estimators of map accuracy. *Remote Sensing of Environment*, 99: 254–270.

Steele, B. M., D. A. Patterson, and R. L. Redmond, 2003. Toward estimation of map accuracy without a probability test sample. *Environmental and Ecological Statistics*, 10: 333–356.

Stehman, S. V., 1992. Comparison of systematic and random sampling for estimating the accuracy of maps generated from remotely sensed data. *Photogrammetric Engineering and Remote Sensing*, 58: 1343–1350.

Stehman, S. V., 1997. Estimating standard errors of accuracy assessment statistics under cluster sampling. *Remote Sensing of Environment*, 60: 258–269.

Stehman, S. V., 1999. Basic probability sampling designs for thematic map accuracy assessment. *International Journal of Remote Sensing*, 20: 2423–2441.

Stehman, S. V., 2000. Practical implications of design-based sampling inference for thematic map accuracy assessment. *Remote Sensing of Environment*, 72: 35–45.

Stehman, S. V., 2001. Statistical rigor and practical utility in thematic map accuracy assessment. *Photogrammetric Engineering and Remote Sensing*, 67: 727–734.

Stehman, S. V., 2004. A critical evaluation of the normalized error matrix in map accuracy assessment. *Photogrammetric Engineering and Remote Sensing*, 70: 743–751.

Stehman, S. V. and R. L. Czaplewski, 1998. Design and analysis for thematic map accuracy assessment: Fundamental principles. *Remote Sensing of Environment*, 64: 331–344.

Stehman, S. V. and J. D. Wickham, 2006. Assessing accuracy of net change derived from land cover maps. *Photogrammetric Engineering and Remote Sensing*, 72: 175–185.

Stehman, S. V., J. D. Wickham, J. H. Smith, J. H., and L. Yang, 2003. Thematic accuracy of the 1992 National Land-Cover Data (NLCD) for the Eastern United States: Statistical methodology and regional results. *Remote Sensing of Environment*, 86: 500–516.

Strahler, A. H., L. Boschetti, G. M. Foody, M. A. Friedl, M. C. Hansen, M. Herold, P. Mayaux, J. T. Morisette, S. V. Stehman, and C. E. Woodcock, 2006. *Global Land Cover Validation: Recommendations for Evaluation and Accuracy Assessment of Global Land Cover Maps*, EUR 22156 EN–DG, Office for Official Publications of the European Communities, Luxembourg, 48 pp.

Van Oort, P. A. J., 2007. Interpreting the change detection error matrix. *Remote Sensing of Environment*, 108: 1–8.

White, R., 2006. Pattern based map comparisons. *Journal of Geographical Systems*, 8: 145–164.

Wickham, J. D., S. V. Stehman, J. H. Smith, and L. Yang, 2004. Thematic accuracy of the 1992 National Land-cover Data for the western United States. *Remote Sensing of Environment*, 91: 452–468.

Wulder, M. A., S. E. Franklin, J. C. White, J. Linke, and S. Magnussen, 2006. An accuracy assessment framework for large-area land cover classification products derived from medium-resolution satellite data. *International Journal of Remote Sensing*, 27: 663–683.

Remote Sensing Applications

A. Lithospheric Sciences

Making Sense of the Third Dimension through Topographic Analysis

Yongxin Deng

Keywords: topographic analysis, DEM, scale, context, error, surface representation, surface parameterization, landform classification.

INTRODUCTION

Prior to the development of digital elevation data, contour lines derived from remote sensing and drawn on paper maps were the main form of elevation data around the world. In many areas, contour lines remain the most common format for elevation data. Replacement of plane-table surveying by aerial photogrammetry in the United States started in the 1940s (USGS 2005). Since then, airborne and later spaceborne remote sensing has become the dominant information source for elevation, using various technologies such as stereoscopic photogrammetry (Giles and Franklin 1996), interferometric radar (Rabus et al. 2003), or laser scanning (French 2003, see also Hodgson et al. 2003). Today, topographic data sets produced from remote sensing exist at a wide range of spatial resolutions (e.g., 0.1 m–5 km or even coarser) and cover large areas of diverse terrain (Lane and Chandler 2003). In fact, it is this availability advantage of topographic data in contrast to other environmental data (e.g., climate, soil, vegetation), together with the rapid development of Geographic Information Systems (GIS) and computer power, that fostered the growth of digital topographic analysis in the past two decades (O'Callaghan and

Mark 1984, Moore et al. 1991, Florinsky 1998a, Wilson and Gallant 2000a, Deng 2007).

The emergence of new data products has recently spawned several new and exciting developments in topographic analysis. The advent of high-resolution elevation data has, for example, encouraged hydrological and geomorphological modeling in lowlands or flatlands (e.g., wetlands, floodplains, and braided river valleys), areas that used to be difficult or impossible to model (Marks and Bates 2000, Lane and Chandler 2003, Lane 2005). Acquisition of digital elevation data over broad regions has facilitated country- or continent-wide environmental modeling (Armstrong and Martz 2003, Wolock et al. 2004, White et al. 2005, Colombo et al. 2007). At the same time, extraction of the elevations that are associated with surface objects such as buildings and trees, which were previously too small to delineate from spaceborne or airborne imagery (Gamba and Houshmand 2000, Fraser et al. 2002), points to the potential for incorporating discrete objects in topographic analysis to support applications such as urban storm-water modeling (Djokic and Maidment 1991). Additionally, the availability of multi-resolution and multi-source data in many settings has prompted the need for (1) data mining and visualization tools

(White et al. 2005, Tate et al. 2007), (2) multi-scale analytical abilities (Gallant and Dowling 2003, MacMillan et al. 2004, Schmidt and Andrew 2005), and (3) understanding of data manipulation effects (Deng and Wilson 2006, Deng et al. 2006).

The strong connections between remote sensing and topographic analysis extend beyond the initial acquisition of elevation data using remote sensing. Environmental information derived from remotely sensed data is often vital for the validation and calibration of topographic analysis procedures and outputs (e.g., Burrough et al. 2001). In more complex cases such as wildfire modeling, there is a need to directly combine remote sensing (e.g., of vegetation) and topographic analysis to safeguard modeling accuracy in mountainous areas, because topography is treated as a part of the 'wildfire environment triangle' (Allgöwer et al. 2003). However, development of present-day fire models often avoids mountainous regions due to the difficulty of incorporating topographic effects (Emilio Chuvieco, personal communication).

Topographic analysis and remote sensing also share several important concepts and tools including: (1) texture analysis on raster images or on digital elevation data (Woodcock and Strahler 1987, Fisher et al. 2004, Lucieer and Stein 2005); (2) semantic object extraction (Schmidt and Dikau 1999, Dehn et al. 2001, Baltsavias 2004); (3) analysis of topographic or image contexts (Chen 1999, MacMillan et al. 2004); (4) image or landform classification methods (MacMillan et al. 2000, Ventura and Irvin 2000, Bruzzone and Prieto 2001, Wang and Jamshidi 2004); and (5) a central concern with spatial scale (Woodcock and Strahler 1987, Band and Moore 1995, Quattrochi and Goodchild 1997, Marceau and Hay 1999, Florinsky and Kuryakova 2000, Deng et al. 2007a, b). These commonalities suggest a shared future for remote sensing and topographic analysis.

Nevertheless, there are areas where the typical focus of topographic analysis is distinguished from that of remote sensing in general. Topographic analysis tends to pay particular attention to terrain-affected biophysical processes and resulting spatial patterns (e.g., Moore and Wilson 1992), to spatially connected topographic surfaces more than discrete cells (e.g., Desmet and Govers 1996, Armstrong and Martz 2003), and to numerous topography-environment relations (e.g., Franklin 1995, Zhu 1997, Deng et al. 2007a, Nelson et al. 2007). For instance, the calculation of incoming solar radiation (e.g., Oliphant et al. 2003) involves not only local slope gradient and azimuth (aspect), which are usually characterized by use of a 3×3 moving window, but also shading and shadowing effects of topography that need the consideration of an irregularly-shaped area defined according to irradiance processes. The calculated insolation may then be linked to other environmental variables (Wilson and Gallant 2000c). The special focuses of topographic analysis suggest a potential opportunity integrating topographic analysis and remote sensing: *Other than reporting elevations, surface objects, and their corresponding accuracies and spatial resolutions, how could the integration of environmental remote sensing and topographic analysis better inform the depiction of topographic surfaces and the study of their biophysical role?*

This chapter provides a comprehensive evaluation of the status quo and future challenges of topographic analysis, so as to assist the understanding of the above question. It is also a forward-looking call for a more surface-oriented topographic remote sensing and for the incorporation of multi-scale topographic contexts in topographic analysis. The chapter is structured in six parts, including this introduction. The next section discusses topographic surface mapping using digital topographic data. The third and fourth sections respectively evaluate topographic surface parameterization and subdivision. The fifth section examines the potential and challenge of incorporating higher-order, multi-scale topographic contexts in topographic analysis. The sixth and final section concludes the chapter by identifying some immediate challenges and opportunities.

SURFACE DEPICTION USING TOPOGRAPHIC DATA

The three most common formats for digital elevation data are: contour lines, triangulated irregular networks (TINs), and regular grids of points, generally termed digital elevation models, or DEMs (Moore et al. 1991, Wilson and Gallant 2000b). Contour lines connect all points of equal elevations and usually employ a uniform contour interval on a map. A TIN is usually constructed by connecting irregularly located surface-specific points into triangular (planar) facets (Peucker 1978, Kumler 1994, Kidner et al. 2000). The most widely used digital elevation format is the DEM, which should be viewed as a representation of topography by 'grid points,' rather than 'areal pixels,' to avoid the appearance of artificial 'step-like' terrain at pixel edges. This conceptualization also distinguishes a DEM and its derivatives from other, 'cell-filled' remote sensing data, such as raster layers of land use and land cover. Because a DEM is a simple format for digital processing and has the most direct linkage with remote sensing, it is much preferred by users. This in turn implies that the potential benefits of improving DEMs are large.

The direct biophysical significance of elevation is rather limited, and is evident mostly in its influence on climatic variables (Geiger 1965, Hutchinson 2007). As a result, the core task of topographic representation and analysis is not to deal with elevation *per se* but to characterize the topographic surface, focusing on how point elevation values of topographic data connect into a surface to impose an influence on the environment (Deng 2007, Hutchinson 2007). In this regard, a mathematical function would be the ideal representation (e.g., Mitasova and Hofierka 1993) if the topographic surface could be taken as smooth and varying only at one scale. But the reality is otherwise (e.g., McClean and Evans 2000), and some kind of discrete spatial sampling has to be employed in an explicit or implicit manner for all topographic data. Schneider (2001a) recognizes this step as *surface discreterization*, which has to be followed by a step of *surface-reconstruction* if the surface needs to be parameterized. This conceptual model with its emphasis on representing and sampling the surface suggests additional criteria – other than 'how to report elevation' – for the evaluation of topographic data formats. It also encourages us to seek new ways of improving the existing data and their formats. To this end, the more inclusive lists of topographic data formats and sources provided by Burrough (1986, p. 40) and Hutchinson (2007) are very useful, since they facilitate efforts to think beyond the 'ordinary.'

The DEM format has obvious weaknesses in reporting the topographic surface (see Moore et al. (1991) for an earlier evaluation): (1) it is essentially a discrete way of depicting a usually continuous surface; (2) it uses a uniform sampling density across the represented surface that is often heterogeneous and may include variable complexities; and (3) it provides no direct information of numerous topographic features (spatial entities) that are environmentally significant, such as peaks, ridges, pits, valleys, hollows, breaklines, etc. Among the three common data formats, in fact, the DEM offers the least support for the existence of a topographic surface, even though it usually comes with the densest sampling and perhaps the most accurate elevation values. Contour lines give realistic depictions of surface shape along the lines, represent surface features such as valleys well, imply surface continuity (or connectivity) between lines, and efficiently depict variations in surface complexity by the varying contour density and curvature (e.g., Moore and Grayson 1991). However, contours are hard to handle digitally. The TIN format, on the other hand, can be based on well-selected surface-specific points, allows adjustments of the vertex density according to the complexity of the surface, and can define important surface features explicitly as triangle vertices, edges, or facets (see Kumler 1994). Most of these strengths would nonetheless disappear if the TIN is poorly constructed.

While impossible at present, it would be desirable if strengths of various data formats could be combined.

Hutchinson (1989), among a few other efforts (see Wilson et al. 2008 for a brief evaluation), proposed a procedure called ANUDEM (see also Hutchinson 2006) to interpolate contour lines into hydrologically correct DEMs using drainage enforcement (e.g., with the assistance of streamlines) and surface-specific points (e.g., high, low, or contour breakpoints, etc.). With sufficient spatial sampling (Wilson et al. 1998), ANUDEM shows more respect to surface reality, at least for analysis of runoff-related processes (Hutchinson and Gallant 2000). As a potential research topic, such strengths of ANUDEM might be combined with a reprocessing procedure (such as filtering or resampling, see the paragraph below) of dense remote sensing elevations to improve or modify future DEM data to support hydrological modeling (e.g., Callow et al. 2007). Following this direction, it is possible to produce process-specific DEMs whose accuracy is high(er) for particular applications. Such DEMs are useful because, when particular biophysical processes are considered, the accuracy of some surface-specific points (e.g., summits as an ecological niche), surface features (e.g., a dike on a floodplain), and particular landforms (e.g., steep slopes associated with gully erosion or landslides) is indeed more important than the accuracy of the remainder of the surface.

In the future, improved DEM 'fitness for use' may imply the need to change (and possibly coarsen) the input data resolution or even the need to employ multiple layers representing different topographic locations with varying details and accuracies in the same analysis. For instance, Hutchinson (2007) indicates that filtering remotely sensed high-resolution DEMs may help identify a coarser but better resolution for surface representations. However, tests have yet to be conducted regarding how a very fine DEM (e.g., 1 m) may be filtered or resampled to improve surface representations at a much coarser resolution (e.g., 10 or 30 m), a process that may incorporate surface-specific points, topographic features, and interpolation procedures such as ANUDEM.

The topographic surface is usually not smooth and is instead covered by various surface objects (e.g., buildings, trees, small water bodies, etc.) and materials (e.g., bare soil, grass, impervious surfaces) that influence its biophysical functions (e.g., in runoff and soil erosion processes). However, most DEM applications and corresponding paradigms – theories, methodologies, and algorithms – represent the spatial variability of 'natural' landscapes within horizontal distances of larger than 0.5–5 km, using DEMs with resolutions of 5–10 m or lower (Wilson and Gallant 2000a). DEMs accordingly depict a 'bare-earth' surface on which objects have been removed or ignored. This situation is perhaps related to some of the traditional DEM generation procedures that are based on the interpolation of contour lines assuming a smooth surface free of objects (Hutchinson and Gallant 2000). This convention of suppressing objects continues for topographic data generation today, despite the often fine spatial resolution of the data that covers both urban and less developed areas. The simultaneous challenge and opportunity are that tools such as airborne lidar can accurately delineate surface objects (e.g., Gamba and Houshmand 2000, Shan and Sampath 2005). In addition, the intentional bare-earth modeling approach at least partly explains our difficulty in combining surface continuity modeling with the representation of discrete objects, as is needed, for example, when modeling hillslope runoff in built environments where buildings and impervious surfaces are key hydrological elements. A 'rough-surface' paradigm is therefore needed that takes advantage of the increasing availability of very high resolution DEM data. As a result, formerly 'sub-scale' terrain objects (e.g., Fraser et al. 2002) and rough and porous surface characteristics (e.g., Butler et al. 1998) can be integrated. Lane (2005), for instance, suggests a re-evaluation of the existing

conceptualization of surface roughness in response to the emergence of very-high resolution DEMs (see also Marks and Bates 2000), which may lead to replacement of conventionally used roughness parameters with distributed, DEM-based porosity values for floodplain inundation modeling.

The issues discussed above focus on representational perspectives, which are related to DEM data uncertainty, a research issue that has attracted numerous efforts in the context of the growth of Geographic Information Science. An up-to-date review and selective bibliography are offered by Fisher and Tate (2006). A central argument since the early efforts has been that DEM error is more of the topographic surface than of the elevation, but the DEM error has usually been reported as elevation error using measures such as sample root mean squared error (RMSE). Therefore, an understanding of the distribution and autocorrelation of error is vital (e.g., Hunter and Goodchild 1997, Carlisle 2005). This suggests the need for other DEM error measurements that complement RMSE. Furthermore, DEM error standards should vary with landscape type, because the same elevation error most likely results in different surface distortions in different landscapes (see Figure 22.1). The primary difficulty in implementing this broader approach

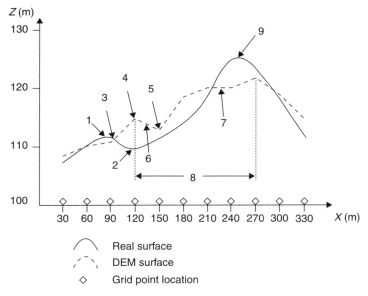

Figure 22.1   Various hypothetical situations comparing elevation vs. surface errors in a DEM representation of a surface profile: (1) loss of real peak or ridge with a small elevation error; (2) loss of real pit or valley; (3) no elevation error but severe surface (shape) error; (4) artificial peak or ridge with a large elevation error; (5) artificial pit or valley with a small elevation error; (6) reversed slope azimuth (also see (3)); (7) reduced slope gradient; (8) displaced catchment area; (9) displaced catchment boundary. The 'Z' axis is elevation and 'X' axis is horizontal distance along a hypothetical transect.

Table 22.1   Surface error types and examples of affected applications

Error Code (as shown in Figure 22.1)	Error type	Influenced applications
1	Topographic object and position	Landmark, soil boundary, mass movement
2	Topographic object and position	Wetland, geohydrology, non-point-source pollution
3	Local shape (correct elevation)	Solar radiation (and air temperature), mass movement direction, wind, vegetation greenness and moisture
4	Topographic object and position	Hydrology, soil erosion potential, road construction
5	Topographic object	Hydrology, soil depth, viewshed
6	Local shape and topographic position	Solar radiation (and air temperature), mass movement direction, wind, vegetation greenness and moisture
7	Local shape (wrong elevation)	Mass and water movement velocity, soil depth, vegetation, geomorphology
8	Topographic surface connection	Hydrology, environmental management
9	Topographic object, position, and surface connectivity	Land resource management, land use planning, hydrology

to error measurement is that there is no 'correct' surface to serve as a standard, even though more accurate elevation values can be obtained from ground control points, or from mapping- or survey-grade GPS measurements. Consequently, various DEM error models have been developed to deal with this lack of 'ground truth.' As evaluated by Liu and Jezek (1999), these error models may include: (1) empirical methods that evaluate DEM error based on point measurements of elevations (e.g., Bolstad and Stowe 1994); (2) analytical approaches that track the process of error propagation (e.g., Heuvelink 1998); and (3) simulation methods that generate a probability surface – according to the 'simulated' structure or autocorrelation of DEM errors – to represent possible error ranges in outputs, using tools such as Monte Carlo simulations (e.g., Lee et al. 1992, Aguilar et al. 2007).

Figure 22.1 and Table 22.1 employ a hypothetical profile to demonstrate: (1) the distinction between surface and elevation errors; (2) the types of surface error; and (3) their non-negligible impacts on applications. It is obvious that the magnitude of surface error may not conform to that of the elevation error, an observation useful for future measurements of surface errors.

SURFACE PARAMETERIZATION

A rough distinction of the analytical tasks in topographic analysis can be drawn between surface parameterization – description of the topographic surface using various attributes that are calculated from topographic data – and surface subdivision (see the next section), even though the two are closely related (Moore et al. 1991, Deng et al. 2007c). Deng (2007) suggested an ontological framework in which topographic analysis tasks can

be compared and linked by identifying the ontological existences (Smith and Mark 2003), or 'things,' that are dealt with. These 'things' include elevation, surface shape, topographic position, topographic context, spatial scale, and landform object. Given a certain topographic data set, most topographic analysis tasks can be linked to one or more of these ontological existences.

In parameterizing the topographic surface, a computational distinction can be made between primary attributes that are directly derived from topographic data (e.g., a DEM) and secondary ones that combine primary attributes for more direct, 'index-like' representations of biophysical processes (Moore et al. 1991, Wilson and Gallant 2000c). This relatively simple, straightforward distinction was later expanded by Florinsky (1998a) who subdivided primary attributes into two groups: local attributes (e.g., slope, aspect, plan curvature, profile curvature) that characterize the surface shape of a small neighborhood area and non-local attributes (e.g., specific catchment area, downslope length along a water flow path) whose definition requires a non-local area identified with reference to a particular biophysical process. A similar division was given by Lindsay (2005), although he added topographic position (e.g., relief) as a separate category. Shary et al. (2002) developed a more sophisticated scheme, accounting for not only local vs. non-local (or regional) differences in morphology, but also physical 'field' contexts (such as 'solar irradiation, gravitational, magnetic, electrical, etc.'; see Shary et al. 2002, p. 5) in which topographic morphology exists. They thus distinguished between field-specific and field-invariant attributes, depending on whether the physical context is explicitly considered in the attribute definition.

According to Deng (2007), the fundamental difference between local and non-local topographic attributes is that the former describes topographic surface shape, whereas the latter addresses the topographic position of a point or object, with or without reference to a biophysical process. Following this interpretation, secondary attributes would be those that combine both 'shape' and 'position' roles of topography in biophysical processes. The most successful secondary attributes include topographic wetness index (Beven and Kirkby 1979, Quinn et al. 1995), topographic transport capacity index (Moore and Wilson 1992), and various solar radiation indices (e.g., Wilson and Gallant 2000c). A list of commonly used primary and secondary attributes, their algorithms, and their possible applications was given by Moore et al. (1991) and also Wilson and Gallant (2000b). These attributes have been widely used across various disciplines in environmental and related sciences (Pike 1988, 2000, Florinsky 1998a, Shary et al. 2002, Evans 2003, Deng et al. 2007c).

Recent research in topographic analysis has highlighted several important issues in paramaterizing topographic surfaces. First, local attributes have been found to be dependent on the data, scale, and algorithm employed (Florinsky 1998b, Jones 1998, Florinsky and Kuryakova 2000, Deng et al. 2007b), whereas non-local attributes may have more absolute real-world meaning, and therefore more solid measure of 'accuracy' (e.g., Fisher 1993, 1996, Wilson et al. 2007). Several tests (e.g., Florinsky 1998b, Schmidt et al. 2003) found the second-order polynomial method of Evans (1972) to be the most robust in deriving attributes for the description of terrain shape, but Florinsky (1998b) cautions that calculated local attributes are only mathematical abstractions, and that the 'real' topographic surface is rarely a smooth mathematical one. Non-local attributes such as relief and specific catchment area are different in this regard: the altitude difference between two points and the possible (runoff) contributing area for a given point – once they are defined precisely – are both certain (at least conceptually), even though we may not be able to measure these variables accurately. This observation signifies the need to develop or refine algorithms for more accurate estimates of non-local attributes, whereas algorithms for local attributes (e.g., slope gradient) should take more flexible forms and employ exploratory 'fitness for use' assessments, even though only a few local attribute algorithms are in wide use. It is also noteworthy that current techniques for routing runoff across the landscape do not agree with one another and are mostly static-state simulations that anticipate a smooth surface with no superimposed objects or porosity for the underlying layer (Wilson et al. 2007, 2008).

A common concern in the topographic analysis literature is that topographic attributes are dependent on the DEM resolution (Zhang and Montgomery 1994, Florinsky and Kuryakova 2000), and this dependency varies with landscape type (Deng et al. 2007b). Moreover, different topographic attributes show rather contrasting sensitivities to scale change (Deng et al. 2007b), thereby casting doubt on the legitimacy of combining these attributes when they are calculated from the same DEM. For instance, topographic wetness index is calculated from slope gradient and specific catchment area, usually from the same DEM, but the scale dependency of these two primary attributes is very different (Deng et al. 2007b). Deng et al. (2007b) observed sharp changes in attribute values as DEM resolution was varied between 5–30 m, indicating the output uncertainty when the analysis was conducted at a single, randomly selected spatial resolution in this popular resolution range. As a consequence, the issue of scale selection is a major challenge in topographic analysis, because many applications in the USA, for example, incorporate USGS (United States Geologic Survey) DEMs (or NED, National Elevation Datasets, see USGS 2006) that are offered at a fixed spatial resolution of 30 m (and 10 m for some places). Using a single-resolution USGS DEM for multiple attributes in an analysis may therefore imply output uncertainty even before other analytical tasks are begun.

On the other hand, local topographic attributes calculated at two remote DEM resolutions, even using the same algorithm, may carry mutually independent meanings (Deng et al. 2007a,b). For instance, a south-facing mountainside (e.g., 10–30 km in size) may contain a north-facing hillside (100–500 m in size), and both may influence the surface temperature of a point on the hillside, just in different ways. In such situations, therefore, it would be impossible to identify which scale is 'the' correct one without specifying the application objective. There is thereby an impending need for both an exploratory attitude and a 'scale-explicit' approach (e.g., not only specifying but also justifying the adopted scale) in topography-based environmental analysis. A related criticism can be made regarding the window size used for local attribute calculations by asking: *Is the 3 × 3 moving window the ideal (or the only) choice for local attributes anyway, especially given the random ways in which DEM resolutions are selected?*

To improve the 'discretization-reconstruction' process in topographic parameterization (Schneider 2001a, see also Wolock and McCabe 2000), Schneider (2001b) suggested developing 'phenomena-based,' or application-specific, topographic surface depictions. This calls for a consideration of the application when selecting data, evaluating error, and calculating topographic derivatives. Fulfilling this goal may require multi-scale topographic characterizations (Deng 2007): Wood (2004), Fisher et al. (2004), and

Deng and Wilson (2007), for example, all defined their own, 'fit-for-use' attributes to map mountain peaks as multi-scale fuzzy entities, and Gallant and Dowling (2003) developed a secondary, multi-scale attribute MRVBF (Multi Resolution Valley Bottom Flatness Index) to map valley bottoms.

Similarly, Schmidt and Dikau (1999) and MacMillan et al. (2004) suggested calculating topographic attributes not only for grid points, but also for defined multi-scale topographic objects (entities), which could be linear (e.g., a flow path) or areal (e.g., a slope unit) on a DEM. This approach is another way of representing 'multi-scale' (and potentially also multi-dimensional) phenomena, depending on how the objects are defined, as will be evaluated later. The 'object attribute' idea should be extended from bare-earth topographic objects (e.g., a peak or slope unit) to other landscape objects on the terrain surface. For example, the slope azimuth and incoming solar radiation for a tree, when parameterized in a 3-D space, may indeed be different from those for the bare-earth grid point near which the tree stands (see Figure 22.1), but the former parameterization is more relevant to tree growth (e.g., in terms of moisture and light conditions).

Using a fractal dimension D (see Goodchild and Mark (1987) for the definition of D) to describe the self-similarity of topographic surfaces across multiple (if not all) scales has attracted numerous efforts and aroused much enthusiasm in the topographic analysis literature, especially during the 1990s (e.g., Yokoya et al. 1989, Ouchi and Matsushita 1992, Zhang et al. 1999). This is understandable given the fact that topographic surfaces in many cases present fractal characteristics, as demonstrated by Huang and Turcotte (1989) and Klinkenberg and Goodchild (1992). An additional merit (as well as a possible danger) of using the fractal dimension is that it is a simplified 'parameterization' approach to a very perplexing scale problem. However, a fundamental challenge of using D seems to be that the very existence of topographic fractals is often proved empirically, but the underlying physical processes that cause such a relationship are not always explained theoretically. In addition, the assumption that a uniform D exists at all scales and landscapes could be problematic (e.g., Mark and Aronson 1984, Gilbert 1989, Yokoya et al. 1989, Klinkenberg and Goodchild 1992).

SURFACE SUBDIVISION: LANDFORM CLASSIFICATION AND OBJECT EXTRACTION

Subdividing and classifying the landscape based entirely (e.g., Burrough et al. 2001) or partly (e.g., Zhu 1997) on topographic properties is an attractive approach to the biophysical environment and its mapping (e.g., Franklin 1995, Hargrove and Hoffman 2004, MacMillan et al. 2004, Wolock et al. 2004). This is because the direct and complete observation of environmental properties across geographic space is often difficult (e.g., vegetation moisture, see Ceccato et al., 2002) or impossible (e.g., soil depth). These classification efforts would cause less confusion if the defined landform classes or objects do exist as *bona fide* features on the topographic surface (Mark and Smith 2004). For example, tracking runoff processes may identify 'real' drainage networks and watersheds that can subdivide the landscape across a large area and/or at multiple spatial scales (e.g., Moore and Grayson 1991, Band et al. 2000, Armstrong and Martz 2003). A viewshed as an areal object is (almost) certain for a point and can be identified if the adopted algorithm and data are reliable enough (Fisher 1993, 1996). However, landform objects with clear and sharp boundaries are uncommon on the continuous topographic surface, and human definitions are often needed for the conceptualization and delineation of landforms (Dehn et al. 2001, Smith and Mark 2003, Fisher et al. 2004, Deng 2007).

Numerous uncertainties and difficulties in subdividing the landscape may arise from multiple sources, including: (1) subjective specification of thresholds or criteria for the thematic delimitation (Dikau 1989); (2) multi-criteria classification, where the output depends on the selection and weight assignments of topographic attributes (Deng and Wilson 2006, Deng et al. 2006); (3) the central concept exists but is vague for efforts of quantitative delimitation (such as peaks, see Fisher 2000); (4) the landform is scale-dependent but the scale-to-scale distinction is fuzzy (e.g., between a mountain and a hill, see Smith and Mark 2003); (5) the landform boundary is non-existent or is gradational (Usery 1996, Gallant and Dowling 2003); and (6) the resultant landform is spatially incongruous (e.g., disjoint cells), so that intact spatial entities cannot be delineated (Burrough et al. 2001).

Landforms are mostly delineated following one of a few common feature-extraction approaches (MacMillan et al. 2004): identifying elemental landform units; classifying (local) topographic attributes; and simulating hydrological processes. The first and second approaches respectively follow so-called 'category theory' and 'set theory' described by Usery (1996), and correspond to two fuzzy classification approaches identified by Burrough and McDonnell (1998, p. 268): semantic importing (SI) and multivariate clustering. Upon evaluating several challenges in topographic analysis, Deng (2007) provided an ontology-based classification of landforms that have been identified in the topographic analysis literature. It was noted that most defined landforms were more or less dependent on subjective control (Deng 2007).

Within the methodological and ontological framework outlined above, several important advances have been made in recent years in classifying landforms. First, fuzzy logic (Zadeh 1965, Burrough and McDonnell 1998, Robinson 2003) has been frequently used to address boundary, scale, and thematic vagueness in landform classifications (Burrough et al. 2001, Fisher et al. 2004, Deng and Wilson 2006). Irvin et al. (1997) and Ventura and Irvin (2000) provided a comparison between crisp and fuzzy sets used in landform classifications. Second, multi-scale approaches in landform delineation have achieved tremendous success in recent years, providing opportunities to compare, link, and integrate multiple spatial scales (e.g., Gallant and Dowling 2003, Fisher et al. 2004, Schmidt and Hewitt 2004, Deng and Wilson 2007). Third, not only cells but also landform objects have been used as basic classification elements (Schmidt and Dikau 1999), which helped with the construction of a hierarchical classification system that may support the integration of topographic contexts (MacMillan et al. 2004).

Two problems persist in the landform classification literature. First, empirically identified landform entities or classes are often insufficiently linked to higher-order spatial contexts (Schmidt and Hewitt 2004). As a result, the classification only has local meaning, cannot be compared to classifications of remote areas, and is difficult to extrapolate even to neighbouring areas (see Burrough et al. 2001). This problem will be addressed in the next section. Second, the strong subjective factors in classification (e.g., threshold definition for SI and attribute selection for clustering) indicate that landform–environment relationships incorporated in the classification outputs can only be implicit (or qualitative) rather than explicit (or quantitative). It is therefore difficult to interpret the classification precisely with specific biophysical meanings (e.g., Ventura and Irvin 2000, Burrough et al. 2001, see also Zhu (1997) for a more successful example). Furthermore, the classification is difficult to improve with the advances in human knowledge and/or data.

A peak delineation algorithm recently developed by Deng and Wilson (2007) shows a possible solution to the above two problems. This algorithm allows multiple semantic meanings of peaks to be imported and translated for application-specific peak definitions and delineations, and it allows topographic contexts to be integrated as higher-order, 'contextual' attributes. To distinguish 'kinds' of topographic locations in terms of the potential for vegetation growth, Deng (in press) recently modified this peak delineation algorithm and used empirically learned topography–NDVI (normalized difference vegetation index)

relationships to guide a fuzzy landform classification procedure.

COARSE SCALE MODELING CHALLENGES: FROM WATERSHEDS TO CONTINENTS

Approximately 80 years ago, Fenneman (1928) published a series of maps dividing the USA into eight major divisions and 25 contiguous physiographic provinces, so as to delineate boundaries between regions that were 'as homogeneous as possible.' Unfortunately, most landform classification and object extraction algorithms today, with the exception of drainage and watershed delineations (e.g., Colombo et al. 2007), are still local activities operating over a limited spatial scale range. Consequently, linkages between landforms (and corresponding processes) cannot be established quantitatively along a spectrum of scales (e.g., from hollow to slope, hill, mountain, and continent). Furthermore, local landform units lack a higher-order context and cannot be compared with remote landform units. This explains why Deng (2007) identified topographic context, or scale-to-scale connections of topography, as a separate topographic existence that has been under-addressed.

Several studies have confronted the challenge of examining local conditions and topographic contexts in a collaborative manner (MacMillan et al. 2000, Schmidt and Hewitt, 2004, Deng and Wilson 2007). Schmidt and Hewitt (2004), for example, subdivided a topographic surface into 15 form element types based on local 'shape' attributes (see Dikau 1989), and then used two larger windows to identify the 'higher scale landscape context' of these form elements, so as to characterize their fuzzy memberships in the class of hills, hillslopes, and/or valleys. This allows two identical form elements to be distinguishable in terms of their topographic contexts (e.g., flat ridge vs. flat valley bottom forms).

A similar approach to Schmidt and Hewitt (2004) may potentially be extended through multiple higher-order scales up to a continental level. This could involve using multiple greatly expanded windows (and perhaps coarsened DEMs) to characterize not only the 'peak-valley' position in a narrow context (e.g., 2–3 km^2), possibly using the algorithms of Gallant and Dowling (2003) and Deng and Wilson (2007), but also the 'mountain-plain' position in one or more much larger contexts (e.g., up to 1×10^6 km^2). A continent-wide topographic reference system may potentially be constructed in this way, in which the third, or topographic, dimension can be established readily. Based on this reference system, other important spatial contexts of landforms, such as social (e.g., a hill in a city) and regional climatic (dry vs. wet mountains) relationships, may also be incorporated in future topographic analysis.

Besides the consideration of topographic contexts, coarse-scale topographic analysis inevitably involves combining spatial resolution and spatial extent in more complex ways. Compared to local models, large-area modeling efforts face a more severe challenge of selecting an appropriate – usually coarser – topographic data resolution for the model, because there are many more data options. Not only are there more types of data sets (e.g., DEMs of various sources and resolutions) available below a coarser spatial resolution, but also there are many methods with which a coarse-resolution DEM can be obtained from a fine-resolution DEM (e.g., Armstrong and Martz 2003). How to match data to the modeled processes and how to meaningfully translate input data (e.g., upscaling) and modeling outputs (e.g., downscaling) between different scales are key challenges. In addition, it is often vital to extrapolate a locally obtained conclusion to a larger area (e.g., Burrough et al. 2001), and it is often necessary to select multiple study areas within a larger area (e.g., a state, a country, or a continent) to obtain more generic conclusions. Kumler (1994) and Klinkenberg and Goodchild (1992) are exemplary in this regard, who selected numerous study areas across the United States based on the physiographic boundaries delineated by Fenneman (1928), and, consequently, provided convincing conclusions regarding elevation data formats and fractals in topographic data respectively.

Other challenges have emerged as well when topographic analysis shifts toward a continental (or very large) scale. First, research questions that need to be addressed at these 'very large' scales may be fundamentally different as a result of the change of dominant processes (e.g., an increased importance of climatic factors, see Colombo et al. (2007)). Second, there is a need to connect (or overlay) topography-based surface subdivisions (e.g., Wolock et al. 2004) with other continent-wide landscape divisions, such as soils, land use and land cover, ecoregions, etc. (e.g., Omernik and Bailey 1997) to refine or produce new environmental information. Third, the applicability of many locally-developed topographic analysis algorithms (e.g., slope, flow-routing) may become questionable at a very coarse resolution and a very large area, and there is a need to reevaluate the algorithm performance at this new scale, for which new algorithms may need to be developed.

CONCLUSIONS

This chapter reviewed topographic analysis and evaluated relevant research efforts in the past 10 to 20 years for the purpose of identifying future research directions and tasks. The focus was on surface representation and analysis, targeting better modeling of the biophysical environment. The review was undertaken in hope that (1) it can help us identify a yardstick useful in evaluating and improving remotely-sensed elevation data, and (2) it can assist the recognition of future collaborative opportunities between the topographic analysis and remote sensing communities, especially when they approach the same biophysical properties for similar modeling purposes. In doing so, several important (remote sensing) topics were not examined in detail in this chapter, such as new remote sensing technologies for topographic data acquisition, and direct integration of topographic and multi-spectral data for anisotropic reflectance correction.

It seems that numerous challenges and opportunities remain, implying topographic analysis may become a more accurate and useful field of scientific research and may need stronger support of remote sensing. In particular, the following research directions hold immediate promise, or are critically needed. First, DEM data needs to be more surface-specific with metadata that reports not only elevation error, but also surface error. A combination of high-resolution topographic data and existing interpolation methods such as ANUDEM may be useful in such endeavors. Second, surface roughness and objects should be included in topographic analysis using high-resolution data. Third, new topographic attributes need to be developed for application-specific analyses, and the analyzed topographic elements should include not only points (or cells), but also multi-scale objects. Fourth, an exploratory approach is needed in handling data, scale, and algorithm uncertainties, as well as in defining landform-environment relationships. Fifth, a multi-scale framework is needed to integrate higher-order topographic contexts for future topographic analysis. In all, addressing these issues will help topographic analysis to better serve future environmental modeling and analysis efforts.

ACKNOWLEDGMENTS

Numerous detailed suggestions provided by Professor Timothy Warner, West Virginia University, and Professor John Wilson, University of Southern California, greatly improved the early drafts of this chapter. Five anonymous reviewers' generous and constructive comments are greatly appreciated. Mr. Shaohan Deng also helped revise the manuscript.

REFERENCES

Aguilar, F. J., M. A. Aguilar, and F. Aguera, 2007. Accuracy assessment of digital elevation models using a non-parametric approach. *International Journal of Geographical Information Science*, 21: 667–686.

Allgöwer, B., J. D. Carlson, and J. W. van Wagtendonk, 2003. Introduction to fire danger rating and remote sensing – Will remote sensing enhance wildland fire danger rating? In E. Chuvieco (ed.), *Wildland Fire Danger Estimation and Mapping: The Role of Remote Sensing Data*. World Scientific Publishing, Singapore. Chapter 1: 1–19.

Armstrong, R. N. and L. W. Martz, 2003. Topographic parameterization in continental hydrology: A study in scale. *Hydrological Processes*, 17: 3763–3781.

Baltsavias, E. P., 2004. Object extraction and revision by image analysis using existing geodata and knowledge: Current status and steps towards operational systems. *ISPRS Journal of Photogrammetry and Remote Sensing*, 58: 129–151.

Band, L. E. and I. D. Moore, 1995. Scale: Landscape attributes and geographical information systems. *Hydrological Processes*, 9: 401–422.

Band, L. E., C. Tague, S. E. Brun, D. E. Tenenbaum, and R. A. Fernandes, 2000. Modelling watersheds as spatial object hierarchies: Structure and dynamics. *Transactions in GIS*, 4: 181–196.

Beven, K. J. and M. J. Kirkby, 1979. A physically based variable contributing area model of basin hydrology. *Hydrology Science Bulletin*, 24: 43–69.

Bolstad, P. V. and T. Stowe, 1994. An evaluation of DEM accuracy: Elevation, slope and aspect. *Photogrammetric Engineering and Remote Sensing*, 23: 387–395.

Bruzzone, L. and D. F. Prieto, 2001. Unsupervised retraining of a maximum likelihood classifier for the analysis of multitemporal remote sensing images. *IEEE Transactions on Geoscience and Remote Sensing*, 39: 456–460.

Burrough, P. A., 1986. *Principles of Geographical Information Systems for Land Resources Assessment*. Oxford Science Publications, Oxford.

Burrough, P. A. and R. A. McDonnell, 1998. *Principles of Geographical Information Systems*. Oxford University Press, Oxford.

Burrough, P. A., J. P. Wilson, P. F. M. van Gaans, and A. J. Hansen, 2001. Fuzzy *k*-means classification of Digital Elevation Models as an aid to forest mapping in the Greater Yellowstone Area, USA. *Landscape Ecology*, 16: 523–546.

Butler, J. B., S. N. Lane, and J. H. Chandler, 1998. Assessment of DEM quality: Characterising surface roughness using close range digital photogrammetry. *Photogrammetric Record*, 16: 271–291.

Callow, J. N., K. P. Van Neil, and G. S. Boggs, 2007. How does modifying a DEM to reflect known hydrology affect subsequent terrain analysis? *Journal of Hydrology*, 332: 30–39.

Carlisle, B. H., 2005. Modelling the spatial distribution of DEM error. *Transactions in GIS*, 9: 521–540.

Ceccato, P., S. Flasse, and J.-M. Gregoire, 2002. Designing a spectral index to estimate vegetation water content from remote sensing data: Part 2. Validation and applications. *Remote Sensing of Environment*, 82: 198–207.

Chen, J. M., 1999. Spatial scaling of a remotely sensed surface parameter by contexture. *Remote Sensing of Environment*, 69: 30–42.

Colombo R., J. V. Vogt, P. Soille, M. L. Paracchini, and A. de Jager, 2007. Deriving river networks and catchments at the European scale from medium resolution digital elevation data. *Catena*, 70: 296–305.

Dehn, M., H. Gärtner, and R. Dikau, 2001. Principles of semantic modeling of landform structures. *Computers and Geosciences*, 27: 1005–1010.

Deng, Y. X., 2007. New trends in digital terrain analysis: Landform definition, representation, and classification. *Progress in Physical Geography* 31: 405–419.

Deng, Y. X., In press. Mapping topographical potentials of vegetation greenness and moisture using fuzzy logic. In F. Columbus (ed), *Mathematical Logic: New Research*. Wova Science Publisher, New York.

Deng, Y. X. and J. P. Wilson, 2006. The role of attribute selection in GIS representations of the biophysical environment. *Annals of the Association of American Geographers*, 96: 47–63.

Deng, Y. X. and J. P. Wilson, 2007. Multi-scale delineation of mountain peaks as fuzzy spatial objects. *International Journal of Geographical Information Science*, 21: DOI: 10.1080/13658810701405623.

Deng, Y. X., J. P. Wilson, and J. Sheng. 2006. Effects of variable attribute weights on landform classification. *Earth Surface Processes and Landforms*, 31: 1452–1462.

Deng, Y. X., X. F. Chen, E. Chuvieco, T. A. Warner, and J. P. Wilson, 2007a. Multi-scale linkages between topographic attributes and vegetation indices in a mountainous landscape. *Remote Sensing of Environment*, 111: 122–134.

Deng, Y. X., J. P. Wilson, and B. O. Bauer, 2007b. DEM resolution dependencies of terrain attributes across a landscape. *International Journal of Geographical Information Science*, 21: 187–213.

Deng, Y. X., J. P. Wilson, and J. C. Gallant, 2007c. Terrain analysis. In: J. P. Wilson and A. S. Fotheringham (eds), *Handbook of Geographic Information Science*. Blackwell, Oxford. Chapter 23: 417–435.

Desmet, P. J. J. and G. Govers, 1996. Comparison of routing algorithms for digital elevation models and their implications for predicting ephemeral gullies. *International Journal of Geographical Information Systems*, 10: 311–331.

Dikau, R., 1989. The application of a digital relief model to landform analysis in geomorphology. In J. Raper (ed.), *Three Dimensional Applications in Geographic Information Systems*. Taylor and Francis, London: 51–77.

Djokic, D. and D. R. Maidment, 1991. Terrain analysis for urban stormwater modeling. *Hydrological Processes*, 5: 115–124.

Evans, I. S., 1972. General geomorphometry, derivatives of altitude, and descriptive statistics. In R. J. Chorley (ed.), *Spatial Analysis in Geomorphology*. Methuen, London: 17–90.

Evans, I. S., 2003. Scale-specific landforms and aspects of the land surface. In I. S. Evans, R. Dikau, E. Tokunaga, H. Ohmori, and M. Hirano (eds), *Concepts and Modelling in Geomorphology: International Perspectives*. Terrapub, Tokyo: 61–84.

Fenneman, N. M., 1928. Physiographic divisions of the United States of America. *Annals of the Association of American Geographers*, 4: 261–353.

Fisher, P. F., 1993. Algorithm and implementation uncertainty in viewshed analysis. *International Journal of Geographical Information Systems*, 7: 331–347.

Fisher, P. F., 1996. Extending the applicability of viewsheds in landscape planning. *Photogrammetric Engineering and Remote Sensing*, 62: 1297–1302.

Fisher, P. F., 2000. Sorites paradox and vague geographies. *Fuzzy Sets and Systems*, 113: 7–18.

Fisher, P. F. and N. J. Tate, 2006. Causes and consequences of error in digital elevation models. *Progress in Physical Geography*, 30: 467–489.

Fisher, P. F., J. Wood, and C. Cheng, 2004. Where is Helvellyn? Fuzziness of multi-scale landscape morphology. *Transactions of the Institute of British Geographers*, 29: 106–128.

Florinsky, I. V., 1998a. Combined analysis of digital terrain models and remotely sensed data in landscape investigations. *Progress in Physical Geography*, 22: 33–60.

Florinsky, I. V., 1998b. Accuracy of local topographic variables derived form digital elevation models. *International Journal of Geographical Information Science*, 12: 47–61.

Florinsky, I. V. and G. A. Kuryakova, 2000. Determination of grid size for digital terrain modelling in landscape investigations: Exemplified by soil moisture distribution at a micro-scale. *International Journal of Geographical Information Science*, 14: 815–832.

Franklin, J., 1995. Predictive vegetation mapping: Geographic modelling of biospatial patterns in relation to environmental gradients. *Progress in Physical Geography*, 19: 474–499.

Fraser, C. S., E. Baltsavias, and A. Gruen, 2002. Processing of Ikonos imagery for submetre 3D positioning and building extraction. *ISPRS Journal of Photogrammetry and Remote Sensing*, 56: 177–194.

French, J. R., 2003. Airborne LIDAR in support of geomorphological and hydraulic modeling. *Earth Surface Processes and Landforms*, 28: 321–335.

Gallant, J. C. and T. I. Dowling, 2003. A multiresolution index of valley bottom flatness for mapping depositional areas. *Water Resources Research*, 39: 1347–1359.

Gamba, P. and B. Houshmand, 2000. Digital surface models and building extraction: A comparison of JFSAR and LIDAR data. *IEEE Transactions on Geoscience and Remote Sensing*, 38: 1959–1968.

Geiger, R., 1965. *The Climate Near the Ground*. Harvard University Press, Cambridge, MA.

Gilbert, L. E., 1989. Are topographic data sets fractal? *Pure and Applied Geophysics*, 131: 241–254.

Giles, P. T. and S. E. Franklin, 1996. Comparison of derivative topographic surfaces of a DEM generated from stereoscopic SPOT images with field measurements. *Photogrammetric Engineering and Remote Sensing*, 62: 1165–1171.

Goodchild, M. F. and D. M. Mark, 1987. The fractal nature of geographic phenomena. *Annals of the Association of American Geographers*, 77: 265–278.

Hargrove, W. W. and F. M. Hoffman, 2004. The potential of multivariate quantitative methods for delineation and visualization of ecoregions. *Environmental Management*, 34: S39–S60.

Heuvelink, G. B. M, 1998. *Error Propagation in Environmental Modelling with GIS*. Taylor and Francis, London.

Hodgson, M. E., J. R. Jensena, L. Schmidtb, S. Schillc, and B. Davis, 2003. An evaluation of LIDAR- and IFSAR-derived digital elevation models in leaf-on conditions with USGS Level 1 and Level 2 DEMs. *Remote Sensing of Environment*, 84: 295–308.

Huang, J. and D. L. Turcotte, 1989. Fractal mapping of digitized images: Application to the topography of Arizona and comparisons with synthetic images. *Journal of Geophysical Research*, 94: 7491–7495.

Hunter, G. J. and M. F. Goodchild, 1997. Modelling the uncertainty of slope and aspect estimates derived from spatial databases. *Geographical Analysis*, 29: 35–49.

Hutchinson, M. F., 1989. A new procedure for gridding elevation and streamline data with automatic removal of spurious pits. *Journal of Hydrology*, 106: 211–232.

Hutchinson, M. F., 2006. ANUDEM Version 5.2. http://cres.anu.edu.au/outputs/anudem.php (last accessed: November 16 2007).

Hutchinson, M. F., 2007. Adding the Z Dimension. In J. P. Wilson, and A. S. Fotheringham (eds), *Handbook of Geographic Information Science*. Blackwell, Oxford. Chapter 8: 144–168.

Hutchinson, M. F. and J. C. Gallant, 2000. Digital elevation models and representation of terrain shape. In: J. P. Wilson, and J. C. Gallant (eds), *Terrain Analysis: Principles and Applications*. John Wiley and Sons, New York. Chapter 2: 29–50.

Irvin, B. J., S. J. Ventura, and B. K. Slater, 1997. Fuzzy and isodata classification of landform elements from digital terrain data in Pleasant Valley, Wisconsin. *Geoderma*, 77: 137–154.

Jones, K. H., 1998. A comparison of algorithms used to compute hill slope as a property of the DEM. *Computers and Geosciences*, 24: 315–323.

Kidner, D. B., J. M. Ware, A. J. Sparks, and C. B. Jones, 2000. Multiscale terrain and topographic modeling with the implicit TIN. *Transactions in GIS*, 4: 379–408.

Klinkenberg, B. and M. F. Goodchild, 1992. The fractal properties of topography – a comparison of methods. *Earth Surface Processes and Landforms*, 17: 217–234.

Kumler, M. P., 1994. An intensive comparison of Triangulated Irregular Networks (TINs) and Digital Elevation Models (DEMs). *Cartographica: The International Journal for Geographic Information and Geovisualization*, 31: 1–99.

Lane, S. N., 2005. Roughness – time for a re-evaluation? *Earth Surface Processes and Landforms*, 30: 251–253.

Lane, S. N. and J. H. Chandler, 2003. Editorial: the generation of high quality topographic data for hydrology and geomorphology: New data sources, new applications and new problems. *Earth Surface Processes and Landforms*, 28: 229–230.

Lee, J., P. K. Snyder, and P. F. Fisher, 1992. Modelling the effect of data errors on feature extraction from digital elevation models. *Photogrammetric Engineering and Remote Sensing*, 58: 1461–1467.

Lindsay, J. B., 2005. The terrain analysis system: A tool for hydro-geomorphic applications. *Hydrological Processes*, 19: 1123–1130.

Liu, H. and K. Jezek, 1999. Investigating DEM error pattern by directional variograms and Fourier analysis. *Geographical Analysis*, 31: 249–266.

Lucieer, A. and A. Stein, 2005. Texture-based landform segmentation of LiDAR imagery. *International Journal of Applied Earth Observation and Geoinformation*, 6: 261–270.

MacMillan, R. A., W. W. Pettapiece, S. C. Nolan, and T. W. Goddard, 2000. A generic procedure for automatically segmenting landforms into landform elements using DEMs, heuristic rules and fuzzy logic. *Fuzzy Sets and Systems*, 113: 81–109.

MacMillan, R. A., R. K. Jones, and D. H. McNabb, 2004. Defining a hierarchy of spatial entities for environmental analysis and modelling using digital elevation models (DEMs). *Computers, Environment and Urban Systems*, 28: 175–200.

Marceau, D. J. and G. J. Hay, 1999. Remote sensing contributions to the scale issue. *Canadian Journal of Remote Sensing*, 25: 357–366.

Mark, D. M. and P. B. Aronson, 1984. Scale-dependent fractal dimensions of topographic surfaces: An empirical investigation, with applications in geomorphology and computer mapping. *Mathematical Geology*, 16: 671–683.

Mark, D. M. and B. Smith, 2004. A science of topography: From qualitative ontology to digital representations. In: M. P. Bishop, and J. F. Shroder (eds), *Geographic Information Science and Mountain Geomorphology*. Springer-Praxis, Chichester: 75–100.

Marks, K. and P. Bates, 2000. Integration of high-resolution topographic data with floodplain flow models. *Hydrological Processes*, 14: 2109–2122.

McClean, C. J. and I. S. Evans, 2000. Apparent fractal dimensions from continental scale digital elevation models using variogram methods. *Transactions in GIS*, 4: 361–378.

Mitasova, H. L. and J. Hofierka, 1993. Interpolation by regularized spline with tension, II: Application to terrain modeling and surface geometry analysis. *Mathematical Geology*, 25: 657–659.

Moore, I. D. and R. B. Grayson, 1991. Terrain-based catchment partitioning and runoff prediction using vector elevation data. *Water Resources Research*, 27: 1177–1191.

Moore, I. D. and J. P. Wilson, 1992. Length-slope factors for the revised Universal Soil Loss Equation: Simplified method of estimation. *Journal of Soil and Water Conservation*, 47: 423–428.

Moore, I. D., R. B. Grayson, and A. R. Ladson, 1991. Digital terrain modelling: A review of hydrological, geomorphological, and biological applications. *Hydrological Processes*, 5: 3–30.

Nelson, A., T. Oberthür, and S. Cook, 2007. Multi-scale correlations between topography and vegetation in a hillside catchment of Honduras. *International Journal of Geographical Information Science*, 21: 145–174.

O'Callaghan, J. F. and D. M. Mark, 1984. The extraction of drainage networks from digital elevation models. *Computer Vision, Graphics and Image Processing*, 28: 323–344.

Oliphant, A. J., R. A. Spronken-Smith, A. P. Sturman, and I. F. Owens, 2003. Spatial variability of surface radiation fluxes in mountainous terrain. *Journal of Applied Meteorology*, 42: 113–128.

Omernik, J. M. and R. G. Bailey, 1997. Distinguishing between watersheds and ecoregions. *Journal of the American Water Resources Association*, 33: 935–949.

Ouchi, S. and M. Matsushita, 1992. Measurement of self-affinity on surfaces as a trial application of fractal geometry to landform analysis. *Geomorphology*, 5: 115–130.

Peucker, T. K., 1978. Data structures for digital terrain models: Discussion and comparison. *Harvard Papers on Geographic Information Systems*.

Pike, R. J., 1988. The geometric signature: Quantifying landslide-terrain types from digital elevation models. *Mathematical Geology*, 20: 491–511.

Pike, R. J., 2000. Geomorphometry – diversity in quantitative surface analysis. *Progress in Physical Geography*, 24: 1–20.

Quattrochi, D. A. and Goodchild, M. F., 1997. *Scale in Remote Sensing and GIS*. CRC Press, London.

Quinn, P. F., K. J. Beven, and R. Lamb, 1995. The ln(α/ tan β) index: How to calculate it and how to use it within the TOPMODEL framework. *Hydrological Processes*, 9: 162–182.

Rabus, B., M. Eineder, A. Roth, and R. Bamler, 2003. The shuttle radar topography mission – a new class of digital elevation models acquired by spaceborne radar. *ISPRS Journal of Photogrammetry and Remote Sensing*, 57: 241–262.

Robinson, V. B., 2003. A perspective on the fundamentals of fuzzy sets and their use in Geographic Information Systems. *Transactions in GIS*, 7: 3–30.

Schmidt, J. and R. Andrew, 2005. Multi-scale landform characterization. *Area*, 37: 341–350.

Schmidt, J. and R. Dikau, 1999. Extracting geomorphometric attributes and objects from digital elevation models – semantics, methods, and future needs. In: R. Dikau, and H. Sourer (eds), *GIS for Earth Surface Systems: Analysis and Modelling of the Natural Environment*. Gebrüder Borntraeger, Berlin: 153–173.

Schmidt, J. and A. Hewitt, 2004. Fuzzy land element classification from DTMs based on geometry and terrain position. *Geoderma*, 121: 243–256.

Schmidt, J., I. S. Evans, and J. Brinkmann, 2003. Comparison of polynomial models for land surface curvature calculation. *International Journal of Geographical Information Science*, 17: 797–814.

Schneider, B., 2001a. On the uncertainty of local shape of lines and surfaces. *Cartography and Geographic Information Science*, 28: 237–247.

Schneider, B., 2001b. Phenomenon-based specification of the digital representation of terrain surfaces. *Transactions in GIS*, 5: 39–52.

Shan, J. and A. Sampath, 2005. Urban DEM generation from raw Lidar data: A labelling algorithm and its performance. *Photogrammetric Engineering and Remote Sensing*, 71: 217–226.

Shary, P. A., L. S. Sharaya, and A. V. Mitusov, 2002. Fundamental quantitative methods of land surface analysis. *Geoderma*, 107: 1–32.

Smith, B. and D. M. Mark, 2003. Do mountains exist? Ontology of landforms and topography. *Environment and Planning B: Planning and Design*, 30: 411–427.

Tate, N. J., P. F. Fisher, and D. J. Martin, 2007. Geographic information systems and surfaces. In: J. P. Wilson and A. S. Fotheringham (eds), *Handbook of Geographic Information Science*. Blackwell, Oxford. Chapter 13: 239–258.

USGS, 2005. Topographic mapping. http://erg.usgs.gov/isb/pubs/booklets/topo/topo.html (last accessed November 16 2007).

USGS, 2006. National Elevation Dataset. http://ned.usgs.gov (last accessed November 16 2007).

Usery, E. L., 1996. A conceptual framework and fuzzy set implementation for geographic features. In P. A. Burrough, and A. U. Frank (eds), *Geographic Objects with Indeterminate Boundaries.* Taylor and Francis, London: 71–86.

Ventura, S. J. and B. J. Irvin, 2000. Automated landform classification for soil-landscape studies. In: J. P. Wilson, and J. C. Gallant (eds), *Terrain Analysis: Principles and Applications.* John Wiley and Sons, New York: 267–294.

Wang, Y. and M. Jamshidi, 2004. Fuzzy logic applied in remote sensing image classification. *IEEE International Conference on Systems, Man and Cybernetics*: 6378–6382.

White, A. B., P. Kumar, and D. Tcheng, 2005. A data mining approach for understanding topographic control on climate-induced inter-annual vegetation variability over the United States. *Remote Sensing of Environment*, 98: 1–20.

Wilson, J. P. and J. C. Gallant, 2000a. *Terrain Analysis: Principles and Applications.* John Wiley and Sons, New York.

Wilson, J. P. and J. C. Gallant, 2000b. Digital terrain analysis. In: J. P. Wilson, and J. C. Gallant (eds), *Terrain Analysis: Principles and Applications.* John Wiley and Sons, New York: 1–27.

Wilson, J. P. and J. C. Gallant, 2000c. Secondary terrain attributes. In: J. P. Wilson, and J. C. Gallant (eds), *Terrain Analysis: Principles and Applications.* John Wiley and Sons, New York: 87–132.

Wilson, J. P., D. J. Spangrud, M. A. Landon, G. A. Nielsen, J. S. Jacobson, and D. A. Tyler, 1998. GPS sampling intensity and pattern effects on computed terrain attributes. *Soil Science Society of America Journal*, 62: 1410–1417.

Wilson, J. P., C. S. Lam, and Y. X. Deng, 2007. Comparison of flow routing algorithms used in Geographic Information Systems. *Hydrological Processes*, 21: 1026–1044.

Wilson, J. P., G. R. Aggett, Y. X. Deng, and C. S. Lam, 2008. Water in the landscape: A review of contemporary flow routing algorithms. In Q. Zhou, B. G. Lees, and G. A. Tang (eds), *Advances in Digital Terrain Analysis.* Lecture Notes in Geoinformation and Cartography, Springer: 213–236.

Wolock, D. M. and G. J. McCabe, 2000. Differences in topographic characteristics computed from 100- and 1000-m resolution digital elevation model data. *Hydrological Processes*, 15: 2223–2236.

Wolock, D. M., T. C. Winter, and G. McMahon, 2004. Delineation and evaluation of hydrologic-landscape regions in the United States using Geographic Information System tools and multivariate statistical analyses. *Environmental Management*, 34: S71–S88.

Wood, J., 2004. A new method for the identification of peaks and summits in surface models. In: *Proceedings of GIScience 2004 – The Third International Conference on Geographic Information Science.* Adelphi, Maryland: 227–230.

Woodcock, C. E. and A. H. Strahler, 1987. The factor of scale in remote sensing. *Remote Sensing of Environment*, 2: 311–332.

Yokoya, N., K. Yamamoto, and N. Funakubo, 1989. Fractal-based analysis and interpolation of 3D natural surface shapes and their application to terrain modeling. *Computer Vision, Graphics, and Image Processing*, 46: 284–302.

Zadeh, L., 1965. Fuzzy sets. *Information and Control*, 8: 338–353.

Zhang, W. H. and D. R. Montgomery, 1994. Digital elevation model grid size, landscape representation, and hydrologic simulations. *Water Resources Research*, 30: 1019–1028.

Zhang, X., N. A. Drake, J. Wainwright, and M. Mulligan, 1999. Comparison of slope estimates from low resolution DEMs: scaling issues and a fractal method for their solution. *Earth Surface Processes and Landforms*, 24: 763–779.

Zhu, A. X., 1997. A similarity model for representing soil spatial information. *Geoderma*, 77: 217–242.

Remote Sensing of Geology

Xianfeng Chen and David J. Campagna

Keywords: mineral mapping, hyperspectral classification, hydrothermal alteration, Planetary geology, hydrocarbon exploration, RADAR interferometry, lidar, landslide hazard assessment, geology.

INTRODUCTION

Earth scientists have employed aerial images for decades. With the advent of satellite data the usefulness of the regional 'synoptic view' for identifying geologic phenomena was more widely recognized (Halbouty 1980). Since then, applications of remote sensing can be found in most of the geological sub-disciplines. Given the multi-disciplinary nature of geologic studies, there are far more applications of the remote sensing method in geology than could conceivably be covered here in any meaningful way. Instead, we offer an account of the current status of mineral mapping using typical hyperspectral imaging systems – for mining and other extractive industries as well as planetary science. For the future directions, we look at landform mapping. This application has been propelled forward by the development of new sensors and new automatic classification techniques that draw upon landform characteristics to extract structural information.

Mineral mapping typically involves the use of imaging spectroscopy (also know as hyperspectral imaging systems), which refers to the acquisition of images with hundreds or more contiguous and narrow spectral bands. Imaging spectroscopy identifies materials through the identification of spectral characteristics related to specific chemical bonds (Goetz and Srivastava 1985, Schaepman, in this volume). These features are then correlated with specific minerals via a variety of analytical methods. Landform mapping, on the other hand,

investigates the topographic expression of the Earth's surface and often relates the landform to specific geological features such as faults. While the analysis of landforms has traditionally been the realm of aerial photography, new instruments that collect topographic data directly, such as lidar (light detection and ranging), also known as airborne laser scanning (ALS, Hyyppä et al. in this volume), have increasingly been applied to map surface landforms in great detail. The following sections will examine some of the technical issues involved in both these areas of geological mapping using remote sensing.

MINERALOGICAL MAPPING

Mineralogical mapping requires significantly more refined data than has been available in conventional broad-band, multispectral remote sensing analysis, and consequently this field has made its greatest advancements following the development of hyperspectral imaging systems. For geologic applications, the spectral range of hyperspectral imaging systems should ideally cover the visible through the near infrared (VNIR) and short-wave infrared (SWIR) to the thermal infrared (TIR) wavelengths. Practically, sensors have been divided into those that sense the VNIR and SWIR, and those that cover TIR wavelengths. Since the first acquisition by the Airborne Imaging Spectrometer (AIS) in 1983 (Vane et al. 1984) VNIR/SWIR hyperspectral instruments have been important tools for mineralogical mapping (Kruse 1988, Crowley et al. 1989, Kruse et al. 1993b, Swayze 1997, Crósta et al. 1998, Dalton et al. 2004).

TIR wavelengths have also been found to be very useful. Numerous researchers have demonstrated the capability of differentiating geologic surface materials such as carbonate, sulfate, and silicate minerals using multispectral TIR instruments (Gillespie et al. 1984, Collins 1991, Hook et al. 1994, Sabine et al. 1994, Rowan and Mars 2003). However, there is a significant limitation in using multispectral TIR for mineralogical mapping: minerals other than quartz cannot be identified uniquely without the benefit of ground information due to the limited spectral differentiation of absorption features (Crowley and Hook 1996). Integrating VNIR/SWIR hyperspectral and TIR multispectral data provides a more promising means of identifying minerals lacking distinct absorption features in the VNIR and SWIR than using these data alone (Chen et al. 2007). Alternatively, hyperspectral TIR data provide a potentially powerful method for mapping silicate minerals (Kirkland et al. 2002, Vaughan et al. 2003).

Spectral features of minerals

Part of the significance of VNIR/SWIR and TIR hyperspectral imagery is that these technologies supply information on inherent physical properties of minerals, namely spectral absorption features, which in turn can be related to mineral composition. This link stems from the characteristic spectral reflectance and emissivity features of minerals (Hook et al. 1994) which are produced by electronic transitions and vibrational processes resulting from the interaction of electromagnetic energy with the atoms and molecules that make up those minerals.

Electronic transitions, including crystal field effects, charge transfer absorptions, conduction bands, and color centers, result from the change of energy states following the absorption of a photon, since isolated atoms and ions have discrete energy states. Crystal field effects determine spectral features of minerals which have transition elements with unfilled electron shells. Charge transfer absorptions involve the absorption of a photon causing an electron to move between ions or between ions and ligands. The absorption features caused by charge transfers occur in the ultraviolet, extending to visible wavelengths, while absorption bands of crystal field transitions are located in the NIR. Conduction bands associated with two energy levels of electrons can cause spectral responses in the visible and NIR. Color centers are caused by irradiation of an imperfect crystal (Clark 1999).

Vibrational processes also cause features in mineral spectra, in this case from the SWIR to beyond the TIR (Hunt 1980). By comparison, electronic transitions tend to dominate mineral spectral features at shorter wavelength regions than vibrational processes (Goetz 1989). Vibrational processes are due to vibrations in the crystal lattices of minerals, and can be understood as an analogue to springs, where the vibration frequency depends on the strength of each spring (the bond in a molecule) as well as the relative masses of each element in the molecule (Clark 1999). Vibrational processes have three modes, including fundamental, overtone, and combination vibrations. Absorption features associated with fundamentals depend on the anion composition, bond strength, and crystal structure of the minerals.

Having explained the background to the interaction of electromagnetic energy and minerals, we will now discuss the spectra of mineral classes (see also van Leeuwen, in this volume). Much of the following discussion comes from Clark (1999), and for a more detailed discussion we direct the reader to that classic work.

Iron-related absorption

A typical electronic process revealed in the spectra of minerals is due to unfilled electron shells of transition elements, of which iron is the most common in minerals. Transition metals have the capacity to combine with other elements using electrons present in more than one shell, and this accounts for several common oxidation states. The energy levels are determined by the valence state of the atom (e.g., Fe^{2+}, Fe^{3+}), its coordination number, and the symmetry of the site it occupies. The crystal field varies with crystal structure from mineral to mineral, thus the amount of splitting varies and the same ion (like Fe^{2+}) produces obviously different absorptions, making specific mineral identification possible from spectroscopy (Figure 23.1). Olivine, for instance, exhibits a strong absorption feature around 1.0 μm related to Fe^{2+} absorptions. Goethite and hematite are good examples of minerals exhibiting Fe^{3+} absorption, as indicated by the presence of diagnostic features around 0.9 μm in their spectra (Figure 23.1).

Iron oxides also have absorption bands caused by charge transfer, interelement transition which involves an electron moving between ions or between ions and ligands. As discussed above, the strength of charge transfers are hundreds to thousands of times greater than those of crystal field transitions. Thus, the absorption band center occurs in the ultraviolet and extends to the visible. These absorption features are, in general, diagnostic. A strong Fe^{3+} absorption of hematite is found in the near ultraviolet and visible wavelength region between 0.5 and 0.6 μm, making hematite appear red in color (Figure 23.1). These features have

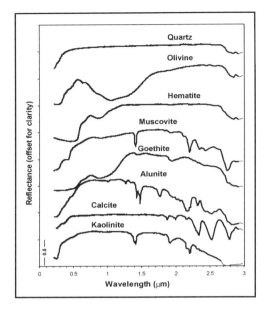

Figure 23.1   Selected mineral reflectance spectra (after Clark et al. 2003b).

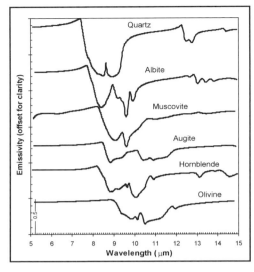

Figure 23.2   Selected silicate mineral emissivity spectra (after Salisbury et al. 1991).

enabled remote sensing of Fe-oxides for a variety of applications.

Silica-related absorption

In the 8–12 μm TIR region, fundamental vibrational processes produce spectral features in silicate spectra, and this spectral region is therefore sometimes known as the 'Si-O stretching region' (Hook et al. 1994). The strong 9-μm Si–O–Si asymmetric stretch results in the distinctive emissivity minimum for quartz. Notably, the wavelength of the Si–O absorption feature varies from 11 to 9 μm, corresponding to minerals with chain, sheet, and framework structures (Hunt 1980) (see Figure 23.2). Thus, both multispectral and hyperspectral TIR remote sensing are potentially very effective for discriminating among silicate minerals.

Carbonate-related absorption

Carbonates show diagnostic vibrational absorption bands due to the CO_3^{2-} ion. There are four fundamental vibrational modes in the free CO_3^{2-} ion: the symmetric stretch, v1: 9.407 μm; the out-of-plane bend, v2: 11.4 μm; the asymmetric stretch, v3: 7.067 μm; and the in-plane bend, v4: 14.7 μm (Clark, 1999). However, in carbonate minerals, only v2, v3, and v4 contribute to the absorption bands. The v3 and v4 bands often appear as a doublet.

There are also CO_3 related absorption regions in the near IR. The two strongest are at 2.50–2.55 and 2.30–2.35 μm. Three weaker absorption bands occur at 2.12–2.16, 1.97–2.00, and 1.85–1.87 μm (Figure 23.1). The band position of each carbonate differs depending on its composition (Hunt and Salisbury 1971; Gaffey et al. 1993). For example, dolomite absorption shifts approximately 0.1 μm to shorter wavelengths than the equivalent calcite absorption (Clark 1999).

Water-related absorption

Water and OH (hydroxyl) contribute particularly diagnostic absorptions in minerals. In particular, the overtone and combination modes form distinct bands in spectra of water-bearing and hydroxyl minerals. The first overtones of the OH stretches contribute the absorption bands at about 1.4 μm, and combinations of the H–O–H bend with OH stretches occur at near 1.9 μm. Therefore, water-bearing mineral will only show a 1.9 μm absorption band, while hydroxyl minerals have both bands at 1.4 and 1.9 μm (e.g., kaolinite). In spectra of OH-bearing minerals, the absorption bands are determined by the OH stretch mode and the ion to which the hydroxyl is attached, typically near 2.7–2.8 μm, but can occur anywhere in the range from about 2.67 to 3.45 μm (Clark et al. 1990, Clark 1999). OH-bearing minerals may have more than one OH related feature because OH commonly occurs in multiple crystallographic sites, typically attached to metal ions. The metal–OH bend, near 10 μm, is not diagnostic as it is usually superimposed on the stronger Si–O fundamental in silicates. In particular, the diagnostic features of clay minerals, at near 2.2–2.3 μm, result from the combination of the metal OH bend and OH stretch (Figure 23.1).

Classification methods

Numerous remote sensing-based geological mapping methods have been developed to exploit the spectral characteristics of rocks and minerals outlined above. These methods vary from statistic-based parametric methods such as minimum distance and maximum likelihood classification, to non-parametric methods such as neural network analysis, to spectral analysis techniques, including spectral feature fitting and linear mixing analysis. Both statistic-based parametric and non-parametric methods rely on training samples to train the classifiers. Therefore, the size of training samples plays a very important role in the classification. Fukunaga (1990) found that the relationship between the required size of the training sample and the dimensionality of data is linear for a linear classifier (e.g., minimum distance classifier) and square for a quadratic classifier (e.g., maximum likelihood classifier). With hyperspectral data, Landgrebe (2000) observed that a quadratic classifier, calculating second-order statistical measures, tends to require excessive numbers of training samples when used with hyperspectral data. For a non-parametric classifier, it has also been estimated that generally as the number of dimensions increases, the number of training samples needs to increase exponentially in order to estimate multivariate densities effectively (Hwang et al. 1994). Additional information on statistical and non-parametric classifiers is given in Jensen et al. (in this volume).

Spectral analysis methods identify minerals focusing on matching image spectra to previously acquired field or laboratory spectra, known as spectral libraries. The spectral library approach is attractive for hyperspectral image analysis because rocks and minerals tend to have distinctive and consistent spectral absorption features, as discussed above. Furthermore, the potential to identify surface materials without any local field data is clearly very attractive. As a result, we will focus on spectral analysis methods hereafter.

Numerous spectral analysis methods have been developed for geological mapping (Mustard and Sunshine 1999), including spectral feature fitting, the spectral angle mapper, and spectral mixture analysis. These spectral analysis methods will be described briefly below.

Spectral feature fitting
Spectral feature fitting (SFF) (Crowley et al. 1989, Clark et al. 1990) compares the absorption

features in the image and reference spectra through calculation of a least-square fit. The SFF algorithm has two steps. First, the continuum, defined as a convex hull fit over the top of each spectrum utilizing straight line-segments to connect local spectrum maxima (Clark et al. 2003a, Kruse et al. 1993b), is removed by dividing the convex hull into the original spectrum. Then, the least-square fit is calculated for the continuum-removed pixel spectrum and reference spectrum at each absorption band. The root mean square error is used to indicate how similar the two spectra are. SFF has an advantage over other methods in that it is sensitive to subtle mineral absorption features, and also minimizes the influence of the effects of variations in mineral grain size and illumination. An advanced example of using SFF is Tetracorder, a rule-based expert system developed by the U.S. Geological Survey (Clark et al. 2003a). The USGS Tetracorder algorithm consists of a set of rules corresponding to absorption features in a reference spectral library. The comparison of image spectra and library spectra in Tetracorder focuses on individual or multiple diagnostic features. This leads to successful discrimination and identification of some minerals with similar spectral features (Clark et al. 2003a, Dalton et al. 2004). The Tetracorder algorithm calculates three parameters: the goodness-of-fit, the apparent depth of an absorption feature, and the product of the fit and the depth. Goodness-of-fit measures how well the image spectra match the library spectra using a modified least-squares approach (Clark et al. 2003b). The apparent depth of the absorption feature relates to the abundance of the material. The product of fit and depth provides a robust estimate of the spectral abundance and overall accuracy of the identification (Swayze 1997).

Spectral angle mapper
The spectral angle mapper (SAM) (Kruse et al. 1993a) calculates the similarity between a pixel spectrum and a reference spectrum in terms of the angle between two n-dimensional vectors, where n is the number of bands of hyperspectral data. SAM is insensitive to solar illumination because it only calculates the hyperspherical direction of the vector, and not the length, which tends to be dominated by solar illumination (Pouch and Campagna 1990). The main advantage of SAM is that it is simple to implement and is quick, particularly for comparing image data with library spectra. It works well for homogenous pixels. SAM is effective at normalizing for topographic illumination variations, but in turn it sacrifices the ability to discriminate rocks with contrasting albedo and relatively flat spectra (Chen et al. 2007).

Spectral unmixing/matched filter

Most image pixels are in reality composed of several different materials due to the limitations of spatial resolution of sensors and the heterogeneity of the feature on the ground. These mixed pixels may result in poor classifications using statistic-based parametric methods. Spectral mixture analysis (SMA) provides a promising way to extract sub-pixel information, based on physical models in which a mixed spectrum is modeled as a combination of a set of pure spectra known as endmembers (Adams et al. 1993). SMA can be modeled as a linear sum of the spectrum of each component weighted by the proportion of the component within the pixels when photons interact with single components within the field of view (Adams et al. 1993). When the scattered photons interact with multiple components, the model should deal with nonlinear effects (Robert et al. 1993). The linear mixing model has been widely used since multiple scattering is assumed to be negligible in most cases (Robert et al. 1998).

The complete linear spectral unmixing model assumes that any image spectrum is the result of a linear combination of the spectra of all endmembers within the scene. This algorithm has the advantage that it is relatively simple and provides a physically meaningful measure of the fraction of each endmember (Robert et al. 1998). One primary limitation of this method is that all endmembers in the scene must be known. Variation in the number of endmembers within the pixels may result in fraction errors. Matched filtering, proposed by Boardman et al. (1995), is a type of linear spectral mixture model in which abundances of only user-defined endmembers are estimated, not all endmembers within the scene. The major advantage of this algorithm is that matched filtering enhances the response of the known endmembers and suppresses the response of the composition of unknown background spectra. However, it may result in false positive fractions for some rare materials.

Case studies

Identifying hydrothermal alteration in Cuprite, Nevada

Cuprite, in western Nevada, is a well-known geological test site, characterized by the exposure of a wide variety of rock and hydrothermal alteration types, with sparse vegetation cover and relatively low topographic relief. The ages of rocks range from Quaternary to Cambrian. Cambrian metasedimentary rocks, including quartzite, phyllite, and limestone, are exposed in western Cuprite (Swayze 1997). Tertiary volcanic rocks and Quaternary alluvial deposits are dominant in the east side of Cuprite (Abrams and Ashely 1980).

The Tertiary volcanic rocks mainly consist of ash-flow and air-fall tuff, conglomerate, and basalt (Ashley 1974, Abrams et al. 1977a, b). Most rocks in the Cuprite area were altered due to now-fossilized hot springs. Three zones of hydrothermal alteration, silicified rocks, opalized rocks, and argillized rocks, have been identified at Cuprite (Abrams and Ashley 1980). The silicified zone, the most intensively altered rocks, contains abundant quartz, some calcite, and minor alunite and kaolinite. The opalized zone, the most widespread alteration zone, surrounds the silicified rocks, and consists of opal and variable amount of alunite and kaolinite. The argillized rocks form the outmost alteration zone, separating the opalized rocks from the unaltered rocks. The major mineral components of the argillized zone include opal, kaolinite, and montmorillonite. The presence of small amount of hematite makes argillized and some opalized rocks appear red in color.

Chen et al. (2007) examined the potential value of integrating VNIR/SWIR hyperspectral and multispectral TIR data for geological mapping in Cuprite. In their research, Airborne Visible/Infrared Imaging Spectrometer (AVIRIS) data and MODIS/ASTER Airborne Simulator (MASTER) multispectral thermal data were used. Two conventional statistic-based classifiers, minimum distance and maximum likelihood classification, and two widely used spectral analysis techniques, SAM and SFF, were applied to VNIR and SWIR AVIRIS data, MASTER TIR data, and the combined data set, respectively. Although the results of SAM and SFF applied to AVIRIS were generally impressive, identifying most rock types with relatively high accuracies, the combined AVIRIS and MASTER TIR data was even more effective for mapping the area (Figure 23.3). SFF applied to the combined data set was able to differentiate silicified alteration from quartzite, both of which exhibit strong distinctive absorption features in the TIR region, but this improvement was at the cost of sacrificing the ability to discriminate rock components with distinct absorption features in the VNIR/SWIR rather than TIR regions, such as argillized and opalized rocks (see Figure 23.3C and D). In general, SAM achieved better performance than SFF in dealing with multiple broad band TIR data, with higher accuracy in discriminating low albedo volcanic rocks and limestone which do not have strong characteristic features in the TIR region (see Figure 23.3A and B).

Mineral mapping with SEBASS hyperspectral TIR data

As discussed above, multispectral TIR instruments that acquire several broad spectral bands have inherent drawbacks in that identifying most surface materials necessarily relies heavily on

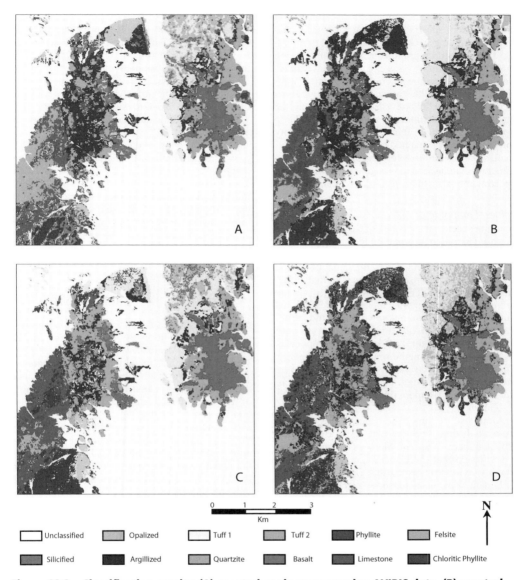

Figure 23.3   Classification results: (A) spectral angle mapper using AVIRIS data, (B) spectral angle mapper using combined AVIRIS and MASTER data, (C) spectral feature fitting using AVIRIS data, (D) spectral feature fitting using the combined data (reprinted from *Remote Sensing of Environment*, volume 110, X. Chen, T. A. Warner, and D. J. Campagna, Integrating visible, near-infrared and short-wave infrared hyperspectral and multispectral thermal imagery for geological mapping at Cuprite, Nevada, page 349, Copyright (2007), with permission from Elsevier). (See the color plate section of this volume for a color version of this figure).

ground truth data. Hyperspectral TIR remote sensing overcomes this problem by using reference spectra from generic spectral library just like its counterpart, hyperspectral VNIR/SWIR . An example of an airborne hyperspectral TIR imaging sensor is the Spatially Enhanced Broadband

Array Spectrograph System (SEBASS), which measures 128 contiguous bands in the 2.5–5.2 and 7.5–13.5 μm wavelength regions (Hackwell et al. 1996). Vaughan et al. (2003) used SEBASS data to identify subtle spectral features associated with silicate and sulfate minerals in an alteration zone, a geothermal site, and a mining district using field and laboratory spectra. Some confusion was found in differentiating between sulfate and clay minerals due to mixtures and fine grain size suppressing the spectral features present. For data calibration, an in-scene atmospheric compensation algorithm (Johnson 1998, Young 1998) was employed to estimate atmospheric parameters such as the atmospheric transmittance and the atmospheric upwelling radiance. Temperature and emissivity information was separated using an emissivity normalization algorithm (Kealy and Hook 1993), and a mineral abundance map was generated with linear spectral unmixing analysis (Adams et al. 1993).

Mineral mapping of Mars

Imaging spectrometry is one of the primary methods used to obtain information regarding planetary geology. Recently, numerous instruments aboard spacecraft have provided very high quality data on nearby planets. For example, the Thermal Emission Spectrometer (TES) was successfully put into orbit around Mars in 1997 aboard the Mars Global Surveyor (TES Team, unknown date), and subsequently collected a vast amount of data up until 2006. The instrument itself has three components that measure incoming infrared and visible energy. The thermal IR spectrometer portion of TES covers the 6–50 μm wavelength with 5 and 10 cm^{-1} spectral sampling, and approximately 10 and 20 cm^{-1} resolution.

A number of important results were obtained from analyzing TES thermal spectra. The first global mineral maps of Mars were created, and it was found that the planet's igneous material is roughly as diverse as Earth's (Bandfield 2002). TES data were used to show that Martian volcanic rocks vary from basaltic to glassy dacite, and that unweathered volcanic minerals dominate the Martian dark regions.

It was also discovered with TES data that carbonate minerals are in the dust, not in surface rocks, as many scientists had predicted (Bandfield et al. 2003). The lack of observed large-scale clay and carbonate mineral deposits suggests Mars has seen much less chemical weathering than Earth. In turn, this indicates Mars' geologic history has been mostly cold and dry. Finally, the identification of gray hematite by TES was influential in determining the location of the later Mars Rover landing (Christensen and Ruff 2004). The hematite absorption feature was mapped in part of the Meridiani Planum, and since the mineral requires the presence of water to form, it was considered a key region for additional surface study.

Hydrocarbon mapping

Another major application of remotely sensed data is found within the energy sector. Early exploration efforts used seeps as indicators of economic deposits of oil and gas. Recently, the detection of marine seeps using radar satellite imagery has been found useful for off-shore exploration. On the sea, oil slicks smooth the surface, reducing the amount of radar backscatter and causing dark anomalies. Significant work has been performed in the Gulf of Mexico to correlate radar anomalies with active natural seeps (De Beukelaer et al. 2003).

On shore delineation of seep evidence has proven to be more difficult and often requires mapping indirect indicators (Schumacher 1996). The surface and near-surface indicators of these microseeps that can be remotely detected are geobotanical anomalies of stressed vegetation and/or mineralogical anomalies associated with the alteration of soils and rocks. The red reflectance of healthy vegetation is dominated by chlorophyll absorption at 0.66 and 0.68 μm (van Leeuwen, in this volume). Changes in chlorophyll concentration produce a shift in the 'red-edge' of chlorophyll absorption near 0.7 μm toward either the blue or red part of the spectrum, the shift being related to a decrease or increase of chlorophyll, respectively. Additionally, hydrocarbons induce changes in soil chemistry, which influences plant vigor and growth. Mineralogical changes that may arise due to hydrocarbon microseeps can include the formation of calcite and pyrite; bleaching of red beds (removal of hematite); clay mineral formation (weathering of feldspars to clays) and alteration (e.g., smectite to kaolinite), etc. With the advent of hyperspectral scanners it is now possible in some situations to indirectly map the changes in reflectance spectra of plants and to detect the anomalous presence of alteration minerals that may be indicative of hydrocarbon microseeps (Schumacher 1996, van der Meer et al. 2002).

An excellent case study for the detection of hydrocarbon microseeps used Probe-1 hyperspectral data over an area in the Ventura Basin of southern California (van der Meer et al. 2002). A tiered image segmentation approach was developed that analyzed both geobotanical and mineralogical anomalies. The data were initially divided into vegetated and non-vegetated (soils/rocks) subsets using NDVI. The vegetated subset was processed to detect the shift in the red-edge position of chlorophyll absorption that may be indicative of the presence of hydrocarbons. The non-vegetated subset was analyzed for the occurrence of areas of calcite enrichment and for the alteration of goethite to hematite, utilizing three different techniques. First, the SAM approach was used to calculate the spectral similarity at each pixel between the test reflectance data and the reference reflectance data; this provides a qualitative estimate of the presence of absorption features that can be related to mineralogy. Next, the Cross Correlogram Spectral Matching algorithm (CCSM; van der Meer and Bakker 1997) was used to match unknown pixel spectra to known laboratory or field spectra; this produces a data layer that indicates the probability of a mineral occurring at a pixel. Finally, the data were subjected to SMA to produce new image layers where each pixel represents the relative abundance of a number of spectral components (calcite, goethite, and hematite). The results were then statistically integrated into a 'predictor map' for the study area to demonstrate that it is possible to identify anomalous spectral responses resulting from hydrocarbon microseeps.

LANDFORM-STRUCTURE MAPPING

Landform and structural analysis has its traditional roots in stereo aerial photography. Visual interpretation was and still is customary for mapping geomorphic features, geologic strata, and structure. Another conventional application is elevation extraction from stereo air photos. This technique has moved from analog to digital methods in the past two decades, greatly enhancing the efficiency of production.

Lidar-derived topography

The need to identify geologic hazards is a major driver for the utilization of advanced remote sensing technologies such as radar interferometry (Kellndorfer and McDonald, in this volume) and lidar (Hyyppä et al. in this volume). The now classic image of the Landers fault displacement in the Mojave Desert of California is an excellent example of a geologic application of radar interferometry (Figure 23.4, Massonnet et al. 1994). This method uses at least two radar images, gathered at different times, and exploiting the interference pattern between the successive dates, the differences in topography is calculated to a centimeter scale. This application is essentially an historical change analysis and can be useful to identify landscape changes caused by geologic events – such as large earthquakes.

While monitoring landscape change is important, the prediction and risk assessment of geologic hazards has been the focus of recent studies. Toward this goal, lidar data has proven to be valuable in areas of dense vegetation where traditional geologic mapping using aerial photographs or space-borne imagery is impractical. Lidar is similar to radar but uses rapid pulses of light pulses instead of radio waves. Lidar systems directly measure the height differences between the instrument platform and a surface, as opposed traditional

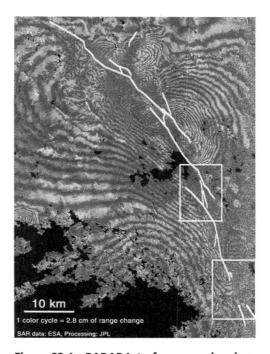

Figure 23.4   RADAR Interferogram showing displacement produced by the Landers earthquake. (Courtesy of the Jet Propulsion Laboratory: Peltzer, G., The Landers earthquake of June 28, 1992 Coseismic study, available at http://www-radar.jpl.nasa.gov/ sect323/InSar4crust/LandersCo.html).

photogrammetric methods of calculating heights from parallax in overlapping aerial photographs. Lidar systems typically consist of a pulse laser scanner that emits multiple (currently up to 200,000 pulses) per second. The time it takes the light energy to reflect back to the instrument from the surface is recorded. The elevation of the reflecting object is then calculated based on the speed of light and the instrument flying height (Hyyppä et al. in this volume).

The significant aspect of lidar for geological applications is that it is potentially possible to obtain laser returns from the ground beneath vegetation. This is due to the fact that even in dense vegetative cover, there are usually enough gaps in the plant cover that some of the light pulses reach the ground. Different quantitative approaches are used to resolve the return time for the ground surface but most utilize a simple ranging to the last detected return. Combining the laser ranging data with knowledge of the mirror scan angle, aircraft height and orientation, and aircraft position from differential Global Positioning System (GPS) techniques and inertial navigation systems, lidar systems calculate geodetically-referenced elevation data with high definition and accuracy.

The emergence of a commercial airborne laser mapping industry has proven to be invaluable to mapping geologic hazards in areas of vegetative cover. Below, we offer two examples that focus on earthquake and landslide hazard assessments using lidar.

Earthquake hazard assessment

In 1996, the Kitsap Public Utility District contracted for a lidar survey on Bainbridge Island, located in the western part of Puget Sound 12 km west of downtown Seattle. While the primary purpose was for hydrologic modeling and water line routing, the data revealed geomorphic features associated with previously unknown fault scarps within the Seattle fault zone (Harding and Berghoff 2000). The impact of this finding motivated the formation of a consortium of federal and local organizations to map the extent of Holocene fault scarps around the Puget Sound for assessing the geologic hazard potential associated with earthquakes.

One result in investigating the earthquake hazard of the Puget Sound area was the mapping and dating of fault scarps in the urban area of Tacoma, Washington (Sherrod et al. 2004). The results of previous geophysical investigations revealed the Tacoma Fault zone as the south bounding structure of the Seattle uplift (Brocher et al. 2001) and help focus the study area for the collection of lidar data. Lidar mapping along the western end of the Tacoma fault zone revealed south-facing fault scarps – some of which are 4 m high – that are mostly indiscernible on aerial photographs. Upon the successful detection of these fault scarps, sites for detailed field investigations were selected. Standard techniques of investigating neotectonic faults, including trenching and dating, were then utilized to discern the last time major earthquakes occurred along the fault system. This study showed that integrating geophysical, lidar, and field data improves our knowledge regarding potential future earthquake hazards in the Seattle–Tacoma metropolitan region.

Landslide hazard assessment

Another application that requires high-resolution elevation data is landslide assessments. Many landslide-prone areas are in regions of high rainfall and rough topography, which is often obscured by dense vegetation. Dense vegetative cover typically precludes the use of aerial photography for detailed morphologic studies. Historical records of landslide occurrence provide a poor general alternative, as this information is only as good as the reporting and record keeping of the local region. However, lidar's capability to detect the ground surface despite vegetation permits detailed mapping of landslide scars for hazard assessments.

Another recent study in Seattle, Washington illustrates just how well lidar is capable of mapping historic landslide scars and other morphological characteristics of landslide-prone areas (Schulz 2004). Seattle landslides generally occur on slopes along Puget Sound or steep, glacially formed ridges. Using a high-resolution (1.8 m horizontal and 0.3 m vertical) DEM derived from the Puget Sound lidar survey, the area was investigated using shaded relief images and slope maps (Figure 23.5). Accuracy of the visual image interpretation was checked in the field and the study found 173 landslides within the study area – nearly four times the number previously estimated.

Not only are landslide hazards a concern in populated areas, they are also a major concern in surface mining operations. The major operation of Lihir gold mine in Papua New Guinea is located within a large caldera with steep slopes surrounding the mine. A recent lidar data acquisition was employed to assess the slope failure risk as the mine expanded toward the slopes (Haneberg et al. 2005). Similar to the Puget Sound study, the lidar-derived DEM data provided key parameters such as slope angle and curvature, as well as roughness (useful to delineate colluvium landslides from rock slumps). Using the semi-empirical method, SMORPH (Slope MORPHology), slope angle class boundaries were integrated with geological units to assess landslide potential. A final risk analysis was generated by qualitatively determining the probability for each hazard class map and the result is used to help plan for future mine operations.

Object-based classification for structural mapping

Remotely sensed imagery contains spatial contextual information such as shape, spatial pattern, and texture. These characteristics are often landform indicators of structure such as faults, and, as we have seen, an important feature in understanding the risk of geologic hazards. Whereas contextual information is essential for visual image interpretation, it is often disregarded in many automatic classification procedures. A new approach, object-based image classification, is a promising technique to combine contextual information and spectral information (Benz et al. 2004, Jensen et al. in this volume). The most important step in object-based classification is image segmentation, which extracts homogenous image objects based on spectral and contextual information. The image objects may directly correspond to objects of interest in the real world. Some geological features might be identified as objects in remotely sensed data, such as volcanoes, intrusive bodies, and faults.

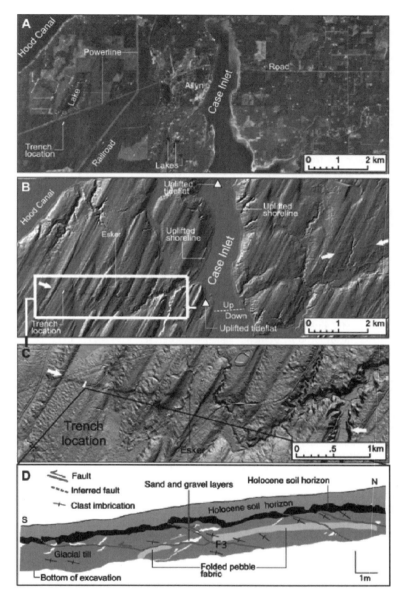

Figure 23.5 A. Aerial photograph of Tacoma Fault zone. B, C. LIDAR shaded relief images. D. Results of trenching and fault profiles. (Courtesy of the USGS, available at http://earthquake.usgs.gov/regional/pacnw/ships/results/tacoma.html).

Gloaguen et al. (2007) investigated the application of object-based classification for extracting normal fault morphology. Their study area, within the Magasi graben, a segment of the East-African rift system, is located in the vicinity of Lake Magadi in Southern Kenya. Within this region are hundreds of recent faults which have well-preserved the morphology due to minimal erosion under arid climatic conditions. Marpu et al. (2006) applied an object-based classification algorithm to an orthorectified and speckle-filtered Radarsat Fine Beam image and Digital Elevation Model (DEM) of study area for extracting image objects that corresponded to faults. This object-based algorithm includes two basic steps: image segmentation and image object classification.

The image segmentation was carried out using the multi-resolution segmentation algorithm in the eCognition (now Definiens) software package, which is based on a bottom-up region-merging approach (Baatz et al. 2004). The classification included a sequence of procedures: feature identification, clustering, threshold calculation, and rule-based classification. The object-based classification was found to have identified individual faults relatively accurately. The fractal dimensions of the classified faults was consistent with the fractal dimension derived directly from the remote sensing images using power spectra and variograms. In summary, this study shows that object-based image classification is a useful potential tool for fault extraction, and the derived fault statistics can give us insight into the intensity and type of deformation of faults.

FUTURE DIRECTIONS AND SUMMARY

The future of geological applications in remote sensing will undoubtedly involve better spectral band width in space or airborne instruments and continued collection of high resolution DEM data sets. The acquisition of routine hyperspectral VNIR and SWIR measurements from space will be particularly important for widespread use of the mineral mapping approaches discussed in this chapter. Unfortunately, acquisition of satellite-based, moderate spatial resolution TIR data is currently limited to the aging experimental ASTER system. The deployment of hyperspectral TIR instruments, and integrated hyperspectral VNIR/SWIR and TIR sensors, currently seem a long way away. In the meantime, therefore, aerial sensors will remain important sources for data acquisition.

Landscape mapping will benefit greatly from repeated observations of high resolution DEMs for monitoring deformation related to earthquakes, landslides, volcanism, and subsidence, to name just a few applications. The acquisition of regional lidar data sets seems particularly important for mapping geologic hazards over wide areas. Placing these data sets in the public domain will facilitate more extensive use in both research and routine mapping.

One of the key features of many of the case studies discussed in this chapter has been the application of a multidisciplinary, integrated approach. Geological remote sensing often employs indirect evidence, and consequently multiple methods can increase the accuracy of interpretations and classifications. Even in the case of direct mineral mapping using hyperspectral techniques, weathering and transported materials may obscure the surface expression of the geology. In such cases, inferences from multiple data sources, and

the use of traditional geological conceptual and mathematical models to extrapolate from areas of higher confidence, can often overcome these limitations.

REFERENCES

Abrams, M. J. and R. P. Ashley, 1980. Alteration mapping using multispectral images – Cuprite Mining District, Esmeralda County, Nevada. *U.S. Geological Survey Open File Report* 80–367.

Abrams, M. J., R. P. Ashley, L. C. Rowan, A. F. H. Goetz, and A. B. Kahle, 1977a. Use of imaging in the 0.46–2.36 μm spectral region for alteration mapping in the Cuprite mining district, Nevada. *U.S. Geological Survey Open-File Report* 77–585, pp. 18.

Abrams, M. J., R. P. Ashley, L. C. Rowan, A. F. H. Goetz, and A. B. Kahle, 1977b. Mapping of hydrothermal alteration in the Cuprite mining district, Nevada, using aircraft scanner imagery for the 0.46–2.36 μm spectral region. *Geology*, 5: 713–718.

Adams, J. B., M. O. Smith, and A. R. Gillespie, 1993. Imaging spectroscopy: Interpretation based on spectral mixture analysis. In C. M. Pieters and P. Englerty, (eds), *Remote Geochemical Analysis: Elemental and Mineralogical Composition*, Cambridge University Press, New York, pp. 145–166.

Ashley, R. P., 1974. Goldfield mining district. In *Guidebook to the Geology of Four Tertiary Volcanic Centers in Central Nevada*. Nevada Bureau of Mines and Geology Report, 19: 49–66.

Baatz, M., U. Benz, S. Dehghani, and M. Heynen, 2004. *eCognition User Guide 4*. Definiens Imagine GmbH, Munchen, Germany.

Bandfield, J. L., 2002. Global mineral distributions on Mars. *J. Geophys. Res.*, 107 (E6), doi:10.1029/2001JE001510.

Bandfield, J. L., T. D. Glotch, and P. R. Christensen, 2003. Spectroscopic identification of carbonate minerals in the Martian dust. *Science*, 301: 1084–1087.

Benz, U., P. Hofmann, G. Willhauck, I. Lingerfelder, and M. Heynen, 2004. Multi-resolution, object-oriented fuzzy analysis of remote sensing data for GIS-ready information. *ISPRS Journal of Photogrammetry and Remote Sensing*, 58: 239–359.

Boardman, J. W., F. A. Kruse, and R. O. Green, 1995. Mapping target signatures via partial unmixing of AVIRIS data. *Summaries of the Fifth JPL Airborne Geoscience Workshop, JPL Publication 95–1*. Pasadena, CA: NASA Jet Propulsion Laboratory, pp. 23–26.

Brocher, T. M., T. Parsons, R. A. Blakely, N. I. Christensen, M. A. Fisher, and R. E. Wells, 2001. Upper crustal structure in Puget Lowland, Washington: Results from 1998 seismic hazards investigation in Puget Sound. *Journal of Geophysical Research*, 106: 13,541–13,564.

Chen, X., T. A. Warner, and D. J. Campagna, 2007. Integrating visible, near-infrared and short-wave infrared hyperspectral and multispectral thermal imagery for geologic mapping at Cuprite, Nevada. *Remote Sensing of Environment*, 110: 344–356. doi:10.1016/j.rse.2007.03.015.

Christensen, P. R. and S. W. Ruff, 2004. Formation of the hematite-bearing unit in Meridiani Planum: Evidence for deposition in standing water. *Journal of Geophysical Research*, 109, E08003, doi:10.1029/2003JE002233.

Clark, R. N., 1999. Spectroscopy of rocks and minerals, and principles of spectroscopy. In A. N. Rencz (ed.), *Remote Sensing for the Earth Sciences: Manual of Remote Sensing*, 3rd edn. John Wiley and Sons, New York, Chapter 1: 3–58.

Clark, R. N., A. J. Gallagher, and G. A. Swayze, 1990. Material absorption band depth mapping of imaging spectrometer data using the complete band shape least-squares algorithm simultaneously fit to multiple spectral features from multiple materials. *Proceedings of the Third Airborne Visible/Infrared Imaging Spectrometer (AVIRIS) Workshop*, JPL Publication 90–54, pp. 176–186.

Clark, R. N., G. A. Swayze, K. E. Livo, R. F. Kokaly, S. J. Sutley, J. B. Dalton, R. R. McDougal, and C. A. Gent, 2003a. Imaging Spectroscopy: earth and planetary remote sensing with the USGS Tetracorder and expert systems. *Journal of Geophysical Research*, 108(E12): 5131, doi:10.1029/2002JE001847.

Clark, R. N., G. A. Swayze, R. Wise, K. E. Livo, T. M. Hoefen, R. F. Kokaly, and S. J. Sutley, 2003b. USGS Digital Spectral Library splib05a, *USGS Open File Report*, 03–395.

Collins, A. H.,1991. Thermal infrared spectra and images of altered volcanic rocks in the Virginia Range, Nevada. *International Journal of Remote Sensing*, 12(7): 1559–1574.

Crósta, A. P., C. Sabine, and J. V. Taranik, 1998. Hydrothermal alteration mapping at Bodie, California, using AVIRIS hyperspectral data. *Remote Sensing of Environment*, 65: 309–319.

Crowley, J. K. and S. J. Hook, 1996. Mapping playa evaporite minerals and associated sediments in Death Valley, California, with multispectral thermal infrared images. *Journal of Geophysical Research*, 101(B1): 643–660.

Crowley, J. K., D. W. Brickey, and L. C. Rowan, 1989. Airborne imaging spectrometer data of the Ruby Mountains, Montana: mineral discrimination using relative absorption band-depth images. *Remote Sensing of Environment*, 29: 121–134.

Dalton, J. B, D. J. Bove, C. S. Mladinich, and B. W. Rockwell, 2004. Identification of spectrally similar materials using USGS Tetracorder algorithm: the calcite-epidote-chlorite problem. *Remote Sensing of Environment*, 89: 455–466.

De Beukelaer, S. M., I. R. MacDonald, N. L. Guinnasso, and J. A. Murray, 2003. Distinct side-scan sonar, RADARSAT SAR, and acoustic profiler signatures of gas and oil seeps on the Gulf of Mexico slope. *Geo-Marine Letters*, 23: 177–186.

Fukunaga, K., 1990. Introduction to Statistical Pattern Recognition. Academic Press, San Diego, California, pp. 592.

Gaffey, S. J., L. A. McFadden, D. Nash, and C. M. Pieters, 1993. Ultraviolet, visible, and near-infrared reflectance spectroscopy: laboratory spectra of geologic materials. In C. M. Pieters, and P. A. J. Englert (eds), *Remote Geochemical Analysis: Elemental and Mineralogical Composition*. Cambridge University Press, Cambridge: pp. 43–78.

Gillespie, A. R., A. B. Kahle, and F. D. Palluconi, 1984. Mapping alluvial fans in Death Valley, California, using multispectral thermal infrared images. *Geophysical Research Letters*, 11: 1153–1156.

Gloaguen, R., P. R. Marpu, and I. Niemeyer, 2007. Automatic extraction of faults and fractal analysis from remote sensing data. *Nonlinear Processes in Geophysics*, 14: 131–138.

Goetz, A. F. H. (1989). Spectral remote sensing in geology. In: G. Asrar (ed.) *Theory and Applications of Optical Remote Sensing*, John Wiley and Sons, Inc., New York, Chapter 12. pp. 491–526.

Goetz, A. F. H. and V. Srivastava, 1985. Mineralogic mapping in the Cuprite Mining District. *Proceedings of the First Airborne Imaging Spectrometer Workshop*. Jet Propulsion Laboratory, Pasadena, California, pp. 22–31.

Hackwell, J. A., D. W. Warren, R. P. Bongiovi, S. J. Hansel, T. L. Hayhurst, D. J. Mabry, M. G. Sivjee, and J. W. Skinner, 1996. LWIR/MWIR imaging hyperspectral sensor for airborne and ground-based remote sensing. *SPIE*, 2819: 102–107.

Halbouty, M. T., 1980, Geologic significance of Landsat data for 15 giant oil and gas fields. *Bulletin of the American Association of Petroleum Geologists*, 64: 8–36.

Haneberg, W. C., A. L. Creighton, E. W. Medley, and D. A. Jonas, 2005, Use of LiDAR to assess slope hazards at the Lihir gold mine, Papua New Guinea. In O. Hungr, R. Fell, R. Couture, and E. Eberhardt (eds), *Landslide Risk Management: Proceedings of International Conference on Landslide Risk Management,* Vancouver, Canada, May 31– June 3, 2005.

Harding, D. J. and G. S. Berghoff, 2000, Fault scarp detection beneath dense vegetation cover: Airborne lidar mapping of the Seattle fault zone, Bainbridge Island, Washington State. *Proceedings of the American Society of Photogrammetry and Remote Sensing Annual Conference*, Washington, D.C., May, 2000, 9.

Hook, S. J., K. E. Karlstrom, C. F. Miller, and K. J. W. McCaffrey, 1994. Mapping the Piute Mountains, California, with thermal infrared multispectral scanner (TIMS) images. *Journal of Geophysical Research*, 99(B8): 15,605–15,622.

Hunt, G. R., 1980. Electromagnetic radiation: the communication link in remote sensing. In B. S. Siegal and A. R. Gillespie (eds), *Remote Sensing in Geology*. Wiley, New York pp. 5–45.

Hunt, G. R. and J. W. Salisbury, 1971. Visible and near infrared spectra of minerals and rocks: II. Carbonates. *Modern Geology*, 2: 195–205.

Hwang, J., S. Lay, and A. Lippman, 1994. Nonparametric multivariate density estimation: a comparative study. *IEEE Transaction on Signal Processing*, 42(10): 2795–2810.

Hyyppä, J., W. Wagner, M. Hollaus, and H. Hyyppä, in this volume. Airborne Laser Scanning. Chapter 14.

Jensen, J. R., J. Im, P. Hardin, and R. R. Jensen, in this volume. Image Classification. Chapter 19.

Johnson, B. R., 1998. In scene atmospheric compensation: Application to SEBASS data collected at the ARM site. Part I. *Aerospace Corporation Technical Report*, ATR-99 (8407)-1.

Kealy, P. S. and S. Hook, 1993. Separating temperatures and emissivity in thermal infrared multispectral scanner data: Implication for recovering land surface temperatures. *IEEE Transactions on Geoscience and Remote Sensing*, 31: 1155–1164.

Kellndorfer, J. and K. McDonald, in this volume. Active and Passive Microwave Systems. Chapter 13.

Kirkland, L. E., K. C. Herr, E. R. Keim, P. M. Adams, J. W. Salisbury, J. A. Hackwell, and A. Treiman, 2002. First use of an airborne thermal infrared hyperspectral scanner for compositional mapping, *Remote Sensing of Environment*, 80: 447–459.

Kruse, F. A., 1988. Use of Airborne Imaging Spectrometer data to map minerals associated with hydrothermally altered rocks in northern Grapevine Mountains, Nevada and California, *Remote Sensing of Environment*, 24(1): 31–51.

Kruse, F. A., A. B. Lefkoff, J. B. Boardman, K.B. Heidebrecht, A. T. Shapiro, P. J. Barloon, and A. F. H. Goetz, 1993a. The spectral image processing system (SIPS) – Interactive visualization and analysis of imaging spectrometer data. *Remote Sensing of Environment*, 44: 145–163.

Kruse, F. A., A. B. Lefkoff, and J. B. Dietz, 1993b. Expert system-based mineral mapping in Northern Death Valley, California/Nevada, Using the Airborne Visible/Infrared Imaging Spectrometer (AVIRIS). *Remote Sensing of Environment*, 44: 309–336.

Landgrebe, D. A., 2000. Information extraction principles and methods for multispectral and hyperspectral image data, In C. H. Chen (ed.). *Information Processing for Remote Sensing*, World Scientific Publishing Co Chapter 1.

Marpu, P. R., I. Niemeyer, and R. Gloaguen, 2006. A procedure for automatic object-based classification, *Proceedings of the 1st International Conference on Object-based Image Analysis*.

Massonnet, D., K. Feigl, M. Rossi, and F. Adragna, 1994. Radar interferometric mapping of deformation in the year after the Landers earthquake. *Nature*, 369: 227–230.

Mustard, J. F. and J. M. Sunshine, 1999. Spectral analysis for earth science: Investigations using remote sensing data. In A. N. Rencz (ed.), *Remote Sensing for the Earth Sciences*: Manual of Remote Sensing, 3rd edn. John Wiley and Sons, New York, Chapter 5 pp. 251–306.

Pouch, G. W. and D. J. Campagna, 1990. Hyperspherical direction cosine transformation for separation of spectral and illumination information in digital scanner data. *Photogrammetric Engineering and Remote Sensing*, 56: 475–479.

Robert, D. A., J. B. Adams, and M. O. Smith, 1993. Discriminating green vegetation, non-photosynthetic vegetation and soils in AVIRIS data. *Remote Sensing of Environment*, 44: 255–270.

Robert, D. A., M. Gardner, R. Church, S. Ustin, G. Scheer, and R. O. Green, 1998. Mapping chaparral in the Santa Monica Mountains using multiple endmember spectral mixture models. *Remote Sensing of Environment*, 65: 257–279.

Rowan, L. C. and J. C. Mars, 2003. Lithologic mapping in the Mountain Pass, California area using Advanced Spaceborne Thermal Emission and Reflection Radiometer (ASTER) data. *Remote Sensing of Environment*, 84: 350–366.

Sabine, C., V. J. Realmuto, and J. V Taranik, 1994. Quantitative estimation of granitoid composition from thermal infrared multispectral scanner (TIMS) data, Desolation Wilderness, northern Sierra Nevada, California. *Journal of Geophysical Research*, 99(B3): 4261–4271.

Salisbury, J. W., L. S. Walter, N. Vergo, and D. M. D'Aria, 1991. *Infrared (2.1–25 micrometers) Spectra of Minerals*. Johns Hopkins University Press, 294 p.

Schaepman, M. E., in this volume. Imaging Spectrometers. Chapter 12.

Schulz, W., 2004. *Landslides Mapped Using LIDAR Imagery, Seattle, Washington*. U.S. Geological Survey Open-File Report 2004–1396, 11 p.

Schumacher, D., 1996. Hydrocarbon Induced Alteration of Soils and Sediments. In: D. Schumacher, and M. A. Abrams, (eds), *Hydrocarbon Migration and its Near-Surface Expression*. AAPG Memoir, 66: 71–89.

Sherrod, B. L., T. M. Brocher, C. S. Weaver, R. C. Bucknam, R. J. Blakely, H. M. Kelsey, A. R. Nelson, and R. Haugerud, 2004. Holocene fault scarps near Tacoma, Washington, USA. *Geology*, 32(1): 9–12.

Swayze, G. A., 1997. The hydrothermal and structural history of the Cuprite Mining District, southwestern Nevada: an integrated geological and geophysical approach, Ph. D. Dissertation, University of Colorado Boulder, pp. 399.

TES Team, unknown date. Mars Global Surveyor TES: Thermal Emission Spectrometer. http://tes.asu.edu/about/index.html. (Last accessed 19 January 2008).

van der Meer, F. and W. Bakker, 1997. CCSM: Cross correlogram spectral matching. *International Journal of Remote Sensing*, 18: 1197–1201.

van der Meer, F., P. van Dijk, H. van der Werff, and H. Yanga, 2002. Remote sensing and petroleum seepage: a review and case study. *Terra Nova*, 14: 1–17.

van Leeuwen, W. J. D., in this volume. Visible, Near-IR and Shortwave IR Spectral Characteristics of Terrestrial Surfaces. Chapter 3.

Vane, G., A. F. H. Geotz, and J. B. Wellman, 1984. Airborne Imaging Spectrometer: a new tool for remote sensing. *IEEE Transactions on Geoscience and Remote Sensing*, 6: 546–549.

Vaughan, R. G., W. M. Calvin, and J. V. Taranik, 2003. SEBASS hyperspectral thermal infrared data: surface emissivity measurement and mineral mapping. *Remote Sensing of Environment*, 85: 48–63.

Young, S. J., 1998. In scene atmospheric compensation: Application to SEBASS data collected at the ARM site. Part II. *Aerospace Corporation Technical Report*, ATR-99 (8407)-1.

24

Remote Sensing of Soils

James B. Campbell

Keywords: soils, remote sensing, inference, SAR, thermal infrared, aerial photography.

INTRODUCTION

The Earth's soils form essential components of natural ecosystems, and the basis for production of food, fuel, and fiber. Use and preservation of these resources requires effective monitoring and management, which in turn depends upon preparation of accurate maps of soil distributions, and monitoring of temporal variations in soil status. It is in this context that remotely sensed imagery forms an essential tool for acquiring data representing soil distribution, and recording both temporal and spatial variation of soil properties.

Soil characteristics

Soil is formed from weathered organic and mineral debris at the Earth's surface, as modified over time by weather and climate. At any specific place, its characteristics are the result of local climate and parent material modified by topography and vegetation over time. Such soil forming factors interact to create soil pedons, three-dimensional soil bodies represented by two-dimensional soil profiles, vertical sequences of distinctive horizons (Buol et al. 2002). Each horizon can be envisioned as a layer formed by biological and physical processes acting in concert over intervals usually measured in several decades to millennia, at a minimum. Local differences in soil forming processes lead to spatial variation in soil characteristics. The soil landscape can be visualized as a mosaic of three-dimensional pedons. Although the usual soil maps represent boundaries between adjacent pedons as sharp edges, in fact they vary greatly in form and shape, characterized by gradations, outliers, inclusions, and transitional zones. Remote sensing of soils therefore must be practiced in the context of complex spatial variation.

With only rare exceptions, remote sensing techniques sense characteristics of the soil surface – even when subsurface properties can be assessed, they may convey little information about the soil profile. Although links between a soil's surface properties and its subsurface horizons can be established, such connections usually depend upon information derived outside the realm of remote sensing. Therefore, it should be recognized that current remote sensing applications usually address characteristics of surface properties, rather than soils in their full pedological context.

One of remote sensing's most effective capabilities is its ability to monitor variation of diurnal or ephemeral soil properties as they change over time, in response to time of day, season, weather, and climate. When observed at broad scale, some of the complexities mentioned above assume less significance, and (for example) regional variations in surface moisture can be tracked from week to week to contribute to understanding the progression of seasonal patterns of crop growth. Although temporal differences in soil properties often complicate applications of remote sensing, they can also reveal soil patterns, as, for example, when soil moisture patterns reveal patterns in surface texture that might not otherwise be observable.

Soil properties of significance for remote sensing

Soils are inherently multivariate, so it is difficult to identify specific properties to characterize soil variability. Any soil is characterized by mixture of mineral and biologic materials, which in combination create a myriad of physical, chemical, and biologic properties that pedologists use to characterize separate soils (National Cooperative Soil Survey). Of these, this chapter highlights four properties that are of broad significance in remote sensing applications.

Texture

Soil texture characterizes the particle size distribution of the mineral component of a soil. The mineral component of soil, by definition, considers particles less than 2 mm in diameter, divided into three size fractions – *clay* (< 0.002 mm in diameter), *silt* (0.002–0.05 mm), and *sand* (0.05–2 mm). The texture of a soil is determined by the proportions (by weight) that fall into each size class. (Particles larger than 2 mm, classified as gravel, are not included in textural classification). For example, a silty clay loam is centered around a proportion of 10% sand, 56% silt, and 34% clay, while a loam is centered around a proportion of 40% sand, 42% silt, and 18% clay. The complete set of designations and definitions is given by the textural triangle (Brady and Weil 2002, Schaetzl and Anderson 2005). Although optical characteristics of varied soil textures have been studied both in the laboratory and in the field (e.g., Stoner and Baumgardner 1981, Baumgardner et al. 1985), texture is so closely interrelated to so many other soil properties that it is difficult to isolate its role from other influences upon remotely sensed field data. Although each soil horizon is characterized by its own texture, in the context of remote sensing, the focus is upon the texture of the surface horizon.

Soil Organic Matter

In humid climates especially, most surface soil accumulates debris from the remains of plants, insects, and animals. As this debris decomposes, it forms the organic component of soil. Soil organic matter forms an important dimension of soil fertility, its moisture-holding capacity, and its color, and therefore is an important variable for many application of remote sensing to the study of soil. Like most other soil properties, its spectral characteristics are closely connected with other properties. Al-Abbas et al. (1972) suggest that organic matter content is most significant as an influence upon soil brightness when its levels exceed 2%, when it dominates the influences of other properties. Others, such as Fox and Sabbagh (2002), have been able to use observations in the red and near infrared regions to detect variations in soil organic matter.

Soil Moisture

Soil moisture forms an important variable for remote sensing applications because of its significance for agriculture, hydrology, climate, and its close relationships with other soil properties. Moisture retained in the soil is usually measured as percent by weight of a soil sample. Although *in situ* moisture measurements can be made using a variety of field-portable instruments, the marked variation of soil moisture from place-to-place and over short time periods renders collection of reliable field data problematic due to practical difficulties of collecting simultaneous observations over large areas. For such reasons, observation of soil moisture and field verification can be much more challenging than might be expected from casual consideration.

Soil properties are inherently interrelated, so that observation of a few properties leads to understanding variation of other properties not directly observed on the imagery. Thus, soils lend themselves to application of inference – the use of qualities observed by remote sensing to map others that cannot be directly observed.

Agricultural Practices

Although this chapter cannot explore the all impact of human cultural practices upon soil properties, it is important to consider the effects of agricultural practices (e.g., Promes et al. 1988, Perfect and Caron 2002, Ding et al. 2002), both upon soil properties themselves and the ability of remote sensing instruments to observe these characteristics.

Applications of remote sensing

Remote sensing can be applied to understand place-to-place variation in soils through either of two contrasting processes. In the first, soil surveyors use optical imagery as part of the soil mapping process, in which a soil surveyor uses imagery partly as a cartographic base to plot field observations, and partly as a source of information about vegetation, drainage, land use, and related characteristics. Using stereoscopic imagery as the source of some topographic information principally enables the soil surveyor to record soils as discrete landscape entities with particular slope and aspect. The soil survey process is integral to preparation of general-purpose soil surveys, which characterize properties common to the landscape unit, but do not attempt to show variation within units – these surveys represent the soil landscape as a series of discrete patterns, as noted above. The soil survey is prepared through application of *surrogates* – observable features, such as vegetation, topography, and land use, to understand patterns that are themselves not directly observable. Because this process addresses fundamental genetic characteristics, it can serve multiple purposes, often including some that were not necessarily envisioned at the time the survey was prepared.

In contrast, a second strategy – *direct sensing* of surface soil properties – records point-to-point variability of surface properties, without regard to underlying profiles (Irons et al. 1989). In comparison to soil survey, direct sensing records variation of single properties, at finer scales, and as continuous variables, sometimes with the ability to record temporal variation of properties such as temperature or moisture. Direct sensing can support preparation of special-purpose soil maps or data, which are highly focused to fulfill a specific mission. Although such data may serve a designated purpose well, they lack the dimensionality that might allow them to be used to serve other purposes.

SOIL SURVEY

Soil scientists use aerial photography to prepare soil surveys that describe spatial variation of soils for applications such as land-use planning, agriculture, forestry, and wildlife management. In the United States and other nations, governmental agencies have used aerial photography in the mapping processes to derive some soil information, and also as the cartographic base to display soil information, largely because the wealth of spatial detail allows users to relate mapped soil boundaries to easily recognizable landscape features. Before preparing a detailed soil survey, soil scientists conduct a reconnaissance survey to understand the general pattern of soils and their relationships with observable landscape features (Soil Survey Division Staff 2007).

The first step is *Mapping*, by which soil scientists delineate soil units, using aerial photography to mark boundaries between principal soil units, applying knowledge gained during the preliminary reconnaissance. Often the mapping process is facilitated by *pre-mapping*, in which pedologists review existing maps and documents to derive an understanding of local soil patterns and problems encountered in previous surveys. Boundaries are marked on field sheets, which record surveyors' interpretations of a soil distribution, usually in greater detail than will be shown on the final map. Likewise, discrete map units or polygons (delineations) are labeled with symbols that designate specific units that will be used in the published version.

In the mapping process, the soil scientist examines aerial photographs, often in stereo, to identify breaks in slope, the boundaries between vegetation classes, and the drainage pattern, to define boundaries between soils. Although soils can seldom be identified from aerial photography alone, the use of aerial photography permits the soil scientist to identify soil boundaries with levels of accuracy and efficiency that that could not be achieved without their use. This practice is possible through application of inference, by which the soil surveyor uses what can be seen on aerial imagery to understand the distribution of qualities that cannot be seen (the soil profile).

During the second step, *Characterization*, which field samples are collected from each prospective map unit, and then subjected to laboratory analysis to determine physical, chemical, and mineralogical properties. These measurements form the basis for *Classification*, the third step, in which soils are assigned to a hierarchy of categories in a national classification system from general (soil order) to specific (soil series) levels (Soil Survey Division Staff 1999). Classification and transect analysis allow estimation of percent composition of soils and miscellaneous land types in each map unit in each survey area. Correlation matches map units within a region to those in ecologically analogous areas. Whereas the other steps may be conducted on a local basis, correlation requires the participation of experienced scientists from adjacent regions, and from national levels in the organization, to provide a broader perspective to understand the relationships between soils of different regions. The final step, *Interpretation*, evaluates each map unit component to assess suitability for prospective agricultural and engineering uses, to provide map users with information concerning the likely suitability of each mapping unit for alternative land uses most common within the region (Soil Survey Division Staff 1993).

The result is a representation of the soil pattern on a map, and a report that describes the soils encountered within each map unit. Each map unit is evaluated with regard to the kinds of uses that might be possible for the region, so that the map serves as a guide to wise use of soil resources. The modern soil survey is possible because aerial photography, or comparable imagery, provides both the detailed base on which soil boundaries can be plotted, and information about the placement of soil boundaries. Soil surveys are possible because soil surveyors understand the critical factors that underlie the place-to-place variation of soils, and how they appear on aerial photographs and related imagery.

DIRECT SENSING OF SOIL PROPERTIES

Alternatively, remote sensing effort can be focused upon direct observation of soil properties. Remote sensing instruments permit direct observation of certain physical characteristics of the soils surface (e.g., Meyers 1975, 1983). By way of example, the following list provides some specifics, without claiming to be exhaustive.

- Surface texture (observed principally in the optical spectrum)
- Soil moisture (observed using optical, mid-IR, and thermal regions)
- Organic matter (observed chiefly in the optical region)
- Surface roughness (sensed using active microwave)
- Mineralogy (most effectively observed using hyperspectral remote sensing)

Although many of these examples represent operational (rather than theoretical) capabilities, it must be emphasized that they are best applied in circumstances that isolate the specific property in question. For example, remote sensing of surface texture can be best accomplished for surfaces that isolate surface texture, perhaps in sandy, barren, surfaces with even topography. Because soil is inherently multidimensional, such artificially constrained situations can be only loosely related to operational applications of remote sensing, in which several properties simultaneously influence the observed spectral response. Direct sensing therefore offers the power of providing first-hand data describing in detail the status of a soil property, but is limited by its narrow applicability. It is for this reason that inference, despite its qualitative character and apparent lack of rigor, endures as an important tool for understanding the nature of soil patterns, even in an era of advanced sensors.

DIGITAL SOIL MAPPING

Digital Soil Mapping (DSM) applies field and laboratory data, manual delineation of soil boundaries, digital terrain data, and geo-referenced imagery to extrapolate soil patterns from known areas to areas where the knowledge of soil distributions are less certain (McBratney et al. 2003, Scull et al. 2003, Lagacherie et al. 2006). DSM relies upon geospatial technologies, such as GPS, and remotely sensed imagery, to gather and register landscape data, and upon analytical capabilities, such as geostatistical interpolation, data mining, and digital elevation models, to derive soil related information within the context of a GIS. Unlike conventional digital techniques, which usually focus upon visualization of observed data, DSM is devoted to extrapolation and interpolation

of soil knowledge to extend knowledge of the soil pattern, and assessment of results for accuracy and uncertainly. Although superficially DSM may seem to apply the methods of direct remote sensing, its essence relies heavily upon inference, and the soil surveyor's understanding of soil distributions.

THE OPTICAL SPECTRUM

Sensors operating in the optical spectrum (van Leeuwen, in this volume) include aerial photography employing varied emulsions, including color, black-and-white, and color infrared. Current capabilities include a variety of aircraft sensors (Stow et al., in this volume), and fine-resolution satellite data (Toutin, in this volume).

Many studies have examined spectral properties of soil surfaces, notably color, texture, moisture, and organic matter. Such studies must isolate the specific properties in question, often through use of laboratory measurements that isolate soils from the multitude of extraneous influence encountered in the field, or by use of artificial materials. Although such studies provide valuable data for the field of remote sensing, it is important to recognize that they usually isolate properties of interest, creating situations that cannot represent soil spectra as encountered in the field, in which numerous properties vary independently. Several authors (e.g., Lee et al. 1988, McBratney et al. 2003) suggest that remotely sensed imagery alone is insufficient as a source of information for soil studies. They recognize that proxies (such as topography, vegetation, drainage patterns) and field observations ultimately form the foundations for derivation of soil information from imagery.

In practical applications, the analyst must encounter soils in the field, in the context of a multiplicity of interrelated soil properties, varied slopes, agricultural practices, and moisture levels. Therefore, the analyst often must rely not only upon information derived by direct sensing, but upon traditional photointerpretation and inference. Such interpretations were originally developed in the context of black-and-white aerial photography at a range of scales, effectively in the context of stereo imagery. However, these practices can be applied to interpretation of color and CIR imagery with equal effectiveness.

Drainage patterns can indicate the general nature of the underlying sediments. For example, a dendritic drainage pattern indicates the presence of uniform surface materials, even slope, and homogeneous drainage. Inspection of cross-sections of stream profiles (as viewed in stereo) provide further clues about the nature of sediments – for example, V-shaped cross-sections suggest the presence of cohesive surface materials, while U-shaped cross-sections are often typical of loess (which occurs as extremely uniform deposits, which tend to form gullies with vertical faces and flat floors). Both V-shaped and U-shaped gullies have sharp angled upper boundaries. Cohesive materials tend to form saucer shaped drainages with round shouldered interfluves.

The image tones of bare fields are often related to microtopography and related variations in local drainage. Restricted drainage in slightly lower regions can be recognized by higher moisture content, and darker tones. Further, lower spots accumulate fine-textured silts and clays washed downslope from higher spots. In contrast, higher spots are often lighter in tone, characterized by lower levels of organic matter, coarser textures, and lower levels of surface moisture. Typically, bare soils characterized by mottled image appearances are characterized by fine textures and poor surface drainage.

An example is presented as Figure 24.1, which depicts a color infrared image of a region of Winneshiek and Fayette Counties, Iowa. Variations in tone and texture in open fields indicate variations in drainage, texture, and organic matter as outlined above. The interfluves are chiefly Mollisols formed in a thin layer of till and loess covering limestone bedrock. Sideslopes are occupied by deep, moderately well-drained soils formed in silty and loamy sediments of underlying glacial till (Typic Hapludolls). Footslopes and drainageways are occupied by deep, somewhat poorly drained Aquic Hapludolls.

These techniques can be applied not only to large-scale black and white aerial photography, but also to imagery acquired by the multiplicity of sensors and platforms currently available. Sullivan et al. (2005) evaluated IKONOS multispectral imagery as a source for soil information at two sites in Tennessee and Alabama. Acquisitions were designed to assess the impact of surface crusting, roughness, and tillage on the ability to depict short-range variation of soil properties. The authors evaluated several interpolation algorithms to investigate the relationship between image tone and soil

Figure 24.1 CIR aerial photograph representing soil patterns, Winneshiek and Fayette Counties, Northeastern Iowa, on Iowan drift, May, 2002. Variations in image tone and texture in open fields indicate local differences in drainage, texture, and organic matter. Interfluves are chiefly Mollisols formed in a thin layer of till and loess covering limestone bedrock. The sideslopes are chiefly occupied by deep, moderately well-drained Typic Hapludolls formed in silty and loamy sediments and the underlying glacial till. Footslopes and drainageways are occupied by deep, somewhat poorly drained Aquic Hapludolls. NAPP photograph, NP0NAPP012985149. (See the color plate section of this volume for a color version of this figure).

properties, especially surface organic matter and clay content.

Ray et al. (2002) employed image brightness as a variable in a regression to assess soil properties such as organic matter, texture, and available nitrogen, phosphorous, and potassium, finding significant relationships in some instances. Simbahan et al. (2006) used satellite imagery as one of several forms of data to improve the quality of fine-resolution maps of soil organic carbon stock assessed within the upper 0.3 m of soil within three large no-till fields (49–65 ha) in Nebraska. They found that relative gains from incorporating secondary information increased with decreasing sampling density. Although such approaches may form the basis for future digital soil mapping, for the immediate future, practical soil mapping must rely upon more established procedures.

In 2005 the United States Department of Agriculture's National Resources Inventory (NRI) program began to use high-resolution IKONOS satellite imagery to prepare soil and natural resource maps of Alaska. In Alaska, satellite imagery offers advantages in providing imagery of remote regions at reasonable cost. The USDA will use a combination of archive and newly tasked IKONOS satellite images to map and apply NRI primary data elements to inventory land use, evaluate loss of farmland to urbanization, to measure the effectiveness of conservation practices, and to detect changes to the landscape from soil erosion.

SHORT-WAVE INFRARED

The short-wavelength infrared (SWIR) region of the spectrum (1.1–2.5 μm) conveys information of significance for understanding soil patterns. Sensors with bands in this region are included in Landsat, ASTER, and AVIRIS instruments, among others. In the context of this present chapter, the SWIR's significance lies chiefly in its ability to record the presence of clay minerals, especially those characterized by presence of Hydroxyl ions (OH^-), common in clays and other altered minerals, and in its sensitivity to surface moisture. Dehaan and Taylor (2004) used hyperspectral imagery to examine mineralogies of surface materials within a semi-arid landscape in New South Wales. Class maps derived from the shortwave infrared part (SWIR) of the spectrum delineated terrain elements distinguished by their clay mineralogy, content, and crystallinity. They propose that their methods can be effective in distinguishing between residual and transported sediments. Bowers and Hanks (1965), and Skidmore et al. (1975) concluded that soil moisture greatly influences the reflection of shortwave radiation from soil surfaces in the SWIR. Lobell and Asner (2002)

described a relatively robust relationship between the degree of saturation and SWIR reflectance, estimating that soil moisture based on a general model of SWIR reflectance contained only half the uncertainty of estimates based in the VNIR.

THERMAL INFRARED

Soils can be observed using sensors sensitive to the thermal infrared spectrum (Quattrochi and Luvall, in this volume). Long-wave infrared radiation conveys information about solar energy that has been absorbed by the Earth's surface as short-wave radiation (with a peak wavelength in the visible region, at about 0.5 μm) then re-radiated by the Earth's surface as long-wave infrared radiation (with a peak wavelength near 10 μm). This spectral region is of interest for the study of soils, as it conveys information about the varied properties of the surface sediments and subsurface geology. Absorption and re-radiation of thermal radiation is a function of thermal properties of surface soil and vegetation cover, controlled by multitude of characteristics, including lithology, moisture levels, depth of surface materials, vegetation cover, land use, slope, and aspect.

Heat capacity (specific heat) measures the ability of a substance to store heat energy. The higher a soil's heat capacity, the more heat it can gain (or lose) per unit change in temperature. Because the heat capacity of dry soil is only about one-fifth the heat capacity of water, moist or saturated soils have high heat capacities relative to comparable dry soils. Therefore, at a given depth, coarse-textured dry soils display larger seasonal swings in temperature than do wet soils. This effect is the consequence of their lower heat capacities, which allow their temperatures to rise or fall more than would comparable wet soils for a specific increase or decrease in heat energy each spring or autumn.

Thermal conductivity indicates the rate at which heat will be transferred by a specific soil for a given temperature gradient. Thermal conductivity tends to increase as soil textures become increasingly fine, with loamy textures having intermediates value between sand and clay, and as moisture content increases. Near the surface, soil temperature varies seasonally as a function of incident solar radiation, rainfall, seasonal swings in overlying air temperature, local vegetation cover, type of soil, and depth. Temperatures near the soil surface vary both daily, as the intensity of solar illumination varies between sunrise and sunset, and annually, as solar angle and length of day vary with latitude and season. As air temperatures warm in spring, soil temperatures increase more slowly and to a lesser extent than do air temperatures, so by summer its maximum temperature is cooler than the maximum air temperature. Likewise, in autumn, soil temperatures cool more slowly and to a lesser extent than air temperatures. A general depiction of a soil's annual temperature pattern is embodied in the way soils in the U.S. are mapped. For example, the mean annual soil temperature of a mesic soil is 8–15°C, and the difference between mean summer and mean winter soil temperatures is more than 6°C, either at a depth of 50 cm from the soil surface or at a densic, lithic, or paralithic contact, whichever is shallower. Other temperature ranges are defined in similar fashion.

Because of soil's greater heat capacity relative to the atmosphere, and the thermal insulation provided by vegetation, seasonal changes in soil temperature diminish with depth, and lag behind seasonal changes in atmospheric temperature. At soil depths below approximately 9 m ($\approx$ 30 ft), soil temperatures are relatively constant across the seasons – this temperature is referred to as the *mean earth temperature*. The amplitude of seasonal changes in soil temperature on either side of the mean earth temperature depends on the type of soil and the depth of the measurement. Thus, spatial variations in thermal responses of surface materials can be assessed by examining temporal variations in thermal behavior of landscape. Thus, the concept of thermal inertia captures the contrasting properties of varied soils to retain and re-radiate solar energy.

As an example, a soil characterized by abundant surface moisture will slowly absorb solar energy during daylight hours, then to retain energy, slowly releasing energy during the night-time. In contrast, neighboring areas characterized by lower moisture content rapidly absorb solar energy during the daylight hours, and re-radiate at higher rates during night-time. Thus, differences in soil texture, moisture capacity, and moisture content will exhibit contrasting thermal behaviors (Burdt et al. 2005).

Observations of thermal inertia are best implemented using a pair of images, one that captures an image of a landscape during the daytime, as it reaches its maximum solar heating, probably just after solar noon. The second member of the image pair records the same landscape during night-time, as the landscape reaches its minimum temperature, just before sunrise. (In practice, such pairs may approximate these times, as pragmatic constraints, especially characteristic of orbital sensors, may not permit observation at optimal times.)

Figure 24.2 presents a schematic illustration of diurnal temperature variation of several broadly defined landscape classes. Two images (images 1 and 2) are acquired at times that permit observation of extremes of temperature, perhaps near noon-time and again just before dawn. These two sets of data permit estimation of the ranges of temperature variation for each class.

Further, it may be useful to consider seasonal image pairs which might permit application of this concept under favorable conditions. Because of effects of atmosphere, length of day, and seasonal vegetation changes, application of March–September pairs (for example) might be preferred rather than January–August pairs, which might seem initially to be an optimum choice. Although seasonal thermal inertia has been applied in interplanetary remote sensing, it apparently has not been used often in the terrestrial case.

Promes et al. (1988) observed furrowed fields with a truck-mounted passive microwave radiometer to evaluate effects of row structure on estimates of soil emissivity. They found that their models could estimate emissivity to within close tolerances for common agricultural practices. Cathcart et al. (2006) investigated applications of thermal infrared data to landmine detection. In this context, they investigated dynamic hyperspectral signature models of various terrain features (e.g., soils) to assess the impact of various environmental factors and processes.

Burdt et al. (2005) examined seasonal soil temperatures and hydrology for plots with several land uses in Southeastern Virginia, and concluded that seasonal temperatures could be related to soil saturation and land use. Verstraeten et al. (2006) estimated soil moisture based upon thermal inertia as observed from meteorological satellite imagery. Ho (1987) modeled heat exchange at the surface satellite temperature data, provided that thermal inertia is known. He proposed a simple inverse model formulated to calculate soil thermal inertia using visible and infrared satellite data. Gauthier and Tabbagh (1994) observed temperatures of hydromorphic soils using airborne thermal remote sensing. Measurements of ground soil thermal properties were used to calculate thermal inertia for each soil and to verify that both deep wet soils with high heat capacities and pebbly soils with high conductivities have higher thermal inertia. Differences in thermal inertia accounted for differences in apparent temperature for fluxes corresponding to those prevailing during the days preceding the flight.

ACTIVE MICROWAVE

As introduced in Kellndorfer and McDonald (in this volume), active microwave sensors, most specifically, synthetic aperture radars (SARs), broadcast microwave energy from an aircraft or satellite to illuminate a narrow strip of the ground surface offset from the flight track of the vehicle. Characteristics of the returned signal can reveal characteristics of the ground surface. Development of capabilities for the study of soil properties using

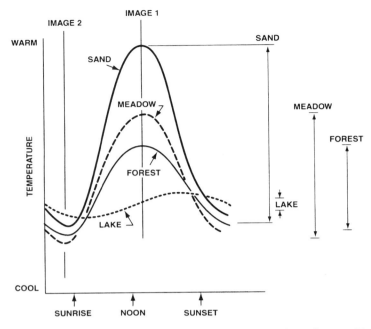

Figure 24.2   Schematic illustration of diurnal temperature variation of several broad classes of land cover. Two images (images 1 and 2) are acquired at times that permit observation of extremes of temperature, perhaps near noontime and again just before dawn. These two sets of data permit estimation of the ranges of temperature variation for each region on the image. Because these variations are determined by the varied thermal behaviors of the several surfaces, they permit interpretation of features represented by the images. (Reproduced with permission from Campbell 2007).

active microwave sensors has been based upon a hierarchy of experiments conducted over the past decades. *In-situ* data can be acquired using microwave antennae mounted on boom trucks for ease of transport to field sites with the desired characteristics.

Concepts developed in this context were implemented using aircraft SAR systems, and then satellite SAR. Although SeaSat (1978) was tailored for observation of the Earth's oceans, it demonstrated the practicality of acquiring Earth resources data of both oceanic and terrestrial surfaces. Within a few years, microwave antennae were carried on the Space Shuttle to implement designs based upon field experiments (Way and Smith 1991). The Shuttle Imaging Radar (SIR), developed by NASA and the Jet Propulsion Laboratory, implemented a series of increasingly sophisticated designs to observe the earth from space, first, through SIR-A (1981, STS-2), then SIR-B (1984, STS-17), and SIR-C/X-SAR (1994, STS-59, STS-68). Successive designs permitted increasingly flexible variations in depression angles and look angles, and multiple frequencies and polarizations that could

be applied and evaluated in varied geographic settings. Proven capabilities have been implemented in operational systems, including RADARSAT, and ERS.

Much of the scattering of a microwave signal is determined by the first surface that intercepts the signal, at least for relatively short wavelengths. For very large portions of the Earth, this surface is likely to be formed from a vegetation canopy. Therefore, to focus our discussion on the effects of soil surfaces upon the backscatter, our discussion focuses upon vegetation-free surfaces as a means of isolating effects determined largely by soil characteristics.

Energy received for a specific cell is described by the *radar equation* (Dobson and Ulaby, 1998):

$$P_r = \frac{P_t G^2 \lambda^2}{(4\pi)^3 R^4} A \sigma_{rt}^0 \qquad (1)$$

where P_t is the power transmitted, P_r is the power returned, G is the antenna gain, λ is the wavelength, σ_{rt}^0 is the backscattering coefficient for specific transmit (t) and receive (r) polarizations, R is the range, and A is the area illuminated at a specific cell.

A discussion of active microwave applications to the examination of soils must focus upon influences upon σ_{rt}^0, the backscattering coefficient (backscattering per unit area). In the context of soils, backscattering is best understood in the context of surface roughness, soil moisture, and surface penetration.

Soil moisture

The microwave signal is sensitive to soil moisture through its influence on the dielectic constant, which measures the ability of a substance to retain an electrical charge. As the dielectric properties of a soil increase, so does the soil contribution to the magnitude of the backscatter. A soil's dielectric properties are influenced by soil moisture, density, texture, mineralogy, and temperature. It has a strong relationship with volumetric soil moisture, and weak relationship to soil texture (Figure 24.3). Although the magnitude of the dielectric constant is weakly related to temperature, frozen soils display a dramatic decrease in dielectric properties relative to that of unfrozen soils.

Drunpob et al. (2005) applied RADARSAT SAR to estimate moisture in a semi-arid coastal watershed. Although their estimates were accurate in portions of the watershed characterized by even terrain, their procedure did not perform well in areas of high relief, where difficulties in geo-referencing created errors. Jackson et al. (1997) observed unvegetated soil surfaces using passive microwave at S and L bands. They found a clear correspondence between the 1 cm volumetric soil moisture and the 2.65 GHz emissivity and between the 3–5 cm soil moisture and the 1.4 GHz emissivity. They confirmed that moisture levels at specific depths control the response at specific frequencies, and that these observations permit monitoring of diurnal variations in soil moisture. Zhang et al. (2003) examined models for estimating dielectric properties of soils at temperatures near freezing, recognizing that cold soils contain a mixture of liquid and frozen moisture.

Surface roughness

Whereas the magnitude of σ^0 is determined largely by the dielectric constant, the form of the reflected

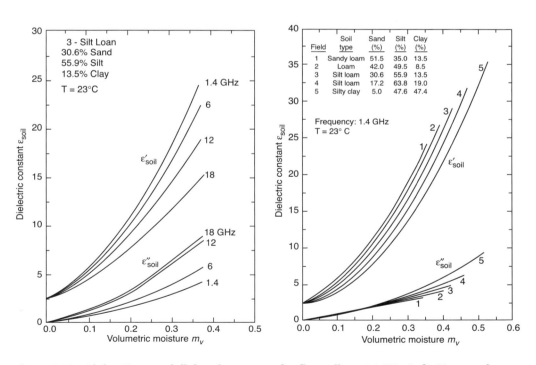

Figure 24.3 Right: Measured dielectric constant for five soils at 1.4 GHz. Left: Measured dielectric constant as a function of volumetric moisture content for a loamy soil at four microwave frequencies. Reproduced by permission from F. T. Ulaby, M. K. Moore, and A. K. Fung. Microwave Remote Sensing: Active and Passive. Norwood, MA: Artech House, Inc. Vol. 3. 1986. © 1986 by Artech House, Inc.

signal is determined largely by the roughness of the surface. If a surface is rough relative to wavelength, energy is re-directed in many directions, creating the *diffuse*, or *isotropic*, behavior characteristic of rough surfaces. Surfaces that are smooth relative to wavelength tend to behave as mirror-like, or *specular* surfaces. For example, an unvegetated gravel surface can behave as a specular surface at X-band, because it is smooth relative to wavelength. On an image, such surfaces would appear in black tones, because they direct energy away from the antenna.

Surface roughness can be characterized by the heights of departures from an arbitrary reference surface. Such a reference surface may take a horizontal, or periodic, form, whereas the roughness can be envisioned as random variations centered on the reference surface. Thus, surface roughness can be characterized as the root means square (rms) of surface roughness. For example, at X-band (3 cm) a surface with facets about 1 cm in height would tend to behave as a rough surface, whereas at L-band (23.5 cm), a much longer wavelength, the same facets would behave as a smooth surface. Thus SAR imagery can characterize roughnesses of irregular surfaces, such as unweathered volcanic surfaces, or tilled fields.

Effects of roughness are related to several variables, including:

- roughness
- azimuth/orientation (especially in tilled surfaces with preferred orientation)
- depression angle
- wavelength
- polarization.

Theoretical and empirical models developed thus far describe actual behavior only under narrow range of circumstances. Sahebi et al. (2002) have proposed procedures, based upon their simulations, for estimating surface roughness from multi-polarization or multi-angular data.

Surface penetration

Under favorable conditions, microwave energy can penetrate the soil surface to be scattered by subsurface features. Depth of penetration depends upon soil properties and characteristics of the microwave signal. It is typically reported as the *skin depth* (penetration depth) – the depth at which the microwave signal has been attenuated to 37% ($1/e$) of its strength at the soil surface.

Penetration is greatest with dry soils, observed at steep depression angles; depth of penetration increases also with increasing wavelength. Therefore, maximum penetration below the soil

surface is likely to occur at long wavelengths and steep depression angles, observed in desert environments.

Scientists in several fields have self-evident interest in understanding penetration below the soil surface, both to detect subsurface features, but also to correctly interpret σ^0, which may record both surface and subsurface effects. Soils are inherently variable, both spatially and vertically, so penetration of the microwave signal offers a double-edged sword – on one hand it offers potential as a tool to understand subsurface characteristics of soils, but on the other, it records highly variable features, with observed effects controlled by a variety of surface and subsurface features. Many of the most successful investigations of subsurface pedologic or geomorphic features focus upon desert environments, characterized by dry soils, and an absence of vegetation cover. In such settings, penetration is likely to be maximized, and in, at least some regions, both spatial and vertical variation are likely to be minimized (Figure 24.4) (Elachi et al. 1984).

In a more recent study, Lasne et al. (2004) examined airborne polarimetric SAR data to investigate L-band penetration of open sand dunes, where buried paleosols apparently acted as subsurface moisture reservoirs. Examination of phase difference between HH and HV data permitted detection at greater depths (5.2 m) than using amplitude difference only.

HYPERSPECTRAL REMOTE SENSING

Conventional remote sensing, as outlined in previous chapters, is based upon use of several rather broadly defined spectral regions. Hyperspectral remote sensing (Schaepman, in this volume) is based upon examination of many narrowly defined spectral channels. Whereas widely used conventional sensor systems may offer perhaps 3 – 7 spectral channels, hyperspectral such as those described below can provide 200 or more channels, each as narrow as 10 nm in width. Hyperspectral sensors implement the concept of 'spectral resolution' to the extreme. Although hyperspectral remote sensing applies the same principles and methods discussed previously, it requires such specialized data sets, instruments, field data, and software that it forms its own field of inquiry.

Some of the principal analytical techniques have included:

- Spectral matching (e.g., Mobley et al. 2005)
- Spectral angle matching (e.g., Sohn et al. 1999)
- Spectral unmixing analysis (e.g, Keshava et al. 2000).

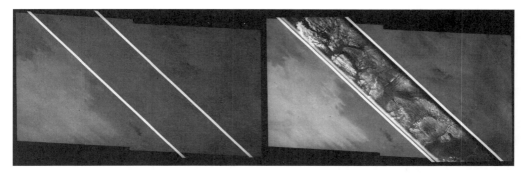

Figure 24.4 **JPL's imaging radar on NASA's second space shuttle flight in November 1981, as reported in the Dec. 3, 1982, edition of Science (McCauley et al., 1982). The experiment, called Shuttle Imaging Radar-A (SIR-A), penetrated cloud cover, varying degrees of vegetation, and dry desert sand to provide new geologic information in poorly surveyed regions. For example, U.S. Geological Survey and NASA-Jet Propulsion Laboratory scientists studying SIR-A data found that the radar had penetrated beneath the extremely dry Selima Sand Sheet dunes and drift sand of the Arba'in desert in the Sudan and Egypt to reveal ancient, buried stream beds, geologic structures, and probable Stone Age habitation sites. (Images courtesy of JPL, Pasadena, CA, and USGS Image Processing Facility, Flagstaff, AZ) (http://pubs.usgs.gov/gip/ deserts/remote/, http://www.jpl.nasa.gov/news/features.cfm?feature=422). (See the color plate section of this volume for a color version of this figure).**

Typical endmembers in unmixing studies in arid regions have included bare soil, water, partially vegetated surfaces, fully vegetated surfaces, and shadows. End members can be investigated in the field to confirm or revise identifications made by computer. Software for hyperspectral analysis often includes provisions for accessing spectral libraries (and for importing additional spectra as acquired in the field or laboratory), and the ability to search for matches with endmembers identified. Although it may not always be possible to uniquely identify matches in spectral libraries, such analyses can narrow the range of alternatives. In some instances mathematical models can assist in defining poorly established end members.

Selige et al. (2006) examine spatial variability of within-field topsoil texture and organic matter using airborne hyperspectral imagery. The percentage sand, clay, organic carbon, and total nitrogen content could be predicted quantitatively and simultaneously by a multivariate calibration approach using either partial least-square regression (PLSR) or multiple linear regression (MLR). The different topsoil parameters are determined simultaneously from the spectral signature contained in the single hyperspectral image, since the various variables were represented by varying combinations of wavebands across the spectra.

Ahn et al. (1999) examined hyperspectral data representing soil patterns using several alternative models. A linear spectral mixing model was effective not only for reducing dimensionality but also for removing vegetation effects for studying soil patterns from a single soil map layer derived from hyperspectral remote sensing data. Block kriging interpolation based on a semivariogram fitted with the isotropic exponential model represented soil patterns very well beyond the limitation of the size of pixel. Fuzzy-c-means clustering analysis showed clear membership patterns and segmented soil patterns effectively, although this is not a soil map in the conventional sense.

Zheng et al. (2005) applied hyperspectral imagery to the study of soil moisture in farmlands near Beijing, using aerial photography, hyperspectral imagery and field data. Correlations between soil moisture and red, green and blue radiance (RGB) values, as well as between soil moisture and hue, saturation and value (HSV) values, were calculated, with observed R^2 values of 0.887 and 0.706, respectively. Secondly, using the combination of RGB and HSV values, another estimation model was established, with an R^2 reaching to 0.900. Finally, they investigated linear regression models using the combinations of the RGB values, HSV values and spectral reflectance data at 1000 nm. Their model achieved an R^2 of 0.903. The result showed that the estimation of soil moisture content by using aerial images and hyperspectral data was rapid and accurate.

Hyperspectral remote sensing is useful for soils studies to the extent that useful soil information is conveyed spectrally. Given that spectral information is usually only a portion of the information necessary to understand soil patterns, its value for soils studies is confined to rather narrow situations in which spectral characteristics of the scene are especially well-defined. Therefore despite the capabilities of hyperspectral remote sensing, its value for soils studies is confined to rather specific situations in which the soil surface is exposed to the sensor.

SUMMARY

Application of remote sensing to the study of soils offers opportunities, but also faces numerous practical challenges. Optical, thermal, hyperspectral, and microwave sensors can characterize specific dimensions of the soil surface including texture, organic matter, moisture content, and surface roughness. Such observations focus upon specific properties, without seeking deeper knowledge about the soil landscape.

Direct observation offer great potential for applications that could have great significance, but also faces many practical obstacles. With only rare exceptions, direct remote sensing applications, by their nature, address characteristics of the soil surface, and provide little direct information concerning the soil profile. Thus, many of the present capabilities of remote sensing address soils in a broad sense of the term, recording characteristics of surface sediments (regolith) rather than soils in their pedologic context. Key issues include:

1 Few, if any, soil properties can be observed independently by remote sensing instruments. Remote sensing instruments collect composite data. Landscapes seldom provide pure signatures of soil surfaces. The soil response is mixed with that of vegetation, tillage, and effects of other agricultural practices. For example, in some instances, sensors convey composite information about several properties mixed together. For example, moisture, texture, and organic matter are closely interwoven in images representing soil color or image tone.

2 Soil properties are inherently multivariate. The variation of any specific property is closely related to that of other properties. This characteristic lends itself to use of inference, as a valuable aid to derivation of soil data from remotely sensed data, but complicates efforts to isolate individual properties.

3 Soils are highly variable over space, and some properties vary markedly over time. These characteristics complicate the effort to collect reference data, and to verify interpretations based upon remotely sensed data.

For the future, remote sensing of soils can be advanced by examining issues such as:

- Coordination of use of multiple sensor systems to the study of soils
- Refining, expanding, and systematization of applications of inference
- Co-orientation of applications of remote sensing with GIS to model soil distributions
- Study of scale of soil variability in relation to levels of detail recorded by sensors
- Development and refining of capabilities to isolate specific properties from composite responses
- Examination of soil patterns over time, to exploit temporal differences in spectral responses of soil surfaces.

ACKNOWLEDGMENTS

The author acknowledges the contributions of Dr. John Galbraith, Dr. James Baker, two anonymous referees, and Dr. Tim Warner.

REFERENCES

Al-Abbas, A. H., P. H. Swain, and M. F. Baumgardner, 1972. Relating organic matter and clay content to the multispectral radiance of soils. *Soil Science*, 114: 477–485.

Ahn, C-W., M. F. Baumgardner, and L. L. Biehl, 1999. Delineation of soil variability using geostatistics and fuzzy clustering analyses of hyperspectral data. *Soil Science Society of America Journal*, 63: 142–150.

Baumgardner, M. F., L. F. Silva, L. L. Biehl, and E. R. Stoner, 1985. Reflectance properties of soils. *Advances in Agronomy*, 38: 1–44.

Bowers, S. A. and A. J. Hanks, 1965. Reflection of radiant energy from soil. *Soil Science*, 100: 130–138.

Buol, S. W., R. J. Southard, R. C. Graham, and P. A. McDaniel, 2002. *Soil Genesis and Classification*. Iowa State University Press, Ames.

Brady, N. C. and R. R. Weil, 2002. *Elements of the Nature and Properties of Soils*. Pearson-Prentice Hall, Upper Saddle River, NJ.

Burdt, A., J. M. Galbraith, and W. L. Daniels, 2005. Season length indicators and land-use effects in southeast Virginia wet flats. *Soil Science Society of America Journal*, 69: 1551–1558.

Campbell, J., 2007. *Introduction to Remote Sensing*, 4th Edition. Guilford Publications, New York.

Cathcart, J. M., R. V. Worrall, and D. P. Cash, 2006. Hyperspectral signature modeling for terrain backgrounds. *Proceedings of SPIE*, Vol. 6239: Article 62390A.

Dehaan, R. L. and G. R. Taylor, 2004. A remote-sensing method of mapping soils and surficial lags from a deeply weathered arid region, near Cobar, NSW, Australia. *Geochemistry: Exploration, Environment, Analysis*, 4: 99–112.

Ding, G., J. M. Novak, D. Amarasiriwardena, P. Hunt, and B. Xing, 2002. Soil organic matter characteristics as affected by tillage management. *Soil Science Society of America Journal*, 66: 421–429.

Dobson, M. C. and F. T. Ulaby, 1998. Mapping soil moisture with imaging radar. In *Manual of Remote Sensing*. (3rd edn, Vol. 2). F. M. Henderson and A. J. Lewis eds. American Society for Photogrammetry and Remote Sensing. John Wiley & Sons. New York. Chapter 8: 407–433.

Drunpob, A., N. B. Chang, M. Beaman, C. Wyatt, and C. Slater, 2005. Seasonal soil moisture variation analysis using RADARSAT-1 satellite image in a semi-arid coastal watershed. *IEEE International Workshop on the Analysis of Multi-Temporal Remote Sensing Images*, pp. 186–190.

Elachi, C., L. E. Roth, and G. G. Schaber, 1984. Spaceborne radar subsurface imaging in hyperarid regions. *IEEE Transactions on Geoscience and Remote Sensing*, GE-22: 383–388.

Fox, G. A. and G. J. Sabbagh, 2002. Estimation of soil organic matter from red and near-infrared remotely sensed data using a soil line Euclidean distance technique. *Soil Science Society of America Journal*, 66: 1922–1929.

Gauthier, F. and A. Tabbagh, 1994. The use of airborne thermal remote sensing for soil mapping: A case study in the Limousin Region (France). *International Journal of Remote Sensing*, 15: 1981–1989.

Ho, D., 1987. A soil thermal model for remote sensing. *IEEE Transactions on Geoscience and Remote Sensing*, GE-25: 221–229.

Irons, J., R. A., Weismiller, and G. W. Petersen, 1989. Soil reflectance. In *Theory and Applications of Optical Remote Sensing* (Ghassem Assrar, ed.).John Wiley & Sons, New York. Chapter 3: 66–106.

Jackson, T. J., P. E. O'Neill, and C. T. Swift, 1997. Passive microwave observation of diurnal surface soil moisture. *IEEE Transactions on Geoscience and Remote Sensing*, 35: 1210–1222.

Kellndorfer, J. and K. McDonald, in this volume. Active and Passive Microwave Systems. Chapter 13.

Keshava, N., J. Kerekes, D. Manolakis, and G. Shaw, 2000. An algorithm taxonomy for hyperspectral unmixing. *SPIE 4049*, pp. 42–63.

Lee, K. S., G. B. Lee, and E. J. Tyler. 1988. Determination of soil characteristics from thematic mapper data of a cropped organic–inorganic soil landscape. *Soil Science Society of America Journal*, 52: 1100–1104.

Lagacherie, P., A. McBratney, and M. Voltz, 2006. *Digital Soil Mapping: an Introductory Perspective*. Elsevier, Amsterdam.

Lasne, Y., P. Paillou, T. August-Bernex, G. Ruffié, and G. Grandjean, 2004. A phase signature for detecting wet subsurface structures using polarimetric L-Band SAR. *IEEE Transactions on Geoscience and Remote Sensing*, 42: 1683–1694.

Lobell, D. B. and G. P. Asner, 2002, Moisture effect on soil reflectance. *Soil Science Society of America Journal*, 66: 722–727.

McBratney, A. B., M. L. Mendonça Santos, and B. Minasny, 2003. On digital soil mapping. *Geoderma*, 117: 3–52.

McCauley, J. F., G. G. Schaber, C. S. Breed, M. J. Grolier, C. V. Haynes, B. Issawi, C. Elachi, and R. Blom, 1982. Subsurface valleys and geoarchaeology of the Eastern Sahara revealed by shuttle radar. *Science*, 218: 1004–1020.

Meyers, V. I., 1975. Crops and soils. in R. G. Reeves (ed), *Manual of Remote Sensing*. American Society of Photogrammetry, Falls Church, VA. Chapter 22: 1715–1813.

Meyers, V. I., 1983. Crops and soils. in R. N. Colwell (ed), *Manual of Remote Sensing* (second edn). American Society of Photogrammetry, Falls Church, Chapter 33: 2111–2228.

Mobley, C. D., L. K. Sundman, C. O. Davis, O. Curtiss, J. H. Bowles, T. V. Downes, R. A. Leathers, M. J. Montes, W. P. Bissett, D. D. R. Kohler, R. P. Reid, E. M. Louchard, and A. Gleason, 2005. Interpretation of hyperspectral remote-sensing imagery by spectrum matching and look-up tables. *Applied Optics*, 44: 3576–3592.

National Cooperative Soil Survey. Soil Survey Manual. (http://soils.usda.gov/technical/manual/print_version/complete.html)

Perfect, E. and J. Caron, 2002. Spectral analysis of tillage-induced differences in soil spatial variability. *Soil Science Society of America Journal*, 66: 1587–1595.

Promes, P. M., T. J. Jackson, and P. E. O'Niell, 1988. Significance of agricultural row structure on the microwave emissivity of soils. *IEEE Transactions on Geoscience and Remote Sensing*, 26: 580–589.

Quattrochi, D. and J. C. Luvall, in this volume. Thermal Remote Sensing in Earth Science Research. Chapter 5.

Ray, S., J. P. Singh, S. Dutta, and S. Panigrahy, 2002. Analysis of within-field variability of crop and soil using field data and spectral information as a pre-cursor to precision crop management. International Archives of Photogrammetry. *Remote Sensing and Spatial Information Systems*, 34(7): 302–307.

Sahebi, Mahmod R., J. Angles, and F. Bonn, 2002. A comparison of multi-polarization and multi-angular approaches for estimating bare soil surface roughness from spaceborne radar data. *Canadian Journal of Remote Sensing*, 28: 641–652.

Schaepman, M. E., in this volume. Imaging Spectrometers. Chapter 12.

Schaetzl, R. and S. Anderson, 2005. *Soils: Genesis and Geomorphology*. Cambridge University Press.

Scull, P., J. Franklin, O. A. Chadwick, and D. McArthur, 2003. Predictive soil mapping-a review. *Progress in Physical Geography*, 27: 171–197.

Selige, T., J. Böhner, and U. Schmidhalter, 2006. High resolution topsoil mapping using hyperspectral image and field data in multivariate regression modeling procedures. *Geoderma*, 36: 234–244.

Simbahan, G. C., A. Dobermann, P. Goovaerts, J. Ping, and M. L. Haddix, 2006. Fine-resolution mapping of soil organic carbon based on multivariate secondary data. *Geoderma*, 132: 471–489.

Skidmore, E. L., J. D. Dickerson, and H. Shimmespfenning, 1975. Evaluating surface-soil water content by measuring reflectance. *Soil Science Society of America Journal*, 39: 238–242.

Sohn, Y., E. Moran, and F. Gurri, 1999. Deforestation in North-Central Yucatan (1985–1995): mapping secondary succession of forest and agricultural land use in Sotuta using the cosine of the angle concept. *Photogrammetric Engineering and Remote Sensing*, 65: 947–958.

Soil Survey Division Staff, 1993. Soil survey manual. *Agr. Handbook* 18. USDA-Natural Resources Conservation Service. Available on-line at: http://soils.usda.gov/technical/manual/ Last verified: May 2 2007.

Soil Survey Division Staff, 1999. Soil taxonomy: A basic system of soil classification for making and interpreting soil surveys (2nd edn) *Agr. Handbook 436*. USDA-Natural Resources Conservation Service. U.S. Govt. Print. Office, Washington, D.C. USA. 869 pp. Available on-line at: http://soils.usda.gov/technical/classification/taxonomy/ Last verified: May 2 2007.

Soil Survey Division Staff, 2007. *National Soil Survey Handbook*. Title 430-VI. USDA-Natural Resources Conservation Service. Available on-line at: http://soils.usda.gov/technical/handbook/ Last verified: May 2 2007.

Stoner, E. R. and M. F. Baumgardner, 1981. Characteristic variations in reflectance of surface soils. *Soil Science Society of America Journal*, 45: 1161–1165.

Stow, D., L. L. Coulter, and C. A. Benkelman, in this volume. Airborne Digital Multispectral Imaging. Chapter 11.

Sullivan, D. G.; J. N. Shaw, and D. Rickman, 2005. IKONOS imagery to estimate surface soil property variability in two Alabama physiographies. *Soil Science Society of America Journal*, 69: 1789–1798.

Toutin, T., in this volume. Fine Spatial Resolution Optical Sensors. Chapter 8.

Ulaby, F. T., M. K. Moore, and A. K. Fung, 1986. *Microwave Remote Sensing: Active and Passive*, Vol. 3. Artech House, Norwood, MA.

van Leeeuwen, W. J. D., in this volume. Visible, Near-IR and Shortwave IR Spectral Characteristics of Terrestrial Surfaces. Chapter 3.

Verstraeten, W. W., F. Veroustraete, C. J. van der Sande, I. Grootaers, and J. Feyen, 2006. Soil moisture retrieval using thermal inertia, determined with visible and thermal spaceborne data, validated for European forests. *Remote Sensing of Environment*, 101: 299–314.

Way, J. and E. A. Smith, 1991. The evolution of synthetic aperture systems and their progression to the EOS SAR. *IEEE Transactions on Geoscience and Remote Sensing*, 29: 962–985.

Zhang, L., J. Shi, Z. Zhang, and K. Zhao, 2003. The Estimation of dielectric constant of frozen soil-water mixture at microwave bands. *IGARSS '03 Proceedings, IEEE International*, 4: 2903–2905.

Zheng, L., M. Li, J. Sun, N. Tang, and X. Zhang, 2005. Estimation of soil moisture with aerial images and hyperspectral data. *International Geoscience and Remote Sensing Symposium (IGARSS 05)* (Article 1525925): 4516–4519.

B. Plant Sciences

Remote Sensing for Studies of Vegetation Condition: Theory and Application

Michael A. Wulder, Joanne C. White, Nicholas C. Coops, and Stephanie Ortlepp

Keywords: vegetation, spectral response, data selection, resolution, change detection.

INTRODUCTION

Remotely sensed data is a proven source of information for detailed characterization of vegetation type (e.g., Gould 2000, Luther et al. 2006), structure (e.g., Gamon et al. 2004, Healey et al. 2006), and condition (e.g., Rossini et al. 2006, Wulder et al. 2006a). The spatial, spectral, and temporal resolution at which the data are acquired is critical in how these vegetation properties are observed, and may ultimately determine the success of a particular remote sensing application. Therefore, when undertaking applications with remotely sensed data, it is imperative to have a clear understanding of the information need that is to be satisfied, thereby allowing for the selection of the most appropriate imagery and analysis methods.

Previous chapters have detailed the capture and characteristics of optical remotely sensed data (Chapters 2, 3, and 7) over a range of spatial and spectral resolutions from both airborne and satellite platforms (Chapters 8–10, and 14). Approaches to image processing (Chapter 15) and applications (Chapters 16–18, and 20) have also been discussed. In this chapter, we discuss the use of remotely sensed data for assessing vegetation conditions at landscape and tree levels, and the considerations that need to be made depending on a given information need. The goal of this chapter is to address the key issues that should be considered when using remotely sensed data to characterize vegetation, and through this, understand how operational applications may be undertaken.

IMAGE RESOLUTIONS AND DATA SELECTION

Remotely sensed data can be characterized by the image spatial resolution (pixel size), spectral resolution (wavelength ranges utilized), temporal resolution (when and how often are images collected), and radiometric resolution (the degree of differentiation within the dynamic range of the sensor). Vegetation is a complex target with a large amount of inherent spectral and spatial variability, and vegetation is typically characterized by strong absorption in the visible wavelengths, particularly the red wavelengths of the electromagnetic spectrum and high reflectance in the near-infrared (NIR) wavelengths. The amount of absorption or reflectance is controlled by vegetation type, amount, density, structure, and vigor. At the leaf scale, pigment concentrations, water content, and structure all contribute to variations in absorption, transmittance, and reflectance. In this section, we discuss the characteristics of remotely sensed data and consider how these various characteristics influence the remote sensing of vegetation.

Of the four types of resolution typically used to characterize remotely sensed data, spatial resolution arguably has the greatest impact on the information content of remotely sensed data, particularly for vegetation targets. Strahler et al. (1986) posit a scene model, based on spatial resolution, for understanding the information content of remotely sensed data. In this model, there are either many objects per pixel (an L-resolution environment) or conversely, many pixels per object (an H-resolution environment) (Figure 25.1). The target objects of interest are therefore important for assessing the utility of a given spatial resolution for a selected application. Table 25.1 provides some examples of commonly used remotely sensed data sources and the type of vegetation information one may expect to extract from these data sources.

Recent advances in the development of satellites with fine to very fine spatial resolution (Table 25.1), combined with the widespread availability of digital camera and scanning technologies, and increasingly sophisticated computer processing

techniques, have contributed to an increase in the use of high spatial resolution imagery to estimate traditional and non-traditional vegetation attributes (Wulder et al. 2004a). However, with increased spatial resolution comes added complexity with respect to defining homogenous vegetation classes. While the increased textural information available in fine or very fine spatial resolution image data allows for improved interpretation based on the shape and texture of ground features, the current techniques used to process and analyze satellite image data, such as the use of standard vegetation indices or per-pixel based classifiers (e.g., maximum likelihood) may not be effective when applied to high spatial resolution image data (Goetz et al. 2003). In this H-resolution environment (Strahler et al. 1986), with many pixels per object (e.g., tree), there will be a large amount of spectral variability associated with individual trees (e.g., pixels representing sunlit crown, shaded crown, and the influence of factors such as branches, cones, and tree morphology). This variability confounds the development of unique spectral signatures for tree or vegetation classification (Culvenor 2003).

Temporal resolution provides an indication of the time it takes for a sensor to return to the same location on the Earth's surface. The revisit time is a function of the satellite orbit, image footprint, and the capacity of the sensor to image off-nadir (e.g., not directly beneath the sensor, but

Figure 25.1   Examples of image spatial resolution over a forested scene with crowns of varying condition. Superimposed pixel sizes range from 30 m (Landsat) (L-resolution model with many objects per pixel) to 4 m (IKONOS multispectral) to 1 m (IKONOS panchromatic) (H-resolution model with many pixels per object). (See the color plate section of this volume for a color version of this figure).

Table 25.1   Example of instrument-related spatial resolution ranges and levels of plant recognition to be expected across a range of image scales (after Wulder 1998). Note, as a heuristic, we categorize images according to their pixel size; very coarse (> 1000m), coarse (100–1000m), moderate (10–100m), fine (1–10m), and very fine (< 1m)

Type or photo scale	Approximate range of spatial resolution (m)	General level of forest vegetation discrimination
Very Coarse Resolution Satellite Images Coarse Spatial Resolution Satellite Images	1000 (AVHRR) 250–1000 (MODIS)	Broad land cover patterns (regional to global mapping)
Moderate Spatial Resolution Satellite Images	30 Coarse Resolution Satellite Images (Landsat) 20 (SPOT multispectral) 10 (SPOT panchromatic)	Separation of extensive masses of evergreen versus deciduous forests (stand level characteristics)
Fine Spatial Resolution Satellite Image	1 (IKONOS panchromatic); 4 (IKONOS multispectral)	Recognition of large individual trees and of broad vegetative types
Very Fine Spatial Resolution Satellite Images (e.g., QuickBird)	0.67 (QuickBird Panchromatic) 2.4 (QuickBird multispectral)	Identification of Individual Trees
Airborne Multispectral Scanners	> 0.3	Initial identification of large individual trees and stand level characteristics
Airborne Video	> 0.04	Identification of individual trees and large shrubs
Digital Frame Camera	> 0.04	Identification of individual trees and large shrubs
1:25,000 to 1:100,000 Photo	0.31–1.24[1]	Recognition of large individual trees and of broad vegetative types
1:10,000 to 1:25,000 Photo	0.12–0.31	Direct identification of major cover types and species occurring in pure stands
1:2,500 to 1:10,000 Photo	0.026–0.12	Identification of individual trees and large shrubs
1:500 to 1:2,500 Photo	0.001–0.026	Identification of individual range plants and grassland types

[1]based upon a typical aerial film and camera configuration utilizing a 150 mm lens

at an angle). The timing of image acquisition should be linked to the target of interest; some disturbance agents may have specific bio-windows (e.g., fire, defoliating or phloem feeding insects) during which imagery must be collected in order to capture the required information (Wulder et al. 2005), while other disturbances may be less time specific (e.g., harvest). For ongoing programs designed to monitor forest change before and after a disturbance event, the acquisition of images should occur in the same season, over a series of years (known as anniversary dates). Anniversary dates are critical to ensure the spectral responses of the vegetation remain relatively consistent over successive years (Lunetta et al. 2004). The reduction in image radiometric quality for off-season imagery resulting from low sun-angles and reduced illumination conditions compromises the ability to capture changes clearly. Selection of scenes captured at the same time each year may reduce issues related to sun angle, shadow, and overall scene brightness.

The temporal characteristics of an imaging system are also important. For some applications, the capacity to incorporate multi-temporal images can be advantageous. For example, analysis of vegetation at both leaf-on and leaf-off periods can

provide important information on the land cover, especially for seasonally variable vegetation, such as deciduous species (Dymond et al. 2002). Aerial-acquisitions are in general more flexible regarding timing than satellite-acquisitions, with the ability to collect images on demand, for example, coincident with insect outbreaks or fires (Stone et al. 2001). For satellite images, there tends to be a trade-off between image spatial resolution and the typical repeat period for image acquisition. Generally, high spatial resolution imagery, including that from satellites such as IKONOS and QuickBird, is acquired from sensors that are able to view off-nadir, and therefore have the potential to revisit a location every 1 to 3.5 days depending on the latitude of the target location. Note however, that shorter revisit times come at the cost of off-nadir viewing. True-nadir image revisit time for QuickBird is 144 days, compared to 16 days for moderate resolution satellites such as Landsat.

Spectral resolution provides an indication of the number and the width of the spectral wavelengths (bands) captured by a particular sensor. Sensors with more bands and narrower spectral widths are described as having a higher spectral resolution. Currently, most operational remote sensing systems are multispectral and have a small number of broad spectral channels: Landsat-7 Enhanced Thematic Mapper Plus (ETM+) data has seven spectral bands in the reflective portion of the electromagnetic spectrum and one band in the thermal-infrared region. Hyperspectral data (e.g., instruments with more than 200 narrow spectral bands (Lefsky and Cohen 2003) are becoming more widely available (Vane and Goetz 1993) both on spaceborne (such as the HYPERION sensor on the EO-1 platform) and airborne platforms such as HyMap (Cocks et al. 1998), CASI (Compact Airborne Spectrographic Imager) (Anger et al. 1994), and the NASA Advanced Airborne Visible/Infrared Imaging Spectrometer (AVIRIS) (Vane et al. 1993). Since phenomena of interest (e.g., foliage discolouration) may manifest in a specific portion of the electromagnetic spectrum, the number, width, and location of a particular sensor's spectral bands along the electromagnetic spectrum will determine whether the data from a given sensor is suitable for characterizing the phenomena.

Radiometric resolution may be interpreted as the number of intensity levels that a sensor can use to record a given signal (Lillesand and Kiefer 2000) and provides an indication of the information content of an image. Most remotely sensed data currently used for vegetation applications (e.g., Landsat, SPOT-5) have a minimum 8-bit radiometric resolution; if a sensor uses eight bits to record data, there are 2^8 or 256 digital values available, ranging from 0 to 255. The finer the radiometric resolution of a sensor the more sensitive it is to detecting small differences in reflected or emitted energy (QuickBird has 11-bit data).

INFORMATION NEEDS AND APPLICATIONS CONSIDERATIONS

The growing number and types of remotely sensed data sources available simplify matching a data source to a particular information need. Since image characteristics determine the nature of the information that may be extracted from an image, this section demonstrates the importance of clearly defining the information need, as a precursor to the selection of appropriate data and analysis approaches. Methodological options may then be considered once the information need is clearly identified. Several logistical issues may emerge when acquiring remotely sensed data to address a specific information need. These issues include the scale at which the target must be measured (e.g., landscape-level or tree-level information); the attributes of interest (change, condition, spatial extent); cost; timeliness; and, repeatability.

The timing of image acquisition can have an impact on the quality of data extracted.

Image acquisition characteristics often require compromise when images are being selected, since non-optimal years or seasons will have an impact on the nature and quality of information captured. Images collected during the winter months (e.g., October–March) can have a lower dynamic range (particularly true at more northerly latitudes), which can reduce spectral overlap between different cover types. Areas of shadow will also be larger, especially in areas of high topography, further reducing the dynamic range. Off-year imagery (that is, imagery from a different year than planned, even when this results in a longer time interval) is typically preferred over off-season imagery, as off-season imagery generally requires more processing, including the need to minimize phenological artifacts, and this will reduce overall mapping quality (Wulder et al. 2004b).

Depending on the spatial extent of the area of interest and the required resolution, more than one scene may be needed to meet the information needs. For example, if the information need requires characterization of the temporal change in insect damage over a large spatial extent, at least two dates of imagery will be required to measure the change, and multiple images will likely be required to fully cover the area of interest. By necessity, the infilling of image areas obscured by clouds and shadows could further increase the number of scenes required to facilitate complete spatial coverage of the area (Homer et al. 1997). Moreover, if the information need requires a high level of detail, the number of images further increases, given that remotely sensed data with a higher spatial resolution typically have a smaller image extent. Multiple images present several image processing challenges such as edge effects, geometric co-registration, image radiometric normalization, phenological and annual differences, and data handling issues.

The costs associated with the acquisition and processing of remotely sensed data are not insignificant. A landscape-level project generally requires the use of multiple scenes, presenting numerous image processing challenges. Tree-level characterization necessitates the use of higher spatial resolution imagery, which is generally much more expensive to acquire than data with a lower spatial resolution. While data acquisition will undoubtedly represent the bulk of the costs, additional costs may be associated with ancillary data processing (e.g., for calibration and validation), data management, and image analysis. Expertise in processing different types and potentially large volumes of remotely sensed and other geographic data, within a geographic information system (GIS) is also often required.

Ancillary data is required for calibration and validation of the analysis methods for most projects. Ancillary data sources are important for generating masks that will restrict the image analysis and aid in vetting the calibration and validation data points. Masks are often used to constrain the variability in spectral response resulting from cover types that are not of interest, such as water or cloud. The use of masks reduces the number of false positives and enables the processing of pixels where there is real change in the object of interest rather than a transition, for example, from cloud or shadow. The set of points remaining enables the interpretation of the results to be made under an assumption that the calibration and validation is not impacted by extraneous conditions. Rogan and Miller (2006) provide a summary of considerations and opportunities for integrating spatial and remotely sensed data to meet applications needs.

APPLICATION OPPORTUNITIES: SPECTRAL CHARACTERISTICS

Different portions of the electromagnetic spectrum may be exploited to satisfy different information requirements. In this section, we present some brief examples of how the visible, near-infrared, and shortwave-infrared portions of the spectrum have been used in a variety of applications, and how these various portions of the electromagnetic spectrum may be combined algorithmically in a vegetation index to further enhance feature discrimination. A more detailed of vegetation response in each portion of the electromagnetic spectrum may be found in Chapter 3.

Visible wavelengths

The visible portion of the electromagnetic spectrum spans 400–700 nm with the spectral reflectance of vegetation in this part of the spectrum heavily influenced by leaf pigmentation, specifically chlorophylls a and b. These pigments reflect highly in the green portion of the spectrum (500 nm) and absorbs the blue (450 nm) and red (670 nm) wavelengths (Hoffer 1978). Other pigments that influence leaf absorption and thus reflectance include carotene, xanthophyll, and anthocyanins (Blackburn 1998). Information on pigments, especially chlorophyll, has been used in applications ranging from agriculture to natural vegetation studies. Pigments are integral in the physiological function of a leaf, and can be used as indicators of its physiological state. For example, the amount of chlorophyll can be used as an indicator of plant productivity as it is linked to the amount of photosynthetically active radiation absorbed by the leaf, and thus to the photosynthetic rate (Gamon and Qiu 1999).

A study by Zarco-Tejada et al. (2005) measured the chlorophyll content of the European grapevine at leaf and canopy levels to determine its physiological condition. For the leaf level, field based measurements of the pigment concentration were made, and a spectrometer was used to measure the reflectance properties of individual leaves. Concurrent canopy level data was acquired by three airborne hyperspectral imaging systems: CASI, ROSIS (Reflective Optics System Imaging Spectrometer), and DAID-7915 (Digital Airborne Imaging Spectrometer). Upon linking the leaf and canopy level data it was found that the best indictors of chlorophyll content used ratios calculated within the 700–750 nm range of the hyperspectral imagery.

Near-infrared wavelengths

The near infrared (NIR) portion of the electromagnetic spectrum ranges from 700 to 1200 nm. Vegetation is characterized by high reflectance in the NIR, controlled primarily by leaf structure; reflectance in these wavelengths occurs at cell walls and at the interfaces between air and water within the leaf (Slaton and Smith 2001). In a typical vegetation spectral response curve, the 'red edge' is the steep portion of the curve in the transition from red to NIR wavelengths. The slope of this curve can indicate plant stress and chlorophyll concentration (Carter and Knapp 2001). The red edge is followed by the NIR plateau, and this portion of the spectrum has been associated with changes at the cellular level, including hydration, health, and arrangement, as well as biomass (Rock et al. 1986). The NIR part of the spectrum has been shown to have a high correlation with hardwood forest biomass (Zheng et al. 2004), and as an indicator of leaf area index. Roberts et al. (1998) correlated leaf age with increasing NIR absorbance in tropical vegetation. Since the NIR can be used to assess vegetation health, it is also an important tool in monitoring tree defoliation due to pests or environmental conditions. For example, the NIR reflectance has been used to detect defoliation levels due to Jack Pine budworm in Wisconsin, USA (Radeloff et al. 1999) and a fungal pathogen in Australia (Coops et al. 2006).

Shortwave infrared wavelengths

The shortwave portion of the spectrum ranges from 1300 to 2400 nm and is strongly influenced by the absorption of water. The moisture contained within a leaf absorbs shortwave infrared radiation, making this range of the spectrum useful in estimating plant water content (Ustin et al. 2004). Vegetation water content is especially important when trying to assess forest fire risk (Maki et al. 2004), and for determining water deficiency in agricultural crops. Tian et al. (2001) used the SWIR reflectance (between 900–1850 and 1700–2500 nm) of wheat leaves to determine moisture stress.

Indices are commonly used with a wide range of remotely sensed data types to integrate multiple wavelength ranges that inform upon vegetation characteristics of interest (Asner et al. 2003). Studies have shown that spectral indices or ratios using the visible wavelengths are sensitive to changes in leaf pigmentation (Blackburn 1998). For example, Chapelle et al. (1992) found that by using a ratio of soybean reflectance spectra and reference spectra, corresponding to the absorption bands of individual pigments, they were able to estimate the concentration of chlorophyll and carotene in the soybean plants. Recent research has resulted in the development of a function for assessing the sensitivity of spectral vegetation indices to biophysical parameters (Lei and Peters 2007).

APPLICATIONS EXAMPLE: SPATIAL CHARACTERISTICS

Based on differing information needs for differing management objectives, digital satellite remote sensing offers a complementary technology for detection and mapping of a range of vegetation related phenomena. Remotely sensed data can facilitate mapping of individual trees or small groups of trees, as well as providing the capacity to cover large spatial areas – ensuring a census, rather than a sample of areas of interest (Wulder et al. 2006b). In addition, remotely sensed data can be easily integrated with other spatial data (such as roads, elevation and climate data) (Dial et al. 2003, Tao et al. 2004) and forest inventory data (Wulder et al. 2005). Furthermore, through the use of more automated processing and interpretation functions, there may be a reduction in interpreter subjective bias (White et al. 2005), which may increase the consistency and reliability of mapping between different areas or dates (Wulder et al. 2006a).

In this section, we present two examples of applications that have used remotely sensed data to map mountain pine beetle (*Dendroctonus ponderosae*) damage to lodgepole pine (*Pinus contorta* Dougl. ex Loud. var. *latifolia* Engelm.) forests in British Columbia, over two different spatial extents. The first example is mapping damage caused by epidemic levels of infestation over very large areas at the landscape level using remotely sensed data with a spatial resolution of 30 m, while the second example is mapping low levels of infestation, at the forest stand level, where individual trees or groups of trees have been infested and killed by the beetle.

Landscape level application example

Over large areas, information on the location, extent, and severity of mountain pine beetle damage is required to determine the resources needed to address the infestation and to allocate those resources effectively. This information is also used for timber supply review, forest inventory update, biodiversity conservation, land use planning, and as baseline information to parameterize and validate the assumptions associated with predictive models. Landscape-level information is also used to direct the location and intensity of more detailed surveys, designed to satisfy operational information needs.

When mountain pine beetles attack and kill a healthy pine tree, the tree's foliage will initially remain green, eventually fading to red (typically within six to eight months after the initial attack). The red coloration of the foliage is characteristic of beetle damage and is termed as the red attack stage; this dramatic change in foliage color enables detection and mapping of beetle damage (Figure 25.1 provides examples of red attack trees). Provided that appropriate imagery is selected to coincide with the manifestation of the red attack stage, the damage can be mapped over large areas in an accurate and timely fashion using Landsat Thematic Mapper (TM) or ETM+ imagery and change detection methods (Skakun et al. 2003, Wulder et al. 2006b).

The creation of a large-area product detailing the location and spatial extent of red attack damage involves the consideration of several logistical issues related to the mapping of a large area, many of which are not often faced in small, research-driven projects. Generating a consistent product over large areas involves the use of multiple scenes collected on different dates and potentially in non-optimal seasons. Other considerations are data availability and quality of available imagery. Imagery that is suitable for the detection and mapping of mountain pine beetle red attack damage must be acquired during the appropriate bio-window for mountain pine beetle (Wulder et al. 2006a).

The Enhanced Wetness Difference Index (EDWI) has been effectively used to detect a range of forest disturbance types (Franklin et al. 2001, 2002, 2003, 2005, Skakun et al. 2003, Wulder et al. 2006b). The EDWI is based on the Tasseled Cap Transformation (TCT) coefficients developed by Huang et al. (2002), which compress Landsat spectral data into a reduced number of bands associated with physical scene characteristics (Crist and Cicone 1984). Though this transformation was originally constructed for agricultural applications, it has been used to reveal some key forest attributes (Cohen et al. 1995). The EDWI is calculated based on a combination of two dates of Landsat images,

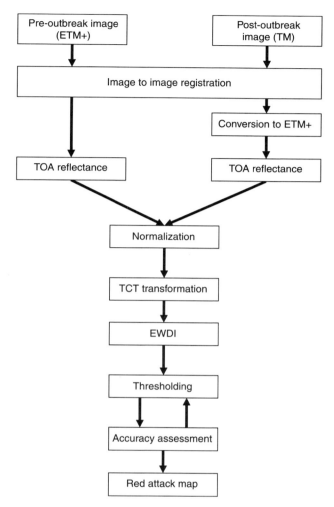

Figure 25.2 Summary of steps included in the processing flow to generate a map of mountain pine beetle red attack damage from two dates of Landsat TM/ETM+ imagery.

making geometric matching of multiple scenes a critical pre-processing step, even with orthorectified imagery, to ensure that the images are properly co-registered.

The mapping approach developed to capture red attack damage involves a sequence of steps including pre-planning, image scene selection, image pre-processing, and analysis (Figure 25.2). The EWDI approach performs best over areas with more homogeneous and extensive attack. Scenes with excessive cloud and haze should be avoided, and any areas of cloud, cloud shadows, haze, topographic shadows, or snow cover should be masked out. Masks must also be generated from forest inventory and harvesting data to identify areas of suitable hosts.

Generally speaking, when applying the EWDI large positive values indicate wetness loss while small EWDI values show no change of wetness, and large negative EWDI values represent wetness gain. Therefore, the areas with large positive values in the EWDI image are likely to be the mountain pine beetle red attacked areas (Skakun et al. 2003). Mountain pine beetle red attack, however, is not the only disturbance that results in a loss of wetness, as other forest management activities will also manifest as decreases in moisture (e.g., forest harvesting).

The EWDI values of the attacked and non-attack pixels can be approximated by a Gaussian distribution, and can be separated by thresholding (Skakun et al. 2003). To determine the EWDI thresholds, calibration and validation data are required for both red attack and non-attacked pine stands. Once the threshold has been set, the accuracy of the output can be verified using reserved validation data. An accuracy assessment provides information on the success of the detection methods used and identifies possible sources of error, and is also valuable for comparing and evaluating different mapping techniques and in the development of new methods.

Tree-level application example

Successful mitigation of mountain pine beetle attack relies on accurate detection of infested trees, and information on the number and location of attacked trees. This detailed information is critical for a range of activities, including sanitation logging, the implementation of silvicultural regimes designed to reduce the susceptibility of host trees, as well as treatments to directly control populations. In all of these cases, information on the location and health status of individual tree crowns is critically important.

The advent of high spatial resolution satellite data, since the launch of the IKONOS satellite in 1999, has resulted in an increased capacity to detect individual trees from space. Airborne digital imagery, such as that obtained by digital cameras sensitive to both the visible and near infrared regions of the spectrum, also provides the capacity to deliver this detailed individual tree information. This spatially driven data is useful for identifying small disturbances focused over a limited spatial extent and can aid as a surrogate for field based measurements (Asner and Warner 2003) or validation efforts (Morisette et al. 2003).

White et al. (2005) use IKONOS 4 m multispectral data to detect mountain pine beetle red attack damage using image classification. In their study, an unsupervised clustering technique was applied to detect red attack damage in forest stands with low and moderate levels of attack, and then compared these estimates to red attack damage estimates generated from air photo interpretation. Results indicate that within a one-pixel buffer (4 m) of identified damage pixels, the accuracy of red attack detection was 70.1% for areas of low infestation (stands with less than 5% of trees damaged) and 92.5% for areas of moderate infestation (stands with between 5 and 20% of trees damaged). Analysis of red attack trees that were missed in the classification of the IKONOS imagery indicated that detection of red attack was most effective for larger tree crowns (diameter >1.5 m) that were <11 m from other red attack trees.

In mountain pine beetle infested forest stands, Kneppeck and Ahern (1989) compared manually derived counts of red attack trees from airborne scanner imagery (with a 1.4 and 3.4 m spatial resolution) to counts estimated by manual interpretation of 1:10,000 air photos. Counts of red attack trees from the 1.4 m resolution imagery were higher (136%) than those derived from the air photo interpretation, while counts from 3.4 m resolution imagery were lower (71%). The results from this study indicated that detailed surveys can benefit from a multi-stage sampling approach where a small sample of ground counts is used to adjust estimates generated from other data sources.

Coops et al. (2006) used helicopter GPS survey measurements of beetle infested pine trees in north-central British Columbia to indicate areas of attack and non-attack stands on QuickBird 2.4 m multispectral data (blue, green, red, near-infrared). The authors tested the ANOVA separability of four classes: sunlit non-attacked crowns, dense red attack crowns, fader crowns, and shadowed crowns, using four QuickBird spectral bands and the NDVI and red/green spectral ratios. Spectral thresholds were used to generate a binary map of red attack and non-attack. The results show that the ratio of the red to green QuickBird spectral bands was the most significant band combination for detecting red attack beetle damage and the information derived from the QuickBird imagery showed good correspondence with both forest health survey data and trends derived from broader spatial resolution Landsat imagery.

CONCLUSION

In this chapter, we have described the capacity of remotely sensed data to characterize and monitor vegetation condition and have demonstrated how information needs influence the choice of remotely sensed data and analysis methods. Regardless of whether the scope of the application is over large areas or at the level of individual trees, there are several logistical issues related to data acquisition and processing that must be addressed. The focus on meeting information needs using remotely sensed data poises the remote sensing community to support sustainable forest management and to play a role in informing policy makers.

The use of remotely sensed data for vegetation characterization has matured rapidly from scientific investigations to operational usage. It is this very success that is producing some of the grand challenges that remain and continue to emerge. The ability to make subtle characterizations of vegetation condition and structure at fine scales has created a demand for this detailed information over large areas. As we described in this chapter, a contradiction in what is desired and what is available may develop, with the typical economies associated with remote sensing being lost if highly detailed characterizations are desired over large areas. A means of addressing this desire for large-area characterizations of finely detailed information is through sampling and modeling; whereby, fine resolution imagery (with high spatial and spectral resolution) is strategically sampled to enable large-area extrapolations with moderate spatial resolution imagery (or other spatial data sources). These and other forms of data integration will aid in meeting increasingly demanding needs for characterizing vegetation over a wide range of scales.

REFERENCES

Anger, C. D., S. Mah, and S. K. Babey, 1994. Technological enhancements to the compact airborne spectrographic imager (*casi*). In *Proceedings of the Second International Airborne Remote Sensing Conference and Exhibition*, September 12–15 Strasbourg, France. Ann Arbor, MI, USA: ERIM International, Inc. pp. 205–214

Asner, G. P. and A. S. Warner, 2003. Canopy shadow in the LBA IKONOS satellite archive: Implications for multispectral studies of tropical forests and savannas. *Remote Sensing of Environment*, 87: 521–533.

Asner, G. P., J. A. Hicke, and D. B. Lobell, 2003. Per-pixel analysis of forest structure: vegetation indices, spectral mixture analysis, and canopy reflectance modelling. In: M. A. Wulder, and S. E. Franklin, (eds), *Remote Sensing of Forest Environments: Concepts and Case Studies*. Kluwer Academic, Boston: pp. 209–254.

Blackburn, G. A., 1998. Spectral indices for estimating photosynthetic pigment concentrations: a test using senescent tree leaves. *International Journal of Remote Sensing*, 19: 657–675.

Carter, G. A. and A. K. Knapp, 2001. Leaf optical properties in higher plants: linking spectral characteristics to stress and chlorophyll concentration. *American Journal of Botany*, 88: 677–684.

Chappelle, E. W., M. S. Kim, and J. E. McMurtrey, 1992. Ratio analysis of reflectance spectra (RARS): an algorithm for the remote estimation of the concentrations of chlorophyll A, chlorophyll B and the carotenoids in soybean leaves. *Remote Sensing of Environment*, 39: 239–247.

Cocks, T., R. Jenssen, and A. Stewart, 1998. The HyMap airborne hyperspectral sensor: The system, calibration and performance. In M. Schaepman et al. (eds), *Proceedings of the 1st EARSeL Workshop on Imaging Spectroscopy*. Remote Sensing Laboratories, Zurich. pp. 37–42.

Cohen, W. B., T. A. Spies, and M. Fiorella, 1995. Estimating the age and structure of forests in a multi-ownership landscape of western Oregon, U.S.A. *International Journal of Remote Sensing*, 16: 721–746.

Coops, N. C., N. Goodwin, C. Stone, and N. Sims, 2006. Assessment of forest plantation canopy condition from high spatial resolution digital imagery. *Canadian Journal of Remote Sensing*, 32: 19–32.

Crist, E. P., and R. C. Cicone, 1984. A physically-based transformation of Thematic Mapper data – the Tasselled Cap, *IEEE Transactions on Geoscience and Remote Sensing*, 22: 256–263.

Culvenor, D. 2003. Extracting individual tree information: A survey of techniques for high spatial resolution imagery. In: M. A. Wulder, and S. E. Franklin, (eds), *Remote Sensing of Forest Environments: Concepts and Case Studies*. Kluwer Academic, Boston: pp. 255–277.

Dial, G., H. Bowen, F. Gerlach, J. Grodecki, and R. Oleszczuk, 2003. IKONOS satellite, imagery, and products. *Remote Sensing of Environment*, 88: 23–36.

Dymond, C. C., D. J. Mladenoff, and V. C. Radeloff, 2002. Phenological differences in Tasseled Cap indices improve deciduous forest classification. *Remote Sensing of Environment*, 80: 460–472.

Franklin, S. E., Lavigne, M. B., Moskal, L. M., Wulder, M. A., and T. M. McCaffrey, 2001. Interpretation of forest harvest conditions in New Brunswick using Landsat TM enhanced wetness difference imagery (EWDI). *Canadian Journal of Remote Sensing*, 27: 118–128.

Franklin, S. E., Lavigne, M., Wulder, M. A., and G. B. Stenhouse, 2002. Change detection and landscape structure mapping using remote sensing. *The Forestry Chronicle*, 78: 618–625.

Franklin, S., M. Wulder, R. Skakun, and A. L. Carroll, 2003. Mountain pine beetle red attack damage classification using stratified Landsat TM data in British Columbia, Canada. *Photogrammetric Engineering and Remote Sensing*, 69: 283–288.

Franklin, S. E., C. B. Jagielko, and M. B. Lavigne, 2005. Sensitivity of the Landsat enhanced wetness difference index (EWDI) to temporal resolution. *Canadian Journal of Remote Sensing*, 31: 149–152.

Gamon, J. A., and H. L. Qiu, 1999. Ecological applications of remote sensing at multiple scales. In: F. I. Pugnaire and F. Valladares, (eds), *Handbook of Functional Plant Ecology*. Marcel Dekker Inc., New York: pp. 805–846.

Gamon, J. A., K. F. Huemmrich, D. R. Peddle, J. Chen, D. Fuentes, F. G. Hall, J. S. Kimball, S. Goetz, J. Gu, K. C. McDonald, J. R. Miller, M. Moghaddam, A. F. Rahman, J.-L. Rougean, E. A. Smith, C. L. Walthall, P. Zarco-Tejada, B. Hu, R. Fernandes, and J. Cihlar, 2004. Remote Sensing in BOREAS: Lessons Learned. *Remote Sensing of Environment*, 30: 139–162.

Goetz, S. J., R. K. Wright, A. J. Smith, E. Zinecker, and E. Schaub, 2003. IKONOS imagery for resource management: Tree cover, impervious surfaces, and riparian buffer analysis in the mid-Atlantic region. *Remote Sensing of Environment*, 88: 195–208.

Gould, W., 2000. Remote sensing of vegetation, plant species richness, and regional biodiversity hotspots. *Ecological Applications*, 10: 1861–1870.

Healy, S. P., Z. Yang, W. B. Cohen, and D. J. Pierce, 2006. Application of two regression-based methods to estimate the effects of partial harvest on forest structure using Landsat data. *Remote Sensing of Environment*, 101: 115–126.

Hoffer, A. M. 1978. Biological and physical considerations in applying computer-aided analysis techniques to remote sensor data. In: P. H. Swain and S. M. Davis (eds),

Remote Sensing: The Quantitative Approach. McGraw-Hill, New York. Chapter 5: 227–289.

Homer, C., R. Ramsey, T. Edwards, and A. Falconer, 1997. Landscape cover-type modeling using a multi-scene thematic mapper mosaic. *Photogrammetric Engineering and Remote Sensing*, 63: 59–67.

Huang, C., B. Wylie, C. Homer, L. Yang, and G. Zylstra, 2002. Derivation of a tasseled cap Transformation based on Landsat-7 at-sensor reflectance. *International Journal of Remote Sensing*, 23: 1741–1748.

Kneppeck, I. D. and F. J. Ahern, 1989. A comparison of images from a pushbroom scanner with normal color aerial photographs for detecting scattered recent conifer mortality. *Photogrammetric Engineering and Remote Sensing*, 55: 333–337.

Lefsky, M. A., and W. B. Cohen, 2003. Selection of remotely sensed data. In: M. A. Wulder, and S. E. Franklin, (eds), Remote Sensing of Forest Environments: Concepts and Case Studies. Kluwer Academic, Boston: pp. 13–46.

Lei, J. and A. J. Peters, 2007. Performance evaluation of spectral vegetation indices using a statistical sensitivity function. *Remote Sensing of Environment*, 106: 59–65.

Lillesand, T. M. and R. W. Kiefer, 2000. *Remote Sensing and Image Interpretation*, 4th edition. John Wiley and Sons, New York, 736 p.

Lunetta, R. S., D. M. Johnson, J. G. Lyon, and J. Crotwell, 2004. Impacts of imagery temporal frequency on land-cover change detection monitoring. *Remote Sensing of Environment*, 89: 444–454.

Luther, J. E., R. A. Fournier, D. E. Piercey, L. Guindon, and R. J. Hall, 2006. Biomass mapping using forest type and structure derived from Landsat TM imagery. *International Journal of Applied Earth Observations and Geoinformation*, 8: 173–187.

Maki, M., M. Ishiara, and M. Tamura, 2004. Estimation of leaf water status to monitor the risk of forest fires by using remotely sensed data, *Remote Sensing Environment*, 90: 441–450.

Morisette, J. T., J. E. Nickeson, P. Davis, Y. Wang, Y. Tian, C. E. Woodcock, N. Shabanov, M. Hansen, W. B. Cohen, D. R. Oetter, and R. E. Kennedy, 2003. High spatial resolution satellite observations for validation of MODIS land products: IKONOS observations acquired under the NASA scientific data purchase. *Remote Sensing of Environment*, 88: 100–110.

Radeloff, V., D. Mladenoff, and M. Boyce, 1999. Detecting Jack Pine budworm defoliation using spectral mixture analysis: separating effects from determinants. *Remote Sensing of Environment*, 69: 156–169.

Roberts, D. A., B. W. Nelson, J. B. Adams, and F. Palmer, 1998. Spectral changes with leaf aging in Amazon caatinga. *Trees*, 12: 315–325.

Rock, B., J. Vogelmann, D. Williams, A. Vogelmann, and T. Hoshizaki, 1986. Remote detection of forest damage. *BioScience*, 36: 439–445.

Rogan, J. and J. Miller, 2006. Integrating GIS and remotely sensed data for mapping forest disturbance and change. Chapter 6 in M. Wulder, and S. Franklin (eds), *Forest Disturbance and Spatial Pattern: Remote Sensing and GIS Approaches.* Taylor and Francis, Boca Raton, Florida, 264 p.

Rossini, M., C. Panigada, M. Meroni, and R. Colombo, 2006. Assessment of oak forest condition based on leaf biochemical variables and chlorophyll fluroescence. T*ree Physiology*, 26: 1487–1496.

Skakun, R. S., M. A. Wulder, and S. E. Franklin, 2003. Sensitivity of the thematic mapper enhanced wetness difference index to detect mountain pine beetle red attack damage. *Remote Sensing Environment*, 86: 433–443.

Slaton, M. R., and W. K. Smith, 2001. Estimating near-infrared leaf reflectance from leaf structural characteristics. *American Journal of Botany*, 88: 278–284.

Stone, C., L. Chisholm, and N. Coops, 2001. Spectral reflectance characteristics of eucalypt foliage damaged by insects. *Australian Journal of Botany*, 49: 687–698.

Strahler, A., C. Woodcock, and J. Smith, 1986. On the nature of models in remote sensing. *Remote Sensing of Environment*, 20: 121–139.

Tao, C. V., Y. Hu, and W. Jiang, 2004. Photogrammetric exploitation of IKONOS imagery for mapping applications. *International Journal of Remote Sensing*, 25: 2833–2853.

Tian, Q., Q. Tong, R. Pu, X. Guo, and C. Zhao, 2001. Spectroscopic determination of wheat water status using 1650–1850 nm spectral absorption features, *International Journal of Remote Sensing*, 10: 2329–2338.

Ustin, S., D. Roberts, J. Gamon, and R. Green, 2004. Using imaging spectroscopy to study ecosystem processes and properties. *BioScience*, 54: 523–534.

Vane, G. and A. Goetz, 1993. Terrestrial imaging spectrometry: current status, future trends. *Remote Sensing of Environment*, 44: 117–126.

Vane, G., R. O. Green, T. G. Chrien, H. T. Enmark, E. G. Hansen, and W. M. Porter, 1993. The airborne visible/infrared imaging spectrometer. R*emote Sensing of the Environment*, 44: 127–143.

White, J. C., M. A. Wulder, D. Brooks, R. Reich, and R. Wheate, 2005. Mapping mountain pine beetle infestation with high spatial resolution satellite imagery. *Remote Sensing of Environment*, 96: 240–251.

Wulder, M. 1998. Optical remote-sening techniques for the assessment of forest inventory and biophysical parameters. *Progress in Physical geography*, 24(4): 449–476.

Wulder, M. A., R. Hall, N. Coops, and S. Franklin, 2004a. High spatial resolution remotely sensed data for ecosystem characterization. *BioScience*, 54: 1–11.

Wulder, M. A., S. Franklin, and J. C. White, 2004b. Sensitivity of hyperclustering and labeling land cover classes to Landsat image acquisition date. *International Journal of Remote Sensing*, 25: 5337–5344.

Wulder, M. A., R. S. Skakun, S. E. Franklin, and J. C. White, 2005. Enhancing forest inventories with mountain pine beetle infestation information. *Forestry Chronicle*, 81: 149–159.

Wulder, M. A., J. C. White, B. Bentz, M. F. Alvarez, and N. C. Coops, 2006a. Estimating the probability of mountain pine beetle red attack damage. *Remote Sensing of Environment*, 101: 150–166.

Wulder, M. A., C. C. Dymond, J. C. White, D. G. Leckie, and Carroll, A. L. 2006b. Surveying mountain pine beetle damage of forests: A review of remote sensing opportunities. *Forest Ecology and Management*, 221: 27–41.

Zarco-Tejada, P. J., A. Berjón, R. López-Lozano, J. R. Miller, P. Martín, V. Cachorro, M. R. González, and A. Frutos, 2005. Assessing Vineyard Condition with Hyperspectral Indices: Leaf and Canopy Reflectance Simulation in a RowStructured Discontinuous Canopy. *Remote Sensing of Environment*, 99: 271–287.

Zheng, D., J. Rademacher, T. Chen, M. Bresee, J. Le Moine, and S. Ryu, 2004. Estimating aboveground biomass using Landsat 7 ETM+ data across a manged landscape in northern Wisconsin, USA. *Remote Sensing of Environment*, 93: 402–411.

Remote Sensing of Cropland Agriculture

M. Duane Nellis, Kevin P. Price,
and Donald Rundquist

Keywords: Crop classification, Crop condition, Crop yield, Crop biophysical characteristics, Crop water management, Crop-related soil characteristics, Precision agriculture, Crop phenology and Nitrogen management.

HISTORY OF REMOTE SENSING IN AGRICULTURE

Remote sensing has long been used in monitoring and analyzing agricultural activities. Well prior to the first coining of the term 'remote sensing' in 1958 by Eveyln Pruitt of the U.S. Office of Naval Research (Estes and Jensen 1998), scientists were using aerial photography to complete soil and crop surveys associated with agricultural areas in the United States and other parts of the world (Goodman 1959). Most of such work in the 1930s involved general crop inventories by the U.S. Department of Agriculture and soil survey mapping as part of the work of the then U.S. Soil Conservation Service. With new developments in infrared photography during World War II, remote sensing techniques evolved that allowed for greater understanding of crop status, water management, and crop-soil condition.

Pioneering work on remote sensing in agriculture was done by Robert Colwell at the University of California in the 1950s, and during the 1960s new laboratories oriented to applications in agriculture, such as the one at Purdue (see Landgrebe 1986), were developed. Crop identification and their areal coverage were early objectives, and Bauer (1985) provides information about projects such as the Corn Blight Watch Experiment and the Crop Identification Technology Assessment for Remote Sensing (CITARS) program.

In the early 1970s, NASA began funding selected universities via its University Affairs Program in an effort to stimulate the use of remote-sensing technologies, and states where agriculture was an important aspect of the economy began applying remote sensing to that sector. Centers and laboratories, such as those at Purdue and Kansas, were early contributors to the evolution of remote-sensing science in agriculture, and the research was important in the ultimate selection of the spectral bands incorporated into future sensor systems.

Subsequent investigations have included many types of sensors, and remote sensing has been proven capable of providing the necessary reliable data on a timely basis for a fraction of the cost of traditional methods of information gathering.

The Large Area Crop Inventory Experiment (LACIE) was the first U.S. government sponsored program aimed at examining the feasibility of using remotely sensed, satellite data, specifically Landsat, to estimate wheat production over large geographic areas. The idea was proposed by the National Research Council in 1960, and with the 1972 launch of the first of the Landsat sensor configuration, the possibility of estimating wheat yield over wide areas became a reality.

The LACIE program was operated jointly under the aegis of NASA, NOAA, and USDA. During 1974–75, the emphasis of the work was on developing both spectral 'signatures' for wheat and the yield-estimation models for the Great Plains of the U.S. Subsequently, the activity was expanded to include Canada and the Soviet Union. The successes of LACIE led to a follow-on project in 1980 called Agriculture and Resources Inventory Surveys Through Aerospace Remote Sensing (AgRISTARS). The goal of this new program was to expand upon LACIE and include monitoring of other crops such as barley, corn, cotton, rice, soybeans, and wheat. An overview of these programs can be found in Rundquist and Samson (1983), Bauer (1985), and Pinter et al. (2003), while details of historical development in remote-sensing science and also agricultural applications are provided by Reeves (1975).

REMOTE SENSING APPLICATIONS IN AGRICULTURE

Crop classification, condition and yield

Remote sensing has played a significant role in crop classification, crop health and yield assessment. Since the earliest stages of crop classification with digital remote sensing data, numerous approaches based on applying supervised and unsupervised classification techniques have been used to map geographic distributions of crops and characterize cropping practices. Depending on geographic area, crop diversity, field size, crop phenology, and soil condition, different band ratios of multispectral data and classifications schemes have been applied. Nellis (1986), for example, used a maximum likelihood classification approach with Landsat data to map irrigated crop area in the U.S. High Plains. Price et al. (1997) further refined such approaches, using a multi-date Landsat Thematic Mapper (TM) dataset in southwest Kansas to map crop distribution and USDA Conservation Reserve Program (CRP) lands in an extensive irrigated area.

Hyperspectral remote sensing has also helped enhance more detailed analysis of crop classification. Thenkabail et al. (2004) performed rigorous analysis of hyperspectral sensors (from 400 to 2500 nm) for crop classification based on data mining techniques consisting of principal components analysis, lambda–lambda models, stepwise Discriminant Analysis and derivative greenness vegetation indices. Through these analyses they established 22 optimal bands that best characterize the agricultural crops. By increasing the number of channels beyond 22 bands, accuracies only increased marginally up to 30 bands and became asymptotic beyond that number. In comparison to Landsat Enhanced Thematic Mapper data and other broadband sensors, these hyperspectral approaches increased accuracy for crop classification from 9 to 43%.

Relative to crop condition, some remote sensing studies have focused on individual physical parameters of the crop system, such as nutrient stress or water availability as variables in analyzing crop health and yield. Other research has focused more on synoptic perspectives of regional crop condition using remote sensing indices. At the same time, some researchers (Seidl et al. 2004) have demonstrated that such approaches can be limited for crop yield and health monitoring given satellite over flight timing in the context of the crop calendar.

The normalized difference vegetation index (NDVI), vegetation condition index (VCI), leaf area index (LAI), General Yield Unified Reference Index (GYURI), and temperature crop index (TCI) are all examples of indices that have been used for mapping and monitoring drought and assessment of vegetation health and productivity (Doraiswamy et al. 2003, Ferencz et al. 2004, Prasad et al. 2006). Wang et al. (2005), for example, used satellite remote sensing of NDVI to provide characterizations of landscape level patterns of net primary productivity within the U.S. Great Plains, and Kogan et al. (2005) used vegetation indices from Advanced Very High Resolution Radiometer (AVHRR) data to model corn yield and early drought warning in China. Hadria et al. (2006) provides an example of developing leaf area indices from four satellite scenarios to estimate distribution of yield and irrigated wheat in semi-arid areas. Zhang et al. (2005) have also modified leaf area indices based on MODIS (MODerate resolution Imaging Spectrometer) using a climate-variability impact index (CVII) related to contributions to monthly anomalies in annual crop growth. Using MODIS the researchers were able to establish the relationship between CVII and LAI to accurately model regional crop forecasts.

In addition, to refine regional crop forecasting, researchers have modified standard NDVI approaches using crop yield masking. This technique involves restricting analysis to a subset region's pixels rather than using all the pixels in the scene. According to work by Kastens et al. (2005) yield correlation masking is shown to have comparable performance to cropland masking across eight major U.S. crop forecasting scenarios. Jensen (2007) provides further examples of the broad range of popular vegetation indices used in remote sensing of agricultural systems.

Recent research has documented radar as a tool for crop monitoring. Chen and Mcnairn (2006) used radar, for example, in rice monitoring within Asia. In their work, they found that backscatter increases significantly during short periods of vegetation growth, which can be used to differentiate rice fields from other land cover.

Recent commercial satellites with fine spatial resolution have also proven of value in mapping crop growth and yield. Yang et al. (2006) used QuickBird satellite imagery for mapping plant growth and yield patterns within grain sorghum fields as compared with airborne multispectral image data. The results suggest QuickBird data and airborne spectral data were equally useful.

Although satellite remotely sensed data have historically been used for assessing specific crop stress parameters, such as indications of nitrogen stress (Reyniers and Vrindts 2006), more detailed insights regarding crop condition are being gained using hyperspectral remote sensing, thermal radiometers, and related devices. Such approaches have also contributed to effective uses of these data in precision agriculture (Yang et al. 2004). Hyperspectral sensors and related techniques can also be used to estimate various other crop biophysical and biochemical parameters, such as leaf nitrogen content, leaf chlorophyll content, and associated factors related to soil moisture (Goel et al. 2003). Remote sensing of soil moisture will be elaborated on later in this chapter. Ye et al. (2006) used hyperspectral images to predict tree crop yield in citrus groves, Vijaya-Kumar and colleagues (2005) used an infrared thermometer and spectral radiometer for screening germplasm and stress in castor beans, while Nicholas (2004) used visible, near infrared, and thermal sensors to assess crop conditions.

Vegetation stages of development (phenology) are influenced by a variety of factors such as available soil moisture, date of planting, air temperature, day length, and soil condition. These factors therefore also influence plant conditions and their productivity. For example, corn crop yields can be negatively impacted if temperatures are too high at the time of pollination. For this reason, knowing the temperature at the time of corn pollination could help forecasters better predict corn yields.

McMaster (2004) has summarized the importance of phenology, asserting that, 'Understanding crop phenology is fundamental to crop management, where timing of management practices is increasingly based on stages of crop development. Simulating canopy development is also critical for crop growth models, whether to predict the appearance of sources and sinks, determining carbon assimilation and transpiration, partitioning carbohydrates and nutrients, or determining critical life cycle events such as anthesis and maturity.'

During the era during which Landsat was temporarily privatized (see also Goward et al., in this volume) after the 1990s, some remote sensing scientists begin experimenting with the use of coarse spatial resolution (1.0 km) imagery (also see Justice et al., in this volume). These data were free, and they allowed for a synoptic view of the conterminous U.S. on a daily to bi-daily temporal frequency. As methods were developed for cloud removal or minimization, new high temporal (weekly to biweekly) cloud-free or near cloud-free datasets were made freely available for a variety of applications and research endeavors. In the early 1980s, Badhwar and Henderson (1981) published a paper in the *Agronomy Journal* describing the use of spectral data for characterizing terrestrial vegetation development. In 1990, Lloyd described the use of a shortwave vegetation index (the Normalized Difference Vegetation Index (NDVI)) for characterizing phenological stages of plant development for terrestrial land cover, and in 1994, Reed et al. described how high temporal resolution NDVI datasets could be used to examine variability in interannual phenology at the continental scale.

Using coarse spatial resolution and high temporal resolution data, for the first time plant response to varying growing conditions could be examined at or near a continental scale, for example, the conterminous US. Since the publication of the manuscripts referenced above, many studies that followed used AVHRR NDVI datasets for such tasks as assessing crop relative condition and making yield forecasts (Steven et al. 1983, Quarmby et al. 1993, Groten 1993, Kastens et al. 2005), characterization of Central Great Plains grass life forms (Reed et al. 1996, Tieszen et al. 1997), studying vegetation response to intra- and interannual climatic variation (Yu et al. 2003, Breshears et al. 2005), and for many other agriculturally related purposes.

Annual NDVI profiles are extracted in operational remote sensing for 12 vegetation phenology metrics (VPMs), and these metrics are used to characterize agricultural vegetation response to varying climatic and land management practices (Reed et al. 1994; Figure 26.1 and Table 26.1).

Crop biophysical characterization

Remote sensing can play an important role in agriculture by providing timely spectral-reflectance information that can be linked to biophysical indicators of plant health. Quantitative techniques can be applied to the spectral data, whether acquired from close-range or by aircraft or satellite-based sensors, in order to estimate crop status/condition. The technology is capable of playing an important role in crop management by providing at least the following types of information:

1 fraction of vegetative cover,
2 chlorophyll content,
3 green leaf area index, and,
4 other measurable biophysical parameters

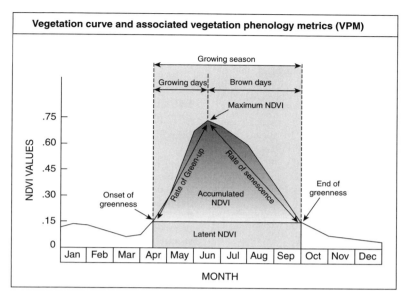

Figure 26.1 A twelve month, hypothetical NDVI temporal response curve for vegetation. Additionally, the vegetation metrics are displayed to show their relation to both NDVI values and time (after Reed et al. 1994).

Table 26.1 Vegetation phenology metrics characterize vegetation phenology and are used to develop summary regional data for research on agro-ecosystem attributes (after Reed et al. 1994)

Type	Metric	Interpretation
Temporal	1 Time of onset of greenness	Beginning of photosynthetic activity
	2 Time of end of greenness	End of photosynthetic activity
	3 Duration of greenness	Length of photosynthetic activity
	4 Time of maximum greenness	Time when photosynthesis at maximum
NDVI-value	5 Value of onset of greenness	Level of photosynthesis at start
	6 Value of end of greenness	Level of photosynthesis at end
	7 Value of maximum NDVI	Level of photosynthesis at maximum
	8 Range of NDVI	Range of measurable photosynthesis
Derived	9 Accumulated NDVI	Net Primary Production (NPP)
	10 Rate of green-up	Acceleration of increasing photosynthetic activity
	11 Rate of senescence	Acceleration of decreasing photosynthetic activity
	12 Mean daily NDVI	Mean daily photosynthetic activity

Fraction of Vegetation Cover

Remote sensing offers, by means of multi-temporal data collection, the capability of monitoring changes in fraction of vegetative cover associated with crop phenology. Details associated with the growth of a corn crop over time were provided by Vina et al. (2004) who used visible atmospherically resistant spectral indices to document a capability for detecting changes in the crop due to biomass accumulation, changes induced by the appearance and development of reproductive structures, and the onset of senescence.

Gitelson et al. (2002), studying wheat and corn, developed spectral indices using only the visible region of the spectrum to estimate vegetation fraction (VF). They found these indices to be more sensitive than NDVI to changes in vegetation fraction at high levels (>60%) of cover, and the error in VF prediction did not exceed 10%.

Chlorophyll Content

One very important and useful indicator of vegetation condition is the content of chlorophyll, a pigment producing the characteristic green color in plants. Variability in chlorophyll content is related to growth stage in the plant's life cycle, photosynthetic capacity/productivity, and stresses (Ustin et al. 1998). The measurement of chlorophyll content is also important with regard to nitrogen management, a key element of variable rate field fertilization technology.

Gitelson et al. (2005) developed a model, based upon field measurements made by means of a hyperspectral radiometer, for non-destructive estimation of chlorophyll in maize and soybean canopies. Separate models were developed for corn and soybeans individually, but the optimum model for evaluating corn and soybeans together, in an effort to obtain a non-species-specific solution, took the form $[(R_{NIR}/R_{720-730}) - 1]$ where R is the reflectance at the specified wavelength in nm. This model allowed estimation ($R^2 = 0.95$) of chlorophyll in the range 0.03–4.33 g/m^2, with an RMSE of less than 0.32 g/m^2 for corn and soybeans considered together.

Numerous other authors have dealt with the issue of remote sensing of chlorophyll content, including Broge and Leblanc (2000), Daughtry et al. (2000), and Broge and Mortensen (2002).

Green Leaf Area Index

The measurement of Leaf Area Index (LAI) is not only important as an estimate of primary production, but also as an input to climate models. Therefore, remote sensing of LAI is often undertaken. Gitelson et al. (2003b) proposed a technique to estimate LAI and green leaf biomass using spectral reflectance either in the green region (around 550 nm) or at the red edge (near 700 nm) along with the near-infrared (beyond 750 nm). Close relationships were found between the spectral indices tested and LAI (ranging from 0 to more than 6) as well as green leaf biomass (ranging from 0 to 3500 kg/ha). Numerous other authors have addressed the topic of remote sensing of LAI including Gilabert et al. (1996), Carlson and Ripley (1997), Broge and Leblanc (2000), and Haboudane et al. (2004).

Other Measurable Biophysical Variables

Other biophysical variables may also be measurable by means of remote sensing. Danson et al. (1992) showed that the first derivatives of reflectance associated with the slopes of the lines near water-absorption bands were highly correlated with leaf water content. Gitelson et al. (2003a) developed a technique based upon remote sensing that accounted for more than 90 percent of the variability in mid-day canopy photosynthesis of irrigated corn. Researchers have also addressed crop yield (e.g., Hatfield 1983, Serrano et al. 2000,

Shanahan et al. 2001, Teal et al. 2006), canopy transpiration (e.g., Inoue and Moran 1997), plant litter (e.g., Nagler et al. 2000, Streck et al. 2002); and phytomass (e.g., Daughtry et al.1992).

THE CHALLENGE: PRACTICAL APPLICATION OF BIOPHYSICAL MEASURES

Using remote sensing, it is possible to infer certain biophysical parameters of cropland vegetation, but in order to make use of the biophysical properties in a practical manner, one must attempt to link them to a set of 'agronomic indicators.' This may mean developing algorithms that incorporate both spectral and agronomic parameters to provide economically viable, practical products (e.g., maps of pigment densities in specific fields, useful descriptors of general crop status and condition as depicted in map form, nutrient management information that can be used effectively and efficiently to improve the application of nitrogen, and biomass measurements used to estimate potential yield in near-real time). It is only through the development of such practical products that remote sensing will gain wide acceptance and use.

Precision farming is an emerging methodology designed to link management actions to site-specific soil and crop conditions, and place inputs of fertilizers, herbicides, and pesticides where they are most needed to maximize farm efficiency and minimize environmental contamination. One of the crucial parts in this system is information on soil and crop conditions at the temporal frequency and spatial resolution required for making crop management decisions. Remote sensing is a viable tool for providing such information.

Crop water management

Information on crop-water demand, water use, soil moisture condition, and related plant growth at different stages of cultivation can be obtained through use of various forms of remote sensing, extending from synoptic views using satellite data to detailed analyses with thermal and hyperspectral sensor systems. Bandara (2003), for example, used NOAA satellite data to assess the performance of three large irrigation projects in Sri Lanka. Within this analysis, estimates using remote sensing of crop-water utilization were compared to actual water availability to determine irrigation efficiency. In a related study, Martin De Santa Olalla et al. (2003) used GIS with NDVI and a hydrological management unit (HMU) to create an Irrigation Advisory Service linked to water requirements of crops and estimates of the volume of irrigation water used.

Linking such approaches to groundwater use has further extended analysis methodologies. Wu et al. (1999), used Landsat Thematic Mapper data to characterize the relationship between land use and groundwater depletion in the fragile U.S. High Plains resource system.

Remote sensing has also been used to evaluate irrigation water distribution at variable scales of the agro-water system. Nellis (1985) used thermal infrared imagery, for example, at the field level to determine parameters related to uniformity of water distribution and conveyance system irregularities in Oregon. In western Turkey, research by Droogers and Kite (2002) used a parametric basin-scale model and a physically based crop-scale model linked to NOAA-AVHRR images to analyze water use for irrigation at the field scale, irrigation scheme scale, and basin scale.

Since early on in digital remote sensing, researchers have used various approaches to estimate crop evapotranspiration, soil moisture, and biomass growth (Bastiaanssen et al. 2001, Ray and Dadhwal 2001). Neale and colleagues (2005) provide an historical perspective on high resolution airborne remote sensing of crop coefficients for obtaining actual crop evapotranspiration. Although most approaches use simple direct correlations between remote sensed digital data and evapotranspiration, some combine various forms of remotely sensed data types. Consoli and colleagues (2006), for example, used IKONOS high resolution satellite along with hyperspectral ground data with agro-meteorological information from orange groves in southern Italy to estimate evapotranspiration fluxes.

Water impact on erosive capacity and sediment yield using GIS and remote sensing has also gained considerable attention by researchers. Khan et al. (2001), for example, used spatial data on land-forms, land cover, and slope to estimate potential erosivity and sediment yield in India. In addition, Wu and colleagues (1997) used spatial data linked to a GIS coupled with Landsat TM land cover data to evaluate USDA Conservation Reserve Program (CRP) lands and related soil properties in the U.S. High Plains.

Clearly remote sensing is playing an ever increasing role in water management of the agricultural system. Such developments have been further enhanced with the evolution of hyperspectral sensors and the ability to link spatially analyzed remote sensing data with other spatial data through GIS and GPS technologies.

Crop and soil characteristics

As has been articulated by Sullivan et al. (2005), numerous studies have demonstrated the utility of remote sensing for distinguishing various soil properties, including erosion prediction, application of agrochemicals to soils for precision management, soil organic carbon, iron oxide content, and soil texture (see also Chapter 24 on soils applications (Campbell, in this volume)). Over 25 years ago, Stoner and Baumgardner's work (1981), for example, showed an increasing level soil carbon was inversely correlated with reflectance in the visible and infrared regions. In contrast, more recent work by Bajwa and Tian (2005) demonstrated the potential of aerial visible/infrared (VIR) hyperspectral imagery for characterizing soil fertility factors in the U.S. Midwest. Soil fertility parameters included pH, organic matter (OM), Ca, Mg, P, K, and soil electrical conductivity. In this analysis, the measured soil fertility characteristics were modeled on first derivatives of the reflectance data using partial least square regression. The model explained a higher degree of variability in Ca (82%), Mg (72%), and OM (66%), for example, and less so for properties such as pH (48%).

Controlled field and laboratory studies have been the basis in remote sensing research related to soil particle size. Such work has often relied on high spectral resolution radiometers (Salisbury and D'Aria 1992, for example). This research has found increasing spectral response with increasing sand content, which is likely associated with a corresponding decrease in water holding capacity of coarser soils.

The refinement in thermal sensing systems and access to hyperspectral sensors (as was earlier noted) have further extended remote sensing applications to soil properties. Salisbury and D'Aria (1992) have used thermal infrared band ratios to estimate quartz content, but varying levels of clay, iron, and soil organic carbon complicated the results.

According to Sullivan et al. (2005), Barnes and Baker (2000), and Russell (2003), spatially and temporally dynamic surface conditions, such as water content, surface roughness, crusting, and crop residue cover, significantly impact spectral response and complicate the remote sensing analysis process. For example, recent studies by Ben-Dor et al. (2003) and Eshel et al. (2004) have shown that crusted soil surfaces of many different types of soils tend to have higher reflective responses due to the simple occurrence of greater quartz exposures when such crusted soils are freshly tilled exposing quartz surfaces.

Ground based sensors or 'on the go' sensors (sensors mounted on a tractor and data mapped with coincident position information) have developed rapidly in recent years. Such efforts provide soil organic matter, electrical conductivity, nitrate content, and compaction (Barnes et al. 2003), and when integrated with other data sources maximize the information for the farm manager.

Numerous remote sensing studies of soil moisture (in addition to earlier references) have focused on the use of thermal and microwave sensing. Casanova et al. (2006) used microwave remote sensing models to improve estimates of soil moisture, and Verstraeten et al. (2006) used a combination of optical and thermal spectral information of METEOSAT imaging to determine thermal inertia relative to soil moisture content.

Also at the interface between water and soil dynamics are processes that lead to soil salinity in irrigated agricultural areas. Again remote sensing has provided important analysis of such applications, as illustrated by the work of Masoud and Koike (2006), in which they used Landsat TM/ETM taken over 16 years coupled with 30 m DEM and field observations in Egypt to successfully document salinity changes in soils and their related land cover.

With increasingly narrower hyperspectral bands in the thermal range of electromagnetic energy and coupled with other spatial data derived from field observations, remote sensing is clearly playing an ever more important role in understanding crop soil characteristics. Such efforts, when linked to GPS, provide promising results in precision agriculture.

Precision Agriculture and 'on-the-go sensors'

Precision Agriculture

Precision farming is an emerging methodology designed to link management actions to site-specific soil and crop conditions, and place inputs of fertilizers, herbicides, and pesticides where they are most needed to maximize farm efficiency and minimize environmental contamination. The core technologies in precision agriculture are GIS, GPS, and remote sensing. The importance of these technologies in agriculture was underscored when NASA (Stennis Space Center), in the early portion of the current century, embarked upon the Ag2020 program in an effort to commercialize the geospatial technologies, develop practical tools for producers, and undertake projects with various types of crops to illustrate the utility of the technologies.

Critical to precision farming is the acquisition of information on soil and crop conditions at the temporal frequency and spatial resolution required for making crop management decisions. Remote sensing is no doubt a viable tool for providing such information. However, for remote sensing to contribute to the management of small parcels, the imagery must be of high spatial resolution. Until relatively recently, such satellite imagery has been unavailable, and though available today, it remains quite expensive. Examples of previous precision

agriculture work done using data acquired by high-resolution satellite sensors include that of Yang et al. (2006) and Bannari et al. (2006). General summaries regarding procedures and issues related to precision agriculture are provided by Barnes et al. (1996), Bramley et al. (1999), Campanella (2000), and Moran (2000).

The Role of Remote Sensing in Precision Agriculture

While 'geospatial technologies,' especially GPS and GIS, are certainly visible, integral components of the movement in agriculture toward precision approaches, remote sensing unfortunately remains poorly understood and less widely used. Today, most combines are equipped with GPS capability, and even planting is often done within a framework of precise geographic location. Similarly, the concept of yield mapping and comparing those results to detailed soils maps have brought GIS to the fore in modern agriculture. Remote sensing, on the other hand, while it has much to offer the agriculturalist, seems less important to producers for several reasons including the high cost of obtaining imagery, the need to register the images to other spatial data, the need for imagery of both high spatial and temporal resolutions, and a learning curve that is rather steep for most on-farm operators and even agronomic consultants. The digital overlay and analysis of spatial, and especially spectral, datasets is perceived as being too difficult for many producers and consultants, and they often are reluctant to acquire the software skills necessary to accomplish the tasks noted previously. The need of most agricultural producers and consultants seems to be useful map products. Another challenge in applying remote sensing to agriculture in general has been the delivery of image data in a timely manner. Pelzmann (1997) indicates that imagery should be delivered within 48 hours of acquisition or less to be truly useful.

Despite the complexities associated with incorporating remote sensing into the day-to-day operations of farming, there are some relatively simple products that can serve to enhance the producer's understanding of his farm as well as facilitate site-specific management. Aerial photographs have potential for delineating management zones (Schepers et al. 2000). Conventional-color aerial photography, acquired at a leaf-off condition, is an important tool in the early stages of preparing a field for site-specific management. Such air photos provide an uncommonly good view of general soil conditions; most specifically, the variability in organic content from place to place (Schepers 2002). Aerial videography has been shown to be a useful tool for monitoring within-field plant-growth variation, as well as establishing management zones for precision farming (Yang et al. 1998).

Nitrogen Management

One of the major environmental issues is the need for nitrogen (N) management, as the groundwater in some areas of intensive agriculture, such as the Midwestern U.S., has become increasingly polluted with nitrates. Because N is relatively cheap and small inputs increase yields, producers have a tendency to apply large amounts of 'insurance nitrogen,' thereby increasing the potential for groundwater pollution (Schepers 2002). Thus, there is a need to develop procedures, including possibly remote sensing, that may have utility for mitigating the problem of N overuse. The appropriate approach seems to be variable-rate technology in order to apply the fertilizer only on plants that are in need and only in portions of the field where that need exists, rather than across the entire field. Studies by Blackmer et al. (1996), Shanahan et al. (2001), and Scharf et al. (2002) provide useful background on the subject of nitrogen management by means of remote sensing.

On-the-go sensors

On-the-go-sensors include implement mounted devices that make measurements of soils, crop canopies, or even individual plants at close-range. The measurements may not necessarily be spectral in nature, but certainly could be.

The essence of the new approaches in site-specific management is highlighted by the fact that combine-mounted yield monitors are commonly used for attempting to assess within-field variation and to delineate crop-management zones (Pinter et. al. 2003). Yet, as Pinter et. al point out, yield maps may not be an accurate portrayal of the extremes of variability, and they do not provide information about yield-reducing stresses because they are acquired at the end of the season. Also, the yield maps only document the spatial distribution; they do not explain the cause of variation (Doerge 1999). This is where remote sensing is capable of making a contribution, but again, high spatial resolution is a requirement.

Systems have been devised for closed-loop, real-time, variable-rate application implements (Adamchuk et al. 2003), including for identifying weeds versus soil or crop residue (e.g., Meyer et al. 1998). Often, the system involves a sprayer, for example to apply herbicide to weeds, which is operated by means of a computer linked to an optical sensor. The sensor provides a computer algorithm with an image of a weed, the analysis proceeds based either on leaf reflectance or leaf shape, and the algorithm makes the decision to activate the spray, thus (hopefully) killing the weed (Schepers 2002). A similar approach could be used to apply nitrogen to plants where leaf color or reflectance indicates a shortage of that particular nutrient. Again, the benefit is that an entire field is spared

from large amounts of fertilizer or pesticides and only the plants in need actually receive the spray. Stamatiadis et al. (2006) used on-the-go multispectral sensing at close-range in a Greek vineyard to demonstrate the value of proximal sensing for optimizing production, improving wine quality, and reducing chemical inputs. Ground-based sensors are also used for monitoring soils. Such instruments provide soil organic matter, electrical conductivity, nitrate content, and compaction (Barnes et al. 2003), and when integrated with other data sources maximize the information for the farm manager.

Specialty crops

Viticulture, the cultivation of grapevines, generally for wine production, is an important agricultural and economic enterprise in many parts of the world, including the United States. Managers of large vineyards know that the extent of grapevine productivity is the result of many topographic, climatic, and edaphic parameters, and the assumption is that these parameters are spatially variable, thus causing differences in vine vigor and yield from place to place in the vineyard. Viticulturalists must monitor vineyard conditions for the purpose of making decisions about irrigation, fertilizer application, canopy management, and when to actually harvest the grapes in order to maximize juice (and thus wine) quality. An additional important objective of ongoing vineyard scrutiny is to detect the presence of stressors including pathogens that affect either the vegetation or the fruit itself. Therefore, remote sensing may be a cost effective tool for providing the viticulturalist with a synoptic view of the vineyard that yields useful practical information.

Several authors have conducted research on the subject of remote sensing of grapevines. General overviews of such work are provided by Lang (1997), Peterson and Johnson (2000), Carothers (2000), and Hall et al. (2002). Examples of 'viticultural sensing' include that by Johnson et al. (2001) who evaluated airborne multispectral imagery for delineating vineyard management zones as a step toward 'precision viticulture' while Lanjeri et al. (2004) used Landsat Thematic Mapper (TM) images and NDVI to assess change over time in Spanish vineyards.

A few investigators have used remote sensing technologies to detect and/or monitor pathogens. Wildman et al. (1983) made use of color infrared aerial photography for detecting and monitoring the spread of Phyloxera, a root louse that can devastate entire grape-growing regions. Johnson et al. (1996) examined airborne CASI multispectral imagery at 5 m spatial resolution for the same purpose.

Work dealing with remote sensing as a means of inferring biophysical characteristics of grapevine canopies includes that of Montero et al. (1999) who used Landsat-TM data and NDVI to measure percent vegetation cover, biomass, plant height, and leaf area index. The authors found positive linear relationships with parameters measured *in-situ* in Spanish vineyards. Lanjeri et al. (2001) also used TM imagery acquired over vineyards in Central Spain to document high positive correlations between NDVI and biomass, plant height, and vegetation cover. Johnson et al. (2003) used IKONOS imagery to convert NDVI to LAI for two commercial vineyards in California, and estimated LAI with an r^2 of 0.72 when compared to *in-situ* measurements. Those authors concluded that remote sensing appears useful for mapping vineyard leaf area in low LAI vineyards. Stamatiadis et al. (2006) made use of ground-based canopy multispectral sensors to estimate biomass production and the quality of Merlot grapes in Northern Greece.

A review of the literature dealing with remote sensing in viticulture leads one to conclude that this application is indeed feasible but that there are also some unique constraints. First, imagery of high spatial resolution, on the order of a few meters, is required in order to make inferences about planted blocks of vines or even individual plants within those blocks (e.g., Hall et al. 2001). Distance between vine rows (often only a meter in California) will, of course, affect the outcome of image analyses. Temporal resolution is important as it relates to the important phenological stages in vine development (e.g., bud break, flowering, veraison, and harvest). Bramley et al. (2003) suggest that veraison +/− two weeks is the optimal time for image acquisition (assuming that imagery is obtained at only one point in time). It seems as though little research has been conducted with hyperspectral sensors, but the previous literature indicates that broad-band sensors are capable of providing useful information. The target background for the sensor imaging rows of grapes (e.g., whether that background is green grass or bare soil) is an important consideration as is the fact that sensor fields of view in vineyards include wood posts and trellis wire, all of which combine to make the search for spectral end-members interesting if not downright difficult.

CHALLENGES AND THE FUTURE

Since the early 1970s, one of the greatest challenges to using satellite remotely sensed data for agricultural applications has been the lack of usable images collected often enough and at consistent intervals over the growing season. This inadequate temporal resolution of nadir-viewing moderate and high spatial resolution satellite acquired imagery can only be resolved by increasing the revisit frequency of the sensors, and this can only be achieved by increasing the number of satellite sensors collecting imagery. Currently, and in the past, the revisit frequency of the moderate resolution sensors like Landsat and SPOT is or has been approximately bimonthly to monthly. Such temporal resolution is inadequate to capture key changing crop conditions (phenological stages) throughout the growing season. This is especially true in areas where there is frequent cloud cover. In the central US, for example, there is often only two or three usable images acquired during the growing season, and due to varying cloud cover conditions, the acquisition dates can significantly vary annually making year-to year comparisons of vegetation conditions very challenging and often impossible.

RapidEye is planning to launch a constellation of five satellites in 2008 with the objective of providing daily geospatial information of high spatial resolution to global customers with an interest in agriculture and forestry, among others (RapidEye 2007). The sensor is to be a five-channel multispectral system operating in the visible and near-infrared. Such an operational capability would no doubt be of great interest to agriculturalists. Without such operational satellite remote sensing system the development of commercial remote sensing products to support agribusiness and the land management decision-making processes would be greatly hampered. Not having an operational system has created a lack of confidence in future availability and consistency in data types and its quality. Insufficient image temporal resolution and the lack of a dependable and consistent remote sensing system are the two most significant factors influencing the use of remotely sensed data in the area of agriculture.

During the early stages of the satellite remote sensing era, most research focused on the use of the data for classification of land cover types with crop types being a major focus among those interested in agricultural applications. Over the past decade or so, the work in agricultural remote sensing has focused more on characterization of plant biophysical properties. As scientists have gained greater access to spectroradiometer data and airborne and space borne hyperspectral data and imagery, a new area of research interest is focusing on the use of near infrared analysis and chemometrics software and models for characterizing the chemical constituencies of plant parts, soils, rocks, etc. Unfortunately, research findings in this area of study will not become operational until high quality hyperspectral satellite imagery become available at an affordable price.

Efforts in the field of precision agricultural are leading to new commercial applications as studies continue to refine the science and instruments used in 'on-the-go' (implement-mounted) sensors. Special interest is now focusing on rapid in-the-field plant leaf nitrogen characterization so that real-time fertilizer applications can be implemented. Similar efforts are being made for other critical nutrients as well. In addition, researchers continue to develop agricultural applications that involve the use of high spatial resolution airborne imagery for field level 'prescription' of fertilizers and herbicides that are administered using 'smart' sprayer applicator technology. Airborne imagery is also being used to defining soil and crop management zones and hyperspectral measurements of plant conditions are being used as an indicator of soil nutrients conditions.

Spectral measurements are also being used to measure plant biophysical properties that can be linked to biogeochemical fluxes, with special interests in CO_2. Remotely sensed data are also being used to map not only land cover, but land use, which has significant influence of CO_2 fluxes. Such data will be very useful for assessing land use practices that influence land owner carbon credit qualifications.

Finally, remote sensing in agriculture is moving toward nano-scale analysis. A new and nontraditional remote sensing application involves the implanting of nano-chips in plant and seed tissue that can be used in near-real time to monitor crop. Clearly, these and other new approaches will reinforce the importance of remote sensing in future analysis of agricultural sciences.

REFERENCES

Adamchuk, V., R. Perk, and J. Schepers, 2003. Applications of remote sensing in site551 specific management. University of Nebraska Cooperative Extension EC03-702.

Badhwar, G. D. and K. E. Henderson. 1981. Estimating Development stages of corn from spectral data – an initial model. Agronomy Journal, 73: 748–755.

Bajwa, S. G. and L. F. Tian, 2005. Soil fertility characterization in agricultural fields using hyperspectral remote sensing. Transactions of the ASAE, 48(6): 2399–2406.

Bandara, K. M. P. S, 2003. Monitoring irrigation performance in Sri Lanka with high560 frequency satellite measurements during the dry season. Agricultural Water Management, 58(2): 159–170.

Bannari, A. Pacheco, K. Staenz, H. McNairn, K. Omari, 2006. Estimating and mapping crop residues cover on agricultural lands using hyperspectral and IKONOS data. Remote Sensing of Environment, 104: 447–459.

Barnes, E. M. and M. G. Baker, 2000. Multispectral data for mapping soil texture: possibilities and limitations. American Society of Agricultural Engineers, 16: 731–741.

Barnes, E., M. Moran, P. Pinter, and T. Clarke, 1996. Mulltispectral remote sensing and site-specific agriculture: examples of current technology and future possibilities. Proceedings, 3rd International Conference on Precision Agriculture, 843–854.

Barnes, E. M., K. A. Sudduth, J. W. Hummel, S. M. Lesch, D. L. Corwin, C. Yang, C. S. T. Daughtry, and W. C. Bausch, 2003. Remote-and ground-based sensor techniques to map soil properties. Photogrammetric Engineering and Remote Sensing, 69(6): 619–630.

Bastiaanssen, W. G. M., R. A. L. Brito, M. G. Bos, R. A. Souza, E. B. Cavalcanti, and M. M. Bakker, 2001. Low cost satellite data for monthly irrigation performance monitoring: benchmarks from Nilo Coelho, Brazil. Irrigation and Drainage Systems, 15(1): 53–79.

Bauer, M. E., 1985. Spectral inputs to crop identification and condition assessment. Proc. IEEE, 73: 1071–1085.

Ben-Dor, E., N. Goldshleger, Y. Benyamini, M. Agassi, and D. G. Blumberg, 2003. The spectral reflectance properties of soil structural crusts in the 1.2- to 2.5-micrometer spectral region. Soil Science Society of America Journal, 67: 289–299.

Blackmer, T., J. Schepers, G. Varvel, and E. Walter-Shea, 1996. Nitrogen deficiency detection using reflected shortwave radiation from irrigated corn canopies. Agronomy Journal, 88: 1–5.

Bramley, R., S. Cook, M. Adams, and R. Corner, 1999. How precision agriculture can help. A Report of the CSIRO Land and Water Division, 30 pp.

Bramley, R., B. Pearse, and P. Chamberlain, 2003. Being profitable precisely – a case study of precision viticulture from Margaret River. The Australian and New Zealand Grapegrower and Winemaker Annual Technical Issue, 84–87.

Breshears, D. D., N. S. Cobb, P. M. Rich, K. Price, C. D. Allen, R. G. Balice, W. H., Romme, J. H. Kastens, M. L. Floyd, J. Belnap, J. J. Anderson, O. B. Myers, and C. W. Meyer, 2005. Regional vegetation die-off in response to global change type drought. Proceedings of the National Academy of Science, 102(42): 15144–15148.

Broge, N. and E. Leblanc, 2000. Comparing prediction power and stability of broadband and hyperspectral vegetation indices for estimation of green leaf area index and canopy chlorophyll density. Remote Sensing of Environment, 76: 156–172.

Broge, N. and J. Mortensen, 2002. Deriving green crop area index and canopy chlorophyll density of winter wheat from spectral reflectance data. Remote Sensing of Environment, 81(1): 45–57.

Campanella, R., 2000. Testing components toward a remote-sensing-based decision support system for cotton production. Photogrammetric Engineering and Remote Sensing, 66(10): 1219–1227.

Campbell, J., in this volume. Remote Sensing of Soils. Chapter 24.

Carlson, T. and D. Ripley, 1997. On the relation between NDVI, fractional vegetation cover, and leaf area index. Remote Sensing of Environment, 62(3): 241–252.

Carothers, J., 2000. Imagery technology meets vineyard management. Practical Winery and Vineyard, 21(1): 54–62.

Casanova, J. J., J. Judge, and J. W. Jones, 2006. Calibration of the CERES-Maize model for linkage with a microwave remote sensing model. Transactions of the American Society of Agricultural and Biological Engineers, 49(3): 783–792.

Chen, C. and H. Mcnairn, 2006. A neural network integrated approach for rice crop monitoring. International Journal of Remote Sensing, 27(7): 1367–1393.

Consoli, S., G. D'Urso, and A. Toscano, 2006. Remote sensing to estimate ET-fluxes and the performance of an irrigation district in southern Italy. *Agricultural Water Management*, 81(3): 295–314.

Danson, F., M. Steven, T. Malthus, and J. Clark, 1992. High-spectral resolution data for determining leaf water content. *International Journal of Remote Sensing*, 13: 461–470.

Daughtry, C., K. Gallo, S. Goward, S. Prince, and W. Kustas, 1992. Spectral estimates of absorbed radiation and phytomass production in corn and soybean canopies. *Remote Sensing of Environment*, 39: 141–152.

Daughtry, C., C. Walthall, M. Kim, E. deColstoun, and J. McMurtrey, 2000. Estimating corn leaf chlorophyll concentration from leaf and canopy reflectance. *Remote Sensing of Environment*, 74: 229–239.

Doerge, T., 1999. Site specific agriculture: yield map interpretation. *Journal of Production Agriculture*, 12(1): 54–61.

Doraiswamy, P. C., S. Moulin, P. W. Cook, and A. Stern, 2003. Crop yield assessment from remote sensing. *Photogrammetric Engineering and Remote Sensing*, 69(6): 665–674.

Droogers, P. and G. Kite, 2002. Remotely sensed data used for modeling at different hydrological scales. *Hydrological Processes*, 16(8): 1543–1556.

Eshel, G. G., J. Levy, and M. J. Singer, 2004. Spectral reflectance properties of crusted soils under solar illumination. *Soil Science Society of America Journal*, 68: 1982–1991.

Estes, J. and J. Jensen, 1998. Development of remote sensing digital image processing systems and raster GIS. In T. Forsman (ed.) *The History of Geographic Information Systems*, Longman, New York, pp. 163–180.

Ferencz, C. S., P. Bognar, J. Lichtenberger, D. Hamar, G. Y. Tarcsai, G. Timar, G. Molnar, S. Z. Pasztor, P. Steinbach, B. Szekely, O. E. Ferencz, and I. Ferencz-Arkos, 2004. Crop yield estimation by satellite remote sensing. *International Journal of Remote Sensing*, 25(20): 4113–4149.

Gilabert, M., S. Gandia, and J. Melia, 1996. Analyses of spectral-biophysical relationships for a corn canopy. *Remote Sensing of Environment*, 55(1): 11–20.

Gitelson, A., Y. Kaufman, R. Stark, and D. Rundquist, 2002. Novel algorithms for remote estimation of vegetation fraction. *Remote Sensing of Environment*, 80: 76–87.

Gitelson, A., S. Verma, A. Vina, D. Rundquist, G. Keydan, B. Leavitt, T. Arkebauer, G. Burba, and A. Suyker, 2003a. Novel technique for remote estimation of CO_2 flux in maize. *Geophysical Research Letters*, 30(9): 1486, doi:10,1029/2002GL016543.

Gitelson, A., A. Vina, T. Arkebauer, D. Rundquist, G. Keydan, and B. Leavitt, 2003b. Remote estimation of leaf area index and green leaf biomass in maize canopies. *Geophysical Research Letters*, 30(5): 1148,doi:10,1029/2002GL016450.

Gitelson, A., A. Vina, V. Ciganda, D. Rundquist, and T. Arkebauer, 2005. Remote estimation of canopy chlorophyll content in crops. *Geophysical Research Letters*, 32, L08403, doi:10.1029/2005GL022688,2005.

Goel, P. K., S. O. Prasher, J. A. Landry, R. M. Patel, A. A. Viau, and J. R. Miller, 2003. Estimation of crop biophysical parameters through airborne and field hyperspectral remote sensing. *Transactions of the American Society of Agricultural Engineers*, 46(4): 1235–1246.

Goodman, M., 1959. A technique for the identification of farm crops on aerial photographs. *Photogrammetric Engineering*, 25: 131–137.

Goward, S. N., T. Arvidson, D. L. Williams, R. Irish, and J. Irons, in this volume. Moderate Spatial Resolution Optical Sensors. Chapter 9.

Groten, S. M. E., 1993. NDVI–crop monitoring and early yield assessment of Burkina Faso. *International Journal of Remote Sensing*, 14(8): 1495–1515.

Haboudane, D., J. Miller, E. Pattey, P. Zarco-Tejada, and I. Strachan, 2004. Hyperspectral vegetation indices and novel algorithms for predicting green LAI of crop canopies: modeling and validation in the context of precision agriculture. *Remote Sensing of Environment*, 90: 337–352.

Hadria, R., B. Duchenin, A. Lahrouni, S. Khabba, S. Er-Raki, G. Dedieu, A. G. Chehbouni, and A. Olioso, 2006. Monitoring of irrigated wheat in a semi-arid climate using crop modeling and remote sensing data: impact of satellite revisit time frequency. *International Journal of Remote Sensing*, 27(6): 1093–1117.

Hall, A. J. P. Louis and D. W. Lamb, 2001. A method for extracting detailed information from high resolution multispectral images of vineyards. *Proceedings of the 6th International Conference on Geocomputation*, ISBN 1864995637, University of Queensland, Brisbane.

Hall, A., D. Lamb, B. Holzapfel, and J. Louis, 2002. Optical remote sensing applications in viticulture – a review. *Australian Journal of Grape and Wine Research*, 8(1): 37–47.

Hatfield, J., 1983. Remote sensing estimators of potential and actual crop yield. *Remote Sensing of Environment*, 13: 301–311.

Inoue, Y. and M. Moran, 1997. A simplified method for remote sensing of daily canopy transpiration – a case study with direct measurements of canopy transpiration in soybean canopies. *International Journal of Remote Sensing*, 18(1): 139–152.

Jensen, J. R., 2007. *Remote Sensing of the Environment, An Earth Resource Perspective*. 2nd Edn. Prentice-Hall, Upper Saddle River, NJ, 544 pp.

Johnson, L., B. Lobitz, R. Armstrong, R. Baldy, E. Weber, J. DeBendictis, and D. Bosch, 1996. Airborne imaging aids vineyard canopy evaluation. *California Agriculture*, 50(4): 14–18.

Johnson, L., D. Bosch, D. Williams, and B. Lobitz, 2001. Remote sensing of vineyard management zones: implications for wine quality. *Applied Engineering in Agriculture*, 17(4): 557–560.

Johnson, L., D. Roczen, S. Youkhana, R. Nemani, and D. Bosch, 2003. Mapping vineyard leaf area with multispectral satellite imagery. *Computers and Electronics in Agriculture*, 38: 33–44.

Justice, C. O. and C. J. Tucker, in this volume. Coarse Spatial Resolution Optical Sensors. Chapter 10.

Kastens, J. H., T. L. Kastens, D. L. A. Kastens, K. P. Price, E. A. Martinko, and R. Lee, 2005. Image masking for crop yield forecasting using time series AVHRR NDVI imagery. *Remote Sensing of Environment*, 99(3): 341–356.

Khan, M. A., V. P. Gupta, and P. C. Moharana, 2001. Watershed prioritization using remote sensing and geographical information system: a case study from Guhiya, India. *Journal of Arid Environments*, 49(3): 465–475.

Kogan, F., B. Yang, W. Guo, Z. Pei, and X. Jiao, 2005. Modelling corn production in China using AVHRR-based vegetation health indices. *International Journal of Remote Sensing*, 26(11): 2325–2336.

Landgrebe, D., 1986. A brief history of the Laboratory for Applications of Remote Sensing (LARS), http://www.lars.purdue.edu/home/LARSHistory.html

Lang, L., 1997. Use of GIS, GPS, and remote sensing: spread to California's winegrowers. *Modern Agriculture*, 1(2): 12–16.

Lanjeri, S., J. Melia, and D. Segarra, 2001. A multi-temporal masking classification for vineyard monitoring in Central Spain. *International Journal of Remote Sensing*, 22(16): 3167–3186.

Lanjeri, S., D. Segarra, and J. Melia, 2004. Interannual vineyard crop variability in the Castilla–LaMancha region during the period 1991–1996 with Landsat Thematic Mapper images. *International Journal of Remote Sensing*, 25(12): 2441–2457.

Lloyd, D., 1990. A phenological classification of terrestrial vegetation cover using shortwave vegetation index imagery. *International Journal of Remote Sensing*, 11: 2269–2279.

Martin De Santa Olalla, F., A. Calera, and A. Dominguez, 2003. Monitoring irrigation water use by combining Irrigation Advisory Service, and remotely sensed data with a geographic information system. *Agricultural Water Management*, 61(2): 111–124.

Masoud, A. A. and K. Koike, 2006. Arid land salinization detected by remotely-sensed landcover changes: a case study in the Siwa region, NW Egypt. *Journal of Arid Environments*, 66(1): 151–167.

McMaster, G. S., 2004. Simulating crop phenology. *4th International Crop Science Congress*. Brisbane, Australia, September 27–October 1, 2004. http://www.cropscience.org.au/icsc2004/poster/2/8/607mcmaster.htm

Meyer, G., T. Mehta, M. Kocher, D. Mortensen, and A. Samal, 1998. Textural imaging and discriminate analysis for distinguishing weeds for spot spraying. *Transactions of the ASAE*, 41(4): 1189–1197.

Montero, F., J. Melia, A. Brasa, D. Segarra, A. Cuesta, and S. Lanjeri, 1999. Assessment of vine development according to available water resources by using remote sensing in LaMancha, Spain. *Agricultural Water Management*, 40: 363–375.

Moran, S., 2000. Image-based remote sensing for agricultural management: perspectives of image providers, research scientists, and users. *Proceedings of the Second International Conference on Geospatial Information in Agriculture and Forestry*, I: 23–I, 29.

Nagler, P., C. Daughtry, and S. Goward, 2000. Plant litter and soil reflectance. *Remote Sensing of Environment*, 71: 207–215.

Neale, C. M. U., H. Jayanthi, and J. L. Wright, 2005. Irrigation water management using high resolution airborne remote sensing. *Irrigation and Drainage Systems*, 19(3–4): 321–336.

Nellis, M. D., 1985. Interpretation of thermal infrared imagery for irrigation water resource management. *Journal of Geography*, 84(1): 11–14.

Nellis, M. D., 1986. Remote sensing for monitoring water demand in Western Kansas. *Papers and Proceedings of the Applied Geography Conferences*, 9: 56–62.

Nicholas, H., 2004. Using remote sensing to determine the date of a fungicide application on winter wheat. *Crop Protection*, 23(9): 853–863.

Pelzmann, R., 1997. Using imagery in field management. *Modern Agriculture*, 1(2): 17- 19.

Peterson, D. and L. Johnson, 2000. The application of Earth science findings to the practical problems of growing winegrapes. *Geographic Information Sciences*, 6(2): 181– 187.

Pinter, P., J. Hatfield, J. Schepers, E. Barnes, S. Moran, C. Daughtry, and D. Upchurch, 2003. Remote sensing for crop management. *Photogrammetric Engineering and Remote Sensing*, 69(6): 647–664.

Pinter, P., J. Ritchie, J. Hatfield, and G. Hart, 2003. The Agricultural Research Service's remote sensing program: an example of interagency collaboration. *Photogrametric Engineering and Remote Sensing*, 69(6): 615–618.

Prasad, A. K., L. Chai, R. P. Singh, and M. Kafatos, 2006. Crop yield estimation model for Iowa using remote sensing and surface parameters. *International Journal of Applied Earth Observation and Geoinformation*, 8(1): 26–33.

Price, K., S. Egbert, R Lee, R. Boyce, and M. D. Nellis, 1997. Mapping land cover in a high plains agro-ecosystem using a multi-date landsat thematic mapper modeling approach. *Transactions of the Kansas Academy of Science*, 100(1–2): 21–33.

Quarmby, N. A., M. Milnes, T. L. Hindle, and N. Silleos, 1993, The use of multitemporal NDVI measurements from AVHRR data for crop yield estimation and prediction. *International Journal of Remote Sensing*, 14(2): 199–210.

RapidEye, 2007. Welcome to RapidEye. http://www.rapideye.de/ (Last accessed January 12 2008).

Ray, S. S. and V. K. Dadhwal, 2001. Estimation of crop evapotranspiration of irrigation command area using remote sensing and GIS. *Agricultural Water Management*, 49(3): 239–249.

Reed, B. C., J. H. F. Brown, D. VanderZee, T. R. Loveland, J. W. Merchant, and D. O. Ohnlen, 1994. Measuring phenological variability from satellite imagery. *Journal of Vegetation Science*, 5: 703–714.

Reed, B. C., T. R. Loveland, and L. L. Tieszen, 1996. An approach for using AVHRR data to monitor U.S. Great Plains Grasslands. *Geocarto International*, 11(3): 13–22.

Reeves, R. (ed.), 1975. *Manual of Remote Sensing*. Falls Church, Virginia: American Society of Photogrammetry.

Reyniers, M. and E. Vrindts, 2006. Measuring wheat nitrogen status from space and ground-based platform. *International Journal of Remote Sensing*, 27(3): 549–567.

Rundquist, D. and S. Samson, 1983. Application of remote sensing in agricultural analysis. Chapter 15 in B. Richason, Jr., *Introduction to Remote Sensing of the Environment*. Kendall/Hunt Publishing Company, Dubuque, pp. 317–337.

Russell, C. A., 2003. Sample preparation and prediction of soil organic matter properties by near-infrared reflectance spectroscopy. *Communications in Soil Science and Plant Analysis*, 34: 1557–1572.

Salisbury, J. W. and D. M. D'Aria, 1992. Infrared (8–14 microm-eters) remote sensing of soil particle size. *Remote Sensing of Environment*, 42: 157–165.

Scharf, P., J. Schmidt, N. Kitchen, S. Sudduth, S. Hong, J. Lory, and J. Davis, 2002. Remote sensing for Nitrogen man-agement. *Journal of Soil and Water Conservation*, 57(6): 518–524.

Schepers, A., 2002. Comparison of GIS approaches that inte-grate soil and crop variables to delineate management zones for precision agriculture. Masters Thesis, Department of Geography, University of Nebraska-Lincoln, 59 pp.

Schepers, J., M. Schlemmer, and R. Ferguson, 2000. Site-specific considerations for managing phosphorus. *Journal of Environmental Quality*, 29: 125–130.

Seidl, M. S., W. D. Batchelor, and J. O. Paz, 2004. Integrating remotely sensed images with a soybean model to improve spatial yield simulation. *Transactions of the American Society of Agricultural Engineers*, 47(6): 2081–2090.

Serrano, L., I. Filella, and J. Penuelas, 2000. Remote sensing of biomass and yield of winter wheat under different Nitrogen supplies. *Crop Science*, 40: 723–731.

Shanahan, J., J. Schepers, D. Francis, G. Varvel, W. Wilhelm, J. Tringe, M. Schlemmer, and D. Major, 2001. Use of remote-sensing imagery to estimate corn grain yield. *Agronomy Journal*, 93: 583–589.

Stamatiadis, S., D. Taskos, C. Tsadilas, C. Christofides, E. Tsadila, and J. Schepers, 2006. Relation of ground-sensor canopy reflectance to biomass production and grape color in two Merlot vineyards. *American Journal of Enology and Viticulture*, 57(4): 415–422.

Steven, M. D., P. V. Biscoe, and K. W. Jaggard, 1983. Esti-mation of sugar beet productivity from reflection in the red and infrared spectral bands. *International Journal of Remote Sensing*, 4: 325–334.

Stoner, E. R. and M. F. Baumgardner, 1981. Characteristic vari-ations in reflectance of surface soils. *Soil Science Society of America Journal*, 45: 1161–1165.

Streck, N., D. Rundquist, and J. Connot, 2002. Estimating resid-ual wheat dry matter from remote sensing measurements. *Photogrammetric Engineering and Remote Sensing*, 68(11): 1193–1201.

Sullivan, D. G., J. N. Shaw, D. Rickman, P. L. Mask, and J. C. Luvall, 2005. Using remote sensing data to evaluate sur-face soil properties in Alabama ultisols. *Soil Science*, 170(12): 954–968.

Teal, R., B. Tubana, K. Girma, K. Freeman, D. Arnall, O. Walsh, and W. Raun, 2006. In season prediction of corn grain yield potential using normalized difference vegetation index. *Agronomy Journal*, 98: 1488–1494.

Thenkabail, P. S., E. A. Enclona, M. S. Ashton, and B. Van Der Meer, 2004. Accuracy assessments of hyperspectral wave-band performance for vegetation analysis applications. *Remote Sensing of Environment*, 91(3–4): 354–376.

Tieszen, L. L., B. C. Reed, N. B. Bliss, B. K. Wylie, and D. D. Dejong, 1997. NDVI C3 and C4 production and distri-butions in Great Plains Grassland cover classes. *Ecological Applications*, 7(1): 59–78.

Ustin, S., M. Smith, S. Jacquemoud, M. Verstraete, and Y. Govaerts, 1998. GeoBotany: Vegetation mapping for Earth sciences, in A. Rencz (ed.), *Manual of Remote Sensing*, Volume 3, John Wiley & Sons, Hoboken, NJ, pp. 189–248.

Verstraeten, W. W., F. Veroustraete, C. J. Van Der Sande, I. Grootaers, and J. Feyen, 2006. Soil moisture retrieval using thermal inertia, determined with visable and thermal space-borne data, validated for European forests. *Remote Sensing of Environment*, 101(3): 299–314.

Vijaya-Kumar, P., Y. S. Ramakrishna, D. V. Bhaskara-Rao, G. Sridhar, G. Srinivasa-Rao, and G. G. S. N. Rao, 2005. Use of remote sensing for drought stress monitoring, yield prediction and varietal evaluation in castor beans (*Ricinus communis* L.). *International Journal of Remote Sensing*, 26(24): 5525–5534.

Vina, A., A. Gitelson, D. Rundquist, G. Keydan, B. Leavitt, and J. Schepers, 2004. Monitoring maize (Zea mays L.) phenology with remote sensing. *Agronomy Journal* 96: 1139–1147.

Wang, J., P. M. Rich, K. P. Price, and W. Dean-Kettle, 2005. Relations between NDVI, grassland production, and crop yield in the central great plains. *Geocarto International*, 20(3): 5–11.

Wildman, W., R. Nagaoka, and L. Lider, 1983. Monitoring spread of grape Phlloxera by color infrared aerial photogra-phy and ground investigation. *American Journal for Enology and Viticulture*, 34(2): 83–94.

Wu, J., M. Ranson, M. D. Nellis, K. Price, and S. Egbert, 1997. Evaluating CRP land soil properties using remote sensing and GIS techniques. *Journal of Soil and Water Conservation* 52(5): 352–358.

Wu, J., M. Ransom, M. D. Nellis, H. Su, and B. Rundquist, 1999. Characterizing the relationships between land use and groundwater for Finney County, Kansas. *Geographical and Environmental Modeling* 3(2): 203–215.

Yang, C., G. Anderson, and J. Everitt, 1998. A view from above: characterizing plant growth with aerial videography. *GPS World*, April, 34–37.

Yang, C., J. H. Everitt, J. M. Bradford, and D. Murden, 2004. Airborne hyperspectral imagery and yield monitor data for mapping cotton yield variability. *Precision Agriculture*, 5(5): 445–461.

Yang, C., J. H. Everitt, and J. M. Bradford, 2006. Compari-son of QuickBird satellite imagery and airborne imagery for mapping grain sorghum yield patterns. *Precision Agriculture*, 7(1): 33–44.

Ye, X., K. Sakai, L. O. Garciano, S. I. Asada, and A. Sasao, 2006. Estimation of citrus yield from airborne hyperspectral images using a neural network model. *Ecological Modelling*, 198(3–4): 426–432.

Yu, F., K. P. Price, J. Ellis, and P. Shi, 2003. Response of seasonal vegetation development to climatic variations in eastern central Asia. *Remote Sensing Environment*, 87: 42–54.

Zhang, P., B. Anderson, B. Tan, D. Huang, and R. Myneni, 2005. Potential monitoring of crop production using a satellite-based Climate-Variability Impact Index. *Agricultural and Forest Meteorology*, 132(3–4): 344–358.

C. Hydrospheric and Cryospheric Sciences

Optical Remote Sensing of the Hydrosphere: From the Open Ocean to Inland Waters

Samantha Lavender

Keywords: ocean colour, marine optics, phytoplankton and satellite oceanography.

INTRODUCTION

This chapter will primarily deal with optical remote sensing, but also mentions the use of sensors that cover the wider breadth of the electromagnetic spectrum as there is an increasing focus on sensor synergy and what all forms of remote sensing can tell us. Optical remote sensing is the detection (primarily from airborne and satellite-mounted sensors, but can also include ship and ground-mounted sensors) of the incident sunlight, in the visible and near infra-red (NIR) parts of the electromagnetic spectrum, reflected from the surface after passive interaction with the water and its constituents. Ocean colour is the spectral variation of water-leaving radiance, L_w (downwelling solar irradiance that penetrates the water surface, interacts with the water body and is backscattered towards the sensor), that can be related to the concentrations of the optically active constituents such as phytoplankton pigments, coloured dissolved organic material (CDOM) and suspended particulate matter (SPM). This chapter introduces the technology of hydrospheric optical remote sensing through a discussion of the sensors commonly used, followed by a description of the associated techniques that have been developed to determine

biogeochemical variables. A short discussion of example applications and a brief summary of the current status and future trends of the field conclude the chapter.

THE PAST, PRESENT AND FUTURE

The history of satellite ocean colour

The first spaceborne ocean colour images were obtained through hand-held cameras on the manned space missions in the 1960s and they clearly demonstrated the potential of satellites for monitoring SPM along the coasts, and the colour variation due to phytoplankton (e.g., Badgley and Childs 1969). The first quantitative work was undertaken by Clarke et al. (1970) who used an airborne spectroradiometer to record the spectral variation of water-leaving radiance above waters with different phytoplankton concentrations. The observations demonstrated a change in the water-leaving radiance spectra with increasing chlorophyll-a concentrations (chl), and that the atmosphere had a notable effect, namely that scattering increased as measurements were recorded at increasing altitude.

Satellite monitoring of the coastal zone was made possible by the Earth Resources Technology Satellite (later renamed Landsat, with Landsat-1 launched in July 1972; Goward et al., in this volume), but the first dedicated marine sensor was the Coastal Zone Color Scanner (CZCS) launched in October 1978 onboard the Nimbus-7 satellite. Ocean colour satellites have generally been launched into polar orbits (following a path that passes close to the North and South Poles) so that almost the entire Earth is covered, with a small amount of overlap at low latitudes and a greater overlap at higher altitudes, as different parts of the Earth's surface are viewed during successive orbits.

CZCS was a proof of concept sensor with a number of design specifications and objectives that included defining requirements for future ocean colour instruments. The sensor recorded the upwelling visible radiance in four wavebands (Table 27.1) with a broad NIR waveband to differentiate sea from land and cloud, and a thermal infra-red band (10.5–12.5 µm) to detect Sea Surface Temperature (SST) variability. CZCS also possessed a tilting mechanism, which pointed the scan plane up to 20 degrees forward or aft of nadir along track, which helped the sensor avoid sunglint (sunlight reflected from the sea-surface without interacting with the components within the water). Finally, the detectors had four gain settings to allow observations at a range of radiance levels and hence sun angles. However, CZCS was operated on an intermittent schedule (due to power demands), the sensor lost the thermal infra-red waveband within the first year, and the other wavebands began degrading from 1981 onwards, although they remained useable to some degree until 1984.

There was then a gap of nearly 20 years before the next ocean colour sensor was launched, and so CZCS has provided a time series which is still used today and has been reprocessed with the latest techniques (Barale et al. 1999 and more recently under the auspices of the NASA Ocean Biology Processing Group (OBPG 2006)). The North Sea Atlas (Holligan et al. 1989) provided an early insight into the CZCS archive by presenting atmospherically corrected ocean colour imagery. Figure 27.1 shows two CZCS colour composites from the North Sea. In the interval of almost 20 years when there were no marine specific ocean colour satellites, the community also developed and used airborne remote sensing together with non-marine specific satellite sensors. An example of a non-marine specific sensor is the Advanced Very High Resolution Radiometer (AVHRR; Justice and Tucker, in this volume). AVHRR has two water-leaving radiance wavebands; waveband 1 detects red wavelengths (580–680 nm) and waveband 2 detects NIR wavelengths (720–1000 nm). AVHRR can be used for

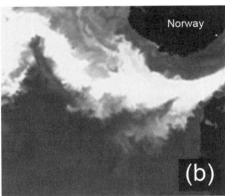

Figure 27.1   CZCS colour composites of (a) 6 September 1979 covering the south eastern North Sea with high levels of suspended particulate matter (yellow to red pixels) and (b) 6 July 1983 with a coccolithophore bloom (white pixels). High chlorophyll concentrations in both images appear as dark blue/grey patches. Wavebands 3 (550 nm), 2 (520 nm) and 1 (440 nm) are displayed as red, green and blue. (See the color plate section of this volume for a color version for this figure).
Source: Images courtesy of Steve Groom at the NERC Earth Observation Data Acquisition and Analysis Service (NEODAAS).

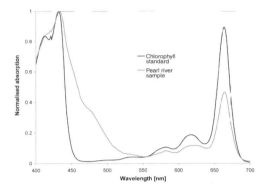

Figure 27.2 Normalized absorption spectrum for a chlorophyll standard and a sample from the Pearl River (China). The horizontal lines at the top of the figure represent the SeaWiFS wavebands.

detecting high levels of SPM or coccolithophore blooms, where there are substantial amounts of backscatter, but has neither the dynamic range nor spectral response to detect subtle changes in ocean colour. Figure 27.2 shows a standard chlorophyll and *in situ* sample normalized absorption spectrum, along with the Sea-viewing Wide Field-of-view Sensor (SeaWiFS) wavebands. See Arnone and Gould (1998), Bowers et al. (1998) and Smyth et al. (2004) for examples of AVHRR ocean colour applications.

Landsat sensors, such as the Thematic Mapper (TM) and Enhanced Thematic Mapper plus (ETM+), which have three visible and one NIR wavebands (450–520, 520–600, 630–690 and 760–900 nm), have also been used for mapping estuaries and near-shore coastal waters as they have a spatial resolution of 30 m. However, the sensors are not optimized for oceanographic applications and so do not have the signal-to-noise ratio (SNR) and level of performance required for water targets which have low reflectances, namely 2–7% compared to 10–50% for a land surface (Hamilton et al. 1993). Landsat sensors also have a large swath (field-of-view, FOV, in the across track direction) with no tilting mechanism to avoid sunglint, acquire data with a poor repeat cycle of 14–17 days and are prioritized for land observations. See Aranuvachapun and LeBlond (1981), Doxaran et al. (2006) and MacFarlane and Robinson (1984) for examples of the use of Landsat sensor data in ocean colour applications. Data is also available via the SPOT (Satellite Pour l'Observation de la Terre, initially launched in 1986) satellites and fine resolution sensors such as IKONOS (the first commercial satellite launched in September 1999; Toutin, in this volume).

The Modular Optoelectronic Scanner (MOS), designed and built by the German Aerospace Research Establishment (DLR), was launched in March 1996 onboard the Indian IRS-P3 spacecraft (Zimmermann and Neumann 1997). It had atmospheric, ocean colour and thermal wavebands, but the mission was pre-operational and had no on-board storage so was not global in coverage. The Ocean Colour Temperature Sensor (OCTS) was launched onboard ADEOS in August 1996, but was lost when ADEOS failed at the end of June 1997 (Gregg 1999). OCTS was an optical radiometer, devoted to the frequent global measurement of ocean colour and SST. ADEOS also carried the Polarization and Directionality of the Earth's Reflectances (POLDER) sensor, which by providing a multiple viewing capability measured several atmospheric paths and so enhanced estimation of atmospheric aerosols but had a coarse spatial resolution (6 m).

The ORBVIEW-2 satellite, which carries the SeaWiFS, was launched in August 1997 (McClain et al. 1998). SeaWiFS can store a limited amount of full resolution (1 km) local area coverage (LAC) data and a complete orbit of the lower resolution (4 km) global area coverage (GAC) data onboard. NASA funded the development of SeaWiFS, and purchased from GeoEye the right to use SeaWiFS data for academic research alongside agencies such as NOAA. From 2005, non-commercial data available through NASA have been limited to just the GAC coverage away from US. SeaWiFS has been the most successful and long lived (it reached its 10 year anniversary in September 2007) ocean colour satellite, and was an improvement on CZCS with seven narrow (20–40 nm) bands in the visible and NIR (see Table 27.1). Figure 27.3 shows the SeaWiFS climatological chlorophyll composite (January 1997–September 2007) derived from GAC imagery.

Since SeaWiFS, over 10 satellite ocean colour sensors have been launched (IOCCG 2007). Here the Moderate Resolution Imaging Spectrometer (MODIS) and Medium Resolution Imaging Spectrometer (MERIS) will be discussed in detail as they are global and the data is freely available to academic users. MODIS sensors were launched onboard both the Terra (EOS AM-1) satellite in December 1999 and the Aqua (EOS PM-1) satellite in May 2002, with the aim of measuring water-leaving radiance in the visible, NIR and thermal infra-red wavebands, in order to generate atmosphere, land and ocean products. MODIS has additional ocean colour wavebands when compared to SeaWiFS, and of particular interest is a narrow waveband for solar stimulated chlorophyll fluorescence and thermal bands for SST. The ENVISAT satellite, launched by the European Space Agency (ESA) in March 2002, carries a total of nine instruments, including

Table 27.1   Spatial resolution and waveband characteristics of selected global ocean colour sensors
Source: Data taken from IOCCG Report 1 (1998)

Sensor	Spatial resolution at nadir (km)	Ocean colour wavebands (nm)
CZCS	0.825	430–450, 510–530, 540–560, 660–680 and 700–800
OCTS	0.700	402–422, 433–453, 480–500, 510–530, 555–575, 655–675, 745–785 and 845–885
SeaWiFS	1.130	402–422, 433–453, 480–500, 500–520, 545–565, 660–680, 745–785 and 845–885
MODIS	1.000	405–420, 438–448, 483–493, 526–536, 546–556, 662–672, 673–683, 743–753 and 862–877
MERIS	1.20/0.300	407.5–417.5, 437.5–447.7, 485–495, 505–515, 555–565, 615–625, 660–670, 677.5–685, 700–710, 750.0–707.5, 758.75–761.25, 770–780, 855–875, 885–895 and 895–905

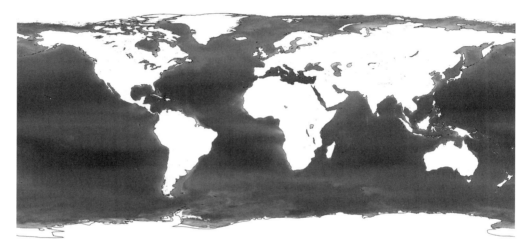

Figure 27.3   SeaWiFS climatological (January 1997–September 2007) global chlorophyll composite with increasing chlorophyll concentration shown by increasing brightness. Courtesy of GeoEYE, the NASA SeaWiFS Project (Code 970.2) and Ocean Biology Processing Group.

MERIS and the Advanced Along-Track Scanning Radiometer (AATSR) for SST. The primary scientific goal of MERIS was originally ocean colour, but the narrow and selectable wavebands (see Table 27.1 for the standard set-up) can also be used for determining atmospheric and land surface information. As with MODIS, MERIS has a solar stimulated chlorophyll fluorescence waveband. A novel result of the mission goals is the production of processed imagery with geophysical values for all pixels, that is each level-2 pixel has one of the three sets of geophysical bands: ocean, land or atmosphere (Merheim-Kealy et al. 1999). MERIS has a full spatial resolution of 300 m for coastal areas and a reduced resolution of 1.2 km with global coverage every 2–3 days (Rast et al. 1999); the full resolution data is available over Europe and where there are receiving stations. Both MODIS and MERIS suffer from sunglint as they

do not tilt and the coverage of MERIS is further reduced by the relatively narrow swath (1150 km as compared to 2330 km for MODIS).

Airborne remote sensing

There are both advantages and disadvantages to using airborne compared to satellite remote sensing. The advantages of airborne remote sensing include the ability to fly at user specific times and with repeat cycles of minutes rather than days and the increased spatial resolutions when compared to satellite sensors (Collins and Pattiaratchi 1984). Aircraft sensors can also be upgraded and regularly calibrated whereas satellite sensors must rely on solar or lunar calibration measurements. The main disadvantages of aircraft sensors are that they are on an unstable platform within the atmosphere,

Table 27.2 Spectral configuration of selected airborne sensors

Sensor	Spectral configuration
ATM	Multispectral (11 wavebands, 20 to 100 nm interval) from 420–13,000 nm.
AVIRIS	Hyperspectral (224 wavebands, 10 nm interval) from 400 to 2400 nm.
CASI	Hyperspectral, but can be flown in a spectral (288 wavebands, 1.9 nm interval) or spatial (8 to 16 wavebands, depends on the model and configuration) mode between 412 and 957 nm.
ROSIS	Hyperspectral (115 waveband, 4 nm interval) between 430 and 820 nm.

which causes additional complexities with both the atmospheric and geometric corrections.

The Fluorescence Line Imager (FLI), built by Moniteq Ltd and ITRES, was one of the first pushbroom systems (Borstadt et al. 1985). Gower and Borstadt (1990) used linear interpolations and bands situated on and each side of the naturally occurring chlorophyll−a fluorescence peak at 685 nm, to calculate the Fluorescence Line Height (FLH). ITRES then went on to develop the Compact Airborne Spectrographic Imager (CASI) which operates over the visible to NIR wavelength range (Table 27.2). The spatial resolution depends on the altitude, ground speed and integration time, but is typically in the 2–10 m range. CASI, as with the FLI, can operate in either a spatially or spectrally optimised mode (Anger et al. 1994). Similar airborne sensors have included the Reflective Optics System Imaging Spectrometer (ROSIS) developed jointly by German industry and research organizations (Kunkel et al. 1991), and Airborne Visible/Infrared Imaging Spectrometer (AVIRIS) with 20 m pixels when flown at an altitude of 65,000 feet (Carder et al. 1993), see Table 27.2.

Airborne simulators of satellite sensors have also been developed. For example, the MODIS airborne simulator has acquired high spatial resolution imagery from the NASA ER-2 high-altitude research aircraft (King et al. 1996). The Daedalus Airborne Thematic Mapper (ATM) has wavebands from the visible to thermal regions of the electromagnetic spectrum (Table 27.2). ATM wavebands 2 (450–520 nm), 3 (520–600 nm), 5 (630–690 nm) and 7 (760–900 nm) are often used as simulations of Landsat-5 TM wavebands 1 to 4.

An important application of active optical sensors is Light Detection and Ranging (lidar, also known as airborne laser scanning). With this technology, a short (laser) pulse of energy is generated by the instrument, scattered and absorbed as it travels through the transmitting medium, and then the backscattered energy that is received

by the sensor, at some later time, is used to determine the round trip time. Hyyppä et al. (in this volume) discuss lidar in more detail, but in the coastal marine environment it is generally used to map bathymetry. Combing lidar and colour aerial photography or multi/hyperspectral data can also greatly assist in the mapping of coastal substrate and coral reefs (e.g., Lubin et al. 2001, Mumby et al. 2001, Mumby and Edwards 2002) where the waters are relatively clear. These optical techniques are constrained by turbidity and, therefore, will have limited application within turbid coastal and inland water bodies. In addition, NASA's Airborne Oceanographic Lidar (AOL) has also been used to acquire individual laser-induced fluorescence spectra for the study of phytoplankton, CDOM, thin oil films and tracer dyes (Hoge et al. 1986). Further applications include oil spill detection (Fingas and Brown 2000) and validation of satellite chlorophyll algorithms (Barbini et al. 2006).

Future sensors

There have been difficulties in obtaining remote sensing imagery for commercial/operational applications, as the data tend to either not be cost-effective or else not available within suitable time-scales. This has partly been addressed with the launch of MODIS (provides unrestricted direct-broadcast data) and MERIS (has a European distribution system aimed at both commercial/operational and scientific uses). The future US operational global ocean colour sensor is the Visible Infrared Imaging Radiometer Suite (VIIRS), which is a multi-spectral scanning radiometer with 22 wavebands between 0.4 and 12 μm. It is planned that VIIRS will initially be launched on the National Polar-orbiting Operational Environmental Satellite System (NPOESS) Preparatory Project (NPP) in 2009, with further NPOESS launches in 2013 and 2016.

Europe plans to launch the Ocean and Land Colour Instrument (OLCI) on Sentinel-3 as part the Global Monitoring for Environment and Security (GMES) initiative. This sensor is of a similar design and specification to MERIS with an additional channel at 1.02 μm to enhance the existing MERIS atmospheric and aerosol correction capabilities. The first Sentinel-3 is planned for launch, with a companion satellite to improve global coverage, in 2012, with follow-up launches to meet observational requirements for robust and continuous data provision.

PROCESSING OF OCEAN COLOUR DATA

Ocean colour remote sensing has developed rapidly since the launch of SeaWiFS, and satellite imagery can now be processed automatically and is being made available via the World Wide Web. However, users should be aware of the underlying assumptions and accuracies as the science is developing and algorithms are continually updated.

Atmospheric correction

Atmospheric correction is very important for both aircraft and satellite sensors as only 5–10% of the total signal originates from the water. It is not feasible to constantly measure all the atmospheric variables necessary, so models have to make broad assumptions, particularly about aerosols. Therefore, an atmospheric correction model requires at least two wavebands in the NIR (IOCCG 1998), preferably situated at around 865–890 and 744–757 nm so as to avoid the atmospheric absorption lines, and a waveband at a longer wavelength is desirable for whitecap (the reflectance generated by breaking waves) removal/estimation and to aid atmospheric correction in coastal waters (Wang and Shi 2005).

The spectral radiance at a set wavelength (λ), $L_s(\lambda)$, received at a remote sensor can be written as a linear sum of the contributions, where: $L_p(\lambda)$ is the path radiance (radiance generated by scattering within the atmosphere); $L_{sky-g}(\lambda)$ is the sky-glitter radiance (irradiance reflected by the water surface); $L_{sun-g}(\lambda)$ is the sun-glitter radiance (or sunglint); $L_{ws}(\lambda)$ is the water-leaving radiance entering the sensor:

$$L_s(\lambda) = L_p(\lambda) + L_{sky-g}(\lambda) + L_{sun-g}(\lambda) + L_{ws}(\lambda) \tag{1}$$

The radiance entering the sensor is related to the water-leaving radiance, $L_w(\lambda)$, where $T_d(\lambda)$ is the diffuse transmittance of the atmosphere (Equation (2)). Diffuse rather than direct transmittance is used as a proportion of the pixel's radiance originates from neighbouring pixels. T_d can be calculated from Equation (3) (Gordon et al. 1983) where $\tau_r(\lambda)$ is the Rayleigh optical depth (caused by the molecules in the atmosphere), $\tau_{o3}(\lambda)$ is the ozone optical depth and θ is the satellite zenith angle. The optical thicknesses of each of the atmospheric constituents define the attenuation of solar irradiance by scattering and absorption as a function of wavelength. In the visible and NIR regions the main constituents are the permanent gases (nitrogen, carbon dioxide, oxygen and argon), aerosols, ozone and water vapour.

$$L_{ws}(\lambda) = L_w(\lambda) \cdot T_d(\lambda) \tag{2}$$

$$T_d(\lambda) = \exp\left(\frac{\tau_r(\lambda)/2 + \tau_{o3}(\lambda)}{\cos\theta}\right) \tag{3}$$

The path radiance can be split into its Rayleigh, $L_r(\lambda)$, and aerosol, $L_a(\lambda)$, scattering components (Equation (4)). This is exact only for a single-scattering atmosphere where there is no Rayleigh–aerosol interaction. In adapting the CZCS atmospheric correction model (Gordon and Castano 1987) for SeaWiFS, the radiance terms were converted to reflectance and a Rayleigh-aerosol interaction term was added to account for multiple scattering (Equation (5)):

$$L_p(\lambda) = L_r(\lambda) + L_a(\lambda) \tag{4}$$

$$R_s(\lambda) = R_r(\lambda) + R_a(\lambda) + R_{ra}(\lambda) + T_d(\lambda) \cdot R_w(\lambda) \tag{5}$$

In the visible and NIR wavelength regions the aerosols are expected to dominate the Rayleigh scattering. The maximum aerosol scattering occurs when the aerosol particle diameter is approximately the same as the radiation's wavelength and is a function of aerosol size, shape and index of refraction. To extrapolate the aerosol contribution we need one waveband to assess the magnitude and a second waveband to assess the wavelength dependence, in addition to rules governing the spectral variation (Gordon and Wang 1994a). If we assume an aerosol power-law type size distribution, the aerosol optical thickness can be calculated from Angstrom's Law (Equation (6)), where B is the Angstrom exponent which is related to the size distribution (Equation (7)), and where n is a fitting parameter, and A is proportional to the total number of aerosol particles (Gregg and Carder 1990).

$$\tau_a(\lambda) = A \cdot \lambda^{-B} \tag{6}$$

$$B = \lambda^n \tag{7}$$

In the original CZCS model, the dark pixel method was used to obtain an Angstrom exponent (Equation (8)) with an extrapolation based on Equation (9). It was assumed that the water-leaving radiance equalled zero and the ratio of aerosol reflectances in the two wavebands, i and j, was constant over an image (even in the presence of spatial inhomogeneities in aerosol concentration). Although the optical depth spectrum does exhibit a strong dependence with wavelength, the ratio of optical depths at different wavelengths may not be constant if large aerosols are present and Equation (9) would then no longer be valid. However, if this were so, the results would still support the use of Equation (10) (Gordon 1978):

$$\varepsilon(\lambda_i, \lambda_j) = \frac{R_a(\lambda_i)}{R_a(\lambda_j)} = \frac{\omega_a(\lambda_i) \cdot \tau_a(\lambda_i) \cdot p_a(\theta, \theta_0, \lambda_i)}{\omega_a(\lambda_j) \cdot \tau_a(\lambda_j) \cdot p_a(\theta, \theta_0, \lambda_j)} \tag{8}$$

$$\varepsilon(\lambda_i, \lambda_j) = \left(\frac{\lambda_i}{\lambda_j}\right)^{-n} \tag{9}$$

$$\varepsilon(\lambda) = \frac{\tau(\lambda)}{\tau(750)} \tag{10}$$

For SeaWiFS the extrapolation (Equation (8)) was not considered valid, as the NIR wavebands are too far away from the visible wavebands (Gordon 1994) and the goal was to recover water-leaving reflectance with an error of no more than 5%. The extrapolation was enhanced by deriving a set of look-up tables based on a combination of several aerosol models: Tropospheric, Maritime, Coastal (mixture of the previous two models), each with four relative humidities (50, 70, 90 and 99%). Multiple scattering is also derived from look-up tables.

The MODIS atmospheric correction algorithm was developed from the SeaWiFS approach and uses MODIS wavebands 15 (743–753 nm) and 16 (862–887 nm) in comparison to SeaWiFS wavebands 7 (745–785 nm) and 8 (845–885 nm). The width of MODIS waveband 16 has been reduced so that the atmospheric correction does not have to account for the oxygen absorption that affects SeaWiFS (Wang 1999). In addition, both MODIS and SeaWiFS employ a whitecap removal algorithm where the normalized whitecap reflectance is calculated from ancillary windspeed information (Gordon and Wang 1994b).

The MERIS atmospheric correction differs from SeaWiFS and MODIS by abandoning the assumption that the Rayleigh and aerosol radiances can be separated (Antoine and Morel 1999). It also accounts for absorbing aerosols by assuming that the water-leaving radiance at 510 nm (the chlorophyll absorption hinge point) is insensitive to changes in chlorophyll concentration, and utilizes a 'bright pixel' component within the atmospheric correction where any significant NIR water-leaving radiance (due to the presence of high concentrations of SPM or coccolithophores which have external calcium carbonate plates) can be estimated and removed before the conventional 'dark pixel' atmospheric correction (Moore et al. 1999). For SeaWiFS, initially an iterative method (Siegel et al. 2000) was developed to correct for non-negligible water reflectance in the NIR arising from moderate to high phytoplankton abundances (chlorophyll concentrations greater than ~ 2 mg m^{-3}) and so independent research was applied to correct for the effects of SPM (e.g., Ruddick et al. 2000, Lavender et al. 2005), but the processing software has now been updated so that it takes non-phytoplankton scattering (Arnone et al. 1998) into account for both MODIS and SeaWiFS (Stumpf et al. 2002). In addition, research has been looking at the MODIS longer wavelengths where the water signal can still be assumed to be negligible (Wang and Shi 2005).

The atmospheric correction of land-orientated satellite sensor systems such as Landsat ETM+ and IKONOS tends to rely on *in situ* measurements of radiance and/or aerosols when they are acquired simultaneously. Software is available in both commercial remote sensing packages or as standalone applications with techniques such as: haze removal by methods such as 'dark object subtraction', which is similar in approach to the ocean colour 'dark pixel' approach, but relies on a small percentage of pixels with the lowest reflectance from deep water and shadow (e.g., Chavez 1988, Kaufman 1989, Vincent et al. 2004); radiative transfer modelling using packages such as the MODerate resolution atmospheric TRANsmission (MODTRAN, Berk et al. 1999) and Second Simulation of the Satellite Signal in the Solar Spectrum (6S, Vermote et al. 1997).

In-water optics

The most appropriate variable to model or measure for the light field just above the water surface is reflectance, $R_{0+}(\lambda)$. This is wavelength dependent and can be determined by the inherent properties (Gould and Arnone 1977) as well as the ratio of upwelling to downwelling light:

$$R_{0+}(\lambda) = \left[\frac{0.176}{Q(\lambda)} \right] * \frac{b_b(\lambda)}{b_b(\lambda) + a(\lambda)} \qquad (11)$$

with 0.176 resulting from an empirical coefficient and approximation of air–water transmission effects and Q as the underwater irradiance-to-radiance ratio $E_u(\lambda)/L_u(\lambda)$, which is only weakly dependent on wavelength. This would be set as π for a Lambertian reflector, but from the work of Bricaud and Morel (1987) the actual value may be closer to 4.5.

The absorption and backscattering coefficients relating to each group of substances present in the water are expressed as a product of their concentrations. Each group is in fact made up of several substances (often present in differing proportions) with slightly different properties and so the specific absorption curve for each group will be subject to a certain amount of variability (Prieur and Sathyendranath 1981). However, it is often sufficiently accurate to consider each group as a routinely measured *in situ* parameter. The chlorophyll concentration is used to represent the phytoplankton, SPM the non-chlorophyllous particles, and Dissolved Organic Matter (DOM) (or more commonly the non-particulate absorption at a wavelength in the ultraviolet to blue wavelength region) the CDOM or gelbstoff.

The total backscattering and absorption coefficients for a body of water can therefore be calculated by summing the contributions from each of the specific in-water constituents:

$$a(\lambda) = a_w(\lambda) + a_{CDOM}(\lambda) \cdot CDOM + a_{chl} + a_{SPM} \cdot SPM \qquad (12)$$

$$b_b(\lambda) = 0.5 \cdot b_w(\lambda) + b_{bchl}(\lambda) + b_{bSPM}(\lambda) \cdot SPM \qquad (13)$$

The factor of 0.5 in the water backscattering term of Equation (13) converts the scattering values to backscattering values, and can be applied because of the symmetry of the volume scattering function (Morel 1974). The chlorophyll absorption and backscattering terms do not have concentration terms in Equations (12) and (13), because they have non-linear relationships with concentration. Equation (13) also describes a monodispersive system with particles of a given refractive index scattering light, so that the scattered light can be assumed to have the same wavelength as the incident light. There is also no multiple scattering and so the particles are independent (i.e., the intensities of the light scattered by each particle can be summed so that the total scattering is proportional to the number of particles).

For SeaWiFS, instead of reflectance, the normalized water-leaving radiance is used (i.e., the water-leaving radiance that would be measured by a sensor placed above the water surface if the Sun was at the zenith and the atmosphere was absent). This can be related to the in-water constituents via the sub-surface reflectance as in Equation (14) (Gordon et al. 1988). This allows different data sets to be compared and can be used instead of reflectance in developing in-water algorithms:

$$L_{wn}(\lambda) = \frac{(1 - \rho(\theta))(1 - \rho)F_0(\lambda, 0) \cdot R_{0-}(\lambda)}{Q \cdot n_w^2(1 - R_d(\lambda) \cdot R_{0-}(\lambda))} \tag{14}$$

In Equation (14), n_w is the ratio of the refracted indices for air and water, and $R_d(\lambda)$ is the water-air reflectance for totally diffuse irradiance $\cong 0.48$. $R_{0-}(\lambda)$ is the sub-surface reflectance just below the surface and $F_0(\lambda)$ is the mean extraterrestrial solar irradiance. The Fresnel reflectance, $\rho(\theta)$, of the sea surface integrated over all the incident angles can be taken as 0.02 (Whitlock et al. 1981). The Fresnel albedo due to the specular reflection of the sun and skylight, ρ, ranges from 0.021 for a nadir sun to 0.064 for a solar zenith angle of 60 degrees, and totally diffuse skylight has a value of 0.066 (Gregg and Carder 1990). Both $\rho(\theta)$ and ρ can be considered independent of wind speed (or sea state), from Austin (1974) and Priesendorfer and Mobley (1986) or Plass et al. (1975) respectively.

Retrieval algorithms and models

Ocean colour measurements can provide information on the concentration of optically active substances present in the surface layer; as ocean colour remote sensing is receiving upwelling radiance from the first optical depth it will underestimate concentrations if there are sub-surface maxima.

The quantitative determination of these substances, such as chl and SPM, can be performed using estimation or retrieval variables, X, by empirical correlation with reflectance $R(\lambda)$ or reflectance ratios $R(\lambda_1)/R(\lambda_2)$ (Tassan 1988):

$$X = f[R(\lambda)] \tag{15}$$

The retrieval variables are usually inserted into a logarithmic expression, the retrieval algorithm (Equation (16)), with parameters A and B being inferred from *in-situ* measurements:

$$\log[\text{chl, SPM}] = A + B \cdot \log[X] \tag{16}$$

The retrieval variable must be highly sensitive to the relevant property and relatively insensitive to other optically active matter, and may utilize two or more wavebands. Retrieval variables often include waveband ratios as they can suppress solar angle and atmospheric effects (Amos and Topliss 1985), and may also cancel out effects caused by the sensor tilt angle and FOV (Morel and Prieur 1977). However, waveband ratios cannot provide information on scattering and absorption and hence discriminate between different absorbing agents, so multispectral methods should be examined (Morel and Prieur 1977).

Waveband combinations are normally chosen by statistical means, such as the highest coefficient of determination and/or the lowest root mean square error (Ritchie and Cooper 1988) or the 'least squares' best-fit equation (Amos and Topliss 1985, Mitchelson et al. 1986). When *in-situ* concentrations are linearly regressed against the retrieval variable there is often a large spread of values around the line of best fit as the retrieval variable will not describe all the variability. The regression technique may be used to determine the line of best fit, but it assumes there is no error in the measurement of the concentration and that the error in the sensor reflectance, $R_s(\lambda)$, is unrelated to the error in the concentration (Whitlock et al. 1982).

Statistical techniques should only be applied under controlled environmental conditions: water depth greater than penetration depth; constant vertical concentration gradient within the penetration depth; *in-situ* data synchronous with the remote sensing; and several validation points. The assumption that a single concentration is the only factor controlling the retrieval variable may not be valid, as the water constituents have non-linear optical relationships and each varies differently with wavelength. The resulting algorithm may only apply to the particular site and also that particular time, inducing large, possibly unacceptable, errors in appreciably different water masses (Tassan 1988). There are also problems matching a point concentration, measured by the vessel, to a sensor value as this requires a geometric correction, because a pixel represents a much larger volume of water. In addition, the concentration may vary rapidly with time. The depth at which the concentration is measured should also be related to the diffuse attenuation coefficient of the remote sensing wavelength, as different wavelengths penetrate (the maximum depth from which light can be detected) to differing depths.

Chlorophyll concentration

Phytoplankton pigment concentrations vary from less than 0.03 mg m^{-3} in oligotrophic waters (poor in nutrients and therefore in phytoplankton) up to over 30 mg m^{-3} in eutrophic waters (nutrient rich waters). The general aim of chlorophyll algorithms is to identify pigment concentrations over the naturally occurring range of 0.01 to 60+ mg m^{-3} using a logarithmic scale. Phytoplankton absorb sunlight, required for photosynthesis, through pigments with 95% of the light absorbed by chlorophyll-a, b, c, photosynthetic carotenoids and photoprotectant carotenoids which are grouped together as the total pigments (Aiken et al. 1995). As the phytoplankton concentration increases the reflectance in the blue decreases (due to the absorption of chlorophyll-like pigments over a broad wavelength band) and the green increases slightly, and a ratio of blue to green water reflectance can be used to derive quantitative estimates of pigment concentration, as discussed by Clarke et al. (1970). At high pigment concentrations the signal at 440 nm becomes too small to be estimated accurately and so algorithms may switch to another ratio that is less sensitive to variations in the pigment concentration.

The current SeaWiFS algorithm is Ocean Colour 4 version 4 (OC4v4), where Equation (18) takes the highest reflectance value from the waveband at 443, 490, or 510 nm, and divides it by the reflectance at 555 nm and inputs this into Equation (17) (O'Reilly et al. 1998):

$$chl = 10^{(0.366-3.067x+1.930x^2+0.649x^3 1.532x^3)} \quad (17)$$

$$x = \log_{10}((R_{rs}443 > R_{rs}490 > R_{rs}510)/R_{rs}555) \quad (18)$$

This type of algorithm is designed for global application where the water is assumed to be predominately Case I (dominated by photosynthetic pigments). In the shelf sea and coastal situations where the water is described as Case II (dominated by CDOM and/or SPM present in the coastal zone) it will become less reliable (IOCCG 2000). Therefore, local algorithms have been developed, e.g., the chlorophyll concentration in the Bay of Biscay has been routinely retrieved from SeaWiFS data using a look-up table as described by Gohin et al. (2002). SeaWiFS and MODIS also have within the SeaWiFS Data

Analysis System (SeaDAS, http://oceancolor.gsfc. nasa.gov/seadas/) several bio-optical models that can be chosen, which will be discussed in more detail in the next section.

The 400–550 nm wavelength region cannot be applied in lakes, mainly because of CDOM absorption. Therefore, algorithms have been developed for the 660–715 nm spectral region. An example algorithm (Equation (19)) is where $L_{700-710}$ and $L_{660-665}$ are upwelling radiances measured with an airborne instrument (Kallio et al. 2003). The empirical parameters (A and B) were obtained by comparing the ratio of the upwelling radiances to *in situ* chl measurements using a least-squares fitting method. MODIS has two wavebands with 250 m spatial resolution (620–670 and 841–876 nm), so these can also be used for relatively large inland water bodies (e.g., Koponen et al. 2004):

$$chl = A \left[\frac{L_{700-710}}{L_{660-665}} \right] + B \quad (19)$$

MODIS and MERIS also have a relatively narrow waveband at around 682.5 nm to detect the solar stimulated chlorophyll fluorescence peak (Rast et al. 1999), which is closely linked to cell physiology and can vary with nutrient status, species composition and growth rate. Gower and Borstadt (1990) demonstrated the possibility of remotely sensing this peak using the airborne FLI as mentioned in the previous airborne remote sensing section. Recent research has extended this fluorescence waveband to applications such as floating sargassum seaweed (Gower et al. 2006).

Beyond chlorophyll

As the derivation of the optically active constituents becomes more complicated in Case II waters either local algorithms or bio-optical models (inverse modelling technique) can be used. SPM will normally enhance reflectance because of particle backscattering that is inversely related to wavelength, being strong in the blue and then decreasing in influence through the spectrum. CDOM reduces the reflectance as it causes significant absorption in the ultraviolet and blue.

Sathyendranath et al. (1989) suggested that the three bio-optical substances could be separated using five well-chosen wavebands, with CDOM being the most difficult variable to estimate (and being highly dependent on a waveband at around 400 nm) followed by the phytoplankton concentration. Implementations for MODIS and SeaWiFS include the Carder model (Carder et al. 2003), GSM01 model (Garver and Siegel 1997, Maritorena et al. 2002) and QAA model (Lee et al. 2002). The inverse modelling approach was also adopted for the MERIS processing in coastal regions (Attema et al. 1998). The approach (Schiller and Doerffer 1999, Doerffer and Schiller 2000) uses an Inverse Radiative Transfer Model-Neural Network (IRTM-NN) to estimate the concentration of algal pigment index 2, yellow substance absorption and SPM using a multiple non-linear regression technique to parameterize the inverse relationship between concentrations and reflectances. A detailed overview of these different approaches is given in IOCCG (2006).

Applications

There are numerous hydrospheric remote sensing applications and although this chapter has concentrated on optical sensors and theory, remote sensing techniques utilize both the optical and microwave regions of the electromagnetic spectrum. This section highlights some example applications from the open ocean to coastal and then inland waters, and highlights some sources of non-optical data.

Remotely sensed measurements of SST can show physical structures and have proved to be a useful tool for commercial fisheries exploitation using the established association of fish schools and thermal fronts (Simpson 1994). SST has historically (with a greater than 20 year time-series) been derived from thermal wavebands, but there has been an increasing usage of passive microwave imagers. Thermal infrared SST measurements (wavebands around $\sim$3.7 μm and/or near 10 μm) are sensitive to the presence of clouds, scattering by aerosols and atmospheric water vapour. Therefore, as for ocean colour imagery, the data requires an atmospheric correction and can only be derived from cloud-free pixels. In the microwave region there is a reduced signal strength, and hence reduced accuracy and resolution for passive microwave SST (primarily near 7 GHz with an additional waveband at 21 GHz for water vapour correction). In addition, these longer wavelengths are largely unaffected by clouds and generally easier to correct for atmospheric effects so can provide an improved frequency of coverage particularly in tropical environments. However, the data quality is influenced by wind-generated surface roughness and precipitation. A detailed review of the SST techniques and microwave remote sensing in general is available within Robinson (2004).

Ocean colour imagery provides additional capability for fisheries through the use of chlorophyll concentration (phytoplankton biomass) and primary production algorithms. Increased chlorophyll levels are frequently found along frontal zones, within upwelling regional and physical features (e.g., eddies where zooplankton and fish populations are known to accumulate for feeding, spawning and early life development).

Primary production has been linked to fish yields (Simpson 1994) and highlights the importance of coastal upwelling regions. Primary production algorithms have progressed from empirical to semi-analytical algorithms, incorporating temperature and light with pigments. The simplest primary production models are depth integrated and based upon the chlorophyll-a concentration or the product of depth-integrated chlorophyll and daily integrated surface Photosynthetically Available Radiation, PAR (Behrenfeld and Falkowski 1997). The input chlorophyll data will cause a large variation in the primary production and so it is essential that the chlorophyll algorithm is accurate.

Harmful algae blooms (commonly called red or brown tides on account of the red or brown coloration given to the water by high concentrations of dinoflagellates) may also cause toxic effects. Shellfish filter large quantities of water and will concentrate the algae in their tissues, which can affect the aquaculture industry directly by mass fish mortalities and also indirectly by shellfish accumulating toxins that can, under certain circumstances, render the shellfish unsafe for human consumption. The monitoring of a bloom (once it has been detected) is important, but it would also be desirable to forecast its occurrence from spectral quality, thermal signature and hydrography (Tester and Stumpf 1998). A study of the local environment will provide information on the factors that influence the initiation of a bloom and its evolution. For example, Raitsos et al. (2006) described coccolithophore bloom size variation in response to the regional environment of the subarctic North Atlantic. The spectral information can be used to identify different species, if they are in abundance and have characteristic absorption/scattering properties, and hence provide an assessment of the potential risks; functional type information is an active area of research (e.g., Sathyendranath et al. 2004, Alvain et al. 2005). An understanding of biological variability may also be enhanced with additional satellite data sources such as wind-stress (e.g., Pradhan et al. 2006, derived from the second European Remote-Sensing Satellite, ERS-2, and NASA QuikSCAT mean wind fields) and radar altimetry Sea Surface Height (SSH) that links to biophysical coupling (e.g., Wilson and Adamec 2002).

Water quality is of paramount importance to managers in nearshore coastal and inland waters (e.g., reservoirs can be the main source of water for drinking, industry and agriculture). The optical remote sensing techniques and products detailed in this chapter apply, but these waters cause particular remote sensing challenges as they are a complex mixture of optically active constituents; exhibit significant heterogeneous patterns; and may only be a few pixels in an ocean colour satellite image with a significant number of 'edge' pixels contaminated by the brighter land signal. Therefore, this community has also relied on land-orientated satellites and airborne remote sensing.

Traditional water quality monitoring depends on *in situ* measurements and the subsequent laboratory analysis of the samples with satellite remote sensing allowing for the integration and extrapolation of these measurements. Since the 1980s, inland water monitoring has tended to use the statistical correlation between broad-waveband reflectances and optically active constituents (chlorophyll and other pigment concentrations such as Phycocyanin within cyanobacterial blooms, Secchi disk depth and total suspended sediments) of the water column (e.g., Kloiber et al. 2002, Vincent et al. 2004). Many studies have also used satellite imagery to map the surface temperature of freshwater inland lakes (e.g., Li et al. 2001, Oesch et al. 2005), and important future variables will also include water levels derived from radar altimetry (e.g., Frappart et al. 2006, ESA 2006).

In nearshore coastal and inland waters, the water depth may be sufficiently shallow that the light penetrates to and returns from the bottom allowing users to map bathymetry, habitats and vegetation. Mumby et al. (2001) demonstrated the use of fine (1 m) spatial resolution airborne multispectral imagery. Lubin et al. (2001) looked at the suitability of existing sensors such as Landsat, and Mumby and Edwards (2002) looked at the cost and accuracy of IKONOS as compared directly to a suite of satellite and airborne instruments including CASI, and Landsat and SPOT sensors.

Future global hyper-spectral satellite sensors (with increased spectral resolution) would aid our ability to remotely sense rapidly changing nearshore coastal environments and under-sampled inland waters. ESA's Compact High Resolution Imaging Spectrometer (CHRIS), launched in October 2001 onboard the Project for On-Board Autonomy small satellite (Barnsley et al. 2004), was not designed as an ocean colour mission. However, its multi-angle (fly-by zenith angles of 0, ±36 and ±55 degrees) and hyperspectral (although the users have defined modes with a selection of 18–62 wavebands and 17–34 m spatial resolution) capabilities have spawned inland and coastal applications.

REFERENCES

Aiken, J., G. F. Moore, C. C. Tree, S. B. Hooker and D. K. Clark, 1995. *The SeaWiFS CZCS-type Pigment Algorithm*. NASA Technical Memorandum 104566, Vol. 29, S. B. Hooker, and E. R. Firestone, NASA Goddard Space Flight Center, Greenbelt, Maryland, USA.

Alvain, S., C. Moulin, Y. Dandonneau and F. M. Breon, 2005. Remote sensing of phytoplankton groups in case 1 waters from global SeaWiFS imagery. *Deep Sea Research Part I*, 52: 1989–2004.

Amos, C. L. and B. J. Topliss, 1985. Discrimination of suspended particulate matter in the Bay of Fundy using the Nimbus 7 Coastal Zone Color Scanner. *Canadian Journal of Remote Sensing*, 11(1): 85–92.

Anger, C. D., S. Mah and S. K. Babey, 1994. Technological enhancements to the Compact Airborne Spectrographic Imager (CASI). *First International Airborne Remote Sensing Conference and Exhibition*, Strasbourg, France.

Antoine, D. and A. Morel, 1999. A multiple scattering algorithm for atmospheric correction of remotely-sensed ocean colour (MERIS instrument): principle and implementation for atmospheres carrying various aerosols including absorbing ones. *International Journal of Remote Sensing*, 20 (9): 1875–1916.

Aranuvachapun, S. and P. H. LeBlond, 1981. Turbidity of coastal water determined from Landsat. *Remote Sensing of Environment*, 84: 113–132.

Arnone, R. A. and R. W. Gould, 1998. Coastal monitoring using ocean color. *Sea Technology*, September: 18–27.

Arnone, R. A., P. Martinolich, R. W. Gould, R. Stumpf, and S. Ladner, 1998. Coastal optical properties using SeaWiFS. In *SPIEE Ocean Optics XIV*, (eds). S. Ackleson and Cambell.

Attema, E., M. Wooding, P. Fletcher and Stuttard, M. 1998. *ENVISAT Mission: Opportunities for Science and Applications*. ESA-ESTEC, Noordwijk, The Netherlands.

Austin, R. W. 1974, The remote sensing of spectral radiance from below the ocean surface. In N. G. Jerlov and E. S. Nielsen (eds), *Optical Aspects of Oceanography*, Academic, pp. 317–344.

Badgley, P. C. and L. F. Childs, 1969. Earth resources surveys from space. In P. C. Badgley, L. Miloy and L. Childs (eds), *Oceans from space*, Gulf Publishing Company, Houston, Texas, USA, pp. 1–28.

Barale, V., D. Larkin, L. Fusco, J. M. Melinotte and G. Pittella, 1999. OCEAN Project: the European archive of CZCS historical data. *International Journal of Remote Sensing*, 20(70): 1201–1218.

Barbini, R., F. Colao, R. Fantoni, L. Fiorani, N,V. Kolodnikova, and A. Palucci, 2006. Laser remote sensing calibration of ocean color satellite data. *Annals of Geophysics*, 49(1): 35–43.

Barnsley, M. J., J. J. Settle, M. Cutter, D. Lobb and F. Teston, 2004. The PROBA-CHRIS mission: A low-cost smallsat for hyperspectral, multi-angle, observations of the Earth surface and atmosphere. *IEEE Transransactions Geoscience Remote Sensing*, 42: 1512–1520.

Behrenfeld, M. J. and P. G. Falkowski, 1997. A consumer's guide to phytoplankton primary production models. *Limnology and Oceanography*, 42(7): 1479–1491.

Berk, A., G. P. Anderson, L. S. Bernstein, P. K. Acharya, H. Dothe, M. W. Mathew, S. M. Adler-Golden, J. H. Chetwynd, S. C. Richtsmeiera, B. Pukallb, C. L. Allredb, L. S. Jeongb and M. L. Hokeb, 1999. MODTRAN4 radiative transfer modeling for atmospheric correction. In A. M. Larar (ed.), *Optical Spectroscopic Techniques and Instrumentation for Atmospheric and Space Research III*, Proceedings of SPIE 3756: 348–353.

Borstadt, G. A., Edel, H. R., Gower, J. F.R. and A. B. Hollinger, 1985. Analysis of test and flight data from the Fluorescence Line Imager. *Canadian Special Publication of Fisheries and Aquatic Sciences*, 83: 38.

Bowers, D. G., S. Boudjelas and G. E. L Harker, 1998. The distribution of fine suspended sediments in the surface waters of the Irish Sea and its relation to tidal stirring. *International Journal of Remote Sensing*, 19(14): 2789–2805.

Bricaud, A. and A. Morel, 1987. Atmospheric corrections and interpretation of marine radiances in CZCS imagery: Use of a reflectance model. *Oceanol. Acta*, 7: 33–50.

Carder, K. L., P. Reinersman, R. F. Chen, F. Muller-Karger, C. O. Davies and M. Hamilton, 1993. AVIRIS calibration and application in coastal oceanic environments. *Remote Sensing of the Environment*, 44: 205–216.

Carder, K. L., F. R. Chen, Z. P. Lee, S. K. Hawes and J. P. Cannizzaro, 2003. *MODIS Ocean Science Team Algorithm Theoretical Basis Document*. ATBD 19, Case 2 Chlorophyll-a. Version 7. 30 January 2003. College of Marine Science, University of South Florida, 67 pp.

Clarke, G. L., G. C. Ewing and C. J. Lorenzen, 1970. Spectra of backscatterd light from the sea obtained from aircraft as a measure of chlorophyll concentration. *Science*, 167: 1119–1121.

Chavez, P. S., 1988. An improved dark-object subtraction technique for atmospheric scattering correction of multi-spectral data. *Remote Sensing of Environment*, 24: 459–479.

Collins, M. and C. Pattiaratchi, 1984. Identification of suspended sediment in coastal waters using airborne thematic mapper data. *International Journal of Remote Sensing*, 5(4): 635–657.

Doerffer, R. and H. Schiller, 2000. Pigment index, sediment and gelbstoff retrieval from directional water leaving reflectances using inverse modelling technique, MERIS ATBD 2.12, Issue 4, Revision 0, 83 pp.

Doxaran, D., P. Castaing and S. J. Lavender, 2006. Monitoring the maximum turbidity zone and detecting fine-scale turbidity features in the Gironde estuary using high spatial resolution satellite sensor (SPOT HRV, Landsat ETM+) data. *International Journal of Remote Sensing*, 27(11): 2303–2321.

ESA, 2006. River and Lake. http://earth.esa.int/riverandlake/ (Last accessed 6 January 2008).

Fingas, M. F. and C. E. Brown, 2000. Oil-spill remote sensing – an update, *Sea Technology*, October: 21–28.

Frappart, F., S. Calmant, M. Cauhopé, F. Seyler and Cazenave, A. 2006. Preliminary results of ENVISAT RA-2-derived water levels validation over the Amazon basin. *Remote Sensing of Environment*, 100(2): 252–264.

Garver, S. A. and D. A. Siegel, 1997. Inherent Optical Property Inversion of Ocean Color Spectra and its Biogeochemical Interpretation: I. Time Series from the Sargasso Sea. *Journal of Geophysical Research*, 102: 18607–18625.

Gohin, F., J. N. Druon and L. Lampert, 2002. A five channel chlorophyll concentration algorithm applied to SeaWiFS data processed by SeaDAS in coastal waters. *International Journal of Remote Sensing*, 23: 1639–1661.

Gordon, H. R., 1978. Removal of atmospheric effects from satellite imagery of the oceans. *Applied Optics*, 17(10): 1631–1636.

Gordon, H. R., 1994. Equivalence of the point and beam spread functions of scattering media: a formal demonstration. *Applied Optics*, 33(6): 1120–1122.

Gordon, H. R. and D. J. Castano, 1987. Coastal Zone Color Scanner atmospheric correction algorithm: multiple scattering effects. *Applied Optics*, 26(11): 2111–2122.

Gordon, H. R. and M. Wang, 1994a. Retrieval of water-leaving radiances and aerosol optical thickness over the oceans with SeaWiFS: a preliminary algorithm. *Applied Optics*, 33: 443–452.

Gordon, H. R. and M. Wang, 1994b. Influence of oceanic whitecaps on atmospheric correction of ocean-color sensors. *Applied Optics*, 33(33): 7754–7763.

Gordon, H. R., D. K. Clark, J. W. Brown, O. B. Brown, R. H. Evans and W. W. Broenkow, 1983. Phytoplankton pigment concentrations in the Middle Atlantic Bight: comparison of ship determinations and CZCS estimates. *Applied Optics*, 22(1): 20–36.

Gordon, H. R., J. W. Brown and R. H. Evans, 1988. Exact Rayleigh scattering calculations for use with the Nimbus-7 Coastal Zone Scanner. *Applied Optics*, 27(5): 862–871.

Gould, R. W. and R. A. Arnone, 1977. Remote sensing estimates of inherent optical properties in a coastal environment, *Remote Sensing of Environment*, 61: 290–301.

Goward, S. N., T. Arvidson, D. L. Williams, R. Irish and J. Irons, in this volume. Moderate Spatial Resolution Optical Sensors. Chapter 9.

Gower. J. F. R. and G. A. Borstadt, 1990. Mapping of phytoplankton by solar-stimulated fluorescence using an imaging spectrometer. *International Journal of Remote Sensing*, 11(2): 313–320.

Gower, J., C. M. Hu, G. Borstad and S. King, 2006. Ocean color satellites show extensive lines of floating sargassum in the Gulf of Mexico. *IEEE Transactions on Geoscience and Remote Sensing*, 44(12): 3619–3625.

Gregg, W. W., 1999. Initial analysis of ocean color data from the ocean color and temperature scanner. *Applied Optics*, 38(3): 476–485.

Gregg, W. W. and K. L. Carder, 1990. A simple spectral solar irradiance model for cloudless maritime atmospheres. *Limnology and Oceanography*, 35(8): 1657–1675.

Hamilton, M. K., C. O. Davis, W. J. Rhea, S. H. Pilorz, and K. L. Carder, 1993. Estimating chlorophyll content and bathymetry of Lake Tahoe using AVIRIS data. *Remote Sensing Environment*, 44: 217–230.

Hoge, F. E., R. N. Swift and J. K. Yungel, 1986. Active–Passive airborne ocean color measurement 2: applications. *Applied Optics*, 25(1): 48–57.

Holligan, P. M., T. Aarup and S. Groom, 1989. The North Sea satellite colour atlas. *Continental Shelf Research*, 9(8): 665–765.

Hyyppä, J., W. Wagner, M. Hollaus and H. Hyyppä, in this volume. Airborne Laser Scanning. Chapter 14.

IOCCG, 1998. *Minimum Requirements for an Operational Ocean-colour Sensor for the Open Ocean*. Report of an IOCCG working group held in Villefranche-sur-Mer, France, October 6–7, 50 pp.

IOCCG, 2000. Remote sensing of ocean colour in coastal, and other optically-complex waters, In S. Sathyendranath, (eds), *Reports of the International Ocean-Colour Coordinating Group*, No. 3. IOCCG. Dartmouth, Canada.

IOCCG, 2006. Remote sensing of inherent optical properties: fundamentals, tests of algorithms, and applications. In Z.-P. Lee (ed.), *Reports of the International Ocean-Colour Coordinating Group*, No. 5, IOCCG, Dartmouth, Canada.

IOCCG, 2007. Ocean colour sensors. http://www.ioccg.org/ sensors_ioccg.html (Last accessed 6 January 2008).

Justice, C. O. and C. J. Tucker, in this volume. Coarse Spatial Resolution Optical Sensors. Chapter 10.

Kallio, K., S. Koponen and J. Pulliainen, 2003. Feasibility of airborne imaging spectrometry for lake monitoring – a case study of spatial chlorophyll a distribution in two meso-eutrophic lakes. *International Journal of Remote Sensing*, 24(19): 3771–3790.

Kaufman, Y. J., 1989. The atmospheric effect on remote sensing and its correction. In G. Asrar, (ed.), *Theory and Applications of Optical Remote Sensing*. John Wiley and Sons, New York, pp. 336–428.

King, M. D., W. P. Menzel, P. S. Grant, J. S. Myers, G. T. Arnold, S. E. Platnick, L. E. Gumley, S. C. Tsay, C. C. Moeller, M. Fitzgerald, K. S. Brown and F. G. Osterwisch, 1996. Airborne scanning spectrometer for remote sensing of cloud, aerosol, water vapour and surface properties. *Journal of Atmospheric Oceanic Technology*, 13: 777–794.

Kloiber, S. M., P. L. Brezonik, L. G. Olmanson and M. E. Bauer, 2002. A procedure for regional lake water clarity assessment using Landsat multispectral data. *Remote Sensing of Environment*, 82: 38–47.

Koponen, S., K. Kallio, J. Pulliainen, J. Vepsäläinen, T. Pyhälahti and M. Hallikainen, 2004. Water quality classification of lakes using 250-m MODIS data. *IEEE Geoscience and Remote Sensing Letters*, 1(4): 287–291.

Kunkel, B., F. Blechinger, D. Viehmann, H. Van der Piepen and R. Doeffer, 1991. ROSIS imaging spectrometer and its potential for ocean parameter measurements (airborne and spaceborne), *International Journal of Remote Sensing*, 12(4): 753–761.

Lavender, S. J., M. H. Pinkerton, G. F. Moore, J Aiken and D. Blondeau-Patissier, 2005. Modification to the Atmospheric Correction of SeaWiFS Ocean Colour Images over Turbid Waters. *Continental Shelf Research*, 25: 539–555.

Lee, Z. P., K. L. Carder and R. Arnone, 2002. Deriving inherent optical properties from water color: A multi-band quasi-analytical algorithm for optically deep waters. *Applied Optics*, 41: 5755–5772.

Li, X., W. Pichel, P. Clemente-Colon, V. Krasnopolsky and J. Sapper, 2001. Validation of coastal sea and lake surface temperature measurements derived from NOAA/AVHRR data. *International Journal of Remote Sensing*, 22: 1285–1303.

Lubin, D., W. Li, P. Dustan, C. H. Mazel and K. Stamnes, 2001. Spectral signatures of coral reefs: features from space. *Remote Sensing of Environment*, 75: 127–137.

MacFarlane, N. and I. S. Robinson, 1984. Atmospheric correction of Landsat MSS data for a multidate suspended sediment algorithm. *International Journal of Remote Sensing*, 5(3): 561–576.

Maritorena, S., D. A. Siegel and A. R. Peterson, 2002. Optimization of Semi-Analytical Ocean Color Model for Global Scale Applications. *Applied Optics*, 41: 2705–2714.

McClain, C. R., M. L. Cleave, G. C. Feldman, W. W. Gregg, S. B. Hooker and N. Kuring, 1998. Science quality SeaWiFS data for global biosphere research. *Sea Technology*, September: 10–15.

Merheim-Kealy, P., J. P. Huot and S. Delwart, 1999. The MERIS ground segment. *International Journal of Remote Sensing*, 20(9): 1703–1712.

Mitchelson, E. G., N. J. Jacob and J. H. Simpson, 1986. Ocean colour algorithms for case 2 waters of the Irish Sea in comparison to algorithms from case 1 waters. *Continental Shelf Research*, 5(3): 403–415.

Moore, G. F., J. Aiken and S. J. Lavender, 1999. The atmospheric correction of water colour and the quantitative retrieval of suspended particulate matter in Case II waters: application to MERIS. *International Journal of Remote Sensing*, 20(9): 1713–1733.

Morel, A., 1974. Optical properties of pure seawater. In N. G. Jerlov, and S. E. Nielsen (eds), *Optical Aspects of Oceanography*. Academic Press, London.

Morel, A. and L. Prieur, 1977. Analysis of variations in ocean colour. *Limnology and Oceanography*, 22: 709–722.

Mumby P. J. and A. J. Edwards, 2002. Mapping marine environments with IKONOS imagery: enhanced spatial resolution can deliver greater thematic accuracy. *Remote Sensing of Environment*, 82(2–3): 248–257.

Mumby, P. J., J. R. M. Chisholm, C. D. Clark, J. D. Hedley and J. Jaubert, 2001. A bird's-eye view of the health of coral reefs. *Nature*, 413: 36.

OBPG, 2006. Ocean color web. http://oceancolor.gsfc.nasa. gov/CZCS/czcs_processing/ (Last accessed 6 January 2008).

O'Reilly, J. E., S. Maritorena, B. G. Mitchell, D. A. Siegel, K. L. Carder, S. A. Garver, M. Kahru and C. McClain, 1998. Ocean color chlorophyll algorithms for SeaWiFS. *Journal of Geophysical Research*, 103(C11): 24937–24953.

Oesch, D. C., J.-M. Jaquet, A. Hauser and S. Wunderle, 2005, Lake surface water temperature retrieval using advanced very high resolution and Moderate Resolution Imaging Spectroradiometer data: Validation and feasibility study. *Journal of Geophysical Research*, 110(C12014): doi:10.1029/2004JC002857.

Plass, G. N., G. W. Kattawar and J. A. Guinn, 1975. Radiative transfer in the Earth's atmosphere: Influence of ocean waves. *Applied Optics*, 14: 1924–1936.

Pradhan, Y., S. J. Lavender, N. J. Hardman-Mountford and J. Aiken, 2006. Seasonal and inter-annual variability of chlorophyll-a concentration in the Mauritanian upwelling: Observation of an anomalous event during 1998–1999. *AMT Deep-Sea Research II Special Issue*, 53: 1548–1559.

Prieur, L. and S. Sathyendrath, 1981. An optical classification of coastal and oceanic waters based on the specific spectral absorption curves of phytoplankton pigments dissolved organic matter and other particulate materials. *Limnology and Oceanography*, 26(4): 671–689.

Priesendorfer, R. W. and C. D. Mobley, 1986. Albedos and glitter patterns of a wind-roughened sea surface. *Journal of Physical Oceanography*, 16: 1293–1316.

Raitsos, D. E., S. J. Lavender, Y. Pradhan, T. Tyrrell, P. C. Reid and M. Edwards, 2006. Coccolithophore bloom size variation in response to the regional environment of the subarctic North Atlantic. *Limnology and Oceanography*, 51: 2122–2130.

Rast, M., J. L. Bezy and S. Bruzzi, 1999. The ESA Medium Resolution Imaging Spectrometer MERIS – a review of the instrument and its mission. *International Journal of Remote Sensing*, 20(9): 1679–1680.

Ritchie, J. C. and C. M. Cooper, 1988. Comparison of measured suspended sediment concentrations with suspended sediment concentrations estimated from Landsat MSS data. *International Journal of Remote Sensing*, 9(3): 379–387.

Robinson, I., 2004. *Measuring the Oceans from Space: The Principles and Methods of Satellite Oceanography*. Springer/Praxis Publishing, Berlin, Germany.

Ruddick, K. G., Ovidio, F. and M. Rijkeboer, 2000. Atmospheric correction of SeaWiFS imagery for turbid and inland waters. *Applied Optics*, 39: 897–912.

Sathyendranath, S., L. Prieur and A. Morel, 1989. A three-component model of ocean colour and its application to remote sensing of phytoplankton pigments in coastal waters. *International Journal of Remote Sensing*, 10(8): 1373–1394.

Sathyendranath, S, L. Watts, E. Devred, T. Platt, C. Caverhill and H. Maass, 2004. Discrimination of diatoms from other phytoplankton using ocean-colour data. *Marine Ecology Progress Series*, 272: 59–68.

Schiller, H. and R. Doerffer, 1999. Neural network for emulation of an inverse model – operational derivation of Case II water properties from MERIS data. *International Journal of Remote Sensing*, 20(9): 1735–1746.

Siegel, D. A., M. Wang, S. Maritorena and W. Robinson, 2000. Atmospheric correction of satellite ocean color imagery: the black pixel assumption. *Applied Optics*, 39: 3582–3591.

Simpson, J. J., 1994, Remote sensing in fisheries: a tool for better management in the utilization of a renewable resource. *Canadian Journal of Fisheries and Aquatic Sciences*, 51: 743–771.

Smyth, T. J., T. Tyrrell and B. Tarrant, 2004. Time series of coccolithophore activity in the Barents Sea, from twenty years of satellite imagery. *Geophysical Research Letters*, 31(11): doi:10.1029/2004GL019735.

Stumpf, R. P., R. A. Arnone, R. W. Gould, P. Martinolich and V. Ransibrahmanakul, 2002. A partially-coupled ocean-atmosphere model for retrieval of water-leaving radiance from SeaWiFS in coastal waters. In S. B. Hooker

and E. R. Firestone (eds), Algorithm Updates for the Fourth SeaWiFS Data Reprocessing, Vol. 22, (NASA Tech. Memo. 2002-206892, NASA Goddard Space Flight Center, Greenbelt, MD, 2003).

Tassan, S., 1988. The effect of dissolved 'yellow substance' on the quantitative retrieval of chlorophyll and total suspended sediment concentrations from remote measurements of water colour. *International Journal of Remote Sensing*, 9(4): 787–797.

Tester, P. A. and R. P. Stumpf, 1998. Phytoplankton blooms and remote sensing: what is the potential for early warning. *Journal of Shellfish Research*, 17(5): 1469–1471.

Toutin, T., in this volume. Fine Spatial Resolution Optical Sensors. Chapter 8.

Vincent, R. K., X. Qin, R. M. L. McKay, J. Miner, K. Czajkowski, J. Savino and T. Bridgeman, 2004. Phycocyanin detection from Landsat TM data for mapping cyanobacterial blooms in Lake Erie. *Remote Sensing of Environment*, 89:381–392.

Vermote, E. F., D. Tanre, J. L. Deuze, M. Herman and J. J. Morcrette, 1997. Second simulation of the satellite signal in the solar spectrum, 6S: an overview. *IEEE Transactions on Geoscience and Remote Sensing*, 35(3): 675–686.

Wang, M., 1999. Validation study of the SeaWiFS oxygen A-band absorption correction: comparing the retrieved cloud optical thicknesses from SeaWiFS measurements. *Applied Optics*, 38(6): 937–944.

Wang, M. and W. Shi, 2005. Estimation of ocean contribution at the MODIS near-infrared wavelengths along the east coast of the U.S.: Two case studies. *Geophysical Research Letters*, 32: L13606.

Whitlock, C. H., L. R. Poole, J. W. Usry, W. M. Houghton, W. G. Witte, W. D. Morris and E. A. Garganus, 1981. Comparison of reflectance with backscatter and absorption parameters for turbid waters. *Applied Optics*, 20(3): 517–522.

Whitlock, C. H., C. Y. Kuo and S. R. LeCroy, 1982. Criteria for the use of regression analysis for remote sensing of sediment and pollutants. *Remote Sensing of Environment*, 12: 151–168.

Wilson, C. and D. Adamec, 2002. A global view of biophysical coupling from SeaWiFS and TOPEX satellite data, 1997–2001, *Geophyical Research Letters*, 29: 10.1029/2001GL014063.

Zimmermann, G. and A. Neumann, 1997. MOS a spaceborne imaging spectrometer for ocean remote sensing. *Backscatter*, 9–13: May.

Remote Sensing of the Cryosphere

Jeff Dozier

Keywords: snow, ice, sea ice, glacier, permafrost.

WHAT IS THE CRYOSPHERE?

The *cryosphere* is the collective term to describe water in the solid form on Earth, including snow, river and lake ice, permafrost, glaciers, ice sheets and ice caps, and sea ice. In Earth's climate, the cryosphere is important because snow and ice comprise cold, wet, bright surfaces that reflect most of the incoming solar radiation back to the atmosphere and to space. Snow and ice therefore significantly affect energy and mass exchange between Earth's surface and atmosphere and are important reservoirs of fresh water.

Study of the cryosphere through remote sensing is addressed through its different components. In this chapter, I consider seasonal snow cover, mountain glaciers and ice caps, continental-scale ice sheets, sea ice, and permafrost. The relationships between the phenomena of interest and the electromagnetic properties, the temporal and scales of variability, and other factors affect the remote sensing strategy.

Seasonal snow cover

Of the seasonal changes that occur on Earth's land surface, perhaps the most profound is the accumulation and melt of the seasonal snow cover. During winter, snow covers 45 million km^2, about 30% of the land surface, affecting climate, weather, and the water balance. Its high albedo reduces the absorbed solar radiation, its low thermal diffusivity insulates the ground, and its large enthalpy of fusion (335,000 J/kg) affects the heat and moisture fluxes. Therefore, snow cover is a huge influence on the hydrologic cycle during the winter and spring for much of Earth's land area. Near many mountain ranges, the seasonal snow cover is the major source of runoff, filling rivers and recharging aquifers that over a billion people, one-sixth of the world's population, depend on for their water resources and that are at risk in a warming climate (Barnett et al. 2005). Snow affects large-scale atmospheric circulation. Early-season snow cover variability in the northern hemisphere, for example, leads to altered circulation patterns, which in turn have implications for climate predictability (Cohen and Entekhabi 1999). For example, winters with greater than normal Eurasian snow cover are usually followed by a weaker Asian monsoon (Liu and Yanai 2002).

Mountain glaciers and ice caps

Occurring in mountain valleys and in high mountain basins at all latitudes, glaciers derive from accumulation of seasonal snow that does not fully melt during the summer. Years of accumulation allow the overlying snow to compress the lower layers until they close off the pore spaces between the snow grains and become glacial ice. As a plastic, the ice flows slowly downhill under the force of gravity. At the lower elevations, snowfall is less and melt rate is greater, so there the accumulation is less than the melt. The difference is made up from the excess ice flowing from the higher elevations, and the integration of accumulation and melt is called the *mass balance* of the glacier.

Currently, most glaciers are shrinking, and this evidence of widespread negative mass balance is one of the more powerful indicators of worldwide climate change. The loss of mass – the transition of ice to water – is a major contributor to sea-level rise in this century, along with thermal expansion of the warming ocean (Meier 1984, Gregory and Oerlemans 1998).

Ice sheets

The large ice sheets of Antarctica and Greenland contain about 2% of Earth's water. With $97\frac{1}{2}\%$ of the water in the oceans, the ice sheets therefore contain about 80% of the fresh water. Of the total mass of water on land in ice, Antarctica has about 90%, Greenland nearly 10%, and the other glaciers the remaining fraction, less than 1%. Ice in East Antarctica is grounded largely above current sea level, whereas West Antarctica is grounded on bedrock below sea level. A plausible risk of a warming climate is for accelerated flow of ice from West Antarctica into the ocean, which would raise sea level dramatically.

The mass balance of the great ice sheets is difficult to measure. Surface measurements are difficult in such harsh environments and are therefore sparse. Much of the understanding we have gained in the last decade about ice sheet mass balance and dynamics has resulted from remote sensing (Bindschadler 2006, Luthcke et al. 2006, *Science* 2006, Velicogna and Wahr 2006).

Sea ice

Sea ice covers much of the polar oceans, 5–8% of the total ocean area, and forms by freezing of sea water. Because ice does not nucleate easily around impurities, salinity of sea ice is 2–5 'practical salinity units' (about equivalent to parts per thousand), whereas ocean water has a salinity of 32–37. When sea ice forms, the ocean water that does not freeze is briny and therefore denser. The seasonal formation of sea ice helps drive the vertical circulation of the ocean. Compared with the unfrozen sea surface, the albedo of sea ice is significantly higher. Sea ice also interrupts energy and mass transfer between the atmospheric boundary layer and the sea surface.

Remote sensing reveals considerable seasonal, regional, and interannual variability in the sea-ice covers of both hemispheres. Along with seasonal snow, sea ice is one of the most variable surface covers. In the southern hemisphere, sea-ice extent varies from 3–4 million km^2 in summer (February) to 17–20 million km^2 in winter. In the northern hemisphere, the confined topography of the Arctic Ocean causes a larger perennial ice area, varying from 7–9 million km^2 to 14–16 million km^2. Arctic sea ice has decreased about 36,000 km^2/yr over the past 30 years, while Antarctic sea ice has decreased about 15,000 km^2/yr (Cavalieri et al. 2003).

Permafrost

Permafrost is generally defined as soil areas that remain perennially frozen, at least for several years, with a thawing active layer during the summer. It is difficult to remotely sense whether the ground is frozen, because the remote sensing signal usually comes from the surface or very near the surface (Duguay et al. 2005). This is especially true in the summer when liquid water is present. The extent of permafrost is important because of its relationship to a changing climate, and the freeze/thaw state of the soil links the terrestrial, energy, and carbon cycle. In the boreal latitudes, the switching on and off of the land–atmosphere carbon exchange coincides with the freeze/thaw transitions. Therefore remote sensing of permafrost and seasonally frozen ground is an important objective.

ELECTROMAGNETIC PROPERTIES OF SNOW AND ICE

Fundamental to understanding how we can remotely sense properties of snow and ice, other than elevation of the ice surface or gravitational estimates of ice mass, is an appreciation of the way snow and ice scatter, absorb, or emit electromagnetic radiation. Snow consists of grains of ice with air and sometimes water in the pore spaces. Glacier ice is snow that has been compressed to eliminate the connectivity between the pores, and thereby consists of pure ice with air bubbles and other impurities. Sea ice is frozen from the ocean water and contains pockets of brine. Glacial ice and sea ice frequently are covered by snow.

Optical properties of pure ice (and water)

The fundamental optical property of a pure bulk material is its complex refractive index $N_\lambda = n_\lambda$. The real part $n_\lambda = c$ is the ratio of the speed of light in vacuum to that at wavelength λ in the medium. The imaginary part k, also called the absorption coefficient, is defined by Equation (1), describing how radiation of intensity I at wavelength λ decays as it passes along distance s through a pure medium. The inclusion of wavelength λ in the denominator allows k to be dimensionless; otherwise it would have dimensions of inverse length. For materials where the refractive index is isotropic (i.e., independent of direction through the molecular structure), the refractive index is the square root of the dielectric constant (which is not a 'constant' but varies with wavelength);

$$\frac{dI_\lambda}{I_\lambda} = -\frac{4\pi k_\lambda}{\lambda}ds \quad \text{so} \quad \frac{I_\lambda(s)}{I_\lambda(0)} = \exp\left(-\frac{4nk_\lambda}{\lambda}s\right) \tag{1}$$

Figure 28.1 shows the absorption coefficients of ice and water as they vary with wavelength. Several features affect the remote sensing signal: from the visible through the infrared wavelengths, ice and water are similarly absorptive. In the microwave part of the spectrum, however, water is much more absorptive than ice by a factor of 80–100. In the visible spectrum, both ice and water are quite transparent, but they both become more absorptive in the near-infrared and especially in the infrared.

Optical properties of snow, glaciers, and sea ice

Starting with the optical properties of a substance, the scattering properties of an element, for example a grain of snow, are calculated with Mie theory. Mie scattering theory has a long, rich history starting with the original paper (Mie 1908) and continuing to efficient, accurate computational methods (Nussenzveig and Wiscombe 1980) and extensions to nonspherical ice particles (Grenfell and Warren 1999). Depending on the size and shape of the scattering element, in comparison to the wavelength of the radiation, and the presence of impurities such as water or dust, one can calculate the scattering properties, generally the single-scattering albedo, the extinction efficiency, and a characterization of the phase function (Bohren and Huffman 1998). Once the properties of an individual element are calculated, one can estimate the reflectivity, absorption, and transparency of a collection of such elements, for example a snow pack, the top layers of a glacier, or sea ice.

The details of doing this, however, depend on the size of the ice grains in relation to the wavelength. In the visible and infrared parts of the spectrum, the snow grains are large compared to the wavelength of the light, so the scattering by one grain does not interfere with the scattering of its neighbor.

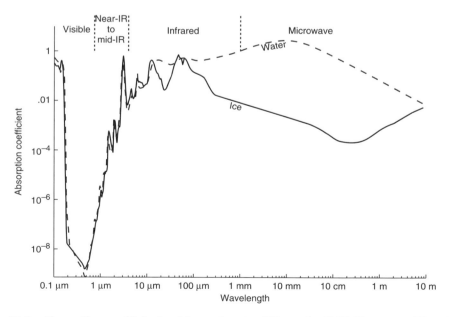

Figure 28.1   Absorption coefficients of ice and water (Wiscombe 2005, Warren and Brandt 2008). From visible through infrared wavelengths, ice and water are similar. In the microwave part of the spectrum, however, water is 80–100 times more absorptive than ice. (See the color plate section of this volume for a color version of this figure).

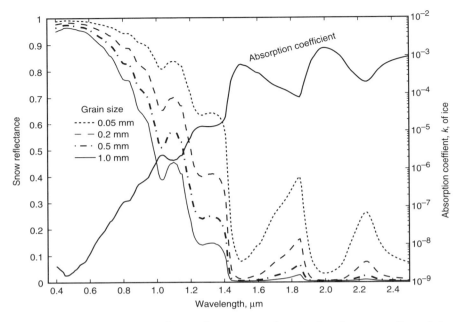

Figure 28.2   Spectral reflectivity of snow for a range of grain sizes from very fine winter snow to coarse spring snow. Also shown, on the right-hand axis, is the absorption coefficient of ice. Where ice is transparent (low absorption coefficient) the reflectivity is high. Where absorption is moderate, snow reflectivity is sensitive to grain size.

Therefore, the grains can be treated as independent scatterers, and multiple scattering theory can be applied for the calculation.

Figure 28.2 shows the reflectivity of clean, deep snow from 0.4–2.5 μm, where solar radiation is the significant energy driver at Earth's surface. In the visible wavelengths where ice is transparent, snow is more reflective than any other surface cover. In the near-infrared ice is more absorptive, so snow is darker and the reflectivity is inversely related to the grain size. In the case of glacier ice, we can reverse the calculation and consider an ice medium with air bubbles as the scatterers. As light penetrates into the snow pack, glacier, or sea ice, many scatterings occur, so the penetration depth is small, ranging from a few mm to a fraction of a meter at most. Because the reflected signal is scattered many times, the reflection is weakly polarized. Moreover, there is only a small contrast between ice and water. Many computer codes exist for multiple scattering calculations; the most widely used and easily available is the discrete-ordinates method (Stamnes et al. 1988).

In the microwave part of the spectrum, however, the relationships between the grain sizes, wavelengths, and spacing between the grains are quite different. Here the grains and the wavelength of the radiation are about the same size, and the grains are packed together relative to the wavelength.

Therefore, the grains cannot be treated as independent scatterers; instead we think of a snowpack or a volume of glacier ice or sea ice as a dense medium itself characterized by extinction per unit volume. Water in the microwave is more absorptive than ice, so wet snow is much more absorptive than dry snow. In dry snow, the combination of the weak absorption by ice and the long wavelength of the radiation in comparison to the grains allow penetration of many meters. In the microwave region, the signal is also strongly polarized. These same properties allow microwave radiation to penetrate through clouds, so remote sensing of the cryosphere with microwave sensors is possible in cloudy weather. However, the microwave signal is affected by grain size, layers in the snowpack, and liquid water, so it is often difficult to untangle all the properties that are part of the signal.

CRYOSPHERIC PRODUCTS AND THEIR INSTRUMENTS

Snow-covered area and albedo

For four decades, satellite remote sensing instruments have measured snow properties from drainage-basin to continental scales (Dozier and

Table 28.1 Spectral and spatial bands used for snow and glacier mapping for a variety of Earth-orbiting satellite sensors

Sensor	Band no.	Spectral range (μm)	Pixel size at nadir (m)	Spectral region
ASTER	1	0.52–0.60	15	VIS
	2	0.63–0.69	15	
nadir	3n/b	0.76–0.86	15	NIR
	4	1.600–1.700	30	
	5	2.145–2.185	30	
	6	2.185–2.225	30	SWIR
	7	2.235–2.285	30	
	8	2.295–2.365	30	
	9	2.360–2.430	30	
Landsat TM	1	0.450–0.515	30	
	2	0.525–0.605	30	VIS
	3	0.630–0.690	30	
	4	0.750–0.900	30	NIR
	5	1.55–1.75	30	SWIR
	7	2.09–2.35	30	
ETM	pan	0.52–0.90	15	VNIR
MODIS	1	0.620–0.670	250	VIS
	2	0.841–0.876	250	NIR
	3	0.459–0.479	500	VIS
	4	0.545–0.565	500	
	5	1.230–1.250	500	
	6	1.628–1.652	500	SWIR
	7	2.105–2.155	500	
AVHRR	1	0.58–0.68	1090	VIS
	2	0.725–1.00	1090	NIR
daytime	3a	1.58–1.64	1090	SWIR
nighttime	3b	3.55–3.93	1090	MWIR

Painter 2004). Snow covered area, derived from multispectral measurements in the visible and near infrared parts of the spectrum, were among the earliest geophysical measurements from satellites (Rango and Itten 1976), and the measurements were soon incorporated into snowmelt-runoff models (Rango and Martinec 1979). The earliest remote sensing of snow properties focused primarily on mapping the snow extent with multispectral sensors, such as the Landsat Multispectral Scanner (MSS, also called the Multispectral Scanning Subsystem), the Landsat Thematic Mapper (TM), and the NOAA Advanced Very High Resolution Radiometer (AVHRR). Going beyond measurements of snow extent, Dozier (1989) proposed a suite of normalized band differences for mapping snow and qualitative grain size with TM data. As a result, most current multispectral schemes for binary mapping of snow cover, by which each pixel is classified as either 'snow' or 'not snow,' are derived from this method. For example, the current scheme in NASA's Earth Observing System (EOS) applies this method to the MODIS (Moderate-Resolution Imaging Spectrometer) instrument for its standard snow map product (Hall et al. 2002).

Table 28.1 shows the seven 'land' bands on MODIS, where the pixel sizes are 250–500 m and the sensor response is such that the bands do not saturate over bright objects, such as snow or clouds, along with similar parameters for ASTER, Landsat TM, and the NOAA AVHRR. A normalized difference snow index (NDSI) is calculated from reflectance R in bands at wavelengths where snow is bright (visible) and where it is dark (shortwave infrared), along with a band used for threshold brightness:

$$\text{NDSI} = \frac{R_{VIS} - R_{SWIR}}{R_{VIS} + R_{SWIR}} \qquad (2)$$

In the MODIS processing, a pixel in a clear area is mapped as snow covered when NSDI > 0.4 and the reflectance in the MODIS near-infrared (band 4) exceeds 0.11. In a forested area, the pixel is mapped as snow when 0.1 < NSDI < 0.4.

MODIS in particular is widely used in hydrological analysis and modeling. The global MODIS snow-cover product (Hall et al. 2002) is produced daily and as an eight-day composite at 500 m spatial resolution. For global climate models, daily snow cover is produced at 0.05° resolution along with monthly global composites. Daily maps are necessary for hydrologic and climate models because snow is dynamic, changing at a slower time scale than atmospheric phenomena, but faster than other surface cover types. The composites are necessary because cloud cover and viewing geometry affect the daily images.

By updating a runoff model with measurements of snow cover, seasonal streamflow forecasts have been improved (McGuire et al. 2006). The satellite observations are able to show the distribution of snow over the topography, which surface measurements do not, and they show considerable snow at the higher elevations after all snow has disappeared from the surface measurement stations.

An additional property measured from MODIS is snow albedo. In the current generation of climate and snowmelt models, snow albedo is typically either prescribed or represented by empirical aging functions, when truly it is a dynamic variable affected by grain growth and light-absorbing impurities. Newer analyses of the snow cover are incorporating the seasonal evolution of both the snow cover and its albedo. In the visible part of the spectrum, clean, deep snow is bright and white, irrespective of the size of the grains. Beyond the visible wavelengths in the near-infrared and shortwave-infrared, however, snow is one of the most 'colorful' substances in nature, its reflectance varying both temporally and spatially. Newly fallen snow usually has a fine grain size, but metamorphism and sintering throughout the winter and spring increase the grain size and reduce reflectance in wavelengths beyond about 0.8 μm (Warren 1982). This behavior of snow is important to the snowpack's energy balance, because the decrease in albedo often occurs during the spring when the incoming solar radiation becomes greater as the solar elevation increases and the days get longer. In the context of hydrologic models, this albedo decay has spatial variability (Molotch et al. 2004). Remotely sensed albedo will typically differ by ±20% from albedo estimated using a common snow-age-based empirical relation applied uniformly across the domain.

A recent development in mapping snow cover and its albedo is 'subpixel' analysis. Snow-covered area in mountainous terrain usually varies at a spatial scale finer than that of the ground instantaneous field-of-view of the remote sensing instrument. This spatial heterogeneity poses a 'mixed-pixel' problem, because the sensor may measure radiance reflected from snow, rock, soil, and vegetation. To use the snow characteristics in hydrologic models, snow must be mapped at subpixel resolution in order to accurately represent its spatial distribution; otherwise, systematic errors may result. For example, in the drier years especially, much of the snow cover is patchy at the lower elevations.

An image classification that identifies each pixel as either snow-covered or not may miss much of this snow.

Mapping of surface constituents at subpixel scale uses a technique called 'spectral mixture analysis,' based on the assumption that the radiance measured at the sensor is a linear combination of radiances reflected from individual surfaces, *endmembers*, whose spectral signatures are unique and well separated above a random image noise level (Sabol et al. 1992, see also Jensen et al., in this volume). A set of simultaneous equations for N endmembers, one for each wavelength band, results. Snow does not have a unique reflectance in each wavelength band, but given its physical characteristics such as grain size and amount and composition of impurities, a snow endmember can be chosen that results in the lowest error in the solution of the simultaneous equations (Painter et al. 2003). The information thereby derived is the fractional snow-covered area for each pixel and the grain size of that snow, from which the albedo can be estimated. Figure 28.3 shows an example for the Sierra Nevada.

Snow water equivalent

Measurements of snow water equivalent (the depth of liquid water a snowpack would produce if melted entirely) from passive microwave sensors constitute one of the longest consistent continental-scale records of an important Earth system variable. The combination of SMMR (Scanning Multichannel Microwave Radiometer), SSM/I (Special Sensor Microwave/Imager) and AMSR-E (Advanced Microwave Scanning Radiometer-EOS) now stretches from 1978 to the present. The principle by which a microwave radiometer estimates snow water equivalent is that the snow attenuates the radiation emitted by the underlying soil, so the microwave brightness temperature is reduced as the snow increases. At the lower frequencies of the microwave band, emission from a dry snow cover is mainly affected by the underlying soil's wetness and roughness. At the higher frequencies, however, the volume scattering by snow particles becomes important and emission is inversely related to snow water equivalence and snow grain size. Therefore, the difference in brightness temperatures between two frequencies correlates with the snow water equivalent (Kelly and Chang 2003). Because of the low amount of energy emitted per unit area at microwave frequencies, the spatial resolution of passive microwave sensors is typically large, e.g., 25×25 km^2 or even larger (Table 28.2).

The equation for snow water equivalent (SWE) is of the form:

$$\text{SWE} = Fc\,(T_{19} - T_{37}) \qquad (3)$$

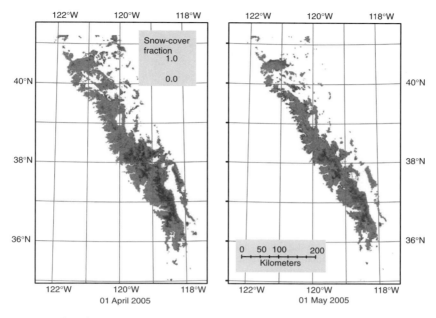

Figure 28.3 **Fractional snow-covered area for the Sierra Nevada from MODIS at the beginning of April and the beginning of May 2005. Not only does the extent of the snow-covered area shrink as the snowmelt seasons progresses, the fractional snow cover in that area also decreases. (See the color plate section of this volume for a color version of this figure).**

Table 28.2   Spatial resolution (instantaneous field-of-view) of passive microwave sensors

Frequency (GHz)	Footprint (km)			
	SSM/R	SSM/I	AMSE-E	AMSR
6.6–6.9	148 × 95		75 × 43	70 × 40
10.7–10.9	91 × 59		51 × 29	46 × 27
18.7–19.4	55 × 41	69 × 43	27 × 16	25 × 14
22–23.8	46 × 30	60 × 40*	32 × 18	29 × 17
36–37	27 × 18	37 × 29	14 × 8	14 × 8
50.3–52.8				10 × 6[1]
85–89		15 × 13	6 × 4	6 × 3

*V-polarization only.

Here F is a factor related to the forest cover fraction and c is a coefficient that depends on a snow class and time of year (a method to correct for grain sizes). T_{19} and T_{37} are brightness temperatures at 19 and 37 GHz, horizontally polarized in most sensors.

Because absorption by liquid water is so much greater than absorption by ice, the presence of liquid water either in the snow itself or in nearby water bodies strongly affects the signal. When snow starts melting, emission significantly increases because of the greater absorption by liquid water, and the observed signals are only from the snow surface. Therefore it is possible to identify melting snow with passive microwave sensors, but once wet, the snow's water equivalent cannot be estimated. A way to improve the estimates of snow water equivalent is to assimilate the passive microwave estimates into a land surface climate model that includes snowmelt and snow albedo feedback (Dong et al. 2007).

Because of the presence of bare ground, vegetation, and water bodies along with snow in a large microwave pixel, and because of the variability of the signal that different snow grain sizes and layered snowpacks cause, there are efforts to put microwave estimates of snow water equivalent on a more solid theoretical basis (Tsang et al. 2000, Jiang et al. 2007). These and other papers use an explicit radiative transfer model for microwave emission from snow, including the soil characteristics and the layering and grain sizes in the snowpack.

Mountain glaciers: extent, equilibrium line, and rate of flow

Study of glacier regimes worldwide reveals widespread wastage since the late 1970s, with a marked acceleration in the late 1980s. Remote sensing is used to document changes in glacier extent – the size of the glacier – and the position of the equilibrium line, which is the elevation on the glacier where winter accumulation is balanced by summer melt (Bindschadler et al. 2001, König et al. 2001). In years when either winter precipitation is greater than normal or summer temperatures are cooler, glacier mass balance is positive, increasing the glacier volume and lowering the equilibrium line. Conversely, lesser amounts of winter accumulation or warmer summers lead to a reduction of glacier volume and a higher equilibrium line.

An example study in the Ak-shirak Range of the central Tien Shan used aerial photographs in the 1940s and 1970s, along with ASTER imagery from 2001, to document a reduction in glacier area of 20% between 1977 and 2001 (Khromova et al. 2003). Similarly, analyses of Landsat TM data indicate accelerated glacier decline worldwide since the 1980s. An inventory of 930 mountain glaciers showed the decline from 1985 to 1999 to represent an annual loss seven times greater than from 1850 to 1973 (Paul et al. 2004).

Because glaciers respond to past and current climatic changes, a globally complete glacier inventory is being developed to keep track of the current extent as well as the rates of change of the world's glaciers. Coordinated by the National Snow and Ice Data Center, the GLIMS project (Global Land Ice Measurements from Space) is using data from ASTER and Landsat Enhanced Thematic Mapper to inventory about 160,000 glaciers worldwide (Raup et al. 2007). Figure 28.4 shows an example of documented glacier retreat in the Garhwal Himalaya over 200+ years, based on field and satellite observations. Analyses with SRTM (Farr et al. 2007) and ASTER data interpreted flow velocity fields using changing surface topography of a Himalayan glacier (Kääb 2005).

A recent advance in radar remote sensing is the application of interferometry (see Kellndorfer and McDonald, in this volume, for background on this technology). A time sequence of repeat-pass images from the ERS-1 and ERS-2 satellite showed surface flow evolution over six years of a glacier surge in Svalbard (Luckman et al. 2002). The analysis showed compressional flow in the upper part of the glacier, followed by extensional flow afterward. Even in the absence of *in situ* data, the interferometric methods can capture the spatial and temporal evolution of a glacier surge.

Ice sheet topography

As *Science* magazine noted in its final 2006 issue, one of that year's major discoveries was the documentation that Greenland and Antarctica are losing ice to the oceans at an accelerating pace

Figure 28.4   ASTER image of the Gangotri Glacier terminus in the Garhwal Himalaya since 1780 (Image from National Snow and Ice Data Center, University of Colorado). (See the color plate section of this volume for a color version of this figure).

(*Science* 2006). It is not clear why the ice sheets are apparently so sensitive to modest ocean warming, but if such rapid shrinking continues, low-elevation coastal areas could be inundated within two centuries.

These discoveries, of accelerating ice loss from Antarctica and Greenland and the importance of ice-sheet dynamics on their mass balances, rest on measurements by a suite of satellite and airborne sensors using novel techniques. These include decades of satellite radar images, laser altimeters on the ICESat mission and on airplanes that precisely measure the elevation of the ice surface, and more recently the GRACE mission that measures ice mass directly through its gravitational pull (Bindschadler 2006).

In 1997, Radarsat completed the first complete radar-based map of Antarctica. Analyses of radar images through the years have enabled detailed measurements of surface velocity, and these in turn have enabled calculation of strain rates and basal shear at the bed. The radar data have also led to the discovery of new ice streams and their velocity variations (Bindschadler et al. 1996). These radar data and declassified satellite data show that ice streams move at variable speeds, in contrast to the older view that ice sheets move at a constant velocity (Bindschadler and Vornberger 1998).

Coupled with the radar images, measurements of surface elevation from laser altimeters (ICESat) and gravity data (from GRACE) now show that both Greenland and Antarctica have been losing ice over the past 5 to 10 years. From 2003 to 2005, Greenland lost about 150 gigaton/year at lower elevations and gained about 50 gigaton/year at higher elevations, with most of the losses occurring during summer (Luthcke et al. 2006). In Antarctica, the gravity data show mass losses of 60–160 gigaton/year (Chen et al. 2006). Most of the loss is from West Antarctica, with East Antarctica in approximate balance (Velicogna and Wahr 2006).

Glaciologists and climatologists have always thought that a warming climate would decrease ice mass, but until recently the focus was on increased melting, partly compensated by greater precipitation at higher elevations. The radar images, however, show that ice velocities have increased over the last decade, with the result that more ice is flowing into the ocean. In West Antarctica, measurements from a synergistic combination of data show that oceanic forcing is a likely cause. Laser altimeter data show thinning of ice near the coastline, radar data show faster flow, Landsat data show retreat of the grounding line, and MODIS data show calving of large icebergs. Warming ocean waters seem to have increased calving of the ice shelves, thereby allowing the ice sheet's outlet glaciers to flow more quickly (Figure 28.5). Glaciers in Greenland have also increased in velocity, perhaps from faster basal slippage.

Sea ice

Sea ice has been monitored with visible and near-infrared data (AVHRR, MODIS), synthetic aperture radar (ERS-1, JERS-1, and Radarsat), and passive microwave data (SMMR, SSM/I, AMSR, and AMSR-E). AVHRR, and MODIS are constrained by persistent cloud cover and the polar night. Radar can characterize sea ice under all weather conditions, but only a small fraction of the entire ice cover can be monitored because of operational and data acquisition constraints and a narrow swath width (100–500 km) that limits spatial and temporal resolution. These data sets have nonetheless been useful for regional studies and for calibration. The most comprehensive and consistent source of global sea ice data has been satellite

passive microwave sensors. Not limited by weather conditions or light levels, they are particularly well suited for monitoring sea ice, because of the strong contrast in microwave emission between open and ice-covered ocean.

The normal product from the passive microwave analyses is sea-ice concentration. This quantity, however, is hard to define precisely because sea ice evolves from open water through very thin 'grease' ice, eventually becoming first-year ice that supports a winter snow cover. Through these stages, the microwave emissivity drops. Thus the retrieved ice concentration can be different even in areas where the true open water fraction is the same.

The basic algorithm for sea-ice concentration stems from the mixture of open water and sea ice in a pixel. The water has a brightness temperature T_W and the ice's brightness temperature is T_I. Setting the sea-ice concentration to f_I, the brightness temperature of the pixel is (Comiso and Steffen 2001):

$$T_B = f_I T_I + (1 - f_I)T_W \quad \text{so} \quad f_I = \frac{T_W - T_B}{T_W - T_I} \tag{4}$$

With no other information, this equation has two unknowns, T_W and T_I. Through measurements acquired in laboratory and field and from aircraft, however, empirical relationships can be found between measurements of water and ice at different microwave frequencies. Then the emissivity of the ice and water and their physical temperatures are known (Comiso 1999). A land mask is applied to ensure that the analysis is carried out only over the ocean. In the Antarctic, this must be updated because of changes in the ice shelves.

Permafrost

As noted in the chapter's introduction, the presence of permafrost is difficult to measure remotely. The chapter on remote sensing of soils (Campbell, in this volume) points out that only the soil moisture near the surface can be directly estimated from a passive or active microwave signal. Soil moisture at depth is estimated from models that use the surface values as input. Because ice is less absorptive than water in the passive microwave region, the brightness temperature will drop when the soil water near the surface freezes, but the freeze/thaw state of the surface is not an indicator of the presence or absence of permafrost. Therefore the approaches to remote sensing of permafrost have mainly involved the use of parameters related to permafrost conditions – topography and surface cover – to indirectly infer permafrost conditions over large areas by using remote-sensing classification algorithms and ground-truth data (Duguay et al. 2005).

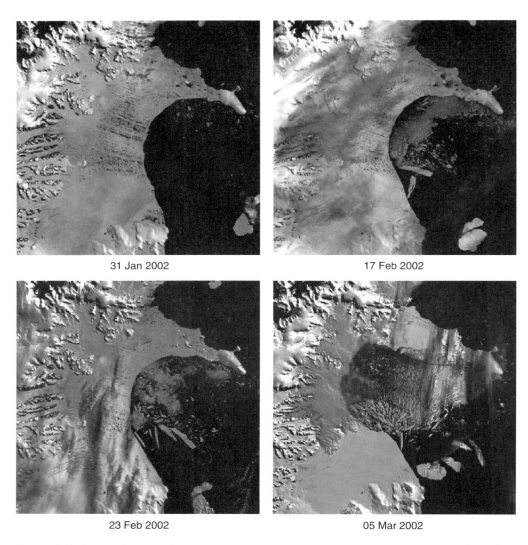

31 Jan 2002

17 Feb 2002

23 Feb 2002

05 Mar 2002

Figure 28.5   MODIS images showing disintegration of the Larsen Ice Shelf in Antarctica. The bottom pair of images show more than 2000 km² of the ice shelf breaking up in two weeks. *Source*: Images from the National Snow and Ice Data Center, University of Colorado.

WHAT IS MISSING FROM THE CURRENT REPERTOIRE?

In snow hydrology, the most important variable is snow water equivalent. While passive microwave sensors can estimate this variable, the large pixel size is useful only where there is limited variability over spatial scales of tens of kilometers. In the mountains, where much of the snow in the mid-latitudes falls, and where over a billion people depend on such snowfall for their water supplies, a passive microwave instrument has too coarse a footprint. Snow in these spatially heterogeneous mountain regions is the source for many of the world's most important rivers. In the western U.S., for example, over 70% of annual streamflow originates as snowmelt, mostly from mountainous areas. Yet currently our knowledge of this resource is inadequate, as the surface measurement network is sparse, does not sample the range of topographic slopes, and does not extend to the highest elevations. Therefore, measurement of the spatial distribution of snow water equivalent is essentially impossible with the network of surface stations. Intriguing results were shown with a multi-frequency, multi-polarization synthetic aperture radar as part of the SIR-C (Shuttle Imaging Radar-C) mission (Shi and Dozier 2000a,b). However, the two SIR-C missions lasted only 10 days, and the results were validated only for a limited field area during these short periods.

A proposed Snow and Cold Land Processes (SCLP) mission is identified as one of 17 high priority needs (National Research Council 2007). The mission's objectives are to measure snow water equivalent, depth, and wetness over land and ice sheets at 100 m spatial resolution at 3–15 day intervals. The configuration will employ a synthetic aperture radar along with a passive microwave radiometer. Another mission described in the NRC report addresses soil moisture and the freeze/thaw state of the soil. SMAP is a proposed combination of a passive microwave radiometer at L-band, along with a radar, also at L-band. The radiometer would estimate soil moisture and freeze/thaw state at 1–3 km resolution, and the radar would provide finer spatial resolution (100 m) over part of the swath.

DATA AVAILABILITY

Cryospheric data products from remote sensing platforms are available from the National Snow and Ice Data Center (http://www.nsidc.org/). The Center also has many interesting data and images from surface sites.

REFERENCES

Barnett, T. P., J. C. Adam, and D. P. Lettenmaier, 2005. Potential impacts of a warming climate on water availability in snow-dominated regions. *Nature,* 438: 303–309, doi: 10.1038/nature04141.

Bindschadler, R., 2006. The environment and evolution of the West Antarctic ice sheet: setting the stage. *Philosophical Transactions of the Royal Society A,* 364: 1583–1605, doi: 10.1098/rsta.2006.1790.

Bindschadler, R. and P. Vornberger, 1998. Changes in the West Antarctic Ice Sheet since 1963 from declassified satellite photography. *Science,* 279: 689–692, doi: 10.1126/science.279.5351.689.

Bindschadler, R., P. Vornberger, D. D. Blankenship, T. Scambos, and R. Jacobel, 1996. Surface velocity and mass balance of Ice Streams D and E, West Antarctica. *Journal of Glaciology,* 42: 461–475.

Bindschadler, R., J. Dowdeswell, D. K. Hall, and J.-G. Winther, 2001. Glaciological applications with Landsat-7 imagery: Early assessments. *Remote Sensing of Environment,* 78: 163–179, doi: 10.1016/S0034-4257(01)00257-7.

Bohren, C. F. and D. R. Huffman, 1998. *Absorption and Scattering of Light by Small Particles,* 530 pp., John Wiley and Sons, New York.

Campbell, J., in this volume. Remote Sensing of Soils. Chapter 24.

Cavalieri, D. J., C. L. Parkinson, and K. Y. Vinnikov, 2003. 30-Year satellite record reveals contrasting Arctic and Antarctic decadal sea ice variability. *Geophysical Research Letters,* 30: 1970, doi: 10.1029/2003GL018031.

Chen, J. L., C. R. Wilson, D. D. Blankenship, and B. D. Tapley, 2006. Antarctic mass rates from GRACE. *Geophysical Research Letters,* 33: L11502, doi: 10.1029/2006GL026369.

Cohen, J. and D. Entekhabi, 1999. Eurasian snow cover variability and Northern Hemisphere climate predictability. *Geophysical Research Letters,* 26: 345–348.

Comiso, J. C., 1999. Algorithm theoretical basis document (ATBD) for the ADEOS/AMSR sea ice algorithm. *AMSR/AMSR-E Documents,* http://sharaku.eorc.jaxa.jp/AMSR/doc/alg/10_alg.pdf (accessed August 19 2007).

Comiso, J. C. and K. Steffen, 2001. Studies of Antarctic sea ice concentrations from satellite data and their applications. *Journal of Geophysical Research,* 106: 31, 361–31,386.

Dong, J., J. P. Walker, P. R. Houser, and C. Sun, 2007. Scanning multichannel microwave radiometer snow water equivalent assimilation. *Journal of Geophysical Research,* 112: D07108, doi: 10.1029/2006JD007209.

Dozier, J., 1989. Spectral signature of alpine snow cover from the Landsat Thematic Mapper. *Remote Sensing of Environment,* 28: 9–22, doi: 10.1016/0034-4257(89)90101-6.

Dozier, J. and T. H. Painter, 2004. Multispectral and hyperspectral remote sensing of alpine snow properties. *Annual Review of Earth and Planetary Sciences,* 32: 465–494, doi: 10.1146/annurev.earth.32.101802.120404.

Duguay, C. R., T. Zhang, D. W. Leverington, and V. E. Romanovsky, 2005. Satellite remote sensing of permafrost and seasonally frozen ground, in *Remote Sensing in Northern Hydrology: Measuring Environmental Change,* edited by C. R. Duguay and A. Pietroniro, Geophysical Monograph Series, 163, American Geophysical Union, Washington, DC pp. 91–118.

Farr, T. G., P. A. Rosen, E. Caro, R. Crippen, R. Duren, S. Hensley, M. Kobrick, M. Paller, E. Rodriguez, L. Roth, D. Seal, S. Shaffer, J. Shimada, J. Umland, M. Werner, M. Oskin, D. Burbank, and D. Alsdorf, 2007. The Shuttle Radar Topography Mission. *Reviews of Geophysics,* 45: RG2004, doi: 10.1029/2005RG000183.

Gregory, J. M. and J. Oerlemans, 1998. Simulated future sea-level rise due to glacier melt based on regionally and seasonally resolved temperature changes. *Nature,* 391: 474–476.

Grenfell, T. C. and S. G. Warren, 1999. Representation of a nonspherical ice particle by a collection of independent spheres for scattering and absorption of radiation. *Journal of Geophysical Research,* 104: 31697–31709.

Hall, D. K., G. A. Riggs, V. V. Salomonson, N. DiGiromamo, and K. J. Bayr, 2002. MODIS snow-cover products. *Remote Sensing of Environment,* 83: 181–194, doi:10.1016/S0034-4257(02)00095-0.

Jensen, J. R., J. Im, P. Hardin, and R. R. Jensen, in this volume. Image Classification. Chapter 19.

Jiang, L., J. Shi, S. Tjuatja, J. Dozier, K. Chen, and L. Zhang, 2007. A parameterized multiple-scattering model for microwave emission from dry snow. *Remote Sensing of Environment*, in press, doi: 10.1016/j.rse.2007.02.034.

Kääb, A., 2005. Combination of SRTM3 and repeat ASTER data for deriving alpine glacier flow velocities in the Bhutan Himalaya. *Remote Sensing of Environment*, 94: 463–474, doi: 10.1016/j.rse.2004.11.003.

Kellndorfer, J. and K. McDonald, in this volume. Active and Passive Microwave Systems. Chapter 13.

Kelly, R. E. J. and A. T. C. Chang, 2003. Development of a passive microwave global snow depth retrieval algorithm for Special Sensor Microwave Imager (SSM/I) and Advanced Microwave Scanning Radiometer-EOS (AMSR-E) data. *Radio Science*, 38: 8076, doi:10.1029/2002RS002648.

Khromova, T. E., M. B. Dyurgerov, and R. G. Barry, 2003. Late-twentieth century changes in glacier extent in the Ak-shirak Range, Central Asia, determined from historical data and ASTER imagery. *Geophysical Research Letters*, 30: 1863, doi: 10.1029/2003GL017233.

König, M., J.-G. Winther, and E. Isaksson, 2001. Measuring snow and glacier ice properties from satellite. *Reviews of Geophysics*, 39: 1–28, doi: 10.1029/1999RG000076.

Liu, X. and M. Yanai, 2002. Influence of Eurasian spring snow cover on Asian summer rainfall. *International Journal of Climatology*, 22: 1075–1089, doi: 10.1002/joc.784.

Luckman, A., T. Murray, and T. Strozzi, 2002. Surface flow evolution throughout a glacier surge measured by satellite radar interferometry. *Geophysical Research Letters*, 29: 2095, doi: 10.1029/2001GL014570.

Luthcke, S. B., H. J. Zwally, W. Abdalati, D. D. Rowlands, R. D. Ray, R. S. Nerem, F. G. Lemoine, J. J. McCarthy, and D. S. Chinn, 2006. Recent Greenland ice mass loss by drainage system from satellite gravity observations. *Science*, 314: 1286–1289, doi: 10.1126/science.1130776.

McGuire, M., A. W. Wood, A. F. Hamlet, and D. P. Lettenmaier, 2006. Use of satellite data for streamflow and reservoir storage forecasts in the Snake River Basin. *Journal of Water Resources Planning and Management*, 132: 97–110, doi: 10.1061/(ASCE)0733-9496(2006)132:2(97).

Meier, M. F., 1984. Contribution of small glaciers to global sea level. *Science*, 226: 1418–1421.

Mie, G., 1908. Beiträge zur Optik trüber Medien, Speziell Kolloidaler Metallösungen. *Annalen der Physik*, 25: 377–445.

Molotch, N. P., T. H. Painter, R. C. Bales, and J. Dozier, 2004. Incorporating remotely sensed snow albedo into spatially distributed snowmelt modeling. *Geophysical Research Letters*, 31: L03501, doi: 10.1029/2003GL019063.

National Research Council, 2007. *Earth Science and Applications from Space: National Imperatives for the Next Decade and Beyond*, 400 pp., National Academies Press, Washington, DC.

Nussenzveig, H. M. and W. J. Wiscombe, 1980. Efficiency factors in Mie scattering. *Physical Review Letters*, 45: 1490–1494.

Painter, T. H., J. Dozier, D. A. Roberts, R. E. Davis, and R. O. Green, 2003. Retrieval of subpixel snow-covered area and grain size from imaging spectrometer data. *Remote Sensing of Environment*, 85: 64–77, doi: 10.1016/S0034-4257(02)00187-6.

Paul, F., A. Kääb, M. Maisch, T. Kellenberger, and W. Haeberli, 2004. Rapid disintegration of alpine glaciers observed with satellite data. *Geophysical Research Letters*, 31: L21402, doi: 10.1029/2004GL020816.

Rango, A. and K. Itten, 1976. Satellite potentials in snow-cover monitoring and runoff prediction. *Nordic Hydrology*, 7: 209–230.

Rango, A. and J. Martinec, 1979. Application of a snowmelt-runoff model using Landsat data. *Nordic Hydrology*, 10: 225–238.

Raup, B., A. Racoviteanu, S. J. S. Khalsa, C. Helm, R. Armstrong, and Y. Arnaud, 2007. The GLIMS geospatial glacier database: A new tool for studying glacier change. *Global and Planetary Change*, 56: 101–110, doi: 10.1016/j.gloplacha.2006.07.018.

Sabol, D. E., J. B. Adams, and M. O. Smith, 1992. Quantitative subpixel spectral detection of targets in multispectral images. *Journal of Geophysical Research*, 97: 2659–2672.

Science, 2006. Breakthrough of the year: the runners-up. 314: 1850–1855, doi: 10.1126/science.314.5807.1850a.

Shi, J. and J. Dozier, 2000a. Estimation of snow water equivalence using SIR-C/X-SAR, Part I: Inferring snow density and subsurface properties. *IEEE Transactions on Geoscience and Remote Sensing*, 38: 2465–2474, doi: 10.1109/36.885195.

Shi, J. and J. Dozier, 2000b. Estimation of snow water equivalence using SIR-C/X-SAR, Part II: Inferring snow depth and grain size. *IEEE Transactions on Geoscience and Remote Sensing*, 38: 2475–2488, doi: 10.1109/36.885196.

Stamnes, K., S.-C. Tsay, W. J. Wiscombe, and K. Jayaweera, 1988. Numerically stable algorithm for discrete-ordinate-method radiative transfer in multiple scattering and emitting layered media. *Applied Optics*, 27: 2502–2509.

Tsang, L., C.-T. Chen, A. T. C. Chang, J. Guo, and K.-H. Ding, 2000. Dense media radiative transfer theory based on quasicrystalline approximation with applications to passive microwave remote sensing of snow. *Radio Science*, 35: 731–750.

Velicogna, I. and J. Wahr, 2006. Measurements of time-variable gravity show mass loss in Antarctica. *Science*, 311: 1754–1756, doi: 10.1126/science.1123785.

Warren, S. G., 1982. Optical properties of snow. *Reviews of Geophysics and Space Physics*, 20: 67–89.

Warren, S. G. and R. E. Brandt, 2008. Optical constants of ice from the ultraviolet to the microwave: A revised compilation, *Journal of Geophysical Research*, 113: D14220, doi: 10.1029/2007JD009744.

Wiscombe, W. J., 2005. Refractive indices of ice and water. ftp://climate1.gsfc.nasa.gov/wiscombe/Refrac_Index/ (accessed August 19 2007).

D. Global Change and Human Environments

29

Remote Sensing for Terrestrial Biogeochemical Modeling

Gregory P. Asner and Scott V. Ollinger

Keywords: carbon cycling, ecosystem modeling, evapotranspiration, hyperspectral imaging, model data assimilation, net primary production, vegetation properties.

INTRODUCTION

Remote sensing and biogeochemical modeling share a highly complementary nature, which has led to a growing number of applications that involve some degree of coupling between the two. Whereas remote sensing represents the only means by which landscape and vegetation properties can be sampled over large and contiguous portions of the Earth's surface, models focus on the underlying biogeochemical processes that regulate the flow and storage of water, carbon and nutrients, often over much longer time scales than can be considered through remote sensing alone. The aims and goals of biogeochemical modeling are wide ranging and include studies of carbon (C) and water cycling, analyses of nitrogen (N) enrichment and leaching to aquatic ecosystems, and trace gas transfers from soils to the atmosphere. Biogeochemical models are also used as part of larger, integrated modeling environments of regional and global biosphere–atmosphere interactions, biogeography, and climate change (e.g., Sellers et al. 1997, Cramer et al. 2001). More recently, biogeochemical modeling has taken a role in decision support for conservation, management, and policy development (Potter et al. 2006).

Remote sensing can provide biogeochemical models with information on vegetation type, leaf area index (LAI), canopy height, the fraction of absorbed photosynthetically active radiation (fPAR), light-use efficiency (LUE), leaf N concentration, pigments and other biochemical compounds to simulate plant growth and mortality. Other remote sensing-related inputs have included temperature, precipitation, solar radiation levels, and soil moisture. In this chapter, we review a number of approaches through which remote sensing data can be applied to the detection of vegetation properties, and discuss the trade-offs of various methods with respect to biogeochemical modeling. Given the breadth of the topic, our goal in preparing this chapter was not to provide a working manual of all remote sensing-model integration methods available. Instead, we sought to summarize important overall strategies for vegetation detection into a framework that involves the types of instruments used, the ecological properties they can be designed to detect, and the manner in which those properties can be utilized by models.

MODELS UTILIZING REMOTE SENSING

This section provides a brief overview of major categories of ecosystem biogeochemistry models that can be driven or guided by remote sensing. In the interest of simplicity, we describe each type of model as being distinct from one another, but readers should be aware that the boundaries between them are often blurry, and hybrid approaches are also available.

Simple empirical models

The simplest type of remote sensing/model linkage consists of a small number of empirically derived algorithms that combine field-based relationships with remotely sensed vegetation properties that correlate strongly with some aspect of ecosystem behavior. For example, Ollinger et al. (2002a) used imaging spectroscopy to detect leaf lignin to nitrogen ratios in temperate forests, which provided a direct connection to decomposition, C:N ratios, and N cycling rates in soils. When such relationships are available, this approach offers a straightforward means of producing estimates that are constrained to known patterns. The resulting accuracy is dependent only on the strength of the observed trends and on the accuracy of the vegetation property estimates. The principal disadvantage is that these approaches include no mechanisms that would allow extrapolation in time or under varying environmental conditions.

Light use efficiency models

Of intermediate complexity are the light-use efficiency (LUE) models, also called production efficiency models (PEM), which use remotely sensed fPAR to estimate maximum carbon C assimilation rates and then adjust for suboptimal climate conditions, using a series of simple climate response algorithms. These models have evolved from the original arguments of Monteith (1972) that the amount of carbon fixed per unit of incident radiation can be used as an organizing principle for estimating overall vegetation productivity. There are now a large number of efficiency models which differ in their details and complexity, but all derive from the idea that knowledge of incident radiation and the light-absorbing properties of the plant canopy can determine the maximum potential photosynthesis for that canopy (Potter et al. 1993, Field et al. 1995, Running et al. 2000). Nearly all applications of efficiency-type methods are based on the idea that the rate of C accumulation by plants (P) depends on environmental and biochemical factors in the following way:

$$P \text{ (NPP OR GPP)} = f\text{PAR} \times \text{PAR} \times \varepsilon^* \times W \times T \tag{1}$$

where PAR is the downwelling photosynthetically active radiation at the top of the canopy, ε^* represents the maximum photochemical conversion efficiency of vegetation foliage under optimal conditions (g MJ^{-1}), and W and T are dimensionless

scalars (0–1) that down-regulate ε^* based on modeled water and temperature stress, respectively. Production efficiency models are particularly appealing for large-scale analyses because of the availability of absorbed PAR estimates from multi-spectral sensors, and because ε^* is both conceptually straightforward and physiologically meaningful. However, a persistent challenge has been the lack of understanding concerning factors controlling variation in ε^* both within and among vegetation types (e.g., Gower et al. 1999).

Ecosystem process models

Of greatest complexity are ecosystem process models, which use remote sensing primarily to initialize important vegetation input variables and then simulate ecological processes – such as photosynthesis, C allocation, respiration, litterfall, decomposition, and water balances – that affect ecosystem behavior. The added complexity in these models allows them to predict a range of additional variables, and to examine responses to environmental factors such as rising CO_2, atmospheric pollution, and physical disturbance. Because they are often designed to be run over longer time scales, they are more suitable for considering changes in ecosystem components such as soil C and nutrient pools that have very long turnover times.

Independent of the application, ecosystem process models often require a large number of parameters, measured in the field or via remote sensing, or from a cumulative knowledge based on the literature (Figure 29.1). Some models, such as Century (Parton et al. 1988, 1998) and PnET-CN (Aber and Driscoll 1997, Ollinger et al. 2002b) require more than 30 input parameters needed to simulate the growth and mortality of plants, and the subsequent accumulation and turnover of soil organic matter (SOM) and nutrients (Table 29.1). Although many of these models were not originally designed to ingest remotely sensed data, they can greatly benefit from the use of remote observations to constrain simulated processes. Other models, such as the Carnegie–Ames–Stanford Approach (CASA; Potter et al. 1993, Field et al. 1995) were designed to be run with fewer parameters, emphasizing those which can be routinely retrieved via remote sensing. Some of these models represent hybrid approaches between efficiency and process-based models, whereby estimates of fPAR and ε^* are linked with algorithms describing more complex biogeochemical processes. To date, these models use remotely sensed data to constrain simulated rates of photosynthesis, plant growth, evapotranspiration, and other plant-related processes. Other more specialized models, such as TerraFlux (Asner et al. 2001), use remotely sensed data to simulate nutrient flows between plants and soils.

Table 29.1 Common biogeochemical models and the types of remote sensing observations currently used to constrain them

Model	Remote sensing constraints	References
CENTURY	Vegetation type	Parton et al. 1988, 1998
Biome-BGC	Vegetation type, LAI, fPAR	Running and Hunt 1993, Running et al. 1994
PnET-II, PnET-CN	Vegetation type, Leaf N	Aber and Driscoll 1997, Ollinger and Smith 2005
CASA-1	Vegetation type, fPAR	Potter et al. 1993, Field et al. 1995
CASA-3D	Vegetation type, fPAR, LAI, Tree height, Crown dimensions, LUE	Huang et al. 2008
SiB	Vegetation type, fPAR, LAI, ET, albedo	Sellers et al. 1986, 1997

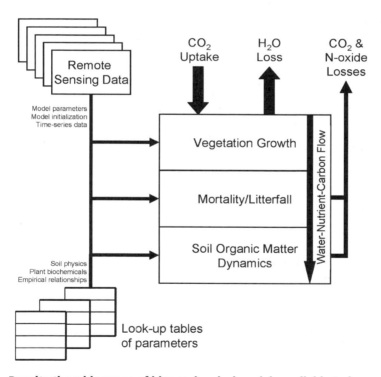

Figure 29.1 Despite the wide range of biogeochemical models available today, nearly all those that provide a spatially explicit rendering of ecosystem processes ultimately require geographic data from remote sensing. Some models use only a few remotely sensed parameters, such as vegetation cover and type, whereas others employ information on physiological, biochemical, and structural properties of vegetation derived from airborne and satellite remote sensing. All models must balance the relative contributions of detailed spatial information from remote sensing with process-oriented data and parameterizations stored in look-up tables.

Table 29.2   Remote sensing observations used to constrain terrestrial biogeochemical models

Measurement	Model control	Typical values	Units
Vegetation type	Many processes related to growth, phenology, litterfall, and decomposition	Broadleaf evergreen forest, shrubs, C3 grasses	Categorical
Fractional PAR absorption	Light interception, GPP, NPP	0 to 100	Percentages
Leaf area index	Light interception, foliar C and nutrient stocks, GPP, NPP, ET, stress	0 to 8+	m^2 foliage/m^2 ground
Leaf pigments	Light-use efficiency, GPP, NPP, stress	Chlorophyll a+b, carotenoid concentrations	$\mu g/cm^2$
Leaf nitrogen	Leaf N concentration, photosynthetic capacity, site fertility, decomposition	0.4 to 5+	Percentages or per leaf area
Vegetation height, Crown dimensions	Aboveground biomass (via field-based allometrics), C stocks, tree mortality, forest succession	0.5 to 40+	Meters

ECOSYSTEM PROPERTIES FROM REMOTE SENSING

Vegetation type

Vegetation type is one of the most basic yet important remote sensing products required by many biogeochemical models (Table 29.2). This is particularly true when the models are run in a spatially-explicit mode at regional to global scales. At a minimum, the models require vegetation classifications such as broadleaf deciduous trees, needleleaf evergreen canopies, and shrub (e.g., Running et al. 1993, 1994, Bonan 1995). This information is critical to setting up model parameterizations related to physiological, biochemical, and structural properties that ultimately affect growth, mortality, and decomposition of plants. For example, the lifespan of temperate deciduous and coniferous trees often differ, and thus the timing and rate of foliage, wood, and root mortality affects C losses to the atmosphere. Similarly, the decomposability of plant tissues varies by vegetation type, and thus a vegetation map is needed to ultimately parameterize decomposition rates spatially in a biogeochemical model. Vegetation type is thus a key entry point for remote sensing into the models.

Plant physiology and growth

Vegetation type is also important for setting up basic model parameters associated with photosynthetic rates, stomatal conductance, light-use efficiency and carbon allocation, all of which directly impact gross and net rates of primary production (GPP, NPP; Table 29.2). For example, variables such as leaf retention time or maximum photosynthesis per unit light intensity might be defined as a function of vegetation type when more explicit information is not available (Ollinger et al. 1998,

Bondeau et al. 1999). These and other vegetation and ecosystem parameters are often derived from a look-up table organized by vegetation type (Sellers et al. 1995).

With the advent of high temporal resolution, global remote sensing instruments such as the Advanced Very High Resolution Radiometer (AVHRR) and the Terra Moderate Resolution Imaging Spectrometer (MODIS), monitoring of vegetation properties directly related to physiology, phenology, and growth became commonplace. The basis for using these sensors rests in the field-based knowledge that fPAR, LAI, and several vegetation indices, such as the NDVI, are positively inter-correlated (Myneni et al. 1997). The ability to detect fPAR in this manner was quickly incorporated in light-use efficiency models to predict GPP and NPP, although the resulting need for estimates of ε^* represents a difficult and ongoing challenge. Some models select a constant value for ε^*, or set it by vegetation type (Potter et al. 1993, Running et al. 2000). Field et al. (1995) argue that a moderately dynamic ε^* caused by climatological stress factors should be more realistic. Other field-based and remote sensing studies have subsequently highlighted the variability of light-use efficiency at local to regional scales (Lobell et al. 2002, Turner et al. 2003). Nonetheless, modeling of GPP and NPP dynamics has remained most closely tied to the temporal variation in fPAR and LAI.

Thus far, the derivation of remotely sensed LAI and fPAR has mostly relied upon optical vegetation indices such as the NDVI, but other methods based on canopy gaps, structure, and the probability of light penetration are gaining momentum. This latter group of methods benefits especially from multi-view angle (MVA) or light detecting and ranging (LiDAR) observations that can determine inter- and intra-crown gaps related to LAI and leaf biomass (Means et al. 1999, Zhang et al. 2000). These approaches are still novel since neither the data nor processing methods are as accessible to non-experts as are optical vegetation indices, but the potential use of these structurally-sensitive remote sensing approaches could change the way biogeochemical and land-surface models are designed and implemented (Asner et al. 1998).

Although the NDVI–*f*PAR–NPP connection has been utilized to explore the spatial and temporal dynamics of plant growth and carbon uptake globally (e.g., Field et al. 1998), other local-to-regional scale studies have demonstrated considerable variation in LUE, prompting a number of more detailed studies. In one example, a meta-analysis by Green et al. (2003) compiled published values of ε^* and a variety of leaf and canopy-level traits from a wide array of C3 plant communities, including deciduous and evergreen tree species, and herbaceous species consisting of grasses, forbs, and legumes. Their results showed that most of the variation in ε^* could be related to mass-based concentrations of N in foliage, suggesting a link with leaf-level photosynthetic capacity and models that include canopy N as determinants of C assimilation. Aber and Driscoll (1997) and Ollinger et al. (2002b) developed and explored linkages between leaf N and photosynthetic capacity in their model PnET. They found leaf N to be one of the most important predictors of growth rates, and eventually incorporated remotely sensed estimates of canopy N to provide a spatially explicit understanding of NPP in temperate forests (Figure 29.2). Their approach focuses on airborne and space-based imaging spectroscopy, along with partial least squares (PLS) regression analysis, to estimate canopy N at fine to moderate spatial resolution. Retrieval of canopy N from sensors such as the NASA Airborne Visible and Infrared Imaging Spectrometer (AVIRIS) has become routine in forested ecosystems, where canopy cover is high, and when shadowing is minimal, although the small scene size and limited data volume have prevented more widespread application of this approach.

Leaf N provides one important window into the physiological functioning of canopies, but leaf pigments are another common approach to remote estimation of plant function. Chlorophyll concentration has been a core remote sensing target in many studies (e.g., Yoder and Waring 1994, Zarco-Tejada et al. 2001) and has been used to estimate annual rates of biomass production and carbon exchange in agricultural systems (Gitelson et al. 2005, 2006), but few biogeochemical models have been designed to explicitly incorporate chlorophyll measurements. Gamon et al. (1990) took a major step toward remote detection of xanthophyll-cycle pigments that may directly express the LUE of plants. They developed the Photochemical Reflectance Index (PRI) to estimate leaf and top-of-canopy LUE, and many have since used the

PRI to understand spatial and inter-specific variations in LUE among plants (reviewed by Ustin et al. 2004). Only recently, however, has the PRI been used in a biogeochemical model. Asner et al. (2004) used Earth Observing-1 Hyperion hyperspectral data to compute the NDVI and PRI of tropical rainforest canopies in the Brazilian Amazon. They found that the NDVI lacked responsiveness to canopy drought conditions that were leading to NPP declines, and this was caused by saturation of the NDVI in the high LAI canopies found in the central Amazon basin. In contrast, the PRI, which expresses top-of-canopy leaf pigment concentrations, accounted for about 20–25% of the field-observed decreases in plant production. Moreover, they found that remote sensing metrics of canopy water content, which penetrate deep into the tropical forest canopies, were highly sensitive to NPP declines. Using the PRI and hyperspectral indices of canopy water, they modeled these NPP decreases with a new version of the CASA model, demonstrating that drought periods could be detected in the tropical forest canopy, and the effect incorporated into a biogeochemical model (Figure 29.3).

Beyond the current hyperspectral metrics for the major pigment groups such as chlorophylls and carotenoids, there is a growing effort to understand how less abundant but functionally important leaf compounds, such as phenolics, might be remotely sensed. Secondary compounds provide information pertinent to plant physiology, leaf herbivory, and a variety of other processes pertinent to biogeochemical models. To date, we know of no models that incorporate remotely sensed (or field-measured) estimates of secondary plant compounds, so this is an area ripe for exploration.

Phenology and litterfall are two additional variables having a major impact on the C, water, and nutrient fluxes of ecosystems. The timing and amount of litterfall is a central determinant of CO_2 emissions via microbial decomposition processes. AVHRR and MODIS timeseries of the NDVI have been used to observe seasonal and inter-annual changes in phenology, and to estimate rates of litterfall production (White et al. 1997). These estimates have subsequently been used to simulate C flows from live to detrital pools in the models (Keyser et al. 2000). Such efforts have shown that phenology exerts major control over the seasonal cycle of CO_2 concentrations in our atmosphere (Sitch et al. 2003).

Carbon and nutrient storage

Growth, phenology, and litterfall are all directly related to the flow of C, nutrients, and water within ecosystems. These processes are also related to C

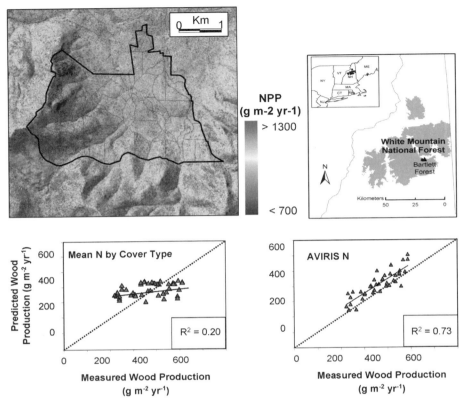

Figure 29.2   Remote sensing of canopy nitrogen provides a spatially-explicit and mechanistic means to constrain biogeochemical model simulations of photosynthesis, net primary production (NPP), and other ecosystem processes. Here, the NASA Airborne Visible and Infrared Imaging Spectrometer (AVIRIS) was used to estimate leaf nitrogen concentrations throughout the Bartlett Forest in New Hampshire, USA. The nitrogen maps were then used to constrain PnET model simulations of photosynthesis and NPP. The resulting NPP maps were far more accurate than those derived simply by parameterizing leaf nitrogen from a look-up table based on vegetation type. (See the color plate section of this volume for a color version of this figure).

Source: Reprinted with permission from *Ecosystems* 8(7), 2005, p. 770, Net primary production and canopy nitrogen in a temperate forest landscape: An analysis using imaging spectrometry, modeling and field data, S. V. Ollinger, and M. L. Smith, Figure 7, Copyright Springer Science and Business Media, Inc., 2005.

and nutrient storage, although indirectly. Storage is a function of inputs and outputs, and thus the flows of materials dictate the standing stock at any given time. Historically, remote sensing has addressed C and nutrient flows in plants more so than the standing stocks. Recently, however, new remote sensing technologies and techniques are gaining access to a more direct mapping of aboveground C and nutrient storage. First, in many environments the fractional cover of live and senescent or dead vegetation is a principal determinant of aboveground C stocks. At the global scale, DeFries

et al. (1999, 2000) used fractional canopy cover from the MODIS Vegetation Continuous Fields (VCF) product to simulate changing C stocks and biochemical fluxes. They showed that fractional cover changes exert a major influence on the C dynamics of terrestrial ecosystems worldwide. At regional scales, detailed mapping of fractional canopy cover can be closely related to aboveground C stocks in shrublands, savannas, and open woodlands (Asner et al. 2005a).

In forested ecosystems without major disturbances or canopy openings, fractional cover does

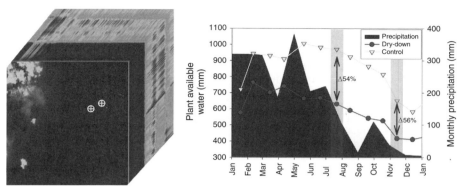

Model and field results	NPP$_{d:c}$(July)	NPP$_{d:c}$(Nov)	NPP$_{d:c}$(2001)
NDVI-NPP model	0.99	0.98	0.98
Hyperion-NPP model	0.81	0.42	0.67
Field NPP data	0.84	0.55	0.73

Figure 29.3   Remote sensing of leaf pigments and canopy water content provide new ways to simulate NPP in terrestrial ecosystems. Here, the NASA Earth Observing-1 Hyperion satellite sensor was used to measure differences in leaf pigments, light-use efficiency, and canopy water content between two 1-ha rainforest stands in the central Amazon (upper left), where seasonal drought occurs (upper right). One 'dry-down' forest stand (d) was treated to remove ~80% of incoming rainfall (red dot, upper left), while a control stand was left unchanged (green dot, upper left; Nepstad et al. 2002). Both stands underwent decreases in plant available water during the dry season, with the dry-down site experiencing a ~50% lower water availability than the control stand (upper right). Hyperion observations of pigments, light-use efficiency, and water captured seasonal and inter-site differences in the ratio of NPP between dry-down and control sites (d:c), whereas the traditional NDVI-based modeling approach failed to do so.
Source: Reprinted from Asner, G. P., D. Nepstad, G. Cardinot, and D. Ray. 2004. Drought stress and carbon uptake in an Amazon forest measured with spaceborne imaging spectroscopy. *Proceedings of the National Academy of Sciences* 101: 6039–6044, Copyright 1998 National Academy of Sciences, USA.

not always provide sufficient information for modeled C and nutrient stocks. Other techniques, such as from LiDAR, fine-scale optical (e.g., IKONOS), synthetic aperture radar (SAR), and multi-view angle passive optical remote sensing (MVA) provide canopy structural information that can be used to estimate C storage (reviewed by Wulder and Franklin 2003). For example, both LiDAR and interferometric SAR provide detailed information on tree canopy height and profile that can be combined with field-based allometric equations to map above-ground biomass and C storage (Lefsky et al. 2002, Treuhaft et al. 2003). To date, only a few models can ingest this information directly (Hurtt et al. 2004), but the straightforward nature of height-based allometrics should make this a more common approach as remotely sensed canopy height estimates become more widely available. Future model developments will make increased use of remotely sensed vegetation structure that is related to C storage.

In parallel with carbon, the nutrient stocks of canopies can be estimated and/or modeled when information pertaining to stand structure, above-ground biomass and tissue nutrient concentrations are available. In situations where nutrient concentrations are relatively invariant within functional groups (as in needleleaf evergreen species or desert shrubs), this can be achieved using a look-up table approach that combines remotely sensed canopy structure (biomass) and vegetation type with known nutrient levels in foliage and wood.

However, there are many cases where greater variation in plant tissue stoichiometry require a more direct means of detecting foliar nutrient concentrations, such as through use of imaging spectroscopy. For example, in temperate deciduous forests, variation in soil type, disturbance history and atmospheric deposition can lead to a wide range of leaf N concentrations both within and among species (Ollinger et al. 2002a, McNeil et al. 2005). Where nitrogen-fixing trees are present, detection of their location and abundance is important because they typically have far higher N:C ratios than do non-fixing species. Finally, the nutrient-to-carbon ratios of foliage among humid tropical tree species are typically more variable than is found in other biomes (Townsend et al. 2007). This translates to much greater uncertainty in biogeochemical modeling studies of tropical forests, unless direct detection of nutrient concentrations in canopy foliage can be achieved.

High spatial resolution data and/or sub-pixel fractional cover algorithms are starting to provide information on tree-fall gaps and dynamics (Asner et al. 2005b) that are important to a class of biogeochemical models that incorporate gap-phase processes in forests (Moorcroft et al. 2001, Huang et al. 2008). These models require spatially explicit (or statistically sound) information on the rate of gap formation, closure, and regrowth to simulate the light environment as well as C fluxes and storage. Progress in sub-pixel unmixing (spectral mixture analysis) with both hyperspectral and multispectral optical imagery is opening doors for mapping forest canopy gap fraction. As the remote sensing technologies and analytical techniques mature over time, the biogeochemical models will follow with more detailed integration of the observations, thereby supplanting the need for statistical parameterizations based purely on field data that is offset geographically or temporally from that of the modeling study.

Evapotranspiration

Plants and soils mediate the hydrology of ecosystems. Plant canopies intercept precipitation, and a fraction of this water is subsequently lost to the atmosphere via evaporation. Soils intercept remaining precipitation, and either lose moisture to evaporation or ground water flow. However, a major fraction of soil water is taken up by the vegetation to support stomatal conductance for photosynthetic function. Water is thus lost via transpiration, and this flux is a major determinant of ecosystem hydrology (Dickenson 1991).

Evapotranspiration (ET) is a function of canopy cover, LAI, roughness, wind speed, energy exchange, and other factors (Table 29.2). ET is also related to the residual energy exchanged between the canopy and atmosphere after taking into account net radiation, sensible heat exchange, and ground heat flux. Both of these perspectives defining ET lead to methods by which this important process is estimated in biogeochemical models. Some models simulate ET based on canopy and micrometeorological parameters (e.g., Sellers et al. 1997). Methods such as the two-source resistance model (Lhomme et al. 1994a,b), utilize LAI, vegetation height, and aerodynamic resistance to simulate sensible and latent heat exchange, and thus evapotranspiration. Other models rely on micrometeorological measurements alone, and do not directly incorporate vegetation structural properties (e.g., Kustas et al. 1990).

For those ET modeling approaches that rely on vegetation properties, canopy cover and LAI can be estimated from optical and other remotely sensed imagery. However, canopy roughness and derived canopy resistance remain difficult to quantify (Shoshany 1993). Roberts et al. (2004) recently estimated canopy rugosity – a metric of roughness – using airborne hyperspectral imagery. Others have used more explicit structural remote sensing technologies such as radar and LiDAR to understand canopy roughness (Lefsky et al. 2002).

Another approach to estimating ET utilizes the visible, near-infrared, and thermal imaging data provided by aircraft and satellite sensors to estimate energy exchanges between the land surface and atmosphere (e.g., Engman 1991, Kustas and Norman 1996). This approach affords a means to bypass, at least to some degree, any need to explicitly estimate vegetation structural properties. For example, Bastiaanssen et al. (1998, 2002) developed the Surface Energy Balance Algorithm for Land (SEBAL) to estimate a complete radiation and energy balance, including resistances for water vapor, heat, and momentum. The primary input to this model is broadband radiance observations from Landsat, MODIS, and similar sensors. Independent of the precise approach, it has long been argued that remote sensing is the only technology that can deliver the suite of radiative and vegetative parameters needed to estimate ET in a consistent manner at regional to global scales (Choudhury 1988).

CHALLENGES IN USING REMOTE SENSING FOR BIOGEOCHEMICAL MODELING

We have highlighted the major remote sensing measurements used in terrestrial biogeochemical models, and we described how these measurements constrain simulations of C, water, and nutrient cycling. The models are often complex, with many parameters that cannot be directly constrained with remote sensing (Figure 29.1). On the other hand, a set of key variables, such as

LAI, fPAR, and nitrogen concentrations, can be remotely estimated, and these vegetation properties have far-reaching control over a range of ecosystem processes (Sellers 1985, 1987).

The challenge remains to improve the remote sensing estimates, but also to find how a broader suite of model parameters co-vary in nature as a way to extend the power of remote sensing controls (or reality checks) in simulations of biogeochemical cycles. Leaf N is a good example of this issue. Leaf N is one of the most important descriptors of canopy function, and it is highly correlated with a range of plant properties such as leaf mass per area, leaf lifespan, and photosynthetic capacity (Reich et al. 1997, Wright et al. 2004). These parameters, in turn, are tightly linked to canopy light capture (e.g., fPAR), plant growth rates, turnover, and decomposition (Aber and Melillo 1991). Leaf N is thus a major target measurement in remote sensing for biogeochemical research (e.g., Wessman et al. 1988, Martin and Aber 1997). However, despite nearly two decades of research, remote sensing of leaf N remains challenging because the physical basis for N retrievals is not clear and because remote sensing data needed to retrieve N are not available over broad spatial scales. N is expressed in chlorophyll in the visible (400–700 nm) region, and in proteins in the shortwave-infrared (1500–2300 nm) range (Curran 1989). A hyperspectral remote sensing signature of a forest canopy is, however, equally or more sensitive to canopy architecture than it is to leaf chlorophyll and proteins (Asner 1998). Nonetheless, leaf N can be estimated remotely using empirical techniques, such as partial least squares (PLS) regression, which does not resolve the contribution of leaf chemistry and canopy structure to the spectral measurement (e.g., Smith et al. 2002, 2003). This fact suggests strong biophysical covariance of leaf and canopy properties in a reflectance spectrum, and this covariance is precisely what is needed to improve strength of remote sensing constraints over modeled processes. In this way, remote sensing and biogeochemical modeling require the same, ecological approach to increase both the accuracy and breadth of the measurements and simulations. Sellers (1985, 1987) was one of the first to express this well in his work on the covariance of physiological processes in plants with respect to remotely sensed signatures. Future work should continue to expand these connections, and to seek broader inter-relationships between remotely sensed and modeled properties of ecosystems.

A second need in remote sensing-modeling research is to address the inherent temporal and spatial mismatch of the observations and the simulated processes. Remotely sensed parameters are often temporally sparse, but spatially rich. In contrast, biogeochemical models require continuous time-series calculations of pertinent fluxes, but the complexity of the models often precludes a pixel-by-pixel simulation approach when high spatial resolution is required. One major research area focuses on the use of remotely sensed data in model-data assimilation (e.g., Bach and Mauser 2003, Rayner et al. 2005). Data assimilation involves a suite of approaches including: (1) model initialization from remote sensing data (e.g., land cover type), (2) update of model state variables through remote sensing (e.g., LAI), (3) remote sensing parameter adjustment through model recalibration (e.g., leaf N concentrations), and (4) estimation of model state variables through model inversion (e.g., fPAR) (Bach and Mauser 2003, Zhang et al. 2006).

Data assimilation also facilitates the integration of multi-temporal remotely sensed parameters from differing spatial sampling schemes (e.g., Landsat land-cover and MODIS phenology) into a biogeochemical and/or land-surface modeling environment. These approaches allow for temporally sparse and spatially detailed observations to be ingested into continuous modeling streams, thereby adjusting model trajectories through time. One problem in doing this, however, is that high frequency changes in vegetation properties (e.g., canopy pigments and physiology) may be missed, since neither the models nor the remote sensing data would capture these changes. On the other hand, model-data assimilation approaches tend to constrain the 'solution space' of current models to a relatively tight degree, thereby decreasing uncertainty of simulated processes over time. Nonetheless, this research area continues to challenge both the remote sensing and modeling communities, and will require further attention as the sensors systems and models evolve in the future.

Finally, it seems that there has been slow recent growth in the area of remote sensing/model integration. Early on, there were rapid advances in the use of remotely sensed data from AVHRR and MODIS-like sensors in land-surface and biogeochemical models (e.g., Potter et al. 1993, Running and Hunt 1993, Sellers et al. 1997). As discussed, vegetation type, cover, LAI and fPAR were readily incorporated into a class of production-efficiency models born during the 1980s–1990s. Since then, however, only a few models have expanded the list of remotely sensed parameters, mainly in the area of 3-D structure (e.g., Hurtt et al. 2004, Huang et al. 2008). However, today we have a range of airborne and space-based observations of canopy properties that have yet to be fully incorporated into the terrestrial models. The use of canopy height information, in particular, represents an underutilized opportunity for modelers, in that making use of these data can require no more than the addition of simple allometric equations relating height to biomass. Such equations are commonly available from field observations and could be added to some models with minimal restructuring. In a sense, the modeling community is not keeping pace with remote sensing developments, or at least has not begun utilizing them fully, and thus a challenge for the modeling community is to develop appropriate constraints in simulations using the newest portfolio of observations now available from remote sensing. Future studies and programs should place emphasis on the co-evolution of these complimentary research areas.

REFERENCES

Aber, J. D. and C. T. Driscoll, 1997. Effects of land use, climate variation, and N deposition on N cycling and C storage in northern hardwood forests. *Global Biogeochemical Cycles*, 11: 639–648.

Aber, J. D. and J. M. Melillo, 1991. *Terrestrial Ecosystems*. Saunders College Publishing, Philadephia.

Asner, G. P., 1998. Biophysical and biochemical sources of variability in canopy reflectance. *Remote Sensing of Environment*, 64: 234–253.

Asner, G. P., B. H. Braswell, D. S. Schimel, and C. A. Wessman, 1998. Ecological research needs from multi-angle remote sensing data. *Remote Sensing of Environment*, 63: 155–165.

Asner, G. P., A. R. Townsend, W. J. Riley, P. A. Matson, J. C. Neff, and C. C. Cleveland. 2001. Physical and bio-geochemical controls over terrestrial ecosystem responses to nitrogen deposition. *Biogeochemistry*, 54: 1–39.

Asner, G. P., D. Nepstad, G. Cardinot, and D. Ray, 2004. Drought stress and carbon uptake in an Amazon forest measured with spaceborne imaging spectroscopy. *Proceedings of the National Academy of Sciences*, 101: 6039–6044.

Asner, G. P., A. J. Elmore, F. R. Hughes, A. S. Warner, and P. M. Vitousek, 2005a. Ecosystem structure along bioclimatic gradients in Hawai from imaging spectroscopy. *Remote Sensing of Environment*, 96: 497–508.

Asner, G. P., D. E. Knapp, A. N. Cooper, M. M. C. Bustamante, and L. P. Olander, 2005b. Ecosystem structure throughout the Brazilian Amazon from Landsat observations and automated spectral unmixing. *Earth Interactions*, 9: 1–31.

Bach, H. and W. Mauser, 2003. Methods and examples for remote sensing data assimilation in land surface process modeling. *IEEE Transactions on Geoscience and Remote Sensing*, 41: 1629–1637.

Bastiaanssen, W. G. M., R. A. Menenti, R. A. Feddes, and A. A. M. Holtslag, 1998. A remote sensing surface energy balance algorithm for land (SEBAL), part 1: formulation. *Journal of Hydrology*, 212/213: 198–212.

Bastiaanssen, W. G. M., M. D. Ahmad, and Y. Chemin, 2002. Satellite surveillance of evaporative depletion across the Indus. *Water Resources Research*, 38: 1–9.

Bonan, G. B., 1995. Land-atmosphere interactions for climate system models: Coupling biophysical, biogeochemical, and ecosystem dynamical processes. *Remote Sensing of Environment*, 51: 57–73.

Bondeau, A., D. W. Kicklighter, and J. Kaduk, 1999. Comparing global models of terrestrial net primary productivity (NPP): importance of vegetation structure on seasonal NPP estimates. *Global Change Biology*, 5: 35–45.

Choudhury, B. J., 1988. Relating Nimbus-7 37 GHz data to global land-surface evaporation, primary productivity and the atmospheric CO_2 concentration. *International Journal of Remote Sensing*, 9: 169–176.

Cramer, W., A. Bondeau, F. I. Woodward, I. C. Prentice, R. A. Betts, V. Brovkin, P. M. Cox, V. Fisher, J. A. Foley, A. D. Friend, C. Kucharik, M. R. Lomas, N. Ramankutty, S. Sitch, B. Smith, A. White, and C. Young-Molling, 2001. Global response of terrestrial ecosystem structure and function to CO_2 and climate change: results from six dynamic global vegetation models. *Global Change Biology*, 7: 357–373.

Curran, P. J., 1989. Remote sensing of foliar chemistry. *Remote Sensing of Environment*, 30: 271–278.

DeFries, R. S., C. B. Field, I. Fung, G. J. Collatz, and L. Bounoua, 1999. Combining satellite data and biogeochemical models to estimate global effects of human-induced land cover change on carbon emissions and primary productivity. *Global Biogeochemical Cycles*, 13: 803–815.

DeFries, R. S., M. C. Hansen, J. R. G. Townshend, A. C. Janetos, and T. R. Loveland, 2000. A new global 1-km dataset of percentage tree cover derived from remote sensing. *Global Change Biology*, 6: 247–254.

Dickenson, R. E., 1991. Global change and terrestrial hydrology – A review. *Tellus*, 43 AB: 176–181.

Engman, E. T., 1991. *Remote Sensing in Hydrology*. Chapman and Hall, London.

Field, C. B., J. T. Randerson, and C. M. Malmström, 1995. Global net primary production: Combining ecology and remote sensing. *Remote Sensing of Environment*, 51: 74–88.

Field, C. B., M. J. Behrenfeld, J. T. Randerson, and P. Falkowski, 1998. Primary production of the biosphere: integrating terrestrial and oceanic components. *Science*, 281: 237–240.

Gamon, J. A., C. B. Field, W. Bilger, A. Björkman, A. L. Fredeen, and J. Peñuelas, 1990. Remote sensing of the xanthophyll cycle and chlorophyll fluorescence in sunflower leaves and canopies. *Oecologia*, 85: 1–7.

Gitelson, A. A., A. Viña, D. C. Rundquist, V. Ciganda, T. J. Arkebauer, 2005. Remote Estimation of Canopy Chlorophyll Content in Crops. *Geophysical Research Letters*, 32: L08403, doi:10.1029/2005GL022688.

Gitelson, A. A., A. Viña, S. B. Verma, D. C. Rundquist, T. J. Arkebauer, G. Keydan, B. Leavitt, V. Ciganda, G. G. Burba, and A. E. Suyker, 2006. Relationship between gross primary production and chlorophyll content in crops: Implications for the synoptic monitoring of vegetation productivity. *Journal of Geophysical Research*, 111: D08S11, doi:10.1029/2005JD006017.

Gower, S. T., C. J. Kucharik, and J. M. Norman, 1999. Direct and indirect estimation of leaf area index, fAPAR, and net primary production of terrestrial ecosystems. *Remote Sensing of Environment*, 70: 29–51.

Green D. S., J. E. Erickson, and E. L. Kruger, 2003. Foliar morphology and canopy nitrogen as predictors of light-use efficiency in terrestrial vegetation. *Agricultural and Forest Meteorology*, 3097: 1–9.

Huang, M., G. P. Asner, M. Keller, and J. A. Berry. 2008. An ecosystem model for tropical forest disturbance and selective logging. Journal of Geophysical Research 113, doi: 10.1029/2007JG000438.

Hurtt, G. C., R. Dubayah, J. Drake, P. R. Moorcroft, S. W. Pacala, J. B. Blair, and M. G. Fearon, 2004. Beyond potential vegetation: Combining LIDAR data and a height-structured model for carbon studies. *Ecological Applications*, 14: 873–883.

Keyser, A. R., J. S. Kimball, R. R. Nemani, and S. W. Running, 2000. Simulating the effects of climate change on the carbon balance of North American high-latitude forests. *Global Change Biology*, 6: 185–195.

Kustas, W. P. and J. M. Norman, 1996. Use of remote sensing for evapotranspiration monitoring over land surfaces. *Hydrological Sciences Journal*, 41: 495–516.

Kustas, W. P., M. S. Moran, R. D. Jackson, L. W. Gay, L. F. W. Duell, K. E. Kunkel, and A. D. Matthias, 1990. Instantaneous and daily values of the surface energy balance over agricultural fields using remote sensing and a reference field in an arid environment. *Remote Sensing of Environment*, 32: 125–141.

Lefsky, M. A., W. B. Cohen, G. G. Parker, and D. J. Harding, 2002. Lidar remote sensing for ecosystem studies. *Bioscience*, 52: 19–30.

Lhomme, J.-P., B. Monteny, and M. Amadou, 1994a. Estimating sensible heat flux from radiometric temperature over sparse millet. *Agricultural Forest Meteorology*, 35: 110–121.

Lhomme, J.-P., B. Monteny, A. Chehbouni, and D. Troufleau, 1994b. Determination of sensible heat flux over Sahelian fallow savannah using infrared thermometry. *Agricultural Forest Meteorology*, 68: 93–105.

Lobell, D. B., J. A. Hicke, G. P. Asner, C. B. Field, C. J. Tucker, and S. O. Los, 2002. Satellite estimates of productivity and light use efficiency in United States agriculture, 1982–98. *Global Change Biology*, 8: 722–735.

Martin, M. E. and J. D. Aber, 1997. High spectral resolution remote sensing of forest canopy lignin, nitrogen, and ecosystem processes. *Ecological Applications*, 7: 431–444.

McNeil, B. E., J. M. Read, and C. T. Driscoll, 2005. Identifying Controls on the Spatial Variability of Foliar Nitrogen in a Large, Complex Ecosystem: the Role of Atmospheric Nitrogen Deposition in the Adirondack Park, NY, USA. *Journal of Agricultural Meteorology*, 60: 1157–1160.

Means, J. E., S. A. Acker, D. J. Harding, J. B. Blair, M. A. Lefsky, W. B. Cohen, M. E. Harmon, and W. A. McKee, 1999. Use of large-footprint scanning airborne lidar to estimate forest stand characteristics in the Western Cascades of Oregon. *Remote Sensing of Environment*, 67: 298–308.

Monteith, J. L., 1972. Solar radiation and productivity in tropical ecosystems. *Applied Ecology*, 9: 747–766.

Moorcroft, P. R., G. C. Hurtt, and S. W. Pacala, 2001. A method for scaling vegetation dynamics: The Ecosystem Demography model (ED). *Ecological Monographs*, 71: 557–585.

Myneni, R. B., R. N. Ramakrishna, and S. W. Running, 1997. Estimation of global leaf area index and absorbed PAR using radiative transfer models. *IEEE Transactions on Geoscience and Remote Sensing*, 35: 1380–1393.

Nepstad, D. C., P. Moutinho, M. B. Dias, E. Davidson, G. Cardinot, D. Markewitz, R. Figueiredo, N. Vianna, J. Chambers, D. Ray, J. B. Guerreiros, P. Lefebvre, L. Sternberg, M. Moreira, L. Barros, F. Y. Ishida, I. Tohlver, E. Belk, K. Kalif, and K. Schwalbe, 2002. The effects of partial throughfall exclusion on canopy processes, aboveground production, and biogeochemistry of an Amazon forest. *Journal of Geophysical Research*, 107: 1–18.

Ollinger, S. V. and M.-L. Smith, 2005. Net primary production and canopy nitrogen in a temperate forest landscape: An analysis using imaging spectroscopy, modeling and field data. *Ecosystems*, 8: 760–778.

Ollinger, S. V., J. D. Aber, and C. A. Federer, 1998. Estimating regional forest productivity and water balances using an ecosystem model linked to a GIS. *Landscape Ecology*, 13(5): 323–334.

Ollinger, S. V., M. L. Smith, M. E. Martin, R. A. Hallett, C. L. Goodale, and J. D. Aber, 2002a. Regional variation in foliar chemistry and soil nitrogen status among forests of diverse history and composition. *Ecology*, 83(2): 339–355.

Ollinger, S. V., J. D. Aber, P. B. Reich and R. Freuder, 2002b. Interactive effects of nitrogen deposition, tropospheric ozone, elevated CO_2 and land use history on the carbon dynamics of northern hardwood forests. *Global Change Biology*, 8: 545–562.

Parton, W. J., A. R. Mosier, and D. S. Schimel, 1988. Dynamics of C, N, P, and S in grassland soils: a model. *Biogeochemistry*, 5: 109–131.

Parton, W. J., M. Hartman, D. Ojima, and D. Schimel, 1998. DAYCENT and its land surface submodel: description and testing. *Global and Planetary Change*, 19: 35–48.

Potter, C., S. Klooster, R. Nemani, V. Genovese, S. Hiatt, M. Fladeland, and P. Gross, 2006. Estimating carbon budgets for U.S. ecosystems. *EOS, Transactions*, 87: 85, 90.

Potter, C. S., J. T. Randerson, C. B. Field, P. A. Matson, P. M. Vitousek, H. A. Mooney, and S. A. Klooster, 1993. Terrestrial Ecosystem Production – a Process Model-Based on Global Satellite and Surface Data. *Global Biogeochemical Cycles*, 7: 811–841.

Rayner, P. J., M. Scholze, W. Knorr, T. Kaminski, R. Giering, and H. Widmann, 2005. Two decades of terrestrial carbon fluxes from a carbon cycle data assimilation system (CCDAS). *Global Biogeochemical Cycles*, 19: 1–20.

Reich, P. B., M. B. Walters, and D. S. Ellsworth, 1997. From tropics to tundra: Global convergence in plant functioning. *Proceedings of the National Academy of Sciences*, 94: 13730–13734.

Roberts, D. A., S. L. Ustin, S. Ogunjemiyo, J. Greenberg, S. Z. Dobrowski, J. Q. Chen, and T. M. Hinckley, 2004. Spectral and structural measures of northwest forest vegetation at leaf to landscape scales. *Ecosystems*, 7: 545–562.

Running, S. W. and E. R. Hunt, 1993. Generalization of a forest ecosystem process model for other biomes, BIOME-BGC, and an application for global-scale models. Pages 141–158 in J. R. Ehleringer and C. B. Field (eds), *Scaling Physiological Processes: Leaf to Globe*. Academic Press, San Diego.

Running, S. W., T. R. Loveland, and L. L. Pierce, 1994. A vegetation classification logic based on remote sensing for use in general biogeochemical models. *Ambio* 23: 77–81.

Running, S. W., P. E. Thornton, R. R. Nemani, and J. M. Glassy, 2000. Global Terrestrial Gross and Net Primary Productivity from the Earth Observing System. In O. Sala, R. Jackson, and

H. Mooney (eds), *Methods in Ecosystem Science*. Springer-Verlag New York.

Sellers, P. J., 1985. Canopy reflectance, photosynthesis and transpiration. *International Journal of Remote Sensing*, 6: 1335–1372.

Sellers, P. J., 1987. Canopy reflectance, photosynthesis, and transpiration. II. The role of biophysics in the linearity of their interdependence. *Remote Sensing of Environment*, 21: 143–183.

Sellers, P. J., Y. Mintz, Y. C. Sud, and A. Dalcher, 1986. A simple biosphere model (SiB) for use within general circulation models. *Journal of the Atmospheric Sciences*, 43: 505–531.

Sellers, P. J., B. W. Meeson, F. G. Hall, G. Asrar, R. E. Murphy, R. A. Schiffer, F. P. Bretherton, R. E. Dickinson, R. G. Ellingson, C. B. Field, K. F. Huemmrich, C. O. Justice, J. M. Melack, N. T. Roulet, D. S. Schimel, and P. D. Try, 1995. Remote-sensing of the land-surface for studies of global change – models, algorithms, experiments. *Remote Sensing of Environment*, 51: 3–26.

Sellers, P. J., R. E. Dickenson, D. A. Randall, A. K. Betts, F. G. Hall, J. A. Berry, G. J. Collatz, A. S. Denning, H. A. Mooney, C. A. Nobre, N. Sato, C. B. Field, and A. Henderson-Sellers, 1997. Modeling the exchanges of energy, water, and carbon between continents and the atmosphere. *Science*, 275: 502–509.

Shoshany, M., 1993. Roughness-reflectance relationship of bare desert terrain: an empirical study. *Remote Sensing of Environment*, 45: 15–27.

Sitch, S., B. Smith, I. C. Prentice, A. Arneth, A. Bondeau, W. Cramer, J. O. Kaplan, S. Levis, W. Lucht, M. T. Sykes, K. Thonicke, and S. Venevsky, 2003. Evaluation of ecosystem dynamics, plant geography and terrestrial carbon cycling in the LPJ dynamic global vegetation model, *Global Change Biology*, 9: 161–185.

Smith, M. L., S. V. Ollinger, M. E. Martin, J. D. Aber, R. A. Hallett, and C. L. Goodale, 2002. Direct estimation of aboveground forest productivity through hyperspectral remote sensing of canopy nitrogen. *Ecological Applications*, 12: 1286–1302.

Smith, M. L., M. E. Martin, L. Plourde, and S. V. Ollinger, 2003. Analysis of hyperspectral data for estimation of temperate forest canopy nitrogen concentration: comparison between an airborne (AVIRIS) and spaceborne (Hyperion) sensor. *IEEE Transactions on Geoscience and Remote Sensing*, 41: 1332–1337.

Townsend, A. R., C. C. Cleveland, G. P. Asner, and M. M. C. Bustamante, 2007. Controls over foliar N:P ratios in tropical rain forests. *Ecology*, 88: 107–118.

Treuhaft, R. N., G. P. Asner, and B. E. Law, 2003. Structure-based forest biomass from fusion of radar and hyperspectral observations. *Geophysical Research Letters* 30: 25–21.

Turner, D. P., S. Urbanski, D. Bremer, S. C. Wofsy, T. Meyers, S. T. Gower, and M. Gregory, 2003. A cross-biome comparison of daily light use efficiency for gross primary production. *Global Change Biology*, 9: 383–395.

Ustin, S. L., D. A. Roberts, J. A. Gamon, G. P. Asner, and R. O. Green, 2004. Using imaging spectroscopy to study ecosystem processes and properties. *Bioscience*, 54: 523–534.

Wessman, C. A., J. D. Aber, D. L. Peterson, and J. M. Melillo, 1988. Remote sensing of canopy chemistry and nitrogen cycling in temperate forest ecosystems. *Nature*, 335: 154–156.

White, M. A., P. E. Thornton, and S. W. Running, 1997. A continental phenology model for monitoring vegetation responses to interannual climatic variability. *Global Biogeochemical Cycles*, 11: 217–235.

Wright, I. J., P. B. Reich, M. Westoby, D. D. Ackerly, Z. Baruch, F. Bongers, J. Cavender-Bares, T. Chapin, J. H. C. Cornelissen, M. Diemer, J. Flexas, E. Garnier, P. K. Groom, J. Gulias, K. Hikosaka, B. B. Lamont, T. Lee, W. Lee, C. Lusk, J. J. Midgley, M.-L/ Navas, Ü. Niinemets, J. Oleksyn, N. Osada, H. Poorter, P. Poot, L. Prior, V. I. Pyankov, C. Roumet, S. C. Thomas, M. G. Tjoelker, E. J. Veneklaas and R. Villar, 2004. The worldwide leaf economics spectrum. *Nature*, 6985: 821–827.

Wulder, M. A., and S. E. Franklin (eds), 2003. *Remote Sensing of Forest Environments*. Kluwer Academic Publishers, Norwell, MA.

Yoder, B. J., and R. H. Waring, 1994. The normalized difference vegetation index of small Douglas-Fir canopies with varying chlorophyll concentrations. *Remote Sensing of Environment*, 49: 81–91.

Zarco-Tejada, P. J., J. R. Miller, T. L. Noland, G. H. Mohammed, and P. H. Sampson, 2001. Scaling-up and model inversion methods with narrowband optical indices for chlorophyll content estimation in closed forest canopies with hyperspectral data. *IEEE Transactions on Geoscience and Remote Sensing*, 39: 1491–1507.

Zhang, Q, X. Xiao, B. H. Braswell, E. Linder, S. V. Ollinger, M-L. Smith, J. P. Jenkins, F. Baret, A. D. Richardson, B. Moore III, and R. Minocha, 2006. Characterization of seasonal variation of forest canopy in a temperate deciduous broadleaf forest, using daily MODIS data. *Remote Sensing of Environment*, 105: 189–203.

Zhang, Y., Y. Tian, Y. Knyazikhin, J. V. Martonchik, D. J. Diner, M. Leroy, and R. B. Myneni, 2000. Prototyping of MISR LAI and FPAR algorithm with POLDER data over Africa. *IEEE Transactions on Geoscience and Remote Sensing*, 38: 2402–2418.

Remote Sensing of Urban Areas

Janet Nichol

Keywords: land use, population, urban environmental quality, urban heat island, air quality.

INTRODUCTION

By the year 2007, more than half of the world's population lived in urban areas (United Nations 2006) and trends suggest the number of urban dwellers will rise to almost 5 billion by 2030, out of a world total of 8.1 billion. This rapid urbanization is a global phenomenon giving rise to a variety of social and environmental problems which affect cities in developed and developing worlds somewhat differently. In the developing world, socio-economic problems resulted in over 1 billion people living in urban slums by 2007 (United Nations 2006), whereas in affluent cities the quality of the urban environment is of greater concern. Common to both is the need for information to assist governments in management and planning, the most basic of which is a count of the population and its trends, as well as how population and activities are distributed spatially in cities. The synoptic view of cities afforded by remote sensing platforms confers great potential for data collection over urban areas although in reality urban areas have been the most challenging to remote sensing for a number of reasons.

The first of these is the nature of the data itself: population and land use are not directly observable either on the ground or from a remote platform, and people are not static. The same is true of most aspects of urban environmental quality, which is largely a subjective concept, and is based on numerous properties which vary at different scales over urban areas. Urban areas are particularly challenging due to their high spatial and spectral variability coupled with the irregular size, shape and orientation of objects. The presence of shadows on images of high-rise areas, and the obstruction of view due to parallax displacement are further drawbacks to the use of fine spatial resolution images, whether airborne or spaceborne.

Additional constraints are imposed by limitations of the spatial and spectral resolution of the sensing systems. For example, individual buildings and streets in urban areas were not detectable on moderate to coarse spatial resolution sensors of the 1970s and 1980s such as Landsat MSS and TM. Since 2000, several fine spatial resolution sensors with resolution in the order of 1–5 m have become available, but paradoxically the increase in spatial and spectral detail has not led to enhanced interpretability using automated techniques. This is because at fine spatial resolutions, spectral variability within objects is often greater than between objects, and methodologies and software for the interpretation of such objects are currently not well developed. The spectral similarity of most man-made surfaces is evident from Figure 30.1, and Figure 30.2 shows that on standard false colour composite images they exhibit a typically steel grey appearance.

In summary, although urban areas pose many challenges to remote sensing, rapid changes in their extent, land use and environment increase the urgency for efficient and timely solutions.

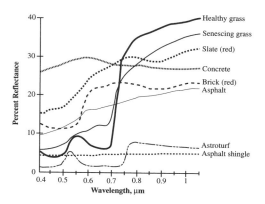

Figure 30.1   Spectral signatures of typical urban surfaces.
Source: Jensen, J. R., Remote Sensing of the Environment: An earth resource perspective. Copyright 2000, p. 417. Reprinted by permission of Pearson Education, Inc., Upper Saddle River, NJ.

REMOTE SENSING FOR LAND USE/LAND COVER MAPPING

Mapping land cover over large areas using satellite sensor data became a reality in the early 1970s with the launch of earth resources satellites. Rapid changes in urban land use especially at the urban periphery necessitate frequent data collection, and planning departments in many modern cities devote considerable resources to the production and updating of land use/land cover maps. Anderson et al. (1976) produced a four-level land use/land cover classification scheme for use with remote sensing data which provides a basic guideline for the level of class detail obtainable using different sensor resolutions. However, in the light of developments in remote sensing systems and methodologies, there is a need for a more appropriate conceptual model relating land use and land cover individually to image resolution and processing routines. The basic premise of the Anderson system, that increased detail in the mapping classes requires finer spatial resolution images, remains applicable only in certain contexts. For example, Welch (1982) observes that aerial photographs with fine resolutions of 0.5–3 m are required for urban land use mapping at Anderson's levels II and III where

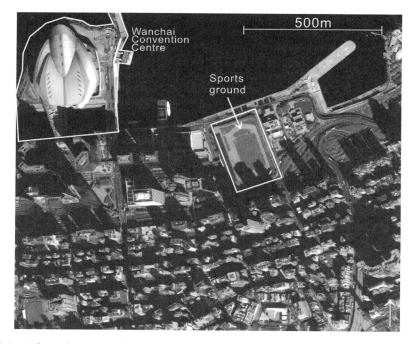

Figure 30.2   False colour, pan-sharpened QuickBird image of the northern shore of Hong Kong island, with the Wanchai Exhibition Centre at top left, and high rise CBD in the lower section. Over the land area, more than half of the ground is obscured by parallax displacement of tall buildings and building shadows. Copyright Digital Globe. (See the color plate section of this volume for a color version of this figure).

actual land uses, including building functions, need to be interpreted. However, in a situation where detection of themes is more important than precise feature identification, a fine resolution is not necessary and may even be a disadvantage. Thus Gao and Skillcorn (1998) were able to identify ten categories of land cover on the rural–urban periphery of Auckland at Anderson's level II using SPOT HRV imagery with 20 m resolution.

Strahler et al.'s (1986) recommendation of two distinct models to represent the relationship between scene elements and image resolution may clarify the above paradox. In Strahler et al.'s H-resolution model, scene elements are larger than the resolution cells and can therefore be directly identified. In the L-resolution model the scene elements of interest are not detectable as they are smaller than the resolution cells.

Mapping of urban land use with moderate resolution sensors: the H-resolution model

The moderate accuracy of *ca.* 75–80% achieved by Gao and Skillcorn (1998), is typical of accuracies achieved by other urban land use studies using the moderate resolution sensors of the 1980s and 1990s, SPOT HRV and Landsat TM (ETM+) (for example Gong and Howarth 1990, Trietz et al. 1992). However, experience has shown that, in the case of automated interpretation, finer spatial resolution does not always provide higher accuracy, and the 20–30 m spatial resolution of these systems may be too detailed to capture the aggregate spectral signature of urban land use types which exhibit high intra-class variability (Forster 1985). For example, the typically high reflectance surfaces of concrete footpaths, parking lots and tile roofs mixed with lower reflecting vegetated gardens, and asphalt roads are all present in the one class 'residential areas' and as a class it may be spectrally multi-modal. By inference, the fine spatial resolution sensors available since 2000 have even more serious limitations in this sense (Aplin et al. 1999). In fact such composite types of land use, as are common in urban areas, are a special case of H-resolution since although the individual components of the land use type are larger than the pixel size and thus identifiable visually, automated methods are unable to utilize all the information available for recognition at a higher level. The QuickBird image (Figure 30.2) shows a convention centre and sports ground between Hong Kong's central business district and harbour. Neither is spectrally homogeneous, and a knowledge-based approach would be required. Barnsley and Barr (1996) define the problem in terms of inappropriate classification methodologies, especially the commonly used

'per pixel' classifiers which assumes that classes are uni-modal. They also discuss several alternative approaches to supplement the spectral information, such as the use of pre-classification smoothing filters (Atkinson et al. 1985), the addition of texture (Gong and Howarth 1990, 1992, Zhang et al. 2002) or ancillary data (Trietz et al. 1992), and the use of contextual information such as image segmentation, in 'per field' classifiers (Aplin et al. 1999, Stuckens et al. 2000).

Thematic mapping at sub-pixel level: The L-resolution model

Broad themes with distinct spectral signatures, such as vegetation or impervious surfaces, are more amenable to automated classification than are land use classes, and finer spatial resolutions usually achieve higher accuracy. However, since in an urban area there will often be more than one cover type within the sensor's IFOV at 20 m or 30 m resolution, hard, 'per-pixel' classifiers do not produce accurate results. Yang et al. (2003) were able to overcome this spatial averageing within the 30 m pixels of Landsat ETM+ for mapping impervious surfaces by modeling the relationship between training sets from the 4 m resolution data of IKONOS and the corresponding Landsat ETM+ data. They obtained an average error of predicted versus actual impervious surfaces ranging from *ca.* 9–11%, and since the method was deemed to be robust, it was subsequently applied over the whole United States. Another technique which is able to address the problem of hard classifiers at L-resolutions, is that of Linear Spectral Unmixing (LSU) since it is able to estimate the proportions of reflective entities (endmembers) within a pixel (Adams et al. 1986). It has been extensively applied to urban land cover mapping (Ridd 1995, Small 2001, 2003, Phinn et al. 2002, Rashed et al. 2003, Nichol and Wong 2007). LSU is a soft classifier which assumes a linear or non-linear (Rashed et al. 2003) combination of land cover elements within a pixel, but since the number of these elements is restricted by the number of image spectral bands, a 3-endmember VIS model of Ridd (1995), based on vegetation, impervious surfaces and soil, or a 4-endmember model of Small (2001, 2003) based on high albedo and low albedo surfaces, vegetation and soil, are commonly used.

Land cover mapping with fine spatial resolution images: Manual or automated?

The ability of fine spatial resolution spaceborne systems such as IKONOS and QuickBird to obtain regular multispectral images calibrated to absolute radiance or reflectance, makes them competitive with aerial photographs for detailed urban mapping (Nichol and Lee 2005). In developed countries airborne digital photography is also becoming a relatively inexpensive digital alternative to standard aerial photographs (Stow et al., in this volume). To obtain fine spatial resolution digital photographs a digital camera is generally flown aboard a low-altitude aircraft equipped with a GPS and a computer for storing data from both the camera and GPS. Akbari et al. (2003) estimated the cost of flying fine resolution (0.3 m) colour aerial photographs over Sacramento, including creation of an orthophoto mosaic, to be US$140 km^{-2}. They cited the very fine spatial resolution, ability to schedule flights as desired, and the reasonable cost as advantages of airborne digital photography over other systems. However, their objective of estimating the areal fractions of different surface types for their impact on the climate and air quality of the city could not be achieved by direct automated classification of the digital images, and visual analysis was required. A similar problem was identified in a survey of road type and condition in Hong Kong, using pan-sharpened QuickBird imagery with 0.86 m resolution (Nichol and Wong 2005b). Figure 30.2 illustrates some problems of interpreting such very fine spatial resolution images in urban areas, with over 50% of the ground surface and most streets obscured due to shadow and parallax displacement of tall buildings. Thus fine spatial resolution images, whether airborne or spaceborne, currently pose challenges to automated interpretation, and further developments in knowledge-based classifiers are required, especially for recognition of relevant ground objects.

APPROACHES TO POPULATION ESTIMATION

In a comparative analysis of remote sensing for population estimation, Lo (1986) distinguished four methods which have provided a contextual framework for subsequent research, namely:

1 counts of individual dwelling units,
2 measurement of the built-up area,
3 measurement of residential land use, and
4 estimations based on pixel radiance.

Only the fourth method was based on automated interpretation. Although developments in remote sensing technology over the last 20 years permit all four to be undertaken using digital images, the first method which counts individual dwelling units still requires visual interpretation. Lo's fourth method, based on pixel radiance is difficult to implement due to the, at best, tenuous relationship between population and pixel brightness, and both methods one and four are difficult to validate since population data are unavailable at the level of individual dwelling units or pixels. For these reasons, moderate spatial resolution sensors such as Landsat TM (ETM+) have become the preferred data source for population estimation, and most recent investigations have utilized methods (1) and (2) which relate population to the extent of urban land cover, and to the extent of residential land cover respectively. Method (3), also known as the Zone-based approach (Lo 2003), first uses digital image classification to identify one or more classes of residential land use, followed by either fieldwork or census data to obtain population density for each residential land use type. If the population of census districts is then computed from the total residential area within each district, the model can be validated. This method presumes a linear or logarithmic relationship between population and residential land area. For US cities, a logarithmic relationship known as the allometric growth model (Tobler 1969) is often suitable (Lo 2003) (i.e., the relative growth rate of population is proportional to the relative growth in residential land area). The model takes the form:

$$\log P = \log a + b \log A \qquad (1)$$

where P is population size, A is residential land area, a is the proportionality coefficient, and b is the scaling factor. When b is larger than 1, positive allometry occurs, where population increases faster than land area, and when less than 1, land area increases faster than population. Lo (2003) found this model better than the linear model for representing the relationship between population and residential land use for census districts in Atlanta, Georgia because as city population increases, people move outwards to occupy more spacious suburban areas. Using a linear model, Lo (2003) regressed population size against area of residential land use for 418 census tracts in Atlanta, and obtained a relationship with $r = 0.66$ that was significant at the 99.9% confidence level. However, r increased to 0.78 using the allometric model.

Further refinement of the zone-based model has been demonstrated by Lu et al. (2006), who noted that within residential areas, more impervious surface represents more buildings and roads and, by implication, more people. They therefore combined classified residential land use areas with an impervious surface fraction image derived from linear spectral unmixing of the same ETM+ image, to derive a parameter, residential impervious surface. A regression developed from block groups with known population, and their percentage of residential impervious surface, gave a fairly high correlation of $R^2 = 0.82$ when applied to the known block groups. Similar accuracies have been reported by Harvey (2002) for a pixel-based approach to population estimation in Australian cities. This approach, which is most similar to Lo's (1986) method (4) based on pixel radiance, uses an initial land cover classification of the image to derive a residential land use class. Then, unlike the zone-based approach, it proceeds to derive regression-based relationships between population and the spectral characteristics of pixels within this class, which permit conversion of pixels to population.

Error in most previous studies has largely been attributed to overestimation of population in low density areas, and underestimation in high density areas (Harvey 2002, Lo 2003, Li and Weng 2005), and prior stratification into high and low density residential areas is commonly recommended for future research. Since both zone-based and pixel-based methods assume that residential land use is synonymous with the low density urban land cover class, the models described above may not be applicable to European or Asian cities where traditionally residential use is mixed with other uses in high density city centres. Furthermore in states with high land prices such as Singapore and Hong Kong, where high-rise residential development is the norm, the allometric model would be difficult to apply.

URBAN ENVIRONMENTAL QUALITY

Urban Environmental Quality (UEQ) is a holistic concept, comprising numerous parameters which affect urbanites synergistically, but which are measured on different scales. Jensen (2000) provides a list of infrastructural features used for quality of life assessment for a house, street or urban complex, but most would require visual interpretation of large scale aerial photographs. On the other hand, UEQ can be considered as a complex, integrated property which varies continuously over the urban landscape. To capture such continuous variability of the urban environment over whole cities, satellite sensor images are the only data source.

However, until recently, satellite sensors were unable to capture data with enough detail to represent the fragmented land cover of urban areas. Indeed, a comparative study of Environmental Quality in Hong Kong (Fung and Siu 2001) used moderate resolution SPOT HRV images to estimate green space at the generalized level of Tertiary Planning Units. A similar scale study to assess the quality of life in Athens Georgia (Lo and Faber 1997) used the normalized difference vegetation index (NDVI) to estimate greenness and temperature from Landsat TM images at 30 m and 60 m resolution, respectively, and data were averaged at

the census district level. The resulting maps at electoral or administrative district level have limited utility for urban and environmental planning, since landscaping and redevelopment in cities takes place at the local scale of individual buildings, blocks or streets, and people experience the environment at this local level.

The availability of fine spatial resolution satellite images, and digital map data representing urban infrastructure down to building level, coupled with the hybrid, raster-vector data handling ability of modern GIS, now permit detailed analysis at the pixel level. The only requirement is the development of appropriate methodologies for integrating data sources. However, due to the different scales of measurement, as well as different measurement units, approaches to integrating the variables are limited. Two general approaches are Principal Component Analysis (PCA) and the GIS overlay approach (Lo and Faber 1997, Nichol and Wong 2005a, 2008).

Nichol and Wong (submitted) tested both techniques for integrating three image-based environmental variables: biomass, temperature and aerosol optical depth (AOD), with other environmental variables, namely noise, building density and building height, for the assessment of UEQ in Kowloon, Hong Kong. The best overall index of UEQ was derived from GIS overlay of the parameters biomass, building density and building height at 4 m resolution (Figure 30.3). A correlation coefficient of 0.79 was obtained, when compared with field questionnaires measuring respondents'

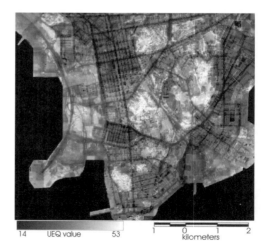

Figure 30.3   Map of UEQ from pixel based overlay of the parameters Vegetation Density, Building Density, and Building Height, at 4 m resolution. See Figure 6 for a false colour composite of the mapped area.

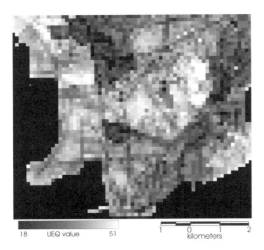

18   UEQ value   51   kilometers

Figure 30.4   Same as Figure 30.3, but at 64 m resolution.

judgment of the environment. This was higher than for maps derived from the PCA approach, and from other resolutions (Figure 30.4), combinations of variables, and mapping units.

MAPPING GREEN SPACE AND URBAN ECOSYSTEMS

Importance of green areas in cities

Green vegetation provides numerous benefits to urban areas, including climate and air quality control, regulation of drainage and runoff, the provision of wildlife habitats, recreational space and aesthetic appeal. For example, Wagrowski and Hites (1997) observe that 45% of polycyclic aromatic hydrocarbons are scavenged and absorbed by urban vegetation in Bloomington, Indiana. In a Landsat sensor based study of Singapore's microclimate, Nichol (1996a) noted the ability of trees to lower air temperatures by approximately 2°C compared with non-vegetated areas. Vegetation moreover is an indicator of pervious, as opposed to impervious surfaces (Yang et al. 2003), the relative proportions of which have been noted as a key environmental, as well as a socio-economic, indicator (Emmanuel 1997). In addition to the mere presence or absence of vegetation, some applications require estimates of the amount of vegetation with a measure of biomass, or recognition of vegetation life form.

Mapping urban vegetation

Until recently, satellite-based studies of urban vegetation have been at a generalized regional level,

using moderate resolution sensors such as SPOT HRV and Landsat TM to map the distribution of green space. Small (2001) recognized the spatial resolution of 30 m from Landsat as being adequate for detecting significant overall variations in green space across the urban area, but too coarse for mapping the precise abundance and distribution of vegetation. Even using the 20 m spatial resolution of the SPOT HRV sensor, Gao and Skillcorn (1998) found that land cover classes having a vegetation component were mapped with lower accuracy than those without (see also Forster 1985). Thus the fragmented distribution of vegetation cover of urban areas, often represented by a single row of street trees or a grassy road verge cannot be characterized using moderate spatial resolution sensors.

For detailed mapping of urban vegetation, traditional methods have combined colour infra-red aerial images with fieldwork. Common variables measured by aerial photograph interpretation include the total green space, and the percentage of tree canopy space as a proportion of the total (Nowak et al. 1996). In Hong Kong, the Environment, Transport and Works Bureau maintains over 500,000 records of urban trees, and the urban and rural planning authorities have a new experimental policy of acquiring large scale, false colour infra-red aerial photographs for environmental monitoring. Although such methods are expensive and time-consuming and require periodic updates to remain valid, they have until recently been the only option.

Recent fine resolution satellite sensors have the ability to obtain accurate measurements of both vegetation cover and biomass. Nichol and Lee (2005) obtained relationships of $R^2 = 0.82$ and $R^2 = 0.81$ between image-derived vegetation cover and vegetation density respectively, and detailed ground plots. Vegetation density was devised as a measure of biomass which is easier to measure in the field than better known indicators such as leaf area index (LAI). The significant results were achieved due to the ability to previously remove shadows cast by tall buildings from the image by radiometric masking. This was possible due to the high (11-bit) radiometric and spatial (4 m) resolution of IKONOS which permitted tree crown shadow to be almost separable spectrally from other cover types in the near-infrared (NIR) band, as well as from building shadow using NDVI (Tucker 1979).

Band combinations for urban vegetation mapping

Since perpendicular vegetation indices such as the Kauth–Thomas greenness index (Kauth and Thomas 1976) are more sensitive to the soil background than ratio-based indices, single bands and band ratios may be more suitable for vegetation mapping in urban areas. This is because the background is not only wet and dry soil, but a wide variety of materials, including concrete, asphalt, brick and tile, which would reduce the ability of perpendicular indices to distinguish differences in vegetation cover. Of the ratio-based indices, the NDVI and chlorophyll index (CI) (Kanemasu 1974) are suitable, the former for its robustness and common usage and the latter because it is a ratio of the green and red wavelengths and represents the intensity of greenness. In some tropical cities such as Hong Kong, where highly reflective and light-toned construction materials such as light asphalt, concrete, brick and tile are used (Akbari 1990), the CI may provide a better measure than NDVI (Nichol and Lee 2005). This is because both the vegetation and background are highly reflecting in the NIR region. Nichol and Lee (2005) give an example of urban Hong Kong in the cool dry season, when urban plantings, though still green, are less vigorous. A decrease in the NIR reflectance is observed as the dry season progresses, followed later by decreased red absorption. Although the NDVI does show higher response for vegetation than for built areas, it is not responsive to vegetation amount, and is unable to distinguish between tree cover and grassy vegetation. Figure 30.5a shows that trees and grass have a similarly light tone on the NDVI image, whereas Figure 30.5b shows that grass is medium toned on the CI image compared to the very bright tone of trees.

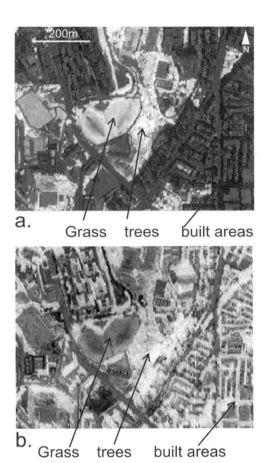

Figure 30.5　IKONOS ratio images showing two cricket pitches with worn, non-vegetated patches in the center, adjacent to a forest plot. a. NDVI showing that grass and trees have similarly high values and are therefore not separable b. Green/red ratio showing that grass and trees are spectrally separable, with grass on the cricket pitch being darker then the trees but brighter than the non-vegetated patches.

Mapping urban vegetation at sub-pixel level

Because of its fragmented nature, urban vegetation has not been accurately mapped by satellite remote sensing systems due to inadequate spatial resolution. Even using the 4 m spatial resolution of IKONOS, 75% of pixels containing vegetation in urban Hong Kong are mixed (Nichol and Wong 2007). The technique of LSU (Adams et al. 1986), which had previously been used to estimate the fractional vegetation amounts for each pixel using moderate resolution sensors (Ridd 1995, Phinn et al. 2002, Small 2003) may be applied effectively at fine spatial resolutions (Nichol and Wong 2007). This is because at fine resolution, pure pixels are available for defining endmembers from the image directly, and also for masking shadows, thereby eliminating their confusion with tree canopies and low albedo surfaces on moderate resolution images. This enables trees and grassy surfaces to constitute distinct endmembers and to be separately mapped in a 4-endmember model containing high and low albedo surfaces, grass and trees.

The resulting fraction images can be conveniently input to spatial analysis models as demonstrated in the designation of wildlife corridors (Nichol and Wong, unpublished). The fraction images are converted to friction surfaces, with the friction values representing the perceived degree of difficulty for wildlife species to traverse the urban area. The friction values, which are allocated according to the fraction of grass, and/or trees in a pixel, may be varied according to species' preferences or tolerance levels. Automated mapping of least cost pathways over different friction

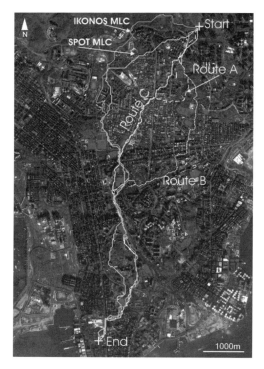

Figure 30.6   False colour IKONOS image of Kowloon Peninsula, Hong Kong, showing routes derived from least cost path analysis, starting at Lion Rock country park forested region at upper right, and ending in Kowloon Park at lower left. Paths A, B and C are derived from IKONOS fraction images resulting from Linear Spectral Unmixing. The longer, MLC paths are derived from the per-pixel, Maximum Likelihood Classifier. (See the color plate section of this volume for a color version of this figure).

surfaces produced different routes A, B and C across the densely urbanized Kowloon Peninsula in Hong Kong (Figure 30.6). For example, Route C is suitable for birds or mammals which prefer trees to urban land cover, but which can also traverse urban areas as a lesser option. It was produced by specifying equal friction weights for all tree fractions. Route B represents the path taken by species which require dense tree cover in order to move across the city, where only tree fractions of >80% were given low friction weights. All three routes derived from the LSU technique are shorter than the two routes derived from the per-pixel maximum likelihood classifier (MLC), since the latter cannot identify green fragments within a pixel.

REMOTE SENSING OF THE URBAN HEAT ISLAND

The urban heat island was first studied empirically by Chandler (1965) working in London, who measured night time air temperatures across the city, and noted increased air temperatures towards the city centre, with a steep temperature gradient on the rural–urban boundary. Urban heat islands are caused in part by replacement of natural evaporative and porous land surfaces by non-evaporative man-made surfaces. These latter surfaces disperse a much greater proportion of energy received into the surrounding atmosphere as sensible heat, as opposed to the predominantly latent heat loss of rural surfaces. Coupled with the generally lower albedo of urban surfaces, these factors result in significantly increased air temperatures in cities compared with their rural surroundings and this difference ($\Delta T(u - r)$), reaches a maximum a few hours after sunset. The usual method of recording urban heat islands is to measure screen-level air temperature by vehicle traverse across the city by night; the main disadvantages being incomplete spatial coverage of such sample data and the lack of time synchronization across the traverse. However, since cities are identifiable on coarse resolution satellite images for their temperature contrasts, as much as for their optical differences with surrounding rural areas, many remote sensing studies have taken place. The first of these were undertaken by satellites launched during the 1970s including NOAA's Television InfraRed Observation Satellite (TIROS) series (Rao 1972), now carrying the Advanced Very High Radiometer (AVHRR) (Carlson et al. 1977) and NASA's Heat Capacity Mapping Mission (HCMM) (Price 1979). Voogt and Oke (2003) summarize recent work using remote sensing to study the urban climate, and give an overview of achievements and limitations. The main advantage of remotely sensed thermal data for studying the urban heat island lies in the ability to provide a time-synchronised dense grid of temperature data over a whole city, but there are numerous constraints, which are discussed below.

The relationship between surface temperature and air temperature

Since the satellite-derived heat island is based on surface radiometric temperature, the optimum usefulness of these data depends on defining their relationship to a more conventional view of the urban heat island such as screen level air temperature at the time of imaging. Although in general, near surface climates are known to be closely related to those of the active surface (Chandler 1965,

Price 1979, Carlson and Boland 1981, Goldreich 1985), Roth et al. (1989) warn against regarding remotely sensed heat islands as synonymous with those based on air temperature measurements near ground level. Carlson and Boland's (1981) work, which demonstrated a close relationship between the conventional heat island measured by screen level air temperature and the surface temperature measured by remote radiometers, confirmed the relevance of satellite-based studies. Nichol (1996a,b) has also demonstrated that for the particular conditions of high-rise housing estates in the equatorial climate of Singapore where wind regimes are low, the surface radiant temperature is significantly related to air temperature. Furthermore, the horizontal (satellite-seen) surfaces were found to be more closely related to air temperature than vertical surfaces, with $r = 0.7$ and 0.49 respectively, for a total of 1524 samples. For an ASTER thermal image obtained on a calm winter night in Hong Kong (Figure 30.7), Nichol (in press) observed a correlation of $r = 0.85$ between image-derived surface temperatures and eighteen

'*in situ*' air temperatures measured on the ground at the image time. In such situations remotely-sensed surface temperature can be accepted as a proxy for ambient air temperatures.

Time of imaging in relation to heat island maximum

Spaceborne thermal sensors such as Landsat's ETM+ and Terra ASTER record mainly during the daytime when densely built, high-rise areas may constitute a heat sink (Nichol 1996a,b). Tropical cities (Nichol 2003) or arid zones in summertime (Carnahan and Larson 1990) may also constitute heat sinks during the day.

Furthermore, the timing of the satellite overpass may not be ideal for detecting temperature differences. Landsat for example, at 9.30 to 10.30 am local time, is near the morning thermal crossover time when minimal thermal contrasts would be expected. Even if marked contrasts are detectable, surface temperature is more variable spatially than air temperature, especially during the day, therefore satellite sensor data may exaggerate the magnitude of the urban heat island.

The few remote sensing studies using night-time images have noted different heat island structure and scales than those based on daytime images. Streutker (2002) for example using AVHRR images of Houston, Texas, noted an inverse relationship between the urban-rural temperature difference ($\Delta T(u - r)$) and rural temperature, which has not been noted on daytime images. Nichol (2005), comparing a Landsat daytime with an ASTER night-time image of Tuen Mun, a town in the western New Territories of Hong Kong (Figure 30.8), noted more pronounced microclimatic variations during the day, with steep temperature gradients over the city. However, apart from barren areas which are warm by day and cool at night, the spatial patterns of temperature on ASTER night-time images (Figure 30.8) are notably similar to those during the day, which supports the relevance of daytime images for heat island studies.

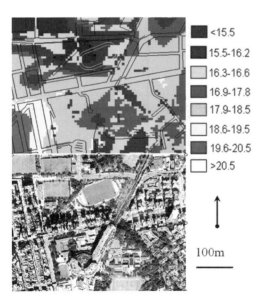

Figure 30.7   Extract from an ASTER night-time image of 31st January 2007 of the Kowloon Peninsula, Hong Kong, and corresponding aerial photograph. The urban heat island core at lower left gives way to lower density development towards the top right. Surface temperatures (°C) are discontinuous, with steep temperature gradients between land cover types. (See the color plate section of this volume for a color version of this figure).

Anisotropy of the satellite view

Satellite sensors record the temperature of horizontal surfaces which may only in rural areas represent the complete radiating surface. The effective (active) surface area of a city, especially in high-rise areas, is much larger than the equivalent countryside of the same size (Roth et al. 1989, Voogt and Oke 2003). In high-rise housing estates areas in Singapore for example, the active surface was found to be 1.7 times greater than the planimetric (satellite seen) surface. Thus nadir views would

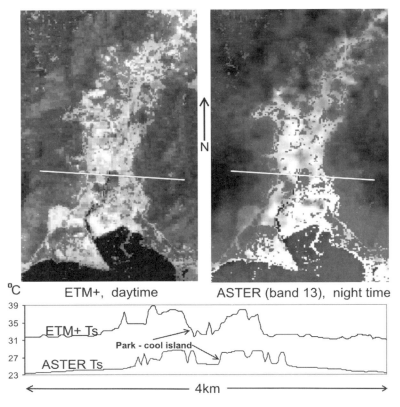

Figure 30.8   East–West traverse of surface temperatures across Tuen Mun town, Hong Kong showing a range of 8°C for the ETM+ daytime image and 6°C for the ASTER night-time image.

be warmer or cooler than off-nadir views depending on the sun position. Voogt and Oke (2003) recommend using ground based observations to construction models for the weighting of temperatures according to area and sun position (see also Nichol 1998).

The need for emissivity and atmospheric correction

Although satellite derived radiance values can readily be converted to equivalent Black Body Temperature (or brightness temperature) using Planck's law (Malaret et al. 1985), this underestimates the surface radiometric temperature if corrections for emissivity differences according to the type of land cover are not carried out (Artis and Carnahan 1982). For example, a clay loam soil of emissivity 0.92, and vegetation of emissivity 0.98, both with a radiometric temperature of 27°C will have brightness temperature (image) values of 20.8 and 25.5°C, respectively. Furthermore, image values can only be considered accurate in clear, dry atmospheres, and a further correction

using atmospheric data in a radiative transfer model such as LOWTRAN (Kneizys et al. 1988) should be made, if absolute temperatures are desirable. In humid atmospheres, energy absorption by atmospheric water vapour may account for brightness temperatures up to 15°C cooler than the surface radiometric temperature (Nichol 1996b).

Spatial resolution of satellite sensors related to scales of urban climate

Oke's (1976) distinction between urban canopy layer and urban boundary layer heat islands may help to define the level of detail and appropriate spatial resolution for satellite-based studies. Thus Oke (1976) states that the canopy layer thermal climate seems to be governed by the individual site character, a situation observable at micro-climatic scale on daytime satellite sensor images. At this scale, the term surface temperature patterns may be more meaningful than surface heat island, since due to the heterogeneous nature of the urban environment, temperatures are discontinuous between structures (Figure 30.7) and

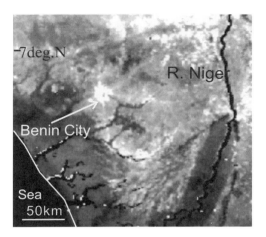

-7deg.N

R. Niger

Benin City

Sea
50km

Figure 30.9   AVHRR thermal (10.5–11.5 μm) band showing a 'classical' heat island situation for Benin city in the tropical forest zone of Nigeria; a distinctly warmer city surrounded by a 'sea' of cooler rural areas.

steep temperature gradients occur during times of solar insolation. No distinct 'island' of warmer temperatures in a 'sea' of cooler ones (as on Figure 30.9) can be observed. Oke's boundary layer heat island on the other hand, defines meteorological conditions overlying the canopy layer, whose characteristics are influenced by the urban area below, and often capped by a temperature inversion, indicating the limit of the urban pollution plume. There exist several thermal satellite sensors with adequate resolution for studies of the urban boundary layer heat island, including MODIS and AVHRR with ~1 km spatial resolution (Figure 30.9). However, for urban canopy layer studies, only Landsat ETM+ and Terra ASTER, with 60 m and 90 m resolution respectively for the thermal bands, possess adequate detail. The strong relationship between surface temperature and NDVI (Gallo et al. 1993, Weng et al. 2004) may provide an alternative for remotely sensed observations of the urban heat island where thermal data are unavailable, or if finer spatial resolution is required.

MONITORING AIR QUALITY IN URBAN AREAS

The gathering of air quality data for urban areas and their source regions is a major challenge because the large areas involved cannot be represented by ground stations. Although recently, satellite sensing systems and methodologies with

adequate spectral and temporal resolution for monitoring aerosols have been developed, it is difficult to undertake such studies at a fine spatial resolution, because the atmospheric signal being sensed is only a small proportion of the total image reflectance. Thus large areas corresponding to large pixels, giving a higher measurable signal compared with the land surface reflectance (noise), are required. The most accessible remotely sensed variable of air quality is Aerosol Optical Depth (AOD). This is a unitless measure of the total amount of aerosol in the atmospheric column and is based on the opacity of the atmosphere in a particular waveband. There is no general algorithm which can estimate aerosol amounts over every kind of surface. Instead, different algorithms have been developed for (i) water, (ii) dark vegetation and (iii) heterogeneous land surfaces respectively. However, techniques for the low reflecting surfaces of water and vegetation are better developed than those over land, because assumptions can be made that the surface reflectance is either zero or near zero. Based on this, Kaufman and Sendra (1987) developed an algorithm which first uses the NDVI to detect dense dark vegetation (DDV) pixels, then uses the short wave infra red (SWIR, 2.1 μm) band, which is not affected by aerosol, to obtain the surface reflectance for the DDV pixels. Then based on the relationship (Kaufman and Sendra 1987):

$$Lsurf_{0.49\mu m} = 0.25 \times Lsurf_{2.1\mu m} \qquad (2)$$

$$Lsurf_{0.66\mu m} = 0.5 \times Lsurf_{2.1\mu m} \qquad (3)$$

the apparent surface reflectance in the blue (0.49 μm) and red (0.66 μm) bands can be obtained. The difference between the actual surface reflectance in these bands, and the observed (top of atmosphere (TOA)) reflectance is assumed to be due to aerosol. This amount is then fitted to a best fit aerosol model, with knowledge of the expected aerosol types in the study area, for example, continental (Lenoble and Brogniez 1984), industrial/urban (Remer et al. 1997), biomass burning (Hao and Liu 1994), to arrive at AOD from the image blue and red wavebands.

From this concept, NASA has developed the MODIS level II AOD product, at 10 km spatial resolution (Kaufman and Tanre 1998) covering the globe. Although the 10 km spatial resolution of this product (MOD04) only provides meaningful depictions on a broad regional scale, it is capable of giving an overview of air quality conditions prevailing over a city's region. Li et al. (2005) have developed a 1 km AOD algorithm, using the same principles as NASA's 10 km DDV algorithm, but using more stringent conditions for the cloud mask and dark vegetation selection at finer resolution. However, since large areas of DDV are generally not available in cities, an assumption must be made that air quality is invariable between the urban, and surrounding rural areas.

Aerosol estimation over complex urban areas has used the heterogeneous land algorithm. This is based on the principle of measurement of the blurring effect between highly contrasting adjacent pixels (Tanre et al. 1979, Mekler and Kaufman 1980), and is often referred to as the 'contrast reduction method'. It has been used by Sifakis and Deschamps (1992), Sifakis et al. (1998) and Retalis et al. (1997) for 'fine' resolution aerosol estimation over Athens using SPOT HRV and Landsat TM images. The major drawback with this method for the derivation of fine resolution aerosol images is that it measures path radiance (aerosol scattering between adjacent image pixels) within a kernel of 15×15 pixels. Thus for Landsat, the 30 m pixel size produces a coarse resolution aerosol image of 450×450 m. A fairly high correlation coefficient ($r = 0.76$) was obtained between Landsat-derived AOD and SO_2 over Athens (Sifakis et al. 1998).

As for AOD, the estimation of other air pollutants from satellite sensor imagery is constrained by the weakness of the signal relative to the total image reflectance. Thus the MOPITT sensor, which measures CO_2 emissions from the Earth surface and TOMS for ozone estimation, have spatial resolutions of 22 km and 39 km at nadir respectively. Since the resolution of these products is too coarse for application at urban scales, and algorithms developed for complex land areas are difficult to apply, the task of deriving accurate aerosol products at urban scales remains challenging.

Over the last 35 years, with the exception of thermal sensors, Earth observation technology has developed the ability to acquire data on urban areas at higher levels of spatial and spectral detail than are actually required for effective urban planning and management. Indeed frequent use of the terms 'multi-' and 'hyper-' suggest imminent shortage of superlatives. The technology for image data acquisition is available and its wider utilisation is partly dependent on the development of data processing algorithms which permit increased automation in data extraction. Specifically, improvements in automated recognition of complex objects on fine spatial resolution images are required to avoid the analysis of large quantities of imagery by 'computer-assisted photograph-interpretation', or on-screen-digitizing. Furthermore, with increasing global concern about the quality of life in cities, algorithms and models for accurate and spatially detailed aerosol, air temperature and population estimates are required. If these barriers can be overcome, the next decade may see costly international space programmes replaced by smaller platforms and specialized sensors run by individual countries, states or large metropolitan regions.

REFERENCES

Adams, J. B., M. O. Smith and P. E Johnson, 1986. Spectral mixture modeling: a new analysis of rock and soil types at the Viking lander. *Journal of Geophysical Research*, 91: 8098–8122.

Akbari, H., 1990. Summer heat islands, urban trees and white surfaces. *ASHRAE Transactions*, 96(1): 1381–1388.

Akbari, H., L. Shea Rose and T. Haider, 2003. Analyzing the land cover of an urban environment using high resolution orthophotos. *Landscape and Urban Planning*, 63: 1–14.

Anderson, J. R., J. E. E. Hardy, J. T. Roach and R. E. Witmer, 1976. A land use and land cover classification system for use with remote sensor data. *Geological Survey Professional Paper 964*. U.S. Government Printing Office, Washington DC.

Aplin, P., P. M. Atkinson and P. J. Curran, 1999. Per-field classification of land use using the forthcoming very fine spatial resolution satellite sensors: problems and potential solutions. In P. Atkinson and N. Tate (eds), *Advances in Remote Sensing and GIS Analysis*. Chichester, John Wiley and Sons, pp. 219–239.

Artis, D. A. and W. H. Carnahan, 1982. Survey of emissivity variability in thermography of urban areas. *Remote Sensing of Environment*, 12: 313–329.

Atkinson, P., J. L. Cushnie, J. G. R. Townsend and A. K. Wilson, 1985. Improving Thematic Mapper land cover classification using filtered data. *International Journal of Remote Sensing*, 6: 955–961.

Barnsley, M. J. and S. L. Barr, 1996. Inferring urban land use from satellite sensor images using kernel-based spatial reclassification. *Photogrammetric Engineering and Remote Sensing*, 62: 949–858.

Carlson, T. N. and F. E. Boland, 1981. Analysis of urban-rural canopy using a surface heat flux temperature model. *Journal of Applied Meteorology*, 17: 998–1013.

Carlson, T. N., A. Augustine and F. E. Boland, 1977. Potential application of satellite temperature measurements in the analyses of land use over urban areas. *Bulletin of American Meteorological Society*, 96: 91–114.

Carnahan, W. H. and C. H. Larson, 1990. An analysis of a heat sink. *Remote Sensing of Environment*, 33: 65–71.

Chandler, T. J., 1965. The Climate of London. Hutchinson, London, p. 292.

Emmanuel, R., 1997. Urban vegetational change as an indicator of demographic trends in cities: the case of Detroit. *Environment and Planning B*, 24: 415–426.

Forster, B. C., 1985. An examination of some problems and solutions in monitoring urban areas from satellite platforms. *International Journal of Remote Sensing*, 6(1): 139–151.

Fung, T. and W. Siu, 2000. Environmental quality and its changes using NDVI *International of Remote Sensing*, 21: 1101–1024.

Fung, T. and W. L. Siu, 2001. A study of green space and its changes in Hong Kong using NDVI. *Geographical and Environmental Modelling*, 5(2): 111–122.

Gallo, K. P., A. L. McNab, T. R. Karl, J. F. Brown, J. J. Hood and J. D. Tarpley, 1993. The use of NOAA Afine resolutionR data for assessment of the urban heat island effect. *Applied Meterorology*, 32(5): 899–908.

Gao, J. and D. Skillcorn, 1998. Capability of SPOT XS data in producing detailed land cover maps at the urban–rural periphery. *International Journal of Remote Sensing*, ´9(15): 2877–2891.

Goldreich,V., 1985. The structure of the ground level heat island in a Central Business District, *Journal of Climatology and Applied Meteorology*, 2: 1237–1244.

Gong, P. and P. J. Howarth, 1990. The use of structural information for improving land cover classification accuracies at the rural–urban fringe. *Photogrammetric Engineering and Remote Sensing*, 56: 67–73.

Hao, W. M. and M. H. Liu, 1994, Spatial and temporal distribution of tropical biomass burning, *Global Biogeochemical Cycles*, 8: 495–503.

Harvey, J. T., 2002. Estimating census district populations from satellite imagery: some approaches and limitations. *International Journal of Remote Sensing*, 23(10): 2071–2095.

Husar, R. B., L. L. Stowe and J. Prospero, 1997. Characterisation of tropospheric aerosols over the oceans. *Journal of Geophysical Research*, 102: 16889–16910.

Jensen, J. R., 2000. Remote Sensing of the Environment: an Earth Resource Perspective. Prentice Hall, New Jersey.

Kanemasu, E. T., 1974. Seasonal canopy reflectance patterns of wheat, sorghum and soyabean. *Remote Sensing of Environment*, 3: 43–47.

Kaufman, Y. J. and C. Sendra, 1987. Algorithm for automatic atmospheric corrections to visible and near-IR satellite imagery, *International Journal of Remote Sensing*, 9(8): 1357–1381.

Kaufman, Y. J. and D. Tanre, 1998. Algorithm for remote sensing of tropospheric aerosol from MODIS, NASA Product ID: MOD04 product report. http://modis.gsfc.nasa.gov/data/atbd.mod3.pdf

Kauth, R. J. and G. S. Thomas, 1976. The Tasseled Cap – a graphic description of the spectral-temporal development of agricultural crops as seen by Landsat. In *Proceedings of a Symposium on Machine Processing of Remotely Sensed Data*, Purdue University, West Lafayette, Indiana, 21 June–1 July, 1976, pp. 4B41–4B51.

Kneizys, F. X., E. P. Shettle, L. W. Abreu, J. H. Chetwynd and G. P. Anderson, 1988. Users guide to LOWTRAN 7. *Air Force Geophysics Laboratory Report*, No. AFGL-TR-88-0177, Hanscom AFB, MA.

Lenoble, J. and C. Brogniez, 1984. A comparative review of radiation aerosol models, *Beitraege zur Physik der Atmosphaere*, 57: 1–20.

Li, C. C., J. T. Mao and K. H. Lau, 2005. Remote Sensing of High Spatial Resolution Aerosol Optical Depth with MODIS Data over Hong Kong, *Chinese Journal of Atmospheric Sciences*, 29(3): 335–342.

Li, G. and Q. Weng, 2005. Using Landsat ETM+ imagery to measure population density in Indianapolis, Indiana, USA. *Photogrammetric Engineering and Remote Sensing*, 71: 947–958.

Lo, C. P., 1986. Applied Remote Sensing, London, Longman.

Lo, C. P., 2003. Zone-based estimation of population and housing units from satellite-generated land use/land cover maps.

In V. Mesev (ed.) *Remotely Sensed Cities*, Taylor and Francis, London, Chapter 7: 157–180.

Lo, C. P. and B. J. Faber, 1997. Integration of Landsat Thematic Mapper and census data for quality of life assessment. *Remote Sensing of Environment*, 62(2): 143–157.

Lu, D., Q. Weng and G. Li, 2006. Residential population estimation using a remote sensing derived impervious surface approach. *International Journal of Remote Sensing*, 27(16): 3553–3570.

Malaret, E., L. A. Bartolucci, D. Fabian Lozano, P. E. Anuta and McGillen, C. D., 1985. Thematic Mapper data quality analysis, *Photogrametric Engineering and Remote Sensing*, 51: 1407–1416.

Mekler, Y. and Y. J. Kaufman, 1980. The effect of the earth atmosphere on contrast reduction for a nonuniform surface albedo and two-halves field, *Journal of Geophysical Research*, 85: 4067–4083.

Nichol, J. E., 1996a. High resolution surface temperature patterns related to urban morphology in a tropical city: a satellite-based study. *Journal of Applied Meteorology*, 35(1): 135–146.

Nichol, J. E., 1996b. Analysis of the urban thermal environment of Singapore using Landsat data. *Environment and Planning B: Planning and Design*, 23: 733–747.

Nichol, J. E., 1998. Visualisation of urban surface temperatures derived from satellite images. *International Journal of Remote Sensing*, 19(9): 1639–1649.

Nichol, J. E., 2003. Heat island studies in the third world cities using GIS and remote sensing. In V. Mesev (ed.) *Remotely Sensed Cities*, Taylor and Francis, Chapter 13.

Nichol, J. E., 2005. Remote sensing of urban heat islands by day and night, *Photogrammetric Engineering and Remote Sensing*, 71(5): 613–621.

Nichol, J. E. and C. M, Lee, 2005. Urban vegetation monitoring in Hong Kong using high resolution multispectral images. *International Journal of Remote Sensing*, 26(5): 903–919.

Nichol, J. E and M. S. Wong, 2005a. Modelling urban environmental quality in a tropical city. *Landscape and Urban Planning*, 73: 49–58.

Nichol, J. E. and M. S. Wong, 2005b. Monitoring road surface type and condition using remote sensing techniques. Technical report to Survey and Mapping Office, Lands Department, the Hong Kong SAR.

Nichol, J. E. and M. S. Wong, 2007. Remote sensing of urban vegetation life form by spectral mixture analysis of IKONOS high resolution satellite images, *International Journal of Remote Sensing*, 28: 985–1000.

Nichol, J. E. and M. S. Wong, 2008. Mapping urban environmental quality using satellite data and multiple parameters. *Environment and Planning B, 35, December 2008, Doi: 10.1068/b34034.*

Nichol, J. E., (in press). An emissivity modulation method for spatial enhancement of thermal satellite images in urban heat island analysis. *Photogrammetric Engineering and Remote Sensing*.

Nowak, D. J., R. A. Rowntree, E. G. Mcpherson, S. M. Sissini, E. R. Kerkmann and J. C. Stevens, 1996. Measuring and analyzing urban tree cover, *Landscape and Urban Planning*, 36: 49–57.

Oke, T. R., 1976. The distinction between canopy and boundary-layer urban heat islands. *Atmosphere*, 14(4): 268–277.

Phinn, S., M. Stanford and P. Scarth, 2002. Monitoring the composition of urban environments based on the vegetation-impervious-surface-soil (VIS) model by sub-pixel analysis techniques. *International Journal of Remote Sensing*, 23: 4131–4153.

Price, J. C., 1979. Assessment of the urban heat island effect through the use of satellite data. *Monthly Weather Review*, 107: 1554–1557.

Rao, P. K., 1972. Remote sensing of urban heat islands from an environmental satellite, *Bulletin American Metorological Society*, 53: 647–648.

Rashed, T., Weeks, J. R., Roberts, D., Rogan, J. and Powell, R., 2003. Measuring the physcial composition of urban morphology using multiple endmember spectral mixture model, *Photogrammetric Engineering and Remote Sensing*, 69(9): 1011–1020.

Remer, L. A., S. Gasso, D. Hegg, Y. J., Kaufman and B. N. Holben, 1997. Urban/industrial aerosols: ground-based, sun/sky radiometer and airborne *in-situ* measurements. *Journal of Geophysical Research*, 102: 16,849–16,860.

Retalis, A., C. Cartalis and E., Athanassiou, 1997. Assessment of the distribution of aerosols in the area of Athens with the use of Landsat Thematic Mapper data, *International Journal of Remote Sensing*, 20(5): 939–945.

Ridd, M., 1995. Exploring a V-I-S (vegetation-impervious-surface-soil) model for urban ecosystem analysis through remote sensing: comparative anatomy for cities. *International Journal of Remote Sensing*, 16: 2165–2185.

Roth, M., T. R. Oke and W. J. Emery, 1989. Satellite derived urban heat islands from three coastal cities and the utilisation of such data in urban climatology. *International Journal of Remote Sensing*, 10(11): 1699–1720.

Sifakis, N. and P. Y. Deschamps, 1992. Mapping of air pollution using SPOT satellite data, *Photogrammetric Engineering and Remote Sensing*, 58: 1433–1437.

Sifakis, N., N. A. Soulakellis and D. K, Paronis, 1998. Quantitative mapping of air pollution density using earth observations: a new processing method and applications to an urban area. *International Journal of Remote Sensing*, 19: 3289–3300.

Small, C., 2001. Estimation of urban vegetation abundance. *International Journal of Remote Sensing*, 22(7): 1305–1334.

Small, C., 2003. High spatial resolution spectral mixture analysis of urban reflectance. *Remote Sensing of Environment*, 88: 170–186.

Stow, D., L. L. Coulter and C. A. Benkelman, in this volume. Airborne Digital Multispectral Imaging. Chapter 11.

Strahler, A. H., C. E. Woodcock and J. A. Smith, 1986. On the nature of models in remote sensing, *Remote Sensing of Environment*, 20: 121–139.

Streutker, D. R., 2002. A remote sensing study of the urban heat island of Houston, Texas, *International Journal of Remote Sensing*, 23: 2595–2608.

Stuckens, J., P. R. Coppin.and M. E. Bauer, 2000. Intrgrating contextual information with per-pixel classification for improved land cover classification. *Remote Sensing of Environment*, 71: 282–296.

Tanre, D., M. Herman, P. Y. Deschamps and A. de Leffe, 1979. Atmospheric modeling for space measurements of ground reflectances, including bidirectional properties, *Applied Optics*, 18: 3587–3594.

Tobler, W., 1969. Satellite confirmation of settlement size coefficients. *Area*, 1: 30–34.

Trietz, P. M., P. J. Howarth and P. Gong, 1992. Application of satellite and GIS technologies for land cover and land use mapping at the rural-urban fringe: a case study. *Photogrammetric Engineering and Remote Sensing*, 58(4): 439–448.

Tucker, C. J., 1979. Red and photographic infra-red linear combinations for monitoring vegetation. *Remote Sensing of Environment*, 8: 127–150.

United Nations, 2006. Urban population to overtake country dwellers for first time, http://www.guardian.co.uk/international/story/0„1798774,00.html (last accessed: 29 November, 2006).

Voogt, J. A. and T. R. Oke, 2003. Thermal remote sensing of urban climates. *Remote Sensing of Environment*, 86: 370–384.

Wagrowski, D. M. and R. A. Hites, 1997. Polycyclic aromatic hydrocarbon accumulation in urban, suburban and rural vegetation. *Environmental Science and Technology*, 31: 279–282.

Welch. R., 1982. Spatial resolution requirements for urban studies. *International Journal of Remote Sensing*, 3: 139–146.

Weng, Q., D. Lu and J. Schubring, 2004. Estimation of land surface temperature-vegetation abundance relationship for urban heat island studies, *International Journal of Remote Sensing*, 89: 467–483.

Yang, L., C. Huang, C. G. Homer, B. K. Wylie and M. J. Coan, 2003. An approach for mapping large-area impervious surfaces: synergistic use of Landsat-7 ETM+ and high spatial resolution imagery. *Canadian Journal of Remote Sensing*, 29(2): 230–240.

Zhang, Q., J. Wang, X. Peng, P. Gong and P. Shi, 2002. Urban built-up land change detection with road density and spectral information from multi-temporal Landsat imagery. *International Journal of Remote Sensing*, 23(15): 3057–3078.

31

Remote Sensing and the Social Sciences

Kelley A. Crews and Stephen J. Walsh

Keywords: ABM (agent-based modeling), CA (cellular automata), discrete-continuous, 'people and pixels', policy, scale-pattern-process.

INTRODUCTION

The challenge in today's practice of remote sensing for the social sciences is that it does not fall neatly into a particular set of sensor systems, does not carry one specific set of approaches, and is carried out globally in diverse systems spanning the natural-anthropogenic divide. In large part this diversity stems from the nature of social science questions to search for answers to more abstract processes that may lack a direct biophysical result (Liverman et al. 1998), necessitating indirect mapping of social processes rather than direct mapping of biophysical phenomena. This legacy starts this chapter's discussion, and motivates the remaining sections that explore how social science applications are at once wholly different from other remote sensing applications, yet function as a microcosm for understanding the accomplishments and challenges facing all remote sensing practicioners. After providing examples of the breadth of social science remote sensing, population–environment interaction studies are examined in more detail as representative of the challenges involved in marrying discrete and continuous data for interdisciplinary analysis of the world's most pressing social and environmental problems. Next, the scale-pattern-process framework as borrowed from landscape ecology and geography is employed as a lens for understanding implications of remote sensing resolutions on data choices and research questions where the assessment of causality and subsequent prediction must face the inherited complexities of mapping often intangible or indirect social processes. Lastly, this discussion is carried into the modeling approaches currently under use and experimentation, with a final discussion about the challenges facing continued use of remote sensing for social science applications.

THE LEGACY OF DIRECT VS. INDIRECT MAPPING

Remote sensing historically was developed for purposes of military reconnaissance and defense, with platforms as diverse as hot air balloons, aircraft, spacecraft, and even pigeons (Jensen 2000). In the 1970s, the development of satellite sensor systems also responded to the needs of the earth science community, particularly with regards to spectral band placement and width as seen in the changes from the Landsat MSS to Landsat Thematic Mapper (TM) and Enhanced Thematic Mapper Plus (ETM+) missions (Jensen 2000). While multiple national governments and private companies continued to expand, refine, or distinguish their own sensor capabilities, in general no single sensor system was built to respond particularly to the needs of the social science community. This legacy continues today, as social scientists work to extract meaningful information from sensor systems designed for other purposes. Yet some would argue that the social science applications of these systems, notably including land change science, have in fact increased the visibility and saliency of remote sensing and related approaches and products, particularly in research applications, more than any other application area (see, e.g., Gutman et al. 2004). At the heart of remote sensing linkages to social science applications lies the juxtaposition of the power and disadvantage of using these data for extracting useful information several orders removed from spectral-based data.

Consider, for example, the fluvial geomorphologist needing to examine changes in landform given perturbations of a flooding regime, or the ecologist wishing to understand vegetation response to shifting climatic events. In each case, the variables of interest (drainage basin patterns, flood extent, vegetative biomass, precipitation pulses) are not directly tied to spectral response but have been shown to be strongly correlated and physiologically related to measurements or indices derived from those spectral measurements. These proxies could be said to be one order removed from the actual spectral measurements taken at sensor. The social sciences, on the other hand, rely upon inferring social processes as manifested by physical phenomenon observable on the landscape via spectral measurements. For example, vegetation indices may be used to capture changes due to management (e.g., agricultural intensification via increased fertilizer, pesticide, herbicide, and irrigation), underlying soil conditions, or climatic conditions (Moran et al. 1994, Walsh et al. 2001, Archer 2004). In this case, the human activity is two orders removed from the at-sensor measurement. More complex (and of greater interest to the social science community) are those phenomena that are even further removed, such as the influence of power structures on household dynamics. In this case, theoretical, physical, and spectral bridges must be built and tested to establish a correlation (or ultimately, a causal connection) between spectral measurements and household dynamics. Further complicating these efforts is the problem of many-to-many relationships, where a given set of household dynamics could result in myriad landscape responses, while simultaneously multiple household scenarios could also possibly lead to a similar landscape (metaphorically equivalent to geomorphology's exposition of the polygenetic continuum). Social science applications therefore represent perhaps the greatest leap of technological, methodological, and theoretical faith faced by remote sensing analysts, further exacerbated by the topical range of their foci. Current application areas include population-environment interactions ranging from migration (e.g., Perz and Skole 2003) to household life cycle assessment (Barbieri et al. 2005), land use management (e.g., Campbell et al. 2005), geographic accessibility (e.g., Verburg et al. 2004), population counts (e.g., Wu and Murray 2007), and landscape ethnography (e.g., Nyerges and Green 2000). Much, though not all, of the literature shows a strong international focus, with projects spanning Asia (Walsh et al. 1999), Africa (Reid et al. 2000), South America (Moran et al. 1994), Central America (Chowdury and Turner 2006) and North America (Walker and Solecki 2004) and covering diverse ecosystems ranging from mountains (Gautam et al. 2004) to desert (Herrmann et al. 2005) to urban (Seto and Fragkias 2005) to Amazonian (Evans et al. 2001). Techniques heavily integrating remotely sensed data include multi-level modeling (Overmars and Verburg 2006), spatially explicit statistical modeling (Geoghegan et al. 2004), model validation (Pontius and Schneider 2001), cellular automata (Yeh and Li 2001), and agent-based modeling (Deadman et al. 2004). As such, the vast and varied applications in the social sciences offer a testing ground for assessing where remote sensing has the greatest potential for societal impact as well as the greatest risk of misinterpretation.

DISPARATE DATA REPRESENTATIONS AND POPULATION–ENVIRONMENT INTERACTIONS

If social science applications indeed represent a test of the growth in utility of remote sensing, then population–environment interactions could arguably be used as the microcosm for understanding both the strengths and weaknesses inherent in, tackled by, and challenging remote sensing. The watershed moment for remote sensing support of population–environment research came in 1998 with the publication of Liverman et al.'s *People and Pixels*. Two years earlier, the US National Research Council brought together researchers involved in NASA-funded research as well as the Human Dimensions of Global Change community to discuss the study of human activities with a strong spatial component, such as land use change, and with an eye toward predictive capabilities for socially significant or policy relevant events and phenomena (Liverman et al. 1998). To this point, *social* spatially explicit data were considered to be primarily the domain of household and community surveys, such as with the US Census files. The apparent divide described in *People and Pixels* lay between the remote sensing specialists, whose work was considered primarily biophysical in nature, and the social scientists, whose work was considered to lack any true biophysical systems understanding. But further underlying this divide was a difference more subtle yet to this day more pervasive: that of data models. Social scientists had in fact been using spatially explicit data, particularly in the form of US Census data and related boundary (TIGER) files, but had done so using a vector data model; remote sensing scientists (with some aerial photogrammetrists notwithstanding) primarily worked within a raster data model. The difference was in many camps considered unbridgeable (McNoleg 2003), but the group convened in 1996 recognized that addressing some of the most compelling social-environmental problems of the day required asking both why and where, and that the inherently interdisciplinary approach to melding social science and remote sensing datasets would ultimately require learning when and how to cross or fill the discrete-continuous chasm. The challenge, though addressed by increasingly sophisticated modeling efforts, remains one of the largest hurdles in social science applications of remotely sensed data today (Rindfuss et al. 2003).

To the remote sensing analyst, a raster view of the world provides simplicity, completeness, and efficient processing: a wall-to-wall representation of a given area is arranged in uniform pixels that not only predefines neighborhood but also makes for streamlined coding. To the social scientist presuming a vector representation of reality, wall-to-wall data are cumbersome when there are areas not of interest within the study area extent, and pixels are only arbitrary units of analysis that hold no meaning for any given phenomenon of interest. Perhaps obviously, two primary options are available for integrating these data in a [mostly] spatially explicit manner: either rasterize the vector, or vectorize the raster. That is, decide which to prioritize: people, or pixels. The options were not new to geographers, particularly those working in GIS, who had been grappling with this type of quandary for some time. What was new, however, were the broader implications of collapsing biophysical data (pixels) to areas of social units (people) or spreading people onto the wall-to-wall landscape (pixels). Embedded in this decision was a still critical dilemma of researchers today: what is the appropriate unit of analysis for population–environment

interaction studies? What were the implications of collapsing watershed processes into political administrative unit boundaries, or of spreading household measures of socioeconomic processes onto the landscape? The answer offered suggested that the most compelling question or process should drive the decision; e.g., assessing the impacts of population density on deforestation might entail spreading people on the landscape beyond where their data were collected, but understanding the impact of declining forest resources on household fertility decisions would mean bringing the forest or pixel information 'to the household.' While it is an overstatement to suggest that the NASA workshop or the Liverman et al. (1998) book started a new movement in social science applications of remote sensing, it is accurate to say it popularized the necessity of interdisciplinary teams to address the question from both sides of the raster-vector or continuous-discrete divide, both 'socializing the pixel' and 'pixelizing the social' (Geoghegan et al. 1998). Such is not to imply that the resultant body of work in land use change and population–environment interactions orbited solely around questions of data integration; rather, these researchers recognized that the key to addressing fundamental questions of social and environmental importance using remote sensing data required theoretically and practically remapping methods of data integration. As such, these fields have led social scientists interested in remote sensing applications in conceptual and material contributions (Walsh and Crews-Meyer 2002, Fox et al. 2003, Gutman et al. 2004, Rindfuss et al. 2004).

SCALE-PATTERN-PROCESS AND DATA RESOLUTION

As diverse as the bodies of theory motivating social science research are, they nonetheless have been enriched by principles borrowed from the natural sciences. Particular emphasis has been placed upon lessons learned from landscape ecology that provide a framework for interpreting landscapes in terms of both composition and configuration. This framework is commonly referred to as 'the paradigm of scale-pattern-process' (Walsh et al. 1999, Turner et al. 2001) in geography and, interestingly, as 'process-pattern-scale' in ecology and biology (Crews-Meyer 2006). The central tenets posit that the scale of analysis (here, especially imagery) chosen impacts the patterns that are [remotely] observable, which are used to infer the processes at work on the landscape (Nagendra et al. 2003, 2004, Walker 2003). Or, taken in reverse, processes drive patterns whose observation is scale-dependent. Typically, this school of thought is operationalized with spatial scale, and particularly with grain or resolution more so than extent (Turner et al. 2001). But in theory and practice, other scales of remotely sensed data are easily as impacted or informed. The grain of temporal scale, frequency of observation or return time, impacts whether observable land use/land cover changes can be disentangled into processes with different temporal signatures, whether long-term (e.g., urbanization), seasonal (e.g., phenological), or even diurnal (e.g., tidal) processes. Again, the extent (here, temporal) may impact choice of research questions or data sources. Spectral resolution, or more broadly stated as choice of sensor system, clearly impacts the ways in which phenomena are conceptualized and tested. For example, the impact of changing market conditions on coffee agriculture will be evidenced differently by sensor systems that provide information on vertical structure, such as assessing age structure or understory/shade species via small footprint LIDAR, versus detecting pest invasions or success of pesticides or irrigation via multi- or hyperspectral optical systems.

Obviously no sensor system can maximize every type of resolution or therefore meet the needs of all application areas given that there are trade-offs between resolutions as a consequence of technological and physical realities such as data transfer speeds and orbital requirements (Jensen 2000, Messina and Crews-Meyer 2000). But guidance for place- and process-based applications of remote sensing (see, e.g., Jensen 2000) can be adapted for social science applications: for example, urban and peri-urban study areas (Wilson et al. 2003; Seto and Fragkias 2005) and population- or cadastral-focused studies (Fox et al. 2003, Schelhas and Sanchez-Azofeifa 2006) tend to require higher spatial resolution (grain), while emergency preparedness and disaster mitigation applications require prioritizing temporal resolution. Clearly, a part of the rich tradition in social science applications of remote sensing has stemmed from operationalizing social science theory into its concomitant physical landscape impacts, whether those connections be theorized *a priori* or explained post-observation. The scale-pattern-process framework offers a means of complementing other social science approaches of positing and assessing causal connections between or among social and biophysical processes.

FRONTIERS OF MODELING

Rindfuss et al. (2004) describe both methods used in and methodological issues arising from the study of population and environment in frontier environments. What is fundamental in their description is the many challenges that emerge from the study of how, where, when, and why land use/land cover changes. These issues generally revolve around questions of time, place, motivations, and implications of land use dynamics that collectively acknowledge the relevance of the human dimension, the complex interplay between people and the environment, the interconnections of endogenous and exogenous forces and factors, and the centrality of maps and models in characterizing landscape patterns and the drivers of change (e.g., Walker and Homma 1996).

While most of the mapping needs of the Land Change Science community are generally addressed through a combination of fine spatial resolution (e.g., QuickBird and Ikonos), hyperspectral (e.g., Hyperion), and multi-spectral (e.g., Landsat Thematic Mapper) remote sensing systems, variables that describe the state and/or condition of the environment and the space–time patterns of land use/land cover are used to initialize, parameterize, calibrate, and validate statistical and spatial simulation models. The intent of these models is to link scale-pattern-process relationships by associating the socio-economic, demographic, biophysical, and geographical drivers of land use/land cover change through variable estimations that involve statistical approaches and spatial simulation models.

Statistical models that are used to describe the variation in land use/land cover patterns tend to emphasize multivariate techniques to estimate the significance and magnitude of the relationships between patterns and processes at the scale of the pixel or at a scale more fundamentally related to the human dimension (e.g., household farms, communities, and provinces). A challenge to conventional statistical methods for land use/land cover modeling, such as regression approaches, is the spatial dependence of land use patterns that imposes biases in the model through the presence of spatial autocorrelation – the ordering of data values as a consequence of location. Spatial autocorrelation is often present as a consequence of the spatial structure or organization of landscape features and land use/land cover types that are mapped by remote sensing systems. Also, spatial autocorrelation is generally present in the cross-sectional and/or longitudinal social surveys that are used to characterize the human dimension as households are often sampled in some clustered form, because of logistical constraints of access in the field and the tendency of human settlements to cluster in space. Further, multivariate statistical models also generally assume spatial stationarity where the nature of the relationship between variables is fixed over space. Human and environmental processes, however, are generally non-stationary in that the relationships depend on where the features occur across the landscape

(Fotheringham et al. 2002). If non-stationarity occurs, global modeling techniques, such as regression approaches, are generally inappropriate or unrealistic as they characterize 'average' effects in place of 'local' conditions. Some global approaches are used that are entirely appropriate and realistic to address the drivers of land use/land cover through econometric models. For example, multi-level models have been used to describe deforestation and agricultural extensification processes on household farms (e.g., Pan and Bilsborrow 2005) in the Ecuadorian Amazon. Multi-level models represent processes across hierarchical spatial scales, for example, the effects of communities on household decision-making about land use patterns on farms (Pan et al. 2004). In addition, spatial lag models are global regression models that generalize the relationships across the region, but take spatial dependence of relationships into consideration through the use of a weighted neighborhood matrix that characterizes neighbors and their spatial interactions (Anselin 2002).

Spatial simulations models are often designed to address a variety of 'What if?' questions through scenarios of land use/land cover change that generally involve the complex interplay between people and environment. Whereas experimentation is often an option for natural science questions, in social sciences experimentation is rarely ethically practical, and simulations provide a means for testing how individuals, households, communities, or even countries would react under a variety of contexts. Using complexity theory as the framework, dynamics systems are examined through Cellular Automata (CA) and Agent-Based Models (ABMs). Complexity theory conceives the world as consisting of self-organized systems, either reproducing their state through negative feedbacks with their environment or moving along trajectories from one state to another as a result of positive feedbacks. The goal is to understand how simple, fundamental processes can be combined to produce complex holistic systems. Land use/land cover simulations that are generated through CA approaches rely on growth or transition rules, neighborhood associations, and a set of initial conditions to assess land use/land cover change patterns for some simulation period (Walsh et al. 2006). ABMs involve the use of autonomous decision-making entities (agents), an environment through which agents interact, rules that define the relationship between agents and their environment, and rules defining the sequence of actions in the model (Parker et al. 2003). Macro-level behaviors 'emerge' from the actions of individual agents as they learn through experiences and change and develop feedbacks with finer scale building blocks as agents. The characterization of the transition or growth rules in the CA and agents in the ABMs often rely on social surveys to describe the human dimension and remote sensing to describe the environment.

THE CHALLENGES AHEAD

Fundamental advantages of remotely sensed data benefiting a variety of application areas do also apply to social science applications as well: a synoptic view, a workaround for inaccessible or unsafe areas, cost-savings compared to field-only collected datasets, and 'seeing' the world beyond the limitations of human visual capability. The question 'What is social?' pinpoints the challenge facing remote sensing for social science pursuits: that few social processes are *directly* manifest on the landscape and thus require using proxies of proxies via indirect mapping. But remote sensing also provides the spatial, environmental, and biophysical context in which social processes occur, without which social science models are left under-specified. With migration, for example, the spatial, environmental, and biophysical contexts clearly change as people migrate, and these changes need to be explicitly included in the model as endogenous factors (Liverman et al. 1998). Understanding how these contexts translate into appropriate choices for selecting remotely sensed data (and methods) is often unclear to social scientists aiming to incorporate remote sensing analysis into their studies, just as remote sensing specialists trained in working with soils and vegetation may not initially understand how remotely sensed data may be best leveraged for studying human populations. To this end, Table 31.1 presents data selection priorities for analysts working to bridge remote sensing and social science applications lacking experience in at least one of these areas.

Ultimately though, just as the social phenomena under investigation, remotely sensed data capture what is often ephemeral, and the snapshot nature of remotely sensed data can impact the certainty with which ephemeral processes or entities are delineated; happily, at least, archival remote sensing does not do the same problems of accurate memory recall that social scientists face when interviewing people for information on past events (Rindfuss et al. 2003). The startup costs of remote sensing projects can be high and pre-startup costs are frequently underfunded (Rindfuss et al. 2004). This is not to say that the costs of other social science primary data collection efforts (such as household surveys) are not sometimes prohibitive as well, but rather to underscore the need for continued improved and subsidized access to remote sensed data such that the entry into remote sensing for social science applications is more easily facilitated.

Table 31.1   Eight tips on data selection for social scientists new to remote sensing, or remote sensing specialists new to social sciences

1	Sometimes cheaper is better; take advantage not only of archived imagery but also of derived products (e.g., SRTM DEMs), particularly when your team lacks a dedicated remote sensing analyst
2	*Caveat emptor:* acquire and read all the metadata, preferably prior to data purchase
3	For slower processes (e.g., urbanization or longer-term migration), target sensor systems with earlier launch dates (e.g., Landsat)
4	For quicker (cyclic or seasonal) change (e.g., short-term displacement or labor migration), prioritize sensors with better return times
5	For areas prone to flooding and/or problematic cloud cover, consider active sensor systems (e.g., radar)
6	Where spatially random processes are occurring or will be modeled (e.g., spontaneous colonization), opt for sensor systems with broader spatial extent
7	For urban, peri-urban, and desert/settlement applications, prioritize high spatial resolution over spectral resolution (e.g., make use of panchromatic bands of multispectral sensor systems)
8	For disaster mitigation (e.g., famine, disease outbreak, natural disasters), opt for broad spatial extent and high return time; also consider higher vertical accuracy of DEMs rather than horizontal resolution (for flooding)

Of paramount importance for all remote sensing applications, not just the social sciences, is the continued development and maintenance of satellite sensor systems designed for broad use. Yet even the workhorse of the US sensor system faces uncertainty, with diminishing funds allocated for earth remote sensing and the Landsat Data Continuity Mission (LDCM) on perpetual standby. The benefits of remote sensing – observation from afar when price, access, or safety prohibit direct observation – are even more important as echoed by recent events ranging in origin from natural (e.g., the December 2005 tsunamis) to anthropogenic (the Iraq War) to in-between (the impacts of Hurricane Katrina). Understanding globally pressing problems such as these requires quality data delivered in a timely fashion at a reasonable price. So what are the implications for scientists and stakeholders that most of the new fine to very fine spatial resolution sensor system development and data provision is becoming increasingly commercialized?

At the same time, data confidentiality is becoming a larger and larger issue (Rindfuss et al. 2004, VanWey et al. 2005). The advent of new data storage and streaming technologies has rendered products and services such as Google Earth nearly mainstream to novice computer users, and GPS technologies have become near omnipresent in many countries through vehicle and cellular phone use in addition to recreational GPS units. But as remotely sensed data become more and more widely available, so too does the ability to link these spatially precise depictions of people without their consent to other data layers. Privacy issues loom large, particularly in a climate of escalated concerns regarding terrorism and security. Yet have or will researchers' noble obligations to privacy and confidentiality become moot in the face of this information available for sale to others around the world with varying motivations and ethics? As the public's expectation of privacy becomes diminished, what is the role of social scientists working with spatially explicit and remotely obtainable data – to act as vanguard to the last bastions of privacy, or to lead the exploration of these data?

On the technological side, there remains the challenge of integrating discrete and continuous data in a way meaningful for social science research. New, multiscale data models are under development, and the rise in popularity of object- or feature-based (as opposed to pixel-based) classification suggests a continued move toward a continuity of data representation. Another answer may lie in better conceptual models of human activity space now that several decades of case studies of spatial patterns of land tenure regimes and impacts on people's movements and activities have been completed, though migration and migratory-based peoples (e.g., pastoralists) continue to stretch the capacity of spatial databases to house temporally dynamic datasets (BurnSilver et al. 2003). Improved temporal and spatial data integration also likely holds the key to unlocking best practices for working with historical datasets in order to understand the landscape legacies of environments and peoples that came before typical remote sensing time series origination (Klepeis and Turner 2001), lest grave mistakes in interpretation of landscapes or cultures be made (Fairhead and Leach 1996).

The boon of complexity theory into social science simulations based on remotely sensed data appears to be a robust path for continued exploration, but modeling efforts require further validation and testing (Pontius et al. 2006). Because the results of spatially explicit simulation models are not, in fact, spatially explicit (or rather, deterministic) in outcome, they remain difficult to interpret and a hard-sell for policymakers. Probability maps (typically with probability of events without regard to actual location) are of little comfort to local stakeholders or decisionmakers, particularly when the conveyance of an understanding of uncertainty is unclear (Pontius et al. 2006), though they may provide useful guidance for regional policymaking. The propagation of error and uncertainty also remains problematic for a variety of applications of remotely sensed data (Brown et al. 2004), though particularly for social scientists.

The real challenge to the interspersion of social science and remote sensing may be how to get social scientists think bigger, conceptually. What can social scientists contribute to or learn from larger scale sensor systems such as GRACE and findings regarding mass water movement and shrinkage of polar ice caps? Typically social scientists employ finer scaled datasets for the bulk of their research, but surely remote sensing of the collective impacts of anthropogenic activity on the Earth's climate or water balance (and the reciprocal impacts as well) should have some bearing on social science pursuits at scales beyond the local and semi-regional. In what other ways can social scientists push the conceptual frontiers of landscape and population assessment: to detect and predict landscapes of fear, of equity, of security? Lastly, should social and remote sensing scientists encourage greater public and policy participation? If so, how can such be done beyond a case study basis and in a more systematic way (Castella et al. 2005)? Ultimately, these questions of data integration, data confidentiality, simulation modeling, error/uncertainty, and conceptual frontiers reveal the strength of the integration of remote sensing and social sciences: social science applications of remote sensing push the boundaries of data usage in unique ways, while remotely sensed data and methods push the boundaries of defining the social and society.

REFERENCES

Anselin, L., 2002. Under the hood: issues in the specification and interpretation of spatial regression models. *Agricultural Economics*, 27(3): 247–267.

Archer, E. R. M., 2004. Beyond the 'climate versus grazing' impasse: Using remote sensing to investigate the effects of grazing system choice on vegetation cover in the eastern Karoo. *Journal of Arid Environments*, 57(3): 381–408.

Barbieri, A. F., R. E. Bilsborrow, and W. K. Pan, 2005. Farm household lifecycles and land use in the Ecuadorian Amazon. *Population and Environment*, 27(1): 1–27.

Brown, D. G., E. A. Addink, J. Duh, and M. A. Bowersox, 2004. Assessing uncertainty in spatial landscape metrics derived from remote sensing data. In R. S. Lunetta and J. G. Lyon (eds), *Remote Sensing and GIS Accuracy Assessment*. CRC Press, Boca Raton. Chapter 16: 221–232.

BurnSilver, S., R. Boone, and K. Galvin, 2003. Linking pastoralists to a heterogeneous landscape: The case of four Maasai group ranches in Kajiado district, Kenya. In J. Fox, R. R. Rindfuss, S. J. Walsh and V. Mishra (eds), *People and the Environment: Approaches for Linking Household and Community Surveys to Remote Sensing and GIS*. Kluwer, Boston. Chapter 6: 173–199.

Campbell, D. J., D. P. Lusch, T. A. Smucker, and E. E. Wangui, 2005. Multiple methods in the study of driving forces of land use and land cover change: A case study of SE Kajiado District, Kenya. *Human Ecology*, 33(6): 763–794.

Castella, J. C., T. N. Trung, and S. Boissau, 2005. Participatory simulation of land-use changes in the northern mountains of Vietnam: the combined use of an agent-based model, a role-playing game, and a geographic information system. *Ecology and Society*, 10(1), article 27.

Chowdhury, R. R. and B. L. Turner, 2006. Reconciling agency and structure in empirical analysis: Smallholder land use in the southern Yucatan, Mexico. *Annals of the Association of American Geographers*, 96(2): 302–322.

Crews-Meyer, K. A., 2006. Temporal extensions of landscape ecology theory and practice: Examples from the Peruvian Amazon. *Professional Geographer*, 58(4): 421–435.

Deadman, P., D. Robinson, E. Moran, and E. Brondizio, 2004. Colonist household decisionmaking and land-use change in the Amazon Rainforest: An agent-based simulation. *Environment and Planning B-Planning and Design* 31(5): 693–709.

Evans, T. P., A. Manire, F. De Castro, E. Brondizio, and S. McCracken, 2001. A dynamic model of household decision-making and parcel level landcover change in the eastern Amazon. *Ecological Modelling*, 143(1–2): 95–113.

Fairhead, J. and M. Leach, 1996. *Misreading the African Landscape: Society and Ecology in a Forest–Savanna Mosaic*. Cambridge University Press, Cambridge, New York.

Fotheringham, A. S., C. Brunsdon, and M. Charlton, 2002. *Geographically Weighted Regression*. Wiley: West Sussex, England, UK.

Fox, J., R. R. Rindfuss, S. J. Walsh, and V. Mishra, 2003. *People and the Environment: Approaches for Linking Household and Community Surveys to Remote Sensing and GIS*. Kluwer, Boston.

Gautam, A. P., G. P. Shivakoti, and E. L. Webb, 2004. Forest cover change, physiography, local economy, and institutions in a mountain watershed in Nepal. *Environmental Management*, 33(1): 48–61.

Geoghegan, J., L. Pritchard, Y. Ogneva-Himmelberger, R. R. Chowdhury, S. Sanderson, and B. L. Turner II, 1998. 'Socializing the Pixel' and 'Pixelizing the Social' in land-use and land-cover change. In D. Liverman, E. Moran, R. R. Rindfuss and P. C. Stern (eds), *People and Pixels: Linking Remote Sensing and Social Science*. National Academy Press, Washington, D.C. Chapter 3: 51–69.

Geoghegan, J., L. Schneider, and C. Vance, 2004. Spatially explicit, statistical land-change models in data sparse conditions. In B. L. Turner, J. Geoghegan, and D. R. Foster (eds), *Integrated Land-Change Science and Tropical Deforestation in the Southern Yucatan: Final Frontiers*. Oxford University Press, New York. Chapter 12: 247–270.

Gutman, G., A. C. Janetos, C. O. Justice, E. F. Moran, J. F. Mustard, R. R. Rindfuss, D. Skole, B. L. Turner II, and M. A. Cochrane, 2004. *Land Change Science: Observing, Monitoring and Understanding Trajectories of Change on the Earth's Surface*. Kluwer Academic Publishers, Dordrecht, The Netherlands.

Herrmann, S. M., A. Anyamba, and C. J. Tucker, 2005. Recent trends in vegetation dynamics in the African Sahel and their relationship to climate. *Global Environmental Change-Part A*, 15(4): 394–404.

Jensen, J. R., 2000. *Remote Sensing of the Environment.* Prentice Hall, Upper Saddle River, New Jersey.

Klepeis, P. and B. L. Turner, 2001. Integrated land history and global change science: The example of the Southern Yucatan Peninsular Region project. *Land Use Policy*, 18(1): 27–39.

Liverman, D., E. Moran, R. R. Rindfuss, and P. C. Stern, 1998. *People and Pixels: Linking Remote Sensing and Social Science.* National Academy Press, Washington, D.C.

McNoleg, O, 2003. An account of the origins of conceptual models of geographic space. *Computers, Environment and Urban Systems*, 27(1): 1–3.

Messina, J. P., and K. A. Crews-Meyer, 2000. A historical perspective on the development of remotely sensed data as applied to medical geography. In D. Albert, W. Gesler, and B. Levergood (eds), *Spatial Analysis, GIS, and Remote Sensing Uses in the Health Sciences.* Ann Arbor Press, Chelsea, Michigan. 147–168.

Moran, E. F., E. Brondizio, P. Mausel, and Y. Wu, 1994. Integrating Amazonian vegetation, land-use, and satellite data. *Bioscience*, 44(5): 329–338.

Nagendra, H., J. Southworth, and C. Tucker, 2003. Accessibility as a determinant of landscape transformation in western Honduras: linking pattern and process. *Landscape Ecology*, 18(2): 141–158.

Nagendra, H., D. K. Munroe, and J. Southworth, 2004. From pattern to process: Landscape fragmentation and the analysis of land use/land cover change. *Agriculture Ecosystems and Environment*, 101(2–3): 111–115.

Nyerges, A. E. and G. M. Green, 2000. The ethnography of landscape: GIS and remote sensing in the study of forest change in west African Guinea savanna. *American Anthropologist*, 102(2): 271–289.

Overmars, K. P. and P. H. Verburg, 2006. Multilevel modelling of land use from field to village level in the Philippines. *Agricultural Systems*, 89(2–3): 435–456.

Pan, W. K. Y. and R. E. Bilsborrow, 2005. The use of a multilevel statistical model to analyze factors influencing land use: a study of the Ecuadorian Amazon. *Global and Planetary Change*, 47(2–4): 232–252.

Pan, W. K. Y., S. J. Walsh, R. E. Bilsborrow, B. G. Frizzelle, C. M. Erlien, and F. D. Baquero, 2004. Farm level models of spatial patterns of land use and land cover dynamics in the Ecuadorian Amazon. *Agriculture, Ecosystems and Environment*, 101: 117–134.

Parker, D. C., S. M. Manson, M. A. Janssen, M. J. Hoffman, and P. Deadman, 2003. Multi-agent systems for the simulation of land use and land cover change: A review. *Annals of the Association of American Geographers*, 93: 314–347.

Perz, S. G. and D. L. Skole, 2003. Social determinants of secondary forests in the Brazilian Amazon. *Social Science Research*, 32(1): 25–60.

Pontius, R. G. and L. C. Schneider, 2001. Land-cover change model validation by an ROC method for the Ipswich watershed, Massachusetts, USA. *Agriculture Ecosystems and Environment*, 85(1–3): 239–248.

Pontius, R. G., A. J. Versluis, and N. R. Malizia, 2006. Visualizing certainty of extrapolations from models of land change. *Landscape Ecology*, 21(7): 1151–1166.

Reid, R. S., R. L. Kruska, N. Muthui, A. Taye, S. Wotton, C. J. Wilson, and W. Mulatu, 2000. Land-use and land-cover dynamics in response to changes in climatic, biological and socio-political forces: the case of southwestern Ethiopia. *Landscape Ecology*, 15(4): 339–355.

Rindfuss, R. R., S. J. Walsh, V. Mishra, J. Fox, and G. Dolcemascolo, 2003. Linking household and remotely sensed data: Methodological and practical problems. In: J. Fox, R. R. Rindfuss, S. J. Walsh and V. Mishra (eds), *People and the Environment: Approaches for Linking Household and Community Surveys to Remote Sensing and GIS.* Kluwer, Boston. Chapter 1: 1–30.

Rindfuss, R. R., S. J. Walsh, B. L. Turner, J. Fox, and V. Mishra, 2004. Developing a science of land change: Challenges and methodological issues. *Proceedings of the National Academy of Sciences of the United States of America*, 101(39): 13976–13981.

Schelhas, J. and G. A. Sanchez-Azofeifa, 2006. Post-frontier forest change adjacent to Braulio Carrillo National Park, Costa Rica. *Human Ecology*, 34(3): 407–431.

Seto, K. C. and M. Fragkias, 2005. Quantifying spatiotemporal patterns of urban land-use change in four cities of China with time series landscape metrics. *Landscape Ecology*, 20(7): 871–888.

Turner, M. G., R. H. Gardner, and R. V. O'Neill, 2001. *Landscape Ecology in Theory and Practice: Pattern and Process.* Springer Press, New York.

VanWey, L. K., R. R. Rindfuss, M. P. Gutmann, B. Entwisle, and D. L. Balk, 2005. Confidentiality and spatially explicit data: Concerns and challenges. *Proceedings of the National Academy of Sciences of the United States of America*, 102(43): 15337–15342, doi:10.1073, pnas. 0507804102.

Verburg, P. H., K. P. Overmars, and N. Witte, 2004. Accessibility and land-use patterns at the forest fringe in the northeastern part of the Philippines. *Geographical Journal*, 170: 238–255.

Walker, R., 2003. Mapping process to pattern in the landscape change of the Amazonian frontier. *Annals of the Association of American Geographers*, 93(2): 376–398.

Walker, R. T. and A. K. O. Homma, 1996. Land use/land cover dynamics in the Brazilian Amazon: An overview. *Ecological Economics*, 18: 67–80.

Walker, R. and W. Solecki, 2004. Theorizing land-cover and land-use change: The case of the Florida Everglades and its degradation. *Annals of the Association of American Geographers*, 94(2): 311–328.

Walsh, S. J. and K. A. Crews-Meyer, 2002. *Linking People, Place, and Policy: A GIScience Approach.* Kluwer Academic Publishers, Boston.

Walsh, S. J., T. P. Evans, W. F. Welsh, B. Entwisle, and R. R. Rindfuss, 1999. Scale-dependent relationships between population and environment in northeastern Thailand. *Photogrammetric Engineering and Remote Sensing*, 65(1): 97–105.

Walsh, S. J., K. A. Crews-Meyer, T. W. Crawford, W. F. Welsh, B. Entwisle, and R. R. Rindfuss, 2001. Separating intra- and inter-annual signals in monsoon-driven Northeast Thailand. In: A. C. Millington, S. J. Walsh, and P. E. Osborne (eds), *Remote Sensing and GIS Applications in Biogeography and Ecology.* Kluwer Academic Publishers, Boston. 91–108.

Walsh, S. J., B. Entwisle, R. R. Rindfuss, and P. H. Page, 2006. Spatial simulation modeling of land use/land cover change scenarios in northeastern Thailand: A cellular automata approach. *Journal of Land Use Science*, 1: 5–28.

Wilson, E. H., J. D. Hurd, D. L. Civco, M. P. Prisloe, and C. Arnold, 2003. Development of a geospatial model to quantify, describe and map urban growth. *Remote Sensing of Environment*, 86(3): 275–285.

Wu, C. S. and A. T. Murray, 2007. Population estimation using Landsat Enhanced Thematic Mapper imagery. *Geographical Analysis*, 39(1): 26–43.

Yeh, A. G. O. and X. Li, 2001. A constrained CA model for the simulation and planning of sustainable urban forms by using GIS. *Environment and Planning B-Planning and Design* 28(5): 733–753.

Hazard Assessment and Disaster Management Using Remote Sensing

Richard Teeuw, Paul Aplin,
Nicholas McWilliam, Toby Wicks,
Matthieu Kervyn and Gerald G. J. Ernst

Keywords: airborne and space-borne sensors, hazard, vulnerability and
risk, disaster cycle, flooding, landslides, volcanic and
biophysical hazards, disaster response.

INTRODUCTION

Disasters occur when two factors – hazard and vulnerability – coincide. The risk of a disaster occurring is proportional to the severity of the hazard and the vulnerability of the affected population. With human population continuing to increase, it is inevitable that more people will be living in hazardous settings, increasing the risk of disaster. With global warming and associated environmental instability, both the frequency and severity of disasters is likely to increase (van Aalst 2006). The United Nations (2005) Hyogo Framework for Action on disaster risk reduction asked governments to: '...*promote the use, application and affordability of recent information and space-based technologies, as well as Earth observation, to support disaster risk reduction.*'

Remote sensing provides many ways of mapping the distribution and severity of hazards and vulnerability – reducing the severity of either of these factors leads to a reduction in the risk of a disaster occurring. For instance, interpreting a Landsat sensor image and producing a geohazard map that can be examined by local communities, increases the awareness of those people and reduces their vulnerability, thereby reducing the risk of disaster.

Remote sensing is a key tool for disaster managers, as it provides a useful means of mapping and monitoring hazards and vulnerable features, before and after destructive events. Remote sensing can play a number of useful roles in management of natural disasters (Cutter 2003, Zeil 2003), with applications during three main stages of the Disaster Cycle:

1 *Pre-disaster* disaster preparedness maps, showing the distribution of vulnerable features (e.g., centres of population, infrastructure) and the relative severities of geohazards, can be used by emergency planners to assess levels of risk and guide disaster reduction strategies. Real-time satellite monitoring is generally of too coarse a spatial resolution to provide effective early warnings of

hazardous events, the notable exception being the use of weather satellites to warn of meteorological hazards.

2 *Event crisis*   once a disaster occurs, fine to coarse spatial resolution satellite imagery (10–1000 m pixels) is useful for assessing the extent and severity of various impacts. With rapid-onset disasters (i.e., earthquakes, eruptions, landslides), aerial photography or fine resolution satellite imagery (0.5–10.0 m pixel), is needed for rapid damage assessments and search operations. The Disaster Monitoring Constellation (DMC; Stevens 2005) has highlighted the value of daily medium to fine spatial resolution satellite imagery for monitoring regions affected by disaster.

3 *Post-disaster*   geomorphological and geo-ecological mapping can assist recovery by highlighting locations with, for example, essential resources (e.g., water, fuel) and reconstruction materials.

For emergency planners, remote sensing can be a direct source of information, such as Digital Elevation Models (DEMs) from the Shuttle Radar Topography Mission (SRTM), as well as being a data source from which thematic maps of hazards, resources and vulnerable features can be produced, via image interpretation or digital image classification. For scientists studying geohazards, remote sensing is one of the most important data sources to feed into models of processes causing hazards and disasters: a greater understanding of those processes improves prediction and mitigation measures. This chapter first examines some of the remote sensing applications in the mapping and monitoring of geohazards, moving from flooding and ground instability to volcanic terrains, before considering biophysical hazards (drought, famine, wildfires, diseases, pollution) and concluding with a section on uses of Earth observation in disaster response.

EXAMPLES OF REMOTE SENSING AND HAZARD ASSESSMENT

Flooding

Floods are the most costly type of natural disaster, globally accounting for about 30% of associated economic losses (Munich Re 1997). Flooding looks set to become even more costly, with global warming leading to sea level rise and increased storminess (Houghton et al. 2001, Lee 2001). Assessing and managing flood risk requires an understanding of the processes and landforms of rivers and coasts, as well as information on vulnerable features, such as population and infrastructure; remote sensing

provides many ways of mapping and monitoring those features.

Coastal and riverine environments contain features that are difficult to map, sample or monitor using conventional fieldwork. Access is, for example, difficult in marshland and mudflats, especially in intertidal areas. Remote sensing data can be collected at a range of scales, from the airborne detection of small features, such as the dimensions of bushes on a floodplain, through to spaceborne observation of regional features, such as a major drainage basin. Summaries of remote sensing applied to hydrology are given by Schultz and Engman (2000), Leuven et al. (2002) and Schmugge et al. (2002).

Remote sensing can play key roles in predicting, detecting, delineating, monitoring and modelling floods. On a continental scale, this involves tracking weather systems and storm clouds using meteorological satellites, such as Meteosat and the Geostationary Operational Environmental Satellite (GOES). On a regional scale, internet-downloadable SRTM DEM data have been used for mapping and modelling of coastal inundation zones (Figure 32.1) – a useful innovation for countries with limited map coverage (Theilen-Willige 2006). Optical satellite sensors' inability to penetrate cloud limits their usage for flood mapping. Radar's ability to operate at night and to 'see' through cloud makes it very useful for mapping and monitoring floods. Imaging radar also allows assessments of catchment soil moisture conditions, river levels, flooded areas and floodwater volumes, enabling early warnings to be given to vulnerable sites downriver (Nico et al. 2000).

For detailed mapping of floodplain geomorphology and land cover, aerial photograph interpretation and photogrammetry remain the most widely-used and well-tested methods. Detailed mapping of floodplains for flood management and modelling involves quantification of: (i) areas where flood water would be contained by elevated land or structures such as dykes; and (ii) the rate at which floodwaters travel over a floodplain, involving evaluations of surface roughness due to features such as vegetation and built structures. In the light of these requirements for hydraulic modelling, detailed mapping and monitoring are needed of features with hydrogeomorphological significance on floodplains. The UK Environment Agency (EA) uses photogrammetry on 1:3,000 scale colour aerial photographs to produce floodplain maps with 15–25 cm contour intervals.

Enhancements in airborne laser altimetry (ALS, also known as lidar; Hyyppä et al., in this volume), allowing levels of micro-relief mapping that rival photogrammetry, as well as lidar's ability to quantify the height and volume of objects on a floodplain, such as buildings and trees, have

Figure 32.1   Simulated 10 m inundation (pale blue) along the NW coast of Costa Rica. Derived from SRTM DEM data, overlain by true-colour Landsat Thematic Mapper (TM) imagery. (See the color plate section of this volume for a color version of this figure).

dramatically improved floodplain hydrodynamic modelling (Cobby et al. 2001, Straatsma and Middelkoop 2006). In the UK, the NextMap airborne Interferometric Synthetic Aperture Radar (InSAR, Kellndorfer and McDonald, in this volume) DEM has been used by insurers to map floodplain boundaries, allowing them to assess the flood risk at a given site, based on its location (Sanders et al. 2005). The UK EA have shown the benefits of multi-sensor aerial surveys, collecting lidar data simultaneously with Compact Airborne Spectrographic Imager (CASI) multispectral data (Brown 2004). The lidar–CASI datasets can be processed by a Geographical Information System (GIS), yielding DEM data for flood modelling and land cover data that can be processed to show the potential cost of flooding, allowing an evaluation of flood risk.

Landslides and ground deformation

For reviews of remote sensing systems that can be used to detect ground instability features, the reader is referred to Mantovani et al. (1996) and Metternicht et al. (2005), as well as Soeters and van

Westen (1996) for examples of remote sensing and GIS applied to landslide hazard assessment. Geo-morphological mapping, based on interpretation of aerial photography and photogrammetric anal-ysis, is a well-established technique for mapping landslides, particularly where steep and unstable slopes make sites too dangerous for walk-over sur-veys. Lidar, both airborne and terrestrial, has been applied to mapping and monitoring slope instabil-ity features, with encouraging results (Adams and Chandler 2002). Weathered rocks and soils rich in shrink-swell clays can be detected using airborne infra-red hyperspectral remote sensing, enabling the mapping of areas susceptible to slope insta-bility (Ben-Dor et al. 2002, Kariuki et al. 2003). Medium spatial resolution spaceborne sensors, such as Landsat Enhanced Thematic Mapper Plus (ETM+), the Advanced Spaceborne Thermal Emission and Reflection Radiometer (ASTER) and Systeme Pour I'Observation de le Terre (SPOT), High Resolution Visible (HRV) have been of lim-ited use in mapping landslides, due to the inher-ently small size of most slope instability features. Successes are reported for the mapping of volcanic debris avalanches, 'cold lahars' and associated col-lapse scars (Francis and Wells 1988, Kerle and van Wyk de Vries 2001, Kervyn et al. in press – see also Figure 32.2). Satellite imagery, particu-larly ASTER, has been used to highlight areas in Ethiopia affected by landslide activity (Temesgen et al. 2001, Andrews Deller 2007) and to map the occurrences of shrink-swell clay minerals in soils (Bourguignon et al. 2007).

InSAR has been used to map ground defor-mation associated with earthquakes (Massonet and Feigl 1998), landslides (Singhroy and Molch 2004), subsidence (Amelung et al. 1999), vol-canoes (Massonnet and Sigmundsson 2000) and glaciers, including ice caps over volcanoes where ice deformation relates to subglacial eruption pro-gression (Gudmundsson et al. 2002). Spaceborne InSAR provides a high spatial sampling density (e.g., the Earth Remote Sensing Satellite (ERS) has pixel sizes of 12–30 m), ~1 cm precision, and visit frequencies of ~1 pass/month, allowing mapping and monitoring of subsidence and/or the action of shrink-swell clays (Boyle et al. 2000). Satellite radar Persistent Scatter Interferometry (PS-InSAR: Ferretti et al. 1999) allows detection of millimetre-scale ground movements, highlighting localized subsidence in urban areas (Riedmann and Haynes 2007). Hooper et al. (2004) used natural persis-tent scatterers, mostly bare rock exposures, in a non-urban environment to solve problems with interferometric coherence.

Volcanic hazards

Volcanic hazards directly concern about 6% of world population and can affect many more

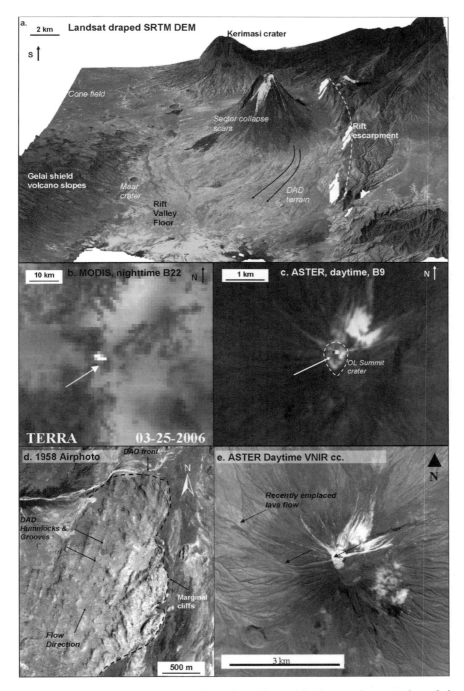

Figure 32.2   Oldoinyo Lengai, Tanzania: the volcano has white lava at its summit and rises 2000 m above a rift valley. Central figure: 3-D view, SRTM DEM with Landsat ETM+ true colour drape. Upper left: night-time Moderate Resolution Imaging Spectrometer (MODIS) band 22 (3.93–3.99 μm) image – white pixels: thermal emission of a volcano-flank lava flow. Upper right: ASTER band 9 (2.36–2.43 μm) scene – thermal anomalies from a summit-crater lava flow. Lower left: aerial photograph of debris avalanche deposits (hummocky relief). Lower right: ASTER Visible and Near-Infra-Red (VNIR) image, acquired a few days after an eruption, which enabled the mapping of lava flows. (See the color plate section of this volume for a color version of this figure).

through atmosphere-climate or socio-economic effects (Marti and Ernst 2005). Remote sensing has revolutionized volcanic hazard management, allowing the derivation of DEMs, measurement of volcanic deformation before, during, and after eruptions, logging of thermal emission changes from craters, lava flows or lava domes, monitoring of eruptions and tracking of emissions. Daily thermal observations at 250–1000 m pixel resolution are acquired automatically on a near-global basis and are freely accessible via the internet within days of data acquisition (Wright et al. 2002). Volcanic clouds can be imaged and tracked for days, on an hourly basis, at 1 km spatial resolution (Rose et al. 2000, Bertrand et al. 2003).

There is a general trends toward usage of imagery with: (i) fine spatial resolutions of 1–5 m (e.g., IKONOS, SPOT High Resolution Geometry (HRG) and the Indian Remote Sensing Satellite (IRS); (ii) more types of multispectral sensors (e.g., ASTER); (iii) high temporal resolution (e.g., Moderate Resolution Imaging Spectrometer (MODIS)); and (iv) near-global coverage of medium to fine spatial resolution archived data (e.g., Landsat Thematic Mapper (TM) and ETM+, ASTER, SRTM), which are crucial for volcanic studies in remote regions (Carn and Oppenheimer 2000, Stevens et al. 2002, Kervyn et al. 2007).

During the 1980 eruption of Mount St Helens, ground-based (C-band) radar data were crucial for forecasting ash dispersal – an acute hazard for international air traffic (Harris and Rose 1983). Most particles in an eruption cloud are fine, whereas the radar return depends on the sixth power of particle size, so that an ash-rich eruption cloud can only be tracked for 10–30 km, unless the presence of low-density ash aggregates increases the effective particle size. Rapid settling of coarse Etna ash is probably why the ash emissions that closed Catania airport ($c.$ 30 km away) could not be detected using the airport radar. However, explosive eruptions often involve water-rich eruption columns, forming ice on the ash particles, increasing their size and scattering, while not increasing their removal velocity proportionally. During the 2000 eruption of Hekla (Iceland), the Reykjavik weather radar (~120 km away) detected the eruption onset within 2 min, tracked the eruption cloud across Iceland and determined eruption end with precision (Lacasse et al. 2004). Such information is crucial to Volcanic Ash Advisory Centres responsible for rapidly issuing forecasts of ash dispersal to civil aviation authorities, who can warn aircraft pilots of the ash hazard.

Volcano degassing studies have benefited from ground-based Differential Optical Absorption Spectrometers (DOAS), which are small, cheap, and provide real-time SO_2 emission data. Instruments are also being developed for determining real-time fluxes of key gases such as HF, HCl and CO_2. This is a huge improvement in volcano monitoring, as the gas data enables better discrimination between eruptive and non-eruptive events. The uses of satellite sensors to monitor volcanic clouds and degassing, mostly utilizing Advanced Very High Resolution Radiometer (AVHRR) data, have been summarized by Rose et al. (2000). This work has increasingly integrated sensor data and ancillary datasets, such as ozone monitoring (e.g., the Total Ozone Mapping Spectrometer (TOMS)), weather balloon data (temperature, moisture, wind directions and speeds), C-band weather radars, atmospheric dispersal models for ash and field maps of ash on the ground (e.g., Rose et al. 2001, 2003, Lacasse et al. 2004). The satellite sensor datasets are now augmented by half-hourly Meteosat Second Generation (MSG) (Pergola et al. 2004) and daily MODIS data (Rose et al. 2003).

The volcanic remote sensing potential of InSAR was illustrated by the 1997 eruption of Okmok, Alaska (Lu et al. 2003). TOPSAR and ERS SAR interferometric and DEM data enabled the study of eruption inflation–deflation phases. InSAR-derived pre-eruptive and post-eruptive DEMs can be used to estimate lava areas, thicknesses and volumes with unprecedented accuracy (15% error). *Differential InSAR* (DInSAR) offers the prospect of monitoring surface deformation to centimetre-scale accuracy (Stevens and Wadge 2004). DInSAR was used to map ground deformation related to seismic unrest at Long Valley caldera, USA, and Nysiros caldera, Greece (Sachpazi et al. 2002). Coherence loss was used to map newly formed lava flows at Kilauea Volcano, Hawaii (Zebker et al. 2000). SAR backscatter contrast was used to map pyroclastic flow deposits at Aniakchak caldera (Rowland et al. 1994).

Landsat TM data were used to detect Short Wave Infra Red (SWIR) radiance from a growing lava dome at Lascar volcano, Chile (Francis and McAllister 1986). Landsat TM or ETM+ and ASTER data can also be used to analyse thermal characteristics of lava flows (Wright et al. 2000), lava domes (Wooster et al. 2000), lava lakes (Harris et al. 1999) and fumarole fields (Harris and Stevenson 1997). High temporal resolution data from sensors such as GOES (4 km pixels), (Along-Track Scanning Radiometer (ATSR): 1 km pixels), AVHRR (1.1 km pixels) and MODIS (0.25–1 km pixels) satisfy the need for operational volcano monitoring. AVHRR data can be acquired at least four times per day: this high sampling frequency reduces the chance of cloud cover and is helpful for examining dynamic volcanic phenomena (Schneider et al. 2000). MODIS provides complete global coverage every 1–2 days, at spatial resolutions of 250, 500 and 1000 m. Its ability to detect SWIR and Thermal Infra-Red (TIR), combined with 200 m precision geolocation of pixels, makes it ideal for automatically detecting and monitoring high-temperature thermal anomalies

(Wright et al. 2002). MODIS has now allowed monitoring of the onset, development and cessation of many eruptive events. Automated MODIS monitoring has recently been extended to the moderate-temperature anomalies of carbonatite volcanism (Kervyn et al., in press). Figure 32.2 shows aspects of the hazardous terrain of Oldoinyo Lengai in Tanzania, using a variety of spaceborne sensors.

Biophysical hazards

'*Droughts are normal climate episodes, yet they are among the most expensive natural disasters...*' (Tadesse et al. 2005: 244). To reduce the impacts of drought, it is necessary to quantify drought occurrence, severity and distribution. Meteorological remote sensing data can provide useful information, enabling prediction and monitoring of general weather conditions, precipitation and so on. Some effort has been made to characterize potential water availability from remotely sensed imagery. Herman et al. (1997) describe a method to estimate accumulated rainfall: this and all similar studies require extensive field measurements (e.g., rain-gauge data) to calibrate image analysis. More commonly, remote sensing is used indirectly to observe drought events, by focusing on effects of water depletion on land cover. In particular, image analysis is conducted to characterize vegetation cover and condition, and soil moisture (McVicar and Jupp 1998). Traditionally, classification techniques and vegetation indices, often the Normalized Difference Vegetation Index (NDVI), were used, but recently drought indices and complex spatial models have been adopted (Vincente-Serrano et al. 2006). Recent approaches have benefited from the emergence of hyperspectral imagery that can enable highly accurate vegetation and soil analysis (Asner et al. 2004).

Much effort has been committed to measuring and monitoring crop stress arising from water scarcity (Tadesse et al. 2005, Clay et al. 2006). This may enable crop yield prediction from early season monitoring (Ferencz et al. 2004, Vincente-Serrano et al. 2006). Crop yield information is extremely valuable in predicting possible food shortages and guiding remediation activities. In extreme cases, drought can lead to failed crops and famine. Remote sensing can provide much useful information related to the extent and severity of drought, and to the distribution and yield of crops. Given that droughts can affect very large areas, the synoptic view from remote sensing can be invaluable. Famine can be avoided or alleviated when this information is used effectively in remediation activities. Soon after the severe famine event in sub-Saharan Africa in the early 1980s, remote sensing was identified as a valuable tool in famine alleviation. Famine early warning systems

were developed that used GIS to integrate remote sensing imagery with social and physical data, to identify areas at risk (Hutchinson 1991, Marsh et al. 1994). Interest continues to be high in monitoring drought conditions and vegetation response in the Sahel region, including analysis into subsistence crop water balances (Senay and Verdin 2003) and vegetation dynamics over the last 25 years (Anyamba and Tucker 2005).

In addition to drought as a driver of failed crops, crop stress and associated food shortages can also arise as a result of crop disease. Remote sensing is used to investigate crop disease using similar approaches to those used for drought-caused crop stress. Much work has focused on measuring the effect of disease on crop health (Qin and Zhang 2005), and projecting crop yield from multitemporal observations. Detailed spatial information on crop disease can be used in precision farming to guide pesticide distribution (West et al. 2003). Another famine-inducing hazard, and one that is also related to water availability, is crop damage caused by migrating locust plagues. Remote sensing is used here to monitor environmental conditions in potential locust habitats, identifying when conditions suit the growth of locust populations, thereby triggering remediation action (Ma et al. 2005). Sivanpillai et al. (2006) used Landsat ETM+ imagery in Kazakhstan to identify reed beds and water, a common habitat for migratory locusts.

The agents that cause diseases and determine their movements are controlled by environmental factors such as climate, water availability and quality, and vegetation. Remotely sensed imagery can be used to identify areas suitable for growth in the population of these agents. Mushinzimana et al. (2006) mention several recent malaria epidemics in east Africa, and outline the use of satellite sensor imagery to monitor the aquatic habitats of mosquito larvae as a first step in controlling the spread of the disease. Information on habitats can be integrated with knowledge on disease transmission to predict the spread of epidemics, demonstrated by Tran and Raffy (2006) for dengue fever. Commonly, GIS integration of remote sensing and other spatial data, enables quick assessment of disease endemic locations, estimation of the human population at risk, prediction of disease distribution in inaccessible areas, and guidance with respect to remediation strategies (Yang et al. 2005).

Wildfire is another environmental hazard that can increase in prevalence as a result of drought conditions. Dry surface conditions can raise the likelihood of fire occurrence and spreading. Remote sensing is used at three stages in wildfire management: (i) fire risk prediction; (ii) real-time fire response; and (iii) post-fire damage assessment. Many investigators have employed satellite sensor imagery to assess conditions of vegetation and other surface features, to estimate fuel load and fire risk (e.g., Chuvieco et al. 2004). Jia et al. (2006) used hyperspectral imagery to determine the abundance of photosynthetically active vegetation as a proxy for fuel. Fire risk maps can be integrated with meteorological and other fire behaviour data to predict fire spread, and this form of analysis can be applied with active fire images to monitor fires as they happen (Gong et al. 2006). The ability to observe fire distribution and predict its movement is obviously a major aid to fire response and control measures. The recent generation of very fine spatial resolution satellite sensors such as IKONOS, QuickBird and OrbView-3 are proving useful for active fire monitoring, owing to their great spatial detail and continuous, frequent coverage (Aplin 2003). After fires occur, multi-temporal imagery can be used to assess the extent and severity of fire damage (Chafer et al. 2004). Fire mapping has become a common remote sensing application in recent years, due in part to a growing interest in accurately estimating atmospheric emissions, given their significance in climate change studies. Various sources of remotely sensed imagery are now routinely processed by suppliers, creating a continuous archive of fire and burn scar maps available for public use (Bradley and Millington 2006).

Whereas drought is noted as one particular causal factor increasing the prevalence of wildfire, climatic conditions are significant in causing both drought and, therefore, wildfire. Large-scale climatic episodes such as the El Nino Southern Oscillation (ENSO), which occurs every few years and is associated with drought conditions (Boyd et al. 2006a), can lead to increased fuel loads and wildfire occurrence (Swap et al. 2003). Investigators are using climatic ENSO information, including remotely sensed imagery, to predict wildfire severity months in advance (Kitzberger 2002).

Although remote sensing may be of limited use in detecting chemical pollution, imagery can be useful for identifying leaks, monitoring damage caused and guiding environmental recovery practices. Perhaps the most well-known disaster of this kind is the 1986 Chernobyl accident, which was observed using the thermal atmospheric sounder capability of the USA's Television Infrared Observation Satellite (Givri 1995). However, due to limitations in the spatial resolution of available instruments and the relatively slow image supply that characterized remote sensing in the 1980s, imagery was used little in the immediate response. Subsequently, considerable use has been made of remote sensing technology to estimate environmental damage and monitor vegetation recovery. Davids and Tyler (2003) examined the health of Scots pine and silver birch a decade after Chernobyl. Boyd et al. (2006b) extended this work, assessing contamination in Belarus with the underlying goal of identifying where formerly high contamination levels have dropped, such that the area no longer poses a health hazard and can be converted back to a more suitable use. Remote sensing has also been used to monitor changes in vegetation spectral properties and soil to identify gas leaks (Smith et al. 2005, van der Werff et al. 2007). Finally, in the aftermath of the September 11, 2001, terrorist attack on the World Trade Center, Cutter (2003) highlighted the value of fine to very fine resolution imagery (orthophotography, lidar) as a key data in spatial decision support systems for rapid disaster response.

DISASTER RESPONSE

The primary role of remote sensing in disaster response is to provide information in support of planning and decision making, to assist emergency agencies in responding effectively to humanitarian needs (e.g., Watkins and Morrow-Jones 1985, Kelly 1998, Bjorgo 2000, Giadia et al. 2003, Kelmelis et al. 2006, Nourbakhsh et al. 2006, Sklaver et al. 2006). In rapid-onset emergencies many of the constraints associated with remote sensing become more pronounced, due to the compressed and unpredictable time scales. Image acquisition, processing, analysis, information extraction and presentation are required in the shortest possible time. For search and rescue purposes – the most time-critical application – the value of remotely sensed information drops rapidly 48 hours after a disaster event (Table 32.1). This section takes information requirements as the driving consideration: remote sensing is treated as essentially a range of data sources within a very diverse geographical information (GI) context, the main vehicle of which is thematic mapping.

The full set of contractual commitments around satellite sensor data distribution is complicated and limits free data flow. The United Nations International Charter on Space and Major Disasters provides a unified system of space data acquisition and delivery to those affected by (or responding to) natural or man-made disasters. An 'Authorised User' can call a single number to request mobilization of space data resources (Radarsat, ERS, ENVISAT, Landsat, SPOT, IRS, NOAA, DMC and others). A 24-hour on-duty operator passes the information to an Emergency Officer, who analyzes the disaster scope and prepares an acquisition plan using available space resources. Data acquisition and delivery takes place on an emergency basis and a Project Manager, experienced in data ordering, handling and application, assists the user. Further discussion of the data policy issues associated with disaster applications is given in Harris

Table 32.1   Remote sensing products for disaster response, ordered by level of processing and delivery time. Terms used for spatial resolution: coarse 100–1000 m; moderate 10–100 m; fine <10 m

Product category	Typical uses	Examples	Typical sources	Typical timescale
Archival imagery	Images provide context and landscape visualization. Medium resolution shows geographical features, but not detailed settlement patterns. May be out-dated	First maps are typically Landsat imagery overlaid with vector data and initial situation information	Landsat ETM+ (near-global); medium to high resolution; SRTM DEM	<12 hours
Near real-time imagery	Initial indication of disaster extent and nature	Regional maps showing extent of damage; or tracking volcanic ash for air traffic safety	MODIS: medium resolution	1–2 days
Specially acquired post-disaster imagery	Visually-derived situational information. May be expensive data	Building damage, landslides, street plans; e.g. Assessing damage in Goma (DRC) after Nyiragongo lava flow disaster	Aerial data, high resolution satellite imagery	Daily for aerial data; 1–2 days for satellite data
Manually traced features	Traced vectors to create map overlay or numerical report	Flood extent, roads, settled areas, oil spill, landslides, eruption impacts	Satellite and aerial sensors	Daily updates, using field reports
Classified images – pixel-based	Automated land cover mapping and measurement Change detection, if time series imagery available	Flood extent, fire extent, agricultural damage	Satellite and aerial sensors	Rapid processing, needs field-work updates
Automated feature extraction	'Intelligent' recognition of significant features	Building and infrastructure damage, roads, fire extent, land-slides. Thermal monitoring of volcanic lakes	Satellite and aerial sensors	Rapid processing, needs field work updates
Photogrammetric modelling and change detection	Detailed measurement of building and infrastructure damage	Building and infrastructure damage, roads, fire extent, land-slides. Monitoring lava domes	Stereo-scopic imagery and DEM data	Slow processing, e.g., 1 day /100 km^2

(this volume). A list of recent activations is given at: www.disasterscharter.org/new_e.html. The Global Monitoring for Environment and Security (GMES) programme aims to provide appropriate data and services to support decision-making on key environmental issues, at local to global scales. GMES adopted a user-driven approach, rather than a technology-push approach, with services classified into three categories:

1 *Reporting/mapping,* including topography or road maps, land-use and harvest, forestry monitoring, mineral and water resources. This service requires extensive Earth surface coverage and periodic data updating.

2 *Operational forecasting,* for marine zones, air quality or crop yields. This service systematically provides data over large regions, allowing modelling and prediction of short, medium or long-term events.

3 *Emergency response,* such as emergency management for disasters, via civil protection organizations responsible for security of people and property. This service concentrates on provision of up-to-date data.

Elements of GMES include PREVIEW and RISK-EOS: networks of service providers delivering GI to support disaster management, via: (i) Asset Mapping: urban areas in zones subject to hazards; (ii) Flood Risk Analysis: based on maps of past flood events, hydraulic simulations and potential damage; (iii) Flash-Flood Early Warning: using real-time precipitation radar and hydrological models; (iv) Burn Scar Mapping: yearly mapping of forest fire areas; (v) Rapid Mapping: in support of the UN disaster charter or for regional-scale crises. Another part of GMES is RESPOND, an alliance of international organizations working with the humanitarian community to improve access to maps, satellite sensor imagery and geographic information. RESPOND makes inputs to all parts of the crisis cycle where GI leads to enhanced deployment of development aid, whether slow onset crises (e.g., famine) or rapid onset crises (e.g., earthquake). GI provision extends beyond spaceborne data to include topographic and thematic map data, as well as information on affected inhabitants and infrastructure provided by relief workers operating in the region. RESPOND offers three types of service, providing a variety of products, support services and training:

1 global base mapping: collation of map data and satellite sensor imagery, from the USA National Geospatial Intelligence Agency (NGA), the United Nations Institute for Training and Research (UNITAR) Operational Satellite Applications Programme (UNOSAT), and internet-accessible databases, such as AlertNet.
2 maintaining an up-to-date GI archive for areas in crisis, with 'Hot spots' defined by users.
3 information on emerging crises: mapping of areas prior to crises, with relevant thematic information (e.g., accessibility, food security information, refugee camp locations, damage mapping).

Finally, disaster response can be assisted in a number of ways by virtual globes, such as Google Earth, Virtual Earth (Microsoft) and World Wind (NASA). These are freely accessible via the internet, are user-friendly and provide near-global moderate to fine-resolution imagery with 3-D perspective views. The integration of internet GIS technology with virtual globes is a promising new frontier of disaster management. For instance, after Hurricane Katrina, fine-resolution satellite imagery posted on Google Earth was utilized by rescue teams (see http://earth.google.com/katrina.html). Some virtual globes allow user-community information, such as photographs or information notices linked to locator coordinates, to be viewed over its image layers. This geowiki approach was utilized soon after the 2006 Kashmir earthquake, where locals phoned overseas relatives, who then posted road access details on Google Earth.

SUMMARY

The application of remote sensing to hazard mapping and disaster management has come a long way over the past half-century. In the 1950s we were largely limited to aerial photography, albeit with near-infra-red capability, requiring specialists in image interpretation and photogrammetry.

The 1960s saw a major boost for the early warning of storm and drought hazards, with the high frequency global coverage provided by meteorological satellites. Since the 1970s, successive generations of optical and radar satellites have produced progressively more detailed regional coverage, with hazardous terrain or post-disaster damage detected by manual interpretation or semi-automated techniques. Over the past decade there has been considerable progress in the automatic detection of some hazardous features, for instance volcanic eruptions or wildfire activity detected using the AVHRR or MODIS sensors. There has also been considerable improvement in the spectral and spatial resolution of recent Earth observation satellites, the prime example being ASTER: its SWIR and thermal bands have aided the detection of areas prone to ground deformation, as well as heatflows associated with volcanic activity.

Looking to the future, real-time imagery should soon be available from the next generation of Radarsat (Singhroy 2006), greatly increasing our ability to assess regional flood risks and possibly giving early warning of terrestrial earthquakes and volcanic eruptions, via InSAR detection of ground deformation. L-band radar data (e.g., the Advanced Land Observing Satellite, ALOS) will probably be used more frequently to map ground deformation in densely vegetated terrain. Meteosat and MODIS look set to provide invaluable datasets for examining and monitoring air pollution and volcanic ash clouds. Airborne lidar will probably be widely used in geohazard assessment, due to its ability to map terrain surfaces beneath vegetation cover. Very fine resolution (sub-metre) satellite imagery looks set to become more widely available, eventually replacing aerial photography for post-disaster damage assessment of urban areas, particularly in developing countries.

Perhaps the biggest need of all is to explore how remote sensing techniques can be shared with emergency planners and disaster managers in developing countries (Ernst et al. in press). Many internet-based data sources have been established, from the daily regional coverage provided by AVHRR, or the archives of Landsat and SRTM data at the Global Land

Cover Facility (http://glcfapp.umiacs.umd.edu:8080/esdi/index.jsp), to geohazard-related GIS layers and geowikis linked to Google Earth. However, we urgently need strategies that enable us to provide developing countries with the expertise, software and hardware for mapping, assessing and monitoring geohazards, vulnerable features and the extent of post-disaster damage.

REFERENCES

Adams, J. C. and J. H. Chandler, 2002. Evaluation of LiDAR and medium-scale photogrammetry for detecting soft-cliff coastal change. *Photogrammetric Record*, 17(99): 405–418.

Amelung, F., D. L. Galloway, J. W. Bell, H. A. Zebker and R. J. Laczniak, 1999. Sensing the ups and downs of Las Vegas: InSAR reveals structural control of land subsidence and aquifer-system deformation, *Geology*, 27: 483–486.

Andrews Deller, M. E., 2007. Space technology for disaster management: data access and its place in the community. In R. M. Teeuw (ed.), *Mapping Hazardous Terrain using remote sensing*. Geological Society of London Special Publication 283: 149–164.

Anyamba, A. and C. J. Tucker, 2005. Analysis of Sahelian vegetation dynamics using NOAA-AVHRR NDVI data from 1981–2003, *Journal of Arid Environments*, 63: 596–614.

Aplin, P., 2003. Remote sensing: base mapping. *Progress in Physical Geography*, 27: 275–283.

Asner, G. P., D. Nepstad, G. Cardinot and D. Ray, 2004. Drought stress and carbon uptake in an Amazon forest measured with spaceborne imaging spectroscopy, *Proceedings of the National Academy of Sciences of the United States of America*, 101: 6039–6044.

Ben-Dor, E., K. Patkin, A. Banin and A. Karnieli, 2002. Mapping several soil properties using DAIS-7915 hyperspectral scanner data – a case study over clayey soils in Israel. *International Journal of Remote Sensing*, 23: 1043–1062.

Bertrand, C., N. Clerbaux, A. Ipe and N. Gonzalez, 2003. Estimation of the 2002 Mount Etna eruption cloud radiative forcing from Meteosat-7 data. *Remote Sensing of Environment*, 87: 257–272.

Bjorgo, E., 2000. Using very high spatial resolution multispectral satellite.sensor imagery to monitor refugee camps. *International Journal of Remote Sensing*, 21: 611–616.

Bourguignon, A., G. Delpont, Chevrel, S. and S. Chabrillat, 2007. Detection and mapping of shrink-swell clays in SW France, using ASTER imagery. In: R. M. Teeuw, (ed.) *Mapping Hazardous Terrain Using Remote Sensing*. Geological Society of London Special Publication, 283: 117–124.

Boyd, D. S., G. M. Foody and P. C. Phipps, 2006a. Dynamics of ENSO drought events on Sabah rainforests observed by NOAA AVHRR. *International Journal of Remote Sensing*, 27: 2197–2219.

Boyd, D. S., J. A. Entwistle, A. G. Flowers, R. P. Armitage and P. C. Goldsmith, 2006b. Remote sensing the radionuclide contaminated Belarusian landscape: a potential for imaging spectrometry? *International Journal of Remote Sensing*, 27: 1865–1874.

Boyle, J., R. Stow and P. Wright, 2000. Imaging of London surface movement for structural damage management and water resource conservation. *Proceedings International Symposium on Operationalization of Remote Sensing*, Enschede, Netherlands, 16–20 August 1999, weblink: http:www.earth.esrin.esa.it/pub/ESA_DOC/gothenburg/381boyle.pdf

Bradley, A. V. and A. C. Millington, 2006. Spatial and temporal scale issues in determining biomass burning regimes in Bolivia and Peru. *International Journal of Remote Sensing*, 27: 2221–2253.

Brown, K., 2004. Increasing classification accuracy of coastal habitats using integrated airborne remote sensing. *EARSeL eProceedings*, 3: 34–42.

Carn, S. A. and C. Oppenheimer, 2000. Remote monitoring of Indonesian volcanoes using satellite data from the Internet. *International Journal of Remote Sensing* 21: 873–910.

Chafer, C. J., M. Noonan and E. Macnaught, 2004. The post-fire measurement of fire severity and intensity in the Christmas 2001 Sydney wildfires. *International Journal of Wildland Fire*, 13: 227–240.

Chuvieco, E., D. Cocero, D. Riano, P. Martin, J. Martinez-Vega, J. de la Riva and F. Perez, 2004. Combining NDVI and surface temperature for the estimation of live fuel moisture content in forest fire danger rating. *Remote Sensing of Environment*, 92: 322–331.

Clay, D. E., K. I. Kim, J. Chang, S. A. Clay and K. Dalsted, 2006. Characterizing water and nitrogen stress in corn using remote sensing. *Agronomy Journal*, 98: 579–587.

Cobby, D. M., D. C. Mason and I. J. Davenport, 2001. Image processing of airborne scanning laser altimetry data for improved river flood modelling. *ISPRS Journal of Photogrammetry and Remote Sensing*, 56(2): 121–138.

Cutter, S. L., 2003. GI Sciences, disasters and emergency management. *Transactions in GIS*, 7(4): 439–445.

Davids, C. and A. N. Tyler, 2003. Detecting contamination-induced tree stress within the Chernobyl exclusion zone, *Remote Sensing of Environment*, 85: 30–38.

Ernst, G. G. J., M. Kervyn and R. M. Teeuw in press. Advances in the remote sensing of volcanoes, eruptions and related hazards. *International Journal of Remote Sensing*.

Ferencz, C., P. Bognar, J. Lichtenberger, D. Hamar, G. Tarscai, G. Timar G. Molnar, S. Pasztor, P. Steinbach, B. Szekely, O. E., Ferencz and I. Ferencz-Arkos, 2004. Crop yield estimation by satellite remote sensing. *International Journal of Remote Sensing*, 25: 4113–4149.

Ferretti, A., F. Rocca and C. Prati, 1999. Non-uniform motion monitoring using the permanent scatterers technique, *Proceedings FRINGE '99: Second ESA International Workshop on ERS SAR interferometry*, 10–12 November 1999. Liège, Belgium, ESA, 1–6.

Francis, P. W. and R. McAllister, 1986. Volcanology from space, using Landsat Thematic Mapper data in the Central Andes. *EOS*, 67: 170–171.

Francis, P. W. and G. L. Wells, 1988, Landsat Thematic Mapper observations of debris avalanches deposits in the central Andes. Bulletin of Volcanology, 50: 258–278.

Giadia, S., D. de Groeve, D. Ehrlich and P. Soille, 2003. Information extraction from very high resolution satellite imagery over Lukole refugee camp, *Tanzania. International Journal of Remote Sensing*, 24(22): 4251–4266.

Givri, J. R., 1995. Satellite remote-sensing data on industrial hazards. *Natural Hazards*, 15: 87–90.

Gong, P., R. L. Pu, Z. Q. Li, J. Scarborough, N. Clinton, and L. M. Levien, 2006. An integrated approach to wildland fire mapping of California, USA using NOAA/AVHRR data. *Photogrammetric Engineering and Remote Sensing*, 72: 139–150.

Gudmundsson, S., H. Gudmundsson, H. Bjornsson, F. Sigmundsson, H. Rott and J. M. Cartensen, 2002. Three-dimensional glacier surface motion maps at the Gjalp eruption site, Iceland, inferred from combining InSAR and other ice displacement data. *Annals of Glaciology*, 34: 315–322.

Harris, A. J. L. and D. A. Stevenson, 1997. Thermal observations of degassing open conduits and fumaroles at Stromboli and Vulcano using remotely sensed data. *Journal of Volcanology and Geothermal Research*, 76: 175–198.

Harris, A. J. L., L. P. Flynn, D. A. Rothery, C. Oppenheimer and S. B. Sherman, 1999. Mass flux measurements at active lava lakes: implications for magma recycling. *Journal of Geophysical Research*, 104(B4): 7117–7136.

Harris, D. M. and W. I. Rose, 1983. Estimating particle sizes, concentrations, and total mass of ash in volcanic clouds using weather radar. *Journal of Geophysical Research*, 88: 10969–10983.

Harris, R, in this volume. Remote Sensing Policy. Chapter 2.

Herman, A., V. B. Kumar, P. A. Arkin and J. V. Kousky, 1997. Objectively determined 10-day African rainfall estimates created for famine early warning systems. *International Journal of Remote Sensing*, 18: 2147–2159.

Hooper, A., H. Zebker, P. Segall and B. Kampes, 2004. A new method for measuring deformation on volcanoes and other natural terrains using InSAR persistent scatterers. *Geophysical Research Letters*, 31.

Houghton, J. T., Y. Ding, D. J. Griggs, M. Nouger, P. J. van der Linden and D. Xiaosu, (eds), 2001. *Climate Change: the Scientific Basis: Contribution of Working Group 1 to the Third Assessment Report of the International Panel on Climate Change*. Cambridge University Press, Cambridge, 881 pp.

Hutchinson, C. F., 1991. Uses of satellite data for famine early warning in sub-Saharan Africa. *International Journal of Remote Sensing*, 12: 1405–1421.

Hyyppä, J., W. Wagner, M. Hollaus and H. Hyyppä, in this volume. Airborne Laser Scanning. Chapter 14.

Jia, G. J., I. C. Burke, A. F. H. Goetz, M. R. Kaufmann and B. C. Kindel, 2006, Assessing spatial patterns of forest fuel using AVIRIS data. *Remote Sensing of Environment*, 102: 318–327.

Kariuki, P. C., F. D. van der Meer and W. Siderius, 2003. Classification of soils based on engineering indices and spectral data. *International Journal of Remote Sensing*, 12: 2567–2574.

Kellndorfer, J. and K. McDonald, in this volume. Active and Passive Microwave Systems. Chapter 13.

Kelly, C., 1998. Bridging the gap: remote sensing and needs assessment – a field experience with displaced populations. *Safety Science*, 30: 123–129.

Kelmelis, J. A., L. Schwartz, C. Christian, M. Crawford and D. King, 2006. Use of geographic information in response to the Sumatra–Andaman earthquake and Indian Ocean tsunami of December 26, 2004. *Photogrammetric Engineering and Remote Sensing*, August 2006.

Kerle, N. and B. van Wyk de Vries, 2001. The 1998 debris avalanche at Casita volcano, Nicaragua – Investigation of structural deformation as the cause of slope instability using remote sensing. *Journal of Volcanology and Geothermal Research*, 105: 49–63

Kervyn M., Kervyn F., Goossens R., Rowland S. K. and Ernst G. G. J., 2007. Mapping volcanic terrain using high-resolution and 3D satellite remote sensing. In R. M. Teeuw, (ed.), *Mapping Hazardous Terrain Using Remote Sensing*. Geological Society of London Special Publication, 283: 5–30.

Kervyn, M., A. J. L. Harris, F. Belton, E. Mbede, P. Jacobs,and G. G. J. Ernst, (in press). Thermal remote sensing of the low-intensity thermal anomalies of Oldoinyo Lengai, Tanzania. *International Journal of Remote Sensing*.

Kervyn, M., J. Klaudius, J. Keller, E. Mbede, P. Jacobs and G. G. J. Ernst (in press). Remote sensing study of sector collapses and debris avalanches deposits at Oldoinyo Lengai and Kerimasi volcanoes, Tanzania. *International Journal of Remote Sensing*.

Kitzberger, T., 2002. ENSO as a forewarning tool of regional fire occurrence in northern Patagonia, Argentina. *International Journal of Wildland Fire*, 11: 33–39.

Lacasse, C., S. Karlsdottir, G. Larsen, H. Soosalu, W. I. Rose and G. G. J. Ernst, 2004. Weather radar observations of the Hekla 2000 eruption cloud, Iceland. *Bulletin of Volcanology*, 66: 457–473.

Lee, E. M., 2001. Living with natural hazards: the costs and management framework. In D. L. Higgett, and E. M. Lee, (eds), *Geomorphological Processes and Landscape Change*. Blackwell, Oxford, 237–268.

Leuven, R. S. E. W., I. Poudevigne and R. M. Teeuw, (eds.) 2002 *Application of Geographic Information Systems and Remote Sensing in River Studies*. Bakhuys Publishers, Leiden, 247 pp.

Lu, Z., E. Fielding, M. R. Patrick and C. M. Trautwein, 2003. Estimating lava volume by precision combination of multiple baseline spaceborne and airborne interfero-metric synthetic aperture radar: The 1997 eruption of Okmok volcano, Alaska. *IEEE Transactions on Geoscience and Remote Sensing*, 41: 1428–1436.

Ma, J. W., X. Z. Han, Hasibagan, C. L. Wang, Y. L. Zhang, J. Y. Tang, Z. Y. Xie and T. Deveson, 2005. Monitoring East Asian migratory locust plagues using remote sensing data and field investigations. *International Journal of Remote Sensing*, 26: 629–634.

Mantovani, F., R. Soeters and C. van Western, 1996. Remote sensing techniques for landslide studies and hazard zonation in Europe. *Geomorphology*, 15: 213–225.

Marsh, S. E., C. F. Hutchinson, E. E. Pfirman, S. A. Desrosiers and C. Vanderharten, 1994. Development of a computer work-station for famine early warning and food security, *Disasters*, 18: 117–129.

Marti, J. and G. G. J. Ernst, (eds), 2005. *Volcanoes and the Environment*. Cambridge, UK: Cambridge University Press.

Massonnet, D. and K. L. Feigl, 1998. Radar interferometry and its application to changes in the Earth's surface. *Reviews of Geophysics*, 36: 441–500.

Massonnet, D. and F. Sigmundsson, 2000. Remote sensing of volcano deformation by radar interferometry from various satellites. In: P.J. Mouginis-Mark, J.A. Crisp, and J.H. Fink, (eds), *Remote Sensing of Active Volcanism. American Geophysical Union Geophysical Monograph*, 116: 207–221.

McVicar, T. R. and D. L. B. Jupp, 1998. The current and potential operational uses of remote sensing to aid decisions on drought exceptional circumstances in Australia: a review. *Agricultural Systems*, 57: 399–468.

Metternicht, G., H. Lorenz and R. Gogu, 2005. Remote sensing of landslides: an analysis of potential contribution to geo-spatial systems for hazard assessment in mountainous environments. *Remote Sensing of Environment*, 98: 284–303.

Munich Re, 1997. *Flooding and Insurance*, Munich Re, Munich.

Mushinzimana, E., S. Munga, N. Minakawa, L. Li, C. C. Feng, L. Brian, U. Kitron, C. Schmidt, L. Beck, G. F. Zhou, A. K. Githeko and G. Y. Yan, 2006. Landscape determinants and remote sensing of anopheline mosquito larval habitats in the western Kenya highlands. *Malaria Journal*, 5: 13.

Nico, G., M. Pappalepore, G. Pasquariello, A. Recife and S. Samarelli, 2000. Comparison of SAR amplitude vs. coherence flood detection methods – a GIS application. *International Journal of Remote Sensing*, 21(8): 1619.

Nourbakhsh, I., R. Sargent, A. Wright, K. Cramer, B. McClendon, and M. Jones, 2006. Mapping disaster zones. *Nature*, 439: 787–788

Pergola, N., V. Tramutoli, F. Marchese, I. Scaffidi and T. Lacava, 2004. Improving volcanic ash cloud detection by a robust satellite technique. *Remote Sensing of Environment*, 90: 1–22.

Qin, Z. H. and M. H. Zhang, 2005. Detection of rice sheath blight for in-season disease management using multispectral remote sensing. *International Journal of Applied Earth Observation and Geoinformation*, 7: 115–128.

Riedmann, M. and M. Haynes, 2007. Developments in synthetic aperture radar interferometry for monitoring geohazards. In: R. M. Teeuw, (ed.), *Mapping Hazardous Terrain Using Remote Sensing*. Geological Society of London Special Publication, 283: 45–51.

Rose, W. I., G. J. S. Bluth and G. G. J. Ernst, 2000. Integrating retrievals of volcanic cloud characteristics from satellite remote sensors: A summary. *Philosophical Transactions of the Royal Society of London*, A358: 1585–1606.

Rose, W. I., G. J. S. Bluth, D. J. Schneider, G. G. J. Ernst, C. M. Riley, L. J. Henderson and R. G. McGimsey, 2001. Observations of volcanic clouds in their first few days of atmospheric residence: The 1992 eruptions of Crater Peak, Mount Spurr volcano, Alaska. *Journal of Geology*, 109: 677–694.

Rose, W. I., Y. Gu, I. M. Watson, T. Yu, G. J. S. Bluth, A. J. Prata, A. J. Krueger, N. Krotkov, S. Carn, M. D. Fromm, D. E. Hunton, A. A. Viggiano, T. M. Miller, J. O. Balletin, G. G. J. Ernst, J. M. Reeves, C. Wilson and B. E. Anderson, 2003. The February–March 2000 eruption of Hekla, Iceland, from a satellite perspective. In A. Robock, and C. Oppenheimer, (eds), *Volcanism and the Earth Atmosphere*. American Geophysical Union Geophysical Monograph 139.

Rowland, S. K., G. A. Smith and P. J. Mouginis-Mark, 1994. Preliminary ERS-1 observations of Alaskan and Aleutian volcanoes. *Remote Sensing of Environment*, 48: 358–369.

Sachpazi, M., C. Kontoes, N. Voulgaris, M. Laigle, G. Vougioukalakis, O. Sikioti, G. Stavrakakis, J. Bastoukas, J. Kalogera, and J. C. Lepine, 2002. Seismological and SAR signature of unrest at Nysiros caldera, Greece. *Journal of Volcanology and Geothermal Research*, 116: 19–33.

Sanders, R., F. Shaw, H. Mackay, H. Galy and M. Foote, 2005. National flood modelling for insurance purposes: using IFSAR for flood risk estimation in Europe. *Hydrology and Earth System Sciences*, 9(4): 449–456.

Schmugge, T. J., W. P. Kustas, J. C. Ritchie, T. J. Jackson and A. Rango, 2002. Remote sensing in hydrology. *Advances in Water Resources*, 25: 1367–1385.

Schneider, D. J., K. G. Dean, J. Dehn, T. P. Miller and V. Y. Kirianov, 2000. Monitoring and analyses of volcanic activity using remote sensing data at the Alaska Volcano Observatory: Case study for Kamchatka, Russia, December 1997. In P. J. Mouginis-Mark, J. A. Crisp and J. H. Fink, (eds), *Remote Sensing of Active Volcanism*. AGU Geophysical Monograph, 116: 65–85.

Schultz, G. A. and E. T. Engman, 2000. *Remote Sensing in Hydrology and Water Management*. Springer, Berlin, 483 pp.

Senay, G. B. and J. Verdin, 2003. Characterization of yield reduction in Ethiopia using a GIS-based crop water balance model, *Canadian Journal of Remote Sensing*, 29: 687–692.

Singhroy, V., 2006. Applications of radar remote sensing. In: *Remote Sensing of Earth Resources: Exploration, Extraction and Environmental Impacts*. Proceedings GRSG2006. At: http://www.grsg.org

Singhroy, V. and K. Molch, 2004. Characterizing and monitoring rockslides from SAR techniques. *Advances in Space Research*, 33: 290–295.

Sivanpillai, R., A. V. Latchinisk, K. L. Driese and V. E. Kambulin, 2006. Mapping locust habitats in River Ili Delta, Kazakhstan, using Landsat imagery, *Agricultural Ecosystems and Environment*, 117: 128–134.

Sklaver, B. A., A. Manangan, S. Bullard, A. Svanberg and T. Handzel, 2006. Rapid imagery through kite aerial photography in a complex humanitarian emergency. *International Journal of Remote Sensing*, 27(21): 4709–4714.

Smith, K. L., J. J. Colls and M. D. Steven, 2005. A facility to investigate effects of elevated soil gas concentration on vegetation. *Water, Air and Soil Pollution*, 161: 75–96.

Soeters, R. and C. J. van Westen, 1996. Slope instability recognition, analysis and zonation. In K. Turner, and R. L. Chuster (eds), *Landslide investigation and Mitigation*. Transportation Research Board, Special Report 247, National Research Council, Washington D. C., pp. 129–177.

Stevens, D., 2005. Space-based technologies for disaster management – making satellite imagery available for emergency response in developing countries. Proceedings IGARSS '05, *IEEE International*, 6: 4366–4369.

Stevens, N. F. and G. Wadge, 2004. Towards operational repeat-pass SAR interferometry at active volcanoes. *Natural Hazards*, 33: 47–76.

Stevens, N. F., V. Manville and D. W. Heron, 2002. The sensitivity of a volcanic flow model to digital elevation model accuracy: Experiments with digitised map contours and interferometric SAR at Ruapehu and Taranaki volcanoes, New Zealand. *Journal of Volcanology and Geothermal Research*, 119: 89–105.

Straatsma, M. W. and H. Middelkoop, 2006. Airborne laser scanning as a tool for lowland floodplain vegetation monitoring. *Hydrobiologia*, 565: 87–103.

Swap, R. J., H. J. Annegarn, J. T. Suttles, M. D. King, S. Platnick, J. L. Privette and R. J. Scholes, 2003. Africa burning: a thematic analysis of the Southern African Regional Science Initiative (SAFARI 2000). *Journal of Geophysical Research – Atmospheres*, 108: 8465.

Tadesse, T., J. F. Brown and M. J. Hayes, 2005. A new approach for predicting drought-related vegetation stress: integrating satellite, climate, and biophysical data over the US central plains, *ISPRS Journal of Photogrammetry and Remote Sensing*, 59: 244–253.

Temesgen, B., M. U. Mohammed and T. Korme, 2001. Natural hazard assessment using GIS and remote sensing methods, with particular reference to the landslides in the Wondogenet Area, Ethiopia. *Physics and Chemistry of the Earth, Part C: Solar, Terrestrial and Planetary Science*, 26: 665–675.

Theilen-Willige, B., 2006. Tsunami risk site selection in Greece, based on remote sensing and GIS methods. *Science of Tsunami Hazards*, 24(1): 35–48.

Tran, A. and M. Raffy, 2006. On the dynamics of dengue epidemics from large-scale information. *Theoretical Population Biology*, 69: 3–12.

United Nations, 2005. Framework for Action 2005–2015: building the resilience of nations and communities to disasters. *Proceedings World Conference on Disaster Reduction*, 18–22 January 2005, Hyogo, Japan. International Strategy for Disaster Reduction, United Nations, A/Conf.206/6.

van Aalst, M. K., 2006. The impacts of climate change on the risk of natural disasters. *Disasters*, 30(1): 5–18.

van der Werff, H. M. A., M. F. Noomen, M. van der Meijde and F. D. van der Meer, 2007. Remote sensing of onshore hydrocarbon seepage: problems and solutions. In R. M. Teeuw, (ed.), *Mapping Hazardous Terrain Using Remote Sensing*. Geological Society of London Special Publication, 283: 125–134.

Vincente-Serrano, S. M., J. M. Cuadrat-Prats and A. Romo, 2006. Early prediction of crop production using drought indices at different time-scales and remote sensing data: application in the Ebro valley (North-East Spain). *International Journal of Remote Sensing*, 27: 511–518.

Watkins, J. F. and H. A. Morrow-Jones, 1985. Small area population estimates using aerial photography imagery to monitor refugee camps. *Photogrammetric Engineering and Remote Sensing*, 51: 1933–1935.

West, J. S., C. Bravo, R. Oberti, D. Lemaire, D. Moshou and H. A. McCartney, 2003. The potential of optical canopy measurement for targeted control of field crop disease. *Annual Review of Phytopathology*, 41: 593–614.

Wooster, M. J., T. Kaneko, S. Nakada and H. Shimizu, 2000. Discrimination of lava dome activity styles using satellite-derived thermal structures. *Journal of Volcanology and Geothermal Research*, 102(1): 97–118.

Wright, R., D. A. Rothery, S. Blake and D. C. Pieri, 2000. Improved remote sensing estimates of lava flow cooling: A case study of the 1991–1993 Mount Etna eruption. *Journal of Geophysical Research*, 105: 23681–23694.

Wright, R., L. Flynn, H. Garbeil, A. Harris and E. Pilger, 2002. Automated volcanic eruption detection using MODIS. *Remote Sensing of Environment*, 82: 135–155.

Yang, G. J., P. Vounastou, X. N. Zhou, J. Utzinger and M. Tanner, 2005. A review of geographic information system and remote sensing with applications to the epidemiology and control of schistosomiasis in China, *Acta Tropica*, 96: 117–129.

Zebker, H. A., F. Amelung, and S. Jonsson, 2000. Remote sensing of volcano surface and internal processes using radar interferometry. In P.J. Mouginis-Mark, J.A. Crisp, and J.H. Fink, (eds), *Remote Sensing of Active Volcanism*. AGU Geophysical Monograph 116: 179–205.

Zeil, P., 2003. Management and prevention of natural disasters – what are the requirements for the effective application of remote sensing? *The International Archives of the Photogrammetry, Remote Sensing and Spatial Information Sciences*, 34(6/W6): 54–56.

Remote Sensing of Land Cover Change

Timothy A. Warner, Abdullah Almutairi,
and Jong Yeol Lee

Keywords: change detection, land use conversion, land use modification, object-oriented analysis, radiometric correction, accuracy assessment.

INTRODUCTION

Environmental change, and especially change that can be related to anthropogenic causes, is recognized as one of the pre-eminent issues of our time (IPCC 2007, Turner et al. 2007). Remote sensing is an important tool for documenting relatively recent land changes, for monitoring current changes, and for calibrating models for predicting future changes (Rindfuss et al. 2004). Remote sensing offers many distinctive strengths for the study of land change:

1 The data are objective, and generally consistent within each scene.
2 Remotely sensed data are spatially explicit, with a defined and relatively uniform spatial scale.
3 Remotely sensed data are spatially comprehensive within the imaged area, although in practice clouds and shadow may limit this characteristic.
4 Each image is acquired in a very short period of time, although images that are composited or mosaicked may integrate information over longer periods.
5 Data can be acquired rapidly over large areas. This characteristic is vital for regional to global scale studies.

6 Data collection is repeatable, a crucial factor for monitoring change. One of the reasons why the Landsat data set has proved so valuable is the very successful plan to acquire repeat seasonal images across the global land masses (Goward et al., in this volume).
7 Image archives represent a source of raw data that can be used to address environmental change issues that may only be identified in the future. Thus, remotely sensed data represent a type of insurance against future information needs.
8 Remotely sensed images are highly persuasive, and can be an effective tool for communicating scientific results to the public (Sohl 1999).

The importance of mapping change is implicit in most of the individual remote sensing applications discussed in earlier chapters in this book. In this chapter, however, we focus on change as a specific application in itself, particularly within the context of land cover change. However, many of the methods and issues discussed are equally applicable to other applications, for example, changes in soils, ice or geomorphology. A discussion of land cover change as a separate chapter in this volume is warranted due to the importance of environmental change, the general consistency in approaches and themes for change analysis across a wide range of applications (Singh 1989, Lunetta and Elvidge 1998, Yuan et al. 1998, Mas 1999, Coppin et al. 2004, Lu et al. 2004, Rogan and Chen 2004, Radke et al. 2005, Wulder and Franklin 2007), and the vast literature on the topic.

This chapter is divided in eight major sections. Following this general introduction, we provide a short discussion of how land cover change has been conceptualized in remote sensing. The subsequent three sections focus on issues in using remote sensing to map change: project design, data selection, and pre-processing. The sixth section, on change analysis methods, provides a survey of the major approaches. Accuracy assessment is addressed in section seven, and the chapter concludes with a discussion of future challenges.

TYPES OF CHANGE

The way change is conceptualized has important implications for the methods chosen for subsequent analysis. Three broad attributes have been used to define change: the categorical nature of the classes, the temporal properties, and the spatial properties.

Nature of the classes

A common approach in conceptualizing change is to focus on whether the classes are discrete or continuous (Coppin et al. 2004). *Conversion* is defined as a change from one discrete class or category to another. For example, a forest stand may be converted into an agricultural field. *Modification* implies a change along a continuum within a single category, such as an increase in biomass in an agricultural crop.

Temporal properties of change

If the change is conceptualized as a shift from one relatively stable state to another, then a single bi-temporal pair of images can be used to map the change. On the other hand, if the change is temporally continuous, and possibly changing in speed, then it would seem that a focus on the change trajectory (Lawrence and Ripple 1999, 2000) may be more appropriate than simply mapping change through a single interval. For predicting future changes, it may be particularly important to understand how the rate of change itself is changing. Monitoring rate of change implies a need for many images.

Spatial properties of change

With the growing interest in the general field of object-based image analysis, in which groups of pixels rather than individual pixels are the focus of analysis, an explicit consideration of the local spatial context of change is rapidly gaining attention (Gamanya et al. 2008).

PROJECT DESIGN

Successful identification of land surface change requires that: (1) the spectral data must be suitable proxies for the change, and (2) the changes that are detected in the change analysis must represent a true change to the scene, and preferably isolate the particular change of interest.

The first step in a remote sensing change analysis project design should be to define the aim of the project (Lunetta 1998). Consideration should be given to the type of change of interest, as discussed in the previous section. After the problem definition step, the desired products and the specifications of those products can be developed, and then the processing methods and the data requirements necessary to achieve the project aims can in turn be identified (Phinn et al. 2000, Wulder and Franklin 2007). Lunetta (1998) recommends separately specifying the required spectral, spatial and temporal resolutions (Warner et al., in this volume) of the data. The desired data properties are then matched against the available data, as well as the costs of both the data and the associated analysis. Exploratory analysis and pilot studies are often used to refine the project methodology. Other important issues in project design are planning the accuracy assessment and communication of results.

In his description of project design, Lunetta (1998) emphasizes the importance of iteration in developing and refining each step of the project based on the constraints, including those of time, money, quality and availability of imagery, and the need for ancillary data. In reality, few projects are likely to proceed in a linear fashion through the steps outlined above.

In many cases, the remote sensing-based change analysis is a component of a larger undertaking. In such cases, it is important to develop the change analysis project design in the context of the overall project. For example, incorporating remote sensing and social science data (Crews and Walsh, in this volume) poses a range of special challenges, including how to link pixels to people and land use, how to overcome the common mismatch in spatial and temporal properties between the image data and the social science data, and how to track and manage error and uncertainty (Rindfuss et al. 2004).

Aspinall (2002) divides the infrastructure necessary to monitor change into the two broad categories of the database and the processing 'toolboxes.' The database comprises both the input data and the derived results, and includes: (1) imagery, (2) supporting maps, (3) field data, (4) derived land cover data sets of individual dates, (5) estimates of uncertainty of the land-cover data, and (6) the output change analysis information. The processing tool boxes include (1) image processing, (2) uncertainty analysis, and (3) change analysis and predictive modeling (Aspinall 2002).

DATA SELECTION

General issues

The issue of data selection is dealt with in detail in the introductory chapter to this volume (Warner et al., in this volume). Within that chapter it was emphasized that it is important to select imagery with appropriate spectral, radiometric, spatial, and temporal properties to allow differentiation of the features of interest. For change detection studies, the temporal properties are particularly important. In the case of a simple bi-temporal analysis, appropriate before and after images should be acquired as close as possible to anniversary dates (i.e., the same day of the year). For more complex studies, involving analysis of seasonal, continuous or episodic change, multiple images may be necessary, greatly increasing the challenge in finding available data that capture the full complexity of the phenomenon of interest. For coarse spatial resolution multi-temporal composites, the period of time encompassed by each image is important in evaluating the potential for discriminating temporal aspects of change, such as the date of spring greenup (Schwartz et al. 2002).

Combining data from different sensors

A special concern for data selection in change detection analysis is the similarity between the successive images to ensure that differences identified from the images are a result of the differences in the scene, and not differences in the image properties. Ideally, the images should all be acquired by the same sensor, from the same viewing geometry, and at the same time of day (Khorram et al. 1999). However, in practice, either because of limited data availability, or the study period exceeds the lifespan of any one sensor, it is often necessary to use data from multiple sensors (Serra et al. 2003).

When using data from different sensors, it is important to consider differences in the spectral, spatial, and radiometric properties of the images. If the spatial properties of the two sensors differ, the higher spatial resolution image should be spatially aggregated to match the lower spatial resolution image (Serra et al. 2003). If the spectral properties of the sensors differ, only the spectral bands that are in common should be used, and the bands should be normalized to the narrower of the two radiometric ranges. Teillet et al. (1997) caution that even slight changes in the spectral range of the bands will result in derived vegetation index values that are notably different. Thus, for some applications, such as monitoring vegetation index changes, even spectral bands that have similar, but not identical, spectral responses cannot be compared directly. On the other hand, Almutairi (2000) found that green, red and near infrared Thematic Mapper (TM) bands could, with little change in accuracy, be substituted for the equivalent MSS bands for the five change methods investigated in a case study in Las Vegas, Nevada.

PREPROCESSING

Two general categories of pre-processing are required: geometric co-registration and spectral radiometric normalization.

Geometric co-registration

Clearly, for any pixel-by-pixel analysis of change, it is necessary for the images to be co-registered spatially, either to a common map projection, or to a common, arbitrary projection. Aerial images and non-nadir views from satellite images tend to have strong topographic-induced distortions due to parallax. For such images, orthocorrection is particularly important, unless the repeat acquisition is planned so that the acquisition locations are precisely replicated (Coulter and Stow 2008, Stow et al. 2008).

Although it has been suggested that a co-registration error of 0.5–1.0 pixels is acceptable, Townshend et al. (1992) found that with a 1.0 pixel error, four of seven test sites evaluated had errors equivalent to 50% of the actual Normalized Difference Vegetation Index (NDVI) change for those locations. They found that, in densely vegetated regions, in order to reduce error in change in NDVI to less than 10%, it was necessary to reduce misregistration to less than 0.2 pixels. This recommendation of a minimum 0.2 pixel co-registration accuracy was supported by subsequent research by Dai and Khorram (1998), although this level of accuracy may be hard to achieve in practice.

The effect of misregistration is dependent on the heterogeneity of the landscape studied. The more spatially heterogeneous the landscape, the greater the number (Bruzzone and Cossu 2003) and total area (Townshend et al. 1992, Dai and Khorram 1998) of false changes that will likely be generated by registration errors. In contrast, false changes are likely to be concentrated along the edges of locally homogeneous image regions, generally resulting in spurious change features that are elongate in shape.

Many methods have been developed for reducing the effects of misregistration error. Most methods attempt to minimize false identification of change by evaluating the local context of change. Amongst the methods suggested are eroding the outer edges of regions identified as changed (Serra et al. 2003), estimating the potential for error artifacts from the local image gradient (Stow 1999), applying a local transform to suppress change observations (Gong et al. 1992, Prenzel and Treitz 2006), applying wavelet transforms to model scale-dependent variability (Carvalho et al. 2001), comparing information in spectral bands that show the change of interest with those that do not show the change (Bruzzone and Serpico 1997), choosing a method least affected by misregistration (Sundaresan et al. 2007) and identifying an optimal smoothing or spatial resolution (Gong et al. 1992, Wang and Ellis 2005).

Radiometric normalization

Unlike geometric co-registration, which is always necessary, radiometric normalization is only needed for some change detection methods. It is therefore crucial to consider the change methods that will be applied before choosing a radiometric normalization method.

Radiometric normalization is not needed for classification in general, as long as the spectral training data are obtained from the image that will be classified (Song et al. 2001). Furthermore, radiometric normalization is not needed for change detection based on linear operations, such as simple image differencing (Song et al. 2001) multi-temporal Kauth and Thomas (1976) transformations, and principal component analysis (PCA) (Collins and Woodcock 1996), as long as empirical data are used to interpret the change images, and the atmospheric effects are dominantly additive.

Radiometric spectral normalization is important, however, for spectral change methods that use *a priori* image properties to identify change (such as image differencing in which a zero digital number (DN) change is assumed to represent no change), or methods that involve ratios between bands (such as NDVI, whether for a single date (Teillet et al. 1997), or for a comparison of multiple dates (Song et al. 2001)). For example, for ratio calculations, the band-dependent additive effect of

aerosol scattering will distort the derived ratio values, if not removed prior to calculation of the ratio (Teillet et al. 1997). However, it is important to recognize that the overall effect of the atmosphere is to reduce the signal-to-noise ratio of the image, and that the information lost cannot be recovered (Song et al. 2001).

Spectral radiometric normalization approaches can be divided into two groups: absolute radiometric correction and relative normalization (Lu et al. 2002; see also Liang, in this volume).

Absolute radiometric correction

Image DNs are arbitrarily-scaled values that only convey information about relative radiance values in the context of the original image. However, if sensor calibration data are available, the DNs can be converted to radiance. The at-sensor radiance can then be scaled to planetary reflectance (also known as apparent, or top of the atmosphere, reflectance), which is a combination of the atmospheric and surface reflectance (NASA 2008).

Reliable and precise radiative transfer models have been developed for calculating the atmospheric transmittance and path radiance. Examples include MODTRAN (Berk et al. 1998) and 6S (Vermote et al. 1997). These physical models require knowledge of the atmospheric conditions which can be obtained from weather balloons, assumed from a small number of default profiles, or estimated from the image itself, using programs such as Atmospheric CORrection Now (ACORN) (Analytical Imaging and Geophysics 2001) and the Fast Line-of-sight Atmospheric Analysis of Spectral Hypercubes (FLAASH) (Alder-Golden et al. 1999).

Since approaches based on radiative transfer models can be difficult to apply in practice, more simple alternative methods have been developed that are based mostly on image properties. The dark object subtraction (DOS) method assumes that there exists at least somewhere in the image a region with close to a zero reflectance (Chavez 1988). If this is true, and there is no additive path radiance, the sensor should detect a least some pixels with zero radiance. Based on this logic, the lowest radiance in any band measured in the scene is an estimate of the path radiance for that band. A number of methods have been built around the DOS concept, including a method that uses the cosine of the solar zenith angle, Tz, (COST, Chavez 1996). Another approach uses 'dense, dark vegetation' (DDV) pixels to estimate haze, because such pixels are observed to have blue and red reflectance that is typically highly correlated with the short wave infared, a region with little expected path radiance (Kaufman et al. 1997, Song and Woodcock 2003).

Lu et al. (2002) found that the relatively simple COST method provided excellent absolute radiometric correction, outperforming more complex methods. Song et al. (2001) found that the best absolute correction approach was DOS with transmittance calculated from a Rayleigh scattering model, although DOS alone was almost as effective. The COST and DDV methods were much less effective, apparently because of incorrect assumptions regarding the true reflectance of dark objects (Song et al. 2001).

An alternative to the conversion of DNs to radiance, and subsequent estimation of solar illumination and reflectance, is to measure the spectral reflectance of representative dark and bright surfaces in the image with a field spectrometer. The measured spectral reflectance is then convolved with sensor spectral band passes, and a regression relationship used to determine the normalization. This approach, termed *empirical line fitting* (Perry et al. 2000), has the benefit of providing a direct link from DN to reflectance. The challenge, however, is to find sufficiently large, homogeneous surfaces that can reliably be characterized with a field spectrometer, which typically have a small field of view (Ben-Dor et al. 1994). In addition, the field measurements should ideally be made at the time of the image acquisition, which clearly is not possible for archived imagery.

Relative normalization

For most change detection algorithms it is only necessary to convert the images of the different dates to the same radiometric scale (Tokola et al. 1999, Yang and Lo 2000, Song et al. 2001), and not necessarily to an absolute scale (Liu et al. 2007). The key to relative normalization methods is to find pixels that can be assumed to have the same spectral reflectance in each of the images. Such pixels are assumed to represent 'pseudo-invariant features' (PIFs) (Schott et al. 1988). PIFs should ideally be at least several pixels in size, and have a relatively homogenous make-up. It is also important that PIFs represent both the dark and bright end of the range of values in each band. Examples of surfaces that have been used as PIFs include deep, clear water, paved surfaces, bare soil and mature forests.

Instead of selecting a number of small PIFs, the entire images can be used to determine the normalization (Yuan and Elvidge 1996, Yang and Lo 2000). Using the entire image is predicated on the assumption that the areas of change are small within the entire image, or that the number of pixels with increased radiance is balanced by those with decreased radiance. A more attractive variant is to exclude pixels that can be observed to have changed, either by manually masking the image, or more commonly, through an initial spectral analysis (Schott et al. 1988, Tokola et al. 1999,

Yang and Lo 2000). For example, in bispectral scattergrams, the unchanged pixels may be identified as a 'central ridge' of values with a high frequency of occurrence (Elvidge et al. 1995, Yuan and Elvidge 1998, Song et al. 2001).

Time-series data normalization

Analysis of change in long time series data, with multiple images, poses a particular challenge for radiometric normalization and quality control. Although the individual images are usually pre-processed to minimize cloud contamination, sensor drift and other artifacts, additional cleaning of the data may be required (Hall-Beyer 2003). For example, Lunetta et al. (2006) applied Fourier analysis to remove anomalous data in Moderate Resolution Imaging Spectroradiometer (MODIS) NDVI time series data, and Galford et al. (2008) used a wavelet analysis to smooth MODIS time series data.

CHANGE ANALYSIS METHODS

In this section, change analysis methods are briefly reviewed and summarized. Although they are discussed in groups, it should be emphasized that this categorization is one of convenience, as there are many different ways to produce such a taxonomy. Figure 33.1 shows a number of change analysis methods applied to a pair of Landsat TM and Enhanced Thematic Mapper Plus (ETM+) images of Las Vegas, Nevada, the city with the fastest growth in the 1990s in the United States (Almutairi 2004, Warner 2005).

The simplest type of change analysis is the identification of the presence of change. A more complex analysis is required to quantify the type of change, whether conversion between classes or modification within a class. Another important distinction is whether the aim is to quantify change, or merely to visualize the change.

Spectral change methods

Spectral change methods are defined here as those methods that are designed to produce an image representing the spectral properties of change (Yuan et al. 1998). Although the original spectral bands are used in many spectral change methods, often a spectral transformation, such as NDVI or the Tasseled Cap (Kauth and Thomas 1976), is applied prior to the spectral change analysis, in order to facilitate the interpretation of the spectral change image (Collins and Woodcock 1996). Thus, many spectral change methods could also be characterized as a type of biophysical modeling.

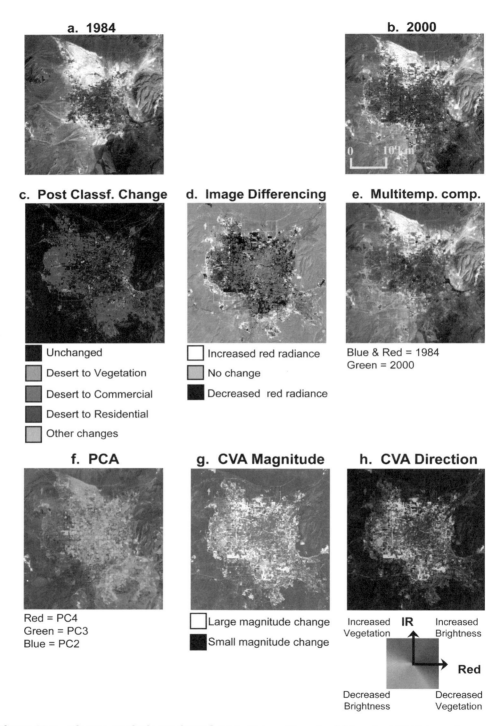

Figure 33.1   Change analysis products for Las Vegas, Nevada. (a) Landsat TM false color composite image, 1986. (b) Landsat ETM+ image, 2000. (c) Post-classification change detection, with selected class transitions shown. (d) Band 3 image differencing. (e) Multi-temporal composite. (f) Multi-date principal component analysis. (g) Change vector analysis, magnitude of change image. (h) Change vector analysis, direction of change image. (See the color plate section of this volume for a color version of this figure).

Spectral change methods are often used as a pre-processing step, prior to applying some type of image classification. For example, thresholding might be applied to the spectral change output data to separate change from no-change pixels. Identifying the appropriate value for this change/no-change threshold is in fact a recurrent concern in spectral change analysis (Yuan et al. 1998). In many cases an assumption is made that change pixels are outliers in the spectral change image, and an arbitrary statistical value, such as plus or minus two standard deviations from the mean, is used as the threshold. Alternatively, since the problem is in essence one of classification, training samples can be used to characterize the change and no-change classes, and also to determine the optimal threshold that minimizes errors (Warner 2005), or provides the best trade-off between missing real change and incorrectly labeling areas as changed (Morisette and Khorram 2000).

Multi-temporal false color composite

Possibly the simplest spectral change method is to display one band from each of two images as a false color composite. For example, the red band from an early image could be displayed in blue and red, and the red band from a subsequent date in green (Figure 33.1e). In the resulting multi-temporal false color composite, unchanged areas are depicted in gray tones, areas of decreased red radiance in magenta, and increased red radiance in green. The method can be extended to a maximum of three dates (Hayes and Sader 2001).

Band differencing and related methods

Another straightforward and common approach is to calculate the *difference* in image DNs between two different dates for one or more spectral bands. In the resulting difference image, the unchanged pixels cluster around zero DN, and the changed pixels have relatively large positive or negative values (Figure 33.1d). By definition, the differencing operation only identifies change, and suppresses information about the original spectral value of the pixels. Therefore it has sometimes been found effective to combine the difference values with the original date DN values as a multiband vector data set, which makes the method equivalent to composite analysis (Cohen and Fiorella 1998). Differencing can be applied to a single image band, or multiple bands. Ridd and Liu (1998) combined multiple band difference values using a *chi-square transformation*, to produce a single, integrated Mahalanobis distance. The advantage of this approach is that the output image can be interpreted as a chi-square statistic, with degrees of freedom equal to the number of bands (Ridd and Liu 1998).

Not surprisingly, it is almost uniformly found that radiometric normalization or correction is necessary prior to image differencing operations (Yuan and Elvidge 1998), unless the results are interpreted with the aid of calibration data. Alternatively, instead of normalization followed by differencing, a *regression* can be calculated directly between the two images, and the residuals used as a measure of change (Ridd and Liu 1998). However, the regression may be biased unless the relationship is developed using only non-changed pixels. *Neighborhood correlation images* (Im et al. 2008) offers yet another interesting alternative, in which change is interpreted from the correlation values generated from pixels over local regions, such as a 3×3 moving window.

Multi-temporal ratios

Ratioing the DN values for individual bands of different dates produces an image conceptually similar to image differencing (Yuan and Elvidge 1998). One important difference, though, is that with ratioing there is an implicit assumption that the significance of the change in DN values is proportional to the original DN value (Liu et al. 2004). The multi-temporal ratio can be expressed as a simple ratio (Date2−Date1), or a normalized ratio (Date2−Date1)/(Date2+Date1) in order to provide a bounded range, with a no-change value in the center of the range.

Principal component analysis (PCA)

The principal component analysis (PCA) transformation generates new, orthogonal (uncorrelated) bands (Richards 1984, Warner 1999, Zhao and Maclean 2000). The PCA transformation is scene-specific, and therefore interpreting the resulting images can be a challenge (Figure 33.1f). An added complexity is that, for multispectral–multi-temporal data, PC bands generally combine both the spectral variations of each individual date, and the temporal change between them. Nevertheless, it has usually been found that the first PC band is related to the overall brightness pattern in the images, and lower order PCs can be interpreted at least in part in terms of change components (Richards 1984). Alternatively, Gong (1993) suggested applying PCA to the results of image differencing, a method that appears to have potential for highlighting just the temporal spectral change.

Multi-temporal Kauth Thomas (MKT)

A particularly rich literature has developed around the Kauth–Thomas (KT) linear transformations of Landsat MSS (Kauth and Thomas 1976) and TM (Crist and Cicone 1984) data that compress the original spectral bands into a small number of transformed bands that can be interpreted as biophysical components, principally Brightness, Greenness, and (for Landsat TM) Wetness. This transformation was extended to incorporate multi-temporal data by Collins and Woodcock (1994, 1996), and has been used in more than 20 change analysis studies (Rogan et al. 2008).

Change vector analysis (CVA)

An approach that is noteworthy for separating the type of change from the amount of change, is change vector analysis (CVA) (Malila 1980). The name of this method points to its underlying conceptualization of each pixel as a vector, with the origin of the vector at the pixel's location in the spectral space at date 1, and the terminus at date 2. The total amount of change is indicated by the vector magnitude (Figure 33.1g), and the type of change by the vector direction (Figure 33.1h).

CVA was originally developed by Malila (1980) using two bands of KT-transformed data. With just two bands, a single angle can be used to specify and interpret the change direction. Subsequent work has extended this method to three bands of KT data using spherical angles (Cohen and Fiorella 1998, Allen and Kupfer 2000). Virag and Colwell (1987) generalized the method further, by modifying the direction component to specify only the sector of change, in other words whether the DN value increased or decreased for each band, thus allowing the differentiation of only broad direction classes.

Hyperspherical direction cosines provide an alternative method of specifying the vector directions, without constraint on the dimensionality of the spectral space (Chen et al. 2003, Warner 2005). Brovolo and Bruzzone (2007) have developed a strong theoretical framework for analyzing change distributions in the polar domain, and shown the relevance of their approach for rigorous analysis of CVA data.

Classification based methods

Post-classification comparison

An independent classification of two images, and subsequent GIS overlay operation, can be used to produce a comprehensive matrix of 'from-to' changes between the two images (Yuan et al. 2005). One benefit of the post-classification change detection approach is that radiometric normalization is not needed, unless a single set of training data is used for classification of both images. Indeed, if there are phenological or other weather-related differences between the images, post-classification change detection would appear to be a particularly good choice of change analysis method, although such data should still be used with caution

(Rogan and Chen 2004). Figure 33.1c shows an example of post-classification comparison in which only the most common change transitions are shown.

The individual classifications can draw on any supervised or unsupervised method, although it is important to maximize the classification accuracy, since the overlay operation tends to compound the errors of the individual dates. If the errors are not correlated between the two classifications (of the same type and at the same locations), the expected change transition error will be at a maximum of the likely range in error, namely the product of the errors of the individual dates (van Oort 2007). However, to the extent that the errors are correlated between the classifications, the expected error is reduced compared to the uncorrelated case, with a minimum error equivalent to the least accurate of the two individual classifications (van Oort 2007).

Composite analysis

An alternative to post-classification analysis is to apply a single classification directly to the multi-temporal composite of both images (Cohen and Fiorella 1998, Lunetta 1998). The challenge with such an approach is that it is hard to identify training areas in a supervised classification for the c^2 change transitions generated by the c classes present. Taking an unsupervised classification approach does not appear to solve the problem either, as change classes may comprise such a small area that they do not generate unique clusters. On the other hand, composite analysis provides a more direct method of identifying change and change sequences (Stow et al. 2008).

Multi-date spectral mixture analysis

Instead of associating a single cover class with each pixel in the individual images, multi-date spectral mixture analysis estimates the proportion of each of the spectral endmembers (Adams et al. 1995) that are assumed to mix linearly to produce the integrated pixel radiance at each date. Multi-date spectral mixture analysis can be conceptualized as a spectral transformation, and therefore could be included under spectral change methods. However, the method is included in this section because the endmembers are typically constrained to sum to unity, thus implying a classification context. The benefit of multi-date spectral mixture analysis is that it potentially can be used to identify conversions between classes, as well as modifications within classes. Rogan et al. (2008) provide a summary table listing over 20 multi-date spectral mixture studies that range from deforestation to urban change and snow cover monitoring applications.

Temporal trajectory methods

The methods discussed so far have focused primarily on bi-temporal image pairs. These methods can in general be adapted to longer time-series data by examining successive image pairs. Such an approach can be useful for looking at how the rate of change varies over time. However, a number of approaches allow a more direct analysis of large layer-stacks of images. For example, PCA can be applied directly to an extended time series of images, and is especially straightforward to interpret if each date has only one associated image spectral feature (e.g., NDVI). This approach has been used to isolate geographical and temporal phenological patterns in vegetation (Hall-Beyer 2003). Lambin and Strahler (1994) used CVA magnitude of change information applied to vegetation index, surface temperature and spatial structure variables to identify subtle temporal changes in West Africa.

A useful theoretical framework for studying landscape evolution over time is provided by paneled metrics, in which the history of a pixel or group of pixels is summarized over multiple images (Crews-Meyer 2002). Temporal trajectories have also been identified through unsupervised classification of time series data. In a study of the recovery of vegetation following the 1980 eruption of Mount St. Helens, Lawrence and Ripple (1999) first estimated vegetation cover from each of eight Landsat TM images from the period 1980–1995. These data were then clustered using unsupervised classification, and the resulting temporal trends interpreted in terms of curves representing different patterns of vegetation recovery (Lawrence and Ripple 2000).

In a conceptually related study, Kennedy et al. (2007) developed four theoretical models of vegetation patterns to describe forest response to disturbance. The TM band 5 reflectance for each pixel in a series of 18 Landsat images covering a 20-year period was analyzed in relation to the theoretical vegetation trajectories, and the year of disturbance, intensity of disturbance, and rate of recovery, estimated (Kennedy et al. 2007).

Object oriented methods

Any of the change analysis methods already discussed, including spectral, classification, and temporal trajectory-based approaches, can potentially be applied to object-oriented change detection. Object-oriented analysis is often presented as being particularly relevant for fine spatial resolution imagery. However, the approach is potentially beneficial whatever the scale, as long as the objects of interest in the image are represented by multiple pixels (i.e., the Woodcock and Strahler (1987) H-resolution case). The benefits of an object-oriented approach for change detection are similar to those of object-oriented classification (Gamanya et al. 2008): a focus on image objects of variable size and shape, rather than arbitrary, fixed-size pixels, a data format that facilitates integration with geographic information systems (GIS), a reduction in fine scale noise in the classification (Im et al. 2008), and a method to overcome the high within-class variability typically associated with fine spatial resolution images.

There are, however, some major challenges in applying an object-oriented approach to change detection. For example, there are at least three options for generating the image segmentation: applying an external segmentation, for example from a cadastral database (Walter 2004), segmenting each image individually, and then overlaying the two segmentations (Gamanya et al. 2008), or generating a single segmentation from the combined data set (Desclée et al. 2006, Stow et al. 2008). Segmentation is a notoriously challenging task because of the wide variety of segmentations can be generated for any one data set (Kim et al. 2008). In general, it is important that the scale of the objects identified provide a balance between too fine a segmentation, which focuses on minor changes in individual pixels, and one that is too coarse, and therefore suppresses meaningful change (Gamanya et al. 2008).

A useful framework for an object based analysis of change is provided by Gamanya et al. (2008), who suggest that the types of spatial change can be categorized in the following groups:

1 existence: stability, appearance, or disappearance.
2 size and shape: expansion, contraction, or deformation.
3 location: displacement or rotation (generally resulting only from misregistration).
4 fragmentation: contraction plus fragmentation.
5 combinations of the above (typical of most real phenomena).

Comparison of methods

The previous sections have detailed a tremendous variety of change detection methods. This raises the obvious question: which methods are best? Unfortunately, there is no definitive answer to this question (Ridd and Liu 1998, Coppin et al. 2004), and many studies have concluded that a single best method cannot be identified.

The first problem in comparing methods is how to define what is meant by the 'best method.' For example, some methods may be slightly less accurate, but have the benefit of ease of implementation, or provide particularly clear visualizations of change. Thus, Sohl (1999) noted that although a modified image differencing provided the greatest overall accuracy, change vector analysis provided rich, qualitative information making it a useful and complementary method.

The second, and more serious problem, is that individual studies have reached opposite conclusions (Coppin et al. 2004). To take just one example, Singh (1989) found image differencing to be more accurate than PCA, and post-classification to be the least accurate. Mas (1999), on the other hand, found post-classification the most accurate, followed by image differencing and PCA. Almutairi (2004) suggested that the explanation for this confusion may lie in the fact that nearly every previous comparative study has been based on just one study site, and that scene properties may be important in determining the relative accuracies of different change detection methods. Thus, in a series of experiments with simulated data, Almutairi (2004) found for example that image differencing produced relatively high accuracies for classes with very good separability, but for classes with poor spectral separability, image differencing was found to have the lowest accuracy.

One way of resolving the complexity of understanding the relative merits of different change analysis methods is through theoretical analyses. Castelli et al. (1998) demonstrated that post-classification comparison is inherently more accurate than image differencing for identifying change, based on an analysis of the probability density functions of classes in the original spectral bands, and the differenced bands. Liu et al. (2004) examined the change/no-change boundary of a range of spectral change methods in bi-temporal space, and found that the relative potential accuracy was first image regression, followed by PCA, image differencing, and lastly, image ratios.

ACCURACY ASSESSMENT

Stehman and Foody (in this volume) provide a comprehensive overview of accuracy assessment issues, including a discussion on change analysis accuracy assessment. Therefore, in this section, only a few major issues will be highlighted. Additional reviews, focused particularly on change detection accuracy assessment, can be found in Biging et al. (1998), Khorram et al. (1999), and Rogan and Miller (2007).

Van Oort (2007) points out that there are three basic types of error matrices reported for change detection products:

1   accuracy assessment of the classification of each independent date.

2   accuracy assessment of the binary change/no-change classification.
3   a complete change transition error matrix, comprising all possible change and no-change classes and their potential confusion with other classes. This matrix has $c^2 \times c^2$ cells, where c is the number of classes in the single date classification.

Although the complete change transition error matrix provides the most useful information, it is only occasionally provided in change studies, because the challenge of collecting sufficient data for so many categories may be prohibitive, and the change categories are often rare, making simple random sampling very inefficient. This latter problem may be addressed by stratification (Stehman and Foody, in this volume), and even potentially by applying different sampling strategies for the change and no-change categories (Khorram et al. 1999). An additional problem with change detection accuracy assessment is that obtaining ground reference data on past land cover can be a problem, although sometimes historical aerial imagery is available, and can be substituted (Khorram et al. 1999).

Error in a change map is caused by both change classification error and spatial misregistration error. As discussed above in the section on georeferencing, the error distribution is likely to vary spatially, with for example, misregistration error concentrated on the edges of relatively homogeneous regions. Salas et al. (2003) found that the perimeter to area ratio of change features provides a useful measure for predicting whether the identified change is an artifact of misregistration.

FUTURE CHALLENGES

It is apparent that despite the vibrant growth in change detection research, there are a number of challenges that still need to be addressed. One of those problems stems from the growing proliferation of the number of sensors with varying spatial, spectral, radiometric, and temporal properties (Warner et al., in this volume). This large range of sensor characteristics suggests new challenges for data continuity, a topic of special importance in change detection. As has been discussed, great care should be taken in combining data from different sensors in change detection studies. To address this concern, new change techniques are being developed that are particularly robust for such situations. For example, Wulder et al. (2008) differenced rank-ordered DN values to identify stand-replacing forest disturbance using combinations of Landsat ETM+, TM, ASTER, and SPOT data. Mercier et al. (2008) developed a method for modeling the relationship between unchanged areas in a comparison of synthetic aperture radar (SAR) and optical imagery. This relationship is then used to predict what local image values would have been, had the sensors been the same.

Advances in image classification open new opportunities in change detection. Object-oriented change detection methods will likely be increasingly important in the future, especially as methods are developed that are able to handle spatial error (Gautama et al. 2006). Potential research in combining fuzzy change detection (Gong 1993, Abuelgasim et al. 1999) and object-oriented methods (Cheng 2002) seems particularly promising.

All remote sensing products have associated error. Summary error statistics, though useful, can be a challenge to interpret, and do not necessarily carry information about the spatial context of the error. One potentially rich area of future research is the development of novel approaches for characterizing and visualizing change detection uncertainty and error.

Perhaps most important of all is the challenge of the development of theoretical and empirical comparisons of change detection methods across a wide range of data sets. The aim of such research should be to provide a comprehensive understanding of the advantages and disadvantages of the current, almost bewildering variety of methods available. In the ideal situation, the user would be able to go through a decision tree and perhaps calculate some simple image metrics, to determine an optimal change detection method.

ACKNOWLEDGMENTS

The authors would like to thank M. Duane Nellis, Steve Yool, John Rogan, and Doug Stow for very helpful suggestions for improving an initial draft of this chapter.

REFERENCES

Abuelgasim, A. A., W. D. Ross, S. Gopal, and C. E. Woodcock, 1999. Change detection using adaptive fuzzy neural networks environmental damage assessment after the Gulf War. *Remote Sensing of Environment*, 70(2): 208–223. DOI:10.1016/S0034-4257(99)00039-5.

Adams, J. B., D. E. Sabol, V. Kapos, R. Almeida-Filho, D. A. Roberts, M. O. Smith, and A. R. Gillespie, 1995. Classification of multispectral images based on fractions of endmembers: application to land-cover change in the Brazilian Amazon. *Remote Sensing of Environment*, 52(2): 137–154.

Alder-Golden, S. M., M. W. Matthew, L. S. Bernstein, R. Y. Levine, A. Berk, S. C. Richtsmeier, P. K. Acharya, G. P. Anderson, G. Felde, J. Gardner, M. Hike, L. S. Jeong,

B. Pukall, J. Mello, A. Ratkowski, and H.-H. Burke, 1999. Atmospheric correction for shortwave spectral imagery based on MODTRAN4. *SPIE Proceedings Imaging Spectrometry*, 3753: 61–69.

Allen, T. R. and J. A. Kupfer, 2000. Application of spherical statistics to change vector analysis of Landsat data: Southern Appalachian spruce–fir forests. *Remote Sensing of Environment*, 74: 482–493.

Almutairi, A., 2000. Monitoring land-cover change detection in an arid urban environment: A comparison of change detection techniques. Unpublished MA thesis, West Virginia University. https://eidr.wvu.edu/eidr/documentdata. eIDR?documentid=1410. (Last date accessed 5/2/2008).

Almutairi, A., 2004. An Investigation of the role of image properties in influencing the accuracy of remote sensing change detection analysis. Unpublished PhD dissertation, West Virginia University. https://etd.wvu.edu/etd/controller.jsp? moduleName=documentdata&jsp%5Fetdld=3596. (last date accessed 5/2/2008).

Analytical Imaging and Geophysics, 2001. *ACORN User's Guide, Stand Alone Version*. Analytical Imaging and Geophysics LLC, Boulder, Colorado. 64 p.

Aspinall, R., 2002. A land-cover data infrastructure for measurement, modeling, and analysis of land-cover change dynamics. *Photogrammetric Engineering and Remote Sensing*, 68(10): 1101–1105.

Ben-Dor, E., F. A. Kruse, A. B. Lefkoff, and A. Banin, 1994. Comparison of three calibration techniques for utilization of GER 63-channel aircraft scanner data of Makhtesh Ramon, Negev, Israel. *Photogrammetric Engineering and Remote Sensing*, 60: 1339–1354.

Berk, A., L. S. Bernstein, G. P. Anderson, P. K. Acharya, D. C. Robertson, J. H. Chetwynd and S. M. Alder-Golden, 1998. MODTRAN Cloud and Multiple Scattering Upgrades with Application to AVIRIS. *Remote Sensing Environment*, 65(3): 367–375.

Biging, G. S., D. R. Colby, and R. G. Congalton, 1998. Sampling systems for change detection accuracy assessment. In R. S. Lunetta and C. D. Elvidge (eds.), *Remote Sensing Change Detection: Environmental Monitoring Methods and Applications*. Ann Arbor Press: Chelsea, Michigan, Chapter 15, pp. 281–308.

Brovolo, F. and L. Bruzzone, 2007. A theoretical framework for unsupervised change detection based on change vector analysis in the polar domain. *IEEE Transactions on Geoscience and Remote Sensing*, 45(1): 218–226, DOI 10.1109/TGRS.2006.885408.

Bruzzone, L. and R. Cossu, 2003. An adaptive approach to reducing registration noise effects in unsupervised change detection. *IEEE Transactions on Geoscience and Remote Sensing*, 41(11): 2455–2465. DOI: 10.1109/TGRS.2003.817268.

Bruzzone, L. and S. B. Serpico, 1997. Detection of changes in remotely-sensed images by the selective use of multi-spectral information. *International Journal of Remote Sensing*, 18(18): 3883–3888.

Carvalho, L. M. T., L. M. G. Fonseca, F. Murtagh, J. G. P. W. Clevers, 2001. Digital change detection with the aid of multiresolution wavelet analysis. *International Journal of Remote Sensing*, 22(18): 3871–3876.

Castelli, V., C. D. Elvidge, C-S. Li, and J. J. Turek, 1998. Classification-based change detection: Theory and applications to the NALC data set. In R. S. Lunetta and C. D. Elvidge (eds), *Remote Sensing Change Detection: Environmental Monitoring Methods and Applications*. Ann Arbor Press: Chelsea, Michigan, Chapter 4, pp. 53–73.

Chavez, P. S., 1988. An improved dark-object subtraction technique for atmospheric scattering correction of multispectral data. *Remote Sensing of Environment*, 24: 459–479.

Chavez, P. S., 1996. Image-based atmospheric corrections – revisited and improved. *Photogrammetric Engineering and Remote Sensing*, 62: 1025–1036.

Chen, J., P. Gong, C. He, R. Pu, and P. Shi, 2003. Land-use/land-cover change detection using improved change-vector analysis. *Photogrammetric Engineering and Remote Sensing*, 69: 369–379.

Cheng, T., 2002. Fuzzy objects: Their changes and uncertainties. *Photogrammetric Engineering and Remote Sensing*, 68: 41–49.

Cohen, W. B. and M. Fiorella, 1998. Comparison of methods for detecting conifer forest change with Thematic Mapper imagery. In R. S. Lunetta and C. D. Elvidge (eds), *Remote Sensing Change Detection: Environmental Monitoring Methods and Applications*. Ann Arbor Press, Chelsea, Michigan, Chapter 6, pp. 89–102.

Collins, J. B. and C. E. Woodcock, 1994. Change detection using the Gramm–Schmidt transformation applied to mapping forest mortality. *Remote Sensing of Environment*, 50: 267–279.

Collins, J. B. and C. E. Woodcock, 1996. An assessment of several linear change detection techniques for mapping forest mortality using multitemporal Landsat TM data. *Remote Sensing of Environment*, 56(1): 66–77.

Coppin, P., I. Jonckheere, K. Nackaerts, B. Muys, and E. Lambin, 2004. Digital change detection methods in ecosystem monitoring: a review. *International Journal of Remote Sensing*, 25(9): 1565–1596.

Coulter, L. and D. Stow, 2008. Assessment of the spatial co-registration of multitemporal imagery from large format digital cameras in the context of detailed change detection. *Sensors*, 8: 2161–2173.

Crist, E. P. and R. C. Cicone, 1984. A physically-based transformation of Thematic-Mapper data – the TM Tasseled Cap. *IEEE Transactions on Geoscience and Remote Sensing*, GE-22(3): 256–263.

Crews, K. and S. J. Walsh, in this volume. Remote Sensing and the Social Sciences. Chapter 31.

Crews-Meyer, K. A., 2002. Characterizing landscape dynamism using paneled-pattern metrics. *Photogrammetric Engineering and Remote Sensing*, 68(10): 1031–1040.

Dai, X. and S. Khorram, 1998. The effects of image misregistration on the accuracy of remotely sensed change detection. *IEEE Transactions on Geoscience and Remote Sensing*, 36(5): 1566–1577.

Desclée, B., P. Bogaert, and P. Defourny, 2006. Forest change detection by statistical object-based method. *Remote Sensing of Environment*, 102: 1–11.

Elvidge, C. D., D. Yuan, R. D. Weerackoon, and R. S. Lunetta, 1995. Relative radiometric normalization of Landsat Multispectral Scanner (MSS) data using and automatic scattergram-controlled regression. *Photogrammetric Engineering and Remote Sensing*, 61(10): 1255–1260.

Galford, G. L., J. F. Mustard, J. Melillo, A. Gendrin, C. C. Cerri, and C. E. P. Cerri, 2008. Wavelet analysis of MODIS time series to detect expansion and intensification of row-crop agriculture in Brazil. *Remote Sensing of Environment*, 112(2): 576–587.

Gamanya, R., P. De Maeyer, and M. De Dapper, 2008. Object-oriented change detection for the city of Harare, Zimbabwe. *Expert Systems with Applications*, 36(1): 571–588. Doi:10.1016/j.eswa.2007.09.067.

Gautama, S., J. D'Haeyer, and W. Philips, 2006. Graph-based change detection in geographic information using VHR satellite images. *International Journal of Remote Sensing*, 27(9): 1809–1824. DOI: 10.1080/01431160612331392545.

Gong, P., 1993. Change detection using principal component analysis and fuzzy set theory. *Canadian Journal of Remote Sensing*, 19(1): 22–29.

Gong, P., E. F. LeDrew, and J. R. Miller, 1992. Registration-noise reduction in difference images for change detection. *International Journal of Remote Sensing*, 13: 773–779.

Goward, S. N., T. Arvidson, D. L. Williams, R. Irish, and J. Irons, in this volume. Moderate Spatial Resolution Optical Sensors. Chapter 9.

Hall-Beyer, M., 2003. Comparison of single-year and multiyear NDVI time series principal components in cold temperate biomes. *IEEE Transactions on Geoscience and Remote Sensing*, 41(11): 2568- 2574. DOI: 10.1109/TGRS.2003.817274.

Hayes, D. J. and S. A. Sader, 2001. Comparison of change-detection techniques for monitoring tropical forest clearing and vegetation regrowth in a time series. *Photogrammetric Engineering and Remote Sensing*, 67(9): 1067–1075.

Im, J., J. R. Jensen, and J. A. Tullis, 2008. Object-based change detection using correlation image analysis and image segmentation. *International Journal of Remote Sensing*, 29(2): 399–423. DOI: 10.1080/01431160601075582.

IPCC, 2007. *Climate Change 2007: Synthesis Report*. Intergovernmental Panel on Climate Change. http://www.ipcc.ch/pdf/assessment-report/ar4/syr/ar4_syr.pdf (last date accessed 5/10/2008).

Kaufman, Y. J., A. Wald, L. A. Remer, B. Gao, R. Li, and L. Flynn, 1997. The MODIS 2.1 μm channel – correlation with visible reflectance for use in remote sensing of aerosol. *IEEE Transactions on Geoscience and Remote Sensing*, 35(5): 1286–1298. DOI: 10.1109/36.628795.

Kauth, R. J. and G. S. Thomas, 1976. The taselled cap – a graphic description of the spectral-temporal development of agricultural crops as seen by Landsat. *Proceedings of the Second Annual Symposium on Machine Processing of Remotely Sensed Data*. Purdue University, West Lafayette, Indiana: pp. 4B-41 to 4B-51.

Kennedy, R. E., W. B. Cohen, and T. A. Schroeder, 2007. Trajectory-based change detection for automated characterization of forest disturbance dynamics. *Remote Sensing of Environment*, 110(3): 370–386.

Khorram, S., G. S. Biging, N. R. Chrisman, D. R. Colby, R. G. Congalton, J. E. Dobson, R. L. Ferguson, M. F. Goodchild, J. R. Jensen, and T. H. Mace, 1999. *Accuracy Assessment of Remote Sensing-Derived Change Detection*. ASPR Monograph Series, ASPRS, Bethesda, Maryland, 64 pp.

Kim, M., M. Madden, and T. Warner, 2008. Estimation of the optimal image object size for the segmentation of forest stands with multispectral IKONOS imagery. In: T. Blaschke, S. Lang, and G. J. Hay (eds), *Object-Based Image Analysis - Spatial Concepts for Knowledge – driven Remote Sensing Applications*. Springer-Verlag, Berlin.

Lambin, E. F. and A. H. Strahler, 1994. Indicators of land-cover change for change-vector analysis in multitemporal space at coarse spatial scales. *International Journal of Remote Sensing*, 15: 2099–2119.

Lawrence, R. L. and W. J. Ripple, 1999. Calculating change curves for multitemporal satellite imagery: Mount St. Helens 1980–1995. *Remote Sensing of Environment* 67: 309–319.

Lawrence, R. L. and W. J. Ripple, 2000. Fifteen years of revegetation of Mount St. Helens: A landscape-scale analysis. *Ecology*, 81: 2742–2752.

Liang, S., in this volume. Quantitative Models and Inversion in Optical Remote Sensing. Chapter 20.

Liu, Y., S. Nishiyama, and T. Yano, 2004. Analysis of four change detection algorithms in bi-temporal space with a case study. *International Journal of Remote Sensing*, 25(11): 2121–2139. DOI: 10.1080/01431160310001606647.

Liu, Y., T. Yano, S. Nishiyama, and R. Kimura, 2007. Radiometric correction for linear change-detection techniques: analysis in bi-temporal space. *International Journal of Remote Sensing*, 28(22): 5143–5157. DOI: 10.1080/01431160701268954.

Lu, D., P. Mausel, E. Brondizio, and E. Moran., 2002. Assessment of atmospheric correction methods for Landsat TM data applicable to Amazon basin LBA research. *International Journal of Remote Sensing*, 23: 2651–2671.

Lu, D., P. Mausel, E. Brondizio, and E. Moran, 2004. Change detection techniques. *International Journal of Remote Sensing*, 25(12): 2365–2407.

Lunetta, R. S., 1998. Applications, project formulation, and analytical approach. In: R. S. Lunetta and C. D. Elvidge (eds), *Remote Sensing Change Detection: Environmental Monitoring Methods and Applications*. Ann Arbor Press: Chelsea, Michigan, Chapter 1, pp. 1–19.

Lunetta, R. S. and C. D. Elvidge (eds), 1998. *Remote Sensing Change Detection: Environmental Monitoring Methods and Applications*. Ann Arbor Press: Chelsea, Michigan.

Lunetta, R. S., J. F. Knight, J. Ediriwickrema, J. G. Lyon, and D. Worthy, 2006. Land-cover change detection using multitemporal MODIS NDVI data. *Remote Sensing of Environment*, 105(2): 142–154. DOI: 10.1016/j.rse.2006.06.018.

Malila, W., 1980. Change vector analysis: an approach for detecting forest changes with Landsat. *Proceedings of the 6th Annual Symposium on Machine Processing of Remotely Sensed Data*, West Lafayette, IN, 3–6 June 1980. Purdue University Press: West Lafayette, IN, pp. 326–335.

Mas, J. F., 1999. Monitoring land-cover changes: a comparison of change detection techniques. *International Journal of Remote Sensing*, 20(1): 139–152.

Mercier, G., G. Moser, and S. B. Serpico, 2008. Conditional copulas for change detection in heterogeneous remote sensing images. *IEEE Transactions on Geoscience and Remote Sensing*, 46(5): 1428–1441. DOI:10.1109/TGRS.2008.916476.

Morisette, J. T. and S. Khorram, 2000. Accuracy assessment curves for satellite-based change detection.

Photogrammetric Engineering and Remote Sensing, 66(7): 875–880.

NASA 2008. *Landast 7 Science Data Users Handbook*. http://landsathandbook.gsfc.nasa.gov/handbook/handbook_toc.html (last accessed 5/4/2008).

Perry, E., T. A. Warner, and H. Foote, 2000. Comparison of atmospheric modeling versus empirical line fitting for mosaicking HYDICE imagery. *International Journal of Remote Sensing*, 21(4): 799–803.

Phinn, S. R., C. Menges, G. J. E. Hill, and M. Stanford, 2000. Optimizing remotely sensed solutions for monitoring, modeling, and managing coastal environments. *Remote Sensing of Environment*, 73(2): 117–132.

Prenzel, B. G. and P. Treitz, 2006. Spectral and spatial filtering for enhanced thematic change analysis of remotely sensed data. *International Journal of Remote Sensing*, 27(5): 835–854. DOI: 10.1080/01431160500300321.

Radke, R. J., S. Andra, O. Al-Kofahi, and B. Roysam, 2005. Image change detection algorithms: A systematic survey. *IEEE Transactions on Image Processing*, 14(3): 294–307. DOI: 10.1109/TIP.2004.838698.

Richards, 1984. Thematic mapping from multitemporal image data using the principal components transformation. *Remote Sensing of Environment*, 16(1): 35–46. DOI:10.1016/0034-4257(84)90025-7

Ridd, M. K. and J. Liu, 1998. A comparison of four algorithms for change detection in an urban environment. *Remote Sensing of Environment*, 63: 95–100.

Rindfuss, R. R., S. J. Walsh, B. L. Turner, II, J. Fox, and V. Mishra, 2004. Developing a science of land change: Challenges and methodological issues. *Proceedings of the National Academies of Science*, 101(39): 13976–13981. DOI:10.1073/pnas.0401545101.

Rogan, J. and D. M. Chen., 2004. Remote sensing technology for mapping and monitoring land-cover and land-use change. *Progress in Planning*, 61(4): 301–325.

Rogan, J. and J. Miller, 2007. Integrating GIS and Remotely Sensed Data for Mapping Forest Disturbance and Change. In: M. A. Wulder and S. E. Franklin (eds), *Understanding Forest Disturbance and Spatial Pattern: GIS and Remote Sensing Approaches*. CRC Press: Boca Rotan, Florida, Chapter 6, pp. 133–172.

Rogan, J., J. Franklin, D. Stow, and D. Roberts, 2009. A comparison of linear change detection methods for mapping multiple types of land-cover change in California. *Photogrammetric Engineering and Remote Sensing* (in press).

Salas, W. A., S. H. Boles, S. Frolking, X. Xiao, and C. Li, 2003. The perimeter/area ratio as an index of misregistration bias in land cover change estimates. *International Journal of Remote Sensing*, 24(5): 1165–1170. DOI: 10.1080/0143116021000044841.

Schott, J. R., C. Salvaggio, and W. J. Volchok, 1988. Radiometric scene normalization using pseudoinvariant features. *Remote Sensing of Environment*, 26: 1–16.

Schwartz, M. D., B. C. Reed, and M. A. White, 2002. Assessing satellite-derived start-of-season measures in the conterminous USA. *International Journal of Climatology*, 22: 1793–1805. DOI: 10.1002/joc.819.

Serra, P., X. Pons, and D. Saurí, 2003. Post-classification change detection with data from different sensors:

some accuracy considerations. *International Journal of Remote Sensing*, (24)16: 3311–3340. DOI: 10.1080/0143116021000021189.

Singh, A., 1989. Digital change detection using remotely sensed data. *International Journal of Remote Sensing*, 10(6): 989–1003.

Sohl, T., 1999. Change analysis in the United Emirate Republics: An investigation of techniques. *Photogrammetric Engineering and Remote Sensing*, 65(4): 475–484.

Song, C. and C. E. Woodcock, 2003. Monitoring forest succession with multitemporal Landsat images: Factors of uncertainty. *IEEE Transactions on Geoscience and Remote Sensing*, 41(11): 2557–2567. DOI:10.1109/TGRS.2003.818367.

Song, C., C. E. Woodcock, K. C. Seto, M. Pax Lenney, and S. A Macomber, 2001. Classification and change detection using Landsat TM data: When and how to correct atmospheric effects? *Remote Sensing of Environment*, 75: 230–244.

Stehman, S. V. and G. Foody, in this volume. Accuracy Assessment. Chapter 21.

Stow, D. A., 1999. Reducing the effects of misregistration on pixel-level change detection. *International Journal of Remote Sensing*, 20(12): 2477–2483.

Stow, D., Y. Hamada, L. Coulter, and Z. Anguelova, 2008. Monitoring shrubland habitat changes through object-based change identification with airborne multi-spectral imagery. *Remote Sensing of Environment*, 112: 1051–1061. DOI: 10.1016/j.rse.2007.07.011.

Sundaresan, A., P. K. Varshney, and M. K. Arora, 2007. Robustness of change detection algorithms in the presence of registration errors. *Photogrammetric Engineering and Remote Sensing*, 73(4): 375–383.

Teillet, P. M., K. Staenz, and D. J. Williams, 1997. Effects of spectral, spatial and radiometric characteristics on remote sensing vegetation indices of forested regions. *Remote Sensing of Environment*, 61: 139–149.

Tokola, T., S. Löfman, and A. Erkkilä, 1999. Relative calibration of multitemporal Landsat data for forest cover change detection. *Remote Sensing of Environment*, 68: 1–11.

Townshend, J. R. G., C. O. Justice, C. Gurney, and J. McManus, 1992. *IEEE Transactions on Geoscience and Remote Sensing*, 30(5): 1054–1060. DOI:10.1109/36.175340.

Turner, B. L., E. F. Lambin, and A. Reenberg, 2007. The emergence of land change science for global environmental change and sustainability. *Proceedings of the National Academy of Sciences*, 104(52): 20666–20671. DOI:10.1073/pnas.0704119104.

van Oort, 2007. Interpreting the change detection error matrix. *Remote Sensing of Environment*, 108: 1–8. DOI:10.1016/j.rse.2006.10.012.

Vermote, E., D. Tanre, J. L. Deuze, M. Herman, and J. J. Morcrette, 1997. Second simulation of the satellite signal in the solar spectrum, 6S: An overview. *IEEE Transactions on Geoscience and Remote Sensing*, 35: 675–686.

Virag, L. A. and J. E. Colwell, 1987. An improved procedure for analysis of change in Thematic Mapper image-pairs. *Proceedings of the Twenty-First International Symposium on Remote Sensing of the Environment*, Ann Arbor, Michigan: ERIM: pp. 1101–1110.

Walter, V., 2004. Object-based classification of remote sensing data for change detection. *ISPRS Journal of Photogrammetry and Remote Sensing*, 58: 225–238.

Wang, H. and E. C. Ellis, 2005. Image misregistration error in change measurements. *Photogrammetric Engineering and Remote Sensing*, 71(9): 1037–1044.

Warner, T. A, 1999. Analysis of spatial patterns in remotely sensed data using multivariate spatial correlation. *Geocarto International*, 14(1): 59–65.

Warner, T. A., 2005. Hyperspherical Direction Cosine Change Vector Analysis. *International Journal of Remote Sensing*, 26(6): 1201–1215.

Warner, T. A., M. D. Nellis, and G. Foody, in this volume. Remote Sensing Data Selection Issues. Chapter 1.

Woodcock, C. E. and A. H. Strahler, 1987. The factor of scale in remote sensing. *Remote Sensing of Environment*, (21): 311–332.

Wulder, M. A. and S. E. Franklin (eds), 2007. *Understanding Forest Disturbance and Spatial Pattern: GIS and Remote Sensing Approaches*. CRC Press: Boca Rotan, Florida.

Wulder, M. A., C. R. Butson, and J. C. White, 2008. Cross-sensor change detection over a forested landscape: Options to enable continuity of medium spatial resolution measures. *Remote Sensing of Environment*, 112: 796–809. DOI:10.1016/j.rse.2007.06.013.

Yang, X. and C. P. Lo 2000. Relative radiometric normalization performance for change detection from multi-data satellite. *Photogrammetric Engineering and Remote Sensing*, 66(8): 967–980.

Yuan, D. and C. Elvidge, 1996. Comparision of relative radiometric normalization techniques. *ISPRS Journal of Photogrammetry and Remote Sensing*, 51: 117–126.

Yuan, D. and C. Elvidge, 1998. NALC land cover change detection pilot study: Washington D.C. area experiments. *Remote Sensing of Environment*, 66: 166–178.

Yuan, D., C. D. Elvidge, and R. S. Lunetta, 1998. Survey of multispectral methods for land cover change analysis. In: R. S. Lunetta and C. D. Elvidge (eds), *Remote Sensing Change Detection: Environmental Monitoring Methods and Applications*. Ann Arbor Press: Chelsea, Michigan, Chapter 2, pp. 21–39.

Yuan, F., K. E. Sawaya, B. C. Loeffelholz, and M. E. Bauer, 2005. Land cover classification and change analysis of the Twin Cities (Minnesota) Metropolitan area by multitemporal Landsat remote sensing. *Remote Sensing of Environment*, 98: 317–328.

Zhao, G. and A. L. Maclean, 2000. Comparison of canonical discriminant analysis and principal component analysis for spectral transformation. *Photogrammetric Engineering and Remote Sensing*, 66(7): 841–847.

Conclusions

34

A Look to the Future

Giles M. Foody, Timothy A. Warner, and
M. Duane Nellis

Keywords: future trends, opportunities, remote sensing policy, applications, technology.

INTRODUCTION

The previous chapters in this book have provided a guide to the current status of remote sensing. The chapters have covered topics ranging from issues connected with the fundamental principles of remote sensing through some major developments in remote sensors to a diverse and growing range of applications. These chapters have highlighted recent trends in the subject, revealing it to be highly dynamic. This is not surprising as remote sensing is an integral part of the general area of geotechnology which has been identified as a major priority area for research and development (Gewin 2004, Mondello et al. 2004, 2006). One consequence of the dynamism of the subject is that major advances beyond the material covered in the book could be made in the near future. Indeed at the time of writing this closing chapter new satellite sensor systems such as CBERS-2B HRC, Deimos-1 and UK-DMC2 have been launched or are scheduled for launch, major documents outlining future plans for aspects of the subject have been produced (e.g., National Science and Technology Council 2007) and the literature has been enhanced greatly by many exciting new publications. Here, however, we wish to consider some possible near future developments to close the book. The overview provided is not meant to be exhaustive but aims to illustrate some likely key future directions for the subject.

Although often considered to be a young subject, remote sensing has, in its photographic origins, a history extending back over some 150 years (Curran 1985, Lillesand et al. 2004). Even the Earth resources satellite systems which are now used widely routinely have also been operational for some four decades (Goward et al., in this volume) and played a major role in studies of key environmental issues such as retreating glaciers and ice sheets (Dozier, in this volume) as well as major land cover related changes such as urban growth, desert fluctuations and deforestation (Lambin 1997, Lepers et al. 2005) providing valuable support and input to environmental research (Wulder et al., in this volume). Similarly, precision agriculture, using both remote images and tractor-mounted 'on-the-go sensors,' has moved from a research environment to an operational environment (Nellis et al., in this volume). Remote sensing, therefore, has a firm and extensive foundation upon which considerable development may be expected. The history of remote sensing has shown that developments, especially in technology, can revolutionize the subject. This has been evident repeatedly with key technical advances such as those associated with the development of infrared photography, digital aerial imaging systems (Stow et al., in this volume), radar (Kellndorfer and McDonald, in this volume), CCDs (Kerekes, in this volume) and lidar (Hyyppä et al., in this volume), amongst others, leading to major advances as well as opening the door for new and perhaps unforeseen applications.

A variety of major possible future trends have been highlighted in the preceding chapters. Throughout, there has been an expectation for normal incremental growth of the subject. One might, for example, expect to see development of new approaches for image classification to derive increasingly more accurate classifications (Jensen et al., in this volume). Methods that have recently become popular, such as support vector machine classifiers (Huang et al. 2002, Melgani and Bruzzone 2004, Pal and Mather 2005) appear to offer many attractions, especially if resources for training the classifier are limited (Foody et al. 2006), and may be the basis for further methodological advances. This may include moves away from conventional thematic mapping practices with effort increasingly directed at tailoring the process to suit the circumstances of a particular application. For example, often interest lies in a sub-set of the classes that occur in the imaged area and by focusing attention on just these classes savings in time, effort and resources may be achieved. This is evident, for example, in the use of one-class classifiers for habitat monitoring in support of conservation activities (Sanchez-Hernandez et al. 2007). Additionally, an increase in the use of soft or fuzzy classifications may help in the study of environmental gradients and transition zones and sub-pixel land cover (Rocchini and Ricotta 2007). The use of soft classifications in a post-classification change detection may also allow a fuller assessment of land cover changes, by revealing modifications as well as conversions of cover. This is particularly valuable as remote sensing based research has focused mainly on conversions, with little attention paid to the severity of change limiting environmental applications and assessment of environmental change (Nepstad et al. 1999, Foody 2001). As well as advances in the techniques for land cover mapping and monitoring applications, major developments in associated activities are anticipated. Thus, for example, topics such as classification accuracy assessment, discussed by Stehman and Foody (in this volume), which many may perceive as a standard and established issue with a defined set of universally accepted methods, have actually been the focus of considerable recent research activity and are recognized as major priority areas for further research as many challenges remain to be addressed (Rindfuss et al. 2004, Strahler et al. 2006). The validation of thematic maps derived via remote sensing is fraught with difficulty but it is central to determining their suitability for use. Furthermore, as map users

become more aware of their actual needs, the ability of remote sensing to provide maps of adequate accuracy may increase (DeFries and Los 1999, Foody 2008) encouraging further use of remote sensing.

Exciting developments that extend well beyond that expected by incremental research are likely. Many of the chapters have flagged potential future activities. For example, the use of currently underused parts of the electromagnetic spectrum, highlighted by Cracknell and Boyd (this volume), may enhance many applications. Indeed exploiting the full information content of existing data sets may enhance the degree of useful information extraction from remotely sensed data. Thus, a move away from analyses based upon simple summary indices such as the NDVI to a fuller use of the data acquired by remote sensors may provide an ability to extract environmental information more accurately (Boyd et al. 1996, Asner et al. 2004, Dash and Curran 2004, Asner and Ollinger, in this volume).

An increase in the use of remote sensing in a diverse array of applications may be expected. For example, an increase in the use of remote sensing for disaster related applications may be expected as the various communities involved in emergency response applications gain a greater appreciation of the potential of remote sensing from its use. The foundations for such applications have been laid, with for example, case studies of remote sensing in support of post-disaster applications discussed by Teeuw et al. (in this volume) and the related data policy issues reviewed by Harris (in this volume), opening the way for increased use of remote sensing in an important application domain. Such applications will also benefit from technological developments which help to enable remote sensing to provide the required data quickly and efficiently (Hutton and Melihen 2006).

One key issue for future research that has been flagged in the preceding chapters is the fundamental issue of radiometric calibration (Liang, in this volume, Schaepman-Strub et al., in this volume, van der Meer et al., in this volume). The importance of calibration for quantitative analyses of optical and microwave remote sensor data sets was highlighted in van Leeuwen (in this volume) and Moghaddam (in this volume), respectively. Appropriately calibrated data sets will also help many applications, including those with a temporal dimension for which data continuity is an important concern. Developing and maintaining long-term archives of well-calibrated remote sensor data are important objectives, especially with the desire to study the Earth's environment within a context of environmental change. Continuity of sensor data sets was highlighted by Justice and Tucker (in this volume) and Goward et al. (in this volume) and is central to research that seeks to aid continuity between satellite systems (e.g., Steven et al. 2003) including future missions (Leimgruber et al. 2005, National Science and Technology Council 2007). There is, in particular, a strong desire to extend the archive of Landsat sensor products into the future since such data have played a major role in establishing remote sensing as a valuable source of environmental information, with a history of use that already spans a period in excess of three decades (Cohen and Goward 2004, Boyd and Danson 2005). Continuity issues, therefore, need consideration in the development of new sensors (Janetos and Justice 2000). Naturally there is also a desire to refine and develop sensors and considerable research using data acquired from a variety of sensors is anticipated (Warner et al. in this volume). Recent and proposed sensors were discussed in Goward et al. (in this volume) and Toutin (in this volume), and often derive from a gradual transition from research to operational systems, something that Schaepman (in this volume) highlights is happening with hyperspectral sensors, as well as a trend for increased commercilization. Developments in sensor technology are likely to result in the acquisition of data with enhanced spatial, spectral and radiometric resolutions as well as from a range of angular geometries (Kerekes, in this volume). Since multi-dimensional data sets can provide a richer characterization of the environment than a conventional image data set this should enhance many applications. Radar systems may, for example, provide information on biophysical variables such as biomass (Moghaddam, in this volume) that can be used to enhance the extraction of information on land cover structure (Bergen et al. 2007). Radar can also be used to derive quantitative soils information, including moisture and roughness (Campbell, in this volume). Similarly, lidar sensors may be used to provide information on forest canopy height and topography that can yield useful structural information (Hyyppä et al., in this volume). The latter may be particularly valuable in some applications. For example, the vertical distribution of a forest canopy can be the most important variable for the accurate prediction of bird species richness and so information on this variable could greatly aid biodiversity studies (Goetz et al. 2007). Lidar also opens new possibilities in geological hazard assessment associated with active tectonics (Chen and Campagna, in this volume). In all cases it is also likely that increased use of ancillary data sets, including field data (Johannsen and Daughtry, in this volume), models (Liang, in this volume), and digital elevation data (Deng, in this volume) will be used, including in areas highlighted in Moghaddam (in this volume), Kerekes (in this volume), Asner and Ollinger (in this volume), and Crews and Walsh (in this volume), as well as other studies (Luoto et al. 2004, Southworth et al. 2006, Leyequien et al. 2007).

It is anticipated that much may be gained by increasing education and training in remote sensing (Merchant and Narumalani, in this volume). Additionally, a greater realization of the opportunities provided by remote sensing may allow the adoption of different perspectives on environmental issues. The latter is an issue highlighted in Crews and Walsh (in this volume) in which the potential of remote sensing to study large areas may require a change of thinking by users and may open the door for new applications.

Although considerable future development is expected for remote sensing as a whole there will be variations in activity within its various component areas. For example, over the vast range of application areas growth may be expected in many but in some the level of activity may remain static or even decline. For example, despite the successes of thermal remote sensing for geological (Chen and Campagna, in this volume), ecological (Quattrochi and Luvall, in this volume), oceanographic (Lavender, in this volume), and urban applications (Nichol, in this volume), few of the planned space-borne sensors incorporate moderate to fine resolution thermal imaging which may limit activity. In contrast, one application area that has seen recent considerable growth in activity is in ecology (Nabiullin, in press). A closer look at some of the recent activity in ecological applications of remote sensing may highlight the type of future growth that might be expected elsewhere in the subject.

Of the many ecological applications, remote sensing has considerable potential as a source of information on major international research priorities such as those linked to biodiversity and its conservation (Wulder et al., in this volume). Remote sensing is especially attractive as a source of information for such applications as it provides an inexpensive means of deriving complete spatial coverage of environmental information for large areas in a consistent and regularly up-datable manner (Muldavin et al. 2001, Luoto et al. 2004, Duro et al. 2007) which is suited to biodiversity assessments (Lassau and Hochuli 2007) of even remote and inaccessible regions (Cayuela et al. 2006). Despite these advantages remote sensing was until recently relatively under-used in biodiversity studies (Innes and Koch 1998, Trisurat et al. 2000). A greater awareness of the potential of remote sensing together with technical advances, notably in sensor spatial resolution, have now enabled remote sensing to be viewed as a repeatable, systematic and spatially exhaustive source of information on variables that impact on biodiversity. The latter includes variables such as productivity, disturbance and land cover (Chowdhury 2006, Duro et al. 2007). The ability to derive information on the latter is critical as land cover changes, notably habitat loss and fragmentation, are the greatest threats to biodiversity (Chapin et al. 2000, Menon et al. 2001, Gaston 2005).

Many of the future trends suggested above can be seen in the history of biodiversity related remote sensing research. For example, the development of fine spatial resolution satellite systems such as IKONOS revolutionized research. While the remote sensing of biodiversity was sometimes viewed as fools errand as features of interest, typically individual organisms such as tress, were smaller than the pixel size (Turner et al. 2003), fine spatial resolution systems (Toutin, in this volume) changed things dramatically. Fine spatial resolution imagery has, for example, allowed some tree species to be mapped accurately from remotely sensed imagery (Martin et al. 1998, Haara and Haarala 2002, Carleer and Wolff 2004, Wang et al. 2004) and it may now be possible to classify some species from satellite sensor data with an accuracy comparable to that from aerial sensor systems (Carleer and Wolff 2004). This is valuable as knowledge of the spatial distribution of tree species can help reveal the underlying processes and enhance knowledge of biodiversity as well as its conservation (Wilson et al. 2004). The opportunities offered by contemporary remote sensing systems has therefore lead to considerable interest in tree species classification research (Sanchez-Azofelfa et al. 2003, Turner et al. 2003, Goodwin et al. 2005) with the fine spatial resolution imagery allowing questions that previously were impractical to study, from space or on the ground, to be addressed. For example, IKONOS data allows studies to be undertaken at the scale of individual tree crowns over large areas (Hurtt et al. 2003, Clark et al. 2004a) providing information on previously poorly known issues such as tree mortality in a tropical rainforest (Clark et al. 2004b).

A major feature of remote sensing is its ability to provide data at a range of scales, especially temporal and spatial. This feature is extremely valuable as many scale dependencies are observed in ecology. While moderate to coarse spatial resolution imagery would normally be inappropriate for applications such as tress species classification they are well suited to other applications. Thus, moderate and coarse spatial resolution imagery have also been used widely for general biodiversity assessments. In particular, through relationships with land cover it has been possible to assess the diversity of species that do not directly affect the remotely sensed response such as insects and birds and assess impacts associated with changes in the habitat mosaic such as fragmentation (e.g., Kerr et al. 2001, Luoto et al. 2002, 2004, Cohen and Goward 2004, Bergen et al. 2007, Fuller et al. 2007, Lassau and Hochuli 2007). Thus, even with relatively coarse spatial resolution, remotely sensed

data may be used to derive useful information on biodiversity (Kerr et al. 2001, Foody and Cutler 2003, Cohen and Goward 2004).

The considerable potential of remote sensing as a source of environmental information is, however, often associated with major challenges. For example, although the provision of complete data coverage for large areas is often seen as a major advantage of remote sensing it does present some problems. For example, it is often assumed that relationships between the biodiversity variable of interest (e.g., species richness) and the remotely sensed response are spatially stationary and hence constant in space, making them transferable between locations. Sometimes, however, relationships may be spatially non-stationary, limiting the generalizabilty of remote sensing methods. Fortunately a range of methods are available to model spatially non-stationary relationships and may help in realization of remote sensing's potential (Osborne et al. 2007). A greater use of field data and models can, therefore, be expected to enhance ecological studies using remote sensing. Similar trends may be expected in other application domains.

To close, this book has aimed to give an overview of the current status of remote sensing. Space constraints have necessarily limited the breadth and depth of coverage but the book gives an overview of the main areas of activity. The snapshot view provided by the book will necessarily date with new developments but this is just a reflection of the dynamism of the subject, an issue that has hopefully been captured in the preceding chapters. Moreover, it is hoped that the book will help drive future work by providing a report on the current status of a broad spectrum of the subject, providing a reference resource for new and established researchers, as well as by illustrating some possible future directions.

REFERENCES

Asner, G. and S. V. Ollinger, in this volume. Remote Sensing for Terrestrial Biogeochemical Modeling. Chapter 29.

Asner G. P., D. Nepstad, G. Cardinot and D. Ray, 2004. Drought stress and carbon uptake in an Amazon forest measured with spaceborne imaging spectroscopy. *Proceedings of the National Academy of Sciences of the United States of America*, 101: 6039–6044.

Bergen, K. M, A. M. Gilboy and D. G. Brown, 2007. Multidimensional vegetation structure in modeling avian habitat. *Ecological Informatics*, 2: 9–22.

Boyd, D. S. and F. M. Danson, 2005. Satellite remote sensing of forest resources: three decades of research development. *Progress in Physical Geography*, 29: 1–26.

Boyd, D. S., G. M. Foody, P. J. Curran, R. M. Lucas and M. Honzak, 1996. An assessment of radiance in Landsat TM middle and thermal infrared wavebands for the detectior of tropical forest regeneration. *International Journal of Remote Sensing*, 17: 249–261.

Campbell, J., in this volume. Remote Sensing of Soils. Chapter 24.

Carleer, A. and E. Wolff, 2004. Exploitation of very high resolution satellite data for tree species identification. *Photogrammetric Engineering and Remote Sensing*, 70: 135–140.

Cayuela, L., J. M. Benayas, A. Justel and J. Salas-Rey, 2006. Modelling tree diversity in a highly fragmented tropical montane landscape. *Global Ecology and Biogeography*, 15: 602–613.

Chapin F. S., E. S. Zavaleta, V. T. Eviner, R. L. Naylor, P. M. Vitousek, H. L. Reynolds, D. U. Hooper, S. Lavorel, O. E. Sala, S. E. Hobbie, M. C. Mack and S. Diaz, 2000. Consequences of changing biodiversity. *Nature*, 405: 234–242.

Chen, X. and D. Campagna, in this volume. Remote Sensing of Geology. Chapter 23.

Chowdhury, R. R., 2006. Landscape change in the Calakmul Biosphere Reserve, Mexico: Modeling the driving forces of smallholder deforestation in land parcels. *Applied Geography*, 26: 129–152.

Clark, D. B., J. M. Read, M. L. Clark, A. M. Cruz, M. F. Dotti and D. A. Clark, 2004a. Application of 1-M and 4-M resolution satellite data to ecological studies of tropical rain forests. *Ecological Applications*, 14: 61–74.

Clark, D. B., C. S. Castro, L. D. A. Alvarado and J. M. Read, 2004b. Quantifying mortality of tropical rain forest trees using high-spatial-resolution satellite data. *Ecology Letters*, 7: 52–59.

Cohen, W. B. and S. N. Goward, 2004. Landsat's role in ecological applications of remote sensing. *Bioscience*, 54: 535–545.

Cracknell, A. P. and D. S. Boyd, in this volume. Interactions of Middle Infrared (3–5 μm) Radiation with the Environment. Chapter 4.

Crews, K. and S. J. Walsh, in this volume. Remote Sensing and the Social Sciences. Chapter 31.

Curran, P. J., 1985. *Principles of Remote Sensing*. Longman, Harlow.

Dash, J. and P. J. Curran, 2004. The MERIS terrestrial chlorophyll index. *International Journal of Remote Sensing*, 25: 5403–5413.

DeFries, R. S. and S. O. Los, 1999. Implications of land-cover misclassification for parameter estimates in global land-surface models: An example from the simple biosphere model (SiB2). *Photogrammetric Engineering and Remote Sensing*, 65: 1083–1088.

Deng, Y., in this volume. Making Sense of the Third Dimension Through Topographic Analysis. Chapter 22.

Dozier, J., in this volume. Remote Sensing of the Cryosphere. Chapter 28.

Duro, D., N. C. Coops, M. A. Wulder, and T. Han, 2007. Development of a large area biodiversity monitoring system driven by remote sensing. *Progress in Physical Geography*, 31: 235–260.

Foody, G. M., 2001. Monitoring the magnitude of land-cover change around the southern limits of the Sahara.

Photogrammetric Engineering and Remote Sensing, 67: 841–847.

Foody, G. M., 2008. Harshness in image classification accuracy assessment. *International Journal of Remote Sensing*, (in press).

Foody, G. M. and M. E. J. Cutler, 2003. Tree biodiversity in protected and logged Bornean tropical rain forests and its measurement by satellite remote sensing. *Journal of Biogeography*, 30: 1053–1066.

Foody, G. M., A. Mathur, C. Sanchez-Hernandez and D. S. Boyd, 2006. Training set size requirements for the classification of a specific class. *Remote Sensing of Environment*, 104: 1–14.

Fuller, R. M., B. J. Devereux, S. Gillings, R. A. Hill and G. S. Amable, 2007. Bird distributions relative to remotely sensed habitats in Great Britain: Towards a framework for national modeling. *Journal of Environmental Management*, 84: 586–605.

Gaston, K. J., 2005. Biodiversity and extinction: species and people. *Progress in Physical Geography*, 29: 239–247.

Gewin, V., 2004. Mapping opportunities. *Nature*, 427: 376–377.

Goetz S., D. Steinberg, R. Dubayah and B. Blair, 2007. Laser remote sensing of canopy habitat heterogeneity as a predictor of bird species richness in an eastern temperate forest, USA. *Remote Sensing of Environment*, 108: 254–263.

Goodwin, N., R. Turner and R. Merton, 2005. Classifying Eucalyptus forests with high spatial and spectral resolution imagery: an investigation of individual species and vegetation communities. *Australian Journal of Botany*, 53: 337–345.

Goward, S. N., T. Arvidson, D. L. Williams, R. Irish and J. Irons, in this volume. Moderate Spatial Resolution Optical Sensors. Chapter 9.

Haara, A. and M. Haarala, 2002. Tree species classification using semi-automatic delineation of trees on aerial images. *Scandinavian Journal of Forest Research*, 17: 556–565.

Harris, R., in this volume. Remote Sensing Policy. Chapter 2.

Huang, C., L. S. Davis and J. R. G. Townshend, 2002. An assessment of support vector machines for land cover classification. *International Journal of Remote Sensing*, 23: 725–749.

Hurtt, G., X. M. Xiao, M. Keller, M. Palace, G. P. Asner, R. Braswell, E. S. Brondizio, M. Cardoso, C. J. R. Carvalho, M. G. Fearon, L. Guild, S. Hagen, S. Hetrick, B. Moore, C. Nobre, J. M. Read, T. Sa, A. Schloss, G. Vourlitis and A. J. Wickel, 2003. IKONOS imagery for the Large Scale Biosphere-Atmosphere Experiment in Amazonia (LBA). *Remote Sensing of Environment*, 88: 111–127.

Hutton, J. and A. Melihen, 2006. Emergency response – remote sensing evolves in the wake of experience. *Photogrammetric Engineering and Remote Sensing*, 72: 977–981.

Hyyppä, J., W. Wagner, M. Hollaus and H. Hyyppä, in this volume. Airborne Laser Scanning. Chapter 14.

Innes, J. L. and Koch, B., 1998. Forest biodiversity and its assessment by remote sensing. *Global Ecology and Biogeography*, 7: 397–419.

Janetos, A. C. and C. O. Justice, 2000. Land cover and global productivity: a measurement strategy for the NASA programme. *International Journal of Remote Sensing*, 21: 1491–1512.

Jensen, J. R., J. Im, P. Hardin, and R. R. Jensen, in this volume. Image Classification. Chapter 19.

Johannsen, C. J. and C. S. T. Daughtry, in this volume. Surface Reference Data Collection. Chapter 17.

Justice, C. O. and C. J. Tucker, in this volume. Coarse Spatial Resolution Optical Sensors. Chapter 10.

Kellndorfer, J. and K. McDonald, in this volume. Active and Passive Microwave Systems. Chapter 13.

Kerekes, J. P., in this volume. Optical Sensor Technology. Chapter 7.

Kerr, J. T., T. R. E. Southwood, and J. Cihlar, 2001. Remotely sensed habitat diversity predicts butterfly species richness and community similarity in Canada. *Proceedings of The National Academy of Sciences of the United States of America*, 98: 11365–11370.

Lambin, E. F., 1997. Modelling and monitoring land-cover change processes in tropical regions. *Progress in Physical Geography*, 21: 375–393.

Lassau, S. A. and D. F. Hochuli, 2007. Associations between wasp communities and forest structure: Do strong local patterns hold across landscapes? *Austral Ecology*, 32: 656–662.

Lavender, S., in this volume. Optical Remote Sensing of the Hydrosphere: From the Open Ocean to Inland Waters. Chapter 27.

Leimgruber, P., C. A. Christen and A. Laborderie, 2005. The impact of Landsat satellite monitoring on conservation biology. *Environmental Monitoring and Assessment*, 106: 81–101.

Lepers E, E. F. Lambin, A. C. Janetos, R. DeFries, F. Achard, N. Ramankutty and R. J. A. Scholes, 2005. A synthesis of information on rapid land-cover change for the period 1981–2000, *Bioscience*, 55: 115–124.

Leyequien, E., J. Verrelst, M. Slot, G. Schaepman-Strub, I. M. A. Heitkonig and A. Skidmore, 2007. Capturing the fugitive: Applying remote sensing to terrestrial animal distribution and diversity. *International Journal of Applied Earth Observation and Geoinformation*, 9: 1–20.

Liang, S., in this volume. Quantitative Models and Inversion in Optical Remote Sensing. Chapter 20.

Lillesand, T. M., R. W. Kiefer and J. W. Chipman, 2004. *Remote Sensing and Image Interpretation*, fifth edition, Wiley, New York.

Luoto, M., M. Kuussaari and T. Toivonen, 2002. Modelling butterfly distribution based on remote sensing data. *Journal of Biogeography*, 29: 1027–1037.

Luoto, M., R. Virkkala, R. K. Heikkinen and K. Rainio, 2004. Predicting bird species richness using remote sensing in boreal agricultural-forest mosaics. *Ecological Applications*, 14: 1946–1962.

Martin, M. E., S. D. Newman, J. D. Aber, and R. G. Congalton, 1998. Determining forest species composition using high spectral resolution remote sensing data. *Remote Sensing of Environment*, 65: 249–254.

Melgani, F. and L. Bruzzone, 2004. Classification of hypersepectral remote sensing images with support vector machines. *IEEE Transactions on Geoscience and Remote Sensing*, 42: 1778–1790.

Menon, S., R. G. Pontius, J. Rose, M. L. Khan, and K. S. Bawa, 2001. Identifying conservation-priority areas in the tropics: a land-use change modeling approach. *Conservation Biology*, 15: 501–512.

Merchant, J. and S. Narumalani, in this volume. Integrating Remote Sensing and Geographic Information systems. Chapter 18.

Moghaddam, M., in this volume. Polarimetric SAR Phenomenology and Inversion Techniques for Vegetated Terrain. Chapter 6.

Mondello, C., G. F. Hepner and R. A. Williamson, 2004. 10-year industry forecast. Phases I–III – study documentation. *Photogrammetric Engineering and Remote Sensing*, 70: 5–58.

Mondello, C., G. F. Hepner and R. A. Williamson, 2006. 10-year remote sensing industry forecast. Phase IV – study documentation. *Photogrammetric Engineering and Remote Sensing*, 72: 985–1000.

Muldavin, E. H., P. Neville, and G. Harper, 2001. Indices of grassland biodiversity in the Chihuahuan Desert ecoregion derived from remote sensing. *Conservation Biology*, 15: 844–855.

Nabiullin, A. A., in press. Measuring remote sensing science: bibliometric analysis of the literature, 1975–2005. *International Journal of Remote Sensing*.

National Science and Technology Council, 2007. *A Plan for a U.S. National Land Imaging Program*, Office of Science and Technology Policy, 120 pp. (available at www.ostp.gov).

Nellis, M. D., K. Price and D. Rundquist, in this volume. Remote Sensing Applications in Agriculture. Chapter 26.

Nepstad, D. C., A. Verissimo, A. Alencar, C. Nobre, E. Lima, P. Lefebvre, P. Schlesinger, C. Potter, P. Moutinho, E. Mendoza, M. Cochrane and V. Brooks, 1999. Large-scale impoverishment of Amazonian forests by logging and fire, *Nature*, 398: 505–508.

Nichol, J., in this volume. Remote Sensing of Urban Areas. Chapter 30.

Osborne, P. E., G. M. Foody and S. Suarez-Seoane, 2007. Non-stationarity and local approaches to modelling the distributions of wildlife. *Diversity and Distributions*, 13: 313–323.

Pal, M. and P. M. Mather, 2005. Support vector machines for classification in remote sensing. *International Journal of Remote Sensing*, 26: 1007–1011.

Quattrochi, D. and J. C. Luvall, in this volume. Thermal Remote Sensing in Earth Science Research. Chapter 5.

Rindfuss, R. R., S. J. Walsh, B. L. Turner, J. Fox and V. Mishra, 2004. Developing a science of land change: Challenges and methodological issues. *Proceedings of the National Academy of Sciences of the United States of America*, 101: 13976–13981.

Rocchini, D. and C. Ricotta, 2007. Are landscapes as crisp as we may think? *Ecological Modelling*, 204: 535–539.

Sanchez-Azofelfa, G. A., K. L. Castro, B. Rivard, M. R. Kalascka and R. C. Harriss, 2003. Remote sensing research priorities in tropical dry forest environments. *Biotropica*, 35: 134–142.

Sanchez-Hernandez, C., D. S. Boyd and G. M. Foody, 2007. One-class classification for mapping a specific land-cover class: SVDD classification of fenland. *IEEE Transactions on Geoscience and Remote Sensing*, 45: 1061–1073.

Schaepman, M. E., in this volume. Imaging Spectrometers. Chapter 12.

Schaepman-Strub, G., M. E. Schaepman, J. V. Martonchik, T. H. Painter, and S. Dangel, in this volume. Radiometry and Reflectance: From Concepts to Measured Quantities. Chapter 15.

Southworth, J., H. Nagendra and D. K. Munroe, 2006. Are parks working? Exploring human–environment tradeoffs in protected area conservation. *Applied Geography*, 26: 87–95.

Stehman, S. V. and G.M. Foody, in this volume. Accuracy Assessment. Chapter 21.

Steven, M. D., T. J. Malthus, F. Baret, H. Xu and M. J. Chopping, 2003. Intercalibration of vegetation indices from different sensor systems. *Remote Sensing of Environment*, 88: 412–422.

Stow, D., L. L. Coulter and C. A. Benkelman, in this volume. Airborne Digital Multispectral Imaging. Chapter 11.

Strahler, A. H., L. Boschetti, G. M. Foody, M. A. Friedl, M. C. Hansen, M. Herold, P. Mayaux, J. T. Morisette, S. V. Stehman and C. E. Woodcock, 2006. *Global Land Cover Validation: Recommendations for Evaluation and Accuracy Assessment of Global Land Cover Maps*, Euopean Commission, Joint Research Centre, Ispra, Italy, EUR 22156 EN, 48 pp.

Teeuw, R., P. Aplin, N. McWilliam, T. Wicks, K. Matthieu, and G. Ernst, in this volume. Hazard Assessment and Disaster Management Using Remote Sensing. Chapter 32.

Toutin, T., in this volume. Fine Spatial Resolution Optical Sensors. Chapter 8.

Trisurat, Y., A. Eiumnoh, S. Murai, M. Z. Hussain, and R. P. Shrestha, 2000. Improvement of tropical vegetation mapping using a remote sensing technique: a case of Khao Yai National Park, Thailand. *International Journal of Remote Sensing*, 21: 2031–2042.

Turner, W., S. Spector, N. Gardiner, M. Fladeland, E. Sterling, and M. Steininger, 2003. Remote sensing for biodiversity science and conservation. *Trends in Ecology and Evolution*, 18: 306–314.

van der Meer, F., H. van der Werff and S. de Jong, in this volume. Pre-processing of Optical Imagery. Chapter 16.

van Leeeuwen, W. J. D., in this volume. Visible, Near-IR, and Shortwave IR Spectral Characteristics of Terrestrial Surfaces. Chapter 3.

Wang, L., W. P. Sousa, P. Gong and G. S. Biging, 2004. Comparison of IKONOS and QuickBird images for mapping mangrove species on the Caribbean coast of Panama. *Remote Sensing of Environment*, 91: 432–440.

Warner, T., M. D. Nellis and G. Foody, in this volume. Remote Sensing Data Selection Issues. Chapter 1.

Wilson, R. J., C. D. Thomas, R. Fox, D. B. Roy and W. E. S. Kunin, 2004. Spatial patterns in species distributions reveal biodiversity change, *Nature*, 432: 393–396.

Wulder, M., J. C. White, S. Ortlepp and N. C. Coops, in this volume. Remote Sensing for Studies of Vegetation Condition: Theory and Application. Chapter 25.

Appendix. Acronyms

1D	One-dimensional
2D	Two-dimensional
3D	Three-dimensional
A/D	Analog-to-digital
AAG	Association of American Geographers
AATSR	Advanced Along-Track Scanning Radiometer
ABBA	Automated Biomass Burning Algorithm
ABM	Agent-based modeling
ACORN	Atmospheric CORrection Now
ADEOS	Advanced Earth Observing Satellite
ADMID	Airborne digital multispectral image data
ADMIS	Airborne digital multispectral imaging system
AgRISTARS	Agriculture and Resources Inventory Surveys Through Aerospace Remote Sensing
AIM	Apparent Image Motion
AirMISR	Airborne Multi-angle Imaging SpectroRadiometer
AIRS	Atmospheric Infrared Sounder
AIRSAR	Airborne Synthetic Aperture Radar (NASA/JPL)
AIS	Airborne Imaging Spectrometer (NASA/JPL)
ALI	Advanced Land Imager
ALOS	Advanced Land Observing Satellite
ALS	Airborne Laser Scanning
ALS-50	Airborne Laser Scanner (Leica)
ALTM	Airborne Laser Terrain Mapper (Optech)
AMSR	Advanced Microwave Scanning Radiometer
AMSR-E	Advanced Microwave Scanning Radiometer for EOS
ANN	Artificial Neural Network
ANOVA	Analysis of Variance
ANZLIC	Australia and New Zealand Land Information Council
AOD	Aerosol Optical Depth
AOL	Airborne Oceanographic Lidar
AOTF	Acousto-Optical Tunable Filters
APEX	Airborne Prism Experiment
ARES	Airborne Reflective Emissive Spectrometer
ASAR	Advanced SAR sensor on Envisat
ASD	Analytical Spectral Devices
ASFMS	Aerial survey flight management system
ASG	Automated SpectroGoniometer
ASPRS	American Society of Photogrammetry and Remote Sensing
ASTER	Advanced Spaceborne Thermal Emission and Reflection Radiometer
AT	Along Track

ATLAS	Advanced Thermal and Land Applications Sensor
ATM	Airborne Thematic Mapper
ATSR	Along Track Scanning Radiometer
AVHRR	Advanced Very High Resolution Radiometer (NOAA meteorological satellites)
AVIRIS	Airborne Visible/Infrared Imaging Spectrometer (NASA JPL)
AVNIR	Advanced Visible and Near Infrared Radiometer
AWiFS	Advanced Wide-Field Sensor
BHR	BiHemispherical Reflectance
BHR_{iso}	BiHemispherical Reflectance under isotropic illumination conditions
BIRD	Bi-spectral Infra-Red Detector
BLIP	Background Limited performance
BNSC	British National Space Centre
bpi	Bits per inch
BRDF	Bidirectional Reflectance Distribution Function
BRF	Bidirectional Reflectance Factor
BTDF	Bidirectional Transmittance Distribution Function
CA	cellular automata
CASA	Carnegie–Ames–Stanford Approach
CASI	Compact Airborne Spectrographic Imager
CBD	Central Business District
CBERS	China–Brazil Earth Resources Satellite
CCD	Charge-Coupled Device
CCRF	Conical-Conical (biconical) Reflectance Factor
CCRS	Canadian Center for Remote Sensing
CCSM	Cross Correlogram Spectral Matching algorithm
CCSP	Climate Change Science Program
CDOM	Colored dissolved organic material
CEO	Centre for Earth Observation of the European Commission.
CEOS	Committee on Earth Observation Satellites
CERES	Clouds and Earth's Radiant Energy System
chl	Chlorophyll-a concentrations
CHM	Canopy Height Model
CHRIS	Compact High Resolution Imaging Spectrometer (SSTL, ESA)
CI	Chlorophyll Index
CIESIN	Center for International Earth Science Information Network
CIR	Color infrared
CLASS	Comprehensive Large Array-data Stewardship System
CNES	Centre National d'Etudes Spatiales, French Space Agency
COFUR	Cost of fulfilling a user request
COST	Cosine of the zenith angle, Tz
COTS	commercial off-the-shelf
CPR	Cloud Profiling Radar
CRP	Conservation Reserve Program
CTIS	Computed Tomography Imaging Spectrometer
CV	Coefficient of Variation
CVA	Change Vector Analysis
CW	Continuous Wave
CZCS	Coastal Zone Color Scanner
DAAC	Distributed Active Archive Center
DAID	Digital Airborne Imaging Spectrometer
dB	Decibel
DDV	Dense Dark Vegetation
DEM	Digital Elevation Model
DESDynI	Deformation, Ecosystem Structure, Dynamics of Ice (NASA)
dGPS	differential Global Positioning System
DHR	Directional-Hemispherical Reflectance
DIMAP	DIctionary Maintenance Programs
DInSAR	Differential SAR Interferometry

DInSAR	Differential InSAR
DLR	Deutsche Forschungsanstalt fuer Luft- und Raumfahrt, German Aerospace Centre
DLR	German Aerospace Research Establishment
DMC	Disaster Monitoring Constellation
DMSP	Defense Meteorological Satellite Program
DN	Digital Number
DOAS	Differential Optical Absorption Spectrometers
DOI	Department of the Interior
DOM	Dissolved Organic Matter
DOQ	Digital Orthophoto Quadrangle
DOS	Dark Object Subtraction
DRC	Democratic Republic of Congo
DSM	Digital Surface Model
DSM	Digital Soil Mapping
DTM	Digital Terrain Model
DVD	Digital Versatile Disc
EBCM	Extended Boundary Condition Method
EDC	EROS Data Center (pre-2005, now obsolete)
EDMSS	Environmental and Disaster Monitoring Satellite System
EDOS	EOS Data and Operations System
EDRS	Environmental Data Records
EIA	Electronics Industries Association
EM	Electromagnetic
EnMAP	Environmental Mapping
ENSO	El Nino Southern Oscillation
ENVISAT	Environmental Satellite
EO	Electro-optical
EO	Earth Observation
EO-1	Earth Observing–1 (NASA mission)
EOC	Earth Observing Camera
EOCAP	Earth Observations Commercialization Applications Program
EOS	Earth Observing System (NASA)
EOSAT	Earth Observation Satellite
EOSDIS	EOS Data and Information System
EPS	Eumetsat Polar System
ERBE	Earth Radiation Budget Experiment
ERIM	Environmental Research Institute of Michigan
EROS	Earth Resources Observation System (pre-2005)
EROS	Earth Resources Observation and Science (post-2005)
ERS	European Remote Sensing (Satellite)
ERS-1	European Remote Sensing satellite – 1
ERS-1/2	Earth Resource Satellite $^1/_2$
ERS-2	European Remote Sensing satellite – 2
ERTS	Earth Resources Technology Satellite
ESA	European Space Agency
ET	Evapotranspiration
ETM+	Enhanced Thematic Mapper Plus (Landsat)
EU	European Union
EUMETSAT	European Organisation for the Exploitation of Meteorological Satellites
EuroSDR	European Spatial Data Research
EWDI	Enhanced Wetness Difference Index
FGDC	Federal Geographic Data Committee
FIGOS	FIeld GOniometer System
FLAASH	Fast Line-of-sight Atmospheric Analysis of Spectral Hypercubes
FLH	Fluorescence Line Height
FLI	Future of Land Imaging
FLI	Fluorescence Line Imager
FLI/PMI	Fluorescence Line Imager/Programmable Multispectral Imager
FMC	Forward motion compensation

FOR	Field of regard
FOV	Field of view
fPAR	Fraction of absorbed photosynthetically active radiation
FSR	Fine spatial resolution
FTS	Fourier Transform Spectrometer
FWHM	Full-width-half-maximum
GAC	Global Area Coverage
GCM	General Circulation Model
GCP	Ground control point
GEMI	Global Environment Monitoring Index
GEMS	Grating electromechanical system
GEO	Group on Earth Observation
GEOSS	Global Earth Observation System of Systems
GeoTIFF	Georeferencing Tagged Image format
GER	Geophysical and Environmental Research Corp.
GI	Geographical Information
GIFOV	Ground instantaneous field of view
GIFTS	Geosynchronous Imaging Fourier Transform Spectrometer
GILS	Government Information Locator Services
GIMSS	Global Inventory Monitoring and Modeling Study
GIQE	General image quality equation
GIS	Geographic Information Systems
GISci	Geographic Information Science
GISTDA	Geo-Informatics and Space Technology Development Agency
GLI	Global Imager
GLIMS	Global Land Ice Measurements from Space
GLONASS	GLObal Navigation Satellite System
GMES	Global Monitoring for Environment and Security
GNI	Gross National Income
GO	Geometric-optical
GOES	Geostationary Operational Environmental Satellite
GPP	Gross Primary Productivity
GPS	Global Positioning System
GPS-AINS	Global Positioning System-aided inertial navigation system
GRACE	Gravity Recovery and Climate Experiment
GRD	Ground resolved distance
GSD	Ground sample distance
GSFC	Goddard Space Flight Center
GTOPO	Global Topographic Data
GYURI	General Yield Unified Reference Index
HCMM	Heat Capacity Mapping Mission
HCRF	Hemispherical-Conical Reflectance Factor
HDF	Hierarchical Data Format
HDRF	Hemispherical-Directional Reflectance Factor
HH	Horizontal transmit, horizontal receive
HIRIS	High Resolution Imaging Spectrometer
HJ	Huan jing
HMU	Hydrological Management Unit
HRG	High Resolution Geometric
HRMSI	High Resolution Multispectral Stereo Imager
HRPT	High Resolution Picture Transmission
HRV	High Resolution Visible (SPOT sensor)
HRVIR	High Resolution Visible Infrared
HSI	Hyperspectral imaging
HSRS	Hot Spot Recognition Sensor System
HSV	Hue, Saturation, Value
HV	Horizontal transmit, vertical receive
HYDICE	HYperspectral Digital Imagery Collection Experiment
HyMap	Hyperspectral Mapper (Integrated Spectronics)

HyspIRI	Hyperspectral Infrared Imager
IAS	Image Assessment System
IC	International Cooperator
ICEsat	Ice, Cloud, and land Elevation Satellite
ICSU	International Council for Science
IEM	Integral Equation Method
IEOS	Integrated Earth Observation System
IFOV	Instantaneous field of view
IFSAR	Interferometric Synthetic Aperture Radar
IGOL	Integrated Global Observations of Land
ILAS	Improved Limb Atmospheric Spectrometer
IMS	Internet Map Service
IMU	Inertial Measurement Unit
INS	Inertial navigation system
InSAR	Interferometric Synthetic Aperture Radar
IPO	Integrated Program Office
IR	Infrared
IRMSS	InfraRed MultiSpectral Scanner
IRS	Indian Remote Sensing (Satellite)
IRTM-NN	Inverse Radiative Transfer Model-Neural Network
ISCCP	International Satellite Cloud Climatology Project
ISPRS	International Society for Photogrammetry and Remote Sensing
JAMI	Japanese Advanced Meteorological Imager
JAXA	Japan Aerospace Exploration Agency
JERS-1	Japanese Earth Resources Satellite – 1
JPL	Jet Propulsion Laboratory (NASA)
JZ	Jian zai
KA	Kirchhoff Approximation
KT	Kauth–Thomas (also called Tasseled Cap Transformation)
KOMPSAT	KOrean MultiPurpose SATellite
LAC	LEISA Atmospheric Corrector
LAC	Local area coverage
LACIE	Large Area Crop Inventory Experiment
LAD	Leaf angle distribution
LAGOS	LAboratory GOniometer System
LAI	Leaf Area Index
Landsat	Land Satellite
LARS	Laboratory for Applications of Remote Sensing
LASER	Light Amplification by Stimulated Emission of Radiation
LCTF	Liquid crystal tunable filter
LDCM	Landsat Data Continuity Mission
LGSOWG	Landsat Ground Station Operations Working Group
Lidar	Light Detection and Ranging
LISS	Linear Imaging Self Scanner
LMS-Q560	Full-Waveform Airborne Laser Scanner (Riegl)
LOWTRAN	Low Resolution Transmission
LPSO	Landsat Project Science Office
LPV	Land Product Validation
LSF	Line spread function
LST	Land surface temperature
LSU	Linear Spectral Unmixing
LTAP	Long-Term Acquisition Plan
LUE	Light Use Efficiency
LUT	Look-up table
LVIS	Laser Vegetation Imaging Sensor (NASA)
LWIR	Longwave infrared
MAD	Mean absolute deviation
MASTER	MODIS/ASTER Airborne Simulator
MAUP	Modifiable areal unit problem
MB	Multiband

MD	Mean deviation
MEMS	Micro electrical and mechanical system
MERIS	Medium Resolution Imaging Spectrometer (ESA)
Meteosat	Meteorological Satellite
MIR	Middle infrared
MISR	Multiangle Imaging SpectroRadiometer
MIVIS	Multispectral Infrared and Visible Imaging Spectrometer
MK-II	Airborne Laser Scanner (Topeye)
MKT	Multi-temporal Kauth Thomas
MLC	Maximum likelihood classifier
MLR	Multiple linear regression
MNF	Minimum noise fraction
MODAPS	MODIS Adaptive Processing System
MODIS	Moderate Resolution Imaging Spectrometer (NASA)
MODTRAN	MODerate resolution atmospheric TRANsmission
MoM	Method of Moments
MOPITT	Measurement of Pollution in the Troposphere
MOS	Modular Optoelectronic Scanner
MRPV	Modified Rahman Pinty Verstraete model
MRVBF	Multi Resolution Valley Bottom Flatness Index
MS	Multispectral
MSG	Meteosat Second Generation
MSI	Multispectral imaging
MSS	Multispectral Scanner (known earlier as Multispectral Scanning Subsystem) (Landsat)
MTF	Modulation transfer function
MTI	Multispectral Thermal Imager
MTSAR-IR	Multi-functional Transport Satellite
MTSAT	Japanese Ministry of Transport geostationary meteorological satellite
MVA	Multiple View Angle
MVISR	Multichannel Visible and IR Scan Radiometer
MWIR	Mid-wave infrared
N	Nitrogen
NAPC	Noise adjusted principal component
NASA	National Aeronautics and Space Administration
NASDA	National Space Development Agency of Japan
NASS	National Agriculture Statistics Service
NBAR	Nadir BRDF Adjusted Reflectance
NCI	Neighborhood correlation images
NDSI	Normalized Difference Snow Index
nDSM	normalized Digital Surface Model
NDVI	Normalized Difference Vegetation Index
NED	National Elevation Datasets
NER	Noise equivalent radiance
NESDIS	NOAA National Environmental Satellite Data and Information Service
NEΔL	Noise equivalent delta radiance
NEΔT	Noise equivalent delta temperature
NE$\Delta\rho$	Noise equivalent delta reflectance
NIH	National Institutes of Health
NIIRS	National image interpretability rating scale
NIR	Near infrared
NIST	National Institute of Standards and Technology (U.S.)
NLCD	National Land Cover Dataset
NLIP	National Land Imaging Program
NMS	National Meteorological Services
NOAA	National Oceanic and Atmospheric Administration
NP	Navigation processor
NPOES	National Polar-orbiting Operational Environmental Satellite System
NPP	NPOESS Preparatory Project (satellite)
NPP	Net primary productivity
NRC	National Research Council

NRCS	Natural Resource Conservation Service
NRI	National Resources Inventory
NSF	National Science Foundation
NSIDC	National Snow and Ice Data Center
NSLRSDA	National Satellite Land Remote Sensing Data Archive
NTIF	New Industry Text Format
NWI	National Wetlands Inventory
OBPG	Ocean Biology Processing Group
OCTS	Ocean Color and Temperature Scanner
OJP	Old Jack Pine
OLCI	Ocean and Land Colour Instrument
OLI	Operational Land Imager
OM	Organic Matter
OMB	Office of Management and Budget
OSTP	Office of Science and Technology Policy
PALSAR	Phased Array L-Band SAR
Pan	Panchromatic
PAR	Photosynthetically active radiation
PARABOLA	Portable Apparatus for Rapid Acquisition of Bidirectional Observation of the Land and Atmosphere
PC	Personal Computer
PCA	Principal component analysis
PEM	Production Efficiency Models
PIF	Pseudo Invariant Feature
Pixel	Picture element
PLSR	Partial least-square regression
PnET-CN	Net photosynthesis evapotranspiration carbon and nitrogen cycling model
POES	Polar Orbiting Environmental Satellites
POLDER	POLarization and Directionality of the Earth's Reflectances
PolInSAR	Polarimetric SAR Interferometry
PRF	Pulse Repetition Frequency
PRI	Photochemical Reflectance Index
PRISM	Panchromatic Remote-sensing Instrument for Stereo Mapping
PROBA	Project for On-board Autonomy (ESA)
PSF	Point spread function
PS-InSAR	Permanent Scatter InSAR
PSU	Primary sampling unit
Radar	Radio Detection and Ranging
Radarsat	Radar Satellite
RAR	Real Aperture Radar
RASCAL	Raster Scanning Airborne Laser Altimeter (NASA)
RBV	Return Beam Vidicon
RCA	Radio Corporation of America
REIS	RapidEye Earth Imaging System
RGB	Red, green, blue
RMS	Root mean square
RMSE	Root mean square error
ROSIS	Reflective Optics System Imaging Spectrometer
RossThickLi-SparseReciprocal	BRDF model of the MODIS BRDF product algorithm
RSI	Remote Sensing Instrument
RST	Rotation, scale, translation
RT	Radiative transfer
SAIL	Scattering by Arbitrarily Inclined Leaves
SAM	Spectral Angle Mapper
SAR	Synthetic Aperture Radar
ScaLARS	Scanning Laser Altimeter of the University of Stuttgart
SCIAMACHY	Scanning Imaging Absorption Spectrometer for Atmospheric Chemistry
SCLP	Snow and Cold Land Processes mission

SeaWiFS	Sea-viewing Wide Field-of-view Sensor
SEBAL	Surface Energy Balance Algorithm for Land
SEBASS	Spatially Enhanced Broadband Array Spectrograph System
SEOSAT	Spanish Earth Observation Satellite
SEVIRI	Spinning Enhanced Visible and Infrared Imager
SFF	Spectral feature fitting
SFSI	Shortwave Infrared (SWIR) Full Spectrum Imager
SGP	Simplified General Perturbations
SI	International System of Units
SI	Semantic Importing
SIR	Shuttle Imaging Radar
SIR-C	Shuttle Imaging Radar – C
SIR-C/X-SAR	Shuttle Imaging Radar, X and C band synthetic aperture radar
SISEX	Shuttle Imaging Spectrometer Experiment
SKLM	Simple Kriging with varying Local Means
SLA	Shuttle Radar Altimeter (NASA)
SLA	Specific Leaf Area
SLICER	Scanning Lidar Imager of Canopies by Echo Recovery (NASA)
SLIM6	Surrey Linear Imager Multispectral 6 Channel
SMA	Spectral mixture analysis
SMAP	Soil Moisture Active/Passive mission
SMMR	Scanning Multichannel Microwave Radiometer
SMORPH	Slope MORPHology
SMOS	Soil Moisture and Ocean Salinity
SNR	Signal-to-noise ratio
SPECTRA	Surface Processes and Ecosystem Changes Through Response Analysis
SPG	SpectroPhotoGoniometer
SPM	Small Perturbation Method
SPM	Suspended particulate matter
SPOT	Satellite Pour l'Observation de la Terre
SRF	Spectral response function
SRS	Simple random sampling
SRTM	Shuttle Radar Topography Mission
SSH	Sea Surface Height
SSM/I	Special Sensor Microwave/Imager
SST	Sea surface temperature
SSU	Secondary sampling unit
STS	Shuttle Transportation System
SVAT	Soil-Vegetation-Atmosphere-Transfer
SVM	Support vector machines
SWE	Snow water equivalent
SWIR	Short-wave infrared
T	Temperature
TBC	To Be Confirmed
TCI	Temperature Crop Index
TCT	Tasseled Cap Transformation
TDI	Time delay integration
TDRS	Tracking and Data Relay Satellite
TDRSS	Tracking and Data Relay Satellite System
TES	Thermal Emission Spectrometer
THEOS	Thailand Earth Observation System
TIGER	Topologically Integrated Geographic Encoding and Referencing system
TIMS	Thermal Infrared Multispectral Scanner
TIN	Triangulated Irregular Network
TIR	Thermal infrared
TIROS	Television Infrared Observation Satellite
TIROS-N	Television Infrared Observation Satellite – NOAA
TLS	Three-line scanners
TM	Thematic Mapper (Landsat)

TOA	Top of Atmosphere
TOMS	Total Ozone Mapping Spectrometer
TOPORAD	TOPOgraphic distribution of solar RADiation (topographic radiation model)
TOVS	TIROS Operational Vertical Sounding
TRMM	Tropical Rainfall Measuring Mission
TRN	Thermal Response Number
U.S.	United States
U2	Utility-2
UAV	Unmanned aerial vehicle
UEQ	Urban Environmental Quality
UHF	Ultra-high Frequency
UHI	Urban heat island
UK	United Kingdom
UK EA	UK Environment Agency
UK-DMC	United Kingdom Disaster Monitoring Constellation
UN	United Nations
UNITAR	United Nations Institute for Training and Research
US	United States
US NGA	National Geospatial Intelligence Agency
USA	United States of America
USDA	United States Department of Agriculture
USGS	United States Geological Survey
USSR	Union of Soviet Socialist Republics
UV	Ultraviolet
V/NIR	Visible and near infrared
VCF	Vegetation Continuous Fields
VCI	Vegetation Condition Index
VF	Vegetation Fraction
VFSR	Very fine spatial resolution
VHF	Very-high Frequency
VIIRS	Visible/Infrared Imager/Radiometer Suite
VIR	Visible / Infrared Imagery
VIRS	Visible and Infrared scanner
VIS	Visible
VIS	Vegetation-Impervious surface-Soil
VNIR	Visible and near infrared
Voxel	Volume element
VPM	Vegetation Phenology Metrics
VV	Vertical send, vertical receive
WASP	Wildfire airborne sensor program
WFI	Wide Field Imager
WGS84	World Geodetic System 1984
WiFS	Wide-Field Sensor
WMO	World Meteorological Organisation
WRS	Worldwide Reference System
XT	Cross Track
ZY	ZiYuan

Index

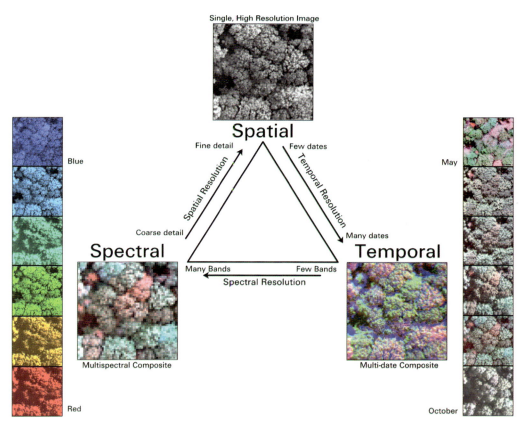

Figure 1.1 Given a limited bandwidth for image acquisition, storage, and communication, trade-offs have to be made regarding the spatial, spectral, and temporal scale of the imagery that can be acquired. Radiometric scale (not shown) is also important.

Source: Figure reproduced from T. Key, T. Warner, J. McGraw, and M. A. Fajvan, 2001. A comparison of multispectral and multitemporal imagery for tree species classification. *Remote Sensing of Environment* 75: 100–112.

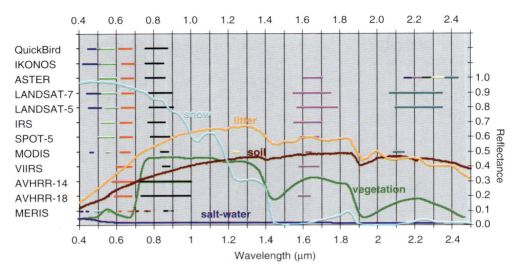

Figure 3.3 Spectral bandwidths and their wavelength positions for a range of frequently used sensors (QuickBird, IKONOS, ASTER, LANDSAT-7, LANDSAT-5, IRS, SPOT-5, MODIS, VIIRS, AVHRR-14, AVHRR-18, and MERIS). The bandwidth is represented by the full width at half-maximum (FWHM) spectral response. Soil, snow, green leaf, and senescent brown leaf spectral signatures are shown to illustrate the importance of the position of the instrument band-passes as they affect the reflectance response of a surface consisting of a mixture of contributing components (e.g., soil, vegetation, water, NPV). The bright soil is an Alfisol soil developed under temperate forests of the humid mid-latitudes.
Source: Based on data from NASA, 1999[1].

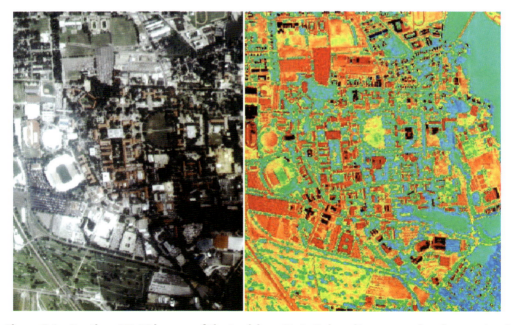

Figure 5.4 Daytime ATLAS images of the Louisiana State University campus showing a natural color image on the left and a thermal image on the right, both at 10 m spatial resolution.

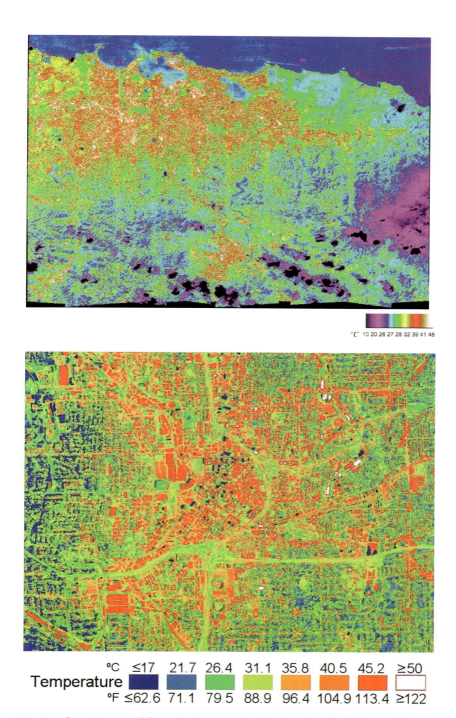

°C 10 20 26 27 28 32 39 41 48

°C	≤17	21.7	26.4	31.1	35.8	40.5	45.2	≥50
Temperature								
°F	≤62.6	71.1	79.5	88.9	96.4	104.9	113.4	≥122

Figure 5.5   Daytime 10 m spatial resolution ATLAS color density sliced thermal images at 10 m spatial resolution of the eastern end of Puerto Rico (top) and of the Atlanta, Georgia central business district (bottom).

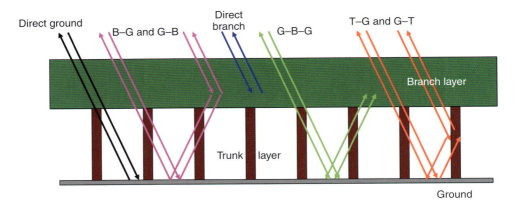

Figure 6.1   Geometry of electromagnetic scattering from vegetated terrain.

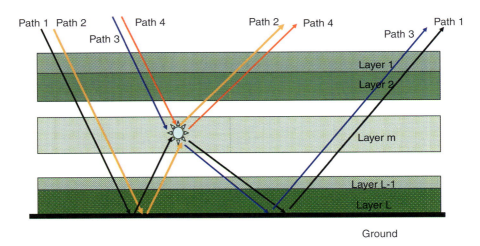

Figure 6.2   Multilayer model geometry used in the Liang et al. (2005) vector RT model, showing the four possible interactions for each layer of vegetation: (1) ground-layer m-ground, (2) ground-layer m, (3) same as two but opposite path, (4) direct from layer m.

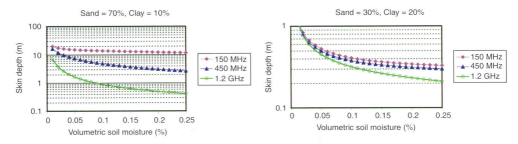

Figure 6.6   Penetration depth at L-band (1.2 GHz), UHF (450 MHz), and VHF (150 MHz) for different soils.

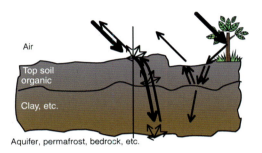

Figure 6.7 Coherent wave scattering paths between trees, surface, and subsurface. These paths contribute significantly to radar backscatter at lower frequencies.

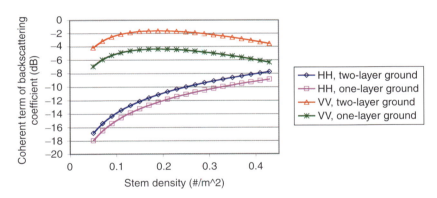

Figure 6.8 Effect of stem-subsurface multiple scattering at VHF, showing the potentially large errors committed if the effect of the subsurface interface is not included in the calculation of the coherent backscattering coefficient. The effect is more pronounced for the VV polarization.

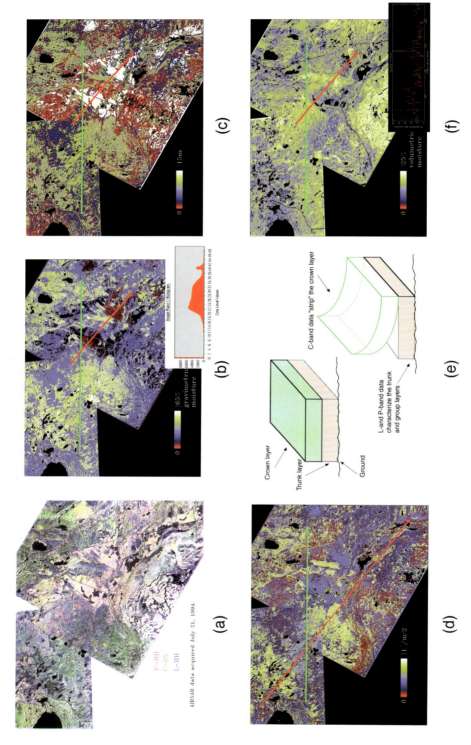

Figure 6.9   Multifrequency multimechanism inversion of forest properties. (a) Three-color overlay of AIRSAR imagery over a boreal forest area with mixed species. (b) Inversion of crown layer moisture from C-band. (c) Inversion of crown layer (and stand) height from C-band. (d) Inversion of stem density from C- and L-band. (e) Conceptual representation of removing the top layer using highest frequency. (f) Inversion of subcanopy soil moisture, after simulation of crown layer backscatter at L- and P-bands, then subtracting from total and using remainder to estimate soil moisture.

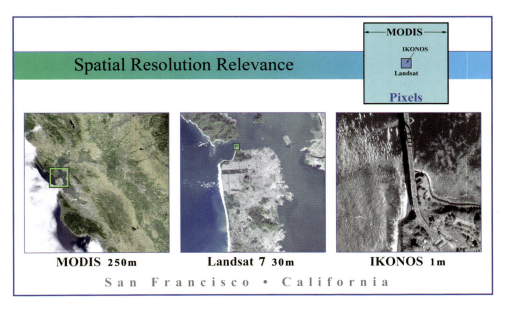

Figure 9.2   Comparative visual examples of coarse-, moderate-, and fine-resolution images of the central California, San Francisco region of the US. Note that the area of the region imaged is the squared product of the sample and line dimensions.
Source: Composition courtesy of Laura Rocchio, Landsat Project Science Office, NASA/GSFC.

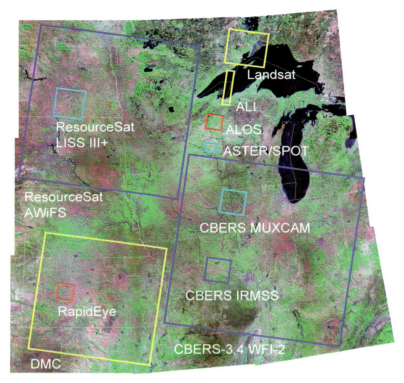

Figure 9.3 Comparative image footprint sizes for a variety of the moderate spatial resolution sensors discussed. Relative placement of the sensor footprints is for illustrative purposes only. The outer margins of the background image are approximately 2000 km across.
Source: Figure supplied by Jim Lacasse, USGS/EROS.

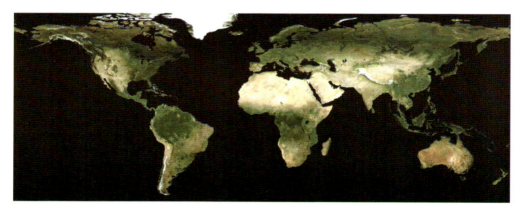

Figure 10.1 A MODIS surface reflectance, simulated true color image derived from the global 8 day surface reflectance products at 500 m (courtesy of E. Vermote and S. Kotchenova of the University of Maryland)

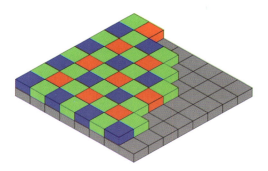

Figure 11.2   Bayer color filter pattern utilized for single-chip color and color infrared (CIR) digital cameras. Pixels shown in blue are sensitive to near infrared radiance in CIR mode.
Source: Image courtesy of Colin M. L. Burnett (http://en.wikipedia.org/wiki/Image:Bayer_pattern_on_sensor.svg).

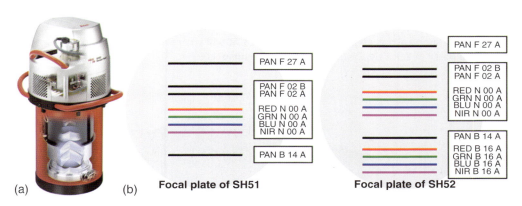

(a)   (b)   **Focal plate of SH51**   **Focal plate of SH52**

Figure 11.3   Leica ADS40 (a) digital sensor and (b) its latest sensor head focal plate configurations. Waveband, forward/nadir/backward/viewing direction, view angle, and sensor (A or B) is indicated on the focal plate diagram.
Source: Courtesy of Leica Geosystems Geospatial Imaging, LLC.

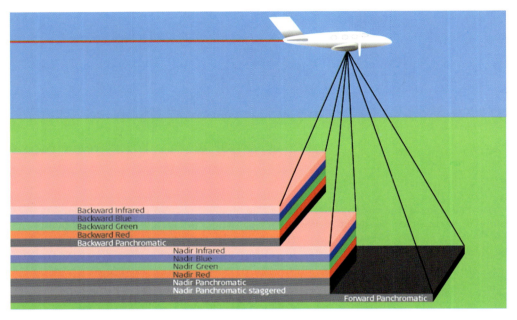

Figure 11.4 Image strips acquired using the Leica ADS40 with SH52 sensor head.
Source: Courtesy of Leica Geosystems Geospatial Imaging, LLC.

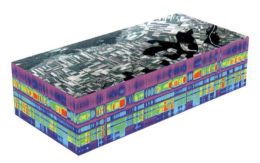

Figure 12.5 Imaging spectrometer data cube acquired by an airborne HyMap system on 26 August 1998 in Switzerland (Limpbach Valley). The sides are color-coded spectra by intensity.

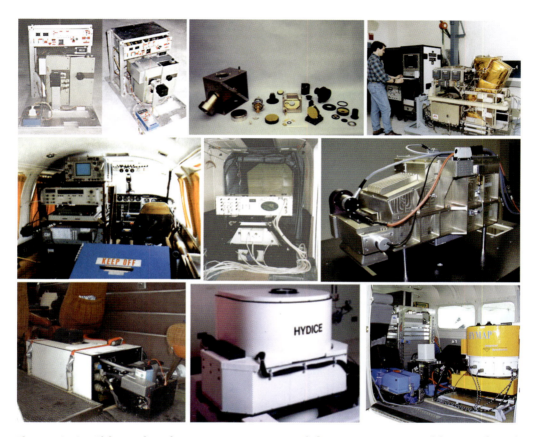

Figure 12.12   Airborne imaging spectrometers. From *left, top*: GER MARK II Airborne Infrared Spectroradiometer, Airborne Imaging Spectrometer (AIS) instrument assembly, Airborne Visible/Infrared Imaging Spectrometer (AVIRIS); *Middle*: Fluorescence Line Imager/Programmable Multispectral Imager (FLI/PMI), Digital Airborne Imaging Spectrometer (DAIS7915), Reflective Optics Imaging Spectrometer (ROSIS); *Bottom*: Shortwave Infrared Full Spectrum Imager (SFSI), Hyperspectral Digital Imagery Collection Experiment (HYDICE) detector assembly, and Hyperspectral Mapper (HyMap).
Source: Photos courtesy of: S.-H. Chang, G. Vane, R. Green, R. Baxton, A. Müller, H. van der Piepen, B. Neville, M. Landers, and M. Schaepman.

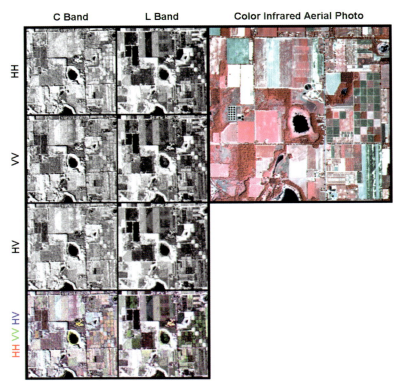

Figure 13.2   Back scatter images of JPL AIRSAR C/L-band channels in hh, vv, and hv polarizations. Kellogg Biological Station in Michigan, USA. Image size is 2.8 by 2.6 sq km.

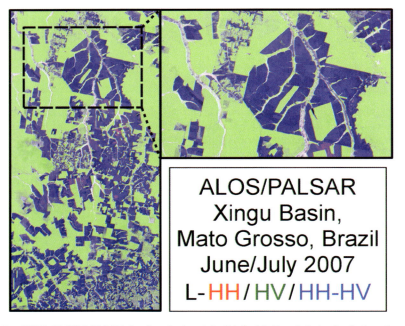

Figure 13.3   JAXA ALOS PALSAR dual-polarimetric (hh/hv) L-Band data depicting Amazon deforestation. Xingu river basin, Brazil. Zoomed image is 70 by 50 sq km.

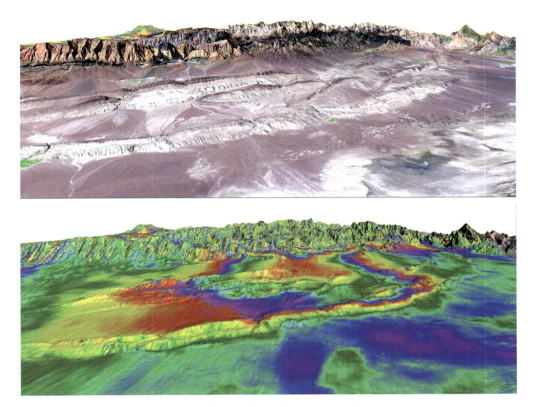

Figure 13.6   Digital elevation data derived from interferometric radar data.
Source: Courtesy NASA/JPL-Caltech.

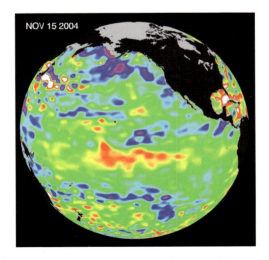

Figure 13.9   Jason radar altimeter derived ocean surface height.
Source: Courtesy NASA/JPL-Caltech.

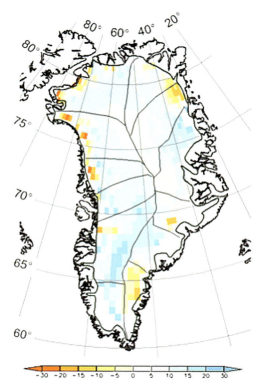

Figure 13.10   Greenland ice thickness derived from ERS-1/2/ radar altimeters (Johannessen et al. 2005).

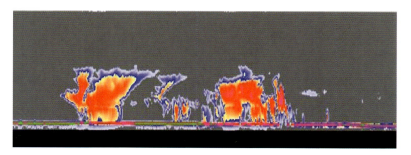

Figure 13.14   Cross-section of tropical clouds and thunderstorm cells from CloudSat profiling radar. (Courtesy of NASA/JPL-Caltech and the Cooperative Institute for Research in the Atmosphere (CIRA), Colorado State University).

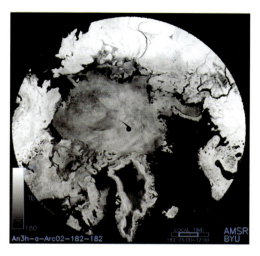

Figure 13.15   North polar view with the AMSR-E radiometer. Data shown are from the H-polarized 19 GHz channel, and were acquired on July 2, 2002. (Courtesy of NASA/JPL-Caltech and the Scatterometer Climate Record Pathfinder project, Microwave Earth Remote Sensing Lab, Brigham Young University).

Figure 14.1   Principle of airborne laser scanning. Airborne laser scanning is a method for acquiring lidar range measurements from an airborne platform and the precise orientation of these measurements. Short laser pluses (4–10 ns) are emitted with a high frequency (e.g., 50–200 kHz) and are continuously deflected in across-flight direction using various scanning methods. The position and rotation of the sensor is continuously recorded along the flight path using a Global Positioning System (GPS) and an Inertial Measurement Unit (IMU). The recorded measurements of sensor position and orientation, beam deflection and range can be converted to a georeferenced three-dimensional (3D) point cloud representing the surface targets that reflected the laser pulses. Courtesy of Hannu Hyyppä.

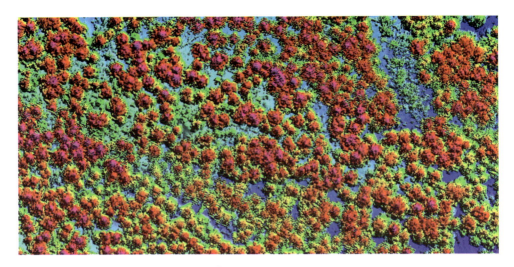

Figure 14.2 Example of lidar-derived digital surface models DSM obtained for forested area. Trees or tree groups are visible. DSMs are typically coded by colors red referring to the highest and blue to the lowest elevations. Courtesy of Hannu Hyyppä.

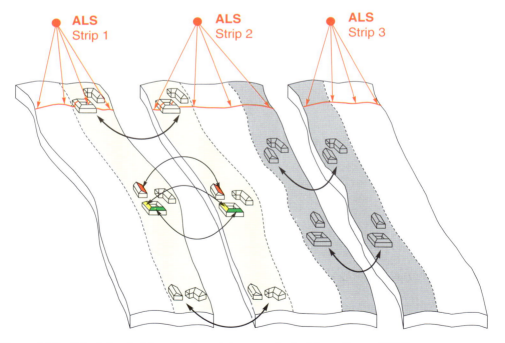

Figure 14.5 Principle of strip adjustment. Image adapted from Kager (2004).

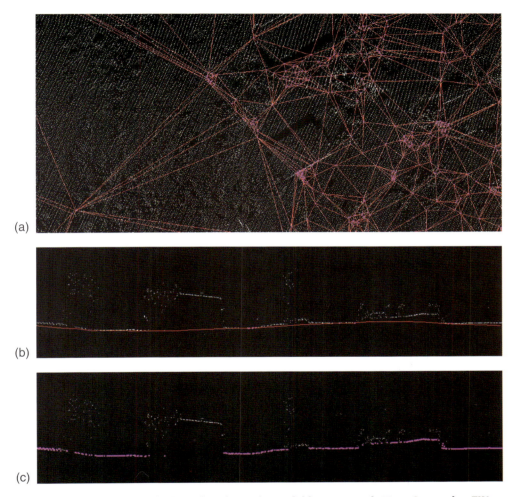

(a)

(b)

(c)

Figure 14.6 The creation of triangulated terrain model in progress in TerraScan using TIN densification method (top). Ground points are partly found and searching *for more* ground points continues (middle). All the ground points are found. Houses and trees are excluded (bottom). Courtesy of Arttu Soininen.

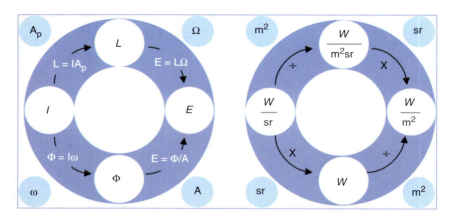

Figure 15.2   Overview of radiometric quantities (left) and corresponding units (right) as used in remote sensing, namely radiant intensity *I*, radiance *L*, irradiance *E*, and radiant flux (or power) Φ.

Incoming/**Reflected**	**Directional**	**Conical**	**Hemispherical**
Directional	Bidirectional Case 1	Directional-conical Case 2	Directional-hemispherical Case 3
Conical	Conical-directional Case 4	Biconical Case 5	Conical-hemispherical Case 6
Hemispherical	Hemispherical-directional Case 7	Hemispherical-conical Case 8	Bihemispherical Case 9

Figure 15.4   Geometrical-optical concepts for the terminology of at-surface reflectance quantities. All quantities including a directional component (i.e., Cases 1–4, 7) are conceptual quantities, whereas measurable quantities (Cases 5, 6, 8, 9) are shaded in gray. (Reprinted from G., Schaepman-Strub, M. E. Schaepman, T. H. Painter, S. Dangel, and J. V. Martonchik, 2006. Reflectance quantities in optical remote sensing–definitions and case studies. *Remote Sensing of Environment*, 103: 27–42).

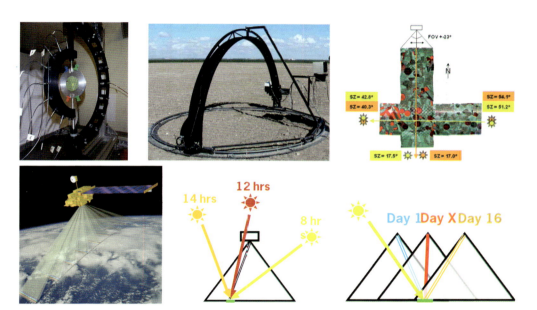

Figure 15.5   Examples of multi-angular sampling principles: (a) laboratory facility to measure leaf optical properties (SPG) (photo courtesy of Stéphane Jacquemoud), (b) field goniometer system (FIGOS), (c) airborne multi-angular sampling during DAISEX'99 using the HyMap sensor (SZ = solar zenith angle (Berger et al. 2001)), (c) spaceborne near-instantaneous multi-angular sampling (MISR with nine cameras), (e) daily composites of geostationary satellite sensors (e.g., Meteosat), (f) multiple-day compositing of polar orbiting satellite sensors (e.g., MODIS 16 days).

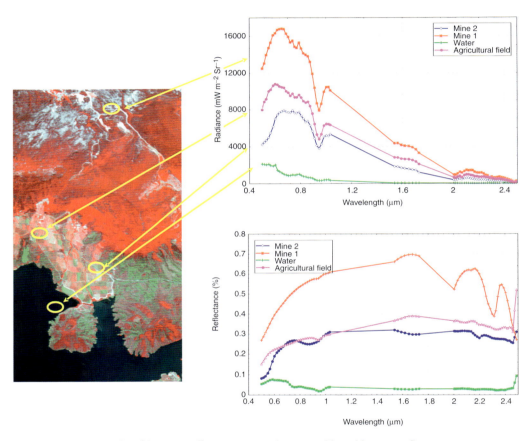

Figure 16.4   Example of image radiance spectra (top graph) and image reflectance spectra (bottom graph). The radiance spectra have the overall appearance of a solar curve, while the reflectance spectra show distinct areas of absorption. Mine 2 is a dolomite mine showing strong absorption at 2.3 μm attributed to Ca-Mg OH, while mine 1 is a bauxite mine showing absorption features in the visible part of the spectrum attributed to iron-oxide.

Image Segmentation and Object-based Classification

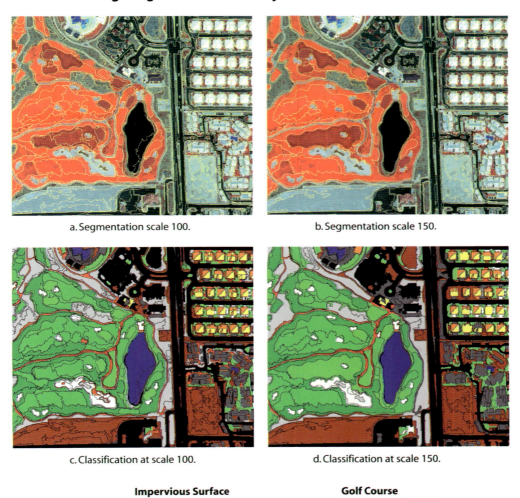

a. Segmentation scale 100.

b. Segmentation scale 150.

c. Classification at scale 100.

d. Classification at scale 150.

	Impervious Surface	Golf Course	
Water	Roof in direct sunlight	Green	Bare soil
Vegetation	Roof oriented away from direct sunlight	Fairway	Cart path
Barrenland	Flat rooftop and other built-up	Rough	Sand
	Asphalt		

Figure 19.1   Example of object-based classification at two different image segmentation scales. The pan-sharpened QuickBird high spatial resolution (61 × 61 cm) multispectral imagery of Las Vegas, NV was acquired on 18 May 2003.

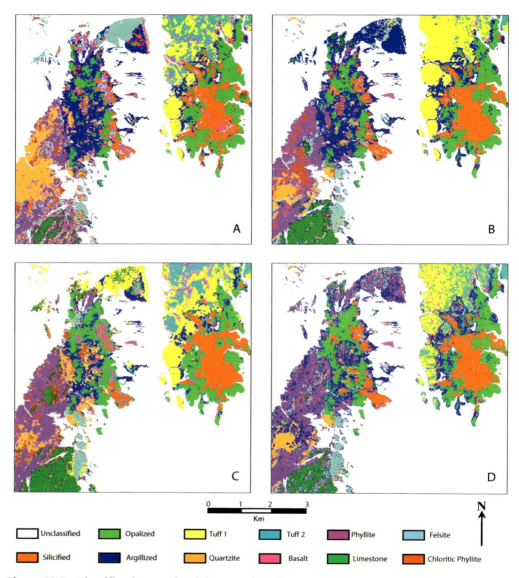

Figure 23.3   Classification results: (A) spectral angle mapper using AVIRIS data, (B) spectral angle mapper using combined AVIRIS and MASTER data, (C) spectral feature fitting using AVIRIS data, (D) spectral feature fitting using the combined data (reprinted from *Remote Sensing of Environment*, volume 110, X. Chen, T. A. Warner, and D. J. Campagna, Integrating visible, near-infrared and short-wave infrared hyperspectral and multispectral thermal imagery for geological mapping at Cuprite, Nevada, page 349, Copyright (2007), with permission from Elsevier).

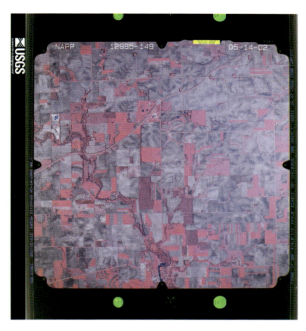

Figure 24.1   CIR aerial photograph representing soil patterns, Winneshiek and Fayette Counties, Northeastern Iowa, on Iowan drift, May, 2002. Variations in image tone and texture in open fields indicate local differences in drainage, texture, and organic matter. Interfluves are chiefly Mollisols formed in a thin layer of till and loess covering limestone bedrock. The sideslopes are chiefly occupied by deep, moderately well-drained Typic Hapludolls formed in silty and loamy sediments and the underlying glacial till. Footslopes and drainageways are occupied by deep, somewhat poorly drained Aquic Hapludolls. NAPP photograph, NP0NAPP012985149.

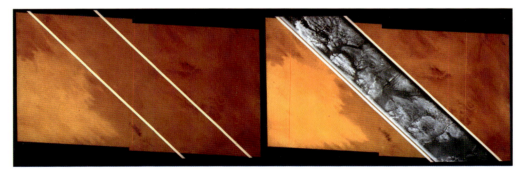

Figure 24.4   JPL's imaging radar on NASA's second space shuttle flight in November 1981, as reported in the Dec. 3, 1982, edition of Science (McCauley et al., 1982). The experiment, called Shuttle Imaging Radar-A (SIR-A), penetrated cloud cover, varying degrees of vegetation, and dry desert sand to provide new geologic information in poorly surveyed regions. For example, U.S. Geological Survey and NASA-Jet Propulsion Laboratory scientists studying SIR-A data found that the radar had penetrated beneath the extremely dry Selima Sand Sheet dunes and drift sand of the Arba'in desert in the Sudan and Egypt to reveal ancient, buried stream beds, geologic structures, and probable Stone Age habitation sites. (Images courtesy of JPL, Pasadena, CA, and USGS Image Processing Facility, Flagstaff, AZ) (http://pubs.usgs.gov/gip/deserts/remote/, http://www.jpl.nasa.gov/news/features.cfm?feature=422).

**Figure 25.1   Examples of image spatial resolution over a forested scene with crowns of
varying condition. Superimposed pixel sizes range from 30 m (Landsat) (L-resolution model
with many objects per pixel) to 4 m (IKONOS multispectral) to 1 m (IKONOS panchromatic)
(H-resolution model with many pixels per object).**

Figure 27.1   CZCS colour composites of (a) 6 September 1979 covering the south eastern North Sea with high levels of suspended particulate matter (yellow to red pixels) and (b) 6 July 1983 with a coccolithophore bloom (white pixels). High chlorophyll concentrations in both images appear as dark blue/grey patches. Wavebands 3 (550 nm), 2 (520 nm) and 1 (440 nm) are displayed as red, green and blue.
Source: Images courtesy of Steve Groom at the NERC Earth Observation Data Acquisition and Analysis Service (NEODAAS).

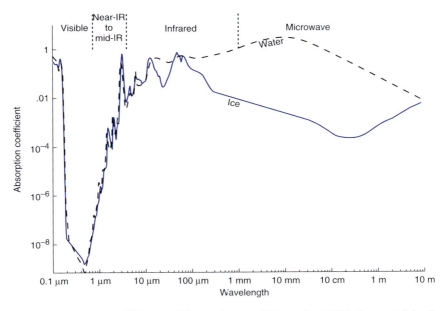

Figure 28.1 Absorption coefficients of ice and water (Wiscombe 2005). From visible through infrared wavelengths, ice and water are similar. In the microwave part of the spectrum, however, water is 80–100 times more absorptive than ice.

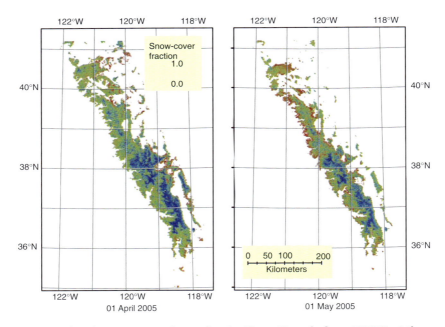

Figure 28.3 Fractional snow-covered area for the Sierra Nevada from MODIS at the beginning of April and the beginning of May 2005. Not only does the extent of the snow-covered area shrink as the snowmelt seasons progresses, the fractional snow cover in that area also decreases.

Figure 28.4   ASTER image of the Gangotri Glacier terminus in the Garhwal Himalaya since 1780 (Image from National Snow and Ice Data Center, University of Colorado).

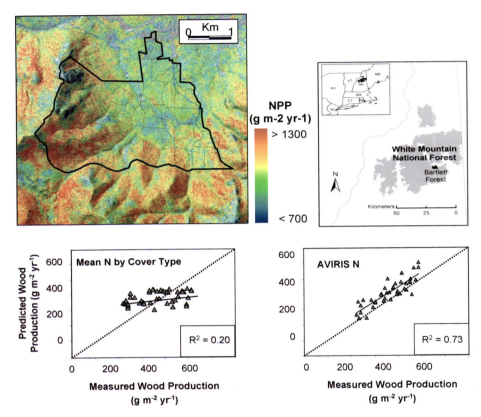

Figure 29.2   Remote sensing of canopy nitrogen provides a spatially-explicit and mechanistic means to constrain biogeochemical model simulations of photosynthesis, net primary production (NPP), and other ecosystem processes. Here, the NASA Airborne Visible and Infrared Imaging Spectrometer (AVIRIS) was used to estimate leaf nitrogen concentrations throughout the Bartlett Forest in New Hampshire, USA. The nitrogen maps were then used to constrain PnET model simulations of photosynthesis and NPP. The resulting NPP maps were far more accurate than those derived simply by parameterizing leaf nitrogen from a look-up table based on vegetation type (Reprinted with permission from *Ecosystems* 8(7), 2005, p. 770, Net primary production and canopy nitrogen in a temperate forest landscape: An analysis using imaging spectrometry, modeling and field data, S. V. Ollinger, and M. L. Smith, Figure 7, Copyright Springer Science and Business Media, Inc., 2005.)

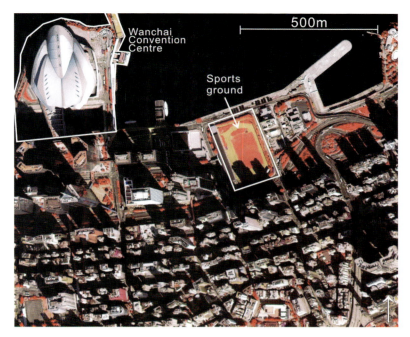

Figure 30.2 False colour, pan-sharpened QuickBird image of the northern shore of Hong Kong island, with the Wanchai Exhibition Centre at top left, and high rise CBD in the lower section. Over the land area, more than half of the ground is obscured by parallax displacement of tall buildings and building shadows. Copyright Digital Globe.

Figure 30.6 False colour IKONOS image of Kowloon Peninsula, Hong Kong, showing routes derived from least cost path analysis, starting at Lion Rock country park forested region at upper right, and ending in Kowloon Park at lower left. Paths A, B and C are derived from IKONOS fraction images resulting from Linear Spectral Unmixing. The longer, MLC paths are derived from the per-pixel, Maximum Likelihood Classifier.

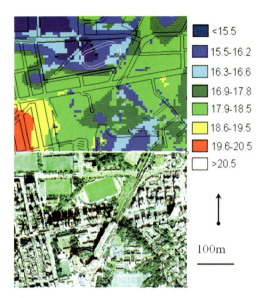

Figure 30.7   Extract from an ASTER night-time image of 31st January 2007 of the Kowloon Peninsula, Hong Kong, and corresponding aerial photograph. The urban heat island core at lower left gives way to lower density development towards the top right. Surface temperatures (°C) are discontinuous, with steep temperature gradients between land cover types.

Figure 32.1   Simulated 10 m inundation (pale blue) along the NW coast of Costa Rica. Derived from SRTM DEM data, overlain by true-colour Landsat Thematic Mapper (TM) imagery.

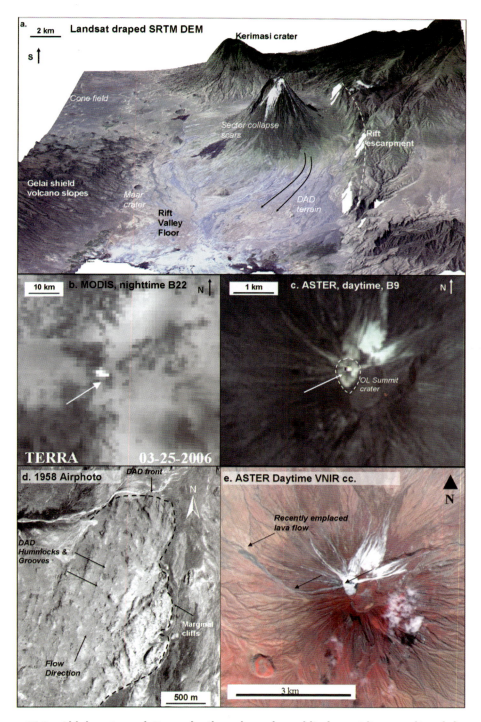

Figure 32.2 Oldoinyo Lengai, Tanzania: the volcano has white lava at its summit and rises 2000 m above a rift valley. Central figure: 3-D view, SRTM DEM with Landsat ETM+ true colour drape. Upper left: night-time Moderate Resolution Imaging Spectrometer (MODIS) band 22 (3.93–3.99 μm) image – white pixels: thermal emission of a volcano-flank lava flow. Upper right: ASTER band 9 (2.36–2.43 μm) scene – thermal anomalies from a summit-crater lava flow. Lower left: aerial photograph of debris avalanche deposits (hummocky relief). Lower right: ASTER Visible and Near-Infra-Red (VNIR) image, acquired a few days after an eruption, which enabled the mapping of lava flows.

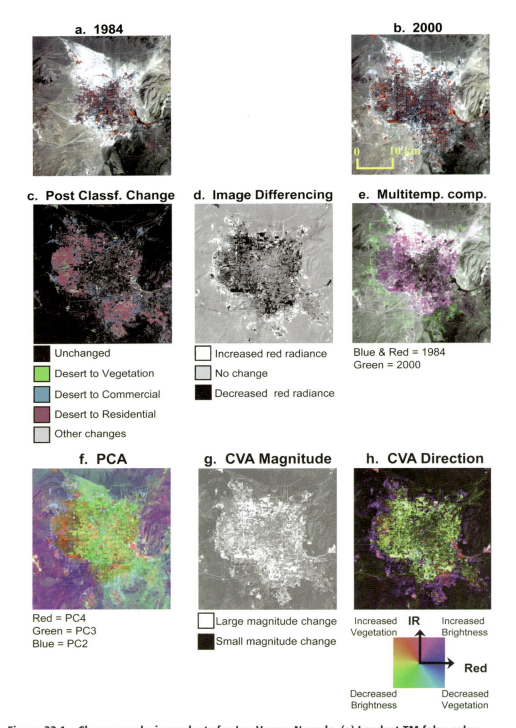

a. 1984

b. 2000

0   10 km

c. Post Classf. Change

■ Unchanged

■ Desert to Vegetation

■ Desert to Commercial

■ Desert to Residential

▢ Other changes

d. Image Differencing

▢ Increased red radiance

▢ No change

■ Decreased red radiance

e. Multitemp. comp.

Blue & Red = 1984
Green = 2000

f. PCA

Red = PC4
Green = PC3
Blue = PC2

g. CVA Magnitude

▢ Large magnitude change

■ Small magnitude change

h. CVA Direction

Increased   **IR**   Increased
Vegetation   Brightness

                                Red

Decreased   Decreased
Brightness   Vegetation

Figure 33.1   Change analysis products for Las Vegas, Nevada. (a) Landsat TM false color composite image, 1986. (b) Landsat ETM+ image, 2000. (c) Post-classification change detection, with selected class transitions shown. (d) Band 3 image differencing. (e) Multi-temporal composite. (f) Multi-date principal component analysis. (g) Change vector analysis, magnitude of change image. (h) Change vector analysis, direction of change image.